INTRODUCTION
TO
APPLIED
MATHEMATICS

GILBERT STRANG
Massachusetts Institute of Technology

WELLESLEY-CAMBRIDGE PRESS
Box 157
Wellesley, Massachusetts 02181

Designed by Michael Michaud, Unicorn Production Services

Printed in the United States of America
4   5   6   7   8   9

Published by Wellesley-Cambridge Press

Box 157   Room 2-240, M.I.T.
Wellesley, MA 02181 USA   Cambridge, MA 02139 USA
(617) 431-8488   (617) 253-4383

Library of Congress Cataloging-in-Publication Data

Strang, Gilbert
  Introduction to applied mathematics.

  Bibliography: p. 738
  Includes index.
  1. Mathematics —   I. Title.
QA37.2.S87   1986   510   84-52450
ISBN 0-9614088-0-4

A manual for instructors is available from the publisher.

Other texts from Wellesley-Cambridge Press

An Analysis of the Finite Element Method, Gilbert Strang and
George Fix (originally Prentice-Hall, ISBN 0-13-032946-0).

Calculus, Gilbert Strang, ISBN 0-9614088-2-0, to be published
in 1990.

CONTENTS

4. ANALYTICAL METHODS

5. NUMERICAL METHODS

6. INITIAL-VALUE PROBLEMS

7. NETWORK FLOWS AND COMBINATORICS

8. OPTIMIZATION

PREFACE

I believe that the teaching of applied mathematics needs a fresh approach. That opinion seems to be widely shared. Many of the textbooks in use today were written a generation ago, and they cannot reflect the ideas (or the algorithms) that have brought such significant change. Certainly there are things that will never be different—but even for the solution of Laplace's equation a great deal is new. In addition to Fourier series and complex variables, it is time to see fast transforms and finite elements. Topics like stability and optimization and matrix methods have earned a more central place—perhaps at the expense of series solutions of differential equations.

Applied mathematics is alive and very vigorous. That ought to be reflected in our teaching. In my own class I became convinced that the textbook is crucial. It must provide a framework into which the applications will fit. A good course has a clear purpose, and you can sense that it is there. It is a pleasure to teach a subject when it is moving forward, and this one is—but the book has to share in that spirit and help to establish it.

The central topics are differential equations and matrix equations—the continuous and discrete. They reinforce each other because they go in parallel. On one side is calculus; on the other is algebra. Differential equations and their transforms are classical, but still beautiful and essential. Nothing is out of date about Fourier! At the same time this course must develop the discrete analogies, in which potential differences rather than derivatives drive the flow. Those analogies are not difficult; they are basic, and I have found that they are welcome. To see the cooperation between calculus and linear algebra is to see one of the best parts of modern applied mathematics.

I am also convinced that these equations are better understood—they are more concrete and useful—when we present at the same time the algorithms that solve them. Those algorithms give support to the theory. And in general, numerical methods belong with the problems they solve. I do not think that the Fast Fourier Transform and difference equations and numerical linear algebra belong only in some uncertain future course. It is *this course* that should recognize what the computer can do, without being dominated by it.

Perhaps I may give special mention to the role of linear algebra. If any subject has become indispensable, that is it. The days of explaining matrix notation in an appendix are gone; this book assumes a working knowledge of matrices. It begins with the solution of $Ax = b$—not by computing A^{-1}, but by recognizing that as elimination goes to an upper triangular matrix, A is being factored into LU or LDL^T. A fuller treatment is given in my textbook *Linear Algebra and Its Applications* (3rd edition, HBJ, 1988). Here we start with the essential facts of linear algebra, and put them to use.

I do not think that the right approach is to model a few isolated examples. The important goal is to find ideas that are shared by a wide range of applications. This is the contribution a book can make, to recognize and explain the underlying pattern. We need to present Kirchhoff's laws, and to see how they lead in the continuous case to the curl and divergence, but we hope to avoid a total and fatal immersion into vector calculus—which has too frequently replaced applied mathematics, and taken all the fun out of it.

It seems natural for the discrete case to come first, but Chapter 1 *should not be too slow*. It is Chapter 2 that points out the triple product $A^T C A$ in the equations of equilibrium, and Chapter 3 finds that framework repeated by differential equations. Applications and examples are given throughout. Where the theory strengthens the understanding, it is provided—but this is not a book about proofs.

After the equations are formulated they need to be solved. Chapter 4 uses Fourier methods and complex variables. Chapter 5 uses numerical methods. It presents scientific computing as an integral part of applied mathematics, with the same successes (especially through Fourier) and the same difficulties (including stability in Chapter 6). Orthogonality is central, as it is for analytical methods.

The book is a text for applied mathematics and advanced calculus and engineering mathematics. It aims to explain what is essential, as far as possible within one book and one year. It also reaches beyond the usual courses, to introduce more recent ideas:

2.5   The Kalman filter
5.3   Iterative methods for $Ax = b$ and $Ax = \lambda x$
5.4   Finite elements
6.2   Chaos and strange attractors
6.6   Shock waves and solitons
7.1   Combinatorial optimization
7.4   Maximal flows and minimal cuts in networks
8.2   Karmarkar's method for linear programming

You will not have time for all of these; I do not. Those that are unfamiliar may stay that way; no objection. Nevertheless they are part of this subject and they fit perfectly into its framework—even Karmarkar's method has a place, whether or not it fulfills everything that has been promised. (It outdoes the simplex method on some problems but by no means on all.) It solves the piecewise linear equations of mathematical programming through a series of linear equations, whose coefficient matrices are again $A^T C A$. Finite elements and recursive filters and Laplace's equation fit this same pattern. In my experience the framework is seen and understood by the reader, and appreciated.

The whole subject is extremely coherent, and sections or chapters which are left for reference are no less valuable on that account. In fact this is also a textbook on numerical analysis (with applications included in the course, not separated) and on optimization (emphasizing the quadratic problems of Chapter 2, the network flows of Chapter 7, and the duality theory of Chapter 8). The book starts with ordinary differential equations and makes the transition to partial differential equations; that is not a great obstacle. It presents the minimum principles of least energy and least action, which are more subtle than equations and sometimes more revealing. But the emphasis must go to equilibrium equations (*boundary-value problems*) and dynamic equations (*initial-value problems*). Those are the central questions, continuous and discrete.

Applied mathematics is a big subject, and this is not a short book. Its goal is to be as useful as possible to the reader, in class and out. Engineering and scientific applications play a larger part than before; infinite series play a smaller part. Those series need to give way, in teaching as they have done in practice, to a more direct approach to the solution. We intend to try—working always with specific examples—to combine the algorithms with the theory. That is the way to work and I believe it is also the way to learn. The effort to teach what students will need and use is absolutely worthwhile.

A personal note before the book begins. These years of writing have been a tremendous pleasure, and one reason was the help offered by friends. It was shamelessly accepted, and what can I do in return? A public acknowledgement is made at the end, of my gratitude for their encouragement. It is what keeps an author going.

In one respect this is a special adventure. I decided that it must be possible to see the book all the way through. I care too much about the subject to mail in a manuscript and say goodbye. Therefore Wellesley-Cambridge Press was created, in order to do it right. The book was printed in the normal way, and it should be handled more efficiently than usual. Bookstores (and individual readers) will need a new address and telephone number, given below. No representative will call! I have to depend on those who enjoy it to say so. More than ever it must stand on its own—if it can. I hope you like the book.

Gilbert Strang

Wellesley-Cambridge Press
Box 157
Wellesley MA 02181
(617) 431-8488

M.I.T.
Room 2-240
Cambridge MA 02139
(617) 253-4383

INTRODUCTION
TO
APPLIED
MATHEMATICS

CHAPTER 1 IN OUTLINE: SYMMETRIC LINEAR SYSTEMS

1.1 Introduction

1.2 Gaussian Elimination—linear equations by rows and by columns
Geometry of the Columns—independent vs. dependent columns
Elimination and Pivots—forward elimination and back substitution
Factorization into $A = LU$—two triangular systems $Lc = b$ and $Ux = c$

1.3 Positive Definite Matrices and $A = LDL^T$—positive pivots in D
Matrices of Order $n = 2$—pivots $d_1 = a$ and $d_2 = c - b^2/a$
The Minimum of a Quadratic—positive definite if $x^T A x > 0$
Test for a Minimum: $n = 2$—positive pivots appear in the squares
Elimination = Completing the Square—each step removes $M = ldl^T$
Elimination = Factorization—$A = l_1 d_1 l_1^T + \cdots + l_n d_n l_n^T = LDL^T$
Matrix Multiplication—rows times columns or columns times rows
Block Matrices—multiplication and inversion by blocks

1.4 Minimum Principles—the minimum of a quadratic $P = \frac{1}{2}x^T A x - x^T b$
Positive Definiteness of $A^T A$ and $A^T C A$—independent columns in A
Least Squares Solution of $Ax = b$—the normal equation $A^T A x = A^T b$
Fitting Measurements by a Straight Line—replacing $Ax = b$ by $Ax = p$
Failure of Independence in the Columns of A—the pseudoinverse A^+
Minimum Potential Energy—$Ax = e$ and $Ce = y$ and $A^T y = f$
The Stiffness Matrix—$K = A^T C A$ is tridiagonal and K^{-1} is positive

1.5 Eigenvalues and Dynamical Systems—$Ax = \lambda x$, $\det(A - \lambda I) = 0$
The Diagonal Form—$S^{-1} A S = \Lambda$ with eigenvectors in S, eigenvalues in Λ
Differential Equations—solutions $e^{\lambda t} x$ for $du/dt = Au$
Second-Order Equations—solutions $e^{i\omega t} x$ for $u_{tt} + Au = 0$
Single Equations of Higher Order—solutions $e^{\lambda t}$ for $u_{tt} + p u_t + q u = 0$
Symmetric Matrices—real eigenvalues, orthogonal eigenvectors, $A = Q \Lambda Q^T$

1.6 A Review of Matrix Theory—$Ax = b$ for b in the column space of A
Incidence Matrices—Kirchhoff's laws and the four subspaces
Factorizations Based on Elimination—$A = LDU$, $PA = LDU$, $PA = LU$
Factorizations Based on Eigenvalues—$A = S\Lambda S^{-1}$ and $A = MJM^{-1}$
Factorizations Based on $A^T A$—$A = QR$, $A = Q_1 \Sigma Q_2^T$, $A = QB$
Review of the Review—rank one matrices $A = vw^T$

1

SYMMETRIC LINEAR SYSTEMS

The simplest model in applied mathematics is a system of linear equations. It is also by far the most important, and we begin this book with an extremely modest example:

$$2x_1 + 4x_2 = 2$$
$$4x_1 + 11x_2 = 1.$$

Such a system, with two equations in two unknowns, is easy to solve. But in discussing this particular problem we have four more general goals in mind, and they can be described in advance:

(1) to establish the method of **Gaussian elimination** for solving a more complicated system, of n equations in n unknowns

(2) to express the solution technique as a **matrix factorization**

(3) to explain the underlying **geometry**

(4) to recognize the associated **minimum principle**.

Part of this will already be familiar. Therefore we mention one thing that is likely to be new: it is the fact that matrix multiplication can be interpreted in four different ways, and that each of those interpretations contributes to the analysis of a linear system. The one which is least known turns out to be surprisingly useful; it multiplies column vectors by row vectors (and produces a full matrix).

The original equations can be written in matrix notation as

$$\begin{bmatrix} 2 & 4 \\ 4 & 11 \end{bmatrix} \begin{bmatrix} x_1 \\ x_2 \end{bmatrix} = \begin{bmatrix} 2 \\ 1 \end{bmatrix}, \quad \text{or} \quad Ax = b,$$

and the solution could be dismissed as $x = A^{-1}b$. To leave it there would be completely wrong. In fact, it is very rarely that the inverse of a matrix needs to be computed. A^{-1} makes the formula easy but it is better to exclude it from the calculations. The real solution to a linear system comes from Gaussian elimination, which has the remarkable effect of splitting A into a product of triangular matrices.

For this example the product is $A = LDL^T$. A lower triangular matrix L multiplies its transpose (which is upper triangular), with a positive diagonal matrix D in between. That is the first of the triple products which dominate this subject. It is the one that comes from the *solution* of a linear problem, while the second comes from the *formulation* of the problem. Perhaps I could also write down the third, which comes from the *dynamics* (when the problem changes with time):

1. LDL^T: triangular L from elimination
   diagonal D from the pivots
2. $A^T CA$: rectangular A from the geometry
   square C from the material properties
3. $Q\Lambda Q^T$: orthogonal Q from the eigenvectors
   diagonal Λ from the eigenvalues

The second one is absolutely crucial. $A^T CA$ is present in all the central problems of equilibrium, whether they are discrete (with matrix equations) or continuous (with differential equations). The first examples come in Section 1.4; then the following chapters present the complete framework. It is the key to the applications.

Later, in Section 1.5, there is the third factorization $A = Q\Lambda Q^T$. It is more delicate, because it comes not from $Ax = b$ but from the eigenvalue problem $Ax = \lambda x$. It depends, as the LDL^T factorization did, on the **symmetry** of A. The same coefficient 4 multiplies x_2 in the first equation and x_1 in the second equation, so that A equals its own transpose. (In matrix notation $A = A^T$.) And the fact that we have not only a symmetric problem but also a minimum problem must be reflected in the positivity of the eigenvalues.

In part this chapter is a summary of linear algebra. It will include matrices that are not symmetric; Section 1.6 describes the changes that come when symmetry is lost. That section also admits rectangular matrices; A can be an m by n matrix of rank r. The list of factorizations is extended beyond the two that come from elimination and eigenvalues, to give a concise but more complete picture of matrix theory.

Most of the chapter is not, however, a review. It moves directly to the special class of matrices, symmetric and positive definite, which are at the center of applied mathematics. Those matrices describe a condition of equilibrium, and they are associated with a minimum of energy—because that is what nature seeks. They also describe oscillations around that equilibrium, when the energy is conserved.

Matrices of this kind are linked directly to minimum principles. They extend into n-dimensional space the familiar test from calculus, that where a function has a minimum its second derivative should be positive. Of course the first derivative should be zero; that is the equation $Ax = b$. Thus the solution to a linear system is also the point at which the energy is minimized. Throughout this book you are faced with that choice, between equations and minimum principles—fortunately it is possible to have both, since they are equivalent statements of the same law:

$$\text{minimize } \tfrac{1}{2}x^T A x - x^T b \text{ or solve } Ax = b.$$

For those who do not want this fast course on matrix theory—or for those who want an even faster course—the main connections to later sections can be given at once. They are

LDL^T: elimination (1.2) leading to direct methods for linear systems (5.1)
$A^T C A$:   least squares (1.4) leading to filters (2.5) and Karmarkar's method (8.2)
$A^T C A$:   elastic springs (1.4) leading to structural mechanics (2.4)
$A^T C A$:   incidence matrices (1.6) leading to networks (2.3) and graphs (7.1)
$Q \Lambda Q^T$:   diagonalization (1.5) leading to the convolution rule (4.2)

These are only the discrete applications. There is a corresponding series of problems in differential equations, in equilibrium and in evolution. They also come from $A^T C A$ or from minimum principles. They are studied analytically by diagonalization (Fourier series) or numerically by discretization (finite differences and finite elements). Therefore the discrete problems come first.

The key steps are indicated by gray shading in the text. Beyond that, a number of textbooks offer a fuller treatment of linear algebra and its applications. One that we are familiar with is mentioned in the list of references.

We start with Gaussian elimination, using examples that are symmetric positive definite. Then the properties of those examples emerge at the same time as the general pattern of elimination (which only cares that the matrix is invertible). We emphasize that although there are other tests for invertibility and other algorithms for solving linear systems—Cramer's rule is the most notorious, based on determinants—it is elimination that is constantly used in applications. For a small matrix that choice probably doesn't matter. For a large matrix it certainly does. For really large problems even elimination is expensive, and it becomes a combinatorial problem to reduce the number of steps; graph theory will help. But in each case the steps of elimination expose part of the inner structure, and the eigenvalues expose another part, and we want to see their relation.

1.2 ■ GAUSSIAN ELIMINATION

We begin with the geometry. One approach is to look at the two equations separately: first $2x_1 + 4x_2 = 2$ and then $4x_1 + 11x_2 = 1$. Each equation is represented by a straight line and those lines intersect at the solution. In our example one line contains every point (x_1, x_2) for which $2x_1 + 4x_2 = 2$. The other line represents $4x_1 + 11x_2 = 1$, its slope is $-4/11$, and it meets the first line at the point $(3, -1)$. Therefore the solution is $x_1 = 3$, $x_2 = -1$. That is the only point on both lines and therefore the only solution to both equations. (There is a dotted line in Fig. 1.1, also going through the solution. We will see that it comes from elimination, which produces a simpler equation but leaves the solution unchanged.)

The problem has a second geometric interpretation, less familiar but equally important. It comes from looking at the *columns* of the system instead of the *rows*. We can rewrite the original problem in vector notation as

$$x_1 \begin{bmatrix} 2 \\ 4 \end{bmatrix} + x_2 \begin{bmatrix} 4 \\ 11 \end{bmatrix} = \begin{bmatrix} 2 \\ 1 \end{bmatrix}, \tag{1}$$

meaning that we want to find the combination of the two vectors on the left which equals the vector on the right. These are vectors in 2-dimensional space, which is a plane, and they are drawn in Fig. 1.1. To multiply a vector by 3, we triple its length:

$$3 \begin{bmatrix} 2 \\ 4 \end{bmatrix} = \begin{bmatrix} 6 \\ 12 \end{bmatrix}.$$

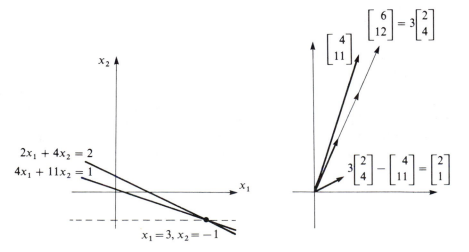

Fig. 1.1. Row picture and column picture of the geometry.

When a vector is multiplied by -1, its direction is reversed:

$$-1\begin{bmatrix} 4 \\ 11 \end{bmatrix} = \begin{bmatrix} -4 \\ -11 \end{bmatrix}.$$

To add vectors, we can describe the result either geometrically by shifting one vector to start where the other one ends, or else algebraically by just adding the corresponding components:

$$3\begin{bmatrix} 2 \\ 4 \end{bmatrix} - 1\begin{bmatrix} 4 \\ 11 \end{bmatrix} = \begin{bmatrix} 6 \\ 12 \end{bmatrix} + \begin{bmatrix} -4 \\ -11 \end{bmatrix} = \begin{bmatrix} 2 \\ 1 \end{bmatrix}.$$

Both methods lead to the same conclusion, that the right weights are $x_1 = 3$, $x_2 = -1$.

Our second example is in four dimensions:

$$\begin{aligned}
2x_1 + x_2 &= 2 \\
x_1 + 2x_2 + x_3 &= 1 \\
x_2 + 2x_3 + x_4 &= 4 \\
x_3 + 2x_4 &= 8
\end{aligned} \qquad (2)$$

The first approach again looks separately at each equation. The graph of $2x_1 + x_2 = 2$ is no longer a line; instead, since x_3 and x_4 can be chosen arbitrarily, it is a three-dimensional surface in four-dimensional space.† It is harder to visualize, but in some four-dimensional sense it must be flat. It is called a *plane*, or sometimes a *hyperplane*, to indicate that it is determined by one linear equation (like an ordinary plane in three dimensions). It has one fewer degree of freedom than the full four-dimensional space. It is perpendicular to the vector $(2, 1, 0, 0)$, and although it is three-dimensional (it would look solid to us) it is a very thin set in four-dimensional space.

The second equation is similar: $x_1 + 2x_2 + x_3 = 1$ describes another hyperplane. It intersects the first in a two-dimensional surface, still flat and still in four dimensions. Then the third hyperplane $x_2 + 2x_3 + x_4 = 4$ intersects that surface in a line. Note that the line does not go through the origin, and is not described by a single equation; in four dimensions, it requires three equations to determine a line. Finally the hyperplane from the fourth equation intersects this line in a point, which is the solution.

That completes one picture of the geometry behind a linear system; it is a family of intersecting hyperplanes. In n dimensions there are n planes, and their intersection is the solution to the linear equations.

---

† The author has tried to draw it, and failed.

Geometry of the Columns: Invertible vs. Singular

The second approach to the geometry, by columns instead of rows, starts from

$$
x_1 \begin{bmatrix} 2 \\ 1 \\ 0 \\ 0 \end{bmatrix} + x_2 \begin{bmatrix} 1 \\ 2 \\ 1 \\ 0 \end{bmatrix} + x_3 \begin{bmatrix} 0 \\ 1 \\ 2 \\ 1 \end{bmatrix} + x_4 \begin{bmatrix} 0 \\ 0 \\ 1 \\ 2 \end{bmatrix} = \begin{bmatrix} 2 \\ 1 \\ 4 \\ 8 \end{bmatrix}.
\tag{3}
$$

This is still $Ax = b$. The left side is a "column at a time" multiplication of A times x:

$$
Ax = \begin{bmatrix} 2 & 1 & 0 & 0 \\ 1 & 2 & 1 & 0 \\ 0 & 1 & 2 & 1 \\ 0 & 0 & 1 & 2 \end{bmatrix} \begin{bmatrix} x_1 \\ x_2 \\ x_3 \\ x_4 \end{bmatrix} = \begin{matrix} x_1 \text{ (column 1)} \\ + x_2 \text{ (column 2)} \\ + x_3 \text{ (column 3)} \\ + x_4 \text{ (column 4)} \end{matrix}.
$$

The product Ax is a ***combination of the four columns of A.***† To solve $Ax = b$ is to find the combination of the four vectors on the left of (3) which produces the vector b on the right. In this example the combination is especially simple; b is the first column plus four times the last column. Thus the solution is $(x_1, x_2, x_3, x_4) = (1, 0, 0, 4)$. It is also the point where the four hyperplanes intersect, in the first approach.

Now we consider a more unpleasant possibility, in which both approaches fail. Suppose the hyperplanes do *not* intersect in a point. Alternatively, suppose the combinations of the columns of A cannot produce the right side b. This is the degenerate case, or ***singular case***. It is typified by a matrix in which the third row is the sum of the first two rows:

$$
Ax = \begin{bmatrix} 1 & 1 & 1 \\ 1 & 2 & 3 \\ 2 & 3 & 4 \end{bmatrix} \begin{bmatrix} x_1 \\ x_2 \\ x_3 \end{bmatrix} = \begin{bmatrix} 0 \\ 1 \\ 5 \end{bmatrix}.
\tag{4}
$$

Each equation represents an ordinary plane in three dimensions. The first plane is $x_1 + x_2 + x_3 = 0$, perpendicular to the vector $(1, 1, 1)$. This plane goes through the origin, since the point $x_1 = x_2 = x_3 = 0$ satisfies the equation. It intersects the second plane $x_1 + 2x_2 + 3x_3 = 1$ in a line. However that line does not touch the third plane—the plane is parallel to the line and they never meet. There is no point on all three planes, and no solution to $Ax = b$.

Algebraically that conclusion is reached by adding the first two equations. We get $2x_1 + 3x_2 + 4x_3 = 1$, which contradicts the third equation. Algorithmically the

---

† I cannot emphasize that too strongly; it is simple but fundamental.

conclusion comes from the failure of elimination, which would lead in a systematic way to $0 = 4$. Here we stay with the geometry, to visualize this singular case.

In the "row at a time" approach the three planes do not meet. The problem is not that two planes are parallel; it is one step more subtle. Viewed from the right direction the planes form a triangle (Fig. 1.2). Since they have no point in common, the equations are inconsistent. In the "column at a time" approach, the three columns of A lie in the same plane. Their combinations do not fill 3-dimensional space, and they cannot produce the right side $b = (0, 1, 5)$. That vector b is not a combination of the columns, and there is no solution.

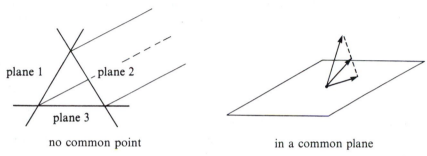

no common point   in a common plane

Fig. 1.2. Dependent rows and dependent columns.

Note one important point: Because there were only 2 independent rows, there were only 2 independent columns. We know that row 3 is the sum of rows 1 and 2. Therefore some combination of the columns will also give zero, and the columns lie in a plane. The combination can be found by solving $Ax = 0$; it is $x = (1, -2, 1)$. The first and third columns in Fig. 1.2 add to twice the second column.

Of course the equations are not always inconsistent. If $b_3 = 5$ is changed to $b_3 = 1$, then the third equation is the sum of the first two. It contains no new information, but it does not destroy the system. In that case there is a whole *line* of solutions lying on all three planes, and the system is **underdetermined**. In the other figure, the right side $b = (0, 1, 1)$ falls into the plane of the columns. It is a combination with weights $x_1 = -1$, $x_2 = 1$, $x_3 = 0$, but we can add any multiple of $1, -2, 1$ to those weights and find another solution.

Now we go back to the nonsingular case, where A is invertible and there is a unique solution $x = A^{-1}b$—which is found not from that formula but from the most basic algorithm in scientific computation.

Elimination and Pivots

Geometry can clarify the underlying problem but it is algebra that solves it. We need a systematic method to compute the solution, and the key idea is simple. The size of the problem is progressively reduced, by eliminating one variable after another, until only the last unknown x_n is left. That unknown is determined by a single equation $d_n x_n = c_n$, which is trivial: $x_n = c_n/d_n$. Then, by going back to each of the eliminated variables, we solve for the remaining unknowns $x_{n-1}, \ldots, x_1$ in reverse order. The final result is a straightforward and efficient algorithm for the solution of n simultaneous linear equations in n unknowns. It is known as **Gaussian elimination**.

For $n = 2$ it is deceptively simple. We start with

$$2x_1 + 4x_2 = 2$$
$$4x_1 + 11x_2 = 1.$$

The upper left entry 2 is called the **pivot**. Some multiple of the pivot will eliminate the entry beneath it, and in this case the multiple is two; we subtract twice equation (1) from equation (2) to arrive at the new system

$$2x_1 + 4x_2 = \quad 2$$
$$3x_2 = -3. \tag{5}$$

The second pivot has now appeared; it is 3, the coefficient of x_2 on the diagonal after elimination. Normally we would use this pivot d_2 to eliminate x_2 from the later equations, but in this case there are no later equations and elimination is complete. The second equation is $d_2 x_2 = c_2$, or $3x_2 = -3$, and it is this equation which corresponds to the dotted line in Fig. 1.1. It gives $x_2 = -1$, and after substituting that answer into the first equation we find $x_1 = 3$. This is a first example of a triangular system, so called from the shape of the left hand side in (5). The process of solving it, from the bottom to the top, is called **back substitution**.

Now consider the example with $n = 4$ equations:

$$Ax = \begin{bmatrix} 2 & 1 & 0 & 0 \\ 1 & 2 & 1 & 0 \\ 0 & 1 & 2 & 1 \\ 0 & 0 & 1 & 2 \end{bmatrix} \begin{bmatrix} x_1 \\ x_2 \\ x_3 \\ x_4 \end{bmatrix} = \begin{bmatrix} 2 \\ 1 \\ 4 \\ 8 \end{bmatrix} = b.$$

For the present we ignore the right side b, the "inhomogeneous term" in the equation. The elimination steps are completely decided by the matrix A, as we reduce it one column at a time to a triangular matrix. The same steps are of course applied to b, to maintain equality, but they can be done afterwards. The immediate goal is to simplify A.

The first pivot is $d_1 = 2$, and we use it to eliminate x_1 from the remaining equations. In this problem, x_1 is already absent from equations (3) and (4). Therefore we subtract from equation (2) the multiple $l_{21} = \frac{1}{2}$ of equation (1):

$$\begin{bmatrix} 2 & 1 & 0 & 0 \\ 1 & 2 & 1 & 0 \\ 0 & 1 & 2 & 1 \\ 0 & 0 & 1 & 2 \end{bmatrix} \rightarrow \begin{bmatrix} 2 & 1 & 0 & 0 \\ 0 & \frac{3}{2} & 1 & 0 \\ 0 & 1 & 2 & 1 \\ 0 & 0 & 1 & 2 \end{bmatrix}.$$

This time the second pivot is $d_2 = \frac{3}{2}$, and we use it to eliminate x_2 from the third equation. The required multiple is $l_{32} = \frac{2}{3}$, and subtraction produces a zero in row 3, column 2:

$$
\begin{bmatrix}
2 & 1 & 0 & 0 \\
0 & \frac{3}{2} & 1 & 0 \\
0 & 0 & \frac{4}{3} & 1 \\
0 & 0 & 1 & 2
\end{bmatrix}.
$$

The final step uses the third pivot $d_3 = \frac{4}{3}$, and multiplies it by $l_{43} = \frac{3}{4}$, in order to annihilate the last nonzero entry below the diagonal. The subtraction $2 - \frac{3}{4} = \frac{5}{4}$ gives the fourth pivot and leaves the final triangular form

$$
U = \begin{bmatrix}
2 & 1 & 0 & 0 \\
0 & \frac{3}{2} & 1 & 0 \\
0 & 0 & \frac{4}{3} & 1 \\
0 & 0 & 0 & \frac{5}{4}
\end{bmatrix}.
$$

This matrix U is upper triangular, with the pivots on its diagonal. Note that the pivots are not the original diagonal entries of A; they are not known until elimination discovers them.

The forward elimination is now completed for A. To keep the four equations correct, we apply the same steps to the right side b:

$$
b = \begin{bmatrix} 2 \\ 1 \\ 4 \\ 8 \end{bmatrix}
\xrightarrow[\text{from row 2}]{\frac{1}{2}\,\text{row 1}}
\begin{bmatrix} 2 \\ 0 \\ 4 \\ 8 \end{bmatrix}
\xrightarrow[\text{from row 3}]{\frac{2}{3}\,\text{row 2}}
\begin{bmatrix} 2 \\ 0 \\ 4 \\ 8 \end{bmatrix}
\xrightarrow[\text{from row 4}]{\frac{3}{4}\,\text{row 3}}
\begin{bmatrix} 2 \\ 0 \\ 4 \\ 5 \end{bmatrix} = c.
\tag{6}
$$

The last step subtracts l_{43} times the third row from the fourth, and produces $8 - \frac{3}{4}(4) = 5$. The new system of equations will be $Ux = c$.

The accident of a zero entry meant that the second step produced no change; that is not very important. What is much more important is to recognize the effect of the zeros in the original matrix A. That matrix is **tridiagonal**. It has three zeros below the diagonal, all outside the "band" of nonzeros. As a result, three of the normal elimination steps were unnecessary; the multipliers l_{31}, l_{41}, and l_{42} vanished since the entries in those locations were already zero (and remained zero). Above the band a similar property holds; the zero entries in A are still there in U. Most large matrices that arise in practice are sparse, and it will be a fundamental question to preserve as much as possible the sparsity of A; in this case it was entirely preserved.

Elimination has taken us from A to U and from b to c. The systems $Ax = b$ and $Ux = c$ have the same solution, since equality was maintained at each step by applying the same operations to A and to b. These operations are completely reversible; we can return from $Ux = c$ to $Ax = b$, so that no solutions are created or destroyed. Therefore the problem is now governed by a triangular matrix,

$$
Ux = \begin{bmatrix} 2 & 1 & 0 & 0 \\ 0 & \frac{3}{2} & 1 & 0 \\ 0 & 0 & \frac{4}{3} & 1 \\ 0 & 0 & 0 & \frac{5}{4} \end{bmatrix} \begin{bmatrix} x_1 \\ x_2 \\ x_3 \\ x_4 \end{bmatrix} = \begin{bmatrix} 2 \\ 0 \\ 4 \\ 5 \end{bmatrix} = c,
$$

and it is this system that back substitution will solve. The last equation is $(\frac{5}{4})x_4 = 5$, and thus $x_4 = 4$. Then the third equation becomes

$$
\tfrac{4}{3}x_3 + x_4 = 4, \quad \text{or} \quad \tfrac{4}{3}x_3 = 0, \quad \text{or} \quad x_3 = 0.
$$

Continuing upwards we have $x_2 = 0$, $x_1 = 1$, and $x = (1, 0, 0, 4)$ as expected.

Remark   Some descriptions of elimination go beyond U, to remove any nonzeros that are above the diagonal. With enough operations the original A can be changed into a diagonal matrix, and even into the identity matrix. This is *Gauss–Jordan* elimination. It is not used in practice, because there is no reason to pay its extra cost. It does lead to the inverse matrix, but normally A^{-1} is not wanted or needed. All we require is $x = A^{-1}b$, which comes from back substitution.

Factorization into $A = LU$

Back substitution solves $Ux = c$. It is easy because U is triangular; that is the object of elimination. It is the second half of the action on the right side of $Ax = b$, and we now look back at the first half—*the step from b to c*. This was "*forward elimination*," when the subtractions that are applied to A on the left side are also applied to b on the right. Those subtractions change A and b to U and c, and the key to elimination is to put them into matrix form.

The steps were listed in equation (6). Multiples l_{ij} of one component were subtracted from a later component. Our whole point is that if these multipliers l_{ij} are put into a matrix L, then **the steps from b to c are exactly the same as solving** $Lc = b$. That is the matrix form of forward elimination, and in this case the multipliers are $\frac{1}{2}$, $\frac{2}{3}$, and $\frac{3}{4}$.

$$
\begin{bmatrix} 1 & & & \\ \frac{1}{2} & 1 & & \\ 0 & \frac{2}{3} & 1 & \\ 0 & 0 & \frac{3}{4} & 1 \end{bmatrix} \begin{bmatrix} 2 \\ 0 \\ 4 \\ 5 \end{bmatrix} = \begin{bmatrix} 2 \\ 1 \\ 4 \\ 8 \end{bmatrix}, \quad \text{or} \quad Lc = b. \tag{7}
$$

We claim that (7) is the same as (6). Suppose the right side b is given and we are solving for c. Certainly the first component $c_1 = 2$ comes from $b_1 = 2$. Look at the second equation. To find $c_2 = 0$ we subtract half of c_1 from $b_2 = 1$; that agrees with the first elimination step. To find c_3 we subtract $\frac{2}{3}$ of c_2 from b_3, and again that is correct. Finally $c_4 = 5$ comes from $b_4 = 8$ by subtracting $\frac{3}{4}c_3 = 3$.

If we turn away from these numbers, and concentrate on the matrix form, the pattern is remarkable. The *forward* half of elimination is governed by the triangular matrix L. The *backward* half is governed by U. The effect of elimination is to break $Ax = b$ into **two triangular systems**, first $Lc = b$ and then $Ux = c$. The matrices L and U come from elimination on the left side of the original equation; L contains the multipliers and U contains the end result. They are the output from A, and then x is the output from b. A good code separates Gaussian elimination into these two subroutines:

> 1. **Factor** (to compute L and U from A)
> 2. **Solve** (to compute x from L and U and b).

This applies to unsymmetric as well as symmetric matrices. There may also be row exchanges, which are never needed for positive definite problems. What is most important is the product of L and U; it turns out to equal the original A.

1A Elimination goes from $Ax = b$ to the triangular system $Ux = c$. The step from b to c is another triangular system $Lc = b$, where L contains the multipliers used in elimination (and 1's on the diagonal). The two equations combine to give

$$Lc = LUx = b, \quad \text{so that} \quad A = LU. \tag{8}$$

Elimination factors A into L times U.

The example illustrates this matrix equation $LU = A$:

$$\begin{bmatrix} 1 & & & \\ \frac{1}{2} & 1 & & \\ & \frac{2}{3} & 1 & \\ & & \frac{3}{4} & 1 \end{bmatrix} \begin{bmatrix} 2 & 1 & & \\ & \frac{3}{2} & 1 & \\ & & \frac{4}{3} & 1 \\ & & & \frac{5}{4} \end{bmatrix} = \begin{bmatrix} 2 & 1 & & \\ 1 & 2 & 1 & \\ & 1 & 2 & 1 \\ & & 1 & 2 \end{bmatrix}.$$

Elimination went from A to U by subtractions. It returns from U to A by undoing those subtractions, and adding back l_{ij} times row j of U to row i. Normally the multipliers fill a lower triangular matrix L, but in the case displayed above L and U and A are band matrices. The factors of a tridiagonal matrix are bidiagonal.

This factorization will be studied again in the next section, to make a special point about symmetry. L is related directly to U, after you divide the rows of U by the pivots. (In other words, multiply the four rows by $\frac{1}{2}, \frac{2}{3}, \frac{3}{4}, \frac{4}{5}$.) Then the diagonal entries are 1 and the upper triangular matrix becomes L^T. *It is the **transpose** of L.*

This creates three factors instead of two, but now they preserve the symmetry:

$$
\begin{bmatrix} 1 & & & \\ \tfrac{1}{2} & 1 & & \\ & \tfrac{2}{3} & 1 & \\ & & \tfrac{3}{4} & 1 \end{bmatrix}
\begin{bmatrix} 2 & & & \\ & \tfrac{3}{2} & & \\ & & \tfrac{4}{3} & \\ & & & \tfrac{5}{4} \end{bmatrix}
\begin{bmatrix} 1 & \tfrac{1}{2} & & \\ & 1 & \tfrac{2}{3} & \\ & & 1 & \tfrac{3}{4} \\ & & & 1 \end{bmatrix}
=
\begin{bmatrix} 2 & 1 & & \\ 1 & 2 & 1 & \\ & 1 & 2 & 1 \\ & & 1 & 2 \end{bmatrix}.
$$

That is the *triple factorization* $A = LDL^T$.

Before ending this section we must emphasize the practical importance of $A = LU$:

> Once L and U are known, every new right side b involves only two triangular systems $Lc = b$ and $Ux = c$ (n^2 steps). The elimination on A is not repeated, and A^{-1} is not needed.

Suppose the components of b are changed to 0, 0, 4, 8. Forward elimination in $Lc = b$ leads to $c = (0, 0, 4, 5)$. Then back substitution gives $x = (0, 0, 0, 4)$. This solution is correct, since 4 times the last column of A does equal b. The essential point is that every vector b can be processed in the same extremely fast way.

In the next section we see how the relation of U to L recovers the symmetry of A, and how positive pivots are linked to a minimization.

EXERCISES

1.2.1 Solve

$$
\begin{bmatrix} 1 & 1 & 1 \\ 1 & 3 & 3 \\ 1 & 3 & 5 \end{bmatrix}
\begin{bmatrix} x_1 \\ x_2 \\ x_3 \end{bmatrix}
=
\begin{bmatrix} 2 \\ 0 \\ 2 \end{bmatrix},
$$

using elimination to reach $Ux = c$ and then back substitution to compute x.

1.2.2 In the preceding problem, describe the graph of the second equation $x_1 + 3x_2 + 3x_3 = 0$. Find its line of intersection with the first surface $x_1 + x_2 + x_3 = 2$, by giving two points on the line.

1.2.3 Solve $Ax = b$ with the same A as in the text, but a different b:

$$
\begin{bmatrix} 2 & 1 & 0 & 0 \\ 1 & 2 & 1 & 0 \\ 0 & 1 & 2 & 1 \\ 0 & 0 & 1 & 2 \end{bmatrix}
\begin{bmatrix} x_1 \\ x_2 \\ x_3 \\ x_4 \end{bmatrix}
=
\begin{bmatrix} 2 \\ 7 \\ 12 \\ 11 \end{bmatrix}.
$$

Without repeating the steps from A to U, apply them only to solve $Lc = b$. Then solve $Ux = c$.

1.2.4 (a) Find the value of q for which elimination fails, in the system

$$\begin{bmatrix} 3 & 6 \\ 6 & q \end{bmatrix} \begin{bmatrix} x_1 \\ x_2 \end{bmatrix} = \begin{bmatrix} 1 \\ 4 \end{bmatrix}.$$

(b) For this value of q, what happens to the first geometrical interpretation (two intersecting lines) in Fig. 1.1?

(c) What happens to the second interpretation, in which b is a combination of the two columns?

(d) What value should replace $b_2 = 4$ to make the system solvable for this q?

1.2.5 For the small angle in Fig. 1.1, find $\cos^2 \theta$ from the inner product formula

$$\cos^2 \theta = \frac{(v^T w)^2}{(v^T v)(w^T w)}, \quad \text{with} \quad v^T w = \begin{bmatrix} 2 & 4 \end{bmatrix} \begin{bmatrix} 4 \\ 11 \end{bmatrix}.$$

1.2.6 By applying elimination to

$$A = \begin{bmatrix} 1 & 2 & 2 \\ 2 & 7 & 7 \\ 2 & 7 & 9 \end{bmatrix},$$

factor it into $A = LU$.

1.2.7 From the multiplication LS show that

$$L = \begin{bmatrix} 1 & & \\ l_{21} & 1 & \\ l_{31} & 0 & 1 \end{bmatrix} \quad \text{is the inverse of} \quad S = \begin{bmatrix} 1 & & \\ -l_{21} & 1 & \\ -l_{31} & 0 & 1 \end{bmatrix}.$$

S subtracts multiples of row 1 and L adds them back.

1.2.8 Unlike the previous exercise, show that

$$L = \begin{bmatrix} 1 & & \\ l_{21} & 1 & \\ l_{31} & l_{32} & 1 \end{bmatrix} \quad \text{is } not \text{ the inverse of} \quad S = \begin{bmatrix} 1 & & \\ -l_{21} & 1 & \\ -l_{31} & -l_{32} & 1 \end{bmatrix}.$$

If S is changed to

$$E = \begin{bmatrix} 1 & & \\ 0 & 1 & \\ 0 & -l_{32} & 1 \end{bmatrix} \begin{bmatrix} 1 & & \\ -l_{21} & 1 & \\ -l_{31} & 0 & 1 \end{bmatrix},$$

show that E is the correct inverse of L. E contains the elimination steps as they are actually done—subtractions of multiples of row 1 followed by subtraction of a multiple of row 2.

1.2.9 Find examples of 2 by 2 matrices such that

(a) $LU \neq UL$
(b) $A^2 = -I$, with real entries in A
(c) $B^2 = 0$, with no zeros in B
(d) $CD = -DC$, not allowing $CD = 0$.

1.2.10 (a) Factor A into LU and solve $Ax = b$ for the 3 right sides:

$$A = \begin{bmatrix} 1 & -1 & 0 \\ -1 & 2 & -1 \\ 0 & -1 & 2 \end{bmatrix}, \quad b = \begin{bmatrix} 1 \\ 0 \\ 0 \end{bmatrix}, \begin{bmatrix} 0 \\ 1 \\ 0 \end{bmatrix}, \begin{bmatrix} 0 \\ 0 \\ 1 \end{bmatrix}.$$

(b) Verify that your solutions x_1, x_2, x_3 are the three columns of A^{-1}. (A times this inverse matrix should give the identity matrix.)

1.2.11 True or false: Every 2 by 2 matrix A can be factored into a lower triangular L times an upper triangular U, with nonzero diagonals. Find L and U (if possible) when

$$A = \begin{bmatrix} a & b \\ c & d \end{bmatrix}.$$

1.2.12 What combination of the vectors

$$v_1 = \begin{bmatrix} 2 \\ 0 \\ 6 \end{bmatrix}, \quad v_2 = \begin{bmatrix} 3 \\ 4 \\ 9 \end{bmatrix}, \quad v_3 = \begin{bmatrix} 2 \\ 0 \\ 7 \end{bmatrix} \quad \text{gives} \quad b = \begin{bmatrix} 2 \\ -8 \\ 7 \end{bmatrix}?$$

1.2.13 What is the intersection point of the three planes $2x_1 + 3x_2 + 2x_3 = 2$, $4x_2 = -8$, $6x_1 + 9x_2 + 7x_3 = 7$?

1.2.14 What are the possibilities other than Fig. 1.2 for a 3 by 3 matrix to be singular? Draw a different figure and find a corresponding matrix A.

1.2.15 Solve by elimination and back-substitution (and row exchange):

$$\begin{bmatrix} 1 & 1 & 1 \\ 3 & 3 & 4 \\ 1 & 2 & 1 \end{bmatrix} \begin{bmatrix} x_1 \\ x_2 \\ x_3 \end{bmatrix} = \begin{bmatrix} 0 \\ 2 \\ -4 \end{bmatrix}.$$

1.2.16 Write down a 3 by 3 matrix with row $1 - 2$ row $2 +$ row $3 = 0$ and find a similar dependence of the columns—a combination that gives zero.

POSITIVE DEFINITE MATRICES AND $A = LDL^T$ ■ 1.3

The previous section went from A to an upper triangular matrix U. It used a sequence of row operations, with multiplying factors l_{ij} chosen to produce zeros below the pivots. This section looks again at those multipliers, and puts them into a matrix L. There is a pattern in the thousands (or millions) of numbers that elimination produces, and when recognized it is beautifully simple. We see it as a factorization of A into triangular matrices: A *equals* L *times* U.

Our plan is to begin by summarizing the main points of the section. It concentrates on symmetric matrices, $A = A^T$, and it takes a major step beyond the mechanics of elimination. In fact it goes one step beyond $A = LU$. The upper triangular U is split into two parts, by dividing each row by its pivot. This is easy to do—we just factor out a diagonal matrix D containing the pivots—but for symmetric matrices it has a remarkable result. The matrix it leaves is exactly L^T, the upper triangular transpose of the lower triangular L. This reappearance of L means that symmetric matrices have a symmetric factorization: A *equals* L *times* D *times* L^T.

One more thing. We have assumed that elimination could proceed without row exchanges; the entries in the pivot positions were never zero. For the most important matrices in this book something extra is true: The pivots are positive. A symmetric matrix with positive pivots is called a *positive definite matrix*, and it is those matrices that will dominate the applications. In the first application, at the end of this section, we concentrate on one of their fundamental properties: They identify the presence of a *minimum*. Thus the main points are

> **(1)**   If the multipliers l_{ij} are put into a lower triangular matrix L, with ones on the main diagonal, the original A is linked to the upper triangular U by $A = LU$
>
> **(2)**   If A is symmetric and the diagonal matrix D containing the pivots is divided out of U, it leaves the transpose of L. Thus U is linked to L by $U = DL^T$
>
> **(3)**   If the pivots are positive then A is positive definite and its *symmetric factorization* is $A = LDL^T$.

These steps will be illustrated for small matrices before going to the general case. We show that factoring A into LDL^T is the same as the method of "completing the square"—which was discovered by Lagrange in 1759, a century before the invention of matrices. It is the key to deciding when a function $F(x_1, \ldots, x_n)$ has a minimum.

We mention that Cholesky went one step further, by taking the square roots of the pivots. He combined $D^{1/2}$ with L into a new triangular matrix $\bar{L} = LD^{1/2}$. Then the other square root $D^{1/2}$ is in the transpose of $\bar{L}$, and the diagonal matrix disappears in

$$A = LD^{1/2}D^{1/2}L^T = \bar{L}\bar{L}^T.$$

This is the ***Cholesky factorization*** of a positive definite A—needing only two matrices, at the cost of computing square roots.

Matrices of Order $n = 2$

The way to understand this section is through a 2 by 2 matrix:

$$A = \begin{bmatrix} a & b \\ b & c \end{bmatrix}.$$

It requires only one elimination step and one multiplier $l_{21} = b/a$. Subtracting this multiple of row 1 from row 2 leaves the upper triangular matrix

$$U = \begin{bmatrix} a & b \\ 0 & c - b^2/a \end{bmatrix}. \tag{1}$$

Thus the pivots are $d_1 = a$ and $d_2 = c - b^2/a$.

Now we construct the lower triangular L, with unit diagonal and $l_{21} = b/a$, and verify each of the steps listed above:

(1) $LU = A$ becomes

$$\begin{bmatrix} 1 & 0 \\ b/a & 1 \end{bmatrix} \begin{bmatrix} a & b \\ 0 & c - b^2/a \end{bmatrix} = \begin{bmatrix} a & b \\ b & c \end{bmatrix}. \tag{2}$$

In other words, we return from U to A by ***adding back*** to row 2 the multiple b/a of row 1. The matrix L gives the inverse of the subtraction step, so that

$$\begin{bmatrix} 1 & 0 \\ b/a & 1 \end{bmatrix} \begin{bmatrix} 1 & 0 \\ -b/a & 1 \end{bmatrix} = \begin{bmatrix} 1 & 0 \\ 0 & 1 \end{bmatrix}, \quad \text{or} \quad LL^{-1} = I.$$

The elimination step applied L^{-1} to A; it subtracted l_{21} times row 1, and reached U. The reverse step applies L to U and recovers A.

(2)   Dividing each row of U by its pivot will factor out a diagonal matrix D:

$$U = \begin{bmatrix} a & b \\ 0 & c - b^2/a \end{bmatrix} = \begin{bmatrix} a & 0 \\ 0 & c - b^2/a \end{bmatrix} \begin{bmatrix} 1 & b/a \\ 0 & 1 \end{bmatrix} = DL^T.$$

The final matrix has ones on the diagonal and it is L^T—the transpose of the earlier L.

(3)   The factorization $A = LU = LDL^T$ becomes explicit:

$$\begin{bmatrix} a & b \\ b & c \end{bmatrix} = \begin{bmatrix} 1 & 0 \\ b/a & 1 \end{bmatrix} \begin{bmatrix} a & 0 \\ 0 & c - b^2/a \end{bmatrix} \begin{bmatrix} 1 & b/a \\ 0 & 1 \end{bmatrix}. \tag{3}$$

The symmetry of A on the left is reflected in the symmetry of the factorization on the right. In fact, the rule for transposing LDL^T guarantees that every such combination is symmetric. The transpose of any product AB is $B^T A^T$, with the transposes in reverse order. Therefore the transpose of a triple product LDL^T has the three individual transposes reversed, but since L^{TT} is L and D^T is D we come back where we started:

$$(LDL^T)^T = L^{TT} D^T L^T = LDL^T. \tag{4}$$

Thus if A has a symmetric factorization $A = LDL^T$, it must be a symmetric matrix: $A = A^T$. The real point is in the opposite direction, to determine how and when a matrix allows such a factorization with $D > 0$.

This is the additional condition on which Cholesky insisted, and we need to emphasize it properly. It is the requirement that the pivots must be positive. Not every symmetric matrix will satisfy this condition. For $n = 2$, positive pivots mean $a > 0$ and $c - b^2/a > 0$. The latter gives $c > 0$, so that both diagonal entries a and c must be positive. But positivity of the main diagonal is only a necessary condition; by itself, *a positive diagonal is not enough* for positive pivots. The condition $c > b^2/a$ also imposes a requirement on b—or on the determinant of the matrix, which is $ac - b^2$. The determinant must be positive too. In other words, it is the *sign* of the diagonal and the *size* of the off-diagonal that are jointly decisive. The matrices

$$\begin{bmatrix} 1 & 5 \\ 5 & 1 \end{bmatrix} \quad \text{and} \quad \begin{bmatrix} -2 & 1 \\ 1 & -2 \end{bmatrix}$$

are not positive definite, although $a > 0$ and $c > 0$ in the first case, and the determinant is positive in the second.

A symmetric matrix with positive pivots is a positive definite matrix. This definition applies to symmetric matrices of any order n, and it is straightforward to check if n is not large. Nevertheless it is not completely satisfactory. The formulas for the pivots become complicated as n increases, and even with $n = 2$ it would take some effort to prove the following simple property: *The sum of two positive definite matrices is positive definite.* Therefore our goal is to complement this definition by another one, equivalent but quite different—so that the second and more basic description can be used when the one based on pivots becomes unworkable.†

The Minimum of a Quadratic

For $n = 2$, the new definition of positive definite matrices will deal with the special function

$$f(x_1, x_2) = ax_1^2 + 2bx_1 x_2 + cx_2^2.$$

---

† Section 1.5 gives a third definition, again different but also equivalent; it requires A to have positive eigenvalues.

This is called a **quadratic form**, since all its terms are of second degree—they involve only squares x_i^2 or products $x_i x_j$. The link between the function f and the matrix A is immediate from a multiplication:

$$f = \begin{bmatrix} x_1 & x_2 \end{bmatrix} \begin{bmatrix} a & b \\ b & c \end{bmatrix} \begin{bmatrix} x_1 \\ x_2 \end{bmatrix}, \quad \text{or} \quad f = x^T A x. \tag{5}$$

Each off-diagonal term contributes $bx_1 x_2$, and the diagonal entries yield the squares ax_1^2 and cx_2^2. There is a complete correspondence between symmetric matrices A and quadratic forms $f = x^T A x$. For the matrix at the beginning of the book,

$$A = \begin{bmatrix} 2 & 4 \\ 4 & 11 \end{bmatrix} \quad \text{corresponds to} \quad f = 2x_1^2 + 8x_1 x_2 + 11x_2^2.$$

The new definition is based on f:

A is positive definite if $f = x^T A x$ is always positive (for $x \neq 0$).

Eventually we show that this is equivalent to positivity of the pivots.

Certainly f is zero when $x_1 = x_2 = 0$; that point is excluded in asking if $f > 0$. Also the first derivatives are zero at $x_1 = x_2 = 0$, since

$$\frac{\partial f}{\partial x_1} = 2ax_1 + 2bx_2$$

$$\frac{\partial f}{\partial x_2} = 2bx_1 + 2cx_2. \tag{6}$$

Everything is contained in the second derivatives, which come directly from the numbers a, b, and c:

$$\frac{\partial^2 f}{\partial x_1^2} = 2a, \quad \frac{\partial^2 f}{\partial x_1 \partial x_2} = 2b, \quad \frac{\partial^2 f}{\partial x_2^2} = 2c.$$

All higher derivatives are zero.

Our question is: **Does f have a minimum at $x_1 = x_2 = 0$?** The function f is zero at the origin, and we ask whether at all other points it is positive:

$$f = ax_1^2 + 2bx_1 x_2 + cx_2^2 > 0. \tag{7}$$

If this is true, its graph will be shaped like a bowl resting at the origin (Fig. 1.3). If not, the bowl might be upside down (when f has a maximum), or it can change into a saddle (when f is positive at some points and negative at others). In every case the origin is a stationary point, where the tangent is horizontal. The first derivatives are zero, and we are extending to two variables or n variables the test that comes from calculus: $f'' > 0$ at a minimum.

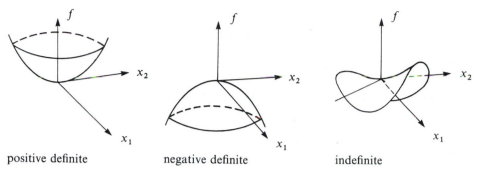

positive definite   negative definite   indefinite

Fig. 1.3. A minimum, a maximum, and a saddle point.

Test for a Minimum: $n = 2$

It is certainly necessary that $a > 0$. Otherwise we look along the x_1 axis, where $x_2 = 0$, and $f = a x_1^2$ is not positive. Similarly we need $c > 0$, or f will not increase in the x_2 direction. In other words the pure second derivatives f_{xx} and f_{yy} must be positive, imitating $f'' > 0$. But there is also a condition on b, coming from the cross derivative f_{xy}, and its sign is not what matters. Each of the quadratics

$$f_1 = x_1^2 + 10x_1 x_2 + x_2^2 \quad \text{and} \quad f_2 = x_1^2 - 10x_1 x_2 + x_2^2$$

has a saddle point rather than a minimum at the origin, even though $a = c = 1$. In the first case, $f_1 = -8$ at $x = (1, -1)$; in the second, $f_2 = -8$ at $x = (1, 1)$. These functions can be both positive and negative.

So we come to the central problem. How is it decided whether a quadratic form is everywhere positive? We answer it first for the particular example $f = 2x_1^2 + 8x_1 x_2 + 11x_2^2$, to see the key idea, and then for a general $f = x^T A x$.

The particular case has $a = 2$ and $c = 11$, both positive. But also the cross term $8x_1 x_2$ must be dominated by the others, or it will produce a saddle point. What is needed is a way of rewriting f so that it is obviously positive. The trick is to include the cross term within a perfect square, and to write f as a *sum of squares*:

$$2x_1^2 + 8x_1 x_2 + 11x_2^2 = 2(x_1 + 2x_2)^2 + 3x_2^2. \tag{8}$$

This is the step of completing the square, and the result is seen to be positive. If the cross term had been too big, say $12x_1 x_2$, the first square would be $2(x_1 + 3x_2)^2$. This uses $18x_2^2$, more than we have to give. The final term is then $(11 - 18)x_2^2$, or $-7x_2^2$. This difference of squares means a saddle point instead of a minimum.

The coefficients on the right side of (8) were not hard to find. By matching the left side they proved it to be positive—which is what we wanted. But to leave it there is to miss a chance of understanding the whole theory. Those same coefficients have *already appeared*, when elimination was applied to A and the result was written as LDL^T:

$$\begin{bmatrix} 2 & 4 \\ 4 & 11 \end{bmatrix} = \begin{bmatrix} 1 & 0 \\ 2 & 1 \end{bmatrix} \begin{bmatrix} 2 & 0 \\ 0 & 3 \end{bmatrix} \begin{bmatrix} 1 & 2 \\ 0 & 1 \end{bmatrix}. \tag{9}$$

The pivots 2 and 3 are the same numbers that are outside the squares in equation (8). The multiplier 2 in L and L^T is the same as the number inside the square. In fact all the entries of L are inside the squares, if they are written out in full:

$$x^T A x = 2(1x_1 + 2x_2)^2 + 3(0x_1 + 1x_2)^2. \tag{10}$$

The first column of L is in the first square (the coefficients 1 and 2) and the second column (0 and 1) produces the second square.

That cannot be an accident. We check first that it always happens in the 2 by 2 case, where the first term ax_1^2 and the cross term $2bx_1x_2$ are matched by the first square,

$$S = a\left(x_1 + \frac{b}{a}x_2\right)^2 = ax_1^2 + 2bx_1x_2 + \frac{b^2}{a}x_2^2.$$

The sign of S is the same as the sign of a. The coefficient of x_2^2 is adjusted in order to equal c, and this correction completes the square:

$$f = a\left(x_1 + \frac{b}{a}x_2\right)^2 + \left(c - \frac{b^2}{a}\right)x_2^2. \tag{11}$$

This identity reveals the conditions for a minimum: $f > 0$ if and only if

$$\boxed{a > 0 \quad \text{and} \quad c - \frac{b^2}{a} > 0.} \tag{12}$$

These conditions are necessary, because if either one is violated we can make the corresponding term in (11) negative, and the other term zero. They are sufficient, because if both conditions are satisfied then everything is positive (except at the origin $x_1 = x_2 = 0$).

You recognize a and $c - b^2/a$. They are the pivots, and therefore the quadratic form $x^T A x$ is positive exactly when the pivots of A are positive. The two definitions of positive definiteness coincide (at least for $n = 2$). **The function f has a minimum when the matrix A is positive definite**. It is this property that we now extend from two variables to n variables, connecting symmetric matrices A to quadratic forms

$$f = x^T A x = \begin{bmatrix} x_1 & x_2 & \cdots & x_n \end{bmatrix} \begin{bmatrix} a_{11} & a_{12} & \cdots & a_{1n} \\ a_{21} & a_{22} & \cdots & a_{2n} \\ \vdots & \vdots & & \vdots \\ a_{n1} & a_{n2} & \cdots & a_{nn} \end{bmatrix} \begin{bmatrix} x_1 \\ x_2 \\ \vdots \\ x_n \end{bmatrix}. \tag{13}$$

The notation will be more complicated but not the underlying idea. The pivots come from one algorithm—*Gaussian elimination*—and $x^T Ax > 0$ is decided by another algorithm—*completing the square*. The key is to show that those two algorithms are the same.

Elimination = Completing the Square

Mistakenly or not, I feel a duty to prove that $A = LDL^T$ for positive definite matrices of all orders. There are several possible proofs. Perhaps the quickest is by induction, from order $n - 1$ to order n; I do not think it is the most illuminating. Two proofs of $A = LU$ are in my text on *Linear Algebra and Its Applications* (the 3rd edition is published by Harcourt Brace Jovanovich, 1988). With symmetry and positive definiteness LU becomes LDL^T. Here I want to try a fresh approach. It looks at each stage of elimination as the removal of a special matrix—one whose rows are all multiples of the same row (the pivot row). That is what Gaussian elimination does. Then the sum of these special matrices, which equals A, can be written as LDL^T by a close look at matrix multiplication.

While expressing elimination as $A = LDL^T$, we are at the same time watching the signs of the pivots—to show they are positive when $f = x^T Ax$ is positive. Perhaps we should point to the main steps in advance, so that the details will be partly optional:

> (1) The matrix removed by each elimination stage has the form ldl^T, column times pivot times row
> (2) The sum of these matrices is LDL^T
> (3) If $x^T Ax > 0$ then the pivots are positive—the first is $d_1 > 0$ and the argument applies again to the remaining matrix that is one order smaller.

You will see everything from an example. The important thing is to think of elimination not only as a method for producing zeros, but as a splitting of A into simpler matrices. The example is

$$A = \begin{bmatrix} 1 & 2 & 2 \\ 2 & 7 & 7 \\ 2 & 7 & 9 \end{bmatrix}.$$

The first stage of elimination produces zeros below the pivot $d_1 = 1$. It multiplies the first row by 2 and subtracts it from the other rows. The effect is to remove the first row and column from the problem, and to leave behind a new matrix in the lower right corner:

$$\begin{bmatrix} 1 & 2 & 2 \\ 2 & 7 & 7 \\ 2 & 7 & 9 \end{bmatrix} = \begin{bmatrix} 1 & 2 & 2 \\ 2 & 4 & 4 \\ 2 & 4 & 4 \end{bmatrix} + \begin{bmatrix} 0 & 0 & 0 \\ 0 & 3 & 3 \\ 0 & 3 & 5 \end{bmatrix}. \tag{14}$$

The matrix that is removed contains multiples of row 1. Then elimination continues on the last matrix. It starts with the next pivot $d_2 = 3$, and the multiple of row 2 to be subtracted is $l_{32} = 1$. This second stage of elimination produces zeros by removing another simple matrix and leaving one that is even simpler:

$$
\begin{bmatrix} 0 & 0 & 0 \\ 0 & 3 & 3 \\ 0 & 3 & 5 \end{bmatrix} = \begin{bmatrix} 0 & 0 & 0 \\ 0 & 3 & 3 \\ 0 & 3 & 3 \end{bmatrix} + \begin{bmatrix} 0 & 0 & 0 \\ 0 & 0 & 0 \\ 0 & 0 & 2 \end{bmatrix}.
\tag{15}
$$

The last pivot is 2. Thus the original A is the sum of *three* matrices, the middle one in (14) and the two on the right side of (15). The first contains multiples of the row 1, 2, 2; the second contains multiples of 0, 1, 1; the third contains multiples of 0, 0, 1. You notice that the same is true of the columns.

We interpret this splitting in two ways. First, it completes the squares in $f = x^T A x$. Second, it gives the factorization $A = LDL^T$. In practice that is the important point. The numbers in LDL^T are a record of elimination, and to solve $Ax = b$ we call up those numbers and apply them to b—which is easy to do for this 3 by 3 matrix. It is only an example, but it illustrates the general rule.

We begin with the sum of squares. The quadratic $f = x^T A x$ is

$$
[x_1 \quad x_2 \quad x_3] \begin{bmatrix} 1 & 2 & 2 \\ 2 & 7 & 7 \\ 2 & 7 & 9 \end{bmatrix} \begin{bmatrix} x_1 \\ x_2 \\ x_3 \end{bmatrix}
$$
$$
= x_1^2 + 4x_1 x_2 + 4x_1 x_3 + 7x_2^2 + 14x_2 x_3 + 9x_3^2.
\tag{16}
$$

The question is whether this quadratic is always positive. The answer is *yes*, because the three matrices in the splitting come from perfect squares. The first matrix matches all terms involving x_1:

$$
[x_1 \quad x_2 \quad x_3] \begin{bmatrix} 1 & 2 & 2 \\ 2 & 4 & 4 \\ 2 & 4 & 4 \end{bmatrix} \begin{bmatrix} x_1 \\ x_2 \\ x_3 \end{bmatrix}
$$
$$
= x_1^2 + 4x_1 x_2 + 4x_1 x_3 + 4x_2^2 + 8x_2 x_3 + 4x_3^2
$$
$$
= (x_1 + 2x_2 + 2x_3)^2.
\tag{17}
$$

The row and column 1, 2, 2 are inside the square, and the pivot $d_1 = 1$ appears outside. The second matrix in the splitting gives a square with $d_2 = 3$ outside and 0, 1, 1 inside,

$$
3x_2^2 + 6x_2 x_3 + 3x_3^2 = 3(x_2 + x_3)^2.
$$

This matches all remaining terms involving x_2. Then the third matrix gives $2x_3^2$. These three squares add up to f in (16), which is therefore positive.

Exactly the same numbers appear in the symmetric factorization $A = LDL^T$:

$$A = \begin{bmatrix} 1 & & \\ 2 & 1 & \\ 2 & 1 & 1 \end{bmatrix} \begin{bmatrix} 1 & & \\ & 3 & \\ & & 2 \end{bmatrix} \begin{bmatrix} 1 & 2 & 2 \\ & 1 & 1 \\ & & 1 \end{bmatrix}.$$

Thus the factorization into LDL^T is essentially the same as the splitting into simple matrices. That is still to be proved in general, but we can settle the question of positive pivots: For any symmetric matrix they signal the presence of a minimum.

1B   The pivots are positive if and only if $x^T A x$ is positive for all nonzero vectors x. In this case the symmetric matrix A is positive definite and $f = x^T A x$ has a minimum at $x = 0$.

The first matrix to be removed from A is M_1. It contains multiples of the first row of A. It uses the first pivot a_{11}, which we call d, and chooses the multiples to match A in the first column. The rows of M_1 are

row 1:	1	times	$\begin{bmatrix} d & a_{12} & \cdots & a_{1n} \end{bmatrix}$
row 2:	$\dfrac{a_{21}}{d}$	times	$\begin{bmatrix} d & a_{12} & \cdots & a_{1n} \end{bmatrix}$
$\vdots$			
row n:	$\dfrac{a_{n1}}{d}$	times	$\begin{bmatrix} d & a_{12} & \cdots & a_{1n} \end{bmatrix}.$

The first column contains $d, a_{21}, \ldots, a_{n1}$ and agrees with the first column of A. But symmetry gives more than that. The column entries are the same as the row entries. If the first column is l, the row in brackets is d times l^T. Therefore we can write M_1 in a most remarkable way, as a column times a row:

$$M_1 = \begin{bmatrix} 1 \\ a_{21}/d \\ \vdots \\ a_{n1}/d \end{bmatrix} \begin{bmatrix} d & a_{12} & \cdots & a_{1n} \end{bmatrix} = ldl^T. \tag{18}$$

Notice above all that *a column times a row is a matrix*. In our example

$$M_1 = \begin{bmatrix} 1 & 2 & 2 \\ 2 & 4 & 4 \\ 2 & 4 & 4 \end{bmatrix} = \begin{bmatrix} 1 \\ 2 \\ 2 \end{bmatrix} \begin{bmatrix} 1 & 2 & 2 \end{bmatrix}$$

In general an n by 1 matrix (a column) times a 1 by n matrix (a row) is an n by n

matrix. If A is not symmetric then the pattern is the same but the row is different from the column. They lead to $A = LU$.

When M_1 is removed from A it leaves a block of order $n - 1$. There are zeros in the first row and column; we call this new matrix B. The next elimination step removes from B another matrix M_2, leaving zeros in two rows and columns and a block of order $n - 2$. This continues until $M_1, M_2, \ldots, M_n$ have been removed and nothing is left. Equation (14) was an example of $A = M_1 + B$, and equation (15) was an example of $B = M_2 + M_3$. Together they split A into $M_1 + M_2 + M_3$.

We now connect this splitting to a sum of squares, to prove 1B. If $x^T A x$ is positive then certainly the first pivot is positive. The choice $x^T = [1 \quad 0 \quad \cdots \quad 0]$ forces $x^T A x = d > 0$. This is the necessary condition seen before, that the main diagonal must be positive. Then to display the whole $x^T A x$ as positive we need n squares, and **the first square is** $x^T M_1 x$. This matches correctly all the terms that involve x_1. It is like $(x_1 + 2x_2 + 2x_3)^2$, which matched the first terms of (16). It does not match the remaining terms, which will be adjusted by the remaining squares. The next square is $x^T M_2 x$, which matches all the terms in x_2.

Removing M_1 from A leaves a smaller problem governed by B. For the block in B, involving only $n - 1$ variables and ignoring the first row and column of zeros, we assume that the result is already proved: Its pivots are positive if and only if $x^T B x$ is positive.† The pivots of this block are the last $n - 1$ pivots of A; elimination continues and these pivots appear. At the same time, $x^T A x$ is positive in dimension n if and only if $x^T B x$ is positive in dimension $n - 1$. (If the latter failed, we could choose x_1 to make the first square zero and then the former would fail.) Therefore the result assumed true for B is also true for A: *The pivots are positive if and only if the function $x^T A x$ is positive.* That completes the proof of 1B.

It follows immediately that if two matrices A_1 and A_2 are positive definite so is $A_1 + A_2$: it passes the test

$$x^T(A_1 + A_2)x = x^T A_1 x + x^T A_2 x > 0.$$

This test succeeds easily, while the pivots are much harder to follow. The pivots of $A_1 + A_2$ are not the pivots of A_1 plus the pivots of A_2 (except for the first one). But they must all be positive.

Elimination = Factorization

The two algorithms, one which eliminated x_1 from $Ax = b$ and the other which eliminated it from $x^T A x$, gave the same matrix B. As they continue, one produces pivots and multipliers and $A = LDL^T$, and the other expresses $x^T A x$ as a sum of squares. We now compare those two end products. It will be like the 2 by 2 and 3 by 3 examples, where the pivots were outside the squares and the multipliers were

---

† This is the "induction hypothesis." The result is already known to be correct for $n = 1$ (and $n = 2$), and we go from each size to the next larger size.

inside. We will be proving that elimination is equivalent to the factorization $A = LU = LDL^T$, and if you stay with it you will be through the longest theoretical section of the book. This is one case in which the theory is basic to numerical applications.

1C   Every symmetric positive definite matrix allows a symmetric factorization

$$A = LDL^T = l_1 d_1 l_1^T + \cdots + l_n d_n l_n^T. \tag{19}$$

The numbers $d_i > 0$ are the pivots in D, and the vectors l_i are the columns of the multiplier matrix L.

Everything rests on one point: The elimination step from A to B is achieved by removing the n by n matrix

$$M_1 = \begin{bmatrix} 1 \\ a_{21}/d \\ \vdots \\ a_{n1}/d \end{bmatrix} d[1 \quad a_{12}/d \quad \cdots \quad a_{1n}/d] = ldl^T.$$

This vector l (or l_1) is the first column of L—including the 1 on the diagonal. The vector dl^T (or $d_1 l_1^T$) is the first row of U. The matrix M_1 is symmetric and removing it leaves a symmetric matrix B. Then the first pivot of B is d_2, its multipliers fill the column l_2, and elimination continues all the way to a block of order one:

$$A \rightarrow \begin{bmatrix} 0 & 0 \\ 0 & * \end{bmatrix} \rightarrow \begin{bmatrix} 0 & 0 \\ 0 & * \end{bmatrix} \rightarrow \cdots \rightarrow \begin{bmatrix} 0 & 0 \\ 0 & d_n \end{bmatrix} = l_n d_n l_n^T.$$
$$\quad\; 1 \quad n-1 \qquad 2 \quad n-2 \qquad\qquad n-1 \quad 1$$

Since $M_j = l_j d_j l_j^T$ is removed at every step, A is decomposed into

$$\boxed{A = l_1 d_1 l_1^T + l_2 d_2 l_2^T + \cdots + l_n d_n l_n^T.}$$

All that remains is to recognize this as the matrix multiplication LDL^T. It involves a sum of columns times rows—the columns of L times the rows of L^T—which is an unusual way to multiply matrices! But it is legal, and it fits perfectly with the steps of Gaussian elimination.

Matrix Multiplication

The usual order of matrix multiplication is "row times column." If the matrices are A and B, then row 1 of A multiplies column 1 of B to give the (1, 1) entry in the corner of AB. This product of vectors is an *inner* product, combining n ordinary multiplications of a_{1j} times b_{j1}. If A and B are n by n, there will be n^2 entries in AB and they normally involve n^3 ordinary multiplications.

However the order of these multiplications is not sacred. Watch a_{11}, when row 1 of A multiplies the columns of B. In these inner products, a_{11} always multiplies a number in the first row of B. Similarly a_{12} multiplies the second row of B. Sooner or later, a_{1n} will multiply the last row of B. The addition of all these products involving row 1 of A gives row 1 of AB:

$$\text{row 1 of } AB = [a_{11} \quad \cdots \quad a_{1n}] \begin{bmatrix} \text{row 1 of } B \\ \vdots \\ \text{row } n \text{ of } B \end{bmatrix}$$

$$= a_{11}(\text{row 1 of } B) + \cdots + a_{1n}(\text{row } n \text{ of } B).$$

This is matrix multiplication "a row at a time." If we write all the rows of AB in this way, and assemble the results, the product AB looks like

$$AB = \begin{bmatrix} a_{11}(\text{row 1 of } B) + \cdots + a_{1n}(\text{row } n \text{ of } B) \\ \vdots \\ a_{n1}(\text{row 1 of } B) + \cdots + a_{nn}(\text{row } n \text{ of } B) \end{bmatrix}.$$

Now you can see another possibility. This expression can be split up in a different way, without changing the final result. The order of multiplication can become

$$AB = \begin{bmatrix} a_{11} \\ \vdots \\ a_{n1} \end{bmatrix} [\text{row 1 of } B] + \cdots + \begin{bmatrix} a_{1n} \\ \vdots \\ a_{nn} \end{bmatrix} [\text{row } n \text{ of } B]. \tag{20}$$

We state this as a general rule and then test it on a specific example.

1D   The matrix AB can be written as a sum of "outer products" of columns times rows:

$$AB = (\text{column 1 of } A)(\text{row 1 of } B) + \cdots + (\text{column } n \text{ of } A)(\text{row } n \text{ of } B). \tag{21}$$

Each product is an n by n matrix of rank one and requires n^2 ordinary multiplications.

EXAMPLE

$$\begin{bmatrix} 1 & 0 \\ 2 & 1 \end{bmatrix} \begin{bmatrix} 2 & 4 \\ 0 & 3 \end{bmatrix} = \begin{bmatrix} 1 \\ 2 \end{bmatrix} [2 \quad 4] + \begin{bmatrix} 0 \\ 1 \end{bmatrix} [0 \quad 3]$$

$$= \begin{bmatrix} 2 & 4 \\ 4 & 8 \end{bmatrix} + \begin{bmatrix} 0 & 0 \\ 0 & 3 \end{bmatrix} = \begin{bmatrix} 2 & 4 \\ 4 & 11 \end{bmatrix}.$$

This is almost $A = LDL^T$. The matrices on the left are L and U. In the middle is the matrix M_1 that was removed by elimination, and the matrix M_2 that was left. The remaining step is to divide out the pivots. Instead of U we want DL^T, instead of LU we want LDL^T, and instead of column times row we want **column** times **pivot** times **row**:

$$\begin{bmatrix} 2 & 4 \\ 4 & 11 \end{bmatrix} = \begin{bmatrix} 1 \\ 2 \end{bmatrix} [2] [1 \quad 2] + \begin{bmatrix} 0 \\ 1 \end{bmatrix} [3] [0 \quad 1]$$

$$= \begin{bmatrix} 1 & 0 \\ 2 & 1 \end{bmatrix} \begin{bmatrix} 2 & 0 \\ 0 & 3 \end{bmatrix} \begin{bmatrix} 1 & 2 \\ 0 & 1 \end{bmatrix}. \tag{22}$$

This is $A = LDL^T$, a sum of the simple matrices $l_j d_j l_j^T$.

Block Matrices

That completes a long section, and we add only two remarks. The first is to mention that matrices can also be multiplied "a column at a time." If the columns of B are $b_1, \ldots, b_n$, then

$$AB = A[b_1 \quad \cdots \quad b_n] = [Ab_1 \quad \cdots \quad Ab_n]. \tag{23}$$

Each column of AB is actually a combination of the columns of A.

The second remark is in the same direction, but more general. In practice, the entries of a matrix might come in blocks. The matrix may even be block diagonal (with square blocks of different sizes) or block triangular (with rectangular blocks off the diagonal). Provided the shapes match, we can carry out **block multiplication**. The blocks can be treated like numbers, except that BA^{-1} replaces b/a and the order of multiplication is important—since $AB \neq BA$ for most matrices. In block elimination of a symmetric matrix, for example,

$$\begin{bmatrix} I & 0 \\ -B^T A^{-1} & I \end{bmatrix} \begin{bmatrix} A & B \\ B^T & C \end{bmatrix} = \begin{bmatrix} A & B \\ 0 & C - B^T A^{-1} B \end{bmatrix}.$$

The block pivots are A and $C - B^T A^{-1} B$, instead of a and $c - b^2/a$. Inverting the first matrix gives a block form of $A = LU = LDL^T$:

$$\begin{bmatrix} A & B \\ B^T & C \end{bmatrix} = \begin{bmatrix} I & 0 \\ B^T A^{-1} & I \end{bmatrix} \begin{bmatrix} A & B \\ 0 & C - B^T A^{-1} B \end{bmatrix}$$

$$= \begin{bmatrix} I & 0 \\ B^T A^{-1} & I \end{bmatrix} \begin{bmatrix} A & 0 \\ 0 & C - B^T A^{-1} B \end{bmatrix} \begin{bmatrix} I & A^{-1} B \\ 0 & I \end{bmatrix}. \tag{24}$$

The inverse in Exercise 1.3.12 is also a 2 by 2 matrix of blocks.

We apologize for all this matrix algebra, but without LDL^T there is no good way to solve symmetric systems.

EXERCISES

1.3.1   Write $A = \begin{bmatrix} 3 & -3 \\ -3 & 5 \end{bmatrix}$ in the forms $A = LDL^T$ and $A = l_1 d_1 l_1^T + l_2 d_2 l_2^T$. Are the pivots positive, so that A is symmetric positive definite? Write $3x_1^2 - 6x_1 x_2 + 5x_2^2$ as a sum of squares.

1.3.2   Factor $A = \begin{bmatrix} 3 & 6 \\ 6 & 8 \end{bmatrix}$ into $A = LDL^T$. Is this matrix positive definite? Write $x^T Ax$ as a combination of two squares.

1.3.3   Find the triangular factors L and U of

$$A = \begin{bmatrix} 1 & 1 & 0 & 0 \\ 1 & 2 & 1 & 0 \\ 0 & 1 & 2 & 1 \\ 0 & 0 & 1 & 2 \end{bmatrix}.$$

In this case U is the same as L^T. What is the pivot matrix D? Solve $Lc = b$ and $Ux = c$ if $b = (1, 0, 0, 0)$.

1.3.4   How do you know from elimination that the rows of L always start with the same zeros as the rows of A? *Note*: Zeros inside the central band of A may be lost by L; this is the "fill-in" that is painful for sparse matrices.

1.3.5   Write $f = x_1^2 + 10x_1 x_2 + x_2^2$ as a difference of squares, and $f = x_1^2 + 10x_1 x_2 + 30x_2^2$ as a sum of squares. What symmetric matrices correspond to these quadratic forms by $f = x^T Ax$?

1.3.6   In the 2 by 2 case, suppose the positive coefficients a and c dominate b in the sense that $a + c > 2b$. Is this enough to guarantee that $ac > b^2$ and the matrix is positive definite? Give a proof or a counterexample.

1.3.7   Decide for or against the positive definiteness of

$$A = \begin{bmatrix} 1 & 1 & 1 \\ 1 & 1 & 1 \\ 1 & 1 & 1 \end{bmatrix} \quad \text{and} \quad A' = \begin{bmatrix} 1 & 1 & 1 \\ 1 & 2 & 2 \\ 1 & 2 & 3 \end{bmatrix}.$$

Write A as $l_1 d_1 l_1^T$ and write A' as LDL^T.

1.3.8   If each diagonal entry a_{ii} is larger than the sum of the absolute values $|a_{ij}|$ along the rest of its row, then the symmetric matrix A is positive definite. How large would c have to be in

$$A = \begin{bmatrix} c & 1 & 1 \\ 1 & c & 1 \\ 1 & 1 & c \end{bmatrix}$$

for this statement to apply? How large does c actually have to be to assure that A is positive definite? Note that

$$x^T A x = (x_1 + x_2 + x_3)^2 + (c - 1)(x_1^2 + x_2^2 + x_3^2);$$

when is this positive?

1.3.9   (i)   The determinant of a triangular matrix is the product of the entries on the diagonal. Thus det $L = 1$ and

$$\det A = \det LDL^T = (\det L)(\det D)(\det L^T) = \det D.$$

The determinant is the *product of the pivots*. Show that det $A > 0$ if A is positive definite.
   (ii)   Give an example with det $A > 0$ in which A is *not* positive definite.
   (iii)   What is the determinant of A in Exercise 1.3.3?

1.3.10   Inverting $A = LDL^T$ gives $A^{-1} = MD^{-1}M^T$, where M is the inverse of L^T. Is M lower triangular or upper triangular? How could you factor A itself, so that the first factor is upper and not lower triangular?

1.3.11   A function $F(x, y)$ has a local minimum at any point where its first derivatives $\partial F / \partial x$ and $\partial F / \partial y$ are zero and the matrix of second derivatives

$$A = \begin{bmatrix} \dfrac{\partial^2 F}{\partial x^2} & \dfrac{\partial^2 F}{\partial x \partial y} \\[3mm] \dfrac{\partial^2 F}{\partial x \partial y} & \dfrac{\partial^2 F}{\partial y^2} \end{bmatrix}$$

is positive definite. Is this true for $F_1 = x^2 - x^2 y^2 + y^2 + y^3$ and $F_2 = \cos x \cos y$ at $x = y = 0$? Does F_1 have a global minimum or can it approach $-\infty$?

1.3.12   Find the inverse of the 2 by 2 symmetric matrix $\begin{bmatrix} a & b \\ b & c \end{bmatrix}$. Verify by direct multiplication that the inverse of a 2 by 2 block symmetric matrix is

$$\begin{bmatrix} A & B \\ B^T & C \end{bmatrix}^{-1} = \begin{bmatrix} A^{-1} + A^{-1} B S B^T A^{-1} & -A^{-1} B S \\ -S B^T A^{-1} & S \end{bmatrix},$$

where $S = (C - B^T A^{-1} B)^{-1}$. A and C are square but B can be rectangular.

1.3.13   For the block quadratic form

$$f = [x^T \quad y^T] \begin{bmatrix} A & B \\ B^T & C \end{bmatrix} \begin{bmatrix} x \\ y \end{bmatrix} = x^T A x + x^T B y + y^T B^T x + y^T C y,$$

find the term that completes the square:

$$f = (x + A^{-1} B y)^T A (x + A^{-1} B y) + y^T (\quad ? \quad) y.$$

The block matrix is positive definite when A and $C - B^T A^{-1} B$ are positive definite.

1.3.14 The rule for block multiplication of AB seems to be: Vertical cuts in A must be matched by horizontal cuts in B, while other cuts (horizontal in A or vertical in B) can be arbitrary. Examples for 3 by 3 matrices are

$$\left[\begin{array}{c|c|c} \times & \times & \times \\ \times & \times & \times \\ \times & \times & \times \end{array}\right] \left[\begin{array}{ccc} \times & \times & \times \\ \times & \times & \times \\ \times & \times & \times \end{array}\right] \qquad \left[\begin{array}{ccc} \times & \times & \times \\ \times & \times & \times \\ \times & \times & \times \end{array}\right] \left[\begin{array}{c|c|c} \times & \times & \times \\ \times & \times & \times \\ \times & \times & \times \end{array}\right]$$

$$\text{column times row} \qquad\qquad \text{row times column}$$

$$\left[\begin{array}{cc|c} \times & \times & \times \\ \hline \times & \times & \times \\ \times & \times & \times \end{array}\right] \left[\begin{array}{cc|c} \times & \times & \times \\ \times & \times & \times \\ \hline \times & \times & \times \end{array}\right]$$

$$\text{matching blocks}$$

Give two more examples and put in numbers to confirm that the multiplication succeeds.

1.3.15 Find the LDL^T factorization, and Cholesky's $\bar{L}\bar{L}^T$ factorization with $\bar{L} = LD^{1/2}$, for the matrix

$$A = \begin{bmatrix} 4 & 12 \\ 12 & 45 \end{bmatrix}.$$

What is the connection to $x^T Ax = (2x_1 + 6x_2)^2 + (3x_2)^2$?

1.3.16 Suppose

$$A = \begin{bmatrix} 1 & & \\ 2 & 1 & \\ 4 & 2 & 1 \end{bmatrix} \begin{bmatrix} 1 & & \\ & 3 & \\ & & 1 \end{bmatrix} \begin{bmatrix} 1 & 2 & 4 \\ & 1 & 2 \\ & & 1 \end{bmatrix} \quad \text{and} \quad b = \begin{bmatrix} 0 \\ 6 \\ 1 \end{bmatrix}.$$

Solve $Ax = b$ by solving two triangular systems. How do you know that A is symmetric positive definite?

1.3.17 If $a_{11} = d$ is the first pivot of A, under what condition is a_{22} the second pivot? Are the pivots of A^{-1} equal to the reciprocals $1/d_i$ of the pivots of A?

1.3.18 Write down in words the sequence of column operations (!) by which the following code computes the usual multipliers l_{ij}. It overwrites a_{ij} with these multipliers for $i > j$, using no extra storage. The notation := signals a definition in terms of existing quantities.

Symmetric factorization LDL^T for positive definite A

For $j = 1, \ldots, n$

    For $p = 1, \ldots, j-1$

        $r_p := d_p a_{jp}$

    $d_j := a_{jj} - \displaystyle\sum_{p=1}^{j-1} a_{jp} r_p$

If $d_j = 0$

> *then* quit

> *else*

> > For $i = j + 1, \ldots, n$

$$a_{ij} := \left(a_{ij} - \sum_{p=1}^{j-1} a_{ip} r_p \right) \Big/ d_j$$

The algorithm requires about $n^3/6$ multiplications.

1.3.19   Explain (and if possible code) the following solution of $Ax = b$ for *positive definite tridiagonal* A. The diagonal of A is originally in $d_1, \ldots, d_n$ and the subdiagonal and superdiagonal in $l_1, \ldots, l_{n-1}$; the solution x overwrites b.

> For $k = 2, \ldots, n$
>
> > $t := l_{k-1}$
> >
> > $l_{k-1} := t/d_{k-1}$
> >
> > $d_k := d_k - t l_{k-1}$
>
> For $k = 2, \ldots, n$
>
> > $b_k := b_k - l_{k-1} b_{k-1}$
>
> For $k = 1, \ldots, n$
>
> > $b_k := b_k/d_k$
>
> For $k = n - 1, \ldots, 1$
>
> > $b_k := b_k - l_k b_{k+1}$

Show how this uses $5n$ multiplications or divisions, and give an example of failure when A is not positive definite.

1.3.20.   If a new row v^T is added to A, what is the change in $A^T A$?

1.4 ■ MINIMUM PRINCIPLES

So far the unknown x has appeared in a linear system $Ax = b$. Now those systems will have competition. They will still be present at the end, when x is actually computed, but the basic statement of the problem will not be a linear equation. Instead, x will be determined by a minimum principle.

We introduce minimum principles through two examples. The first starts with equations that have no solution; the original matrix A is rectangular, and there are more equations than unknowns. In that case it is generally impossible to solve $Ax = b$. Nevertheless such systems constantly appear in applications, and the unknown x has to be determined; it will be chosen to make Ax as close as possible to b. It minimizes the quadratic $\| Ax - b \|^2$, and the choice of x which makes this error a minimum is the *least squares solution* to $Ax = b$.

The second example is chosen from a thousand possibilities in engineering and science. It is based on a fundamental principle of mechanics, that nature acts so as to *minimize energy*. The system looks for a point of *equilibrium*, and an equilibrium is successful only if it is stable. It should require the input of energy to move away from equilibrium; otherwise, if a movement releases rather than consumes energy, that movement will grow.† It is like a ball on a sphere (or a child on a slide). Balanced at the top, the ball is unstable; the steady state is at the bottom, where the potential energy is a minimum.

Fortunately, the energy is often a quadratic. Its minimum occurs where the derivative is zero, and the derivative of a quadratic function is linear. Therefore the equilibrium point is determined by a system of linear equations. The scalar case, with only one degree of freedom, is simple but typical: the energy is described by a parabola $P(x)$, so that

$$P(x) = \tfrac{1}{2}Ax^2 - bx \quad \text{and} \quad \frac{dP}{dx} = Ax - b.$$

The derivative is zero at $x = A^{-1}b$. Provided A is positive this equilibrium is stable; the parabola opens upward. In more dimensions x becomes a vector, A is a symmetric matrix, and the parabola becomes a paraboloid (Fig. 1.4). But the minimum still occurs where $Ax = b$:

1E   If A is positive definite, then the quadratic $P(x) = \tfrac{1}{2}x^T Ax - x^T b$ is minimized at the point where $Ax = b$. The minimum value is $P(A^{-1}b) = -\tfrac{1}{2}b^T A^{-1}b$.

Proof   Suppose x is the solution to $Ax = b$. We want to show that at any other point y, $P(y)$ is larger than $P(x)$. Their difference is

---

† I think this also applies to us.

$$P(y) - P(x) = \tfrac{1}{2} y^T A y - y^T b - \tfrac{1}{2} x^T A x + x^T b$$
$$= \tfrac{1}{2} y^T A y - y^T A x + \tfrac{1}{2} x^T A x$$
$$= \tfrac{1}{2}(y - x)^T A (y - x). \tag{1}$$

Since A is positive definite, the last expression can never be negative; it is zero only if $y - x = 0$. Therefore $P(y)$ is larger than $P(x)$, and the minimum occurs at $x = A^{-1}b$. At that point

$$P_{\min} = \tfrac{1}{2}(A^{-1}b)^T A (A^{-1}b) - (A^{-1}b)^T b = -\tfrac{1}{2} b^T A^{-1} b, \tag{2}$$

completing the proof.

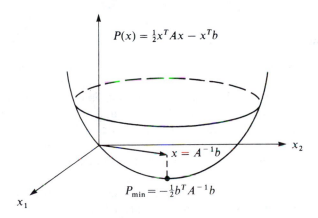

Fig. 1.4. The minimum of a quadratic function $P(x)$.

The equilibrium point can also be determined from calculus. The first derivatives $\partial P / \partial x_i$ must be zero, and we take the typical example

$$P = \tfrac{1}{2}[x_1 \ \ x_2] \begin{bmatrix} 2 & 3 \\ 3 & 5 \end{bmatrix} \begin{bmatrix} x_1 \\ x_2 \end{bmatrix} - [x_1 \ \ x_2] \begin{bmatrix} 7 \\ 8 \end{bmatrix}$$
$$= \tfrac{1}{2}(2x_1^2 + 6x_1 x_2 + 5x_2^2) - 7x_1 - 8x_2$$
$$\frac{\partial P}{\partial x_1} = 2x_1 + 3x_2 - 7 = 0$$
$$\frac{\partial P}{\partial x_2} = 3x_1 + 5x_2 - 8 = 0$$

The vanishing of the derivatives has produced $Ax = b$. Then to verify that this x gives a minimum, calculus looks also at the second derivatives. It is here that we use the positive definiteness of A, and the work of the previous section. The diagonal entries 2 and 5 are positive, and so is the determinant. For $n = 2$ this guarantees a minimum.

A third very simple proof of 1E comes from rewriting P as

$$\tfrac{1}{2}x^T A x - x^T b = \tfrac{1}{2}(x - A^{-1}b)^T A (x - A^{-1}b) - \tfrac{1}{2}b^T A^{-1}b.$$

This immediately puts the minimum at $x = A^{-1}b$, to bring the first term on the right side down to zero. That leaves only the constant term $P_{min} = -\tfrac{1}{2}b^T A^{-1}b$. It is interesting that the minimum is never positive.

The geometry is very straightforward. For $b = 0$ the minimum occurs at the origin $x = 0$. For other vectors b, the graph of P is shifted until it is centered at $A^{-1}b$, and it is lowered so that the value of P at equilibrium (the minimum energy) is P_{min}. The graph of $P(x)$ goes through $P(0) = 0$, so the minimum is not above zero. The cross section of the bowl in Fig. 1.4 is an ellipse.

Positive Definiteness of $A^T A$ and $A^T C A$

The examples share one feature which we have not been brave enough to mention. They do not start with a positive definite matrix A! In fact the underlying matrix, in these examples and throughout the book, is very frequently *rectangular*. It might give the connection between the nodes and edges of a network, and the number of nodes is different from the number of edges. But when everything is put together, and we come to an equation of equilibrium, we do get back to n equations in n unknowns—with a square coefficient matrix. That matrix is not the original A. Instead it is $A^T A$, or more often $A^T C A$.

You must recognize the difference between this section and the last one. There we started with a square matrix and factored it into LDL^T. The equations were given, and the problem was to solve them. That is one side of applied mathematics. The other side—which is probably more important and harder—is to find the equations in the first place. We are given physical laws that connect voltage with current, or pressure with flow. The geometry adds laws of its own, determined by the connections between nodes and edges. From these laws, with the physics represented by C and the geometry by A, come the equations of equilibrium and the minimization of energy. The coefficient matrix gets assembled as $A^T C A$, in formulating the equations, and then it gets disassembled into LDL^T, in solving them.

The matrices $A^T A$ and $A^T C A$ are always symmetric (if C is) and we need to know when they are positive definite.

1F (i) If A has linearly independent columns—it can be square or rectangular—then the product $A^T A$ is symmetric positive definite.
  (ii) If also C is symmetric positive definite, so is the triple product $A^T C A$.

The first is a special case of the second, with $C = I$. (The identity matrix is certainly positive definite: its pivots are all 1, its eigenvalues are all 1, and $x^T I x = x_1^2$

$+ \cdots + x_n^2$ is positive except at $x = 0$.) If the matrix A is m by n, then linear independence of the columns forces $n \leq m$. It cannot have more columns than rows, or there would be no chance that the columns were independent. (The number of independent columns equals the number of independent rows, by the fundamental theorem of linear algebra.) A typical example is

$$A^T C A = \begin{bmatrix} 1 & 1 & 0 \\ 2 & 2 & 1 \end{bmatrix} \begin{bmatrix} 1 & 0 & 0 \\ 0 & 1 & 0 \\ 0 & 0 & 3 \end{bmatrix} \begin{bmatrix} 1 & 2 \\ 1 & 2 \\ 0 & 1 \end{bmatrix} = \begin{bmatrix} 2 & 4 \\ 4 & 11 \end{bmatrix},$$

with $m = 3$ rows, $n = 2$ columns (independent!), and an m by m positive definite matrix C in the middle. The final product $A^T C A$ is the n by n matrix at the start of this book.

To prove that $A^T A$ is positive definite we test its quadratic form:

$$\boxed{x^T A^T A x \text{ is the same as } (Ax)^T(Ax) \text{ or } \|Ax\|^2.}$$

This is the square of the length of Ax. It is positive unless $Ax = 0$. But now enters the independence of the columns: The combination Ax of the columns is never zero except in the unavoidable case $x = 0$. Thus $x^T A^T A x$ is positive except if $x = 0$. That makes $A^T A$ positive definite.†

For $A^T C A$, the quadratic form is again positive: $x^T A^T C A x = (Ax)^T C(Ax)$ and C is assumed positive definite. The only way to obtain zero is from $Ax = 0$. Appealing as above to the independence of the columns of A, $Ax = 0$ happens only if $x = 0$. Thus $A^T C A$ is positive definite.

We now apply these results to the examples.

EXAMPLE 1:   *Least Squares Solution of $Ax = b$*

We are given m equations in n unknowns, with $m > n$. The problem is overdetermined, and this system is not likely to have a solution. It is easy to describe the right hand sides for which $Ax = b$ can be solved; the equation asks that they be combinations of the columns of A, since the matrix-vector multiplication Ax gives

$$x_1(\text{column } 1) + \cdots + x_n(\text{column } n) = b.$$

The vectors b that can be obtained this way form an n-dimensional subspace—it is the **column space** of A. This subspace lies within m-dimensional space; all the columns have m components. Thus $Ax = b$ has a solution only if b lies in this n-dimensional subspace of m-dimensional space—and with $n < m$, that is unlikely. A subspace is very thin.

---

† Its near relative AA^T is *not* positive definite (Exercise 1.4.9).

For example, suppose the system $Ax = b$ is

$$
\begin{aligned}
C + t_1 D &= b_1 \\
C + t_2 D &= b_2 \\
C + t_3 D &= b_3 \\
C + t_4 D &= b_4
\end{aligned}
\quad \text{or} \quad
\begin{bmatrix} 1 & 0 \\ 1 & 1 \\ 1 & 3 \\ 1 & 4 \end{bmatrix}
\begin{bmatrix} C \\ D \end{bmatrix}
=
\begin{bmatrix} b_1 \\ b_2 \\ b_3 \\ b_4 \end{bmatrix}.
\tag{3}
$$

These equations ask for a line $y = C + Dt$ that passes through four given points; it tries for $y = b_1$ at $t = 0$, $y = b_2$ at $t = 1$, $y = b_3$ at $t = 3$, and $y = b_4$ at $t = 4$. Normally that is impossible (Fig. 1.5). The four equations can be solved only if the four points happen to fall on a line. Then the vector b is a combination of the columns of A. The b_i might be equal, giving a multiple of the first column and a horizontal line $y = C$; they might be proportional to the second column 0, 1, 3, 4, corresponding to a special line $y = Dt$; or they might be a combination of the two. The choice $b = (0, 8, 8, 20)$ is none of these, and **there will be an error $e = b - Ax$.†**

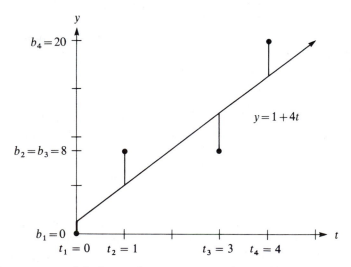

Fig. 1.5. The nearest straight line to the measurements 0, 8, 8, 20.

The best line is the one that makes e as small as possible. To minimize the error is to minimize $\| Ax - b \|^2 = (Ax - b)^T (Ax - b)$, if we measure a vector in the usual way: The square of its length is

$$
\| e \|^2 = e^T e = e_1^2 + e_2^2 + \cdots + e_n^2.
\tag{4}
$$

For a right triangle this is just Pythagoras' formula for the hypotenuse. In n

---

† The usual notation in statistics is $e = Y - X\beta$. We emphasize that the errors e_i are the vertical distances to the line, not the shortest distances.

dimensions there are n perpendicular sides, and $\|e\|$ is the length of the diagonal in an n-dimensional box. An n-dimensional unit cube has a diagonal $(1, 1, ..., 1)$ of length $\sqrt{n}$.

Which vector x will minimize $(Ax - b)^T(Ax - b) = x^TA^TAx - x^TA^Tb - b^TAx + b^Tb$? This is a quadratic, and we can answer that question by 1E. The constant term b^Tb will not affect the minimization and can be ignored. The cross term b^TAx is the same as x^TA^Tb, since for real vectors like b and Ax the order in the inner product is irrelevant. If we divide by two, we reach the minimization of $P = \frac{1}{2}x^TA^TAx - x^TA^Tb$.† That is the problem solved earlier for $\frac{1}{2}x^TAx - x^Tb$ and we can copy that solution—except that the matrix is no longer A but A^TA, and the vector multiplying x is no longer b but A^Tb. This changes $Ax = b$ into $A^TAx = A^Tb$:

1G   The vector x that minimizes $\|Ax - b\|^2$ is the solution to the **normal equations**

$$A^TAx = A^Tb. \tag{5}$$

This vector $x = (A^TA)^{-1}A^Tb$ is the **least squares solution** to $Ax = b$.

The error $e = b - Ax$ will not be zero, but its inner product with every column of A is zero. That gives the normal equations $A^Te = 0$, or $A^TAx = A^Tb$. The error e is perpendicular to the whole column space of A, and b is decomposed into

$$b = Ax + (b - Ax) = \text{closest point in column space} + \text{error}.$$

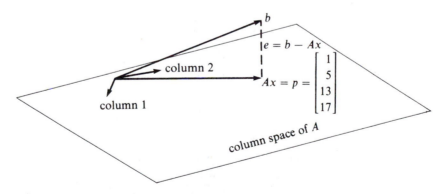

Fig. 1.6. The nearest point to b is $p = $ column $1 + 4$ (column 2).

This reflects what is clear geometrically, that if Ax is the closest point to b in the column space, then the line from b to Ax is perpendicular to that space (Fig. 1.6). It

---

† The factor $\frac{1}{2}$ has no special place in least squares. For physical problems it goes naturally into the definition of energy.

is natural to call this closest point p the **projection** of b onto the column space. Algebraically, the rule is easy to summarize:

> if $Ax = b$ has no solution, multiply by A^T and solve $A^T Ax = A^T b$.

Fitting Measurements by a Straight Line

We return to the four points $(0, 0)$, $(1, 8)$, $(3, 8)$, $(4, 20)$, which are not in a straight line. Equation (3) looks for such a line, but from Fig. 1.5 it is hopeless; the four equations have no solution. There is no choice $y = C + Dt$ which will fit all four points exactly. Therefore we choose the values of C and D which make the error a minimum.

These optimal values $x = (C, D)$ come from the normal equations. For those equations we need $A^T A$ and $A^T b$:

$$A^T A = \begin{bmatrix} 1 & 1 & 1 & 1 \\ 0 & 1 & 3 & 4 \end{bmatrix} \begin{bmatrix} 1 & 0 \\ 1 & 1 \\ 1 & 3 \\ 1 & 4 \end{bmatrix} = \begin{bmatrix} 4 & 8 \\ 8 & 26 \end{bmatrix}$$

$$A^T b = \begin{bmatrix} 1 & 1 & 1 & 1 \\ 0 & 1 & 3 & 4 \end{bmatrix} \begin{bmatrix} 0 \\ 8 \\ 8 \\ 20 \end{bmatrix} = \begin{bmatrix} 36 \\ 112 \end{bmatrix}.$$

Then the system to be solved is $A^T Ax = A^T b$:

$$4C + 8D = 36$$
$$8C + 26D = 112.$$

It gives $C = 1$, $D = 4$, and **the best straight line is** $y = 1 + 4t$.

On the closest line, the values of y at times $t = 0, 1, 3, 4$ are $1, 5, 13, 17$. This is the projection p, the first column of A plus 4 times the second column. It is the closest point to the vector b with components $0, 8, 8, 20$. The error $b - Ax$ agrees with the vertical distances $-1, 3, -5, 3$ in Fig. 1.5, and it should be perpendicular to Ax:

$$\begin{bmatrix} -1 & 3 & -5 & 3 \end{bmatrix} \begin{bmatrix} 1 \\ 5 \\ 13 \\ 17 \end{bmatrix} = 0.$$

Notice how b and p and the four errors appear in both figures! In the 4-dimensional figure we see the perpendicularity. In the plane figure we see the actual line.

The sum of the individual errors is zero: $-1+3-5+3=0$. This is always true since the first row of A^T is $[1 \quad 1 \quad 1 \quad 1]$. The first equation in $A^TAx=A^Tb$ is $[1 \quad 1 \quad 1 \quad 1]e=0$, and therefore the sum of the e_i is zero. Geometrically, this can be interpreted in terms of the "center point" at $\bar{t}=(0+1+3+4)/4$, $\bar{b}=(0+8+8+20)/4$. *The best straight line goes through this center point:* $C+D\bar{t}=\bar{b}$ or $1+4\cdot2=9$.

This application of a minimum principle has solved one form of a basic problem in statistics, the problem of **linear regression**:

1H   Given the measurements $b_1, \ldots, b_m$ at times $t_1, \ldots, t_m$, the line $y=C+Dt$ which minimizes the error $\|e\|^2$ is determined by

$$A^TAx=A^Tb \quad \text{or} \quad \begin{bmatrix} m & \Sigma t_i \\ \Sigma t_i & \Sigma t_i^2 \end{bmatrix}\begin{bmatrix} C \\ D \end{bmatrix} = \begin{bmatrix} \Sigma b_i \\ \Sigma t_i b_i \end{bmatrix}. \tag{6}$$

The best line is $y=\bar{b}+D(t-\bar{t})$, with $D=\dfrac{\Sigma(t_i-\bar{t})b_i}{\Sigma(t_i-\bar{t})^2}$.

Failure of Independence in the Columns of A

We could take a second measurement b_5 at one of the same times, say at $t_5=t_4=4$, without any difficulty. Repeating a row of A does not spoil the independence of its columns, and the best line tries to fit both b_4 and b_5. But we could *not* repeat one of the columns, and look for the best line $y=C+Dt+E$, without serious trouble. In this case C and E cannot be distinguished, and the new third column of A would be identical to the first:

$$A = \begin{bmatrix} 1 & 0 & 1 \\ 1 & 1 & 1 \\ 1 & 3 & 1 \\ 1 & 4 & 1 \end{bmatrix}.$$

These columns are not independent and A^TA will not be positive definite:

$$A^TA = \begin{bmatrix} 4 & 8 & 4 \\ 8 & 26 & 8 \\ 4 & 8 & 4 \end{bmatrix}.$$

There is a repeated column and a repeated row, and $x=(1, 0, -1)$ gives $x^TA^TAx=0$. The quadratic form is zero at other points than the origin.

It is true that $A^T A$ is *semidefinite*: $x^T A^T A x$ is never negative, since it still equals $\|Ax\|^2$. But the matrix is singular and we cannot expect a unique solution to the normal equations. It is for exactly this situation that the **pseudoinverse** was invented. It chooses one particular solution, in this case $C = E = \frac{1}{2}$ and $D = 4$, which leaves the best line at $y = 1 + 4t$. This allows the method of least squares to continue even with dependent columns. The optimal Ax is still the projection p— the nearest point to b in the column space. But the formula $x = (A^T A)^{-1} A^T b$ breaks down and the pseudoinverse has to take the place of the inverse.

In almost all applications, the columns will be independent.

EXAMPLE 2: *Minimum Potential Energy*

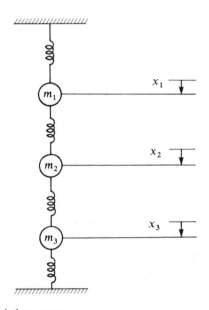

Fig. 1.7. Four springs and three masses.

We are given four springs in a line, with masses between them. The ends at the top and bottom are fixed (Fig. 1.7). The problem is to find the displacements x_1, x_2, x_3 of the three masses. It seems simple, and a little old-fashioned, but it will be important for an enormous class of applications. Therefore we write down everything.

There are four vectors involved. Two of them are associated with the springs, and the other two are defined at the nodes (where the masses are). They are

(1) the vector x of nodal displacements x_1, x_2, x_3
(2) the vector e of elongations of the springs
(3) the vector y of forces in the springs (the springs exert forces $-y$ on the nodes)
(4) the vector f of external forces on the nodes.

Each vector is connected to the one above it by a matrix equation:

$$Ax = e \quad \text{and} \quad Ce = y \quad \text{and} \quad A^T y = f.$$

The first step is to find those matrices A and C and A^T.

(1 to 2)   The elongation in the ith spring is $e_i = x_i - x_{i-1}$. One end of the spring moves a distance x_i, the other end moves x_{i-1}, and the difference is the elongation. Since the boundary displacements are zero—fixed endpoints means $x_0 = x_4 = 0$— we have four equations in three unknowns:

$$e = Ax \quad \text{or} \quad \begin{bmatrix} e_1 \\ e_2 \\ e_3 \\ e_4 \end{bmatrix} = \begin{bmatrix} 1 & & \\ -1 & 1 & \\ & -1 & 1 \\ & & -1 \end{bmatrix} \begin{bmatrix} x_1 \\ x_2 \\ x_3 \end{bmatrix}. \tag{7}$$

A positive e_i describes a spring in tension; $e_4 < 0$ for the lowest spring, in compression.

(2 to 3)   The force in the ith spring is proportional to its elongation, $y_i = c_i e_i$. This comes from Hooke's law for a linearly elastic material. It is the "constitutive law," determining the spring constants c_i. It leads to a diagonal (but positive definite) matrix C:

$$y = Ce \quad \text{or} \quad \begin{bmatrix} y_1 \\ y_2 \\ y_3 \\ y_4 \end{bmatrix} = \begin{bmatrix} c_1 & & & \\ & c_2 & & \\ & & c_3 & \\ & & & c_4 \end{bmatrix} \begin{bmatrix} e_1 \\ e_2 \\ e_3 \\ e_4 \end{bmatrix} \tag{8}$$

(3 to 4)   The net force at each node is zero: $f_i = y_i - y_{i+1}$. This is the equilibrium equation balancing the internal forces from the springs against the external forces f acting at the nodes. The force f_i (positive downwards) acts to stretch spring i and compress spring $i + 1$. Therefore the balance of forces is

$$f = A^T y \quad \text{or} \quad \begin{bmatrix} f_1 \\ f_2 \\ f_3 \end{bmatrix} = \begin{bmatrix} 1 & -1 & & \\ & 1 & -1 & \\ & & 1 & -1 \end{bmatrix} \begin{bmatrix} y_1 \\ y_2 \\ y_3 \\ y_4 \end{bmatrix}. \tag{9}$$

It is no accident that the matrix connecting x to e is the transpose of the matrix connecting y to f. That fact appears automatically in the fundamental variational principles of physics. It means that the internal work of stretching the springs equals the external work done at the nodes:

$$y^T e = y^T Ax = f^T x. \tag{10}$$

In this case the external forces f_1, f_2, f_3 are $m_1 g, m_2 g$, and $m_3 g$, the masses multiplied by the gravitational constant.

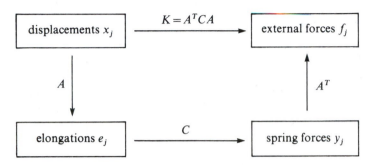

Fig. 1.8. The basic framework of mechanics.

Now we assemble the three laws into one. This is simple but crucial. The equations $e = Ax$ and $y = Ce$ combine to give $y = CAx$; e has been eliminated. Then CAx is substituted for y in the third law $A^T y = f$, to give the fundamental equation:

1I The nodal displacements x are determined from the nodal forces f by the equations

$$\left.\begin{array}{c} Ax = e \\ Ce = y \\ A^T y = f \end{array}\right\} \quad \text{or} \quad A^T C A x = f. \tag{11}$$

The matrix $K = A^T C A$ is symmetric positive definite.

You see here the fundamental equation of mechanics, for a system in equilibrium.

The Stiffness Matrix

The coefficient matrix $A^T C A$ is one step beyond the $A^T A$ of least squares. It combines all three laws, and it is a positive definite matrix—known in computational mechanics as the **stiffness matrix**, and denoted by K.

The same matrix $K = A^T C A$ appears in the underlying minimum principle. The springs search for the position in which the total potential energy, of masses and springs together, is a minimum. The potential energy of the masses is negative,

$$-m_1 g x_1 - m_2 g x_2 - m_3 g x_3 = -x^T f,$$

since it requires that much work to lift them back to their original positions. The potential energy of the springs is positive,

$$\tfrac{1}{2}c_1e_1^2 + \tfrac{1}{2}c_2e_2^2 + \tfrac{1}{2}c_3e_3^2 + \tfrac{1}{2}c_4e_4^2 = \tfrac{1}{2}e^TCe = \tfrac{1}{2}x^TA^TCAx.$$

The springs are under strain and will give back energy if they are released.† Thus the total potential energy, for any displacements x, is the quadratic

$$\boxed{P(x) = \tfrac{1}{2}x^TA^TCAx - x^Tf.}$$

The vector that minimizes $P(x)$ is the one that solves equation (11): $A^TCAx = f$.

The structure of A^TCA is important. If A is m by n, then there are m springs and C is a square matrix of order m. The vectors e and y have m components, and x and f have n (one for each node). In our case $m = 4$ and $n = 3$; multiplying the three matrices gives

$$\underset{\substack{n \text{ by } m \ \ m \text{ by } m \ \ m \text{ by } n}}{A^T \quad C \quad A} = \begin{bmatrix} c_1 + c_2 & -c_2 & 0 \\ -c_2 & c_2 + c_3 & -c_3 \\ 0 & -c_3 & c_3 + c_4 \end{bmatrix}.$$

Off the diagonal, the nonzero entries reflect the connections between nodes. Since node 1 is not connected to node 3, the $(1, 3)$ and $(3, 1)$ entries are zero. Whenever the springs are connected along a line, their matrix A^TCA will be *tridiagonal*: there are only three nonzero diagonals, and the (i, j) entry is zero if $|i - j| > 1$.

Suppose the springs are identical, with spring constants $c_i = 1$. Then the stiffness matrix A^TCA has the special form

$$A^TCA = \begin{bmatrix} 2 & -1 & 0 \\ -1 & 2 & -1 \\ 0 & -1 & 2 \end{bmatrix}.$$

It is a "second difference" matrix, since the equation $A^TCAx = f$ involves the differences $-x_{i+1} + 2x_i - x_{i-1} = f_i$, with $x_0 = x_4 = 0$. At a typical node like the second, the forces are $x_1 - x_2$ from the spring above, $x_3 - x_2$ from the spring below, and f_2 from gravity. We will see how the second differences become second derivatives, for a continuum of springs.

Remark 1   There is a further property of A^TCA which appears after elimination. The factorization into LDL^T is

$$\begin{bmatrix} 2 & -1 & 0 \\ -1 & 2 & -1 \\ 0 & -1 & 2 \end{bmatrix} = \begin{bmatrix} 1 & & \\ -\tfrac{1}{2} & 1 & \\ 0 & -\tfrac{2}{3} & 1 \end{bmatrix} \begin{bmatrix} 2 & & \\ & \tfrac{3}{2} & \\ & & \tfrac{4}{3} \end{bmatrix} \begin{bmatrix} 1 & -\tfrac{1}{2} & 0 \\ & 1 & -\tfrac{2}{3} \\ & & 1 \end{bmatrix}. \qquad (12)$$

---

† At elongation e, the force is ce and the work in a further stretching is $ce\,de$. Integrating from $e = 0$, the total energy stored in the spring is $\tfrac{1}{2}ce^2$.

We knew the pivots would be positive, because $A^T C A$ is positive definite. As in the earlier example of elimination (the 1, 2, 1 tridiagonal matrix in Section 1.2, with $+1$ instead of -1 off the diagonal), the factors L and L^T are bidiagonal; their structure is inherited from $A^T C A$. The new property is the *positivity* of the inverse:

$$L^{-1} = \begin{bmatrix} 1 & 0 & 0 \\ \frac{1}{2} & 1 & 0 \\ \frac{1}{3} & \frac{2}{3} & 1 \end{bmatrix} \quad \text{and} \quad (A^T C A)^{-1} = L^{-T} D^{-1} L^{-1} = \frac{1}{4} \begin{bmatrix} 3 & 2 & 1 \\ 2 & 4 & 2 \\ 1 & 2 & 3 \end{bmatrix}. \tag{13}$$

The inverse of the stiffness matrix is positive. It is also positive definite, but you should not confuse those two properties: neither one implies the other.

Positivity means that when all the forces f go in one direction, so do all the displacements x. Multiplying a positive f by a positive matrix gives a positive x $= (A^T C A)^{-1} f$. In the continuous case we will find the same property for a membrane; when all forces act downwards, the displacement is everywhere down. (It is not true for a beam, governed by fourth derivatives or fourth differences.) The key is in the negative entries off the diagonal of $A^T C A$.† They guarantee $K^{-1} > 0$.

Remark 2 A computer would solve this problem at the speed of light, but there is a crucial point to be made about the computation. *The inverse is a full matrix*, without the band structure and the zeros of $A^T C A$. In this case we actually lose if we allow K^{-1} into the computation; multiplying f by the inverse takes much longer (n^2 steps for an n by n matrix) than ordinary elimination. The solution should be found as $x = L^{-T} D^{-1} L^{-1} f$, but not by combining those three matrices into a single inverse. Instead each separate step $L^{-1} f$, $D^{-1}(L^{-1} f)$, and $L^{-T}(D^{-1} L^{-1} f)$ is a simple back substitution—or in the case of $c = L^{-1} f$ it is a forward substitution:

$$Lc = f \quad \text{is} \quad \begin{bmatrix} 1 & 0 & 0 \\ -\frac{1}{2} & 1 & 0 \\ 0 & -\frac{2}{3} & 1 \end{bmatrix} \begin{bmatrix} c_1 \\ c_2 \\ c_3 \end{bmatrix} = \begin{bmatrix} f_1 \\ f_2 \\ f_3 \end{bmatrix},$$

and c_1, c_2, c_3 are found in that order. For an n by n matrix this needs only $n - 1$ multiplications, D^{-1} needs n more, and back substitution uses $n - 1$—a total of $3n - 2$, which is much less than n^2. The solution of tridiagonal systems, or any system with a narrow band matrix, is tremendously fast—provided the inverse never appears.

---

†A matrix with non-positive off-diagonal entries is an *M-matrix* if its inverse is nonnegative. No less than 40 equivalent descriptions have been given without assuming symmetry: all pivots are positive, all real eigenvalues are positive, and 38 others. With symmetry this means it is positive definite.

EXERCISES

1.4.1 Find the minimum value (and the minimizing x) for

(i) $P(x) = \frac{1}{2}(x_1^2 + x_2^2) - x_1 b_1 - x_2 b_2$

(ii) $P(x) = \frac{1}{2}(x_1^2 + 2x_1 x_2 + 2x_2^2) - x_1 + x_2$.

1.4.2 What equations determine the minimizing x for

(i) $P = \frac{1}{2} x^T A x - x^T b$

(ii) $P = \frac{1}{2} x^T A^T A x - x^T A^T b$

(iii) $E = \| A x - b \|^2$?

1.4.3 Find the best least squares solutions to all three problems

$$\begin{bmatrix} 2 & -1 \\ 2 & 2 \\ -1 & 2 \end{bmatrix} \begin{bmatrix} x_1 \\ x_2 \end{bmatrix} = \begin{bmatrix} 1 \\ 1 \\ 1 \end{bmatrix} \quad \text{and} \quad \begin{bmatrix} 2 \\ -1 \\ 2 \end{bmatrix} \quad \text{and} \quad \begin{bmatrix} 2 \\ 2 \\ -1 \end{bmatrix}.$$

1.4.4 Find the best least squares solution to

$$A x = \begin{bmatrix} 1 & 1 \\ 1 & 2 \\ 1 & 3 \end{bmatrix} \begin{bmatrix} C \\ D \end{bmatrix} = \begin{bmatrix} 0 \\ 4 \\ 2 \end{bmatrix} = b.$$

Graph the measurements $0, 4, 2$ at times $t = 1, 2, 3$, and the best straight line.

1.4.5 The best fit to b_1, b_2, b_3, b_4 by a horizontal line (a constant function $y = C$) is their average $C = (b_1 + b_2 + b_3 + b_4)/4$. Confirm this by least squares solution of

$$A x = \begin{bmatrix} 1 \\ 1 \\ 1 \\ 1 \end{bmatrix} [C] = \begin{bmatrix} b_1 \\ b_2 \\ b_3 \\ b_4 \end{bmatrix}.$$

From calculus, which C minimizes the error $E = (b_1 - C)^2 + \cdots + (b_4 - c)^2$?

1.4.6 Which three equations are to be solved by least squares, if we look for the line $y = Dt$ through the origin that best fits the data $y = 4$ at $t = 1$, $y = 5$ at $t = 2$, $y = 8$ at $t = 3$? What is the least squares solution $\bar{D}$ that gives the best line?

1.4.7 For the three measurements $b = 0, 3, 12$ at times $t = 0, 1, 2$, find

(i) the best horizontal line $y = C$

(ii) the best straight line $y = C + Dt$

(iii) the best parabola $y = C + Dt + Et^2$.

1.4.8 Multiple regression fits two-dimensional data by a plane $y = C + Dt + Ez$, instead of one-dimensional data by a line. If we are given $b_1 = 2$ at $t = z = 0$; $b_2 = 2$ at $t = 0$, $z = 1$; $b_3 = 1$ at $t = 1$, $z = 0$; and $b_4 = 5$ at $t = z = 1$, write down the 4 equations in the 3 unknowns C, D, E. What is the least squares solution from the normal equations?

1.4.9   For a matrix with more columns than rows, like

$$A = \begin{bmatrix} 1 & 1 & 0 \\ 2 & 2 & 1 \end{bmatrix} \quad \text{or even} \quad A = [1 \quad 2],$$

the matrix $A^T A$ is not positive definite. Why is it impossible for these columns to be independent? Compute $A^T A$ in both cases and check that it is singular.

1.4.10   In a system with three springs and two forces and displacements write out the equations $e = Ax$, $y = Ce$, and $A^T y = f$. For unit forces and spring constants, what are the displacements?

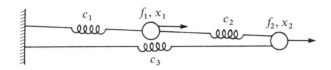

1.4.11   Suppose the lowest spring in Fig. 1.7 is removed, leaving masses m_1, m_2, m_3 hanging from the three remaining springs. The equation $e = Ax$ becomes

$$\begin{bmatrix} e_1 \\ e_2 \\ e_3 \end{bmatrix} = \begin{bmatrix} 1 & 0 & 0 \\ -1 & 1 & 0 \\ 0 & -1 & 1 \end{bmatrix} \begin{bmatrix} x_1 \\ x_2 \\ x_3 \end{bmatrix}.$$

Find the corresponding equations $y = Ce$ and $A^T y = f$, and solve the last equation for y. This is the **determinate** case, with square matrices, when the factors in $A^T CA$ can be inverted separately and y can be found before x.

1.4.12   For the same 3 by 3 problem find $K = A^T CA$ and A^{-1} and K^{-1}. If the forces f_1, f_2, f_3 are all positive, acting in the same direction, how do you know that the displacements x_1, x_2, x_3 are also positive?

1.4.13   If the matrix A is a column of 1's as in problem 5 above, can you invent a four spring-one displacement system which produces this matrix? With spring constants c_1, c_2, c_3, c_4 and force f, what is the displacement x?

1.4.14   If two springs with constants c_1 and c_2 hang (i) from separate supports, or (ii) with one attached below the other, what masses m_1 and m_2 in each case will produce equal displacements x_1 and x_2?

1.4.15   Suppose the three equations are changed to $e = Ax - b$, $y = Ce$, and $A^T y = 0$. Eliminate e and y to find a single equation for x.

EIGENVALUES AND DYNAMICAL SYSTEMS ■ 1.5

There is another way to study a symmetric matrix, more subtle than $A = LDL^T$ and less mechanical. It involves the eigenvalues and eigenvectors of A. For any square matrix, we can look for vectors x that are in the same direction as Ax; those are the eigenvectors. Multiplication by A normally changes the direction of a vector, but *for certain exceptional vectors Ax is a multiple of* x. That is the basic equation $Ax = \lambda x$. The number λ is the eigenvalue, which tells whether x is stretched or contracted or reversed by A.

For the identity matrix nothing changes direction. Every vector is an eigenvector and the eigenvalue is $\lambda = 1$. Of course this is special. It is much more normal to have n different eigenvalues, and they are easy to recognize when the matrix is diagonal:

$$A = \begin{bmatrix} 3 & 0 & 0 \\ 0 & 2 & 0 \\ 0 & 0 & 0 \end{bmatrix} \text{ has } \lambda_1 = 3 \text{ with } x_1 = \begin{bmatrix} 1 \\ 0 \\ 0 \end{bmatrix}, \lambda_2 = 2 \text{ with } x_2 = \begin{bmatrix} 0 \\ 1 \\ 0 \end{bmatrix},$$

$$\lambda_3 = 0 \text{ with } x_3 = \begin{bmatrix} 0 \\ 0 \\ 1 \end{bmatrix}.$$

On each eigenvector the matrix acts like a multiple of the identity, but the multiples 3, 2, 0 are different in the three directions. A vector like $x = (2, 4, 6)$ is a mixture $2x_1 + 4x_2 + 6x_3$ of eigenvectors, and when A multiplies x it gives

$$Ax = 2\lambda_1 x_1 + 4\lambda_2 x_2 + 6\lambda_3 x_3 = \begin{bmatrix} 6 \\ 8 \\ 0 \end{bmatrix}.$$

This was a typical vector x—not an eigenvector—but the action of A was still determined by its eigenvalues and eigenvectors.

Notice that there was nothing exceptional about $\lambda_3 = 0$. Like every other number, zero might be an eigenvalue and it might not. If it is, then its eigenvector satisfies $Ax = 0x$. Thus x is in the nullspace of A; a zero eigenvalue signals that A has linearly dependent columns (and rows). Its determinant is zero. Invertible matrices have $\lambda \neq 0$, while singular matrices include zero among their eigenvalues.

Diagonal matrices are certainly the simplest. The eigenvalues are the diagonal entries and the eigenvectors are in the coordinate directions. For other matrices we find the eigenvalues first, before the eigenvectors, by the following test:

If $Ax = \lambda x$ then $(A - \lambda I)x = 0$ and $A - \lambda I$ has dependent columns. Therefore the determinant of $A - \lambda I$ must be zero.

In other words, shifting the matrix by λI makes it singular. We look for those exceptional numbers λ, and the eigenvector is in the nullspace of $A - \lambda I$.† Actually it is not correct to speak about *the* eigenvector, since if x is doubled or tripled or multiplied by -1 it is still an eigenvector. The equation $Ax = \lambda x$ still holds after rescaling x, and it is the direction of x that is important.

EXAMPLE 1. $A = \begin{bmatrix} 2 & 1 \\ 1 & 2 \end{bmatrix}$ and $A - \lambda I = \begin{bmatrix} 2 - \lambda & 1 \\ 1 & 2 - \lambda \end{bmatrix}.$

The eigenvalues of A are the numbers that make the determinant of $A - \lambda I$ zero:

$$\det(A - \lambda I) = (2 - \lambda)^2 - 1 = \lambda^2 - 4\lambda + 3 = 0.$$

This factors into $(\lambda - 1)(\lambda - 3) = 0$, and the eigenvalues are 1 and 3. For each eigenvalue the equation $(A - \lambda I)x = 0$ yields the corresponding eigenvector:

$$\lambda_1 = 1 \quad A - \lambda_1 I = \begin{bmatrix} 1 & 1 \\ 1 & 1 \end{bmatrix} \quad x_1 = \begin{bmatrix} 1 \\ -1 \end{bmatrix}$$

$$\lambda_2 = 3 \quad A - \lambda_2 I = \begin{bmatrix} -1 & 1 \\ 1 & -1 \end{bmatrix} \quad x_2 = \begin{bmatrix} 1 \\ 1 \end{bmatrix}.$$

You see that the shifted matrix $A - \lambda I$ has dependent columns and $(A - \lambda I)x = 0$ in each case. That is equivalent to $Ax = \lambda x$. If the calculation is done in this order, the equation $Ax = \lambda x$ can be used to check the algebra.

This example is partly special and partly typical. For a 2 by 2 matrix, the determinant of $A - \lambda I$ will always be a quadratic polynomial; its first term is λ^2. The eigenvalues may not be integers like 1 and 3. They can easily turn out to be complex numbers in the unsymmetric case

$$A = \begin{bmatrix} a & b \\ b' & c \end{bmatrix} \quad \text{and} \quad \det \begin{bmatrix} a - \lambda & b \\ b' & c - \lambda \end{bmatrix} = \lambda^2 - (a + c)\lambda + ac - bb'.$$

This polynomial, whose roots are the eigenvalues, is the ***characteristic polynomial*** of A. For an n by n matrix it has degree n and begins with $(-\lambda)^n$. There are n roots and therefore n eigenvalues. For each eigenvalue there is a solution of $(A - \lambda I)x = 0$—or possibly several independent eigenvectors x, if λ is a repeated eigenvalue.

In the 2 by 2 case the formula for the roots of a quadratic gives the eigenvalues exactly. When $n = 3$ and $n = 4$ there are still formulas for the roots, although they are seldom remembered and rarely used. For $n > 4$ there is not and never will be such a formula. Most eigenvalues have to be found numerically, and that is not done by computing the determinant of $A - \lambda I$. Instead the matrix is gradually brought into

---

† We use I to keep vectors and matrices straight; $(A - \lambda)x = 0$ is shorter but mixed up.

diagonal form (see Chapter 5) and the eigenvalues begin to appear on the main diagonal. The computation takes longer than elimination—perhaps 20 times longer, if it achieves good accuracy—but that is not unreasonable. The algorithms for eigenvalues depend on the size and structure of A, and they are a major part of numerical linear algebra.

The Diagonal Form

It is fair to say that *diagonalization* is the great contribution of the eigenvectors. Suppose the columns of S are the eigenvectors of A. Then we want to show that $S^{-1}AS$ is a diagonal matrix. Furthermore it is extremely special; it contains the eigenvalues. It is denoted by Λ (capital *lambda*) as a way of remembering that $\lambda_1, \dots, \lambda_n$ are the entries on the diagonal.

1J Suppose the n by n matrix A has n linearly independent eigenvectors. Then if these vectors are the columns of S, it follows that

$$S^{-1}AS = \Lambda = \begin{bmatrix} \lambda_1 & & & \\ & \lambda_2 & & \\ & & \ddots & \\ & & & \lambda_n \end{bmatrix}. \tag{1}$$

This is so important that we do it twice. First the eigenvectors x_i go into S and we multiply a column at a time. Since they are eigenvectors,

$$AS = A \begin{bmatrix} x_1 & x_2 & \dots & x_n \end{bmatrix} = \begin{bmatrix} \lambda_1 x_1 & \lambda_2 x_2 & \dots & \lambda_n x_n \end{bmatrix}.$$

Then the trick is to split this last matrix into

$$AS = \begin{bmatrix} \lambda_1 x_1 & \lambda_2 x_2 & \dots & \lambda_n x_n \end{bmatrix} = \begin{bmatrix} x_1 & x_2 & \dots & x_n \end{bmatrix} \begin{bmatrix} \lambda_1 & & & \\ & \lambda_2 & & \\ & & \ddots & \\ & & & \lambda_n \end{bmatrix}.$$

That is S times Λ. Ordinary matrix multiplication has produced

$$\boxed{AS = S\Lambda \quad \text{or} \quad S^{-1}AS = \Lambda \quad \text{or} \quad A = S\Lambda S^{-1}.} \tag{2}$$

The matrix S is invertible, because its columns (the eigenvectors) were linearly independent. If the eigenvalues are distinct—the matrix A has no repeated

eigenvalues—then automatically the eigenvectors are independent and diagonalization succeeds.

For some matrices with repeated eigenvalues, the matrix Λ can still be reached.
This is clear if A is diagonal, because it is *already* equal to Λ; the eigenvectors form
the identity matrix $S = I$ and $S^{-1}AS = \Lambda$ is trivial. We show below that for some
other examples with repeated eigenvalues no diagonalization is possible; the matrix
is defective and it has too few eigenvectors. But the most important matrices—
including all that are symmetric or skew-symmetric or orthogonal—always have a
full set of eigenvectors and can be diagonalized. For those matrices it makes no
difference whether any eigenvalues are repeated.

The $S^{-1}AS$ or SAS^{-1} calculation can be displayed for the 2 by 2 example

$$A = \begin{bmatrix} 2 & 1 \\ 1 & 2 \end{bmatrix}, \quad \text{with } \lambda_1 = 1 \text{ and } \lambda_2 = 3.$$

Its eigenvectors are $(1, -1)$ and $(1, 1)$ and they go into the columns of S:

$$S = \begin{bmatrix} 1 & 1 \\ -1 & 1 \end{bmatrix} \quad \text{and} \quad S^{-1} = \begin{bmatrix} \frac{1}{2} & -\frac{1}{2} \\ \frac{1}{2} & \frac{1}{2} \end{bmatrix}. \tag{3}$$

Then the triple factorization, based on eigenvalues instead of pivots, is

$$A = S\Lambda S^{-1} = \begin{bmatrix} 1 & 1 \\ -1 & 1 \end{bmatrix} \begin{bmatrix} 1 & \\ & 3 \end{bmatrix} \begin{bmatrix} \frac{1}{2} & -\frac{1}{2} \\ \frac{1}{2} & \frac{1}{2} \end{bmatrix}.$$

This looks very much like $A = LDL^T$. In fact the resemblance will be closer if we
divide S (and multiply S^{-1}) by the square root of 2. That yields an equally good
eigenvector matrix S—the columns can be scaled in any way we choose—and the
scaling by $\sqrt{2}$ gives a special matrix that we call Q:

$$A = Q\Lambda Q^{-1} = \begin{bmatrix} \dfrac{1}{\sqrt{2}} & \dfrac{1}{\sqrt{2}} \\ -\dfrac{1}{\sqrt{2}} & \dfrac{1}{\sqrt{2}} \end{bmatrix} \begin{bmatrix} 1 & \\ & 3 \end{bmatrix} \begin{bmatrix} \dfrac{1}{\sqrt{2}} & -\dfrac{1}{\sqrt{2}} \\ \dfrac{1}{\sqrt{2}} & \dfrac{1}{\sqrt{2}} \end{bmatrix}. \tag{4}$$

The third matrix is now the transpose of the first; Q^{-1} *is the same as* Q^T. The
columns are unit vectors, and they are perpendicular.

Thus our 2 by 2 matrix has become $A = Q\Lambda Q^T$. This special form is possible
when A is symmetric; the eigenvector matrix S in $A = S\Lambda S^{-1}$ becomes an
orthogonal matrix Q. Perhaps we can postpone the proof and concentrate on
drawing out everything we see in this example.

One important point: The pivots are not the same as the eigenvalues! The eigenvalues 1 and 3 are in Λ, whereas the pivots are in the middle of LDL^T:

$$\begin{bmatrix} 2 & 1 \\ 1 & 2 \end{bmatrix} = \begin{bmatrix} 1 & 0 \\ \frac{1}{2} & 1 \end{bmatrix} \begin{bmatrix} 2 & \\ & \frac{3}{2} \end{bmatrix} \begin{bmatrix} 1 & \frac{1}{2} \\ 0 & 1 \end{bmatrix}.$$

It is true that all pivots and eigenvalues are positive; that is because A is positive definite.† It is also true that the product of the pivots equals the product of the eigenvalues: $(2)(\frac{3}{2}) = (1)(3)$. Both sides equal the determinant. And one more number can be found directly from the matrix, to check that the eigenvalues are correct. It is the **sum** of the eigenvalues (different from the sum of the pivots) and it is revealed by looking at the main diagonal of A.

1K The sum of the diagonal entries is the **trace** of A. It equals the sum of the eigenvalues. The product of the eigenvalues equals the **determinant** of A.

The trace in the example is $2 + 2 = 4$, which agrees with $1 + 3$. That sum stays constant no matter how we change the off-diagonal entries. Of course the eigenvalues are not constant—if an off-diagonal entry disappears the eigenvalues become 2 and 2—but their sum is always 4.

We prove trace $A = $ trace $\Lambda = \lambda_1 + \cdots + \lambda_n$ in two steps:

(1) trace $CD = $ trace DC for any matrices, by checking their diagonals
(2) the choices $C = S$ and $D = \Lambda S^{-1}$ give trace $S\Lambda S^{-1} = $ trace Λ.

That completes the proof for every diagonalized matrix $A = S\Lambda S^{-1}$, symmetric or not. A similar argument for the determinant leads to det $A = $ det $\Lambda = \lambda_1 \lambda_2 \cdots \lambda_n$:

(1) (det C)(det D) = det CD for any matrices
(2) (det S)(det S^{-1}) = 1 by choosing $C = S$ and $D = S^{-1}$
(3) det $A = $ (det S)(det Λ)(det S^{-1}) = det Λ.

The last point (3) is what we wanted to prove. For matrices that cannot be diagonalized there is an alternative proof in the exercises.

The rest of this section has two goals:

(1) to show how the eigenvalues and eigenvectors lead to the solution of differential equations
(2) to show that all symmetric matrices can be factored into $A = Q\Lambda Q^T$, with positive Λ corresponding to positive definite A.

The first is the most important application and the second is the most important theorem.

---

† If $A = A^T$ then the pivots and eigenvalues always have the same signs—the number of positive pivots indicates the number of positive eigenvalues.

Differential Equations

We emphasize from the start that $Ax = \lambda x$ is fundamentally different from $Ax = b$. Previously the problems were static; now they are dynamic. The springs and masses are no longer stationary. Instead they oscillate around the point of equilibrium, governed by Newton's law $F = ma$:

$$M\frac{d^2u}{dt^2} + Ku = 0. \tag{5}$$

The first term is mass times acceleration, for several masses at once. M is the **mass matrix**. The second term is $-F$ (again for several masses); it contains the forces from the springs. The spring displacements are denoted by u instead of x, for this and all future differential equations, because x is needed for other purposes.

This system of differential equations, linear with constant coefficients, has a set of special solutions. Each of those solutions is a **pure oscillation**, with all masses moving at the same frequency, and the general solution is a combination of these "normal modes." Our goal is to connect the frequencies to the eigenvalues, and the amplitudes of oscillation to the eigenvectors.

The difficulty is to do that much and stop. Later in the book we go further, since dynamical systems are of enormous importance. Here we aim at the most fundamental facts for equations of first and second order:

$$\frac{du}{dt} = Au \quad \text{and} \quad \frac{d^2u}{dt^2} + Au = 0.$$

It is natural to let A be symmetric.

The first example is our 2 by 2 matrix with eigenvalues 1 and 3:

$$\frac{du}{dt} = \begin{bmatrix} 2 & 1 \\ 1 & 2 \end{bmatrix} u \quad \text{or} \quad \begin{aligned} \frac{du_1}{dt} &= 2u_1 + u_2 \\ \frac{du_2}{dt} &= u_1 + 2u_2. \end{aligned} \tag{6}$$

Suppose $u_1 = 5$ and $u_2 = 0$ at the starting time $t = 0$. These are the **initial values**. The differential equation should determine u_1 and u_2 at all later times, but the difficulty is that the two equations are coupled. Without knowing u_2 we cannot solve the first equation for u_1—and vice versa. If u_2 were always zero then $du_1/dt = 2u_1$ would be easy. Even though this is not the right equation—it is only a 1 by 1 problem—we solve it first:

$$\text{if} \quad \frac{du_1}{dt} = 2u_1 \quad \text{and} \quad u_1(0) = 5 \quad \text{then} \quad u_1(t) = 5e^{2t}.$$

The solution is a pure exponential. The exponent 2 comes from the differential equation and the coefficient 5 comes from the initial condition. The solution $5e^{2t}$ satisfies both requirements. This 1 by 1 system is the prototype for all constant coefficient problems, and the way to solve two or more differential equations is to recover as much as possible of this simple case.†

There are two ways to proceed with the system (6). First, we could look for combinations of u_1 and u_2 that "uncouple" the equations. In this case they can be found by trial and error. If we subtract the equations we have

$$\frac{d}{dt}(u_1 - u_2) = u_1 - u_2 \quad \text{with} \quad u_1 - u_2 = 5 \quad \text{at} \quad t = 0.$$

If we add the equations then

$$\frac{d}{dt}(u_1 + u_2) = 3(u_1 + u_2) \quad \text{with} \quad u_1 + u_2 = 5 \quad \text{at} \quad t = 0.$$

These are each 1 by 1 and their solutions are pure exponentials:

$$u_1 - u_2 = 5e^t \quad \text{and} \quad u_1 + u_2 = 5e^{3t}. \tag{7}$$

You notice the eigenvalues 1 and 3 in the exponents. To recover u_1 and u_2 we add and subtract and divide by 2, reversing the steps that led to (7):

$$u_1 = \tfrac{5}{2}e^t + \tfrac{5}{2}e^{3t} \quad \text{and} \quad u_2 = -\tfrac{5}{2}e^t + \tfrac{5}{2}e^{3t}. \tag{8}$$

These have the correct initial values 5 and 0. It was the eigenvectors of A that held the key to the right combinations $u_1 \pm u_2$, and we need a general rule. It is exceptionally easy:

> The pure exponential solutions to $du/dt = Au$ combine the eigenvalues and eigenvectors into $u(t) = e^{\lambda t}x$.

All n of the components grow or decay or oscillate together. Their evolution is controlled by λ, and their relative size (and sign) is controlled by x. It is $Ax = \lambda x$ that is responsible for $du/dt = Au$:

$$\frac{d}{dt}(e^{\lambda t}x) = \lambda e^{\lambda t}x \quad \text{and} \quad A(e^{\lambda t}x) = \lambda e^{\lambda t}x. \tag{9}$$

Thus $u = e^{\lambda t}x$ is a solution. So is any multiple $ce^{\lambda t}x$. This is true for each pair λ_i and

---

† We come back to this all-important solution $e^{at}u_0$ in studying initial-value problems. There will be a forcing term f, and stability if a is negative.

x_i (with $Ax_i = \lambda_i x_i$) and it is true for any combination

$$u(t) = c_1 e^{\lambda_1 t} x_1 + c_2 e^{\lambda_2 t} x_2 + \cdots + c_n e^{\lambda_n t} x_n. \tag{10}$$

If A has a full set of n independent eigenvectors—it is diagonalizable—then this is the **general solution** to $du/dt = Au$. The right choice of the n constants c_i will match the n initial conditions in the vector $u = u_0$. The pure exponentials are the complete key to the differential equation.

The example had two independent eigenvectors $(1, -1)$ and $(1, 1)$:

$$u(t) = c_1 e^t \begin{bmatrix} 1 \\ -1 \end{bmatrix} + c_2 e^{3t} \begin{bmatrix} 1 \\ 1 \end{bmatrix}.$$

At $t = 0$ the exponentials are $e^0 = 1$ and the initial conditions give

$$c_1 \begin{bmatrix} 1 \\ -1 \end{bmatrix} + c_2 \begin{bmatrix} 1 \\ 1 \end{bmatrix} = u_0, \quad \text{or} \quad \begin{bmatrix} 1 & 1 \\ -1 & 1 \end{bmatrix} \begin{bmatrix} c_1 \\ c_2 \end{bmatrix} = u_0.$$

With $u_0 = (5, 0)$ we found $c = (5/2, 5/2)$.

Notice the matrix in this equation. It is S, formed from the eigenvectors. It always appears when we set $t = 0$ in the general solution (10):

$$c_1 x_1 + \cdots + c_n x_n = u_0 \quad \text{or} \quad Sc = \begin{bmatrix} x_1 & \cdots & x_n \end{bmatrix} \begin{bmatrix} c_1 \\ \vdots \\ c_n \end{bmatrix} = u_0. \tag{11}$$

Thus the right constants are $c = S^{-1} u_0$.

1L   The solution to $du/dt = Au$ is the combination (10) of pure exponentials (if A has n independent eigenvectors) with constants given by $c = S^{-1} u_0$.

Note   A change of unknowns from u to $v = S^{-1} u$ uncouples the n equations:

$$\frac{du}{dt} = Au \quad \text{becomes} \quad S \frac{dv}{dt} = ASv. \tag{12}$$

Since $S^{-1} AS = \Lambda$ and $v_0 = S^{-1} u_0 = c$, this is a diagonal problem:

$$\frac{dv}{dt} = \Lambda v \quad \text{with solution} \quad v = e^{\Lambda t} c.$$

Translating back to u, that is our sum of pure exponentials:

$$u = Sv = Se^{\Lambda t}c = \begin{bmatrix} x_1 & \cdots & x_n \end{bmatrix} \begin{bmatrix} e^{\lambda_1 t} & & \\ & \ddots & \\ & & e^{\lambda_n t} \end{bmatrix} \begin{bmatrix} c_1 \\ \vdots \\ c_n \end{bmatrix}$$

$$= c_1 e^{\lambda_1 t} x_1 + \cdots + c_n e^{\lambda_n t} x_n.$$

The example is an equation of "cooperation," with growth in u_2 helping growth in u_1—since $du_1/dt = 2u_1 + u_2$. Similarly u_1 helps u_2; they might be Canada and America. Alone they would grow like e^{2t}. Together they grow like e^{3t}, unless their initial values are exactly opposite. In that exceptional case the first exponential $e^t x_1$ is the solution, and it grows only like e^t.

Second-Order Equations

A law of competition, like cooperation, would lead to a first-order system $du/dt = Au$. The off-diagonal entries would be negative instead of positive. This is common in chemistry or biology—with no reflection on chemists or biologists. The columns of A add to zero and the total $u_1 + \cdots + u_n$ is fixed. It is also common in the matrices of physics, where Newton's law $F = ma$ is a second-order equation. The acceleration is a second derivative, and our goal is to understand how this "inertial term" alters the behavior of the solution.

The algebra stays almost the same when d^2u/dt^2 replaces du/dt. We still look for pure exponential solutions, and they still come from the eigenvectors. The only change is in $e^{\lambda t}$. This was the solution to $du/dt = \lambda u$, but now the 1 by 1 equation is

$$\frac{d^2u}{dt^2} + \lambda u = 0. \tag{13}$$

It has two initial conditions: $u = u_0$ and $du/dt = u_0'$. The displacement and velocity are both given at $t = 0$, because the equation is second-order in time. Its solution is

$$u(t) = u_0 \cos \omega t + \frac{u_0'}{\omega} \sin \omega t. \tag{14}$$

At $t = 0$ the cosine equals one, the sine equals zero, and u equals u_0. Similarly $du/dt = u_0'$; the initial conditions are right. However it remains to determine ω. The equation (13) will be satisfied by $\cos \omega t$ (and also by $\sin \omega t$ and any combination of sine and cosine) if

$$\frac{d^2}{dt^2} \cos \omega t + \lambda \cos \omega t = 0.$$

Since the second derivative of the cosine is $-\omega^2 \cos \omega t$, we need

$$\boxed{\lambda = \omega^2.} \tag{15}$$

In second-order equations the frequency ω (omega) is the *square root* of the eigenvalue.

To go from this model problem with one unknown to the vector problem $d^2u/dt^2 = -Au$, we use the eigenvalues and eigenvectors of A. Suppose that $Ax = \lambda x$. Then there is a solution which is a pure multiple of x. The dependence on time is still a combination of sine and cosine, with the same frequency $\omega = \sqrt{\lambda}$:

$$u(t) = [a \cos \omega t + b \sin \omega t]x.$$

This satisfies $u_{tt} = -Au$. The two derivatives produce $-\omega^2$; multiplication by A produces λ. These cancel because $\lambda = \omega^2$. Of course this one solution cannot match all $2n$ initial conditions in u_0 and u_0', but if A has n eigenvectors then there are n pure oscillations. Each has two free constants a and b, and the general solution is

$$u = (a_1 \cos \omega_1 t + b_1 \sin \omega_1 t)x_1 + \cdots + (a_n \cos \omega_n t + b_n \sin \omega_n t)x_n. \tag{16}$$

The initial displacement u_0 is easily distinguished from the velocity u_0'. At $t = 0$ the sines are zero and the cosines are 1, so u_0 is matched by the a's:

$$a_1 x_1 + \cdots + a_n x_n = u_0, \quad \text{or} \quad Sa = u_0, \quad \text{or} \quad a = S^{-1}u_0.$$

These are the same constants as in $u_t = Au$. The derivative at $t = 0$ is

$$b_1 \omega_1 x_1 + \cdots + b_n \omega_n x_n = u_0',$$

which determines the b's and completes the solution.

EXAMPLE $\quad \dfrac{d^2u}{dt^2} + \begin{bmatrix} 2 & -1 \\ -1 & 2 \end{bmatrix} u = 0.$

The eigenvalues are again 1 and 3, even with minus signs (competition rather than cooperation) off the diagonal. The eigenvectors are $x_1 = (1, 1)$ and $x_2 = (1, -1)$, the same as before but in reverse order:

$$\begin{bmatrix} 2 & -1 \\ -1 & 2 \end{bmatrix}\begin{bmatrix} 1 \\ 1 \end{bmatrix} = 1\begin{bmatrix} 1 \\ 1 \end{bmatrix}, \quad \begin{bmatrix} 2 & -1 \\ -1 & 2 \end{bmatrix}\begin{bmatrix} 1 \\ -1 \end{bmatrix} = 3\begin{bmatrix} 1 \\ -1 \end{bmatrix}.$$

The general solution to the second-order equation is

$$u = (a_1 \cos t + b_1 \sin t)\begin{bmatrix} 1 \\ 1 \end{bmatrix} + (a_2 \cos \sqrt{3}t + b_2 \sin \sqrt{3}t)\begin{bmatrix} 1 \\ -1 \end{bmatrix}.$$

This is an interesting oscillation. It is a mixture of two pure oscillations, with frequencies 1 and $\sqrt{3}$. When u_0' is zero, the system starts from rest and the sines disappear. If the first mass is pulled away from equilibrium, so that the system is started at $u_0 = (1, 0)$ and left free to oscillate, the solution is

$$u(t) = \tfrac{1}{2} \cos t \begin{bmatrix} 1 \\ 1 \end{bmatrix} + \tfrac{1}{2} \cos \sqrt{3}t \begin{bmatrix} 1 \\ -1 \end{bmatrix}.$$

Thus the initial conditions are satisfied by an equal mixture of the two oscillations.

We can interpret this solution physically. The system represents two unit masses, connected to each other and to stationary walls by three identical springs (Fig. 1.9a). The first mass is pulled down, the second mass is held in place, and at $t = 0$ we let go. Their motion $u(t)$ becomes an average of two pure oscillations, corresponding to the two eigenvectors. In the first mode, the masses move exactly in unison and the spring in the middle is never stretched (Fig. 1.9b). The frequency $\omega_1 = 1$ is the same as for a single spring and a single mass. In the faster mode $x_2 = (1, -1)$, with components of opposite sign and with frequency $\sqrt{3}$, the masses move in opposite directions (Fig. 1.9c). The general solution is a combination of these two normal modes, and our particular solution is half of each.

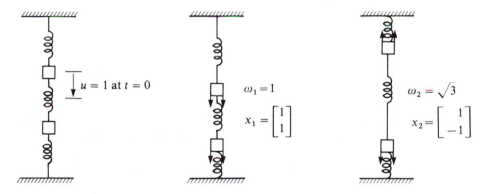

Fig. 1.9. Oscillations around equilibrium.

As time goes on, the motion is "almost periodic." It never returns exactly to the initial conditions, since both cosines are 1 only if t and $\sqrt{3}t$ are multiples of π. Their ratio would have to be a fraction—but $\sqrt{3}$ is not a fraction. The best we can say is that the masses come *arbitrarily close* to reproducing the initial situation. It is like a billiard ball on a perfectly smooth table, hit at an angle that does not bring it back to the start. The energy is fixed, and sooner or later the system gets near to every state with this energy. The special cases when the ball crosses the table and comes straight back correspond to pure oscillations, when only $\cos t$ or $\cos \sqrt{3}t$ is present. Then the motion is exactly periodic.

Remark Everything is changed if the eigenvalues of A are not real and positive. Then $\omega = \sqrt{\lambda}$ will produce complex frequencies (the marginal case $\lambda = 0$ is in

Exercise 1.5.8). The sine and cosine of ωt become less convenient than the equivalent choice $e^{i\omega t}$ and $e^{-i\omega t}$. If $\lambda = -1$, for example, the frequencies are $\omega = \pm i$. Then the exponentials are e^t and e^{-t}, which grow or decay instead of oscillating. In the good case A is symmetric positive definite and $\lambda > 0$ and ω is real.

Single Equations of High Order

There is another application which uses one of the ideas from this section—the exponentials $e^{\lambda t}$. However it does not use the eigenvectors. The unknown is only a scalar, as it was in the model problems $du/dt = \lambda u$ and $d^2 u/dt^2 + \lambda u = 0$. Here we consider the more general case

$$\frac{d^2 u}{dt^2} + p\frac{du}{dt} + qu = 0, \tag{17}$$

and it would be easy to allow derivatives of any order.

The key step is the first one, to search for exponential solutions. Substituting $e^{\lambda t}$ into the equation, it succeeds if

$$\lambda^2 e^{\lambda t} + p\lambda e^{\lambda t} + qe^{\lambda t} = 0.$$

In other words λ must satisfy

$$\lambda^2 + p\lambda + q = 0. \tag{18}$$

This quadratic has two roots λ_1 and λ_2. Unless they are equal they give two independent solutions to (17), and the complete solution is a combination

$$u = c_1 e^{\lambda_1 t} + c_2 e^{\lambda_2 t}. \tag{19}$$

The constants c_1 and c_2 allow us to match initial values u_0 and u'_0.

EXAMPLE $u'' + 6u' + 8u = 0$ with $u_0 = 3$ and $u'_0 = -8$.

The equation for λ becomes $\lambda^2 + 6\lambda + 8 = 0$, or $(\lambda + 2)(\lambda + 4) = 0$. The roots are -2 and -4, so that $u = c_1 e^{-2t} + c_2 e^{-4t}$. The initial conditions require $c_1 + c_2 = 3$ and $-2c_1 - 4c_2 = -8$ (from the derivative at $t = 0$). The constants must be $c_1 = 2$ and $c_2 = 1$.

It is important to say that this problem—and all others like it—could be converted to a first-order system $dU/dt = AU$. The vector U contains the displacement and velocity:

$$U = \begin{bmatrix} u \\ du/dt \end{bmatrix} \quad \text{with} \quad U_0 = \begin{bmatrix} u_0 \\ u'_0 \end{bmatrix}.$$

Now we need two differential equations. One is the trivial statement $du/dt = du/dt$, which is the same as $dU_1/dt = U_2$. The other is the real equation $u'' + pu' + qu = 0$, or $U_2' + pU_2 + qU_1 = 0$. In vector form these equations are

$$\frac{d}{dt}\begin{bmatrix} U_1 \\ U_2 \end{bmatrix} = \begin{bmatrix} 0 & 1 \\ -q & -p \end{bmatrix}\begin{bmatrix} U_1 \\ U_2 \end{bmatrix}. \tag{20}$$

This can be solved in the normal way, starting with $U = e^{\lambda t}x$.

The point to notice is the equation for the eigenvalues:

$$A - \lambda I = \begin{bmatrix} -\lambda & 1 \\ -q & -p-\lambda \end{bmatrix} \quad \text{must have determinant} \quad \lambda^2 + p\lambda + q = 0.$$

That is the same as (18). The eigenvalues of A are identical with the exponents λ_1 and λ_2 in the original problem. In the example they are -2 and -4. As long as these exponents are not equal the problem is easy, whether as a second-order equation or a first-order system.

Unfortunately, a root λ is sometimes repeated:

$$u'' - 6u' + 9u = 0 \quad \text{leads to} \quad \lambda^2 - 6\lambda + 9 = 0.$$

This factors into $(\lambda - 3)^2$, and 3 is a repeated root. We are temporarily in trouble; the only pure exponential solution is e^{3t}. There must be a second solution, but it has a different form. It involves the same exponential *multiplied by* t:

$$u = te^{3t} \quad \text{is also a solution.}$$

You can verify that this satisfies the equation. It can be found systematically from the theory of differential equations. It can also be found by a trick that turns up when formulas become degenerate: If the usual solutions are $u(\lambda_1)$ and $u(\lambda_2)$, then at $\lambda_1 = \lambda_2$ they become u and $du/d\lambda$. In our case u is $e^{\lambda t}$ and $du/d\lambda$ yields the mysterious factor t.

The complete solution is $c_1 e^{3t} + c_2 te^{3t}$. The breakdown of a repeated root produced this exceptional second term, and we close by studying the same breakdown in the matrix case:

$$A = \begin{bmatrix} 0 & 1 \\ -q & -p \end{bmatrix} \quad \text{becomes} \quad A = \begin{bmatrix} 0 & 1 \\ -9 & 6 \end{bmatrix}.$$

This matrix looks harmless. Its characteristic equation $\det(A - \lambda I) = 0$ is $\lambda^2 - 6\lambda + 9 = 0$ (of course). The eigenvalue $\lambda = 3$ is repeated. The sum of eigenvalues is the trace 6, and the product is the determinant 9. But *there is only one eigenvector*:

$$(A - 3I)x = 0 \quad \text{gives} \quad \begin{bmatrix} -3 & 1 \\ -9 & 3 \end{bmatrix}\begin{bmatrix} x_1 \\ x_2 \end{bmatrix} = \begin{bmatrix} 0 \\ 0 \end{bmatrix} \quad \text{or} \quad x = c\begin{bmatrix} 1 \\ 3 \end{bmatrix}.$$

In the language of linear algebra, the nullspace is only one-dimensional. We hoped for two eigenvectors but found only one. Therefore A is not diagonalizable; there is no S such that $A = SAS^{-1}$. The eigenvalue matrix would be $\Lambda = 3I$, and then $S\Lambda S^{-1}$ would collapse to $3I$—which does not agree with A.

This matrix needs a *generalized eigenvector y*. It is connected to the true eigenvector x by $(A - 3I)y = x$, and therefore fails the eigenvector test $(A - 3I)y = 0$. However it comes close; multiplying again by $A - 3I$ gives

$$(A - 3I)^2 y = (A - 3I)x = 0.$$

We may take $x = (1, 3)$ and $y = (0, 1)$.

The best we can do—or the best Jordan could do—is to put x and y into the columns of S. Then the diagonal matrix Λ is replaced by the **Jordan form** J, in which an extra entry 1 appears above the repeated eigenvalue:

$$J = \begin{bmatrix} 3 & 1 \\ 0 & 3 \end{bmatrix} \quad \text{and} \quad A = SJS^{-1}.$$

The system $du/dt = Au$ could not be decoupled, but at least it can be made triangular. Then back substitution leads to the term te^{3t} which completes the solution.

Symmetric Matrices

The solutions of $Ax = \lambda x$ give the solution to $du/dt = Au$. That is the meeting point of two major theories, linear algebra and differential equations. They intersect in the eigenvalue problem and this part of the book is about that common ground. We have forgotten $Ax = b$ and we have not met $du/dt = Au + f$; the problem of forced oscillations is one of the first in Chapter 6. Here we stay with a limited but fundamental goal—to recognize how eigenvalues and eigenvectors solve differential equations, and how they reflect the properties of A. They carry the critical information about a matrix, the part that decides the behavior of u, and we look now for the special properties of λ and x when A is symmetric.

In a word, *symmetry means real eigenvalues and perpendicular eigenvectors*. A matrix with these properties (we mean a full set of eigenvectors) must be symmetric; every symmetric matrix has these properties. Diagonal matrices are included as a special case; they are obviously symmetric. Their eigenvalues are on the main diagonal and their eigenvectors lie along the coordinate axes. Those directions are certainly orthogonal. For other symmetric matrices the eigenvectors point in other directions, but the key property remains true: The eigenvectors are *perpendicular*. Every symmetric matrix is diagonal when the axes are rotated by the right matrix Q.

Earlier we had $S^{-1}AS = \Lambda$. The columns of S were the eigenvectors of A, not necessarily perpendicular. Now we change S to Q, as a way to emphasize what happens when A is symmetric. The columns of Q are still the eigenvectors, but here they are perpendicular and scaled to have length one. The result is that Q is an *orthogonal matrix*. Its inverse equals its transpose, as we see by multiplying

$$Q^TQ = \begin{bmatrix} x_1^T \\ \vdots \\ x_n^T \end{bmatrix} \begin{bmatrix} x_1 & \cdots & x_n \end{bmatrix} = \begin{bmatrix} 1 & & \\ & \ddots & \\ & & 1 \end{bmatrix} = I.$$

The diagonal entries are $x_i^T x_i = 1$ because of the scaling, and the off-diagonal entries are $x_i^T x_j = 0$ because of perpendicularity. Thus $Q^{-1} = Q^T$, and either of these symbols takes the place of S^{-1} in the main theorem about symmetric matrices.

1M If A is real and symmetric then

  (1) its eigenvalues are all real
  (2) its eigenvectors can be chosen orthonormal (perpendicular with length one)
  (3) $A = Q\Lambda Q^{-1} = Q\Lambda Q^T$.

If A is also positive definite then

  (4) its eigenvalues are not only real but positive.

Before the proof we look at four examples:

$$A_1 = \begin{bmatrix} 0 & 1 \\ -1 & 0 \end{bmatrix} \quad A_2 = \begin{bmatrix} 0 & 1 \\ 1 & 0 \end{bmatrix} \quad A_3 = \begin{bmatrix} 1 & 2 \\ -2 & 3 \end{bmatrix} \quad A_4 = \begin{bmatrix} 2 & 2 \\ 2 & 4 \end{bmatrix}.$$

The first is *skew*-symmetric and its eigenvalues are not real. The determinant of $A_1 - \lambda I$ is $\lambda^2 + 1 = 0$ and the eigenvalues are i and $-i$. (The eigenvectors are still perpendicular.) The second matrix is symmetric, with real eigenvalues 1 and -1; it is not positive definite. The third matrix has no special properties; its eigenvalues are $2 \pm \sqrt{3}\,i$. (Complex eigenvalues come in pairs $a \pm ib$, if A is real.) The fourth matrix has everything; it is symmetric positive definite with $\lambda = 3 \pm \sqrt{5}$. Its eigenvectors have to be perpendicular.

We now suppose that λ and x could be complex, and prove otherwise. The "conjugate" of $\lambda = a + ib$ is $\bar{\lambda} = a - ib$, and taking conjugates throughout $Ax = \lambda x$ would give $\bar{A}\bar{x} = \bar{\lambda}\bar{x}$. The transpose is $\bar{x}^T \bar{A}^T = \bar{x}^T \bar{\lambda}$, but in our symmetric case $\bar{A}^T$ is the same as A. Now only one idea is needed—to multiply by the right vectors:

$$\text{from } Ax = \lambda x \quad \text{we have} \quad \bar{x}^T A x = \bar{x}^T \lambda x$$

$$\text{from } \bar{x}^T A = \bar{x}^T \bar{\lambda} \quad \text{we have} \quad \bar{x}^T A x = \bar{x}^T \bar{\lambda} x.$$

Comparing the results, $\lambda = a + ib$ is the same as $\bar{\lambda} = a - ib$. Thus b is zero and λ is real. The eigenvector, which is found from $(A - \lambda I)x = 0$, can also be chosen real.

Remark   The proof depended entirely on $\bar{A}^T = A$. For a real matrix this is the same as symmetry: $A^T = A$. For a complex matrix it means that A is ***Hermitian***—each entry a_{ij} is the conjugate of a_{ji}. The diagonal is real, and three examples of Hermitian matrices are

$$\begin{bmatrix} 1 & 2+3i \\ 2-3i & 4 \end{bmatrix}, \quad \begin{bmatrix} 0 & i \\ -i & 0 \end{bmatrix}, \quad \begin{bmatrix} 1 & 2 \\ 2 & 3 \end{bmatrix}.$$

All their eigenvalues are real.

For eigenvectors that come from different eigenvalues, say $Ax = \lambda_1 x$ and $Ay = \lambda_2 y$, the same idea of taking inner products gives

$$y^T A x = y^T \lambda_1 x \quad \text{and} \quad y^T A^T x = y^T \lambda_2 x.$$

Since $A = A^T$ and $\lambda_2 \neq \lambda_1$, the common factor $y^T x$ must be zero: the eigenvectors are perpendicular. Except for one remaining possibility, this completes the theory. If A happens to have repeated eigenvalues, as in $A = I$, a slight perturbation is needed to make them distinct. The eigenvectors of the perturbed symmetric matrix are orthogonal, and they remain so as we approach A, which therefore has *a full set of n orthonormal eigenvectors*: $x_i^T x_j = 0$ for $i \neq j$. The eigenvectors go into the columns of Q (as they went into S). Since they are perpendicular and can be scaled to unit length, Q is an orthogonal matrix: $A = Q \Lambda Q^{-1} = Q \Lambda Q^T$.

Finally $\lambda > 0$ if A is positive definite. Again the idea is the same: the product of $Ax = \lambda x$ with x^T is $x^T A x = \lambda x^T x$. But $x^T x > 0$ and $x^T A x > 0$ (since A is positive definite) so λ must be positive.

The converse of 1M is equally true. If $A = Q \Lambda Q^T$ with real eigenvalues in Λ, then A is necessarily symmetric. If $A = Q \Lambda Q^T$ with positive eigenvalues in Λ, then A is necessarily positive definite.

EXAMPLE $A = \begin{bmatrix} 2 & -1 & 0 \\ -1 & 2 & -1 \\ 0 & -1 & 2 \end{bmatrix}$ $\det(A - \lambda I) = -\lambda^3 + 6\lambda^2 - 10\lambda + 4.$

The eigenvalues are $\lambda_1 = 2 - \sqrt{2}$, $\lambda_2 = 2$, $\lambda_3 = 2 + \sqrt{2}$. All are positive so A is positive definite. Their sum is 6 (the trace of A) and their product is 4 (the determinant).

This matrix comes from three oscillating masses, and the eigenvectors describe the three normal modes of oscillation. They give the special solutions ($a \cos \sqrt{\lambda}t + b \sin \sqrt{\lambda}t)x$ to the second-order equation $u_{tt} + Au = 0$. Comparing the three eigenvectors, the first (with smallest λ_1) moves all three masses in the same direction—the one in the center moves $\sqrt{2}$ further. The zero in the second eigenvector fixes the mass at the center; the others move oppositely. The third eigenvector (with highest frequency) moves masses 1 and 3 together but the mass between them is opposed (Fig. 1.10). For N masses we would find $0, 1, \ldots, N-1$ changes of sign in the N eigenvectors.

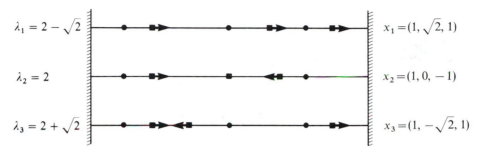

Fig. 1.10. Three normal modes of oscillation.

We close this section with comments on the different (but equivalent) conditions for a matrix to be positive definite:

(1) $x^T A x > 0$ for all nonzero vectors x
(2) all pivots are positive
(3) all eigenvalues are positive.

The three tests are useful at different times. The quadratic $x^T A x$ enters in equilibrium problems, where we look for a minimum. Eigenvalues enter in dynamic problems. Test (1) is convenient for a sum of matrices and test (3) for matrix powers:

if A and B are positive definite, so is $A + B$

if A is positive definite, so are A^2, A^{-1}, and all A^n

The former comes directly from $x^T(A + B)x = x^T A x + x^T B x$. The latter comes from the eigenvalues; they are λ_i^2 or λ_i^{-1} or λ_i^n, all positive. Even fractional powers are allowed, like the positive definite square root $A^{1/2}$ (Exercise 1.5.23) with eigenvalues $\lambda_i^{1/2}$. The zeroth power is the identity matrix, $A^0 = I$ with eigenvalues $\lambda_i^0 = 1$.

The second test enters in elimination, where the pivots decide if a *particular numerical matrix* is positive definite. Remember that positivity of the determinant is not a sufficient test. Certainly the determinant will be positive if the pivots are (it is their product). But it is also positive when two pivots are negative, as in

$$A = \begin{bmatrix} -1 & 0 \\ 0 & -1 \end{bmatrix} \text{ with determinant } +1.$$

The complete determinant test applies not just to A but to all the submatrices $A^{(1)}$, $A^{(2)}$, $\ldots$, $A^{(n)} = A$ in the upper left corner. The submatrix $A^{(k)}$ is k by k, and its determinant is the product of the first k pivots. When the pivots are positive these determinants are positive. And conversely, if a pivot $d_k \leq 0$ appears then the sequence of positive determinants is broken. Thus we have a fourth test for positive definiteness, very close to the second one:

| (4)   All the submatrices $A^{(k)}$ have positive determinants.

This completes the matrix theory.

EXERCISES

1.5.1   Find the eigenvalues of $A = \begin{bmatrix} a & b \\ b & a \end{bmatrix}$ and show that $(1, -1)$ and $(1, 1)$ are always eigenvectors. Confirm that $\lambda_1 + \lambda_2$ equals the sum of diagonal entries (the trace) and $\lambda_1 \lambda_2$ equals the determinant. Under what conditions on a and b is this matrix positive definite?

1.5.2   Write the preceding matrix in the form $A = S\Lambda S^{-1} = Q\Lambda Q^T$.

1.5.3   Find all eigenvalues and all eigenvectors (there are more than usual) of

$$A = \begin{bmatrix} 1 & 1 & 1 \\ 1 & 1 & 1 \\ 1 & 1 & 1 \end{bmatrix}.$$

1.5.4   Find the eigenvalues and eigenvectors of

$$A_1 = \begin{bmatrix} 3 & 4 \\ 4 & -3 \end{bmatrix} \quad \text{and} \quad A_2 = \begin{bmatrix} 1 & -1 & 0 \\ -1 & 2 & -1 \\ 0 & -1 & 1 \end{bmatrix}.$$

Check the trace and determinant.

1.5.5   Solve the first-order system

$$\frac{du}{dt} = \begin{bmatrix} 3 & 4 \\ 4 & -3 \end{bmatrix} u \quad \text{with} \quad u_0 = \begin{bmatrix} 0 \\ 6 \end{bmatrix}.$$

1.5.6   Solve the second-order system

$$\frac{d^2 u}{dt^2} + \begin{bmatrix} 1 & -1 & 0 \\ -1 & 2 & -1 \\ 0 & -1 & 1 \end{bmatrix} u = 0 \quad \text{with} \quad u_0 = \begin{bmatrix} 2 \\ -1 \\ -1 \end{bmatrix} \quad \text{and} \quad u_0' = \begin{bmatrix} 0 \\ 0 \\ 0 \end{bmatrix}.$$

These initial conditions do not activate the zero eigenvalue (see the following exercises).

1.5.7   Suppose each column of A adds to zero, as in

$$A = \begin{bmatrix} 3 & -1 & 0 \\ -2 & 2 & -1 \\ -1 & -1 & 1 \end{bmatrix}.$$

(a) Prove that zero is an eigenvalue and A is singular, by showing that the vector of ones is an eigenvector of A^T. (A and A^T have the same eigenvalues, but not the same eigenvectors.)

(b) Find the other eigenvalues of this matrix A, and all three eigenvectors.

1.5.8 With this 3 by 3 matrix, add the three equations $du/dt = Au$ to show that $u_1 + u_2 + u_3$ is a constant. What is the general solution (as in equation (10)) for this example? *Note:* When $\omega = 0$, $\sin \omega t$ is replaced by t in the general solution—just as the 1 by 1 model problem $d^2u/dt^2 = 0$ is solved by $u = a + bt$.

1.5.9 The $x - y$ axes are rotated through an angle θ by

$$Q = \begin{bmatrix} \cos \theta & \sin \theta \\ -\sin \theta & \cos \theta \end{bmatrix}.$$

(1) Verify that $Q^T = Q^{-1}$, so that Q is orthogonal.

(2) The rotated vector Qx is never in the same direction as x, so Q has no real eigenvalues. Find the (complex) eigenvalues and eigenvectors.

1.5.10 (a) Find the eigenvectors of $A = \begin{bmatrix} 0 & 1 \\ -1 & 0 \end{bmatrix}$ and show that they are perpendicular—remembering that the inner product of complex vectors is $x_1^H x_2 = \bar{x}_1^T x_2$ instead of $x_1^T x_2$.

(b) Solve the system $du/dt = Au$ with $u_0 = (3, 4)$.

1.5.11 Why is the sum of entries on the diagonal of AB equal to the sum along the diagonal of BA? In other words, what terms contribute to the trace of AB?

1.5.12 Show that the determinant equals the product of the eigenvalues by imagining that the characteristic polynomial is factored into

$$\det(A - \lambda I) = (\lambda_1 - \lambda)(\lambda_2 - \lambda) \cdots (\lambda_n - \lambda), \tag{*}$$

and making a clever choice of λ.

1.5.13 Show that the trace equals the sum of the eigenvalues, in two steps. First, find the coefficient of $(-\lambda)^{n-1}$ on the right side of (*). Next, look for all the terms in

$$\det(A - \lambda I) = \det \begin{bmatrix} a_{11} - \lambda & a_{12} & \cdots & a_{1n} \\ a_{21} & a_{22} - \lambda & \cdots & a_{2n} \\ \vdots & \vdots & & \vdots \\ a_{n1} & a_{n2} & \cdots & a_{nn} - \lambda \end{bmatrix}$$

which involve $(-\lambda)^{n-1}$. Explain why they all come from the main diagonal, and find the coefficient of $(-\lambda)^{n-1}$ on the left side of (*). Compare.

1.5.14   (a)   Show how the equation $u' = Au$ becomes $v' = Jv$ if $A = SJS^{-1}$ and $v = S^{-1}u$.

(b)   By back substitution (second equation first) solve

$$\begin{array}{l} dv_1/dt = 3v_1 + v_2 \\ dv_2/dt = \qquad 3v_2 \end{array} \quad \text{with} \quad v(0) = \begin{bmatrix} 1 \\ 1 \end{bmatrix}.$$

The term involving te^{3t} enters because of the repeated eigenvalue.

1.5.15   With masses $m_1 = m_2 = 1$ and spring constants $c_1 = c_3 = 4$, $c_2 = 6$, the differential equation is

$$u_{tt} + Ku = 0 \quad \text{with} \quad K = \begin{bmatrix} 10 & -6 \\ -6 & 10 \end{bmatrix}.$$

Find its natural frequencies ω_1 and ω_2, and from the eigenvectors find its two pure oscillations $u = (a \cos \omega t + b \sin \omega t)x$.

1.5.16   If the first mass starts at equilibrium and the second is displaced to $u_2 = 6$, with initial velocities $v_1 = v_2 = 0$, find their motions u_1 and u_2.

1.5.17   If $Kx = \omega^2 x$, show that $u = (ce^{i\omega t} + de^{-i\omega t})x$ solves the differential equation $u_{tt} + Ku = 0$. This exponential form is an alternative to the trigonometric form $u = (a \cos \omega t + b \sin \omega t)x$.

1.5.18   Solve the example in the text, with frequencies $\omega_1 = 1$ and $\omega_2 = \sqrt{3}$ as in (15), if the masses start at $u_1 = u_2 = 0$ with velocities $v_1 = 1 - \sqrt{3}$ and $v_2 = 1 + \sqrt{3}$. Show that $u(t)$ is never again zero.

1.5.19   If K is negative instead of positive, and $u_{tt} = u$ instead of $u_{tt} + u = 0$, solutions will grow or decay rather than oscillating. Solve $u_{tt} = u$ with $u(0) = 2$, $du/dt(0) = 0$.

1.5.20   Suppose there is a damping term proportional to velocity in

$$M \frac{d^2 u}{dt^2} + F \frac{du}{dt} + Ku = 0.$$

When will $u = e^{\lambda t}x$ be a solution?

1.5.21   If $A = \begin{bmatrix} a & b \\ b & c \end{bmatrix}$ then the determinant of $A - \lambda I$ is $(a - \lambda)(c - \lambda) - b^2$. From the formula for the roots of a quadratic, show that both eigenvalues are real.

1.5.22   Multiplying columns times rows, $A = Q\Lambda Q^T$ is

$$A = x_1 \lambda_1 x_1^T + x_2 \lambda_2 x_2^T + \cdots + x_n \lambda_n x_n^T.$$

This is the *spectral theorem*: Every symmetric matrix is a combination with weights λ of projections xx^T onto the eigenvectors. Write out this combination after rescaling to unit length the eigenvectors in the two text examples

$$A = \begin{bmatrix} 2 & 1 \\ 1 & 2 \end{bmatrix} \quad \text{and} \quad A = \begin{bmatrix} 2 & -1 & 0 \\ -1 & 2 & -1 \\ 0 & -1 & 2 \end{bmatrix}.$$

1.5.23 (*Positive definite square root*) Suppose A is positive definite: $A = Q\Lambda Q^T$ with $\lambda_i > 0$. Let $A^{1/2} = Q\Lambda^{1/2}Q^T$ be the matrix with the same eigenvectors in Q and with eigenvalues $\lambda_i^{1/2}$. Explain why this $A^{1/2}$ is symmetric positive definite, and its square is A. Find $A^{1/2}$ if

$$A = \begin{bmatrix} 10 & -6 \\ -6 & 10 \end{bmatrix}.$$

1.5.24 If K and M are positive definite and $Kx = \lambda Mx$, prove that λ is positive. This is the *generalized eigenvalue problem*, with two matrices. Find the two eigenvalues when

$$M = \begin{bmatrix} 1 & 0 \\ 0 & 2 \end{bmatrix} \quad \text{and} \quad K = \begin{bmatrix} 2 & -1 \\ -1 & 2 \end{bmatrix}.$$

1.5.25 For the same matrices M and K, coming from masses $m_1 = 1$ and $m_2 = 2$ and spring constants $c_1 = c_2 = c_3 = 1$, find the two pure oscillations $u = (a \cos \omega t + b \sin \omega t)x$ of the system $Mu_{tt} + Ku = 0$. Since $M^{-1}K$ is no longer symmetric, its eigenvectors x_1 and x_2 are no longer perpendicular; verify that now $x_1^T M x_2 = 0$.

1.5.26 Suppose a single mass m is between two springs with constants c_1 and c_2; their other ends are fixed. Write down (1) the second-order equation (Newton's law) for the displacement u of the mass, and (2) the frequency ω in the solution $u = a \cos \omega t + b \sin \omega t$.

1.6 ■ A REVIEW OF MATRIX THEORY

We close this chapter by admitting matrices that are not positive definite, not symmetric, not invertible, and in some cases not square. A will be an m by n matrix of rank r. The most important results can be summarized in a few pages, and it is amazing how closely they follow the direction set by $A = LDL^T$ and $A = Q\Lambda Q^T$. These equations will be replaced by others, but the essential facts are still captured by matrix factorizations—and the new ones still involve the interplay of triangular matrices, orthogonal matrices, and diagonal matrices.

Our review accepts $x^T y = 0$ as the test for two vectors to be perpendicular. It goes directly to the columns and rows and subspaces associated with A. If the matrix has five columns each lying in 7-dimensional space, or seven rows in 5-dimensional space, the key is to visualize their combinations. A subspace is what you get by taking *all linear combinations* of n given vectors. And another subspace, orthogonal to the first, contains all vectors that are perpendicular to the n given vectors. Those are the two ways to describe a plane or a line, and they extend to vector spaces of any dimension.

We begin with the most basic problem of linear algebra—to understand $Ax = b$. We want to know when this system has a solution. Numerically that is decided by elimination, but there is a more direct approach. The product Ax is always a *combination of the columns of* A, as in

$$Ax = \begin{bmatrix} 1 & -1 & 0 \\ 0 & 1 & -1 \\ 1 & 0 & -1 \end{bmatrix} \begin{bmatrix} x_1 \\ x_2 \\ x_3 \end{bmatrix} = x_1 \begin{bmatrix} 1 \\ 0 \\ 1 \end{bmatrix} + x_2 \begin{bmatrix} -1 \\ 1 \\ 0 \end{bmatrix} + x_3 \begin{bmatrix} 0 \\ -1 \\ -1 \end{bmatrix}. \tag{1}$$

Everything flows from that simple observation: to solve $Ax = b$ is to find a combination of the columns that gives b. We consider all possible combinations Ax, coming from all choices of x. Those products Ax form the **column space** of the matrix—the set of all vectors that are combinations of the columns. In the example the columns lie in 3-dimensional space. Their combinations fill out a plane. The plane goes through the origin, since one of the combinations has weights $x_1 = x_2 = x_3 = 0$. Some vectors b do not lie on the plane, and for them $Ax = b$ cannot be solved.

For an m by n matrix the columns lie in m-dimensional space. If there are r linearly independent columns then the column space has dimension r; it contains all combinations of those r columns, and the other $n - r$ columns add nothing new. (The example has $r = 2$, and the third column is a combination of the first two.) The column space is the whole of m-dimensional space when there are m independent columns, and $r = m$. In that case there is at least one solution for every b.

In general r is called the **rank** of the matrix, and the column space is denoted by $\mathscr{R}(A)$. The main point is:

1N Ax is always a combination of the columns of A; it is in the column space $\mathscr{R}(A)$. The system $Ax = b$ has a solution exactly when the right side b is also in the column space.

$Ax = b$ asks for a combination that produces b, and it exists when b is in the column space.

The 3 by 3 example has rank $r = 2$. The third column is not independent of the others, and the column space is only a plane. It is easy to describe that plane. In every column the sum of the first two components equals the third component. All combinations of the columns will inherit this property; if b is in $\mathscr{R}(A)$ then $b_1 + b_2 - b_3 = 0$. This is a plane through the origin and it is perpendicular to $y = (1, 1, -1)$. A random vector like $b = (2, 3, 4)$ is not in the column space and $Ax = b$ has no solution, since b misses the plane and $y^T b$ is not zero:

$$b_1 + b_2 - b_3 = y^T b = \begin{bmatrix} 1 & 1 & -1 \end{bmatrix} \begin{bmatrix} 2 \\ 3 \\ 4 \end{bmatrix} \neq 0.$$

The vector $b = (2, 3, 5)$ does lie in the plane, since $b_1 + b_2 - b_3 = 0$, and it must be a combination of the columns. In that case $Ax = b$ can be solved:

$$\begin{bmatrix} 2 \\ 3 \\ 5 \end{bmatrix} = 5 \begin{bmatrix} 1 \\ 0 \\ 1 \end{bmatrix} + 3 \begin{bmatrix} -1 \\ 1 \\ 0 \end{bmatrix} \quad \text{so that} \quad x = \begin{bmatrix} 5 \\ 3 \\ 0 \end{bmatrix}.$$

However there are other solutions for this b! If the components of x are 6, 4, 1 instead of 5, 3, 0, the combination Ax is unchanged. Adding the components 1, 1, 1 has no effect because that combination of the columns is zero:

$$A \begin{bmatrix} 1 \\ 1 \\ 1 \end{bmatrix} = \begin{bmatrix} 1 & -1 & 0 \\ 0 & 1 & -1 \\ 1 & 0 & -1 \end{bmatrix} \begin{bmatrix} 1 \\ 1 \\ 1 \end{bmatrix} = \begin{bmatrix} 0 \\ 0 \\ 0 \end{bmatrix}.$$

The vector with components 1, 1, 1 is in the **nullspace** of A; it satisfies $Ax = 0$. The same is true of $x = (3, 3, 3)$ and $x = (0, 0, 0)$. The nullspace of this matrix is a line through the origin.

For an m by n matrix the nullspace lies in n-dimensional space. It has dimension $n - r$ (in this case $3 - 2$). It is denoted by $\mathscr{N}(A)$ and contains all solutions to $Ax = 0$. Of course the nullspace always contains the vector $x = 0$; the question is whether it contains any nonzero vectors—in which case the columns are not independent.

Just as the column space was the key to existence of solutions, the nullspace is the key to uniqueness. The test is to see whether the columns are linearly independent—in other words whether the only combination that gives $Ax = 0$ is $x = 0$.

10   If the matrix A has linearly independent columns then

(1)   The nullspace contains only the point $x = 0$
(2)   The solution to $Ax = b$ (if there is one) is unique
(3)   The rank is $r = n$.

In general any two solutions to $Ax = b$ differ by a vector in the nullspace.

If $Ax_1 = b$ and $Ax_2 = b$ then $A(x_1 - x_2) = 0$ and the difference $x_1 - x_2$ lies in the nullspace.

Incidence Matrices

For the 3 by 3 example, this linear algebra is given a natural interpretation by the following graph:

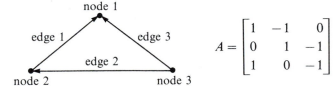

$$A = \begin{bmatrix} 1 & -1 & 0 \\ 0 & 1 & -1 \\ 1 & 0 & -1 \end{bmatrix}$$

The edges correspond to rows of the matrix; the nodes correspond to columns. The first edge goes from node 2 to node 1 and its row (the first row) has -1 in column 2 and $+1$ in column 1. The other edges (other rows) are constructed in the same way. Each row indicates the node it leaves by -1 and the node it enters by $+1$. The arrows are responsible for these signs but the flow can go either way. Then A is the *edge-node incidence matrix* of the graph, and the system $Ax = b$ is

$$x_1 - x_2 = b_1$$
$$x_2 - x_3 = b_2 \tag{2}$$
$$x_1 - x_3 = b_3.$$

Again we see the requirement $b_1 + b_2 - b_3 = 0$. On the left side of (2), the sum of the first two rows equals the third row. Therefore the same must be true on the right side:

$$(x_1 - x_2) + (x_2 - x_3) = (x_1 - x_3) \quad \text{so} \quad b_1 + b_2 = b_3. \tag{3}$$

Otherwise there will be no solution to $Ax = b$ and the vector b will lie outside the column space. All combinations of the columns satisfy (3) and are therefore perpendicular ($y^T b = 0$) to $y = (1, 1, -1)$.

This can also be seen on the graph. If x_1, x_2, x_3 are the potentials at the three nodes of the graph, then b_1, b_2, b_3 are the potential differences. The requirement (3) becomes one of the fundamental laws of circuit theory. It is **Kirchhoff's voltage law**: the sum of potential drops around a closed loop is zero. Since the loop goes against the arrow on edge 3, it is b_1 plus b_2 *minus* b_3 that must be zero.

Graph theory extends also to the nullspace. Equation (2) asks us to determine the potentials from the potential differences, and that is impossible. We can add a constant to all three potentials without changing their differences. In other words the vector $x = (1, 1, 1)$ is in the nullspace. Any multiple $cx = (c, c, c)$ can be added to any solution of (2). You can expect that this vector of 1's is in the nullspace of every incidence matrix.

There is another important thing to be learned from this example. It comes from the *transpose* of A, in which the rows are exchanged with the columns. In general A^T is n by m; it is the *node-edge* incidence matrix; the nodes now correspond to rows. Here A^T is square and we look at $A^T y = f$:

$$\begin{bmatrix} 1 & 0 & 1 \\ -1 & 1 & 0 \\ 0 & -1 & -1 \end{bmatrix} \begin{bmatrix} y_1 \\ y_2 \\ y_3 \end{bmatrix} = \begin{bmatrix} y_1 + y_3 \\ -y_1 + y_2 \\ -y_2 - y_3 \end{bmatrix} = \begin{bmatrix} f_1 \\ f_2 \\ f_3 \end{bmatrix}. \tag{4}$$

This is another system of linear equations and the basic questions are the same as for $Ax = b$:

(1)   Which right sides f allow the system to be solved?
(2)   Are the solutions to $A^T y = f$ unique? For this we solve $A^T y = 0$.

Those questions are easy for such a small matrix, but they will throw light on all matrices. Our first step is to answer them directly:

(1)   The sum of the left sides in $A^T y = f$ is zero, and the same must be true on the right side:

$$(y_1 + y_3) + (-y_1 + y_2) + (-y_2 - y_3) = 0 \quad \text{so} \quad f_1 + f_2 + f_3 = 0. \tag{5}$$

No net flow can come from outside the network. The column space of A^T contains only the vectors f whose components add to zero. In other words, all combinations of the columns are perpendicular to $x = (1, 1, 1)$.

(2)   The solutions to $A^T y = f$ are not unique, since there is a nonzero solution to $A^T y = 0$. It is the vector y with components $1, 1, -1$:

$$A^T y = \begin{bmatrix} 1 & 0 & 1 \\ -1 & 1 & 0 \\ 0 & -1 & -1 \end{bmatrix} \begin{bmatrix} 1 \\ 1 \\ -1 \end{bmatrix} = \begin{bmatrix} 0 \\ 0 \\ 0 \end{bmatrix}.$$

Any multiple of this y can be added to any solution of $A^T y = f$.

You cannot look at those answers without recognizing x and y. The vector $x = (1, 1, 1)$ that is perpendicular to the columns of A^T is the same vector that was in the nullspace of A. And the vector $y = (1, 1, -1)$ that was perpendicular to the columns of A has now appeared in the nullspace of A^T.

That is a beautiful pattern, and it never fails. Every matrix produces *four fundamental subspaces*, two from $Ax = b$ and two from $A^T y = f$. First come the column space $\mathscr{R}(A)$ and the nullspace $\mathscr{N}(A)$—the former in m-dimensional space R^m and the latter in n-dimensional space R^n. Then come the column space and nullspace of A^T, which are in R^n and R^m. These two subspaces are denoted by $\mathscr{R}(A^T)$ and $\mathscr{N}(A^T)$, and they are the *row space* and *left nullspace* of A. The dimensions and the perpendicularity of these four spaces make up the *fundamental theorem of linear algebra*:

1P   The row space is perpendicular to the nullspace and their dimensions add to n:

$$\dim \mathscr{R}(A^T) + \dim \mathscr{N}(A) = r + (n - r) = n.$$

The column space is perpendicular to the left nullspace and their dimensions add to m:

$$\dim \mathscr{R}(A) + \dim \mathscr{N}(A^T) = r + (m - r) = m.$$

The row space and column space have the same dimension r, equal to the rank of A. The number of linearly independent rows equals the number of linearly independent columns.

Thus every x in the nullspace is perpendicular to every f in the row space:

$$\text{if } A^T y = f \quad \text{and} \quad Ax = 0 \quad \text{then} \quad x^T f = x^T A^T y = 0^T y = 0.$$

It also comes directly from the equation $Ax = 0$ that puts x in the nullspace:

$$Ax = \begin{bmatrix} \text{row } 1 \\ \text{row } 2 \\ \cdots \\ \text{row } m \end{bmatrix} \begin{bmatrix} \\ x \\ \\ \end{bmatrix} = \begin{bmatrix} 0 \\ 0 \\ \cdot \\ 0 \end{bmatrix}.$$

Each equation makes x perpendicular to a row of A. Similarly all columns are perpendicular to each y in the left nullspace:

$$A^T y = \begin{bmatrix} \text{column } 1 \\ \text{column } 2 \\ \cdots \\ \text{column } n \end{bmatrix} \begin{bmatrix} \\ y \\ \\ \end{bmatrix} = \begin{bmatrix} 0 \\ 0 \\ \cdot \\ 0 \end{bmatrix}.$$

The two pairs of subspaces and the action of A are illustrated in Fig. 1.11—which contains all the information in the fundamental theorem.

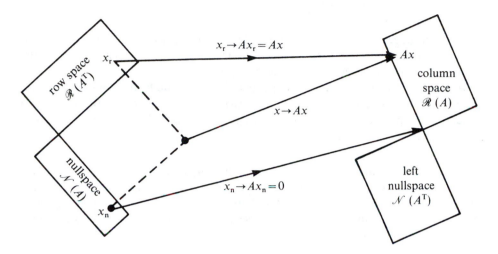

Fig. 1.11. The four subspaces and the action of A.

The rank and the four dimensions can be found for every incidence matrix. Consider the rectangular A associated with five edges and four nodes:

$$A = \begin{bmatrix} -1 & 1 & 0 & 0 \\ -1 & 0 & 1 & 0 \\ 0 & -1 & 1 & 0 \\ 0 & -1 & 0 & 1 \\ 0 & 0 & -1 & 1 \end{bmatrix}$$

The nullspace of A is still a line. It contains the vector x with components 1, 1, 1, 1. The product Ax is the sum of the four columns of A, and it is zero. In other words x is perpendicular to each of the rows.

The nullspace of A^T is now 2-dimensional, since there are two independent loops in the graph. Their edges lead to

$$y_1 = \begin{bmatrix} 1 \\ -1 \\ 1 \\ 0 \\ 0 \end{bmatrix} \quad \text{and} \quad y_2 = \begin{bmatrix} 0 \\ 0 \\ -1 \\ 1 \\ -1 \end{bmatrix}$$

if we watch the arrows as we go around the loops. These vectors satisfy $A^T y_1 = 0$ and $A^T y_2 = 0$. The loop around the whole square comes from $y_1 + y_2$ and is not independent of the smaller loops. The rank is $r = 3$, which gives the dimension of the column space and row space.

Column space: Any 3 columns are independent but the sum of all four is zero. $Ax = b$ can be solved only when b is perpendicular to y_1 and y_2. This is Kirchhoff's *voltage law* for the potential differences b.

Row space: Any 3 rows are independent if they come from edges that contain no loops (for example rows 1, 2, 4). The *current law* $A^T y = f$ can be solved only when f is perpendicular to $x = (1, 1, 1, 1)$. Then $f_1 + f_2 + f_3 + f_4 = 0$ and the total current entering from outside is zero.

Kirchhoff's laws will be studied more systematically in Chapter 2; this is a preview.

The pattern is the same for every incidence matrix, provided the graph is *connected*. Between every two nodes there must be a path of edges (ignoring arrows). This is the important case, because a disconnected graph with two or more pieces has an incidence matrix like

$$A = \begin{bmatrix} A_1 & 0 \\ 0 & A_2 \end{bmatrix}.$$

Each block is an incidence matrix for its own piece of the graph. Every graph is built out of connected graphs, and the incidence matrix of a connected graph can be completely understood.

1Q For a connected graph with m edges and n nodes the incidence matrix has rank $r = n - 1$. The nullspace of A is the line through the vector $(1, 1, ..., 1)$. The nullspace of A^T has dimension $m - r = m - n + 1$ and it contains vectors y_i from $m - n + 1$ independent loops of the graph.

The equation $Ax = b$ can be solved when b is perpendicular to all y_i (Kirchhoff's voltage law). The current law $A^T y = f$ can be solved when f is perpendicular to $(1, 1, ..., 1)$.

The rank is $n - 1$ because the nullspace is 1-dimensional. If $Ax = 0$ then x has the form $(c, c, ..., c)$ by Exercise 1.6.5, and x is a multiple of $(1, 1, ..., 1)$.

That completes a quick introduction to linear algebra. It covers the first part of the course, about subspaces and transposes and $Ax = b$. Next comes the advanced course, in which each step is compressed into a matrix factorization. We admit all matrices, not only incidence matrices.

The list of factorizations will be partly a restatement of what you know and partly a reference for the future. It is not expected that they will all be presented in class.

Factorizations Based on Elimination

The first two factorizations on the list extend the triple product $A = LDL^T$ to matrices that are square and invertible, but not necessarily symmetric or positive definite. Then the third admits rectangular matrices.

1. $A = LDU$:   **Triangular factorization without row exchanges.**
L is lower triangular, U is upper triangular, and both have ones on the diagonal. D is the diagonal matrix of nonzero pivots. The factors L, D, U come directly from elimination as in Section 1.3; the only difference is that without symmetry we have $U \neq L^T$. The factorization succeeds **only if the pivots are not zero.**

This is the basic calculation in scientific computing. The first stage of elimination changes the $n^2 - n$ entries below the first row. After all stages the total operation count on the left side of $Ax = b$ is

$$(n^2 - n) + ((n-1)^2 - (n-1)) + \cdots + (1^2 - 1) = \frac{n^3 - n}{3}.$$

Thus elimination uses about $n^3/3$ multiplications and subtractions for a full matrix of order n. It splits $Ax = b$ into two triangular systems $Lc = b$ and $DUx = c$. On the right side of the equations it needs only n^2 steps, half in forward elimination and half during back-substitution.

2. $PA = LDU$:   **Triangular factorization with row exchanges.**
For every invertible A there is a permutation matrix P that will reorder the rows to achieve nonzero pivots. Thus PA has an LDU factorization. In particular P is needed if $a_{11} = 0$; the first row is exchanged for a later row before elimination begins.

In practice a row exchange is allowed at every step of elimination, to put the largest entry of the column into the pivot position. This is "partial pivoting." For positive definite matrices it is not required and $P = I$.

3. $PA = LU$:   **Reduction to echelon form.**
Every rectangular matrix A can be changed by row operations into a matrix U that has zeros below its pivots $d_1, \ldots, d_r$. The last $m - r$ rows of U are entirely zero. In this factorization L is square as before, while D and U are combined into a single rectangular matrix with the same shape as A. For the sake of confusion it is again named U.

EXAMPLES:

$$A = \begin{bmatrix} 1 & 1 \\ 2 & 7 \end{bmatrix} = \begin{bmatrix} 1 & 0 \\ 2 & 1 \end{bmatrix} \begin{bmatrix} 1 & 0 \\ 0 & 5 \end{bmatrix} \begin{bmatrix} 1 & 1 \\ 0 & 1 \end{bmatrix} = LDU$$

$$A = \begin{bmatrix} 0 & 1 \\ 1 & 6 \end{bmatrix}, \quad P = \begin{bmatrix} 0 & 1 \\ 1 & 0 \end{bmatrix}, \quad PA = \begin{bmatrix} 1 & 6 \\ 0 & 1 \end{bmatrix} = LDU \quad \text{with} \quad L = D = I$$

$$A = \begin{bmatrix} 1 & 0 \\ 1 & 1 \\ 0 & -1 \end{bmatrix} = \begin{bmatrix} 1 & 0 & 0 \\ 1 & 1 & 0 \\ 0 & -1 & 1 \end{bmatrix} \begin{bmatrix} 1 & 0 \\ 0 & 1 \\ 0 & 0 \end{bmatrix} = LU \quad \text{with} \quad P = I.$$

Each example takes the usual elimination steps from A to U, and keeps a record of those steps in L. Below the diagonal, l_{ij} is the multiple of row j that is subtracted from row i. For square matrices the *determinant* is the product of the pivots, with sign reversed when the number of row exchanges is odd. In the first example det $A = 5$; the second matrix has det $A = -2$.

Factorizations Based on Eigenvalues

The next factorizations are associated with the eigenvalue problem, and therefore with square matrices. If A is symmetric we established in Section 1.5 that $A = Q\Lambda Q^T$; Λ is the diagonal matrix of eigenvalues and Q is the orthogonal matrix of eigenvectors. Without symmetry this is no longer possible; almost every matrix can be diagonalized but the eigenvectors are not normally orthogonal. The exceptions, which are not diagonalizable, are covered by the Jordan form.

   4. $A = S\Lambda S^{-1}$:   **Diagonalization of A.**
A can be diagonalized if it has a full set of n linearly independent eigenvectors. They are the columns of S, and $S^{-1}AS$ is the diagonal matrix Λ of eigenvalues. In the best case, with $A^T A = AA^T$, the eigenvectors can be chosen orthonormal and S becomes Q. This includes symmetric matrices (whose eigenvalues are real), skew-symmetric matrices ($A = -A^T$ and the eigenvalues are imaginary), and orthogonal matrices ($A^T = A^{-1}$ and all $|\lambda_i| = 1$).

   5. $A = MJM^{-1}$:   **Jordan form.**
Every square matrix A is similar to a Jordan matrix J, with the eigenvalues $J_{ii} = \lambda_i$ on the diagonal. J has a diagonal block of the form

$$\begin{bmatrix} \lambda & 1 & & \\ & \cdot & \cdot & \\ & & \cdot & 1 \\ & & & \lambda \end{bmatrix}$$

for each independent eigenvector. The sizes of the blocks can vary; if A is diagonalizable then they are all 1 by 1 and J is diagonal. Thus if the eigenvalues are distinct or if repeated eigenvalues admit a full set of eigenvectors, J is Λ and M is S; the Jordan form then reduces to the diagonalization $A = S\Lambda S^{-1}$. In general A may have $l < n$ eigenvectors, and the sizes of the l blocks in J (with $n - l$ off-diagonal 1's) are determined by A.

EXAMPLES:

$A = \begin{bmatrix} 1 & 2 \\ 0 & 3 \end{bmatrix}$ has eigenvalues 1 and 3 with eigenvectors $x_1 = \begin{bmatrix} 1 \\ 0 \end{bmatrix}$, $x_2 = \begin{bmatrix} 1 \\ 1 \end{bmatrix}$.

Therefore $A = S\Lambda S^{-1} = \begin{bmatrix} 1 & 1 \\ 0 & 1 \end{bmatrix} \begin{bmatrix} 1 & \\ & 3 \end{bmatrix} \begin{bmatrix} 1 & -1 \\ 0 & 1 \end{bmatrix}$.

$A = \begin{bmatrix} 3 & 2 \\ 0 & 3 \end{bmatrix}$ has eigenvalues $\lambda_1 = \lambda_2 = 3$ and only one eigenvector $x_1 = \begin{bmatrix} 1 \\ 0 \end{bmatrix}$.

Therefore A is not diagonalizable and its Jordan form has a single block:

$$J = M^{-1}AM = \begin{bmatrix} 1 & 0 \\ 0 & 2 \end{bmatrix} \begin{bmatrix} 3 & 2 \\ 0 & 3 \end{bmatrix} \begin{bmatrix} 1 & 0 \\ 0 & \frac{1}{2} \end{bmatrix} = \begin{bmatrix} 3 & 1 \\ 0 & 3 \end{bmatrix}.$$

In every case the product of the eigenvalues is the determinant.

Factorizations Based on $A^T A$

The last three factorizations apply to m by n matrices A but they involve the square matrix $A^T A$. This product is always symmetric. Furthermore $A^T A$ is at least *positive semidefinite*. A quadratic like $x^T A^T A x$ can never be negative because it is the square of a length:

$$x^T A^T A x = (Ax)^T (Ax) = \|Ax\|^2 \geq 0. \tag{6}$$

The eigenvalues of $A^T A$ are never negative; the question is whether any eigenvalue is zero. It is again equation (6) that gives the answer: Everything is strictly positive except when $Ax = 0$. From this simple observation come the key properties of $A^T A$:

1R   If the columns of A are linearly independent then $A^T A$ is symmetric positive definite. If the columns are linearly dependent then $A^T A$ is not invertible; at least one of its eigenvalues is zero. In general A and $A^T A$ have the same nullspace, the same row space, and the same rank.

The whole point is that Ax and $A^T Ax$ and $x^T A^T Ax$ are either all zero or all nonzero.

A good example is the 3 by 3 matrix that came earlier from a triangular graph. We now call it A_0 to emphasize that its columns are dependent:

$$A_0 = \begin{bmatrix} 1 & -1 & 0 \\ 0 & 1 & -1 \\ 1 & 0 & -1 \end{bmatrix} \quad \text{and} \quad A_0^T A_0 = \begin{bmatrix} 2 & -1 & -1 \\ -1 & 2 & -1 \\ -1 & -1 & 2 \end{bmatrix}. \tag{7}$$

$A_0^T A_0$ is symmetric and only semidefinite; its eigenvalues are 3, 3, and 0. Each matrix has the vector x with components 1, 1, 1 in its nullspace. Both matrices have rank 2, and their row spaces contain all vectors perpendicular to $(1, 1, 1)$.

Now remove the last column of A_0. That leaves independent columns and it removes the last row and column from $A_0^T A_0$:

$$A = \begin{bmatrix} 1 & -1 \\ 0 & 1 \\ 1 & 0 \end{bmatrix} \quad \text{and} \quad A^T A = \begin{bmatrix} 2 & -1 \\ -1 & 2 \end{bmatrix}. \tag{8}$$

With this change $A^T A$ becomes symmetric positive definite. Its eigenvalues are 3 and 1, and they fall between the earlier eigenvalues 3, 3, 0. Notice that A^T still has dependent columns and AA^T (which is not the transpose of $A^T A$!) is *semidefinite*:

$$A^T = \begin{bmatrix} 1 & 0 & 1 \\ -1 & 1 & 0 \end{bmatrix} \quad \text{and} \quad AA^T = \begin{bmatrix} 2 & -1 & 1 \\ -1 & 1 & 0 \\ 1 & 0 & 1 \end{bmatrix}. \tag{9}$$

The eigenvalues of AA^T are 3, 1, 0, agreeing with $A^T A$ except for the zero eigenvalue. The eigenvector associated with $\lambda = 0$ has components 1, 1, -1 and we have seen it before; it is in the nullspace of A^T and A_0^T and AA^T and $A_0 A_0^T$. It came from the loop in the triangular graph.

Finally we describe the last three factorizations. All are important but the first is supremely important; orthogonalization is central to numerical linear algebra. We summarize it briefly here, and return to Gram-Schmidt in Chapter 5.

6. $A = QR$: Orthogonalization of the columns of A.
A must have independent columns $a_1, \ldots, a_n$. Q has **orthonormal** columns $q_1, \ldots, q_n$. (Orthonormal means that $q_i^T q_j = 0$ if $i \neq j$ and $q_i^T q_i = 1$—in other words $Q^T Q = I$.) The link between A and Q is an upper triangular matrix R:

$$A = \begin{bmatrix} a_1 & a_2 & a_3 \end{bmatrix} = \begin{bmatrix} q_1 & q_2 & q_3 \end{bmatrix} \begin{bmatrix} q_1^T a_1 & q_1^T a_2 & q_1^T a_3 \\ 0 & q_2^T a_2 & q_2^T a_3 \\ 0 & 0 & q_3^T a_3 \end{bmatrix} = QR.$$

Gram-Schmidt chooses q_1 in the direction of a_1, and q_2 in the plane of a_1 and a_2. It subtracts multiples of earlier columns from later columns, to produce orthogonal columns:

$$A = \begin{bmatrix} 1 & 0 & 0 \\ -1 & 1 & 0 \\ 0 & -1 & 1 \\ 0 & 0 & -1 \end{bmatrix} \rightarrow \begin{bmatrix} 1 & \frac{1}{2} & 0 \\ -1 & \frac{1}{2} & 0 \\ 0 & -1 & 1 \\ 0 & 0 & -1 \end{bmatrix} \rightarrow \begin{bmatrix} 1 & \frac{1}{2} & \frac{1}{3} \\ -1 & \frac{1}{2} & \frac{1}{3} \\ 0 & -1 & \frac{1}{3} \\ 0 & 0 & -1 \end{bmatrix}.$$

Dividing those columns by their lengths gives Q. Then $R^T R$ is $A^T A$.

7. $A = Q_1 \Sigma Q_2^T$: Singular Value Decomposition.
A is m by n of rank r. Its "singular values" are the square roots of the r positive eigenvalues of $A^T A$, which are also the r positive eigenvalues of AA^T. These square

roots $\sigma_1, \ldots, \sigma_r$ are the only nonzero entries in the m by n diagonal matrix Σ. Q_1 and Q_2 are orthogonal matrices of orders m and n, and their columns contain the eigenvectors of AA^T and A^TA.

In case A is symmetric positive definite, its singular value decomposition is just $Q\Lambda Q^T$; Σ is the eigenvalue matrix and $Q_1 = Q_2 = Q$. For general rectangular matrices the *SVD* is new and useful; it is a stable way to display the rank in Σ, the column space and left nullspace in Q_1, and the row space and nullspace in Q_2^T. It leads to the *pseudoinverse* when A is not invertible.†

8. $A = QB$: **Polar decomposition.**
The polar form of an invertible matrix is the analogue of $z = re^{i\theta}$ for a nonzero complex number. It splits A into an orthogonal matrix Q (like $e^{i\theta}$) and a symmetric positive definite matrix B (like r). In continuum mechanics it separates the rotation from the stretching.

B is the positive definite square root of A^TA, just as r is the positive square root of $\bar{z}z$. Then $Q = AB^{-1}$ automatically has orthonormal columns:

$$Q^TQ = B^{-1}A^TAB^{-1} = B^{-1}B^2B^{-1} = I.$$

There is another polar factorization $A = B'Q'$ with the matrices in the opposite order, and we can also have $A = QB$ with rectangular A and Q.

The examples use the same incidence matrix A as above, with two independent columns taken from A_0. All the factorizations 6–8 involve square roots of A^TA and you will see the final results; there is no way to preserve whole numbers.

Orthogonalization: $A = \begin{bmatrix} 1 & -1 \\ 0 & 1 \\ 1 & 0 \end{bmatrix} = \begin{bmatrix} 1/\sqrt{2} & -1/\sqrt{6} \\ 0 & 2/\sqrt{6} \\ 1/\sqrt{2} & 1/\sqrt{6} \end{bmatrix}$

$$\times \begin{bmatrix} \sqrt{2} & -1/\sqrt{2} \\ 0 & \sqrt{6}/2 \end{bmatrix} = QR$$

Singular Value
Decomposition: $A = \begin{bmatrix} 0 & 2/\sqrt{6} & 1/\sqrt{3} \\ 1/\sqrt{2} & -1/\sqrt{6} & 1/\sqrt{3} \\ 1/\sqrt{2} & 1/\sqrt{6} & -1/\sqrt{3} \end{bmatrix} \begin{bmatrix} 1 & \\ & \sqrt{3} \\ & \end{bmatrix}$

$$\times \begin{bmatrix} 1/\sqrt{2} & 1/\sqrt{2} \\ 1/\sqrt{2} & -1/\sqrt{2} \end{bmatrix} = Q_1 \Sigma Q_2^T$$

---

† This is developed in my textbook *Linear Algebra and Its Applications* (Academic Press, 1980) and in other references.

$$\text{Polar form: } A = \begin{bmatrix} 1/\sqrt{3} & -1/\sqrt{3} \\ \dfrac{-1+\sqrt{3}}{2\sqrt{3}} & \dfrac{1+\sqrt{3}}{2\sqrt{3}} \\ \dfrac{1+\sqrt{3}}{2\sqrt{3}} & \dfrac{-1+\sqrt{3}}{2\sqrt{3}} \end{bmatrix} \begin{bmatrix} \dfrac{1+\sqrt{3}}{2} & \dfrac{1-\sqrt{3}}{2} \\ \dfrac{1-\sqrt{3}}{2} & \dfrac{1+\sqrt{3}}{2} \end{bmatrix} = QB.$$

That completes the list of factorizations. Every combination of orthogonal, triangular, diagonal, and positive definite matrices seems to be important. Several may be familiar but in a different disguise; Gram-Schmidt is often discussed in textbooks but not written as QR. It stabilizes an enormous range of calculations. LDU is the key to a linear system, while $S\Lambda S^{-1}$ diagonalizes a system of differential equations. When A is symmetric these become the LDL^T and $Q\Lambda Q^T$ on which this chapter was built.

Review of the Review

Our examples so far have been incidence matrices. The nullspace of A_0 was a line through $(1, 1, ..., 1)$ and the nullspace of A_0^T came from $m - n + 1$ loops of the graph. The rank was $n - 1$. It is hard to find better examples, but by going to the other extreme we can find matrices that are simpler. Instead of rank $n - 1$, they will be matrices of **rank one**.

If the rank is one then the column space is 1-dimensional. All columns are multiples of the same column, as in

$$A = \begin{bmatrix} 1 & 2 & 1 \\ 3 & 6 & 3 \\ 2 & 4 & 2 \end{bmatrix}.$$

Notice that the row space is also 1-dimensional (as it must be). All rows are multiples of the same row vector w^T, and a rank-one matrix is the product of *a column times a row*:

$$A = vw^T = \begin{bmatrix} 1 \\ 3 \\ 2 \end{bmatrix} \begin{bmatrix} 1 & 2 & 1 \end{bmatrix} = \begin{bmatrix} 1 & 2 & 1 \\ 3 & 6 & 3 \\ 2 & 4 & 2 \end{bmatrix}.$$

For a matrix like this we can find the nullspace and left nullspace and eigenvalues and eigenvectors. The determinant is especially easy; it is zero, and the matrix is not invertible.

Here are the four fundamental spaces for the rank-one matrix $A = vw^T$:

(1) The column space contains all multiples of v
(2) The row space contains all multiples of w

(3)   The nullspace contains all vectors orthogonal to w
(4)   The left nullspace contains all vectors orthogonal to v.

Of course the matrix is wiped out in one step of elimination. The 3 by 3 example factors into

$$A = \begin{bmatrix} 1 & 2 & 1 \\ 3 & 6 & 3 \\ 2 & 4 & 2 \end{bmatrix} = \begin{bmatrix} 1 & & \\ 3 & 1 & \\ 2 & 0 & 1 \end{bmatrix} \begin{bmatrix} 1 & 2 & 1 \\ & 0 & 0 \\ & & 0 \end{bmatrix}$$

and there are no pivots after the first.

What are the eigenvalues and eigenvectors? Each vector in the nullspace is an eigenvector corresponding to $\lambda = 0$; it satisfies $Ax = 0x$. In our case the nullspace is a plane and $\lambda = 0$ is a double eigenvalue. In general $\lambda = 0$ is repeated $n - 1$ times— the nullspace has dimension $n - 1$—and only one eigenvalue remains to be found. Its eigenvector is v itself:

$$Av = vw^T v \quad \text{with eigenvalue} \quad \lambda = w^T v. \tag{10}$$

That product $vw^T v$ deserves another look. It began as a matrix vw^T times a vector v. It ended as a vector v times a number $w^T v$. Without changing the order of the matrices, that last multiplication can be done first. This happens also for $A^2 = (vw^T)(vw^T)$, where the number $\lambda = w^T v$ can be extracted from the middle to leave $A^2 = \lambda A$. It happens again in the exercises. In our example $Av = \lambda v$ is

$$Av = \begin{bmatrix} 1 & 2 & 1 \\ 3 & 6 & 3 \\ 2 & 4 & 2 \end{bmatrix} \begin{bmatrix} 1 \\ 3 \\ 2 \end{bmatrix} = 9 \begin{bmatrix} 1 \\ 3 \\ 2 \end{bmatrix}.$$

You recognize 9 as the trace of A. The sum of the diagonal entries equals the sum of the eigenvalues, and for a rank one matrix the eigenvalues are $0, \ldots, 0$, and $w^T v$.

I did not expect it, but rank one matrices are also excellent illustrations of the Jordan form. Normally this form is diagonal; all we need is a full set of independent eigenvectors. The nullspace produces $n - 1$ eigenvectors, each with $\lambda = 0$, and the other eigenvector is supposed to be v. Therefore *if v is not in the nullspace* the matrix can be diagonalized. The eigenvectors go into the columns of the diagonalizing matrix S:

$$S = \begin{bmatrix} 1 & 1 & 1 \\ 3 & -1 & 0 \\ 2 & 1 & -1 \end{bmatrix} \quad \text{and} \quad S^{-1} A S = \Lambda = \begin{bmatrix} 9 & & \\ & 0 & \\ & & 0 \end{bmatrix}. \tag{11}$$

The first column of S is v and its eigenvalue $\lambda = 9$ is in Λ. The other columns of S are perpendicular to w and their eigenvalues are $\lambda = 0$. We could check (11) by computing S^{-1} and multiplying out $S^{-1} A S$, but linear algebra can be trusted.

In this case the Jordan form was $J = \Lambda$. But there is an exception, when $A = vw^T$ cannot be diagonalized. It occurs when $w^T v = 0$, so that v happens to fall into the nullspace. The eigenvalue $\lambda = w^T v$ is zero like the others. Then the diagonal matrix would be $\Lambda = 0$, and we would be trying for $A = S\Lambda S^{-1} = 0$—which is not possible. Instead there are only $n - 1$ independent eigenvectors, all in the nullspace, and we have to accept an off-diagonal 1 in the Jordan form. Here is an example:

$$
v = \begin{bmatrix} 1 \\ 3 \\ 2 \end{bmatrix}, \quad w = \begin{bmatrix} 1 \\ -1 \\ 1 \end{bmatrix}, \quad A = vw^T = \begin{bmatrix} 1 & -1 & 1 \\ 3 & -3 & 3 \\ 2 & -2 & 2 \end{bmatrix}, \quad J = \begin{bmatrix} 0 & 1 & 0 \\ 0 & 0 & 0 \\ 0 & 0 & 0 \end{bmatrix}.
$$

The trace of A is zero, $v^T w$ is zero, and all eigenvalues are zero. But there are only two independent eigenvectors.

We mention two applications of rank one matrices. They are present whenever you change a row or a column. The original matrix need not have rank one, but the change does. This happens in least squares, when a new measurement arrives. It brings a rank one change in the LU factorization, or the QR factorization, or the SVD—and this updating is a key step in numerical linear algebra. The exercises ask about the inverse. The difference between A^{-1} and $(A - vw^T)^{-1}$ also has rank one, and its "Sherman-Morrison formula" is important and not obvious.

In the other application vw^T is symmetric; v is the same as w. This vector has length one and we call it u:

$$
A = uu^T \quad \text{with eigenvalue} \quad u^T u = 1.
$$

A is the **projection matrix** onto the line through u. For example,

$$
u = \begin{bmatrix} .6 \\ .8 \end{bmatrix} \quad \text{gives} \quad A = uu^T = \begin{bmatrix} .36 & .48 \\ .48 & .64 \end{bmatrix}.
$$

The trace is $.36 + .64 = 1$. When A multiplies b it gives the projection $uu^T b$—the multiple of u that is closest to b. Like all projections this matrix satisfies $A^2 = A$.

That completes our review. We did not use the determinant, which leads to a test for invertibility and to Cramer's formula for $x = A^{-1}b$. And we did not define "dimension" or "basis." The dimension is the largest number of linearly independent vectors in the given space. Such a largest set of vectors is a basis, and their combinations produce all other vectors in the space. The n columns of an invertible matrix give one basis for the whole space R^n, and (unless the matrix is symmetric) the rows give a different basis. Despite these omissions, and others, I hope you have seen something new.

The exercises range over much of linear algebra, and they also touch one topic outside it: Exercise 1.6.1 asks how to decide whether a graph is connected. In principle you could compute the incidence matrix and check its rank. If possible, can you start from a list of edges and find a more efficient algorithm? It is a first contact with the combinatorial problems of Chapter 7—which are matrix problems of an important and special kind.

EXERCISES

1.6.1   Suppose you are given a list of the edges of a graph; for edge i you are given its two node numbers $a(i)$ and $b(i)$. Describe the steps of an algorithm (or write a code) to decide whether or not the graph is connected.

1.6.2   Write down the incidence matrices A_1 and A_2 for the following graphs:

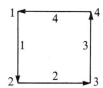
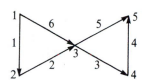

For which right sides does $A_1 x = b$ have a solution? Which vectors are in the nullspace of A_1^T?

1.6.3   The previous matrix A_2 should have $n - 1$ independent rows; which are they? There should also be $m - n + 1$ independent vectors in the nullspace of A_2^T, one from each loop; which are they?

1.6.4   If those two graphs are the disconnected pieces of a single graph with 9 nodes and 10 edges, what is the rank of its incidence matrix? With n nodes, m edges, and p unconnected pieces, find the number of independent solutions to $Ax = 0$ and $A^T y = 0$.

1.6.5   If A is the incidence matrix of a connected graph and $Ax = 0$, show that $x_1 = x_2 = \cdots = x_n$. Each row of $Ax = 0$ is an equation $x_j - x_k = 0$; how do you prove that $x_j = x_k$ even when no edge goes from node j to node k?

1.6.6   In a graph with N nodes and N edges show that there must be a loop.

1.6.7   For electrical networks x represents potentials, Ax represents potential differences, y represents currents, and $A^T y = 0$ is Kirchhoff's current law (Section 2.3). Tellegen's theorem says that Ax is perpendicular to y. How does this follow from the fundamental theorem of linear algebra?

1.6.8   By transposing $AA^{-1} = I$, show that the transpose of A^{-1} is the inverse of A^T. Verify $(A^{-1})^T = (A^T)^{-1}$ for the 2 by 2 matrix

$$A = \begin{bmatrix} 1 & 0 \\ 4 & 1 \end{bmatrix}.$$

1.6.9   Describe the vectors in the column space, nullspace, row space, and left nullspace of

$$A = \begin{bmatrix} 0 & 1 & 0 & 0 \\ 0 & 0 & 1 & 0 \\ 0 & 0 & 0 & 0 \end{bmatrix}.$$

1.6.10   In the list below, which classes of matrices contain

$$A = \begin{bmatrix} 0 & 1 & 0 & 0 \\ 0 & 0 & 1 & 0 \\ 0 & 0 & 0 & 1 \\ 1 & 0 & 0 & 0 \end{bmatrix} \quad \text{and} \quad B = \begin{bmatrix} \frac{1}{3} & \frac{1}{3} & \frac{1}{3} \\ \frac{1}{3} & \frac{1}{3} & \frac{1}{3} \\ \frac{1}{3} & \frac{1}{3} & \frac{1}{3} \end{bmatrix}?$$

Symmetric, orthogonal, triangular, invertible, projection, permutation, Jordan form, diagonalizable. Find the eigenvalues of A and B.

1.6.11   Find the LU factorizations of A_1 and PA_2:

$$A_1 = \begin{bmatrix} 1 & -1 & 0 \\ 0 & 1 & -1 \\ 1 & 0 & -1 \end{bmatrix}, \quad A_2 = \begin{bmatrix} -1 & 1 & 0 & 0 \\ -1 & 0 & 1 & 0 \\ 0 & -1 & 1 & 0 \\ 0 & -1 & 0 & 1 \\ 0 & 0 & -1 & 1 \end{bmatrix}, \quad P = \begin{bmatrix} 1 & 0 & 0 & 0 & 0 \\ 0 & 1 & 0 & 0 & 0 \\ 0 & 0 & 0 & 1 & 0 \\ 0 & 0 & 1 & 0 & 0 \\ 0 & 0 & 0 & 0 & 1 \end{bmatrix}.$$

L contains the multipliers l_{ij}, with $l_{ii} = 1$, and the permutation P exchanges rows 3 and 4 to avoid a zero in the pivot position.

1.6.12   Invert $PA = LDU$ to find a formula for A^{-1}. It allows the inverse to be computed with n^3 multiplications—or about $n^3/2$ when A is symmetric.

1.6.13   Factor A into LU and find $A^{-1} = U^{-1}L^{-1}$ if

$$A = \begin{bmatrix} 1 & -1 & 0 & 0 \\ -1 & 2 & -1 & 0 \\ 0 & -1 & 2 & -1 \\ 0 & 0 & -1 & 2 \end{bmatrix}.$$

Solve $Ax = b$ if b has components $1, 0, 0, 0$.

1.6.14   (a)   Find two vectors that are orthogonal to $(1, 1, 1, 0)$ and $(0, 1, 1, 1)$, by making these the rows of A and solving $Ax = 0$.
   (b)   Find 2 equations in 4 unknowns whose solutions are the linear combinations of $(1, 1, 1, 0)$ and $(0, 1, 1, 1)$.

1.6.15   The fundamental theorem says that either b is in the column space or it is not orthogonal to the left nullspace: Either (1) $Ax = b$ for some x, or (2) $A^T y = 0$, $y^T b \neq 0$ for some y. Show directly that (1) and (2) cannot both be true.

1.6.16   Find matrices A for which the number of solutions to $Ax = b$ is

(1)   0 or 1 depending on b
(2) ∞ for every b
(3)   0 or ∞ depending on b
(4)   1 for every b.

How is the rank r related to m and n in each of your examples?

1.6.17   Why is the plane $x + 2y + 3z = 0$ perpendicular to the vector with components 1, 2, 3? Give the equation of a plane parallel to that one but containing the point $x = y = z = 1$.

1.6.18   Given 4 vectors in 5-dimensional space, how could a computer decide if they are linearly independent?

1.6.19   True or false: **1.**   There is no matrix whose row space contains $[1 \quad 2 \quad 1 \quad 1]^T$ and whose nullspace contains $[1 \quad -2 \quad 1 \quad 1]^T$.
   2.   Exactly one vector is in both the row space and the nullspace.
   3.   If A and B have rank 3 then $A + B$ has rank at most 6.
   4.   The rank of the matrix with every $a_{ij} = 1$ is $r = 1$.
   5.   The rank of the n by n matrix with $a_{ij} = i + j$ is $r = n$.

1.6.20   From $A = S \Lambda S^{-1}$ construct the matrix that has eigenvalues 3 and 4 with eigenvectors $(2, 0)$ and $(1, 1)$.

1.6.21   Let $v = \begin{bmatrix} 1 \\ 1 \end{bmatrix}$ and $w = \begin{bmatrix} 1 \\ -1 \end{bmatrix}$ so that $v^T w = 0$.

   (a)   What is $A = v w^T$?
   (b)   What are the eigenvalues of A?
   (c)   What are its eigenvectors?
   (d)   What is its Jordan form?

1.6.22   If $A = S \Lambda S^{-1}$ show by transposing that A and A^T have the same eigenvalues. What matrix contains the eigenvectors of A^T? They agree with the eigenvectors of A only if $A A^T = A^T A$.

1.6.23   If the pivots of A are $d_1 = 2$ and $d_2 = 3$ (without a row exchange) what can you say about the eigenvalues? What if you also know that the trace is $a_{11} + a_{22} = 6$?

1.6.24   Show that $A^T A$ can never have a negative eigenvalue: if $A^T A x = \lambda x$ then $\lambda = \|Ax\|^2 / \|x\|^2$.

1.6.25   Prove (using the trick of multiplying by y^T) that if $A A^T y = 0$ then $A^T y = 0$. The matrices $A A^T$ and A^T have the same nullspace, the same row space, and the same rank. Find all three if

$$A = \begin{bmatrix} 1 & 1 \\ 0 & -1 \\ -1 & 0 \end{bmatrix}.$$

Note   The following three exercises all involve the matrix

$$A = \begin{bmatrix} 3 & 0 \\ 4 & 5 \end{bmatrix}.$$

1.6.26   Factor A into a product QR of an orthogonal matrix times an upper triangular matrix.

1.6.27 Factor A into its polar form $A = QB$.

1.6.28 Find the singular value decomposition $A = Q_1 \Sigma Q_2^T$.

1.6.29 If u has components $\frac{1}{2}, \frac{1}{2}, -\frac{1}{2}, -\frac{1}{2}$, find

(a) the rank-one projection matrix $A = uu^T$
(b) the projection of $b = (1, 0, 0, 0)$ onto the line through u
(c) the eigenvalues of A
(d) A^2.

1.6.30 If u has components $\frac{2}{3}, \frac{2}{3}, \frac{1}{3}$, find

(a) the rank-two projection matrix $P = I - uu^T$
(b) the projection Pb of $b = (1, 0, 0)$ onto the plane perpendicular to u
(c) the eigenvalues of P
(d) P^2.

1.6.31 If u has components $\frac{2}{3}, \frac{2}{3}, \frac{1}{3}$, find

(a) the Householder matrix $H = I - 2uu^T$
(b) the eigenvalues of H
(c) H^2.

1.6.32 Describe each of the four fundamental subspaces (lines or planes?) of

$$A = \begin{bmatrix} 1 \\ 1 \end{bmatrix} \begin{bmatrix} 2 & 1 & 1 \end{bmatrix} = \begin{bmatrix} 2 & 1 & 1 \\ 2 & 1 & 1 \end{bmatrix}.$$

1.6.33 For a rank-one matrix $A = vw^T$ show that $A^T A$ is a multiple of ww^T, and that it has the eigenvalue $\|v\|^2 \|w\|^2$ with eigenvector w. (The square root $\|v\| \|w\|$ is the nonzero singular value of A.)

1.6.34 The inverse of $B = I - vw^T$ has the form $B^{-1} = I - cvw^T$. By multiplication find the number c. Under what condition on v and w is B not invertible?

1.6.35 (a) If A is invertible then the inverse of $B = A - vw^T$ has the form $B^{-1} = A^{-1} - cA^{-1}vw^T A^{-1}$. By multiplication (watching for a scalar inside $vw^T A^{-1}vw^T A^{-1}$) find the number c.

(b) If you subtract 1 from the first entry a_{11} of A, what matrix is subtracted from A^{-1}? In A^{-1} let q be the first column, r^T be the first row, and s be the first entry.

2

EQUILIBRIUM EQUATIONS AND MINIMUM PRINCIPLES

A FRAMEWORK FOR THE APPLICATIONS ■ 2.1

This chapter begins a series of problems in applied mathematics. Each one is fundamental in its own field—electrical networks, mechanics, statistics, fluids, signal processing, heat flow, splines, and finite elements. That list could be made much longer; virtually every subject starts with a linear equation of equilibrium. (Chemistry, economics, and geodesy will also appear.) What makes it possible to study such a sequence of different applications is that mathematically they fit into a single framework. The theories are separate but parallel.†

A part of that framework you have already seen. The least squares problem in Section 1.4 minimized $(Ax-b)^T(Ax-b)$, and led to the normal equations $A^TAx = A^Tb$. The line of springs minimized potential energy and led to $A^TCAx=f$. A combination that allows for both b and f, and for the positive definite C as well as the rectangular A, is all we need for the general case. What is new is the presence of a *second unknown*, found at the same time as the first. A full solution gives currents as well as voltages, stresses as well as displacements, and flow rates as well as pressures. These new unknowns y satisfy minimum principles and constraint equations of their own, based on the same inputs b, f, C, and A. Of course they are

---

† The resistance of such a long series of problems would normally be too great; but their resistance in parallel...

not independent; to solve for x is to solve for y. But more than that, the laws they satisfy are connected by duality.

We begin with a network, to make clear the difference between x and y. The vector x represents the **nodal variables**: it starts with components $x_1, \ldots, x_N$, one for each of the nodes. Every problem assigns its own meaning to the x_j, and uses its own language, but if we choose a single description it is this: x_j is the **potential** at node j. It will be one of the primary variables throughout this book. The other variable enters when two neighboring nodes are at different potentials and something happens on the edge between them. That produces the dual unknown, the vector which represents the **edge variables**. It has m components $y_1, \ldots, y_m$, one for every edge, and normally there are more edges than nodes. (The figure has $m = 6$ and $N = 4$.) Again each different application gives its own interpretation to the vector y, but there is a single word that is frequently right: y_i is the **flow** on edge i. Therefore it is also called a flow variable, or a through variable.

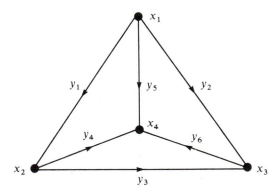

Fig. 2.1. Four nodal variables and six edge variables.

The flows y_i are determined by two things: the **potential differences** across the edges and the **physical properties** of the edges. The latter are represented in an m by m matrix C, which in most cases is especially simple. Often it is a diagonal matrix; its diagonal entries $c_1, \ldots, c_m$ are positive, and they give the "material constants" for each edge. If there is coupling between edges, or between experimental observations in statistics, then C may also have off-diagonal terms—but still this matrix is symmetric and positive definite. It is the other determining factor, the difference between potentials, that connects the edge variables y to the nodal variables x.

This connection is made by a rectangular matrix A_0. It is an m by N matrix, and every entry is 0, 1, or -1. It is called an "incidence matrix", with a **row for every edge** and a **column for every node**. This matrix A_0 records the geometry of the network. When applied to the potential x it yields the potential difference across each edge—and this is the quantity that produces flow.

To establish the signs of the flow we need a direction on each edge. It is indicated by an arrow, which can be fixed in any convenient way—say from the lower numbered node to the higher. Then edge 1 in the figure goes from node 1 to node 2, and the potential difference is $x_2 - x_1$. If this difference is positive the flow will go in the opposite direction—it *leaves* the node which is at a higher potential. In order that $x_2 - x_1$ will be the first component of $A_0 x$, we put $+1$ in column 2 and -1 in column 1; these go into the first row of A_0. The other rows are similar. If edge i goes from node j to node k, then row i has -1 in column j and $+1$ in column k. For the network in Fig. 2.1, the matrix is

$$A_0 = \begin{bmatrix} -1 & 1 & 0 & 0 \\ -1 & 0 & 1 & 0 \\ 0 & -1 & 1 & 0 \\ 0 & -1 & 0 & 1 \\ -1 & 0 & 0 & 1 \\ 0 & 0 & -1 & 1 \end{bmatrix} \begin{matrix} 1 \\ 2 \\ 3 \\ 4 \\ 5 \\ 6 \end{matrix} \quad edges$$

$$\qquad\qquad 1 \quad 2 \quad 3 \quad 4 \qquad nodes$$

This is the **edge-node incidence matrix**, or the **connectivity matrix**, or the **topology matrix**.

The six rows describe how the six edges (or branches) meet the four nodes (or vertices). Each row contains a $+1$ and a -1, to give the direction of the arrow. Each column tells which edges are incident on the corresponding node.

Notice one important point: Every row adds to zero. The columns of A_0 add up to the zero column. Therefore A_0 fails to have linearly independent columns. We cannot determine all four potentials, even if we know everything about the flows. The underlying reason is clear. It is only *differences* in potential that appear, so that x_1, x_2, x_3, x_4 can all be increased or decreased by the same amount and nothing is changed. To remove this useless degree of freedom, we may arbitrarily fix one of the potentials, say $x_4 = 0$. (In the language of electrical engineering, the last node is **grounded**.) The effect is to remove the last column of A_0, and to leave an incidence matrix A which does have independent columns:

$$A = \begin{bmatrix} -1 & 1 & 0 \\ -1 & 0 & 1 \\ 0 & -1 & 1 \\ 0 & -1 & 0 \\ -1 & 0 & 0 \\ 0 & 0 & -1 \end{bmatrix}.$$

It multiplies the unknown potentials x_1, x_2, x_3 to give the potential differences

$$Ax = (x_2 - x_1, x_3 - x_1, x_3 - x_2, -x_2, -x_1, -x_3)$$

across the six edges. It is this matrix that enters the calculations.

We will denote the number of independent nodal variables by n. In this case

$n = 3$. For electrical networks n is $N - 1$, with one node grounded. For mechanical networks n is the total number of degrees of freedom at the nodes, not counting any nodes which are fixed. And in some applications (of which statistics is one) the original matrix is not the incidence matrix of a network. The columns of A_0 are already independent, and in that case A_0 is A—without the removal of the last column. Always the matrix A will be m by n.

It remains to introduce the vectors b and f. Between them they make something happen; if they are both zero then $x = y = 0$. They provide the external forces, or the input data, or the *voltage and current sources*, to which the system reacts. The vector b has m components $b_1, \ldots, b_m$, one for each edge. It combines with the potential differences to give the vector $e = b - Ax$. Then *this edge vector e produces flow*. Ohm's law or Hooke's law or somebody's law is always $y = Ce$. The geometry in A, the inputs in b, and the material properties in C connect the potentials to the flows, by the equation $y = C(b - Ax)$. This is one fundamental equation, and to find the two unknown vectors x and y we need one more.

The second fundamental equation is $A^Ty = f$. It expresses a balance or an equilibrium at the nodes. In mechanics it is a balance of internal forces with external forces. In statistics the external forces are usually absent and the system $A^Ty = 0$ or $A^TC(b - Ax) = 0$ leads to the normal equations. In an electrical circuit $A^Ty = 0$ is **Kirchhoff's current law**, requiring that the net flow into each node is zero. The key point, decisive for the mathematics as well as the applications, is that the matrix in this equation is the same as before. It is the reappearance of A, or rather the appearance of its transpose, which gives symmetry to Fig. 2.2 and to the equations of equilibrium.†

We can verify that A^Ty gives the flows into the nodes:

$$A^Ty = \begin{bmatrix} -1 & -1 & 0 & 0 & -1 & 0 \\ 1 & 0 & -1 & -1 & 0 & 0 \\ 0 & 1 & 1 & 0 & 0 & -1 \end{bmatrix} \begin{bmatrix} y_1 \\ y_2 \\ y_3 \\ y_4 \\ y_5 \\ y_6 \end{bmatrix} = \begin{bmatrix} -y_1 - y_2 - y_5 \\ y_1 - y_3 - y_4 \\ y_2 + y_3 - y_6 \end{bmatrix}.$$

From Fig. 2.1 the flow into node 3 is $y_2 + y_3 - y_6$, agreeing with the third component of A^Ty. Each row of A^T, like each column of A, corresponds to one particular node. It has entries $+1$ for all edges that enter the node and -1 for all edges that leave. Therefore A^Ty adds the flows on these edges, with the correct signs to give the total flow into the node. This balances the external vector f.

To summarize this section, we first rewrite the separate equations from the three steps:

$$\boxed{e = b - Ax, \quad y = Ce, \quad A^Ty = f.}$$
(1)

---

† The transpose is the node-edge incidence matrix.

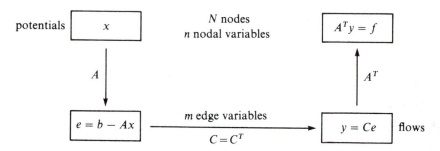

Fig. 2.2. The framework for the equilibrium equations.

Eliminating the intermediate vector, which is $e = C^{-1}y$, leaves two equations for y and x:

2A   The flows y and the potentials x satisfy the equilibrium equations

$$C^{-1}y + Ax = b \quad \text{and} \quad A^Ty = f. \tag{2}$$

Solving the first equation for $y = C(b - Ax)$ and substituting into $A^Ty = f$ gives a single equation for x:

$$A^TC(b - Ax) = f, \quad \text{or} \quad A^TCAx = A^TCb - f. \tag{3}$$

Its coefficient matrix A^TCA is symmetric positive definite.

This last equation combines the three steps of the diagram into one. Starting from x we reach $b - Ax$ at the first step, $C(b - Ax)$ at the second step, and then $A^TC(b - Ax)$ at the last step. To maintain equilibrium, this vector is balanced by f.

Note that usually $A^Ty = f$ cannot be solved directly for y. Our example has $n = 3$ equations in $m = 6$ unknowns, and y is found only with the help of x. The determinate case $m = n$ does sometimes arise in practice; in that exceptional situation f alone is enough to determine y. But ordinarily equation (3) must be solved for x—the potentials come first—and then $y = C(b - Ax)$ gives the flows.

The solution of (3) is the essential problem. It is difficult and expensive when n is large, and a major part of numerical analysis is devoted to finding an efficient algorithm. That will be the subject of Chapter 5. We can say in advance that there are poor algorithms and better algorithms, but no absolute winner. It is not even clear whether to work with the system (2) or the single equation (3); the explicit construction of $K = A^TCA$ may or may not be a mistake. What is clear is the need to understand the background as fully as possible, and for that reason we will reach the same equation by a different approach—from minimum principles.

Remarks on the Equilibrium Equations

1. We regard the pair of equations (2) as the *fundamental equations of equilibrium*. They fit into a symmetric system that is easy to remember:

$$\begin{bmatrix} C^{-1} & A \\ A^T & 0 \end{bmatrix} \begin{bmatrix} y \\ x \end{bmatrix} = \begin{bmatrix} b \\ f \end{bmatrix}. \tag{4}$$

This block matrix is not positive definite! It is true that C^{-1} is positive definite, so the first m pivots (which depend only on C^{-1}) are all positive. But then the remaining n pivots are negative. In the language of the next section, the minimization is over and the maximization has begun.

2. The triple product $A^T C A$ appears halfway through elimination. The first m steps produce zeros in the lower left block, when the first equation is multiplied by $A^T C$ and subtracted from the second. That leaves

$$\begin{bmatrix} C^{-1} & A \\ 0 & -A^T C A \end{bmatrix} \begin{bmatrix} y \\ x \end{bmatrix} = \begin{bmatrix} b \\ f - A^T C b \end{bmatrix}.$$

Then the equation for x is $-A^T C A x = f - A^T C b$, exactly as we found in (3). The elimination here and the substitution there amount to the same thing. The coefficient matrix is $-A^T C A$, and the minus sign makes it negative definite. After solving for x, back-substitution gives the flows $y = C(b - Ax)$.

3. There is a neat formula for the matrices $A^T A$ and $A^T C A$, in the network case. Away from the main diagonal $(A^T A)_{jk}$ is either -1 or 0—depending whether or not the network has an edge from node j to node k. This comes directly from the matrix multiplication $A^T A$; row j of A^T (which is column j of A) multiplies column k of A. The two columns can overlap only at one edge, the one between j and k, and there the product of 1 and -1 gives $(A^T A)_{jk} = -1$. If that edge is absent the entry is zero. In $A^T C A$ the diagonal matrix C contributes a constant c_i for each edge, and that extra factor gives

$$(A^T C A)_{jk} = -c_i, \text{ if edge } i \text{ goes between nodes } j \text{ and } k. \tag{5}$$

On the main diagonal the terms are positive: $(A^T A)_{kk}$ has $+1$ for every edge in or out of node k. Again C contributes an extra factor to each of these terms and

$$(A^T C A)_{kk} = \text{sum of all } c_i, \text{ for edges meeting node } k. \tag{6}$$

Frequently this diagonal entry matches the sum along the rest of the row; that will always be true for $A_0^T C A_0$. It is only if node k is connected to the grounded node that there is a difference. Then the diagonal contains an extra c_i for that edge to ground, which puts it ahead of the off-diagonal sum. For Fig. 2.1, with six edges and three ungrounded nodes,

$$A^T C A = \begin{bmatrix} c_1 + c_2 + c_5 & -c_1 & -c_2 \\ -c_1 & c_1 + c_3 + c_4 & -c_3 \\ -c_2 & -c_3 & c_2 + c_3 + c_6 \end{bmatrix}. \tag{7}$$

The matrices $A^T A$ and $A^T C A$ are symmetric positive definite "M-matrices," mentioned in an earlier footnote (Section 1.4). Apart from zeros, they are negative off the diagonal and ***all entries in their inverses are positive***.

4.   Every area of applied mathematics has its own interpretation of A, C, b, and f. Some problems also have their own choice of sign; each section will translate between the specific application and the common framework. In certain cases (for example structures) the sign of x is reversed, and so is the sign of A:

$$\begin{bmatrix} C^{-1} & -A \\ A^T & 0 \end{bmatrix} \begin{bmatrix} y \\ -x \end{bmatrix} = \begin{bmatrix} b \\ f \end{bmatrix}. \tag{8}$$

It is now $+A^T C A$ which appears halfway through elimination, and therefore all the pivots will be positive. But the matrix in (8) is no longer symmetric.

5.   Strictly speaking Fig. 2.1 represents a ***directed graph***. It is a ***network*** when a number c_i is assigned to every edge.

6.   It is remarkable how well the framework extends to problems that are *continuous* rather than discrete. Instead of a finite number of edges, the flow may fill a region in the plane. The unknowns change from vectors to functions, and the potential differences change to derivatives; they still produce the flow. It is governed by differential equations instead of matrix equations. Nevertheless the pattern in Fig. 2.2 still leads to $A^T C A$, and describes the equilibrium of the system—this is the theme of Chapter 3.

EXERCISES

2.1.1   For a graph with edges around a square and across one diagonal ($N = 4$ and $m = 5$), number the nodes and edges and write down the incidence matrix A_0. What is $A_0^T A_0$?

The next three exercises refer to the network with four nodes and six edges at the beginning of the section. The edge constants are $c_1, \ldots, c_6$.

2.1.2   (a)   Compute the 4 by 4 matrices $A_0^T A_0$ and $A_0^T C A_0$ for the network in Fig. 2.1. Notice that like the original A_0, its columns add up to the zero column.

(b)   Verify that removing the last row and column of $A_0^T C A_0$ leaves $A^T C A$ in equation (7). What is $A^T A$?

(c)   Show that this $A^T A$ is positive definite by applying one of the tests in Chapter 1 (for example, compute the determinants or the pivots).

2.1.3   For the triangular network in Fig. 2.1, let $f_1 = f_2 = f_3 = 1$ and $f_4 = -3$. With $C = I$ and $b = 0$, solve the equilibrium equation $-A^T C A X = f$. (Note that f_4 and x_4 do not enter, because $x_4 = 0$ and the last column of A_0 was removed.) Solve also for y, and describe the flows through the network.

2.1.4   Suppose there are "batteries" $b_4 = b_5 = b_6 = 1$ on the inner edges of the network in Fig. 2.1. With $f = b_1 = b_2 = b_3 = 0$ and $C = I$, write down the 9 by 9 equilibrium system (4). Show that *there is no flow*: the solution has $y = 0$ (but not $x = 0$).

2.1.5   For a network with only $m = 3$ edges and $N = 3$ nodes, at the vertices of a triangle with arrows clockwise, write down A_0 and A. With $f = 0$, $b_1 = b_2 = b_3 = 1$, and $C = I$, find x and y.

2.1.6   Suppose a network has N nodes and every pair is connected by an edge. Find m, the number of edges.

2.1.7   Imagine an R by R network in the plane, with nodes at the $N = R^2$ points with integer coordinates between 1 and R.

(a)   If horizontal and vertical edges make it into a network of unit squares find the number of edges. Show that the approximate ratio of m to N is 2 to 1.

(b)   If the network also includes the diagonals of slope $+1$ in each square, adding $(R - 1)^2$ new edges, show that the approximate ratio of m to N for this triangular mesh is 3 to 1.

2.1.8   Show that the particular matrix

$$M = \begin{bmatrix} C^{-1} & A \\ A^T & 0 \end{bmatrix} = \begin{bmatrix} 1 & 1 \\ 1 & 0 \end{bmatrix}$$

is neither positive definite nor negative definite, by finding its pivots (and also its eigenvalues). Does

$$\begin{bmatrix} x_1 & x_2 \end{bmatrix} \begin{bmatrix} 1 & 1 \\ 1 & 0 \end{bmatrix} \begin{bmatrix} x_1 \\ x_2 \end{bmatrix} = x_1^2 + 2x_1 x_2$$

have a minimum or maximum or saddle point at $x_1 = x_2 = 0$?

2.1.9   Given a network and its incidence matrices (ungrounded and grounded):

$$A_0 = \begin{bmatrix} -1 & 1 & 0 & 0 \\ -1 & 0 & 1 & 0 \\ 0 & -1 & 1 & 0 \\ 0 & -1 & 0 & 1 \\ 0 & 0 & -1 & 1 \end{bmatrix} \qquad A = \begin{bmatrix} -1 & 1 & 0 \\ -1 & 0 & 1 \\ 0 & -1 & 1 \\ 0 & -1 & 0 \\ 0 & 0 & -1 \end{bmatrix}$$

(i)   find $A^T C A$ from equations (5) and (6), with $-c_1, -c_2, -c_3$ appearing off the diagonal

(ii)   find $A^T C A$ from "column-row" multiplications, where the first column of A^T and the first row of A give

$$
\begin{bmatrix} -1 \\ 1 \\ 0 \end{bmatrix} [c_1][-1 \quad 1 \quad 0] = \begin{bmatrix} c_1 & -c_1 & 0 \\ -c_1 & c_1 & 0 \\ 0 & 0 & 0 \end{bmatrix}.
$$

Add up the five products of this kind (one for each edge).

2.1.10   Eliminate y to find the equations for x in the systems

$$
\begin{bmatrix} C^{-1} & A \\ A^T & D \end{bmatrix} \begin{bmatrix} y \\ x \end{bmatrix} = \begin{bmatrix} 0 \\ f \end{bmatrix} \quad \text{and} \quad \begin{bmatrix} C_1^{-1} & 0 & A_1 \\ 0 & C_2^{-1} & A_2 \\ A_1^T & A_2^T & 0 \end{bmatrix} \begin{bmatrix} y_1 \\ y_2 \\ x \end{bmatrix} = \begin{bmatrix} 0 \\ 0 \\ f \end{bmatrix}.
$$

2.1.11   What is the equation for the vector x that minimizes $P(x) = \frac{1}{2}(b - Ax)^T C(b - Ax)$?

2.1.12   Draw a network with no loops (a *tree*). Check that with one node grounded the incidence matrix A is square, and find A^{-1}. All entries of the inverse are 1, -1, or 0.

2.1.13   Suppose A_0 is a 12 by 9 incidence matrix from a connected graph. Its exact form is unknown but it has 12 edges and 9 nodes, none grounded.
(a) How many columns of A_0 are independent?
(b) How many rows are independent, and what do the corresponding edges look like on the graph?
(c) What condition on f makes $A_0^T y = f$ solvable?
(d) How many independent solutions are there to $A_0^T y = 0$, and how can they be found from the graph? (See page 74.)
(e) For which vectors b does $A_0 x = b$ have a solution?

2.2 ■ CONSTRAINTS AND LAGRANGE MULTIPLIERS

This section introduces minimization subject to constraints. Up to now, we have computed the absolute minimum of a quadratic. When it was $P(x) = \frac{1}{2}x^T Ax - x^T b$, that minimum was at the point where $Ax = b$ (page 33). Now the quadratic is $Q(y)$ $= \frac{1}{2}y^T C^{-1} y - b^T y$, and its derivatives $\partial Q / \partial y_i$ are zero when $C^{-1}y = b$. But if the problem includes constraints—*extra conditions that must be satisfied by y*—this equation is no longer correct. Its solution $y = Cb$ may violate the constraints. The correct equation for the constrained minimizer will have a new term, and our goal in this section is to find that term and understand it.

The first approach is by example, to see how a constraint can enter. The function Q will be $\frac{1}{2}(y_1^2 + y_2^2)$, so that minimizing Q is essentially the same as minimizing the length of the vector y. Here C is the identity matrix, and b is zero. If there is no constraint, the zero vector is the best y! But we will require y to lie on the line $2y_1 - y_2 = 5$. That is the constraint equation (later written $A^T y = f$). The problem is **to find the point on this line which is nearest to the origin.**

In Fig. 2.3, the nearest y is on the perpendicular ray. As you move the point y along its line, the distance to the origin goes down and then up again. It hits bottom at the point $(2, -1)$. The second figure shows that more clearly. The graph of Q is the large bowl sitting on the origin, and the constraint cuts through that bowl. *Since y is restricted to the dashed line, Q is restricted to the dashed parabola above that line.* It is the lowest point on the parabola that we want to compute.

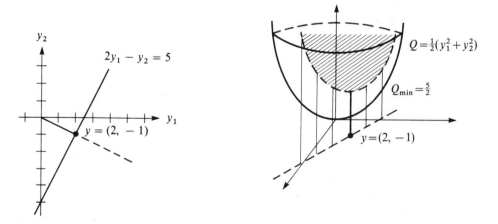

Fig. 2.3. The distance to a line: Minimization with a constraint.

May I do the computation twice? The first method will be easy in this example, but difficult or impossible in general. The second method (which introduces *Lagrange multipliers*) will look harder—but once you get the knack it is a better approach. It leads to a lot of beautiful (and practical) mathematics.

Method 1. Solve $2y_1 - y_2 = 5$ for y_2, substitute into Q, and minimize:

$$Q = \tfrac{1}{2}(y_1^2 + y_2^2) = \tfrac{1}{2}(y_1^2 + (2y_1 - 5)^2) \text{ has } Q' = y_1 + 2(2y_1 - 5).$$

The derivative Q' is zero at $y_1 = 2$, and then $y_2 = 2y_1 - 5 = -1$.

Method 2. Build the constraint into a new function L:

$$L = Q + x_1(2y_1 - y_2 - 5) = \tfrac{1}{2}(y_1^2 + y_2^2) + x_1(2y_1 - y_2 - 5).$$

Where Method 1 removed the unknown y_2, this method adds a new unknown x_1—called the Lagrange multiplier.† It avoids the step of solving the constraint equation for y_2 (not hard here, but in other problems that step could wipe out the first method). The best y is found, by calculus, where the derivatives of L are zero:

$$
\begin{array}{llll}
\partial L/\partial y_1 = 0 & y_1 & + 2x_1 = 0 & \\
\partial L/\partial y_2 = 0 & y_2 & - x_1 = 0 & \text{(1)} \\
\partial L/\partial x_1 = 0 & 2y_1 - y_2 & = 5. &
\end{array}
$$

The first equation gives $y_1 = -2x_1$, the second equation is $y_2 = x_1$, and then the third is $-5x_1 = 5$. The Lagrange multiplier is $x_1 = -1$, and the nearest point y is $(y_1, y_2) = (2, -1)$. Note that the derivative with respect to x_1 brought back the constraint; L was constructed to make that happen. At present, the multiplier x_1 has no meaning of its own—but that will change.

You may recognize the system (1). I hope you do! The last equation is $A^T y = f$, with $A = [2 \quad -1]$. The first equations are $C^{-1}y + Ax = b$, where $C^{-1} = I$ and $b = 0$ because Q is $\tfrac{1}{2}y^T I y$. **We have recovered the two fundamental equations of equilibrium.** More than that, the matrix $A^T C A$ is here too; it is the number 5. When we reached $-5x_1 = 5$ by eliminating the y's, that was $-A^T C A x = f$. We want to do that again more systematically, for the general case with n constraints and m unknown y's.

Minimization of $Q = \tfrac{1}{2}y^T C^{-1}y - b^T y$ subject to the constraint $A^T y = f$.

1. Introduce multipliers $x_1, \ldots, x_n$ for the constraints (the rows of $A^T y = f$).
2. Build the constraints into a function of $m + n$ variables,

$$L(x, y) = Q + x^T(A^T y - f) = \tfrac{1}{2}y^T C^{-1}y - b^T y + (Ax)^T y - x^T f. \qquad \text{(2)}$$

---

† Some authors denote Lagrange multipliers by λ instead of x, but in this book λ is reserved for eigenvalues.

3. Solve the $m + n$ equations $\partial L/\partial y_i = 0$ and $\partial L/\partial x_j = 0$, which are

$$\begin{array}{ll} C^{-1}y - b + Ax = 0 \\ A^T y - f = 0 \end{array} \quad \text{or} \quad \begin{bmatrix} C^{-1} & A \\ A^T & 0 \end{bmatrix} \begin{bmatrix} y \\ x \end{bmatrix} = \begin{bmatrix} b \\ f \end{bmatrix}. \tag{3}$$

The minimizing equations are identical to the equilibrium equations. They can be derived from the three steps in the framework ($e = b - Ax$, $y = Ce$, and the balance law $A^T y = f$), or they can be derived from a minimum principle. In a moment we find a further possibility—a minimum principle in x alone—which also leads to equilibrium and to $A^T C A$. It comes from eliminating y.

When there are several Lagrange multipliers $x_1, \ldots, x_n$, and each one multiplies a constraint equation from $A^T y = f$, the function L builds in those n terms. In vector notation they add to $x^T(A^T y - f)$. The steps are straightforward and they give the equations we want, but it remains to explain why. For that we go back to the relation between Methods 1 and 2—now allowing the constraint to be nonlinear. We will make the theory brief.

Problem: Minimize $Q(y_1, y_2)$ subject to $A(y_1, y_2) = f$.

Method 1 solves the constraint equation for y_2 (if possible), and substitutes into Q. That produces a function R of y_1 alone, which we minimize by setting its derivative to zero:

compute $R(y_1) = Q(y_1, y_2(y_1))$ and solve $R' = 0$.

Suppose that can be done—what do we get? The derivative of R is

$$R' = \frac{\partial Q}{\partial y_1} + \frac{\partial Q}{\partial y_2} \frac{\partial y_2}{\partial y_1} = 0. \tag{4}$$

The derivative of $A(y_1, y_2(y_1)) = f$ (a constant!) is

$$A' = \frac{\partial A}{\partial y_1} + \frac{\partial A}{\partial y_2} \frac{\partial y_2}{\partial y_1} = 0. \tag{5}$$

Then $\partial Q/\partial y_1$ divided by $\partial A/\partial y_1$ equals $\partial Q/\partial y_2$ divided by $\partial A/\partial y_2$. Call that ratio $-x$, so

$$\frac{\partial Q}{\partial y_1} + x \frac{\partial A}{\partial y_1} = 0, \quad \frac{\partial Q}{\partial y_2} + x \frac{\partial A}{\partial y_2} = 0. \tag{6}$$

We have reached the same result as Method 2. *Equation* (6) *says that the y derivatives of* $L = Q(y) + x(A(y) - f)$ *are zero. The x derivative is also zero, bringing back the* constraint $A(y) = f$. Because of Lagrange, all derivatives are zero at the solution— and one of the central ideas of calculus is saved.

Notice that $L = Q$ when the constraint is satisfied. The extra term becomes zero and disappears. Therefore we do have $\partial Q/\partial y = 0$ at the minimum, but **only in the directions permitted by the constraints.** In other directions, the derivative of Q is not zero. For the example with y on the line $2y_1 - y_2 = 5$, the allowed variations all satisfied $2dy_1 - dy_2 = 0$. The corresponding variations in $Q = \frac{1}{2}(y_1^2 + y_2^2)$ were $y_1 dy_1 + y_2 dy_2$. If the Q-variation is zero whenever the A-variation is zero, then (y_1, y_2) is proportional to $(2, -1)$. The proportionality constant is the multiplier $-x$, exactly as in equation (6). In the example it is 1, and $y = (2, -1)$.

EXAMPLE 2 (nonlinear, with symmetric K) *Minimize* $Q = y^T K y$ *subject to* $y^T y = 1$

Differentiating Q gives $2Ky$. Differentiating the constraint gives $2y$. If those are proportional (with proportionality constant $-x$) then $2Ky = -2xy$. The minimizing y is an *eigenvector* of K, chosen with unit length. The value of Q is $y^T(-xy) = -x$. **The minimum of Q is the smallest eigenvalue of K**—and the maximum of Q is the largest eigenvalue.

An equivalent approach is to minimize a ratio of quadratics, the *Rayleigh quotient* $y^T K y/y^T y$. Again the minimum is λ_{min}, reached at the eigenvector. And a third approach is to work with $L = y^T K y + x(y^T y - 1)$, whose derivatives again give $Ky = -xy$. Question: What happens when the constraint is $y^T M y = 1$? This corresponds to the quotient $y^T K y/y^T M y$, and leads to the generalized eigenvalue problem $Ky = -xMy$. Often K is the stiffness matrix, and M is the mass matrix.

Lagrange multipliers lead to the equations for a constrained minimum. The method works, and it does not require us to solve $A(y) = f$. (If $y_1 e^{y_1 y_2} = 2$, we couldn't find y_1.) That Lagrange method was our primary goal, and you can solve examples without reading any further—but there is more to be said.

Those equilibrium equations describe a *saddle point* for L, in which x is just as important as y. The Lagrangian $L(x, y)$ is being minimized with respect to y, and at the same time it is *maximized* with respect to x. When x is removed we come back to our **primal problem**, the constrained minimum of Q. In the opposite direction— eliminating y and keeping x—we arrive at the **dual problem**. The dual uses the same data A, C, b, and f, but they go into a different function $P(x)$. It is constructed so that the two problems have, in some fundamental sense, the same answer: **The minimum of $Q(y)$ equals the maximum of $-P(x)$.**

The rest of the section is devoted to the duality between P and Q—the connection between two complementary minimum principles. It is applied to structures in Section 2.4, and ultimately this duality will be the justification for Lagrange multipliers in Chapter 8. We try to develop it gradually, maintaining at every step the link to equilibrium equations—which directly identify the minimizing y and the maximizing x.

An Example of Duality

Duality is more than a formal equivalence between a minimization and a maximization. If the minimum problem has a geometric meaning, or if it has a physical application, then so does the maximum problem. The best example I know is in three-dimensional space. It is the *distance to a line*, and there are two ways to find it. The equality between those two is our first statement of duality:

the minimum distance to the points on a line

equals

the maximum distance to planes through that line.

All distances are measured from the origin.

That example is worth thinking about, to visualize the farthest plane. You could add a third dimension to Fig. 2.3, coming out of the book, and keep the same line. Planes through the line are those that cut the paper along $2y_1 - y_2 = 5$. If the plane makes a small angle with the paper, it will pass close to the origin; the distance is small. If the plane is the same as the paper, it contains the origin and the distance is zero. If the plane is **perpendicular** to the paper, then its nearest point to the origin is the one marked y in the figure. This perpendicular plane is at a maximum distance from the origin, equal to the minimum distance to the line. That is duality.

Notice this one point: Hiding inside the maximum problem is a minimum problem that has to be solved first. We have to know the distance to the plane! That is much easier than the distance to a line, and it illustrates perfectly the object of Lagrange multipliers—to exchange the original harder problem (with constraints) for two easy problems.

The Dual Problem in Higher Dimensions

Our goal is to discover the function $P(x)$ in the dual problem. We stay with linear equations, going beyond three dimensions and beyond the simple distance function $Q = \frac{1}{2}y^T y$. In other applications Q involves a "material matrix" C in the quadratic part, and a source term b in the linear part:

PRIMAL   *Minimize* $Q(y) = \frac{1}{2}y^T C^{-1} y - b^T y$ *subject to* $A^T y = f$.

A typical constraint is Kirchhoff's current law; the net flow into each node is zero. The Lagrange multipliers $x_1, \ldots, x_n$ build the n constraints into L:

$$L = Q + x^T(A^T y - f) = \frac{1}{2}y^T C^{-1} y - (b - Ax)^T y - x^T f \tag{7}$$

Then the step to the dual problem is *to eliminate y and leave only x*. We minimize L with respect to y, and this is an ordinary *unconstrained* minimization—because the constraints are already built in. The minimum of the quadratic occurs when $C^{-1}y = b - Ax$, in other words at $y = C(b - Ax)$. Substituting into (7) and simplifying yields the minimum value—which we call $-P(x)$:

$$\min L(x, y) = -\tfrac{1}{2}(b - Ax)^T C(b - Ax) - x^T f = -P(x). \tag{8}$$

This is the function that enters the dual problem. Notice that it involves the matrix C, where Q involved C^{-1}. It also involves the special combination $b - Ax$ that was previously denoted by e; thus P is $\tfrac{1}{2}e^T Ce + x^T f$. In many applications P represents the *total potential energy*. That energy is minimized at equilibrium, which means that $-P(x)$ is maximized—and that is the unconstrained problem we are looking for:

DUAL   *Maximize* $- P(x) = -\tfrac{1}{2}(Ax - b)^T C(Ax - b) - x^T f.$

We took the liberty of reversing signs (twice!) in $Ax - b$, so that the quadratic part of the energy would come first. That part is $\tfrac{1}{2}x^T A^T C Ax$, which brings back the triple product $A^T CA$. Remember that $A^T CA$ entered the equilibrium equations in the same way it has just done here—by eliminating y.

The final step is to connect the dual to the primal. We have to show that *minimum = maximum*. (Those are not the minimum and maximum of the same thing, or it would be ridiculous! Duality is $\min(Q) = \max(-P)$.) It turns out that half of the proof, namely *minimum $\geqslant$ maximum*, is easy. That is known as **weak duality**:

2B   For any admissible x and y (meaning $A^T y = f$) we have $Q(y) \geq -P(x)$.

This comes directly from the construction. Remember that $-P$ was the minimum value of L, so that $L(x, y) \geqslant -P(x)$ for every y. But when the constraint $A^T y = f$ is satisfied, L is the same as Q: the built-in term disappears. We conclude that $Q(y) \geq -P(x)$, which is weak duality.

Note: To make the picture really right, Q should be the maximum of $L(x, y)$ over all x—just as $-P$ was the minimum over all y. When $L = Q + x^T(A^T y - f)$ is maximized over x, the result is remarkable. If $A^T y = f$ the maximum is Q, since the other term is zero. If $A^T y \neq f$, the maximum is $+\infty$. (We can choose x to be extremely large with the right signs.) Thus the true formula for Q is the usual $\tfrac{1}{2}y^T C^{-1}y - b^T y$ when the constraint is satisfied, and $+\infty$ when $A^T y \neq f$. This trick of using $+\infty$ in Q makes the inequalities look much neater: **for all x and y**

$$Q(y) \geq L(x, y) \geq -P(x). \tag{9}$$

A Proof of Duality

We have to show that equality can hold in (9). Q can come down and $-P$ can come up, to a point where they meet. That will be a saddle point of L, and it will solve both the primal problem and the dual. It will be the minimum of Q, which by weak duality can never go below $-P$. It will be the maximum of $-P$, which can never go above Q.

We know where that saddle point is. It is given by the equilibrium equations. This section started by discovering the condition $C^{-1}y + Ax = b$, for a minimization constrained by $A^T y = f$. Now the analysis is coming full circle—after introducing the dual problem—to see that the same equations solve the maximization. The theory of duality could provide a very general proof, but for quadratic functions and linear constraints we can appeal directly to matrix algebra. To find the point where $Q = -P$, we write out $Q + P$ and substitute $A^T y$ for f:

$$Q + P = \tfrac{1}{2} y^T C^{-1} y - b^T y + \tfrac{1}{2}(Ax - b)^T C(Ax - b) + x^T f$$
$$= \tfrac{1}{2}(C^{-1} y + Ax - b)^T C(C^{-1} y + Ax - b). \tag{10}$$

The algebra has produced $\tfrac{1}{2} z^T C z$, with $z = C^{-1} y + Ax - b$. Since C is positive definite, this confirms that $Q + P \geq 0$. (That was weak duality.) It also indicates how to reach full duality: *z must be zero.* That is exactly the equation $C^{-1} y + Ax = b$. We summarize the theory before going to an optional application:

2C The constrained minimum of $Q(y)$ equals the maximum of $-P(x)$. The minimizing y and maximizing x (the saddle point of L, where $Q + P = 0$) satisfy
$$C^{-1} y + Ax = b \quad \text{and} \quad A^T y = f.$$
The term Ax corrects $C^{-1} y = b$ to account for the constraint, leaving $m + n$ equations in $m + n$ unknowns—the original $y_1, \ldots, y_m$ and the Lagrange multipliers $x_1, \ldots, x_n$.

That was pretty condensed, and I would like to go back to an earlier example before any new applications. The example was the line of springs with internal forces y, external forces f, and material constants c. The energy in the springs is

$$Q = \tfrac{1}{2} y_1^2/c_1 + \tfrac{1}{2} y_2^2/c_2 + \tfrac{1}{2} y_3^2/c_3 = \tfrac{1}{2} y^T C^{-1} y.$$

The force balance equations $A^T y = f$ are the constraints on y: $y_1 - y_2 = f_1$ and $y_2 - y_3 = f_2$. The springs minimize Q subject to $A^T y = f$, and this constraint is handled by Lagrange multipliers:

$$L = Q - x_1(y_1 - y_2 - f_1) - x_2(y_2 - y_3 - f_2).$$

The derivatives of L give the equations of equilibrium:

$$
\begin{array}{lll}
\partial L/\partial y_1: & y_1/c_1 - x_1 = 0 & \\
\partial L/\partial y_2: & y_2/c_2 + x_1 - x_2 = 0 & C^{-1} y - Ax = 0 \\
\partial L/\partial y_3: & y_3/c_3 + x_2 = 0 & \\
\partial L/\partial x_1: & y_1 - y_2 = f_1 & A^T y = f \\
\partial L/\partial x_2: & y_2 - y_3 = f_2 &
\end{array}
$$

Solving for y produces $y = CAx$. Substituting into $A^T y = f$ gives $A^T CAx = f$. If we prefer minimum principles (but who does?), then $y = CAx$ is substituted into L. The result is $-P$:

$$L = \tfrac{1}{2}y^T C^{-1}y - x^T(A^T y - f)$$
$$= \tfrac{1}{2}x^T A^T C C^{-1}CAx - x^T(A^T CAx - f)$$
$$= -\tfrac{1}{2}x^T A^T CAx + x^T f = -P.$$

Then maximizing $-P$ or minimizing P leads back to the linear system $A^T CAx = f$.

The reader may have noticed that the sign of x was everywhere reversed. That brings us in line with the sign convention in mechanical engineering, where $P = \tfrac{1}{2}e^T Ce - x^T f$ is the total potential energy (and is minimized!). You will see it again for *trusses* in Section 2.4—which could come before or after this section. A truss is a *two-dimensional* system of springs and masses (like an old Army cot), and that extra dimension will make life interesting.

An Application to Geometry

Our goal is to apply duality theory to problems of equilibrium, when the underlying equations are independent of time. In mechanics P will be the energy and Q will be the complementary energy. In a resistor Q corresponds to $I^2 R$ and P corresponds to V^2/R (plus linear terms that make $-P = Q$). For nonlinear problems this idea still applies; duality is present even if P and Q are not quadratic, or if there are inequality constraints—as in the combinatorial optimization of Chapter 7 and the linear programming of Chapter 8. Here we stay with linear equations, and begin with a "pure" example from geometry. It is a chance to see how projections appear in matrix algebra.

Suppose $a_1, \ldots, a_n$ are vectors in m-dimensional space. As always, their combinations $x_1 a_1 + \cdots + x_n a_n$ fill out a subspace S. It is the space "spanned" by the given vectors, and if they are linearly independent (as we assume) they will be a basis for S. If the vectors were $a_1 = (1, 0, 0)$ and $a_2 = (0, 1, 0)$, their combinations would be $(x_1, x_2, 0)$ and S would be a horizontal plane in 3-dimensional space.

Orthogonal to S is another subspace T. It contains every vector y that is perpendicular to all the a_i. In the example where S is a horizontal plane, T must be the vertical axis: $y = (0, 0, y_3)$ is a typical vector in T. In general T will have dimension $m - n$, and all vectors in T are perpendicular to all vectors in S.

Now consider any vector b in m-dimensional space, and look at its projections on S and T. The figure shows S and T as perpendicular lines in the plane ($m = 2$ and $n = 1$). If our intuition is right, and if Pythagoras is right, then

$$\text{(distance to } S)^2 + \text{(distance to } T)^2 = \text{(length of } b)^2. \tag{11}$$

We want to show that this is exactly the great duality theorem $\min(Q) = \max(-P)$, and that the equilibrium equations locate the projections.

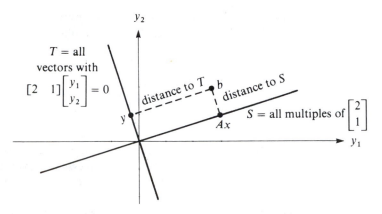

Fig. 2.4. Projection of b onto orthogonal subspaces S and T.

In this application C is the identity matrix and $f = 0$; we need only A and b. A is the matrix whose columns are the vectors $a_1, \dots, a_n$. Their combinations $x_1 a_1 + \dots + x_n a_n$ are exactly the vectors Ax, by matrix multiplication. The subspace S contains all these combinations, so it is the *column space* of A. The distance from b to S is the distance from b to the nearest Ax:

$$\text{(distance to } S)^2 = \min_x (Ax - b)^T (Ax - b).$$

This is the problem of least squares, the minimization of $P(x)$.

Next we turn to T. It contains all vectors y that are perpendicular to the a_i; that translates into $A^T y = 0$. Again this is matrix multiplication, but now a row at a time. When T was the vertical axis containing all $y = (0, 0, y_3)$, we had

$$A^T y = \begin{bmatrix} 1 & 0 & 0 \\ 0 & 1 & 0 \end{bmatrix} \begin{bmatrix} y_1 \\ y_2 \\ y_3 \end{bmatrix} = \begin{bmatrix} 0 \\ 0 \end{bmatrix}.$$

Thus T is the *nullspace* of A^T. It is automatic that every Ax in S is at right angles to every y in T: $(Ax)^T y = x^T A^T y = 0$. The column space of A and the nullspace of A^T are orthogonal. The distance from b to T is a minimization subject to $A^T y = 0$:

$$\text{(distance to } T)^2 = \min_{A^T y = 0} \|b - y\|^2 = \min_{A^T y = 0} y^T y - 2b^T y + b^T b.$$

This is the dual to least squares; it is the minimization of Q.

Substituting these distances into Pythagoras' law and cancelling $b^T b = $ (length of $b)^2$, (11) becomes

$$\min_x (Ax - b)^T (Ax - b) + \min_{A^T y = 0} (y^T y - 2b^T y) = 0.$$

Dividing by 2 and remembering that $C = I$ and $f = 0$, we have duality:

$$\boxed{\min P + \min Q = 0.\dagger}$$

(12)

In other applications those perpendicular spaces S and T are still present. For networks of resistors and batteries, the vector of currents is orthogonal to the vector of potential differences. One is in the nullspace of A^T (the current law) and the other is in the column space of A (the voltage law). In mechanics the self-stresses are orthogonal to the strains. But it would be impossible to find a clearer example than this one from geometry, and we emphasize the key points:

1. Projection onto S leads to x
2. Projection onto T leads to y
3. Duality is equivalent to Pythagoras' law.

The intuition is correct about right angles, in any number of dimensions, and the projections onto S and T produce a rectangle. The equilibrium equation is $y + Ax = b$, and one look at the figure shows that it is satisfied. The two components add up to b; no other points on S and T can do so. The equilibrium equation (with $C = I$) solves both optimization problems, and it matches the geometry.

Projections

We complete that example with an optional section. It finds an explicit formula for the projections Ax and y, starting from

$$\begin{bmatrix} I & A \\ A^T & 0 \end{bmatrix} \begin{bmatrix} y \\ x \end{bmatrix} = \begin{bmatrix} b \\ 0 \end{bmatrix}.$$

(13)

Combining $y = b - Ax$ with $A^T y = 0$ gives the normal equation

$$A^T Ax = A^T b, \quad \text{or} \quad x = (A^T A)^{-1} A^T b.$$

Therefore Ax, the projection of b onto S, is

$$Ax = A(A^T A)^{-1} A^T b.$$

(14)

That is important. Geometrically, Ax is the point on S closest to b. Algebraically, it comes from multiplying b by the m by m **projection matrix**

$$\boxed{P = A(A^T A)^{-1} A^T.}$$

(15)

---

† Duality was originally $\min Q = \max (-P)$. But always $\max (-P) = -\min (P)$; making $+P$ as small as possible makes $-P$ as large as possible. Thus duality is also $\min P + \min Q = 0$.

P is a symmetric matrix. It takes every point b to its projection Pb, and if it is applied again, that point does not move. In other words,

$$P^T = P \quad \text{and} \quad P^2 = P. \tag{16}$$

Any matrix with these properties is a projection. The special case $P = I$ is a projection onto the whole space; in that case $m = n$, and the nearest point to b is b itself. The special case $P = 0$ is a projection onto the origin (a zero-dimensional subspace). In between come the typical projections, which annihilate all vectors in T and leave only the component Pb in S. In these cases $A^T A$ is n by n and positive definite, while P and A and A^T and AA^T are not invertible. Formula (15) gives the projection onto the column space of A, and we cannot split apart the matrix $(A^T A)^{-1}$.

For the line in Fig. 2.4, P is easy to compute. The matrix A has just one column— A^T is $[2 \quad 1]$—and S contains all multiples of this column. Therefore

$$A^T A = 5 \quad \text{and} \quad P = \frac{AA^T}{5} = \frac{1}{5}\begin{bmatrix} 2 \\ 1 \end{bmatrix}[2 \quad 1] = \begin{bmatrix} 4/5 & 2/5 \\ 2/5 & 1/5 \end{bmatrix}.$$

This matrix is symmetric, and equals its own square: $P^2 = P$. It is a singular matrix, with determinant zero and no inverse, because every vector on T is projected to zero:

$$b = \begin{bmatrix} -1 \\ 2 \end{bmatrix} \text{ is on } T, \text{ and } Pb = \begin{bmatrix} 4/5 & 2/5 \\ 2/5 & 1/5 \end{bmatrix}\begin{bmatrix} -1 \\ 2 \end{bmatrix} = \begin{bmatrix} 0 \\ 0 \end{bmatrix}.$$

Of course the projection onto T is just the opposite. It leads to the point marked y in Fig. 2.4, which is $b - Ax$ or $b - Pb$ or $(I - P)b$. Therefore the projection matrix which gives y is $I - P$:

$$I - P = \begin{bmatrix} 1 & 0 \\ 0 & 1 \end{bmatrix} - \begin{bmatrix} 4/5 & 2/5 \\ 2/5 & 1/5 \end{bmatrix} = \begin{bmatrix} 1/5 & -2/5 \\ -2/5 & 4/5 \end{bmatrix}.$$

Again this is symmetric and equal to its own square: $(I - P)^2 = I - 2P + P = I - P$. If it is multiplied by P, it gives zero, since one projection followed by a perpendicular projection leads to the origin.

What is the distance from $b = (5, 5)$ to T? From the figure it is the length of Ax, and we have a formula for every step:

$$x = \frac{A^T b}{A^T A} = \frac{[2 \quad 1]\begin{bmatrix} 5 \\ 5 \end{bmatrix}}{[2 \quad 1]\begin{bmatrix} 2 \\ 1 \end{bmatrix}} = \frac{15}{5} = 3$$

$$Ax = 3\begin{bmatrix} 2 \\ 1 \end{bmatrix} \quad \text{agrees with } Pb = \begin{bmatrix} 4/5 & 2/5 \\ 2/5 & 1/5 \end{bmatrix}\begin{bmatrix} 5 \\ 5 \end{bmatrix} = \begin{bmatrix} 6 \\ 3 \end{bmatrix}.$$

Its length squared is $6^2 + 3^2 = 45$. The other side of the rectangle is $y = (-1, 2)$, whose length squared is $(-1)^2 + 2^2 = 5$. Then $45 + 5$ agrees with $\|b\|^2 = 5^2 + 5^2 = 50$, confirming duality.

2D The column space S of a rectangular matrix A is orthogonal to the nullspace T of A^T. The projection of any vector b onto S is

$$Ax = A(A^TA)^{-1}A^Tb = Pb. \tag{17}$$

The projection of b onto T is $y = b - Ax$. The duality between them is equivalent to Pythagoras' law

$$\|Ax\|^2 + \|y\|^2 = \|b\|^2. \tag{18}$$

EXERCISES

2.2.1 Minimize $Q = \frac{1}{2}(y_1^2 + \frac{1}{3}y_2^2)$ subject to $y_1 + y_2 = 8$ in two ways:

(a) Solve $\partial L/\partial y = 0$, $\partial L/\partial x = 0$ for the Lagrangian $L = Q + x_1(y_1 + y_2 - 8)$.
(b) Solve the equilibrium equations (with $b = 0$) for x and y.
What is the optimal y, and what is the minimum of Q? What is the dual quadratic $-P(x)$, and where is it maximized?

2.2.2 Find the nearest point to the origin on the plane $y_1 + y_2 + \ldots + y_m = 1$ by solving for y_1, substituting into $Q = \frac{1}{2}(y_1^2 + \ldots + y_m^2)$, and minimizing with respect to the other y's. Then solve the same problem with Lagrange multipliers.

Note We could try to solve $A^Ty = f$ for the first n y's in terms of $y_{n+1}, \ldots, y_m$. Then substitution in Q leaves only these $m - n$ unknowns. The method fails if the flows $y_1, \ldots, y_n$ contain a loop—which is avoided in Section 2.3 but is sometimes difficult.

2.2.3 Find the minimum by Lagrange multipliers of

(a) $Q = \frac{1}{2}(y_1^2 + y_2^2)$ subject to $y_1 - y_2 = 1$

(b) $Q = \frac{1}{2}(y_1^2 + y_2^2 + y_3^2)$ subject to $y_1 - y_2 = 1$, $y_2 - y_3 = 2$ (use x_1 and x_2)

(c) $Q = y_1^2 + y_1y_2 + y_2^2 + y_2y_3 + y_3^2 - y_3$ subject to $y_1 + y_2 = 2$

(d) $Q = \frac{1}{2}(y_1^2 + 2y_1y_2) - y_2$ subject to $y_1 + y_2 = 0$ (watch for maximum).

Find the corresponding $P(x)$ in parts (a) and (b), and maximize $-P(x)$.

2.2.4 Find the rectangle with corners at points $(\pm y_1, \pm y_2)$ on the ellipse $y_1^2 + 4y_2^2 = 1$, such that the perimeter $4y_1 + 4y_2$ is as large as possible.

2.2.5 In three dimensions (and without any formulas) how far is the origin from the line $y_2 = 1$, $y_3 = 1$? What plane through this line is farthest from the origin? Where is the saddle point y—the nearest point on the line and on this farthest plane?

2.2.6 The minimum distance to the surface $A^T y = f$ equals the maximum distance to the hyperplanes which _____
Complete this statement of duality.

2.2.7 How far is it from the origin $(0, 0, 0)$ to the plane $y_1 + 2y_2 + 2y_3 = 18$? Write this constraint as $A^T y = 18$, and solve for y in

$$\begin{bmatrix} I & A \\ A^T & 0 \end{bmatrix} \begin{bmatrix} y \\ x \end{bmatrix} = \begin{bmatrix} 0 \\ 18 \end{bmatrix}.$$

2.2.8 The previous question brings together several parts of mathematics if you answer it more than once:

(i) The vector y to the nearest point on the plane must be on the perpendicular ray. Therefore y must be a multiple of $(1, 2, 2)$. What multiple lies on the plane $y_1 + 2y_2 + 2y_3 = 18$? What is the length of this y?
(ii) Since $A^T = [1\ 2\ 2]$ has length $(1 + 4 + 4)^{1/2} = 3$, the Schwarz inequality for inner products gives

$$A^T y \leq \|A\|\ \|y\| \quad \text{or} \quad 18 \leq 3\|y\|.$$

What is the minimum possible length $\|y\|$? Conclusion: The distance to the plane $A^T y = f$ is $|f|/\|A\|$.

2.2.9 In the first example of duality—"the minimum distance to points equals the maximum distance to planes"—how do you know immediately that maximum $\leq$ minimum? In other words explain *weak duality*: The distance to any plane through the line is not greater than the distance to any point on the line.

2.2.10 If $b = (15, 10)$ in the geometry example of Fig. 2.4, what are the optimal Ax and y and what are the lengths in $\|Ax\|^2 + \|y\|^2 = \|b\|^2$?

2.2.11 Find the projection matrix P onto the $45°$ line $x_1 = x_2$, which is the column space of $A = \begin{bmatrix} 1 \\ 1 \end{bmatrix}$. Verify that $P^2 = P = P^T$. What is the projection of the point $b = (0, 1)$ onto this line? What are the eigenvalues of P?

2.2.12 If Ax is on S and y is on T but $z = y + Ax - b$ is not zero, show that Ax and y miss duality by $z^T z$ exactly as in (4). In other words, if $A^T y = 0$ verify the "quadrilateral law"

$$\|b - Ax\|^2 + \|b - y\|^2 = \|b\|^2 + \|z\|^2.$$

2.2.13 Suppose $A = \begin{bmatrix} 1 & 1 \\ 1 & 2 \\ 1 & 3 \end{bmatrix}$ and $b = \begin{bmatrix} 6 \\ 0 \\ 0 \end{bmatrix}$.

Compute $x = (A^T A)^{-1} A^T b$, $P = A(A^T A)^{-1} A^T$, and $Ax = Pb$. What is the dimension of T (the subspace $A^T y = 0$) and how far away is b?

2.2.14 For any projection matrix P, with $P = P^T = P^2$, verify that

$$\|Pb\|^2 + \|(I - P)b\|^2 = \|b\|^2.$$

2.2.15 Suppose S is an n-dimensional subspace of R^m, and P is the matrix that projects onto S. From the geometry rather than the formula $P = A(A^TA)^{-1}A^T$, find the vectors whose direction is the same after projection:

(a) Why is every vector in S (and every vector orthogonal to S) an eigenvector of P?
(b) What are the corresponding eigenvalues?
(c) What is the column space of P (the set of all possible Pb)?
(d) What is the rank of P (the dimension of the column space)?
(e) What is the determinant of P?

2.2.16. In m dimensions, how far is it from the origin to the hyperplane $x_1 + x_2 + \cdots + x_m = 1$? Which point on the plane is nearest to the origin?

2.3 ■ ELECTRICAL NETWORKS

The laws governing a resistive network are justly famous: Kirchhoff's current law, Kirchhoff's voltage law, and Ohm's law. They are completely straightforward, but I never found it easy to keep them in good order. It is tempting, especially for a small network, to compute one current or voltage without getting the others straight—and often without getting the signs straight. Our hope is to combine the three laws into a systematic procedure, so that even if the computer does more calculations than we might do, it does them correctly.

You will guess that it is Kirchhoff's two laws, KCL and KVL, which lead to A^T and A. These laws involve only the "topology" of the network. They start with A_0, a rectangular incidence matrix with dimensionless entries 1, 0, and -1; it records the connections between the nodes. KCL requires zero net current entering each node; there is no buildup of charge, with only resistors in the circuit. KVL requires the sum of the voltage drops around each closed loop to be zero. The simplest way to apply this law is to assign a potential x_j to each node. Then since a loop comes back to the same node and the same potential it started from, the potential differences around the loop must add to zero. It is a sum like $(x_2 - x_1) + (x_3 - x_2) + (x_1 - x_3)$, in which everything cancels.

The remaining law reflects, as always, the properties of the material—in this case it gives the *conductances* c_i. These are the diagonal entries of the matrix C. (The

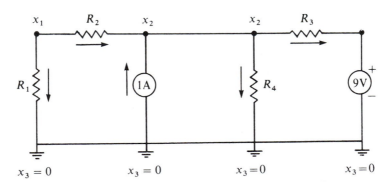

Fig. 2.5. Resistive network with voltage source and current source.

resistances $R_i = 1/c_i$ are on the diagonal of C^{-1}.) Ohm's law is $V = IR$ on each edge, or in vector notation $y = Ce$, connecting the voltage drops to the currents. We choose an example (Fig. 2.5) with four resistors and three nodes, plus a voltage source and a current source. The conventions are straightforward:

(i)   One node is chosen as ground, with potential $x_3 = 0$.
(ii)   The potentials, currents, and voltage sources may be positive or negative—depending on the arrows.
(iii)   Current in the direction of the arrow is positive, and $b > 0$ if the voltage

source contributes to positive current; the direction from $-$ to $+$ across the battery then agrees with the arrow. (In the figure the voltage source is $b = -9$.)

(iv)   The potential decreases in the direction of the current. Across R_3, for example, Ohm's law gives

$$I_3 = \frac{V_3}{R_3} = \frac{1}{R_3}(b + x_{start} - x_{end}) = \frac{1}{R_3}(-9 + x_2).$$

We write V_1, V_2, V_3, V_4 for the voltage drops across the resistors; they are the components of e. The components of y will be the currents I_1, I_2, I_3, I_4.

The vector x of node potentials and the vector y of currents through the resistors should fit the framework of the previous sections. For circuits the equilibrium equations are simpler than minimum or maximum principles, and we write them down directly:

1. $x \to e$:   The voltage drops across the resistors are

$$V_1 = x_1 - x_3, \; V_2 = x_1 - x_2, \; V_3 = -9 + x_2 - x_3, \; V_4 = x_2 - x_3.$$

In vector notation this is

$$e = b - A_0 x = \begin{bmatrix} 0 \\ 0 \\ -9 \\ 0 \end{bmatrix} - \begin{bmatrix} -1 & 0 & 1 \\ -1 & 1 & 0 \\ 0 & -1 & 1 \\ 0 & -1 & 1 \end{bmatrix} \begin{bmatrix} x_1 \\ x_2 \\ x_3 \end{bmatrix}$$

2. $e \to y$. The currents in the resistors satisfy Ohm's law:

$$y = \begin{bmatrix} I_1 \\ I_2 \\ I_3 \\ I_4 \end{bmatrix} = \begin{bmatrix} 1/R_1 & & & \\ & 1/R_2 & & \\ & & 1/R_3 & \\ & & & 1/R_4 \end{bmatrix} \begin{bmatrix} V_1 \\ V_2 \\ V_3 \\ V_4 \end{bmatrix} = Ce.$$

3. $y \to f$. Kirchhoff's current law at the three nodes gives

$$-I_1 - I_2 = 0, \quad I_2 - I_3 - I_4 = -1, \quad I_1 + I_3 + I_4 = 1.$$

The coefficient matrix in these equations is the transpose of A_0:

$$\begin{bmatrix} -1 & -1 & 0 & 0 \\ 0 & 1 & -1 & -1 \\ 1 & 0 & 1 & 1 \end{bmatrix} \begin{bmatrix} I_1 \\ I_2 \\ I_3 \\ I_4 \end{bmatrix} = \begin{bmatrix} 0 \\ -1 \\ 1 \end{bmatrix}, \quad \text{or} \quad A_0^T y = f.$$

Thus the equilibrium equations are $y = Ce = C(b - A_0 x)$ and $A_0^T y = f$. They will combine to give the key equation. But first there is a part of A_0 that has to be amputated.

Each row of the incidence matrix A_0 adds to zero, which means trouble. It is just at this point that grounding node 3 is crucial; by fixing $x_3 = 0$ as a boundary condition the last column of A_0 (which multiplies x_3) can be dropped. The same is true of the last row of A_0^T, for a similar (in fact dual) reason. The three equations in Kirchhoff's current law add to $0 = 0$, so the third one contains no independent information; it is a consequence of the other two. After grounding node 3 and removing its column, the final matrix A is 4 by 2. It has two independent columns, and the final x and f have only $n = 2$ components.

Now we summarize the equations of resistive circuit theory.

2E Ohm's law $y = C(b - Ax)$ and Kirchhoff's current law $A^T y = f$ combine to give the network equation

$$A^T C A x = A^T C b - f. \tag{1}$$

The solution x gives the potentials at the nodes and y gives the currents.

In solving the example, we start with the current law $A^T y = f$ and Ohm's law (rewritten as $C^{-1} y + Ax = b$) together in a single large system. Then elimination will lead to equation (1) for x:

$$\begin{bmatrix} C^{-1} & A \\ A^T & 0 \end{bmatrix} \begin{bmatrix} y \\ x \end{bmatrix} = \left[\begin{array}{cccc|cc} R_1 & 0 & 0 & 0 & -1 & 0 \\ 0 & R_2 & 0 & 0 & -1 & 1 \\ 0 & 0 & R_3 & 0 & 0 & -1 \\ 0 & 0 & 0 & R_4 & 0 & -1 \\ \hline -1 & -1 & 0 & 0 & 0 & 0 \\ 0 & 1 & -1 & -1 & 0 & 0 \end{array} \right] \begin{bmatrix} I_1 \\ I_2 \\ I_3 \\ I_4 \\ \hline x_1 \\ x_2 \end{bmatrix} = \begin{bmatrix} 0 \\ 0 \\ -9 \\ 0 \\ \hline 0 \\ -1 \end{bmatrix}$$

$$= \begin{bmatrix} b \\ f \end{bmatrix}. \tag{2}$$

Note that x_3 and f_3 are removed. If all resistances are $R_i = 1$, then $C = I$ and elimination leads to $-A^T A x = f - A^T b$. This is the equation we actually solve. It appears in the lower right corner, where zeros are now:

$$\begin{bmatrix} -2 & 1 \\ 1 & -3 \end{bmatrix} \begin{bmatrix} x_1 \\ x_2 \end{bmatrix} = \begin{bmatrix} 0 \\ -1 \end{bmatrix} - \begin{bmatrix} 0 \\ 9 \end{bmatrix}.$$

The potentials are $x_1 = 2$, $x_2 = 4$, and the currents are

$$y = C(b - Ax) = \begin{bmatrix} 2 \\ -2 \\ -5 \\ 4 \end{bmatrix}.$$

These potentials and currents (with all $R_i = 1$) are illustrated in Fig. 2.6, together with the basic framework for resistive networks.

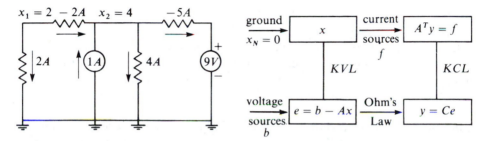

Fig. 2.6. Currents and voltages in the example and in general.

This framework gives a systematic approach to circuit theory. It can allow other possibilities which were missed by the example:

1. Nodes 1 and 3 might be directly connected by a voltage source of strength V. That fixes $x_1 = V$ as well as $x_3 = 0$. In this case x_1 is also removed from the vector x of unknown potentials, leaving only x_2. Of course x_1 does not disappear without a trace; the voltage drops are

$$e = b - Ax = \begin{bmatrix} 0 \\ 0 \\ -9 \\ 0 \end{bmatrix} - \begin{bmatrix} -1 & 0 \\ -1 & 1 \\ 0 & -1 \\ 0 & -1 \end{bmatrix} \begin{bmatrix} x_1 \\ x_2 \end{bmatrix} = \begin{bmatrix} V \\ V \\ -9 \\ 0 \end{bmatrix} - \begin{bmatrix} 0 \\ 1 \\ -1 \\ -1 \end{bmatrix} [x_2]$$

This new source effectively contributes to b, just as the 9-volt battery did. When some potential is fixed, it multiplies a column of A and appears in the forcing term b. If the only active source is this voltage V, then the total current I between terminals 1 and 3 gives the *network resistance* V/I between those two nodes.

2. Instead of fixing the voltage $x_1 = V$ we could fix the current $y_1 = I$. This adds a new and final arrow to Fig. 2.6b, pointing into the box labeled y. It gives a current source which is in *series* with a resistor, and it effectively contributes to f. Like the

voltage V, the current I appears in the matrix multiplication; Kirchhoff's current law $A^T y = f$ becomes

$$
\begin{bmatrix} -1 & -1 & 0 & 0 \\ 0 & 1 & -1 & -1 \end{bmatrix} \begin{bmatrix} I \\ I_2 \\ I_3 \\ I_4 \end{bmatrix} = \begin{bmatrix} 0 \\ -1 \end{bmatrix} \quad \text{or}
$$

$$
\begin{bmatrix} -1 & 0 & 0 \\ 1 & -1 & -1 \end{bmatrix} \begin{bmatrix} I_2 \\ I_3 \\ I_4 \end{bmatrix} = \begin{bmatrix} I \\ -1 \end{bmatrix}.
$$

The right side is the new f.

3.   Along with batteries we could have **transistors**, which are nonlinear voltage sources. Their strength is not fixed; it depends on the voltages at their terminals. In other words, a transistor is a "voltage-dependent" voltage source. It makes Kirchhoff's laws nonlinear (and a diode makes Ohm's law nonlinear).

A fourth possibility is to allow **capacitors**, for which current is proportional not to voltage but to its rate of change: $I = C \, dV/dt$. A capacitor stores electrical energy, rather than dissipating it like a resistor. Complementary to it is an **inductor**, for which voltage is proportional to the rate of change of current: $V = L \, dI/dt$. It stores magnetic energy. With direct currents that are independent of time, the capacitor acts like an infinite resistance and forces $I = 0$ (an open circuit). The inductor acts like an infinite conductance and forces $V = 0$ (a short circuit). Their real applications are in time-dependent problems with *alternating currents*, but first we make a table to connect the general framework of this chapter and the specific networks of this section.

Linear Resistive Circuits: Unknowns and Equations

x:	potentials at the nodes
$x_N = 0$:	boundary condition grounding the last node
A:	incidence matrix with grounded column removed
Ax:	potential differences between nodes
b:	voltage sources
$e = b - Ax$:	potential drops V_j across the resistors
C:	diagonal matrix of conductances $c_j = 1/R_j$
y:	currents I_j through the resistors
f:	current sources at the nodes
$A^T y = f$:	Kirchhoff's current law
$C^{-1} y + Ax = b$:	Ohm's law

RLC Circuits

There are linear devices other than resistors. The standard *RLC* circuit also allows capacitors and inductors, leading to differential equations instead of algebraic equations for x and y. One possibility is to have a pure initial value problem, as in Chapter 6; the currents decay with time as the resistors dissipate energy. The other possibility, of greater importance in many applications, is a sinusoidal forcing term with fixed frequency ω. In this case each voltage can be written as $v \cos \omega t = Re(ve^{i\omega t})$, the real part of a complex number $v \cos \omega t + iv \sin \omega t$. These alternating voltages produce alternating currents. Therefore the problem is time-varying, but the dependence on time of every current has the same convenient form: it is the real part of $ye^{i\omega t}$. The edge unknowns are now complex numbers $y_1, \ldots, y_m$.

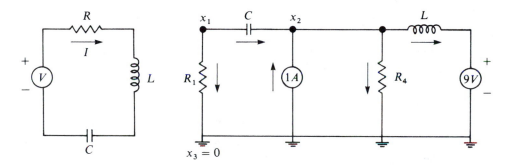

Fig. 2.7. An *RLC* loop and an *RLC* circuit.

The circuit becomes as simple to analyze as before, after changing from real resistances R to **complex impedances** z. To find z we look at the equation for a basic *RLC* loop (Fig. 2.7). If the inductance is L and the capacitance is C, then $V = L \, dI/dt$ for the first and $I = C \, dV/dt$ (or $V = \int I \, dt/C$) for the second. These voltages in inductance, resistance, and capacitance combine to give

$$L\frac{dI}{dt} + RI + \frac{1}{C}\int I dt = \text{applied voltage} = Re(ve^{i\omega t}). \tag{3}$$

Substituting $I = Re(ye^{i\omega t})$ for the current produces a factor $i\omega$ in the derivative and integral, and our equation is the real part of

$$\left(i\omega L + R + \frac{1}{i\omega C}\right) ye^{i\omega t} = ve^{i\omega t}.$$

Cancelling the common factor $e^{i\omega t}$—a simple but crucial step, which is possible because all sources have the common frequency ω—we reach the key equation

$$v = yz \text{ with impedance } z = i\omega L + R + \frac{1}{i\omega C}. \tag{4}$$

This is Ohm's law $V = IR$, with R made complex. For the basic loop in Fig. 2.7 there is nothing more to do; y is v/z, and the current is the real part of $(v/z)e^{i\omega t}$. The change from R to z added the imaginary terms $i\omega L$ and $1/i\omega C = -i/\omega C$, and it produced an impedance of magnitude

$$|z| = \left| R + i\left(\omega L - \frac{1}{\omega C}\right) \right| = \left(R^2 + \left(\omega L - \frac{1}{\omega C}\right)^2 \right)^{1/2}. \tag{5}$$

This is larger than R, so the magnitude $|y| = |v/z|$ of the current is below v/R. In other words, the capacitor and inductor have reduced the flow.

Notice the one special frequency at which the inductor cancels the capacitor. That happens if

$$\omega L = \frac{1}{\omega C}, \quad \text{or} \quad \omega = \sqrt{\frac{1}{LC}}. \tag{6}$$

Then the imaginary part of the impedance disappears and the circuit is back to $z = R$. That is what happens when you tune a radio—you are adjusting the capacitance C to match the frequency you want. At that resonant frequency, given by (6), the impedance is smallest and you can hear the signal (after it is amplified). Other frequencies will be received at the same time, but the imaginary part of their impedance z is much larger and their signal is inaudible.

The capacitor and inductor not only reduce the flow by increasing $|z|$, they also change its *phase*. If the complex number z is $|z|e^{i\theta}$, making an angle θ with the real axis, then the current is

$$I = Re\left(\frac{V}{z}e^{i\omega t}\right) = Re\left(\frac{v}{|z|}e^{i(\omega t - \theta)}\right). \tag{7}$$

Therefore terms that used to involve $\cos \omega t$ now involve $\cos(\omega t - \theta)$; the cycles are advanced or retarded depending on θ. If the circuit contains only resistors then $\theta = 0$; the imaginary part is absent and there is no phase change. If the circuit contains *no* resistors, then z is a pure imaginary number (from $i\omega L$ and $-i/\omega C$). Its phase angle is $\pi/2$ or $3\pi/2$. In that case the current is completely out of phase with the voltage; if the voltage is proportional to $\cos \omega t$ then the current involves $\sin \omega t$.

For a more general RLC circuit, we go back to the basic framework. Each voltage source and current source, together with each potential and current, is represented by a complex number. These are the components of b, f, x, and y. At the end, after the equilibrium equations are solved, the actual time-varying currents are found from the real parts of $ye^{i\omega t}$. Except for complex coefficients nothing is changed, and n complex equations only mean $2n$ real equations. The coefficient matrix is $A^T C A$, with the same A and with a complex diagonal matrix C. Its entries are the "admittances," the reciprocals $1/z$ of the impedances.

EXAMPLE Suppose the resistances R_2 and R_3 in Fig. 2.5 are changed to a capacitance C and an inductance L (Fig. 2.7). The connectivity matrix A is unchanged. Therefore the equilibrium equation is exactly as before, except that the diagonal of C^{-1} contains impedances instead of resistances:

$$
\begin{bmatrix} C^{-1} & A \\ A^T & 0 \end{bmatrix} \begin{bmatrix} y \\ x \end{bmatrix} =
\left[
\begin{array}{cccc:cc}
R_1 & & & & -1 & 0 \\
& (i\omega C)^{-1} & & & -1 & 1 \\
& & i\omega L & & 0 & -1 \\
& & & R_4 & 0 & -1 \\
\hdashline
-1 & -1 & 0 & 0 & 0 & 0 \\
0 & 1 & -1 & -1 & 0 & 0
\end{array}
\right]
\begin{bmatrix} y_1 \\ y_2 \\ y_3 \\ y_4 \\ x_1 \\ x_2 \end{bmatrix}
= \begin{bmatrix} b \\ f \end{bmatrix} \quad (8)
$$

The solution gives the currents and potentials, $Re(ye^{i\omega t})$ and $Re(xe^{i\omega t})$.

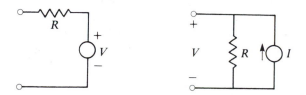

Fig. 2.8. Thévenin and Norton equivalents.

In many applications there is a "box" whose internal circuit we want to ignore. All that matters to the rest of the circuit is the total impedance of the box. If it contains a family of resistors, in parallel or in series, then it has an equivalent resistance R_{box}.† If the box also contains inductors and capacitors, it has an equivalent impedance z_{box}. Once this equivalent is known there is no reason to look inside the box.

One additional possibility: The box may contain voltage or current sources. In that case it has a "Thévenin equivalent"—a single voltage source in series with an

---

† Resistors in parallel combine according to $1/R = \Sigma\, 1/R_i$, and resistors in series by $R = \Sigma\, R_i$. The same is true of impedances, and a succession of these steps leads to the resistance R_{box} or the impedance z_{box}.

impedance. Alternatively it can be replaced by its "Norton equivalent"—a current source in parallel with an impedance (Fig. 2.8). I hope it is intuitive that these equivalents can be found. Mathematically, they mean that a diagonal block of A^TCA is connected to the rest of the matrix at only one edge, and elimination can begin within this block. The block is reduced to a single entry, which is all that the rest of the circuit will know.

Minimum Principles and Loop Currents

In closing, we mention two other approaches to networks and their currents. One is to base the equations on minimum principles. The duality theory of the previous section fits perfectly, but I am sorry to say that it is not much used. It is so straightforward to write down the equations of Ohm and Kirchhoff that quadratics are never missed. Certainly a quadratic is present; it is the **power** VI or V^2/R or I^2R. It gives the rate at which electrical energy is converted into heat by the resistor (or into other energy by a light bulb or loudspeaker). The minimum principles are so beautiful that they deserve to be seen:

(1)   If a current source connects two nodes, then current flows through the network so as to minimize the heat loss $\Sigma\, I_j^2 R_j = y^T C^{-1} y$ subject to Kirchhoff's law.

(2)   If a voltage source is applied between two nodes, then the potentials at the other nodes adjust so as to minimize the heat loss $\Sigma\, V_j^2/R_j = e^T C e$.

The power is in one case $2Q(y)$ and in the other $2P(x)$; the factor $\frac{1}{2}$ which appears in other applications does not appear here.

Notice what the second principle says. The voltage source produces a vector b, given by the strength of the battery. Then the potentials in the network try to achieve an Ax which is as close as possible to b. **Nature solves a least squares problem**, in its effort to save heat.

The last approach to network equations, one which is more often used, is to reduce the number of unknown currents from m to $m - n$. To do that we use the n equations in Kirchhoff's current law. The only difficulty is that the n currents to be eliminated, and the $m - n$ currents to be kept, cannot be chosen at random.

There is a simple rule that always works: Choose n edges that form a **tree**.† A tree is a graph of the simplest possible kind. It has N nodes and only $n = N - 1$ edges; all edges are needed just to connect the nodes. Between any two nodes there is exactly one path, and **there are no loops** (Fig. 2.9a). With $N = 4$, for example, the tree could have its three edges coming out of one node, or they could form three sides of a square. Such a tree is easy to construct; just add edges and avoid loops until all nodes are connected.

Pick any spanning tree in the network. Each of the remaining $m - n$ edges, when

---

† Its full name is **spanning tree**, to indicate that it reaches all N nodes. Kirchhoff found an amazing formula for the number of spanning trees in a network: it is the determinant of $A^T A$.

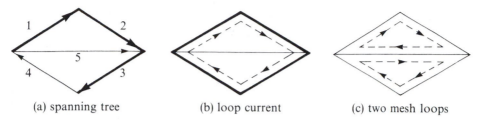

(a) spanning tree   (b) loop current   (c) two mesh loops

Fig. 2.9. Spanning trees and loops.

it is added to the tree, will close a loop (Fig. 2.9b). We can imagine that the current on that extra edge goes around the loop. Certainly each of these loop currents satisfies Kirchhoff's law; for current flowing around a loop there is no buildup of charge. In matrix terms, ***each loop current is a solution to $A^T y = 0$.*** It is easy to check that the total flow in the network is the sum of the $m - n$ loop currents (each accounting for one edge outside the tree, and the currents within the tree being automatically controlled by Kirchhoff's law). Therefore the loop currents give a complete set of independent unknowns.

The second loop current would come from adding the horizontal edge to the tree; the loop goes around the top. This is a case in which the special properties of incidence matrices A made it possible to do part of the elimination by hand. We have eliminated three of the five unknown currents by using the three equations $A^T y = 0$. Normally the nullspace of a matrix is not easy to find, but here any y that gives current flowing around a loop immediately satisfies $A^T y = 0$. The loop in Fig. 2.9b corresponds to a vector y with components $I_1 = I_2 = I_3 = I_4 = 1$ and $I_5 = 0$.

For graphs that lie in a plane, as this one does, there is another way to find $m - n$ independent loop currents. Just look at the network; it has $m - n = 2$ "mesh loops." Given a current around each loop in Fig. 2.9c, addition produces a complete set of currents in the network—and this is the general solution to $A^T y = 0$. The same idea will appear for fluid flows in Chapter 3, where Kirchhoff's law becomes a differential equation $\partial w_1 / \partial x + \partial w_2 / \partial y = 0$. There the general solution comes from a stream function s, with $w_1 = \partial s / \partial y$ and $w_2 = - \partial s / \partial x$. On a discrete network the analogue of s is the set of currents on the mesh loops.

Either method uses an edge-loop matrix B, instead of the edge-node matrix A. B has a column for each loop, indicating which edges that loop contains (and in which direction):

$$B = \begin{bmatrix} 1 & 1 \\ 1 & 1 \\ 1 & 0 \\ 1 & 0 \\ 0 & -1 \end{bmatrix} \quad \text{and} \quad A = \begin{bmatrix} -1 & 1 & 0 \\ 0 & -1 & 1 \\ 0 & 0 & -1 \\ 1 & 0 & 0 \\ -1 & 0 & 1 \end{bmatrix}.$$

The first column of B comes from the loop around the outside; the second is from the top triangle. (Their difference would produce the lower triangle. Remember that only two loops can be independent.) Both columns satisfy $A^T y = 0$, and any other solution is a combination $y = Bz$ of these two columns.† The equilibrium equation is now

$$C^{-1}y + Ax = b, \quad \text{or} \quad C^{-1}Bz + Ax = b, \quad \text{or} \quad B^T C^{-1}Bz = B^T b. \tag{9}$$

Instead of $n = 3$ unknown potentials and the matrix $A^T C^{-1}A$, we now have $m - n = 2$ unknowns z_j and the matrix $B^T C^{-1}B$. The structure is the same, and for n close to m the problem has been reduced.

EXERCISES

2.3.1 Circuit (a) has a current source of strength I. Write down the 2 by 2 incidence matrix A_0 and remove its second column to form A. Verify that $A^T C A = \dfrac{1}{R_1} + \dfrac{1}{R_2}$. (This is the conductance of two resistors in parallel.) What is the full 3 by 3 system, corresponding to (2) in the text, for I_1, I_2, x_1? What is its solution?

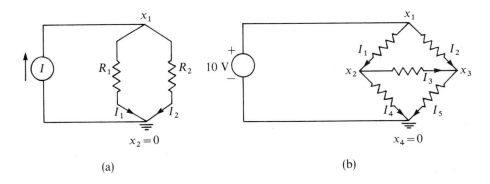

(a)   (b)

2.3.2 For circuit (b) in the figure the voltage source fixes $x_1 = 10$. Write down the five equations $e = -A_0 x$, and set $x_1 = 10$, $x_4 = 0$ to reach $e = b - Ax$; A has two columns multiplying x_2 and x_3. Similarly $A^T y = f$ has only the two equations $I_1 - I_3 - I_4 = 0$ and $I_2 + I_3 - I_5 = 0$, at nodes 2 and 3. With unit resistors solve the equilibrium equations for y and x. What is the total current $I = I_4 + I_5$ into node 4 and what is the network resistance V/I to the voltage $V = 10$?

2.3.3 What would happen to the potentials and currents, in the example based on Fig. 2.5, if instead of grounding node 3 we had fixed $x_3 = 5$? What if we had grounded node 1 instead of node 3?

---

† Note that $A^T B = 0$, or by transposing $B^T A = 0$.

2.3.4 The text discusses an applied voltage V across x_1 and x_3 in Fig. 2.5, which determines the network resistance $R = V/I$ between those nodes. With all $R_i = 1$, find R in two ways:

1. Use $1/r = 1/R_3 + 1/R_4$ to find the resistance of R_3 and R_4 in parallel. Then R_2 is in series with r, and that combination is in parallel with R_1. What is the equivalent resistance R?

2. Let $x_1 = V$, ignore the battery and current source in the figure, and solve the equilibrium equations for x_2 and y. If I is the current into node 3, then $R = V/I$.

2.3.5 (a) If $K = A^T CA$ is symmetric and invertible, how do you know that K^{-1} is also symmetric? (Do something to $KK^{-1} = I$.) (b) The potential at node i due to a unit current source from node j to ground is $(K^{-1})_{ij}$, since $x = K^{-1}f$. How does this compare with the potential at node j due to a unit current source from node i to ground?

2.3.6 How large is the current around the basic RLC loop (Fig. 2.7) if $R = 3$, $\omega L = 5$, $\omega C = 1$? Put the voltage $V = \cos \omega t$ and the resulting current on a (sinusoidal) graph; what is their difference in phase?

2.3.7 With $R_1 = C = L = R_4 = 1$ in the circuit of Fig. 2.7, set up (but don't solve) the equilibrium system (8) for the currents and potentials. Can you get two equations for the potentials x_1 and x_2?

2.3.8 Suppose the n simultaneous equations $Ax = b$ involve complex matrices and vectors: $A = A_1 + iA_2, b = b_1 + ib_2$, with solution $x = x_1 + ix_2$. By taking the real and imaginary parts of $Ax = b$, find $2n$ real equations for the $2n$ real unknowns x_1 and x_2.

2.3.9 How much heat is dissipated by the resistors in Fig. 2.5, with the given battery and current source?

2.3.10 For a planar graph, as in Fig. 2.9c, why are there $m - n$ mesh loops?

2.3.11 (a) Find all eight spanning trees in Fig. 2.9. (b) Why does $A^T B = $ (node-edge matrix) (edge-loop matrix) $= 0$?

2.3.12 In fluid problems, flow rate is the analogue of current. What is the analogue of potential?

2.3.13 Consider the graph below with four nodes and all six possible edges.
(a) Count its spanning trees (each with three edges and no loops)
(b) Form A and $A^T A$ (3 by 3) and find its determinant
(c) Compare with the formula N^{N-2}
(d) How many independent loops in this graph?

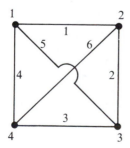

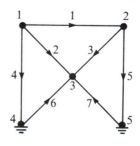

2.3.14   (a)   Write down the 7 by 5 incidence matrix A_0 for the network above.   (b)   Which vectors satisfy $A_0^T y = 0$?   (c)   Ground both of the indicated nodes and find 10 equations for the 3 potentials and 7 currents, if every edge has unit resistance and $f_3 = 24$ amps enters node 3 (current source) and flows to ground.   (d)   Solve this system for the potentials and currents.

STRUCTURES IN EQUILIBRIUM ■ 2.4

The problem of statics is to find the internal stresses and the displacements, when loads are applied to a structure. This is older than the problem of voltages and currents, but still very much alive. It is the first half of mechanics; the other half is dynamics, when the movement depends on time. Here there is just enough displacement to create internal forces that balance the applied forces. Then the structure—a bridge or a building or an airplane—rests comfortably in equilibrium.

Our structures will be built from elastic bars. Each bar resists a change in its length, and that resisting force permits the structure to withstand a load. The nodal connections are by pin joints, or hinges, which can transmit forces only *along* the bars. The pin joints allow the bars to point in any direction they choose; each bar as a whole can turn. Unlike beams, however, the bars do not bend; unlike plates, they are one-dimensional; unlike shells, they are simple and straight. Those others are the building blocks of more complicated structures, leading toward the fully three-dimensional theory of continuum mechanics. Each step of that theory can fit into our pattern, as long as the displacements depend linearly on the forces.

This section takes the first step, for a framework that is constructed entirely from bars. We will restrict it to lie in a plane, and call it a *truss*.

Three trusses are shown in Fig. 2.10 together with their supports. The fixed supports provide two reactive forces, resisting movement in each direction but allowing rotation. (The roller that comes later in Fig. 2.12d only prevents vertical displacement.) At every unsupported node *two external forces* f_j are applied; they will be given by the problem, and some of them may be zero. At such a node, which is free to move, *two displacements* x_j are to be determined.

Suppose the truss has m bars and N nodes. Then there are initially $2N$ nodal displacements, in two independent directions (normally horizontal and vertical) at each node. The supports will fix certain displacements, two at each fixed support and one at each roller; we denote by r the number of known displacements. (In the electrical problem only one node was grounded, so $r = 1$.) There remain $n = 2N - r$ degrees of freedom in the displacements, and we refer to this vector of n unknowns as x. Thus the r fixed displacements, often zero, disappear from the vector x.

In the same way there are $2N$ nodal forces f_j, in two directions at each node. The r forces from the supports are unknown reactions, supplying whatever force is required to maintain the r fixed displacements. The other $n = 2N - r$ forces are prescribed, and they make up the vector f of *applied forces*. It is exactly like the line of four springs in Section 1.4, which initially had $N = 5$ nodes. Fixing both ends of the line gave $r = 2$, and left $N - r = 3$ nodal unknowns—the displacements of the masses between the springs. In that one-dimensional case N replaced $2N$; for a space truss, with three independent directions at each node, it would be $3N$. The trusses in Fig. 2.10 have $n = 4$ unknown displacements and applied forces at the upper nodes, and $r = 4$ fixed displacements and unknown reactions at the supports.

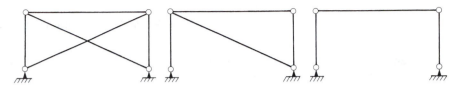

Fig. 2.10. Trusses with $m > n$ (indeterminate), $m = n$ (determinate), $m < n$ (unstable).

But m (the number of bars) goes from 5 to 4 to 3, and the trusses behave very differently.†

So far the analogies are clear. The displacements correspond to potentials, and some are fixed. The forces correspond to current sources, and the balance at each node is like Kirchhoff's current law. There is, however, one important difference: There are no arrows on the edges! They are not needed, since $y_i > 0$ always means that bar i is in **tension**, and $y_i < 0$ means it is in **compression**. In mechanics the arrows are not on the edges but at the nodes. They decide the directions of positive displacement. As before, these directions can be fixed in any definite way—a negative x_j simply means that the node moves in the direction opposite to the arrow.

There is one other difference: we also have to fix the signs of the external forces f. It is natural to use the same arrows as for the displacements—a positive force is one in the direction of a positive displacement. This convention leads to $e = + Ax$ (rather than $e = - Ax$) and this is the usual rule in mechanics. The vector b is ordinarily zero.

To find A we begin as always with A_0. It still has m rows, one for every bar. But instead of two nonzero entries in each row, $+1$ and -1, we now expect four. They will be $\pm\cos\theta$ and $\pm\sin\theta$, where θ is the slope angle of the bar. (The row still adds to zero.) For a horizontal or vertical bar we are back to ± 1, since $\sin\theta$ or $\cos\theta$ is zero, but for other slopes the nodes at the ends of the bar each contribute two terms. We show next how these terms appear, both in the rule $e = Ax$ for elongation of the bars and in the transposed rule $A^T y = f$ for force balance at the nodes.

Suppose the ends of a bar are moved; how much does the bar stretch? If its original length is L and its angle is θ, then before stretching it extends $L\cos\theta$ horizontally and $L\sin\theta$ vertically (Fig. 2.11a). When its ends are moved these components change; for the stretched bar Pythagoras gives

$$l^2 = (L\cos\theta + x_1 - x_3)^2 + (L\sin\theta + x_2 - x_4)^2.$$

This is the new length. Computing the squares and then the square root,

$$l^2 = L^2 + 2L(x_1 \cos\theta - x_3 \cos\theta + x_2 \sin\theta - x_4 \sin\theta) + O(x_j^2)$$
$$l = L + (x_1 \cos\theta - x_3 \cos\theta + x_2 \sin\theta - x_4 \sin\theta) \quad + O(x_j^2/L).$$

---

†It is the appearance of $2N$ that brings new possibilities. Previously m was never below n $= N - 1$; now the relation of m to n (and the shape of A) is not so certain. And even $m \geq n$ will not guarantee stability.

The difference $l - L$ is the elongation, denoted by e. Assuming small deformations (much smaller than in the Figure!) and ignoring the x_j^2/L corrections, the expression in parentheses is a typical term in $A_0 x$. The nonzero entries in each row of A_0 are $\pm \cos \theta$ and $\pm \sin \theta$. It is like an incidence matrix, but now it describes a genuinely two-dimensional geometry. In three dimensions the entries will be $\pm \cos \theta_1$, $\pm \cos \theta_2$, $\pm \cos \theta_3$—the three cosines that give the direction of the bar.

As it stands this matrix A_0 is m by $2N$, but we drop the r columns corresponding to fixed displacements. This leaves $2N - r = n$ columns, one for every unknown displacement x_j. The resulting matrix A is m by n. The relation $e = Ax$ between elongations of the bars and displacements of the nodes is the **compatibility equation**. It is linear, by our assumption of small deformations.

The transpose of this same matrix must appear in the **equilibrium equation**. That is the balance $A^T y = f$ between internal forces y in the bars and applied forces f at the nodes. Since each node is in equilibrium, the net force on it—both horizontal and vertical—must be zero. In Fig. 2.11b the horizontal applied force is f_H, and the balance of horizontal forces gives

$$-y_1 \cos \theta_1 - y_2 \cos \theta_2 - y_3 \cos \theta_3 = f_H. \tag{1}$$

This is a typical equation in the system $A^T y = f$; there will be another one for the vertical components, with sines instead of cosines and f_V instead of f_H.†

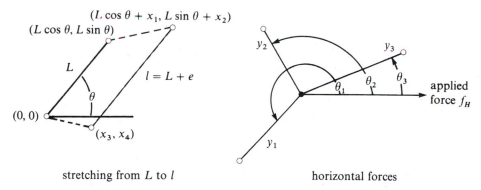

stretching from L to l horizontal forces

Fig. 2.11. The elongation $e = Ax$ and the force balance $A^T y = f$.

We check the signs in (1). A positive internal force, $y_1 > 0$, means that bar 1 is in tension; it is pulling on the hinge. The slope angle of bar 1 is $\theta = \theta_1 - \pi$, so the term $-y_1 \cos \theta_1$ is $y_1 \cos \theta$; it comes from the same matrix entry $+\cos \theta$ as the term $x_1 \cos \theta$ in the compatibility equation. Similarly the coefficient of y_3 is $-\cos \theta_3$, like the coefficient of x_3 in $l - L$. The difference is that in $e = Ax$ we looked at each

---

† Some find A from this force balance, others prefer $e = Ax$ from displacements. Note that $a_{ij} > 0$ when a positive displacement $x_j > 0$ *stretches* bar i (and when $f_j > 0$ produces tension). That keeps track of the signs.

bar, and now we look at each node. In that case we found the rows of A; in this case we go down its columns.

Finally there is $y = Ce$, the constitutive law, connecting the force in each bar with its elongation. C is a diagonal matrix, m by m, with the elastic constants of the bars on its diagonal. This is the "material matrix," and Hooke's law $y = Ce$ completes the path from the displacements x to the applied forces f:

$$f = A^T y = A^T Ce = A^T CAx, \quad \text{or} \quad f = Kx. \tag{2}$$

These are the n equations which are actually solved. The symmetric coefficient matrix $K = A^T CA$ is the **stiffness matrix** of the structure.

We collect these steps in a single table:

Structural equilibrium

x:	displacements at the free nodes ($n = 2N - r$ degrees of freedom)
$e = Ax$:	elongations of the m bars (compatibility equation)
b:	zero
C:	diagonal matrix of elastic constants of the bars
$y = Ce$:	internal forces in the bars (Hooke's law)
f:	external forces applied at the free nodes
$A^T y = f$:	force balance at the nodes (equilibrium equation)
$K = A^T CA$:	stiffness matrix of the structure

One further note: Equilibrium also holds at the supports. It has the same form as $A^T y = f$, except that it comes from the r columns of the matrix A_0 that were dropped to produce A, and from the r reactive forces that were dropped to produce f. In other words, the typical equilibrium equation (1) remains true at the supports. It is solved *after* we know y. Then the r extra equations in $A_0^T y = f$ immediately give the r components of f (the reactions at fixed nodes) which had been unknown.

Four Types of Trusses

We apply the theory to the simple trusses in Fig. 2.12. It is amazing how many possibilities appear with only a few nodes. In fact there are four basic types of trusses, and the figure contains one of each type.

EXAMPLE 1   There are $m = 2$ bars, $N = 3$ nodes, and $r = 4$ reactions (two at each support). Therefore $n = 2N - r = 2$, and the matrix A is 2 by 2. To compute it we write down the equilibrium equations at the free node. If the angles θ are $\pm 45°$ then $\cos \theta$ and $\sin \theta$ are $\pm 1/\sqrt{2}$:

$$
\begin{aligned}
\frac{1}{\sqrt{2}} y_1 - \frac{1}{\sqrt{2}} y_2 &= f_H \\
\frac{1}{\sqrt{2}} y_1 + \frac{1}{\sqrt{2}} y_2 &= f_V.
\end{aligned}
\tag{3}
$$

This coefficient matrix is A^T. A itself will appear in the compatibility equation, if the node is displaced horizontally by x_H and vertically by x_V. That stretches the bars by

$$e_1 = \frac{1}{\sqrt{2}} x_H + \frac{1}{\sqrt{2}} x_V$$

$$e_2 = -\frac{1}{\sqrt{2}} x_H + \frac{1}{\sqrt{2}} x_V.$$

Then the stiffness matrix is

$$K = A^T C A = \begin{bmatrix} 1/\sqrt{2} & -1/\sqrt{2} \\ 1/\sqrt{2} & 1/\sqrt{2} \end{bmatrix} \begin{bmatrix} c_1 & 0 \\ 0 & c_2 \end{bmatrix} \begin{bmatrix} 1/\sqrt{2} & 1/\sqrt{2} \\ -1/\sqrt{2} & 1/\sqrt{2} \end{bmatrix}$$

$$= \frac{1}{2} \begin{bmatrix} c_1 + c_2 & c_1 - c_2 \\ c_1 - c_2 & c_1 + c_2 \end{bmatrix}.$$

This example has a special property: the matrix A is *square* rather than rectangular, and the equation $A^T y = f$ alone is enough to find the forces in the bars. In other words, we can solve directly for y_1 and y_2. Then we go backwards: Hooke's law gives $e = C^{-1} y$ for the elongations, and the displacements come from $x = A^{-1} e$. The triple product $K = A^T C A$ can be inverted one matrix at a time, which is impossible when A is rectangular.

A truss with this special property is called **statically determinate**: The matrix A is square and nonsingular. This requires that $m = n = 2N - r$. In that case the equilibrium equation $A^T y = f$ can be solved immediately for y.

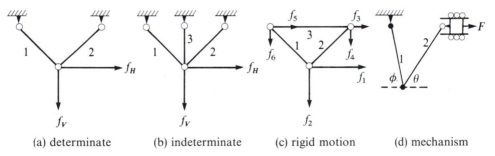

(a) determinate   (b) indeterminate   (c) rigid motion   (d) mechanism

Fig. 2.12. The four types of truss: two stable and two unstable.

EXAMPLE 2 With a third bar we have $m = 3$, $N = 4$, and $r = 6$. Thus $n = 2N - r$ again equals 2. The matrix A will be 3 by 2, and now the truss is **statically indeterminate**. It has $m > n$ with n independent columns. The equilibrium equations include a vertical force y_3 from the new bar, with $\cos \theta_3 = 0$ and $\sin \theta_3 = 1$:

$$\frac{1}{\sqrt{2}} y_1 - \frac{1}{\sqrt{2}} y_2 \qquad = f_H$$

$$\frac{1}{\sqrt{2}} y_1 + \frac{1}{\sqrt{2}} y_2 + y_3 = f_V. \tag{4}$$

Since we cannot solve for a unique y, the displacements come first from $Kx = f$:

$$K = \begin{bmatrix} 1/\sqrt{2} & -1/\sqrt{2} & 0 \\ 1/\sqrt{2} & 1/\sqrt{2} & 1 \end{bmatrix} \begin{bmatrix} c_1 & & \\ & c_2 & \\ & & c_3 \end{bmatrix} \begin{bmatrix} 1/\sqrt{2} & 1/\sqrt{2} \\ -1/\sqrt{2} & 1/\sqrt{2} \\ 0 & 1 \end{bmatrix}$$

$$= \begin{bmatrix} \dfrac{c_1 + c_2}{2} & \dfrac{c_1 - c_2}{2} \\ \dfrac{c_1 - c_2}{2} & \dfrac{c_1 + c_2}{2} + c_3 \end{bmatrix}.$$

Then $e = Ax$ and $y = Ce$, which is the right y among the solutions to $A^T y = f$.

EXAMPLE 3   The supports are gone in Fig. 2.12c, leaving $m = 3$ bars, $N = 3$ nodes, and $r = 0$. Therefore $n = 2N - r = 6$, and the matrix A has six columns. This gives six equations of equilibrium to be satisfied, and only three bar forces y_i to do it with: $A^T y = f$ is

$$\frac{1}{\sqrt{2}} y_1 - \frac{1}{\sqrt{2}} y_2 \qquad = f_1$$

$$\frac{1}{\sqrt{2}} y_1 + \frac{1}{\sqrt{2}} y_2 \qquad = f_2$$

$$\frac{1}{\sqrt{2}} y_2 + y_3 = f_3$$

$$-\frac{1}{\sqrt{2}} y_2 \qquad = f_4$$

$$-\frac{1}{\sqrt{2}} y_1 \qquad -y_3 = f_5$$

$$-\frac{1}{\sqrt{2}} y_1 \qquad = f_6.$$

In general these are impossible to solve, and a typical set of six applied forces will move the truss right off the page. It may also spin around, but the truss will not change form; it will undergo a *rigid motion*.

The natural question is: When is the truss in equilibrium? For some forces it must be possible to solve the six equations, and balance $f_1, ..., f_6$ by forces y_1, y_2, y_3 in the bars. Then the truss stays where it is. Certainly we can select y_1, y_2, y_3 and find such a set of forces; the equations will be solvable because the solution was picked first. In matrix terms, $A^T y = f$ is solvable if f is a combination of the columns of A^T. Most vectors f with six components will not lie in the three-dimensional column space, but some will.

A neater answer comes from a little more engineering, or more thought. The six forces leave the triangle in place if their combined horizontal and vertical components are zero,

$$f_1 + f_3 + f_5 = 0 \tag{5a}$$
$$f_2 + f_4 + f_6 = 0, \tag{5b}$$

and if their *moment* around a typical point is zero. Take that point to be the upper left node. It is connected to a sloping bar of length $\sqrt{2}$ and a horizontal bar of length 2 (since the angle is 45°). Therefore the "lever arm" of f_4 is twice that of f_1 and f_2, and there is no rotation if

$$f_1 - f_2 - 2f_4 = 0. \tag{5c}$$

These are the three conditions on f to avoid rigid motion. When they are satisfied, the six equations can be solved. An algebraist would derive them directly from those six equations: equations 1, 3, 5 add to (5a), equations 2, 4, 6 add to (5b), and

equation 1 − equation 2 − 2 (equation 4) gives $f_1 - f_2 - 2f_4 = 0$.

Whenever $r < 3$ there will have to be conditions of this kind on f, to keep the truss from moving. It has too few supports.

EXAMPLE 4   With a roller support and a fixed support, there are $r = 3$ reactions and they prevent rigid motion. The truss in Fig. 2.12d has $m = 2$ bars, $N = 3$ nodes, and $n = 2N - r = 3$ equilibrium equations. Again there are more equations than unknowns, $m < n$, in $A^T y = f$:

$$y_1 \cos \varphi - y_2 \cos \theta = 0$$
$$y_1 \sin \varphi + y_2 \sin \theta = 0 \tag{6}$$
$$y_2 \cos \theta = F.$$

If the applied force F at the roller (or slider) is zero, then $y_1 = y_2 = 0$. Otherwise these equations have no solution; the truss must move. And when it moves, it does not stay rigid. There are enough supports ($r = 3$) to prevent it from leaving, but it will *change shape*. As F pulls to the right, the bars become horizontal and the lower node moves up. This deformation of the truss is a ***mechanism***.

A mechanism is a bad thing for a bridge, but it is not always bad in a car. In fact this example is a "crank and slider," converting translation into rotation—just as the straight motion of cylinders is converted to rotation of the wheels. In practice the force F alternates in direction, moving the slider back and forth while bar 1 goes around its support, and the problem becomes dynamic.

Staying with statics, we have ***instability*** (large motion of the truss from small forces) whenever A has ***dependent columns***. There will be displacements x that satisfy $Ax = 0$, so the truss can move without stretching ($e = 0$). For most forces there will be no solution to $A^T y = f$, so the truss cannot get itself into equilibrium. This is where a fundamental fact in statics meets a fundamental theorem in linear algebra—that $Ax = 0$ has many solutions exactly when $A^T y = f$ has none.

2F If A does not have independent columns then its rank is below n and

(1) there are nonzero solutions to $Ax = 0$ (displacements without strain)
(2) there are forces for which $A^T y = f$ has no solution (instability).

If A has fewer than n independent columns, so has A^T. The combinations $A^T y$ fill less than the whole of n-dimensional space, and all forces f that are not among these combinations cannot be equilibrated by the bars. The truss may move rigidly, or it may have a mechanism and deform. A mechanism can be eliminated by adding more bars. And a rigid motion can be eliminated by adding more supports. We summarize the four possibilities:

> Stable truss (the columns of A are linearly independent):
> **(1)** Statically determinate: $m = n$
> **(2)** Statically indeterminate: $m > n$
> Unstable truss (the columns of A are linearly dependent):
> **(3)** Rigid motion: too few supports
> **(4)** Mechanism: the truss deforms

The distinction between 3 and 4 is in the solutions to $Ax = 0$. If all solutions correspond to translation or rotation of the whole truss, we are entirely in case 3. If $Ax = 0$ has nonzero solutions but rigid motions are not among them (there are enough supports) we are entirely in case 4. In general an unstable truss could move rigidly and also deform, if it is really unsafe.

Note that if $m < n$ there are too few bars for stability. We are certainly in case 3 or 4 or both. The n columns cannot be independent if there are only $m < n$ rows. But a truss with more bars, $m > n$, could still be unstable. It might have plenty of edges in one region, but resting on an unsafe bridge like the one in Fig. 2.10c (without diagonal braces) it can still collapse.

Remark 1. If there are only three reactive forces, $r = 3$, they can be calculated directly from the applied forces f—without knowing the bar forces y. The reactions must combine with f to prevent rigid motions, which gives three equations: zero horizontal force, zero vertical force, and zero moment.

Remark 2. In continuum mechanics the strain ε is dimensionless and the stress σ is force per unit area. These match our e and y after dividing by the length and cross-sectional area of the bar:

$$\varepsilon = \frac{e}{L} \quad \text{and} \quad \sigma = \frac{y}{A}.$$

Then the constant in Hooke's law $\sigma = E\varepsilon$ depends only on the material and not on the shape of the bar. E is *Young's modulus*, with the units of stress. For each separate bar it gives the elastic constant in $y = ce$, from

$$c = \frac{y}{e} = \frac{\sigma A}{\varepsilon L} = \frac{EA}{L} \tag{7}$$

Remark 3. The vector b, which has remained at zero, will enter if a bar is stretched before it goes into the truss. Such a truss is *prestressed*, and the initial stretching b is added to the usual deformation: $e = Ax + b$. It is done to maintain tension $(y > 0)$ and avoid compression, since a bar is in much greater danger of buckling than of being pulled too hard. (On the other hand concrete bridges depend completely on being in compression. Concrete falls apart in tension, and we need iron to reinforce it or steel cables to support the bridge.) This happens every day, but the most beautiful examples of trusses are the "tensegrity structures" of Buckminster Fuller. Every bar is in tension, and a mechanism is barely prevented.

Interpretation of Lagrange Multipliers

I am often asked in class about the physical meaning of Lagrange multipliers. It is not an easy question. Part of the answer is that the multipliers can represent the potential or the displacement. This indicates that also in other cases they have an importance of their own. They are not just numbers in the middle of an optimization problem, and we can give more examples of their significance.

The simplest constraints are the grounding of a node and the fixing of a support. The potential or displacement is set to zero, say $x_N = 0$, and column N is removed from A_0. That leaves a matrix with independent columns, and the constraint is immediately forgotten; it was so easy to enforce. I believe that Lagrange would have complained. The equation $x_N = 0$ is a constraint like any other, and should have a multiplier. Since it is a constraint on x, the multiplier will be y—the opposite situation from the constraints $A^T y = f$. The Lagrangian $L = P(x) + yx_N$

builds in the constraint $x_N = 0$. Then $\partial L/\partial x_N = 0$ is $\partial P/\partial x_N + y = 0$ and it reveals what y is:

> The Lagrange multiplier is the correction needed to satisfy the Nth equilibrium equation. In statics it is the reaction force supplied by the support. In a network it is the current into the grounded node.

Equilibrium $A^T y = f$ was not enforced at the support, whose column is not even present in A. But equilibrium must hold, and the multiplier is the extra force (computable from $A_0^T y$) that is required from the support.

Chapter 8 will add one more sentence to this interpretation. The multiplier is the *price you pay for the constraint*. If a movement to $x_N = \delta$ costs w in work done by the support, then the multiplier is w/δ — and work divided by displacement brings us back to the force supplied by the support.

Energy and Virtual Work

For a truss we reached the stiffness equation directly from compatibility $e = Ax$, Hooke's law $y = Ce$, and equilibrium $A^T y = f$. Together they gave $A^T C A x = f$, or $Kx = f$. For more complicated structures (and for continuum mechanics in general) the great theorems are the principles of minimum energy and the principle of virtual work. They lead to the same equations, just as the minimum principles of Section 2.2 reproduced the network equations of Section 2.1. It takes only a few lines and it is worth the effort.

The fundamental fact of equilibrium is that internal forces balance external forces; there is no net force on any particle or it would move. We can express this fact in a more subtle way. The quantity called "work" is the product of force and displacement (counting only displacement that is parallel to the force). Since the net forces are zero at equilibrium, their *virtual work* in any imaginary infinitesimal displacement from equilibrium will be zero. This virtual work is partly external and partly internal. If the nodal displacements are $x_1, \ldots, x_n$ and the applied forces are $f_1, \ldots, f_n$, the external work is easy to compute: it is

$$W_{\text{ext}} = f_1 x_1 + \cdots + f_n x_n = f^T x. \tag{8}$$

We might have written δx, and later δe, to emphasize that the virtual displacements are infinitesimal. For the internal work we look within each bar: the bar forces are $y_1, \ldots, y_m$ and the displacements produce elongations $e_1, \ldots, e_m$. Since a tension $y_i > 0$ acts against an elongation $e_i > 0$, the work in the bars is

$$W_{\text{int}} = -y_1 e_1 - \cdots - y_m e_m = -y^T e. \tag{9}$$

Therefore the *principle of virtual work* becomes $f^T x = y^T e$. But the bar elongations can be determined from the nodal displacements, $e = Ax$, so we actually have

$$f^T x = y^T A x = (A^T y)^T x. \tag{10}$$

Because x was arbitrary, the principle of virtual work requires the equilibrium equation $f = A^T y$. This is the first main point.

Note that we ignored the r reactive forces, since they are at supports which prevent displacement. The virtual displacements have to be "kinematically admissible." With zero displacements the reactive forces do no work.

One more step will bring us to the energy. Within each bar, the work of internal forces is stored exactly as in a spring. It becomes **strain energy**, ready to be recovered; an elastic system is conservative. From Hooke's law the force is $y = ce$ and the work done is converted into

$$\text{energy in bar} = \int_0^e y \, de = \int_0^e ce \, de = \tfrac{1}{2} ce^2.$$

Summing over all m bars, the strain energy in matrix notation is $U = \Sigma \tfrac{1}{2} c_i e_i^2 = \tfrac{1}{2} e^T C e$. This is internal potential energy. The external part is $-f^T x$, force times distance. Therefore the **total potential energy** is

$$\boxed{P(x) = \tfrac{1}{2} e^T C e - f^T x = \tfrac{1}{2} x^T A^T C A x - f^T x.} \tag{11}$$

You can guess what is coming:

2G   At equilibrium, the displacement vector x minimizes the potential energy P.

The minimum of the quadratic P occurs at $A^T C A x = f$, which is the stiffness equation. Thus the minimum principle has reproduced the equation of equilibrium. It can be interpreted as $f_j = \partial U / \partial x_j$: the external forces balance the internal forces, which are the derivatives of the strain energy. Note that P has no minimum when $A^T C A$ fails to be positive definite; A then has dependent columns and the structure can move with zero strain energy. The stability of the truss corresponds to the positive definiteness of $K = A^T C A$.

There must be a dual principle in terms of y. It begins with virtual forces in the bars instead of virtual displacements at the nodes. These changes in bar forces have to respect the constraints $A^T y = f$—they must be "statically admissible," maintaining equilibrium rather than compatibility. Again the virtual work is zero and a quadratic is minimized, but this time it is the **complementary energy**—expressed in terms of y rather than x. For a single bar this energy is

$$\int_0^y e \, dy = \int_0^y \frac{y}{c} \, dy = \frac{1}{2} \frac{y^2}{c}.$$

Summing over all bars, the complementary energy is the quadratic Q we met earlier:

$$Q(y) = \sum \frac{1}{2} \frac{y_i^2}{c_i} = \frac{1}{2} y^T C^{-1} y. \tag{12}$$

Therefore the energy principle discovered by Castigliano becomes exactly our theorem of duality. And the saddle point problem for $L = Q + x^T(A^T y - f)$ is known as the Hellinger–Reissner principle.

2H At equilibrium, the bar forces y minimize the complementary energy $Q(y)$ subject to $A^T y = f$. Furthermore the minimum values of P and Q satisfy $P_{min} = -Q_{min}$.

This is identical to the main result of Section 2.2 (with $b = 0$) after a sign change in x. There the quadratic was $P = \frac{1}{2} x^T A^T C A x + x^T f$; changing x to $-x$ reverses the linear term and produces the potential energy of a truss. The equations also reverse sign: mechanics has elongations $e = Ax$ instead of $e = -Ax$, $y = CAx$ instead of $y = -CAx$, and $f = A^T C A x$ instead of $f = -A^T C A x$. But the minimum of P is unchanged since $-x$ is as admissible as x.

The two principles are in perfect duality, but in practice one completely dominates the other. The **displacement method** (which minimizes P) is in constant use; the **force method** (which minimizes Q) is comparatively dormant. The reason can be found in the principles themselves. In the first, kinematic constraints like $x_j = 0$ are easy to impose. In the complementary principle we have to obey $A^T y = f$, and that is harder to do. It asks us to identify all the "redundancies," which are the solutions to $A^T y = 0$; they are the $m - n$ degrees of freedom in minimizing Q. For a small truss these self-stresses can be computed and added to a particular solution of $A^T y = f$. Codes for the nullspace are beginning to appear. But for a large truss or a discrete approximation to a continuous structure—as in the finite element method of Section 5.4, where thousands of unknowns are quite common—the displacement method seems to win.

EXERCISES

2.4.1 Write down $m, N, r,$ and n for the three trusses in Fig. 2.10, and establish which is statically determinate, which is statically indeterminate, and which one has a mechanism. Describe the mechanism (the uncontrolled deformation).

2.4.2 With horizontal forces f_H^1 and f_H^2 pulling the upper nodes in Fig. 2.10a to the right, and vertical forces f_V^1 and f_V^2 pulling them up, write down the four equilibrium equations $A^T y = f$. Assuming the diagonal is at $30°$ and all $c_i = 1$, form the stiffness matrix $K = A^T C A = A^T A$.

2.4.3 With a single horizontal force f_H^1 applied to the upper left node in Fig. 2.10b, and the diagonal still at 30°, find the four equations $A^T y = f$. Since A is square solve directly for y. What reactive forces are supplied by the supports?

2.4.4 For the truss in Fig. 2.10c, write down the equations $A^T y = f$ in three unknowns y_1, y_2, y_3 to balance the four external forces $f_H^1, f_H^2, f_V^1, f_V^2$. Under what condition on these forces will the equations have a solution (allowing the truss to avoid collapse)?

2.4.5 For example 1 in the text, from Fig. 2.12a, equilibrium at the left support gives

$$-\frac{1}{\sqrt{2}} y_1 = f_H^1 \qquad \text{(horizontal reaction)}$$

$$-\frac{1}{\sqrt{2}} y_1 = f_V^1 \qquad \text{(vertical reaction)}$$

What are the corresponding equations at the second support? These four equations correspond to the four columns of A_0 eliminated by the fixed displacements $x_H^1 = x_V^1 = x_H^2 = x_V^2 = 0$ at the supports.

2.4.6 For example 3 (Fig. 2.12c) let the forces be $f_1 = f_2 = f_4 = f_6 = 0, f_3 = 1, f_5 = -1$. These satisfy the conditions for no rigid motion. Write down directly the solution to the 6 equations in the text for y_1, y_2, y_3.

2.4.7 In example 4 with a mechanism, what forces f_H and f_V at the lower node would make it possible to solve the three equations $A^T y = f$? F still acts horizontally at the roller.

2.4.8 With the bridge in Fig. 2.10a on top of the one in Fig. 2.10c (the supports remain only at the bottom) show that $m = n = 8$ but there is still a mechanism. What force would make this ladder collapse?

2.4.9 Sketch a six-sided truss with fixed supports at two opposite vertices. Will one diagonal crossbar between free nodes make it stable, or what is the mechanism? What are m and n? What if a second crossbar is added?

2.4.10 If we create a new node in Fig. 2.10a where the diagonals cross, is the resulting truss statically determinate or indeterminate?

2.4.11 In continuum mechanics, work is the product of stress and strain integrated over the structure: $W = \int \sigma \varepsilon \, dV$. If a bar has uniform stress $\sigma = y/A$ and uniform strain $\varepsilon = e/L$, show by integrating over the volume of the bar that $W = ye$. Then the sum over all bars is $W_{\text{total}} = y^T e$; show that this equals $f^T x$.

2.4.12 At the equilibrium $x = K^{-1} f$, show that the strain energy U (the quadratic term in P) equals $-P_{\min}$, and therefore $U = Q_{\min}$.

2.4.13 The "stiffness coefficients" k_{ij} in K give the forces f_i corresponding to a unit displacement $x_j = 1$, since $Kx = f$. What are the "flexibility coefficients" that give the displacements x_i caused by a unit force $f_j = 1$?

2.4.14   At equilibrium, where $x = K^{-1}f$, the terms in the potential energy $P(x)$ are $\frac{1}{2}x^T Kx$ $= \frac{1}{2}f^T K^{-1}f$ and $f^T x = f^T K^{-1}f$. The internal strain energy and the external potential energy are not equal! Why not?

Note   The point of virtual work is that, starting from x and making a small change v, the *changes* in internal and external terms are equal.

2.4.15   (a)   Turn the square network of Exercise 2.3.14 into a truss. With the usual pin supports at the two nodes that were grounded, write down the 7 by 6 matrix in $e = Ax$.
   (b)   Which of the four types of truss is this?
   (c)   What is the rank of A and what are the solutions to $Ax = 0$?
   (d)   What are the solutions to $A^T y = 0$?

2.4.16   Suppose a truss consists of *one bar* at an angle θ with the horizontal. Sketch forces f_1 and f_2 at the upper end, acting in the positive x and y directions, and corresponding forces f_3 and f_4 at the lower end. Write down the 1 by 4 matrix A_0, the 4 by 1 matrix A_0^T, and the 4 by 4 matrix $A_0^T CA_0$. For which forces can the equation $A_0^T y = f$ be solved?

2.4.17   For *networks*, a typical row of $A_0^T CA_0$ (say row 1) is described on page 92: The diagonal entry is Σc_i, including all edges into node 1, and each $-c_i$ appears along the row. It is in column k if edge i connects nodes 1 and k. ($A^T CA$ is the same with the grounded row and column removed.) The problem is to describe $A_0^T CA_0$ for *trusses*, and the idea is to put together the special $A_0^T CA_0$ found in the previous exercise (*a 4 by 4 matrix for each bar*).
   (a) Suppose bar i goes at angle θ_i from node 1 to node k. By assembling the $A_0^T CA_0$ for each bar, show how the 2 by 2 upper left corner of $A_0^T CA_0$ contains

$$\begin{bmatrix} \Sigma c_i \cos^2 \theta_i & \Sigma c_i \cos \theta_i \sin \theta_i \\ \Sigma c_i \cos \theta_i \sin \theta_i & \Sigma c_i \sin^2 \theta_i \end{bmatrix}$$

   (b) Where do those terms appear (with minus signs) in the first two rows? All rows of $A_0^T CA_0$ add to zero.

2.4.18   This is another approach to $A_0^T CA_0$ for trusses. The first column of A_0 contain $\cos \theta_i$ in row k, if bar i goes at angle θ_i from node 1 to node k. The second column contains $\sin \theta_i$. Multiply out $A_0^T CA_0$ to find its 2 by 2 upper left corner.

2.4.19   Sketch a square truss with horizontal forces f_1, f_3, f_5, f_7 and vertical forces f_2, f_4, f_6, f_8 at the nodes, numbered clockwise.
   (a) Write down A_0 and $A_0^T CA_0$.
   (b) There should be $8 - 4 = 4$ independent solutions of $A_0 x = 0$. Describe or draw four movements x of the truss (rigid motion or mechanism?) that produce no stretching.
   (c) From combinations of the 8 equations $A_0^T y = f$, show that $x^T f$ must be zero for the four movements x of part (b). *For equilibrium, the force f must not activate the instabilities x.*

LEAST SQUARES ESTIMATION AND THE KALMAN FILTER   ■ 2.5

Least squares carries us directly to a quadratic, the square of the error. When there is no solution to $Ax = b$, the best possible x is the one that minimizes $\|Ax - b\|^2$. But in what sense is this the best? Is there something better, if the measurements b_i are not equally reliable? It seems reasonable to attach more weight to the reliable measurements, but there has to be a systematic way.

This section begins with weighted least squares. If the first observation is trusted more than the second—it comes from a larger sample or a more careful measurement—then we minimize $w_1^2(Ax - b)_1^2 + w_2^2(Ax - b)_2^2 + \cdots$ with $w_1 > w_2$. The first observation is weighted more heavily. It is also possible that the measurements are coupled, and in the extreme case b_2 might be an exact copy of b_1; the vote for Vice-President is not independent of the vote for President. Such a correlation can be accounted for by off-diagonal entries in the weighting matrix W.

Our first step will be to find the best solution x for a given W. Then the next step is to decide on the right weighting matrix, or equivalently the right $C = W^T W$, if there is statistical information about the errors. We expect that the measurement b_i contains an error e_i; we do not know what it is. Nevertheless it is often reasonable to suppose that we know the average size (or expected value) of e_i and e_i^2 and $e_i e_j$. This will determine an optimal W.

There is a third step that enters when measurements continue to arrive. Suppose we are fitting the data by a straight line, and the experiment produces a measurement every second. Each new data point means a change in the best line, and we will fall hopelessly behind by trying every time to solve a new normal equation $A^T A x = A^T b$. Instead, we only want the *change* in x. What is needed for computation in real time is **recursive least squares**. Therefore we look for the change in $(A^T A)^{-1}$ or $(A^T C A)^{-1}$ produced by a new row of A.

Finally there is the possibility that the model itself is not stationary; the experiment is not standing still. In that case a different quantity x_i is estimated after each time step. If the state x_{i+1} is unrelated to x_i, the earlier computations will be useless—but that is not likely in practice. It is much more common to have a linear relation $x_{i+1} = F_i x_i$, with known F_i and unknown x_i. Each step brings new information through a measurement b_i of $A_i x_i$. But there is error in the measurement and error in the linear relation:

$$b_i = A_i x_i + e_i$$
$$x_{i+1} = F_i x_i + \varepsilon_i \tag{1}$$

The problem is to separate the signal from the noise.

The solution to this nonstationary problem is given by the **Kalman filter**. It is a digital filter (the process comes in discrete steps) and it is recursive: the estimate of x_{i+1} is computed from the previous estimate of x_i and from one new measurement. Its discovery produced a revolution in signal processing. Earlier filters assumed a stationary process and amplified the right band of frequencies; Fourier analysis takes the signals x_i, in state space, into the frequency domain. That remains extremely important.† But new applications and the technology of digital integrated circuits have produced a new theory of optimal filtering.

Thus we study the **linear estimation problem**, going from weighted least squares to recursive filters. The first part is fundamental for statistics; the weights come from the variance and covariance. The latter part is fundamental for signal processing, orbit determination, smoothing, and all applications that separate information from noise. Those are outside the traditional engineering mathematics courses, and many students will not need them immediately, but for a large group of readers they should be here. Fortunately they fit into the framework of $A^T C A$.

Weighted Least Squares

In ordinary least squares we are given a system $Ax = b$ with no solution. A has m rows and n columns, with $m > n$; there are more measurements $b_1, \ldots, b_m$ than free parameters $x_1, \ldots, x_n$. The best choice (we will call it $\hat{x}$) is the one that minimizes the length of the error vector $e = b - A\hat{x}$. If we measure this length in the usual way, so that $\|e\|^2 = (b - Ax)^T(b - Ax)$ is the sum of squares of the m separate errors, minimizing this quadratic gives the normal equations of Section 1.4:

$$A^T A\hat{x} = A^T b, \quad \text{or} \quad \hat{x} = (A^T A)^{-1} A^T b. \tag{2}$$

Geometrically, $A\hat{x}$ is a projection of b; it is the closest vector to b among all possible vectors Ax. These candidates fill out the column space of A, and the least squares choice

$$A\hat{x} = A(A^T A)^{-1} A^T b = Pb$$

is the projection of b onto this column space.

Algebraically, we reach the normal equations just by *multiplying both sides of the unsolvable equation $Ax = b$ by A^T*. That produces a square coefficient matrix $A^T A$; there are now n equations for the unknowns $x_1, \ldots, x_n$. We will assume that A has "full rank"—its columns are independent—so that $A^T A$ is invertible and $\hat{x}$ is completely determined.††

Now suppose the m measurements are not equally reliable. If we trust b_1 more than b_2, we try harder to make e_1 small. There is a neat way to do that, by attaching weights to the errors and minimizing $\|We\|^2 = w_1^2 e_1^2 + w_2^2 e_2^2 + \cdots$. In other words, we find the best least squares solution to $WAx = Wb$. When the measurements are independent the weighting matrix W is diagonal; in ordinary least squares it is $W = I$. When the measurements are correlated, so that the error in b_1 is not independent of the error in b_2, there should also be an off-diagonal weight w_{12}

---

† Dolby stereo uses a very successful nonlinear filter.

†† Otherwise a further condition is needed to specify $\hat{x}$. If $A^T Ax = A^T b$ has multiple solutions the pseudoinverse A^+ chooses the smallest one. The formula (2) in which $A^T A$ is no longer invertible is replaced by $\hat{x} = A^+ b$.

$(=w_{21})$. We will determine the w that is statistically correct, but first comes the computation of the new x.

The formula is easy. A and b are replaced by WA and Wb in the minimization, and therefore the same is true in the normal equations:

$$(WA)^T WA\hat{x} = (WA)^T Wb. \tag{3}$$

In other words, $A^T W^T WA\hat{x} = A^T W^T Wb$. Notice that only the combination $C = W^T W$ appears on both sides; we have exactly the familiar equilibrium equation $A^T CA\hat{x} = A^T Cb$. Its solution $\hat{x}$ (or $\hat{x}_W$, if we emphasize the dependence on W) is

$$\hat{x} = (A^T CA)^{-1} A^T Cb = Lb. \tag{4}$$

The weighted least squares solution to an overdetermined system $Ax = b$ still depends linearly on b. If $C = I$ we are back to ordinary least squares. And if A is invertible we are back to the only reasonable solution $x = A^{-1}b$; in that case $L = A^{-1}C^{-1}A^{-T}A^T C$, or $L = A^{-1}$.

EXAMPLE   Suppose a patient's pulse is measured at $x = 70$ beats per minute, then at $x = 80$, then at $x = 120$. The least squares solution of these three inconsistent equations minimizes $(x - 70)^2 + (x - 80)^2 + (x - 120)^2$. Differentiation gives $(x - 70) + (x - 80) + (x - 120) = 0$ at $x = \hat{x}$, and so does the formula $A^T A\hat{x} = A^T b$:

$$[1 \quad 1 \quad 1] \begin{bmatrix} 1 \\ 1 \\ 1 \end{bmatrix} [\hat{x}] = [1 \quad 1 \quad 1] \begin{bmatrix} 70 \\ 80 \\ 120 \end{bmatrix} \quad \text{or} \quad 3\hat{x} = 270.$$

The best answer is the average value $\hat{x} = 90$. (The value 95 would be best in minimizing the maximum error—which is then 25—but extreme cases dominate the answer. Or the value 80 would minimize the sum of the $|e_i|$, which keeps outliers like 120 under control. But it is the *squares* of the errors that lead to the best theory.) Now suppose the third reading was taken quickly, when the patient and doctor were nervous. It deserves less weight. If we assign weights $w_1 = w_2 = 1$ and $w_3 < 1$, the minimization of $(x - 70)^2 + (x - 80)^2 + w_3^2(x - 120)^2$ gives the weighted normal equations $A^T W^T WAx = A^T W^T Wb$:

$$[1 \quad 1 \quad 1] \begin{bmatrix} 1 & & \\ & 1 & \\ & & w_3^2 \end{bmatrix} \begin{bmatrix} 1 \\ 1 \\ 1 \end{bmatrix} [\hat{x}] = [1 \quad 1 \quad 1] \begin{bmatrix} 1 & & \\ & 1 & \\ & & w_3^2 \end{bmatrix} \begin{bmatrix} 70 \\ 80 \\ 120 \end{bmatrix}$$

$$\text{or} \quad \hat{x} = \frac{70 + 80 + 120w_3^2}{1 + 1 + w_3^2}.$$

This is a weighted average of the readings. With $w_3 = 1$ it is the ordinary average, but with $w_3 = \frac{1}{2}$ we find $\hat{x} = 80$. The high reading is balanced by the lower weight.

In every case we reach the triple product $A^T C A$. Therefore it must be related to the other sections of this chapter and to the three equations in our basic framework:

$$e = b - Ax, \quad y = Ce, \quad \text{and} \quad A^T y = 0. \tag{5}$$

Eliminating y and e leaves $A^T C A x = A^T C b$. These equations continue to govern the problem, even though A is no longer the incidence matrix of a network. I don't know whether the dual unknown y has a natural interpretation in statistics.

The Choice of $C = W^T W$

The next step is to choose the matrix C. In electrical or mechanical networks it was given by the physical properties of the material. Here it has to come from the statistical properties of the experiment. To compute C we first assume that the errors $e = b - Ax$ are not biased to one side or the other. Their expected value is $E[e_i] = 0$. That does not imply (although it sounds like it) that we actually expect the errors to be zero. It does imply that the average error is zero, when each e_i is weighted by the probability that it will occur. The error has **zero mean**.† For this to make sense, we think of the errors as coming from a population of possible errors. We do not know exactly which errors occurred in our particular experiment— otherwise we could remove them—but we do know their probabilities.

Here are three examples of populations and their probabilities:

1.   *Discrete*: A child's age is estimated in years, and the errors $e_i = -1, 0, 1$ occur with equal probabilities $p_i = \frac{1}{3}$.

2.   *Uniform*: The moon's velocity is measured in feet per second, and after rounding to an integer the errors are uniformly distributed between $-\frac{1}{2}$ and $\frac{1}{2}$. In this continuous distribution, the probability of any specific error like $e = \frac{1}{3}$ is zero! We can measure two things:
  (1)   the probability $F(x)$ of an error less than x
  (2)   the probability $p(x)dx$ of an error between x and $x + dx$
F is the *cumulative* probability; it allows all errors up to x. Its derivative p is the probability *distribution*—the frequency of errors near x, even if the chance of hitting x exactly is zero. Errors below x have probability $F(x)$ and errors below $x + dx$ have probability $F(x + dx)$. Errors in between have probability

$$F(x + dx) - F(x) = p(x)dx, \quad \text{and thus} \quad p = \frac{dF}{dx}.$$

---

† This is not a real restriction, since otherwise the mean can just be subtracted from b.

In our uniform case the length of an interval decides the probability of errors falling into that interval. The whole interval from $-\frac{1}{2}$ to $\frac{1}{2}$ has length 1 and probability 1. The interval from $-\frac{1}{2}$ to x has length $x + \frac{1}{2}$ and probability $F(x) = x + \frac{1}{2}$. The interval from x to $x + dx$ has length dx and probability $p(x)dx = dx$. Thus $p = 1 = dF/dx$.

3.   *Normal distribution*: Your quiz score is compared to the class average, and the difference (positive or negative) is distributed in an exponentially decreasing way: The chance that the difference falls between x and $x + dx$ is

$$p(x)dx = \frac{1}{\sqrt{2\pi}} e^{-x^2/2} dx, \text{ with total probability } \int_{-\infty}^{\infty} p(x)dx = 1. \qquad (6)$$

This $p(x)$ is called the **normal** (or **Gaussian**) distribution. The factor $\sqrt{2\pi}$ ensures that the total probability equals one, and the graph of $e^{-x^2/2}$ is the famous bell-shaped curve in Fig. 2.13.

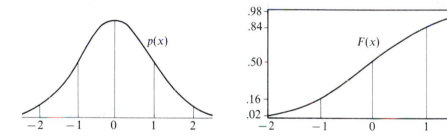

Fig. 2.13. The normal distribution and its integral.

This normal distribution would be more reasonable for physical measurements than for quizzes. Perhaps scores on the SAT are Gaussian. Even with non-Gaussian probabilities, the bell-shaped curve begins to appear when we add a large number of independent samples. Combining the ages of many children or the velocities of many moons, the central limit theorem says that probabilities like $p = \frac{1}{3}$ fade into a normal distribution. Only the scaling of the axes can vary; x might be rescaled to x/σ, changing the probabilities to

$$p(x) = \frac{1}{\sqrt{2\pi}\,\sigma} e^{-x^2/2\sigma^2}, \text{ still with } \int_{-\infty}^{\infty} p(x)dx = 1. \qquad (7)$$

Values of x near $\pm\infty$ have almost no chance of occurring (and may actually be ruled out, if quiz scores are between 0 and 100). Even though it looks awkward, and its integrals are not easy to evaluate, this normal distribution is by far the most common throughout probability theory. The limit theorems for repeated experiments make it natural and unavoidable.

The second figure is for a normal distribution with mean zero and $\sigma^2 = 1$. It shows the probability that a sample value will fall below x. It is the cumulative probability $F(x)$, the integral of p from $-\infty$ to x. The probability of a negative x is $F(0) = \frac{1}{2}$. The probability of falling into the range $-1 \leq x \leq 1$ is the difference $F(1) - F(-1) \sim .68$. Only about one sample in 22 lies as far out as $|x| = 2$ and about one in 400 is beyond $|x| = 3$.

All three examples have zero mean, since the errors e and $-e$ are equally likely. Thus the expected value (the sum of all possible errors in the population, multiplied by their probabilities) is zero:

1. $E[e] = \frac{1}{3}(-1) + \frac{1}{3}(0) + \frac{1}{3}(1) = 0$

2. $E[e] = \displaystyle\int_{-1/2}^{1/2} x \, dx = 0$

3. $E[e] = \displaystyle\int_{-\infty}^{\infty} x \, e^{-x^2/2} \, dx/\sqrt{2\pi} = 0.$

The quantity which is *not* expected to be zero is e^2, the square of the error. Its expected value is called the **variance**. The variance depends only on the magnitude of the error, and ignores its sign. For these populations the variances are

1. $E[e^2] = \frac{1}{3}(-1)^2 + \frac{1}{3}(0)^2 + \frac{1}{3}(1)^2 = \frac{2}{3}$

2. $E[e^2] = \displaystyle\int_{-1/2}^{1/2} x^2 dx = \frac{1}{12}$

3. $E[e^2] = \displaystyle\int_{-\infty}^{\infty} x^2 \, e^{-x^2/2} \, dx/\sqrt{2\pi} = 1.$

When the normal distribution is rescaled as in (7), the variance becomes $E[e^2] = \sigma^2$. The wilder the experiment, the wider will be the bell-shaped curve and the larger the variance. This notation "sigma squared" is used for the variance of all populations, Gaussian or not, so the three cases can be written $\sigma^2 = \frac{2}{3}$, $\sigma^2 = \frac{1}{12}$, $\sigma^2 = 1$. We summarize the formulas in the case of zero mean:

$$\text{mean} = E[e] = \int xp(x)dx \text{ and variance} = E[e^2] = \int x^2 p(x)dx.$$

If the mean $E[e] = \mu$ were not zero, the variance would be computed around the mean and it would be $\sigma^2 = E[(e - \mu)^2] = E[e^2] - \mu^2$.

The square root of the variance is σ, the **standard deviation**. Only this number will be needed to determine the proper weight for least squares; the full probability distribution $p(x)$ is not required. But for two measurements we also need to know the mutual dependence of the errors, which is measured by their **covariance**. If the errors are independent—they are two completely separate samples from the population of all possible errors—then their covariance is zero. But if a pulse is taken twice in the same visit, with the same stethoscope, the error covariance is probably not zero. In general it is the expected value of the product $e_i e_j$ of the two errors:

$$covariance = E[e_i e_j] = \int \int (e_i)(e_j)(joint\ probability\ of\ e_i\ and\ e_j).$$

In the independent case, by far the most common in practice, this joint probability is the product of the separate probabilities and the integral splits into

$$E[e_i e_j] = \int (e_i)(probability\ of\ e_i) \int (e_j)(probability\ of\ e_j) = 0 \cdot 0 = 0.$$

In that independent case *the right weights are* $w_i = 1/\sigma_i$. A small variance means a more reliable observation and a larger w. Thus for independent observations, least squares minimizes

$$\| W(b - Ax) \|^2 = \| We \|^2 = \frac{e_1^2}{\sigma_1^2} + \cdots + \frac{e_m^2}{\sigma_m^2}. \tag{8}$$

The matrices W and $C = W^T W$ are then diagonal; C contains the numbers $1/\sigma_i^2$. When the variances are all equal, they have no effect and we return to ordinary least squares.

To show why this is the right choice, and to determine C when the observations fail to be independent, we need a theorem. It can be expressed in matrix notation, if all the variances and covariances of the errors $e_1, \ldots, e_m$ are included in a single matrix. That is the **covariance matrix** V. Its diagonal entries V_{ii} are the variances $E[e_i^2]$ and its off-diagonal entries V_{ij} are the covariances $E[e_i e_j]$. Since all these products e_i^2 and $e_i e_j$ appear in the m by m matrix ee^T (a column times a row), we can abbreviate the covariance matrix as the expected value of ee^T:

$$V = E[ee^T]. \tag{9}$$

V is symmetric positive definite, except for an unusual semidefinite case below.

Now comes the choice of $C = W^T W$. All we know about the errors, except for their mean $E[e] = E[b - Ax] = 0$, is contained in the matrix V. What we require of any rule $\hat{x} = Lb$, which estimates the true but unknown parameters x from the measurements b, is that it be *linear* and *unbiased*. It is certainly linear if L is a matrix. It is unbiased if the expected error $x - \hat{x}$—the error in the estimate, not in the observations—is also zero:

$$E[x - \hat{x}] = E[x - Lb] = E[x - LAx - Le] = E[(I - LA)x] = 0. \tag{10}$$

Thus L is unbiased if it is a left inverse of the rectangular matrix A: $LA = I$.† Under

---

† $LA = I$ just means: Whenever b can be fitted exactly by some x (so that $Ax = b$) our choice $\hat{x}$ should agree with x: $\hat{x} = Lb = LAx = x$. If the data lies on a straight line the best fit should be that line.

this restriction, Gauss picked out the best L (we call it L_0) and the best C in the following way: **The matrix C should be the inverse of the covariance matrix V.**

His rule leads to $L_0 = (A^T V^{-1} A)^{-1} A^T V^{-1}$, which does satisfy $L_0 A = I$, and for completeness we include a proof that this choice is optimal.

2I The best linear unbiased estimate (BLUE) is the one with $C = V^{-1}$. The optimal estimate $\hat{x}$ and the optimal matrix L_0 are then

$$\hat{x} = (A^T V^{-1} A)^{-1} A^T V^{-1} b = L_0 b. \tag{11}$$

This choice minimizes the expected error in the estimate, measured by the covariance matrix $P = E[(x - \hat{x})(x - \hat{x})^T]$.

Proof The matrix to be minimized is

$$P = E[(x - \hat{x})(x - \hat{x})^T] = E[(x - Lb)(x - Lb)^T]$$
$$= E[(x - LAx - Le)(x - LAx - Le)^T].$$

Since $x = LAx$ and L is linear, this is just

$$P = E[(Le)(Le)^T] = LE[ee^T]L^T = LVL^T.$$

Thus it is LVL^T that we minimize, subject to $LA = I$. To show that L_0 is the optimal choice, write any L as $L_0 + (L - L_0)$ and substitute into $P = LVL^T$:

$$P = L_0 V L_0^T + (L - L_0) V L_0^T + L_0 V (L - L_0)^T + (L - L_0) V (L - L_0)^T. \tag{12}$$

The middle terms are transposes of one another, and they are zero:

$$(L - L_0) V L_0^T = (L - L_0) V V^{-1} A (A^T V^{-1} A)^{-1} = 0$$

because $V V^{-1}$ is the identity and $(L - L_0) A = I - I = 0$. Furthermore the last term in (12) is symmetric and at least positive semidefinite. It is smallest when $L = L_0$, which is therefore the minimizing choice; the proof is complete.

The matrix P for this optimal estimate x comes out neatly when we simplify

$$P = L_0 V L_0^T = (A^T V^{-1} A)^{-1} A^T V^{-1} V V^{-1} A (A^T V^{-1} A)^{-1}.$$

Cancelling V and $A^T V^{-1} A$ with their inverses gives the key formula for P:

2J The covariance matrix P for the error in $\hat{x}$, when $\hat{x}$ is computed from the optimal rule $A^T V^{-1} A \hat{x} = A^T V^{-1} b$, is

$$P = (A^T V^{-1} A)^{-1}. \tag{13}$$

P gives the expected errors in $\hat{x}$ just as V gave the expected errors in b. These matrices average over all experiments—they do not depend on the particular measurement b that led to the particular estimate $\hat{x}$. We emphasize that $P = E[(x - \hat{x})(x - \hat{x})^T]$ is the fundamental matrix in filtering theory. Its inverse $A^T V^{-1} A$ is the *information matrix*, and it is exactly the triple product $A^T C A$. It measures the information content of the experiment. It goes up as the variance V goes down, since $C = V^{-1}$. It also goes up as the experiment continues; every new row in A makes it larger.

In the most important case, with independent errors and unit variances and $C = I$, we are back to "white noise" and ordinary least squares. Then the information matrix is $A^T A$.

Remark 1.   We can always obtain $C = I$ from $C = V^{-1}$ by a change of variables. Factor the matrix V^{-1} into $W^T W$ and introduce $\bar{e} = We$. These normalized errors $\bar{e} = W(b - Ax)$ still have mean zero, and their covariance matrix is

$$E[(We)(We)^T] = WE[ee^T]W^T = WVW^T = I.$$

The weighting matrix W returns us to white noise—a unit covariance problem, with simpler theory and simpler computations.†

Remark 2.   If one of the variances is zero, say $\sigma_1^2 = 0$, then the first measurement is exact. The first row and column of V are zero, and V is not positive definite or even invertible. (The weighting matrix has $(V^{-1})_{11} = \infty$.) This just means that the first equation in $Ax = b$ should be given infinite weight and be solved exactly. If this were true of all m measurements we would have to solve all m equations $Ax = b$, without the help of least squares; but with exact measurements that would be possible.

EXAMPLE   Suppose the doctor takes m different readings of the patient's pulse. If they are equally reliable, his least squares estimate $\hat{x}$ will be their average. We can estimate the error in this average, knowing the expected errors of the individual readings. The underlying question is: How much does it help to repeat an experiment? If each error $(x - x_i)^2$ has expected value σ^2, what is the expected value of $(x - \hat{x})^2$? The answer is given by P, the error covariance matrix. Our formula becomes

$$P^{-1} = A^T V^{-1} A = \begin{bmatrix} 1 & 1 & \cdots & 1 \end{bmatrix} \begin{bmatrix} \sigma^{-2} & & \\ & \ddots & \\ & & \sigma^{-2} \end{bmatrix} \begin{bmatrix} 1 \\ 1 \\ \vdots \\ 1 \end{bmatrix} = \frac{m}{\sigma^2}.$$

By taking m measurements, the variance goes from $V = \sigma^2$ to $P = \sigma^2/m$.

---

† Throughout mathematics there is this choice between a change of variable or a change in the definition of length. We introduce $X = Wx$ or we measure the original x by $\|x\|^2 = x^T V^{-1} x$. It is really Hobson's choice; it makes no difference.

Recursive Least Squares

Suppose we have estimated the vector x as well as possible from a first set of measurements b_0. We write this estimate as x_0 (rather than $\hat{x}_0$). The underlying equation is $A_0 x = b_0$, probably overdetermined and unsolvable. Its least squares solution, weighted as it should be by the covariance matrix V_0 of the errors in b_0, is

$$x_0 = (A_0^T V_0^{-1} A_0)^{-1} A_0^T V_0^{-1} b_0. \tag{14}$$

The error $x - x_0$ has zero mean and its covariance matrix is

$$P_0 = E[(x - x_0)(x - x_0)^T] = (A_0^T V_0^{-1} A_0)^{-1}. \tag{15}$$

Question   If more data arrives, can the best estimate for the combined system $A_0 x = b_0$, $A_1 x = b_1$ be computed from x_0 and b_1 *without restarting the calculation from b_0?*

You can imagine how this situation will arise. We are fitting a straight line through points that come from radar. At any moment we can stop and compute the best line. When the next point misses that line, the estimate has to be changed. If the error in the new point is independent of the earlier error there must be a chance to work recursively. We want to find x_1 from the previous estimate x_0 and the new data b_1.

The final result should be the same as computing x_1 from scratch. For that the right choice of C is V^{-1}, where

$$V = \begin{bmatrix} V_0 & 0 \\ 0 & V_1 \end{bmatrix} \text{ is the covariance matrix of the errors } \begin{bmatrix} e_0 \\ e_1 \end{bmatrix}.$$

V is block-diagonal because e_1 is independent of e_0. Therefore the coefficient matrix $A^T C A$ in the equation for x_1 is

$$P_1^{-1} = \begin{bmatrix} A_0 \\ A_1 \end{bmatrix}^T \begin{bmatrix} V_0 & 0 \\ 0 & V_1 \end{bmatrix}^{-1} \begin{bmatrix} A_0 \\ A_1 \end{bmatrix} = A_0^T V_0^{-1} A_0 + A_1^T V_1^{-1} A_1. \tag{16}$$

Remember that x_1 is not based only on b_1; it is best for the combined system $A_0 x = b_0$, $A_1 x = b_1$. Since the normal equations are $A^T V^{-1} A x = A^T V^{-1} b$, and the matrix on the left is exactly P^{-1}, the optimal x_1 is

$$x_1 = P_1 \begin{bmatrix} A_0 \\ A_1 \end{bmatrix}^T V^{-1} \begin{bmatrix} b_0 \\ b_1 \end{bmatrix} = P_1 (A_0^T V_0^{-1} b_0 + A_1^T V_1^{-1} b_1). \tag{17}$$

This is the solution we hope to find recursively, by using the already computed x_0 in place of b_0 in (17). The difficulty is that the b_0 term is multiplied by P_1. Therefore *we update the matrix P along with the estimate* of x, using (16):

$$P_1^{-1} = P_0^{-1} + A_1^T V_1^{-1} A_1. \tag{18}$$

This gives the increase in information from the second measurements. It is a decrease in P, which measures uncertainty. Note again that (18) does not depend on the particular measurements b_0 or b_1; it uses only their statistical properties, and gives the statistical properties of x_1.

Of course the actual estimate x_1 has to be based on the actual b_0 and b_1. It is given by (17), and the whole subject of recursive least squares depends on rewriting that formula:

$$\begin{aligned} x_1 &= P_1(P_0^{-1}x_0 + A_1^T V_1^{-1} b_1) \\ &= P_1(P_1^{-1}x_0 - A_1^T V_1^{-1} A_1 x_0 + A_1^T V_1^{-1} b_1) \\ &= x_0 + K_1(b_1 - A_1 x_0). \end{aligned} \tag{19}$$

The matrix $K_1 = P_1 A_1^T V_1^{-1}$ is called the **gain matrix**. With this manipulation the formula has become recursive; it uses x_0 instead of b_0. This is the result we want, and it is easy to interpret.

Suppose the new measurement is exactly consistent with the original x_0, so that $b_1 = A_1 x_0$. Then there is no reason to change our estimate of x. The best guess is still $x_1 = x_0$, whenever the new points b_1 fall on the line determined by the old points b_0. In general this will not happen. There will be **prediction error** $b_1 - A_1 x_0$, called an **innovation**. It is the unexpected part of b_1, and (19) amplifies it by the gain matrix K_1 to give the correction to x_0. The new P_1 and x_1 contain everything we know, and we are ready for another measurement b_2.

When b_2 comes, we appreciate the formulas (18) and (19). The algebra that produced them applies equally well to the new situation. The subscripts 0 and 1 are replaced by 1 and 2, and later by $i-1$ and i, to give the fundamental theorem of recursive least squares:

2K   The least squares estimate of x, and its variance P, are given recursively by

$$\begin{aligned} P_i^{-1} &= P_{i-1}^{-1} + A_i^T V_i^{-1} A_i \\ x_i &= x_{i-1} + K_i(b_i - A_i x_{i-1}) \quad \text{with} \quad K_i = P_i A_i^T V_i^{-1}. \end{aligned} \tag{20}$$

EXAMPLE   We continue taking the pulse of the same patient. To keep the algebra simple, suppose each reading has variance σ^2. We computed the best estimate x_m after m readings and its variance P_m:

$$x_m = \frac{b_1 + \cdots + b_m}{m} \quad \text{and} \quad P_m = \frac{\sigma^2}{m} = E[(x - x_m)^2]. \tag{21}$$

How does this come from the recursive formulas?

The last measurement is $x = b_m$, almost sure to be inconsistent with the previous readings. This equation has coefficient matrix $A_m = [1]$ and covariance matrix $V_m = [\sigma^2]$. Therefore the recursion gives

$$P_m^{-1} = P_{m-1}^{-1} + \frac{1}{\sigma^2} \quad \text{and} \quad x_m = x_{m-1} + K_m(b_m - x_{m-1}).$$

The first equation says that $1/\sigma^2$ is added to P^{-1} at every step, so P_m^{-1} is m/σ^2. Thus $P_m = \sigma^2/m$ is correct. The second equation has gain matrix $K_m = P_m A_m^T V_m^{-1} = [1/m]$, which gives the right formula for x_m:

$$x_m = x_{m-1} + \frac{1}{m}(b_m - x_{m-1})$$

$$= \frac{1}{m} b_m + \frac{m-1}{m}\left(\frac{b_1 + b_2 + \cdots + b_{m-1}}{m-1}\right) = \frac{b_1 + b_2 + \cdots + b_m}{m}.$$

For this simple example the recursion succeeds. Of course the real importance of the idea appears when b and x are vectors and V and P are matrices. Then (20) is enormously simpler than a recalculation of the whole estimate at every step, and it becomes possible to understand an experiment as it happens.

The Kalman Filter

We can hardly hope, in a few lines, to explain one of the most crucial advances in forecasting and filtering. Its applications are very wide but there is a common thread: signals are coming in and have to be processed to remove noise.[†] When these signals are all measurements of the same quantity we have seen what to do; the recursion (20) gives the correct filter. But if the quantity being estimated is changing, so that something different is measured at every step, then the filter must account for this change. The optimal filtering problem becomes dynamic.

To estimate the unknown vector x_i at each step we have two equations:

(1)   Measurements $b_0, \ldots, b_m$ have been taken up to time $t = m$. If these were exact they would be related to the true values by $b_i = A_i x_i$. In general the measurements are not exact, and the matrices A_i are rectangular.

(2)   A known law $x_{i+1} = F_i x_i$ governs the *change in state* as time goes forward. Again we allow different matrices F_i at each step, and again there may be error. This time it is an error ε_i in the model instead of an error e_i in the measurement. Thus our equations are of two kinds:

$$A_i x_i = b_i \qquad \text{(with error } e_i)$$

$$-F_i x_i + x_{i+1} = 0 \quad \text{(with error } \varepsilon_i).$$

---

† Airplanes land on automatic control while a Kalman filter is fitting speed and position to the laws of motion.

We want to combine these equations into a single system. Suppose we know b_0, b_1, b_2; we are ready to estimate x_2. The equations up to that point go together into

$$
\begin{bmatrix}
A_0 & & \\
-F_0 & I & \\
& A_1 & \\
& -F_1 & I \\
& & A_2
\end{bmatrix}
\begin{bmatrix}
x_0 \\
x_1 \\
x_2
\end{bmatrix}
=
\begin{bmatrix}
b_0 \\
0 \\
b_1 \\
0 \\
b_2
\end{bmatrix}.
\tag{22}
$$

The whole object of filtering is to find the best solution to this system! According to Gauss, we should solve it by least squares (with his weighting matrix). And according to Kalman, we should solve it recursively. The work done here should not repeat the previous step, and it should be put to use at the next step.

That would be easy if the earlier estimates stayed the same at each new step. If x_0 and x_1 were already decided in equation (22), the only unknown would be x_2. But that is not the case; the whole solution of (22) is affected by the measurement b_2. We are computing the best estimates for x_0 and x_1 as well as x_2, based on all the data available up to $t = 2$. The improvement on earlier estimates is called **smoothing**, and the estimation of x_2 is called **filtering**. We write these estimates—the least squares solution of (22)—as $x_{0/2}$, $x_{1/2}$, and $x_{2/2}$.

Most applications concentrate on the new value $x_{2/2}$, which predicts $x_{3/2} = F_2 x_{2/2}$ at the following time. Then b_3 will correct this predicted value to a filtered value, after one more step. It is only in the unlikely case that b_3 exactly matches the prediction $A_3 x_{3/2}$—the innovation (the difference between the two) is then zero—that the filtered $x_{3/3}$ will agree with the predicted $x_{3/2}$.

This problem is a direct extension of recursive least squares, where we continually estimated the same vector x. The system was not evolving, F_i was the identity matrix, and the dynamic equation was just $x_{i+1} = x_i$. Furthermore there was no error ε_i in this equation; it was satisfied exactly. Then $x_0 = x_1 = x_2$ in the system (22), which reduces to

$$
\begin{bmatrix}
A_0 \\
A_1 \\
A_2
\end{bmatrix}
[x] =
\begin{bmatrix}
b_0 \\
b_1 \\
b_2
\end{bmatrix}.
\tag{23}
$$

It is this system that was solved a few pages ago by recursive least squares, adding the block A_2 and measurements b_2.

In the Kalman filter, it is unreasonable to suppose that $x_{i+1} = F_i x_i$ is exact. The model is almost never that good. It is also unreasonable to assume that the errors ε_i and e_i have the same size, when they are not even measured in the same units. Frequently the e_i are independent with variance σ^2 and the ε_i are independent with another variance σ^2/c^2. Then the correct weighting for the equations in (22) is easy to write down. The rows involving A and b are *divided by* σ, and the rows involving F

and $-I$ are *divided by* σ/c. (These rows could also be allowed a right hand side, an external control in $x_{i+1} = F_i x_i + u_j$.) The effect of the weighting, in which σ appears everywhere and can be factored out, is to change (22) into

$$
\begin{bmatrix}
A_0 & & \\
-cF_0 & cI & \\
& A_1 & \\
& -cF_1 & cI \\
& & A_2
\end{bmatrix}
\begin{bmatrix}
x_0 \\
x_1 \\
x_2
\end{bmatrix}
=
\begin{bmatrix}
b_0 \\
0 \\
b_1 \\
0 \\
b_2
\end{bmatrix}.
\tag{24}
$$

This is the system $Ax = b$ which the filter solves by ordinary least squares.

The direct and foolish approach is to start each step from scratch, with $A^T A x = A^T b$. The recursive approach is to use as much as possible from the previous step. At that time the last two rows of the block matrix were absent; the matrix ended at A_1. In the present step, the new block $-cF_1$ affects the 2, 2 entry of $A^T A$, and the third row and column are completely new. That is the general rule: $A^T A$ is a block tridiagonal matrix which changes its lower right block and adds an extra row and column at every step.

The algebraic problem is to find the effect of these changes on the estimate $x = (A^T A)^{-1} A^T b$ and its covariance matrix $P = (A^T A)^{-1}$. The formulas are straightforward, but they have to be a little complicated. (Most books on the Kalman filter are rather discouraging.) In fact, an unthinking use of these formulas is numerically unstable. It is much better to update the factorization $A^T A = LDL^T$ than the inverse matrix. That is called a **square root filter**; it avoids the "square" $A^T A$. It is a second stage in developing Kalman's beautiful idea, that the problem could be solved sequentially and that a dynamic model $x_{i+1} = F_i x_i$ could be included. We emphasize the importance of computing P along with the estimate of x. In fact, since P does not depend on the data b, it can be computed by itself—before the experiment! This has become an important design tool in simulating the accuracy of an estimation and improving it before the event occurs.

There is also a third stage, which we summarize briefly as a guide to numerical filtering. Instead of the LDL^T factorization it works with $A = QR$. This is the **orthogonalization** of the columns of A, to be discussed more completely in Chapter 5. It puts $Ax = b$ into an upper triangular form

$$
\begin{bmatrix} R \\ 0 \end{bmatrix} [x] = \begin{bmatrix} c \\ d \end{bmatrix} \text{ with } R = \begin{bmatrix} R_0 & R_{0,1} & 0 \\ 0 & R_1 & R_{1,2} \\ 0 & 0 & R_2 \end{bmatrix}.
\tag{25}
$$

$Ax = b$ has been multiplied by an orthogonal matrix without changing its least squares solution. In this form the best x is clear; nothing can be done about $0x = d$, so the right thing is to solve $Rx = c$. Furthermore the covariance matrix P, which was $(A^T A)^{-1}$, is now equal to $(R^T R)^{-1}$. The part we want most, the covariance matrix P_2 for the last component $x_{2/2}$, is in the lower right corner of $(R^T R)^{-1}$—and that corner is just $R_2^{-1}(R_2^{-1})^T$. All the crucial information is in R, which can be computed recursively. The only change from the previous step, apart from the two nonzero blocks in the last column, is in R_1. What is most important is its extreme stability against roundoff error; it is connected to A by a set of mutually orthogonal unit vectors, and nothing is more stable than that.

EXAMPLE We can compare filtering with ordinary least squares on a problem where they might be expected to give the same answer—but they don't. It is known as a **steady model**, since all observations $b_1, \ldots, b_m$ measure the same thing. Furthermore all errors are independent with variance $\sigma^2 = 1$. The two problems look identical:

(1) The least squares solution of the m equations $x = b_i$ is

$$\hat{x} = \frac{b_1 + b_2 + \cdots + b_m}{m} \quad \text{with variance} \quad P = \frac{1}{m}.$$

This was checked earlier by the formulas of recursive least squares.

(2) The filtered solution comes from the m equations $x_i = b_i$ together with the steady model $x_{i+1} = x_i$.

Nevertheless there is a difference. It comes from the presence of errors ε_i in the equation $x_{i+1} = x_i$. We are at the same time assuming that x does not change, and allowing for the possibility that it does. When it drifts away from x_0, the *latest* x_i is the predicted value of x_{i+1}. Then we also measure b_{i+1}, and the best estimate is a combination of prediction and measurement. As in least squares, this must use all the measurements—but the recent b_i are weighted more heavily. In ordinary least squares, which does not allow for drift, the b_i were weighted equally.

The difference can be seen after two measurements and again after three. The equations (22) that combine $x_i = b_i$ with $x_{i+1} = x_i$ are

$$
\begin{bmatrix} 1 & \\ -1 & 1 \\ & 1 \end{bmatrix}
\begin{bmatrix} x_0 \\ x_1 \end{bmatrix} =
\begin{bmatrix} b_0 \\ 0 \\ b_1 \end{bmatrix}
\quad \text{and} \quad
\begin{bmatrix} 1 & & \\ -1 & 1 & \\ & 1 & \\ & -1 & 1 \\ & & 1 \end{bmatrix}
\begin{bmatrix} x_0 \\ x_1 \\ x_2 \end{bmatrix} =
\begin{bmatrix} b_0 \\ 0 \\ b_1 \\ 0 \\ b_2 \end{bmatrix}.
\tag{26}
$$

If we force $x_{i+1} = x_i$ to hold exactly, we are back to ordinary least squares. But if we allow drift errors ε_i, also of variance one, then it is the equations in (26) that are solved by least squares. Forming the normal equations $A^T A x = A^T b$ in the two cases gives

$$
\begin{bmatrix} 2 & -1 \\ -1 & 2 \end{bmatrix}
\begin{bmatrix} x_{0/1} \\ x_{1/1} \end{bmatrix} =
\begin{bmatrix} b_0 \\ b_1 \end{bmatrix}
\quad \text{and} \quad
\begin{bmatrix} 2 & -1 & 0 \\ -1 & 3 & -1 \\ 0 & -1 & 2 \end{bmatrix}
\begin{bmatrix} x_{0/2} \\ x_{1/2} \\ x_{2/2} \end{bmatrix} =
\begin{bmatrix} b_0 \\ b_1 \\ b_2 \end{bmatrix}.
\tag{27}
$$

You see that only one entry is changed (from 2 to 3) as the new row and column are added. This pattern would continue: $A^T A$ has 3's along its diagonal, except at the top and bottom. The factors in LDL^T or in QR change only their last entries— which is the reason the Kalman filter works.

We can see what happens without complicated formulas. The inverses in (27) are

$$(A^T A)^{-1} = \frac{1}{3}\begin{bmatrix} 2 & 1 \\ 1 & 2 \end{bmatrix} \quad \text{and} \quad (A^T A)^{-1} = \frac{1}{8}\begin{bmatrix} 5 & 2 & 1 \\ 2 & 4 & 2 \\ 1 & 2 & 5 \end{bmatrix}. \tag{28}$$

After the measurements b_0 and b_1, the best estimates come from the first equation in (27). Multiplying by $(A^T A)^{-1}$ gives, not the ordinary average, but

$$x_{0/1} = \frac{2b_0 + b_1}{3}, \quad x_{1/1} = \frac{b_0 + 2b_1}{3}.$$

The new data b_1 has a heavier weight $2/3$ in the filtered estimate of x_1. And b_1 also appears in the smoothed estimate $x_{0/1}$. There it is weighted less heavily, but still we know more about x_0 after measuring b_1.

With b_2, the filtered and smoothed estimates change to

$$x_{0/2} = \frac{5b_0 + 2b_1 + b_2}{8}, \quad x_{1/2} = \frac{2b_0 + 4b_1 + 2b_2}{8}, \quad x_{2/2} = \frac{b_0 + 2b_1 + 5b_2}{8}.$$

The last is the most important. It is the best estimate of x_2, using b_0 and b_1 but emphasizing b_2. The possibility of drift produces, as we continue, an exponential decay of the weight attached to old measurements.

The other half of the Kalman filter computes the matrices P_i. They give the reliability of the estimates of x. For the filtered estimates $x_{i/i}$, P_i appears as the last entry of $(A^T A)^{-1}$: $P_1 = 2/3$ and $P_2 = 5/8$. From b_0 alone we would have had $P_0 = 1$. Thus the estimation errors are steadily decreasing. The reciprocals $1, 3/2, 8/5$ are the information matrices (this chapter is closing with $A^T C A$!) and they increase with every new measurement of the model.

EXERCISES

2.5.1 What happens to the weighted average of three pulse readings b_1, b_2, b_3 as the third weight w_3 approaches zero? What is the limit if this reading is *more* reliable than the others, and $w_3 \to \infty$? The other weights are $w_1 = w_2 = 1$.

2.5.2 From m independent measurements $x = b_i$ of the pulse rate, with different variances $\sigma_1^2, \ldots, \sigma_m^2$, what is the best estimate $\hat{x}$?

2.5.3 In the previous exercise, what is the variance P in the best estimate?

2.5.4   Produce the formula $P = \sigma^2/m$ in (21), assuming equal variances σ^2 in each of the m independent measurements, from

$$P = E[(x - \hat{x})^2] = E\left[\left(\frac{x - b_1}{m} + \cdots + \frac{x - b_m}{m}\right)^2\right]$$

$$= E\left[\left(\frac{x - b_1}{m}\right)^2\right] + \cdots + E\left[\left(\frac{x - b_m}{m}\right)^2\right].$$

2.5.5   (a)   If two measurement errors have variances $E[e_1^2] = 3$ and $E[e_2^2] = 2$, with covariance $E[e_1 e_2] = -1$, what is their covariance matrix V?
(b)   If the measurements are $b_1 = 0$ and $b_2 = 5$, what is the best estimate $\hat{x}$—the weighted least squares solution of $Ax = b$ with $A = \begin{bmatrix} 1 \\ 1 \end{bmatrix}$?
(c)   What is $P = (A^T V^{-1} A)^{-1}$ for this experiment?

2.5.6   Suppose the errors $e_i = -2, -1, 0, 1, 2$ occur with probabilities $p_i = 1/12, 1/6, 1/2,$ $1/6, 1/12$. Show that the mean is $E[e] = 0$ and find the variance $\sigma^2 = E[e^2]$.

2.5.7   Where are the points of inflection, at which $d^2 p/dx^2 = 0$, in the normal probability distribution $p = e^{-x^2/2}/\sqrt{2\pi}$?

2.5.8   From Fig. 2.13b what is the approximate probability that the sample value x lies above the standard deviation $\sigma = 1$? This is the area under the bell-shaped curve between $x = 1$ and $x = \infty$.

2.5.9   Suppose that for IQ's the theoretical mean is 100 and the standard deviation is $\sigma = 16$. What is the probability that your IQ exceeds 132 (if you were random and IQ's were Gaussian)? This is all statistically doubtful.

2.5.10   Find the connection between the integral of $p(x)$ and the error function:

$$\text{Prob}(x \le X) = \frac{1}{\sqrt{2\pi}} \int_0^X e^{-x^2/2}\, dx \quad \text{and} \quad \text{erf } X = \frac{2}{\sqrt{\pi}} \int_0^X e^{-x^2}\, dx.$$

2.5.11   What are the probabilities $p_2, \ldots, p_{12}$ of rolling the numbers $2, \ldots, 12$ with a pair of dice? Compute the variance $\sigma^2 = \Sigma\, p_k(k - 7)^2$. Note that the variance is taken around the mean value 7.

2.5.12   If an event occurs with probability p in each trial, what is the probability that in n trials it will occur x times? This is the *binomial distribution* of x.

2.5.13   If the values $x_n = 0, 1, 2, \ldots$ occur with Poisson probabilities $p_n = (n!e)^{-1} = e^{-1}, e^{-1},$ $e^{-1}/2, e^{-1}/6, \ldots$ show that the total probability is 1 (remembering the series for e). Find the mean $E[x] = x_0 p_0 + x_1 p_1 + \cdots = \Sigma\, ne^{-1}/n!$

2.5.14   Show that the Gauss normal equation $A^T V^{-1} A x = A^T V^{-1} b$ does give a formula $x = L_0 b$ that satisfies $L_0 A = I$.

2.5.15   If a child's age is estimated as b_1 by his mother with error probabilities $e_{-1} = .1$, $e_0 = .8, e_1 = .1$ and b_2 by his father with error probabilities $e_{-1} = .2, e_0 = .6, e_1 = .2$, find the variances σ_1^2 and σ_2^2 and the best weighted solution x to the two equations $x = b_1, x = b_2$.

2.5.16 From one measurement $x = b_0$ with the variance σ_0^2, the best estimate is $x_0 = b_0$ and *its* variance is $P_0 = \sigma_0^2$. Use the recursion formulas (19–20) to find the new estimate x_1 and its variance P_1 after a second measurement b_1 with variance σ_1^2.

2.5.17 In the Kalman filter, suppose the equations $x_{i+1} = F_i x_i$ hold exactly (the errors ε_1 are zero). Show that (22) reduces to a form similar to (23),

$$
\begin{bmatrix} A_0 \\ A_1 F_0 \\ A_2 F_1 F_0 \end{bmatrix} [x_0] = \begin{bmatrix} b_0 \\ b_1 \\ b_2 \end{bmatrix}.
$$

2.5.18 For the steady model, extend equations (26–27–28) to the next measurement b_3 and find the information matrix that follows 1, 3/2, 8/5.

CHAPTER 2 IN OUTLINE: EQUILIBRIUM EQUATIONS

2.1 A Framework for the Applications—input A, C, b, f, output x and y
Remarks on the Equilibrium Equations—$A^T C A x = A^T C b - f$

2.2 Constraints and Lagrange Multipliers—minimum of $Q(y)$ with $A^T y = f$
Duality—shortest distance to line = longest distance to planes
Duality in Higher Dimensions—minimum of Q = maximum of $-P$
Lagrange Multipliers—saddle point $-P(x) \le L(x,y) \le Q(y)$
An Application to Geometry—perpendicular subspaces and Pythagoras' law
Projections—the projection matrix $A(A^T A)^{-1} A^T$

2.3 Resistive Networks—Kirchhoff's law and Ohm's law
RLC circuits—capacitors and inductors and complex impedances
Minimum Principles and Loop Currents—each loop solves $A^T y = 0$

2.4 Structures in Equilibrium—the matrices A and C and $K = A^T C A$
Four Types of Trusses—determinate, indeterminate, rigid motion, mechanism
Interpretation of Lagrange Multipliers—reactive forces or cost of constraints
Energy and Virtual Work—potential energy P and complementary energy Q

2.5 Least Squares and the Kalman Filter—the normal equations solved recursively
Weighted Least Squares—measurements that are not equally reliable
The Choice of $C = W^T W$—linear unbiased estimates and covariance matrix
Recursive Least Squares—updating the estimate and variance of x
The Kalman Filter—nonstationary model and its information matrix

3

EQUILIBRIUM IN THE CONTINUOUS CASE

INTRODUCTION: ONE-DIMENSIONAL PROBLEMS ■ 3.1

This chapter takes the jump from matrix equations to differential equations. Instead of a finite number of resistors or springs, the physical system covers a whole region—an interval like $0 \leq x \leq 1$ or a plane domain or a part of three-dimensional space. The material is no longer lumped; it is distributed over the region. Its density and conductance and elasticity can vary from one point to the next. Therefore the potential differences between individual nodes must be replaced by derivatives.

To take this step we copy the main idea of calculus. The slope of a curve comes from the average slope over increasingly small intervals: $(u(x + \Delta x) - u(x))/\Delta x$ approaches the slope du/dx. The average velocity of a car comes from reading the odometer at times t and $t + \Delta t$, and its instantaneous velocity appears as $\Delta t \to 0$. Physical laws are derived in the same way, as the limiting cases of finite equilibrium equations.

Instead of a line of springs, consider a continuous elastic bar. Hang it vertically, so its top is fixed; its displacement is $u = 0$ at $x = 0$. Elsewhere along the bar, the point originally at distance x from the top is displaced down to $x + u(x)$. The stretching at any point is measured by the derivative $e = du/dx$. This corresponds to the elongation of a spring; it is the *strain* in the bar. Where u is constant, the rod is unstretched and the strain is zero. Otherwise the stretching produces an internal force proportional to the strain and depending on the elastic constant c. This force is $w = c \, du/dx$. The equilibrium of our small piece requires the difference in internal forces at its two ends to balance the external force of gravity (Fig. 3.1):

$$\left(c\frac{du}{dx}\right)_{x+\Delta x} - \left(c\frac{du}{dx}\right)_x + f\Delta x = 0.$$

The external force per unit length is $f = \rho a g$, density times cross-sectional area times the gravitational constant. Dividing by Δx, we approach the differential equation of equilibrium:

$$-\frac{d}{dx}\left(c\frac{du}{dx}\right) = f. \tag{1}$$

This is the matrix equation $A^T C A x = f$, in the continuum limit.

You can already see the analogies between the discrete case and the continuous case. We list them in a table, and then discuss them individually.

DISCRETE	CONTINUOUS
nodal unknown x	displacement or potential u
elongation $e = Ax$	strain $e = du/dx$
spring forces $y = Ce$	bar forces $w = ce$
equilibrium $A^T y = f$	equilibrium $-dw/dx = f$
incidence matrix A	differential operator d/dx
diagonal matrix C	multiplication by $c(x)$
transposed matrix A^T	transposed operator $-d/dx$
matrix equation $A^T C A x = f$	differential equation $-\dfrac{d}{dx}\left(c\dfrac{du}{dx}\right) = f$
fixed (or grounded) node	displacement boundary condition $u(0) = 0$
unstretched spring	stress boundary condition $w(1) = 0$

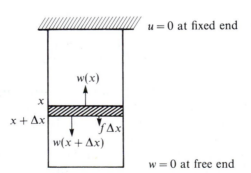

Fig. 3.1. Equilibrium of an elastic bar.

Clearly the basic framework is not changed. The simplest step comes from C, which is decided by the bar itself. Multiplication by a diagonal matrix, which is $y_i = c_i e_i$ in the discrete case, becomes multiplication by a function, $w(x) = c(x)e(x)$. Similarly A and its transpose reflect the geometry:

$$A = \frac{d}{dx} \quad \text{and} \quad A^T = -\frac{d}{dx}. \tag{2}$$

The support at $x = 0$ again "grounds one node." To specify u at this point is called an **essential** (or *kinematic* or *Dirichlet*) **boundary condition**:

$$u(0) = 0. \tag{3}$$

The other end is free, and that introduces something new. There is no internal force at that end, since it is in the air, and therefore

$$w(1) = c(1)\frac{du}{dx}(1) = 0. \tag{4}$$

This is a **natural** (or *dynamic* or *Neumann*) **boundary condition**. It comes directly from the physics, and it will come again from the minimum principle for u. But before that, we need to look more closely at the step from matrices to derivatives.

A to A^T: Integration by Parts

Equation (2) says that the transpose of d/dx is $-d/dx$. Thus **differentiation is skew-symmetric**. Taking its transpose reverses the sign, as it does with the matrix that comes from a discrete approximation to the derivative. That matrix is

$$A = \frac{1}{2\Delta x} \begin{bmatrix} 0 & 1 & & \\ -1 & 0 & 1 & \\ & -1 & 0 & 1 \\ & & \cdot & \cdot \end{bmatrix} = -A^T. \tag{5}$$

When A is applied to the values $u_1, u_2, u_3, \ldots$ of the function u, taken at points spaced by Δx, it gives the difference quotient that approximates du/dx:

$$A \begin{bmatrix} u_1 \\ u_2 \\ u_3 \\ \cdot \end{bmatrix} = \frac{1}{2\Delta x} \begin{bmatrix} u_2 \\ u_3 - u_1 \\ u_4 - u_2 \\ \cdot \end{bmatrix} \quad \text{or} \quad \frac{u(x + \Delta x) - u(x - \Delta x)}{2\Delta x} \approx \frac{du}{dx}.$$

A is a rescaled incidence matrix, and d/dx is its continuous limit. (You can even see the boundary condition $u_0 = 0$.) It is obvious that $A^T = -A$, but to complete the analogy we have to give a meaning to $(d/dx)^T$—the transpose in the continuous case.

In transposing a derivative, there are no rows and columns to exchange. Transposing a matrix is too easy; the underlying rule is mostly forgotten. That rule is **to keep the inner product** $x^T(A^T y)$ **equal to** $(Ax)^T y$. The matrix formula $A_{ij}^T = A_{ji}$ has exactly that property; both sides equal $x^T A^T y$. To apply the rule in the continuous case we need an **inner product of functions**.

The continuous analogue of a sum $e^T w = e_1 w_1 + \dots + e_m w_m$ is an integral. Since e and w are defined over the length of the bar, from $x = 0$ to $x = 1$, the natural choice for their inner product is

$$
e^T w = (e,w) = \int_0^1 e(x)w(x)dx. \tag{6}
$$

The notation (e,w) is more common than $e^T w$ when functions are involved; quantum mechanics would use $\langle e,w \rangle$. On an electrical conductor it is an integral of voltage times current; in fluids it will be velocity times flow rate; here it is strain times stress. Then the key to $(d/dx)^T$ lies in an integration by parts:†

$$
\int_0^1 \frac{du}{dx} w\, dx = - \int_0^1 u \frac{dw}{dx} dx + [uw]_{x=0}^{x=1}. \tag{7}
$$

If the last term (the boundary term) is zero, this equation becomes exactly $(Au, w) = (u, A^T w)$. A is the derivative d/dx applied to u, and A^T is $-d/dx$ applied to w. Formally, the rule for transposes is satisfied and we have $(d/dx)^T = -d/dx$.

You see that everything depends on the conditions at $x = 0$ and $x = 1$. In fact, **a differential operator comes with boundary conditions**. They are an essential part of its definition. That was equally true for the incidence matrices A_0 and A, whose only difference was in the grounding of a node. In that same analogy, A_0 is d/dx with no boundary conditions and A is d/dx with $u(0) = 0$. This condition guarantees that $uw = 0$ at $x = 0$. It removes one part of the boundary term $[uw]_0^1$, but it leaves the end $x = 1$ undecided. Therefore the boundary condition that comes with $(d/dx)^T$ must make this other term zero.

The condition we want is $w(1) = 0$. It is needed both physically and mathematically. Physically it describes the free end of the bar, where there is no force. Mathematically it produces zero for the integrated term, and it leaves $(Au,w) = (u, A^T w)$:

$$
\int_0^1 \frac{du}{dx} w\, dx = - \int_0^1 u \frac{dw}{dx} dx.
$$

This completes the definition of A^T:

$$
A^T w = - \frac{dw}{dx} \text{ subject to the condition } w(1) = 0.
$$

---

† That is the key to most of advanced calculus.

If there had been no boundary conditions on u, then A^T would have two conditions $w(0) = w(1) = 0$. When there are two conditions on u, there are none on w. A combination $w + ku$ is also possible at either end. But the example above is special, because the differential equation $A^T C A u = f$ can be solved directly:

The equation is $\quad -\dfrac{d}{dx}\left(c\,\dfrac{du}{dx} \right) = f.$ $\hspace{3cm}$ (8)

One integration with $w(1) = 0$ gives

$$w = c\,\frac{du}{dx} = \int_x^1 f\, dx. \hspace{2cm} (9)$$

Dividing by c, a second integration produces u.

EXAMPLE 1   Uniform bar with $f = 1$ and $c = 1$

Equation (9) gives $w = 1 - x$ and its integral is

$$u = \int_0^x (1 - x)dx = x - \tfrac{1}{2}x^2. \hspace{2cm} (10)$$

This satisfies $-u'' = 1$ and $u(0) = 0$ and $w(1) = 0$.

It is unusual that we could solve for w on the way to u. That corresponds in the discrete problem to *square* matrices, when we can separately invert each factor of $A^T C A$. The first step gives $w = (A^T)^{-1}f$; then $e = C^{-1}w$ (division by c, as above); and finally $u = A^{-1}e = (A^T C A)^{-1}f$. The inverses of the differentiations A^T and A are the integrations (9) and (10). It is like inverting a triangular matrix and then inverting its transpose.

Thus $A = d/dx$ with one boundary condition is invertible. If both ends are fixed, as they were for the line of springs in Chapter 1, then A specifies both $u(0)$ and $u(1)$. That corresponds to $m > n$ for matrices, and A^T has no boundary conditions. It is the statically indeterminate case, in which A is rectangular and u has to be found before w. The triple product $A^T C A$ is still square and invertible and symmetric positive definite, and $A^T C A u = f$. The framework for one-dimensional differential equations (Fig. 3.2) looks exactly like the framework for matrix equations.

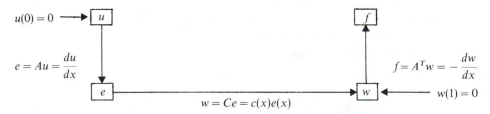

Fig. 3.2. The framework in the continuous case.

EXAMPLE 2 $$-\frac{d^2u}{dx^2} = 1 \text{ with } u(0) = 0 \text{ and } u(1) = b \qquad (11)$$

Since the equation makes u'' constant, u must be a quadratic $p_0 + p_1 x + p_2 x^2$. Its three coefficients p_0, p_1, p_2 are determined by the conditions in (11):

$$\frac{d^2u}{dx^2} = 2p_2 = -1, \; u(0) = p_0 = 0, \; u(1) = p_0 + p_1 + p_2 = b.$$

The quadratic that satisfies all three conditions is $u(x) = bx + \frac{1}{2}(x - x^2)$. It is the sum of two terms: a uniform stretching bx produced by fixing the end $x = 1$ at $u = b$, and a multiple of $x - x^2$ that vanishes at both ends. The stress in the bar (with $c = 1$) can be found as soon as u is known:

$$w = \frac{du}{dx} = b + (\tfrac{1}{2} - x). \qquad (12)$$

Again one part comes from the boundary condition and the other part from the force. This example illustrates that **boundary conditions need not be homogeneous** (i.e., zero). Just as a voltage source can fix the potential at a node to be $x = 10$, so we can specify the displacement at the end of the bar to be $u = 10$. And just as a current source can specify y, so a weight attached to the bottom of the bar would determine $w(1)$. (Here we had no weight.) The key is to find one condition at each boundary point, and those not associated with A must come with A^T.

If we are true mathematicians, we should make a small change in the inner products (Au, w) and (u, A^Tw). Two pages earlier, those were integrals from 0 to 1. But since A comes with a boundary condition, Au *is really the pair du/dx and $u(0)$.* Then

$$(Au, w) \text{ should be } \int_0^1 \frac{du}{dx} w \; dx + u(0)w(0).$$

Similarly A^Tw tells us both $-dw/dx$ and $w(1)$, and

$$(u, A^Tw) \text{ should be } \int_0^1 u\left(-\frac{dw}{dx}\right) dx + u(1)w(1).$$

With this correction, integration by parts always achieves $(Au, w) = (u, A^Tw)$—even if the numbers $u(0)$ and $w(1)$ are not zero. If A had told us du/dx and $u(0)$ and $u(1)$, then $u(0)w(0)$ and $-u(1)w(1)$ would both have gone into (Au, w) and there would be no boundary term with A^T.

The earlier method was OK if it is understood in this way: We learn what goes with A^T by deciding what will guarantee $[uw]_0^1 = 0$ *if* the conditions are homogeneous. Note that $w(0) = 0$, a second condition at the same end, would have failed.

Remark   The equation $-(cu')' = f$ has other applications. If u is the *temperature* then c is the diffusivity or thermal conductivity. The heat source is f, the condition $u(0) = T$ fixes the temperature at one end, and $w(1) = 0$ insulates the other end—through which there is no flow.

Sturm-Liouville Problems

New applications can produce an extra term in the differential equation:

$$-\frac{d}{dx}\left(c\,\frac{du}{dx}\right) + qu = f. \tag{13}$$

Fortunately qu does not spoil the symmetry, and fits completely into the same framework. It corresponds to the appearance of a second term in the matrix equation; instead of $A^TCAu = f$ we have $(A_1^TC_1A_1 + A_2^TC_2A_2)u = f$. This presents no difficulty, if we set

$$C = \begin{bmatrix} C_1 & 0 \\ 0 & C_2 \end{bmatrix} \quad \text{and} \quad A = \begin{bmatrix} A_1 \\ A_2 \end{bmatrix}.$$

Then the matrix equation with the extra term is identical with the single equation $A^TCAu = f$, and we are back to normal. For the differential equation it is exactly the same: C_1 is multiplication by $c(x)$, C_2 is multiplication by $q(x)$, and $A_2 = I$. Therefore A^TCA is

$$\begin{bmatrix} -d/dx & I \end{bmatrix} \begin{bmatrix} c(x) & 0 \\ 0 & q(x) \end{bmatrix} \begin{bmatrix} d/dx \\ I \end{bmatrix} = -\frac{d}{dx}\left(c\,\frac{d}{dx}\right) + q.$$

Applied to u, it gives f. This is the equation (13), called a **Sturm-Liouville problem** if it has boundary conditions of the right kind. As before, these conditions must remove the term $[cu\,du/dx]_0^1$ which comes from integration by parts. Of course the combination $u(0) = 0$ and $c\,du/dx\,(1) = 0$ still succeeds.

This is the basic one-dimensional equation, symmetric and second-order, and it appears everywhere. In quantum theory it is Schrödinger's equation. For the oscillations of a drum it is Bessel's equation. On a sphere it is Legendre's equation. In many applications the right side is λu, not f, and (13) becomes an eigenvalue problem. However the starting point must be the **constant-coefficient equation** and its **exponential solutions**:

$$\boxed{\begin{array}{l} -cu'' + qu = 0 \text{ is solved by} \\ u = e^{x\sqrt{q/c}} \quad \text{and} \quad u = e^{-x\sqrt{q/c}}. \end{array}}$$

That has to be checked. The two derivatives in u'' bring out the constant q/c. In one case each derivative yields $\sqrt{q/c}$; the other produces $-\sqrt{q/c}$. In both cases the equation is satisfied, because c times q/c equals q, and therefore cu'' matches qu.

All constant-coefficient equations have exponential solutions e^{ax}. The systematic approach is to substitute $u = e^{ax}$ into the differential equation:

$$-ca^2e^{ax} + qe^{ax} = 0 \quad \text{or} \quad ca^2 = q.$$

Then $a = \pm\sqrt{q/c}$, as anticipated. The complete solution is a combination of these two exponentials:

$$u = d_1 e^{x\sqrt{q/c}} + d_2 e^{-x\sqrt{q/c}}. \tag{14}$$

The constants d_1 and d_2 are determined here by the *boundary* conditions, while in an initial-value problem they are determined by the initial conditions. There is one boundary condition at each end, instead of u and u' at $t = 0$. In either problem there are two conditions and two constants.

Now we put this solution to use. It will be $u = 0$ unless either the boundary conditions are nonzero or the right side of the equation becomes nonzero. Those are the fundamental cases and we take them separately:

(1) $-cu'' + qu = 0$ with $u(0) = A$ and $u(1) = B$. The constants d_1 and d_2 are determined by $d_1 + d_2 = A$ and $d_1 e^{\sqrt{q/c}} + d_2 e^{-\sqrt{q/c}} = B$.

(2) $-cu'' + qu = f_0$ with $u(0) = u(1) = 0$. For constant f_0 there is a *particular solution* $u = f_0/q$. It is added to the *homogeneous solution* in (14), and their sum is made to satisfy the boundary conditions:

$$d_1 + d_2 + \frac{f_0}{q} = 0 \quad \text{and} \quad d_1 e^{\sqrt{q/c}} + d_2 e^{-\sqrt{q/c}} + \frac{f_0}{q} = 0.$$

In both cases we arrive at the numbers d_1 and d_2.

That is the answer for constant-coefficient problems, in principle. In practice there is more to say. If the right side of the equation is $f_0 e^{bx}$ instead of a constant, the particular solution is also a multiple of e^{bx}. If the right side is $f_0 \cos bx$ or $f_0 \sin bx$, the particular solution has the same form. Those are the three important cases—*constants, exponentials,* and *sinusoids*—and the only difficulties come for special values of b, c, and q. We postpone that unpleasantness, in order to put the solution with constant $f_0 = 1$ to one extra use—to illustrate the difference between a *regular perturbation* (small q) and a *singular perturbation* (small c).

Limiting Cases and Perturbations

Normally a higher derivative dominates a lower derivative. Differential equations are classified almost without looking at their low-order terms. This approach is justified for terms that produce only a small change in the solution. That will be the case if we take $c = f_0 = 1$ and compare the solutions of

$$-\frac{d^2 u}{dx^2} + qu = 1 \quad \text{and} \quad -\frac{d^2 U}{dx^2} = 1 \tag{15}$$

for small q. At $q = 0$ the problems are identical; their solution with zero boundary conditions is $U = \frac{1}{2}(x - x^2)$. Then the test is to see whether or not that U almost solves the problem $-u'' + qu = 1$. In this example it does. The error term qU is small compared to 1 and the boundary conditions are met. Therefore the perturbation is completely "regular" and the problem can be studied in two ways:

(1)   Fit the general solution $u = d_1 e^{\sqrt{q}x} + d_2 e^{-\sqrt{q}x} + \dfrac{1}{q}$ to the boundary conditions. The particular solution $1/q$ satisfies the equation $-u'' + qu = 1$, the exponentials satisfy $-u'' + qu = 0$, and with the right choice of d_1 and d_2 their sum solves the perturbed problem.

(2)   Starting with U at $q = 0$, look for a solution of the form $u = U + qV + q^2W + \cdots$. The first correction term qV is determined by substituting into the differential equation:

$$-U'' - qV'' - \cdots + qU + q^2V + \cdots = 1.$$

We already have $-U'' = 1$. Matching coefficients of q gives

$$V'' = U = \tfrac{1}{2}(x - x^2) \quad \text{or} \quad V = \frac{x^3}{12} - \frac{x^4}{24} - \frac{x}{24} \tag{16}$$

with the integration constant $-1/24$ chosen to make $V(0) = V(1) = 0$.

Both methods give the same solution. The first is exact but its behavior is not so clear for small q. The second form $U + qV + \cdots$ clarifies that behavior but is not satisfactory for large q. Between them we know almost everything.
   The solution for negative q is also interesting. The exact solution is exact only as far as $q = -\pi^2$. At that point there is a solution to $-u'' - \pi^2 u = 0$ which satisfies $u(0) = u(1) = 0$ but is not identically zero. It is $\sin \pi x$, and any multiple of this function can be added to any solution u (if there is one). Thus at $q = -\pi^2$ the problem stops being positive definite; the second derivative no longer dominates the low-order term. The equation describes the kind of rope that boxers and children skip over, and they keep q near $-\pi^2$ so the rope will turn with a small force f. Mathematically it will turn with zero force, $\sin \pi x$ is in the nullspace, and the differential equation breaks down.
   At the opposite extreme, when c is small and $q = 1$, we compare $-cu'' + u = 1$ with its limit $U = 1$. Already there is something exceptional; boundary conditions cannot be imposed on U. There is not even a differential equation. Nevertheless the solution to the first problem stays extremely close to $U = 1$, **except near the boundaries**. Again there are two approaches:

(1)   Fit the general solution $u = d_1 e^{x/\sqrt{c}} + d_2 e^{-x/\sqrt{c}} + 1$ to $u(0) = u(1) = 0$.
(2)   Near each end find a solution that connects the boundary value $u = 0$ to the interior value $U = 1$.

Again method (1) is exact; it is among the exercises. Method (2) may not be exact but you could never see the difference in Fig. 3.3. There is a **boundary layer** at each end in which all the action occurs. The layer reaches approximately to $x = 8\sqrt{c}$, which is enough for the special solution $u = 1 - e^{-x/\sqrt{c}}$ to climb from $u(0) = 0$ to $u = 1 - e^{-8}$. At that point it has virtually met the interior solution $U = 1$. Then a similar exponential at the other end connects $U = 1$ back to $u = 0$, in another boundary layer.

The perturbation is singular because the unperturbed solution $U = 1$ completely misses the boundary conditions. The leading term $-cu''$ is disappearing as c goes to zero, but it remains powerful inside the layer. Elsewhere the problem is calm.

(a) small q (b) small c

Fig. 3.3. A regular perturbation and a singular perturbation: $-cu'' + qu = 1$.

Note finally that a first derivative du/dx standing alone in (13) would have destroyed the whole framework. It corresponds to adding a skew-symmetric matrix to the existing $A^T C A$. Such a term does appear in fluid dynamics, and it illustrates the difference between diffusion and convection. Diffusion is symmetric and convection is not.

EXERCISES

3.1.1 For a bar with constant c but with decreasing $f = 1 - x$, find $w(x)$ and $u(x)$ as in equations (8–10).

3.1.2 For a hanging bar with constant f but weakening elasticity $c(x) = 1 - x$, find the displacement $u(x)$. The first step $w = (1 - x)f$ is the same as in (9), but there will be stretching even at $x = 1$ where there is no force. (The condition is $w = c\, du/dx = 0$ at the free end, and $c = 0$ allows $du/dx \neq 0$.)

3.1.3 Suppose a bar is free at both ends: $w(0) = w(1) = 0$. This allows rigid motion. Show that if $u(x)$ satisfies the differential equation and these boundary conditions, so does $u(x) + C$ for any constant C.

3.1.4 With the bar still free at both ends, what is the condition on the external force f in order that $-\dfrac{dw}{dx} = f(x)$, $w(0) = w(1) = 0$ has a solution? (Integrate both sides of the equation from 0 to 1.) This corresponds in the discrete case to solving $A_0^T y = f$; there is no solution for most f, because the left sides of the equations add to zero.

3.1.5   Find the displacement for an exponential force, $-u'' = e^x$ with $u(0) = u(1) = 0$.

Note that $A + Bx$ is the general solution to $-u'' = 0$; it can be added to any particular solution for the given f, and A and B can be adjusted to fit the boundary conditions.

3.1.6   Suppose the force f is constant but the elastic constant c jumps from $c = 1$ for $x \leq \frac{1}{2}$ to $c = 2$ for $x > \frac{1}{2}$. Solve $-dw/dx = f$ with $w(1) = 0$ as before, and then solve $c \, du/dx = w$ with $u(0) = 0$. Even if c jumps, the combination $w = c \, du/dx$ remains smooth.

3.1.7   Find the next term $W(x)$ in $u = \frac{1}{2}(x - x^2) + q(\frac{1}{12}x^3 - \frac{1}{24}x^4 - \frac{1}{24}x) + q^2 W + \cdots$. Choose W to match the q^2 terms in $-u'' + qu = 1$ and to satisfy $W(0) = W(1) = 0$.

3.1.8   For the negative value $q = -1$ show that $u = d_1 \cos x + d_2 \sin x - 1$ satisfies the differential equation $-u'' - u = 1$. The exponentials are e^{ix} and e^{-ix}, and they can be replaced by the sine and cosine.

3.1.9   If the condition at $x = 1$ were $u'(1) = 0$, why would no boundary layer be needed in Figure 3.3?

3.1.10   Verify that $u = d_1 e^{x/\sqrt{c}} + d_2 e^{-x/\sqrt{c}} + 1$ is an exact solution to $-cu'' + u = 1$. The condition $u = 0$ at $x = 0$ gives $d_1 + d_2 + 1 = 0$; find a similar equation from $u(1) = 0$ and solve for d_2. We expect $d_2 \approx -1$ to produce the boundary layer at $x = 0$.

3.1.11   What is the general solution to the constant-coefficient equation $-u'' + pu' = 0$? Try exponentials $u = e^{ax}$.

3.1.12   For $-u'' + pu' = 1$ with small p, find the regular perturbation pV by substituting $u = \frac{1}{2}(x - x^2) + pV$ and keeping the terms that are linear in p.

3.1.13   The solution to $-cu'' + u' = 1$ is $u = d_1 + d_2 e^{x/c} + x$. Find d_1 and d_2 if $u(0) = u(1) = 0$, and find their limits as $c \to 0$. The limit of u should satisfy $U' = 1$; which boundary condition does it keep and which end has a boundary layer?

3.1.14   Find the exponentials $u = e^{ax}$ that satisfy $-u'' + 5u' - 4u = 0$ and the combination that has $u(0) = 4$ and $u(1) = 4e$.

3.1.15   Solve the equation $-u'' = f$ with $u(0) = 0$ and $u'(1) = 0$ when f is a *delta function* at $x = \frac{1}{2}$. The impulse f is zero (and u is linear, $u = Ax + B$) except at $\frac{1}{2}$, where u' has a unit step down. The bar is stretched above $x = \frac{1}{2}$, then free.

3.1.16   Solve the same problem with $u(0) = u(1) = 0$, leading to the *Green's function* of page 351. The solution to $-u'' = \delta$ is again piecewise linear.

3.1.17   My class thinks that w in equation (9) should be $\int_0^x f \, dx + C$. But what constant of integration makes $w(1) = 0$?

Notes on the Dirac delta function (δ = unit impulse at $x = 0$)
Its integral from $-\infty$ to x is a *step function*: jump from 0 to 1 at $x = 0$
Second integral is a *ramp function* ($= x$ for $x > 0$; solution to $u'' = \delta$)
Third integral is a *quadratic spline* ($= \frac{1}{2}x^2$ for $x > 0$; jump in second derivative)
Fourth integral is a *cubic spline* ($= \frac{1}{6}x^3$ for $x > 0$; solution to $u'''' = \delta$, p. 177)
Its derivative δ' is a *doublet* (p. 327)
Delta function $\delta(x)\delta(y)$ in two dimensions: $\iint f(x, y) \, \delta(x)\delta(y) \, dxdy = f(0, 0)$
Defining property: $\int v(x)\delta dx = v(0)$ for every smooth function v

3.2 ■ DIFFERENTIAL EQUATIONS OF EQUILIBRIUM

The correspondence between discrete and continuous equilibrium problems is almost complete. We identified A as d/dx and A^T as $-d/dx$, with boundary conditions. The matrix equation $A^T C A x = f$ became a differential equation $-d/dx(c\,du/dx) = f$ for the displacement (or potential). Now it remains to find the quadratics $P(u)$ and $Q(w)$ that replace

$$P(x) = \tfrac{1}{2}(Ax)^T C(Ax) - x^T f \quad \text{and} \quad Q(y) = \tfrac{1}{2} y^T C^{-1} y.$$

Then the differential equations of equilibrium will come from minimum principles.

Of course those minimum principles must lead to $-(cu')' = f$. But how does that happen? Chapter 1 confirmed that a quadratic like $P = \tfrac{1}{2} x^T K x - x^T b$ is minimized at the point where $K x = b$. In finite dimensions calculus is sufficient; the partial derivatives of P are zero. In the continuous case, minimizing $P(u)$ requires "the derivative of P with respect to a function." That is something new, and we write it as $\delta P/\delta u$—taking advantage of the Greek alphabet to use another form of the same letter delta. To make sense of this expression, and to find the differential equation that determines the minimizing u, is the main goal of the *calculus of variations*.

We certainly do not intend an exhaustive treatment of that subject; the main points are in Section 3.6. But the basic idea is so natural—to perturb u and find the resulting change in $P(u)$—that we can begin to explain it here. We can also apply it to functionals that involve the second derivative of u; they lead to fourth-order differential equations. And we anticipate two-dimensional problems, in which u is $u(x,y)$ and w is $w(x,y)$ and the minimum principles involve partial derivatives like $\partial u/\partial x$ and $\partial u/\partial y$. The equations that result from $\delta P/\delta u = 0$ are then *partial* differential equations. They express nature's preference for minimum energy, or minimum time, or least action.

Those equations for the minimizing functions are known as the *Euler equations*. They give the differential form of the problem, while minimizing P or Q gives the variational form. Intermediate between the energy and the Euler equation is a third statement of absolutely central importance: it is the *equation of virtual work*. We will meet it, fortunately and unavoidably, in the transition from P to $\delta P/\delta u$. It is also known as the *weak form* of the equation, and in Section 5.4 on finite elements it is the key to numerical methods for continuous equilibrium problems.

The fourth-order equation leads naturally to third-degree polynomials (the solutions of $d^4 u/dx^4 = 0$) and to one of the great successes in the approximation of functions. The section ends on that high (but optional) note, the interpolation of prescribed values $u_0, u_1, \ldots, u_n$ by a *cubic spline*. The spline is a third-degree polynomial between the interpolation points, where it has the right values and continuous second derivatives.

Minimum Principles

The minimum principles for u and w are easy to guess. In finite dimensions, the vector x minimized the potential energy P (or maximized $-P$), and this energy was

$$P(x) = \tfrac{1}{2}(Ax)^T C(Ax) - f^T x = \tfrac{1}{2}\sum_1^m c_i(Ax)_i^2 - \sum_1^n f_j x_j \quad \text{with } x_0 = 0.$$

In the continuous case, an elastic bar is fixed at the top; its displacement there is $u(0) = 0$. The point originally at a distance x down the bar is displaced by an additional distance $u(x)$. By analogy with the sum in the equation above, the total potential energy is an integral

$$P(u) = \int_0^1 \left[\frac{c}{2}\left(\frac{du}{dx}\right)^2 - f(x)u(x) \right] dx \quad \text{with } u(0) = 0. \tag{1}$$

The inner product of f and u is now $\int fu\, dx$, and the differences Ax have become derivatives du/dx, but the problem is still to minimize P.

The other minimum principle also changes from a sum to an integral:

$$Q(y) = \tfrac{1}{2}y^T C^{-1}y = \sum y_i^2/2c_i \text{ with } A^T y = f$$

$$Q(w) = \int_0^1 \frac{1}{2c}w^2 dx \text{ with } -\frac{dw}{dx} = f(x) \text{ and } w(1) = 0. \tag{2}$$

The constraint $-dw/dx = f$ is the continuous form of $A^T y = f$. The goal is to find the minimizing u and w and to show that the two problems are dual. We start with P, as our first problem in the calculus of variations.

3A   The function that minimizes $P(u)$ satisfies the differential equation

$$-\frac{d}{dx}\left(c\,\frac{du}{dx}\right) = f(x).$$

There is one essential boundary condition and one natural boundary condition:

$$u(0) = 0 \quad \text{and} \quad c(1)\frac{du}{dx}(1) = 0.$$

Proof   If u is minimizing, and if it is compared with $u + v$, the energy cannot decrease: $P(u) \le P(u + v)$. By expanding $P(u + v)$, this simple statement will lead to the fundamental condition on u:

$$P(u + v) = \int_0^1 \left[\frac{c}{2} \left(\frac{du}{dx} + \frac{dv}{dx} \right)^2 - f(u + v) \right] dx$$

$$= P(u) + \int_0^1 \left[c \frac{du}{dx} \frac{dv}{dx} - fv \right] dx + \int_0^1 \frac{c}{2} \left(\frac{dv}{dx} \right)^2 dx.$$

Since $P(u + v)$ cannot be less than $P(u)$, their difference is not negative:

$$\int_0^1 \left[c \frac{du}{dx} \frac{dv}{dx} - fv \right] dx + \int_0^1 \frac{c}{2} \left(\frac{dv}{dx} \right)^2 dx \geq 0 \text{ for every } v. \tag{3}$$

We have found the terms that depend on v, and it is the first integral that will be decisive. That integral is $\delta P/\delta u$, the **first variation** of P in the direction of v. It corresponds to the first-order term in the expansion $F(x + \Delta x) = F(x) + \Delta x F'(x) + \cdots$ of an ordinary function. The second integral in (3) corresponds to the second-order term $\frac{1}{2}(\Delta x)^2 F''$, and it is sure to be positive—but the first term locates the minimum.

At a minimum, **the first variation must be zero**. Otherwise, if it is not zero for some particular v, then either for that v or for $-v$ it will be negative. (It changes sign with v.) Then if v is multiplied by a small number ε, the first variation is multiplied by ε and the second by the smaller number ε^2. Eventually the first term dominates. If that linear term is negative, it will force the left side of (3) to be negative—producing a disaster. The function $u + \varepsilon v$ would give a smaller value of P than u itself, which is impossible if u is minimizing.

Just as the first derivative dF/dx is zero at an ordinary minimum of F, so the first variation $\delta P/\delta u$ must be zero at a minimum of P:

If u is the function that minimizes P, then

$$\int_0^1 c \frac{du}{dx} \frac{dv}{dx} dx = \int_0^1 f(x)v(x)dx \quad \text{for every test function } v. \tag{4}$$

This is the **equation of virtual work**. It is also the **weak form** of the differential equation $-(cu')' = f$.

The form (4) is called weak because it depends on the test functions v. Those are the "virtual displacements." Their work (force times displacement and stress times strain) balances to zero. Therefore the forces themselves must balance, and in a moment that will give us the strong form—which is $-(cu')' = f$, without any v.

There is a boundary condition on the test functions, since $P(u + v)$ can only be compared to $P(u)$ when $u + v$ satisfies the same conditions as u:

$$u(0) = 0 \quad \text{and} \quad u(0) + v(0) = 0 \quad \text{lead to} \quad v(0) = 0. \tag{5}$$

For nonzero conditions, $u(0) = a$ and $u(0) + v(0) = a$ again give $v(0) = 0$. Thus v is subject to every condition that is imposed on u, but these conditions are made homogeneous; their right hand sides are zero.

Now comes the final step, from the weak form to the strong form. The key is to integrate the left side of equation (4) by parts:

$$\int_0^1 c \frac{du}{dx} \frac{dv}{dx} \, dx = - \int_0^1 \frac{d}{dx} \left(c \frac{du}{dx} \right) v \, dx + \left[c \frac{du}{dx} v \right]_{x=0}^{x=1}. \tag{6}$$

By (4), this has to equal $\int fv \, dx$ whenever $v(0) = 0$. Suppose v is a delta function concentrated at a point x between 0 and 1. Then $\int fv \, dx$ is the value $f(x)$ at that point. It must agree with the right side of (6) when $v = \delta$ is an impulse:

$$-\frac{d}{dx} \left(c \frac{du}{dx} \right) = f(x) \text{ at every } x. \tag{7}$$

That is the Euler equation.† It is totally expected, and it ensures that the integral on the right side of (6) does match $\int fv \, dx$. It leaves untouched the final term in (6), which must be zero on its own. Since we control $v(0) = 0$ but there is no condition at the other end $x = 1$, the only way to ensure $[cu'v]_{x=0}^{x=1} = 0$ is to require

$$c(1)u'(1) = 0, \quad \text{or} \quad w(1) = 0. \tag{8}$$

This is the boundary condition that comes **naturally** from the minimum principle, and it completes the proof of 3A. The minimum of P occurs where $-(cu')' = f$ and $u(0) = 0$ and $w(1) = 0$.

That is a typical argument in the calculus of variations.

Remark 1 Suppose the coefficient $c(x)$ jumps from c_- to c_+ at the point $x = x_0$, because a new material begins at that point. Then the integration by parts in (6) is more safely done in two pieces, from 0 to x_0 and from x_0 to 1. The first piece produces an extra term $c_-u'_-v_-$ and the second produces $c_+u'_+v_+$—one by approaching x_0 from the left and the other from the right. If the test function v is constant around x_0, so that $v_- = v_+$, then these two new terms cancel if and only if

$$c_-u'_- = c_+u'_+, \quad \text{or} \quad w_- = w_+. \tag{9}$$

The jump in c must be compensated by a jump in du/dx. That is the natural boundary condition at the jump. The product $w = c \, du/dx$ is continuous at x_0; otherwise equation (7) would have a delta function on the left side.

---

† This differential equation for the minimizing u is also called the Euler-Lagrange equation—but Lagrange is with us enough already.

Remark 2   Continuing that sentence, a concentrated force at x_0 (a delta function in f) requires that $w = c \, du/dx$ **must** have a jump. The size of the jump matches the size of the concentrated force. Thus an instantaneous change in the internal force w balances the concentrated external force.

EXAMPLE   Suppose there is a **concentrated force** $f = 4\delta(x_0)$ at the midpoint $x_0 = \frac{1}{2}$ of the bar. Then the bar (with $c = 1$) is stretched down to that point and unstretched below it:

$$u(x) = 4x \text{ for } x \le \tfrac{1}{2}, \, u(x) = 2 \text{ for } x \ge \tfrac{1}{2}. \tag{10}$$

This function satisfies equation (7), which becomes $0 = 0$ everywhere except at x_0. At that point u' goes from 4 to 0; the jump in $-w$ equals 4 and balances f. The weak form is also satisfied:

$$\int_0^1 c \frac{du}{dx} \frac{dv}{dx} \, dx = \int_0^1 fv \, dx \quad \text{becomes} \quad \int_0^{1/2} 4 \frac{dv}{dx} \, dx = 4v \left(\frac{1}{2} \right),$$

which is true for every v since $v(0) = 0$.

Remark 3   (A mathematical point) We have spoken about "every function v" but that was an exaggeration. The **admissible functions** are those for which the total energy $P(v)$ is finite. The dominant term in P is the internal **strain energy** that integrates $(dv/dx)^2$ times $c/2$. In particular, $v(x)$ cannot have a jump. If it did, then v' would include a delta function and the energy would be infinite. Exercise 3.2.13 indicates why the square of the delta function has $\int \delta^2 \, dx = \infty$.

This theoretical point is important for two reasons. First, choosing v to be a delta function in (6) was completely improper. We should have used functions v which were more and more concentrated at the point x, but still had finite energy. In the limit the conclusion is the same: The strong form holds for any reasonable functions $f(x)$ and $c(x)$. The second consequence of requiring finite energy is very practical. For numerical solution, the **finite element method** chooses n trial functions $T_1(x), \dots, T_n(x)$. The approximate solution $U = U_1 T_1 + \cdots + U_n T_n$ is the combination that satisfies the equation (4) of virtual work, when the test functions are the v's:

$$\int_0^1 c \, U'v' dx = \int_0^1 fv \, dx \quad \text{for each} \quad v = T_1, \dots, T_n. \tag{11}$$

That gives n equations for the coefficients $U_1, \dots, U_n$. But it requires that each $T_j(x)$ is admissible, with finite energy. Therefore we may not legally use trial functions that have jump discontinuities.†

---

† The simplest finite element is a series of connected linear pieces. There are jumps in T' but $\int (T')^2 dx$ is safely finite. The real difficulties come for fourth-order equations, in which the slopes must also match.

Complementary Minimum Principle for *w*

We turn to the other quadratic $Q(w)$. It is the **complementary energy** and it completes the duality—although the details can be omitted in class:

$$Q(w) = \int_0^1 \frac{1}{2c} w^2 dx \quad \text{with} \quad -\frac{dw}{dx} = f(x) \quad \text{and} \quad w(1) = 0. \tag{12}$$

In this example the constraints are remarkable; they completely determine w. Integrating $-w' = f$ and fixing the integration constant by $w(1) = 0$, *only one function is admissible* in the variational principle. The internal forces w can be found directly from the external forces f and $w(1)$, without looking at the displacements.

When the constraints determine w, the problem is statically determinate. It corresponds to a square matrix A and a unique solution to $A^T y = f$. If the elastic bar were fixed at both ends, $u(0) = u(1) = 0$, the situation would be different. One degree of freedom is removed from u and given to w, and the constant of integration $w(1)$ is not known. To find it we need to use the minimum principle (12). Since this is the first time that a differential equation has appeared in the constraint, it is important to know what to do.

In the discrete case we introduced a Lagrange multiplier. Therefore we do the same here, recognizing that the multiplier will be a function $u(x)$ instead of a vector. There is a constraint at every point, rather than an equation at a finite number of nodes. Multiplying the constraint $-w' = f$ by $u(x)$ and building it into the function, (12) changes into

$$L(u,w) = \int_0^1 \left[\frac{1}{2c} w^2 + u \left(\frac{dw}{dx} + f \right) \right] dx. \tag{13}$$

This leads to a new problem in the calculus of variations, similar to the minimization of $P(u)$. We want the condition on the optimal w.

The method is the same: Perturb w by a "virtual force" V, and compute the difference produced by $w + V$ in (13). When w is minimizing, the new terms involving V can only increase the energy. Those terms are

$$\int_0^1 \left[\frac{1}{2c} (2wV + V^2) + u \frac{dV}{dx} \right] dx \geq 0. \tag{14}$$

Again everything depends on the first variation, the part that is linear in V. If that is not zero then (14) will fail; the V^2 term will not save it. Therefore the first variation must vanish, and

$$\frac{\delta L}{\delta w} = \int_0^1 \left[\frac{1}{c} wV + u \frac{dV}{dx} \right] dx = 0 \quad \text{for every } V. \tag{15}$$

Integrating by parts, this "weak form" becomes

$$\int_0^1 \left[\frac{w}{c} - \frac{du}{dx} \right] V \, dx + [uV]_0^1 = 0. \tag{16}$$

When V is a delta function we reach the strong form $w/c = du/dx$. This Euler equation is now the link between w and u. At the same time it confirms that the Lagrange multiplier u is identical to the displacement! At least they satisfy the same equation, since $w = cu'$ and $-w' = f$ lead to $-(cu')' = f$. Only the boundary conditions remain to be checked.

The condition on V comes from the essential condition on w. If $w(1) = 0$ is required for admissibility, then $(w + V)(1) = 0$ is also required and $V(1) = 0$. That eliminates $[uV]$ at the end where $x = 1$. To eliminate it at the other end $x = 0$, where V is completely arbitrary, we need $u(0) = 0$. That is the **natural condition** when $Q(w)$ is minimized; it was the **essential condition** when $P(u)$ was minimized. It confirms that the multiplier satisfies the same condition as the displacement, and the two are identical.

We summarize these steps. They took us from a constrained Q to the Lagrangian L (in which the constraints are built in). If we substitute $c \, du/dx$ for w, the expression (13) changes into $-P(u)$. These are exactly the steps that governed duality in the matrix case, and they govern it here too.

3B (Duality in the continuous case) The minimum value of $Q(w)$ equals the maximum value of $-P(u)$, and the optimal u and w are a saddle point for L:

$$\min_w Q(w) = \min_w (\max_u L(u,w)) = \max_u (\min_w L(u,w)) = \max_u -P(u). \tag{17}$$

The displacements are Lagrange multipliers for the force balance $-dw/dx = f$.

Maximizing $-P$ corresponds to minimizing P, and the optimal u is given by the Euler equation $-(cu')' = f$. It will be an exercise to show that always $Q + P \geq 0$ (weak duality), and that the optimal u and w achieve $Q + P = 0$ (strong duality).

Fourth-order Equations

We turn from an elastic rod to a beam. The difference is that **the beam bends**. It becomes curved, and the bending produces internal forces that try to straighten it. The restoring force is no longer governed by the stretching, but by the **curvature**. The displacement u, which was previously measured in the direction of the rod, is now perpendicular to the beam.

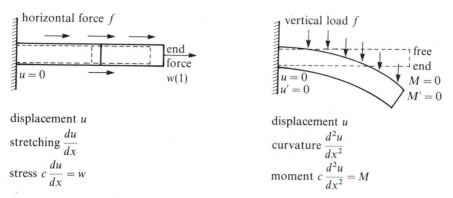

displacement u

stretching $\dfrac{du}{dx}$

stress $c\dfrac{du}{dx} = w$

displacement u

curvature $\dfrac{d^2u}{dx^2}$

moment $c\dfrac{d^2u}{dx^2} = M$

Fig. 3.4. A rod versus a beam.

Mathematically the difference is in A. For stretching we had $Au = du/dx$; for bending it will be $Au = d^2u/dx^2$. This is the leading term in the curvature, and if it is zero then u is linear and the beam is straight. If it is not zero then there is a *bending moment* $M = CAu$, or $M = cu''$. The balance between the restoring force and the applied force f will give the equation of equilibrium. You recognize that M is taking the place of w; the material constant c is the bending stiffness.

We might make a similar comparison in two dimensions between a plate and an elastic membrane. When the plate is bent, it fights back. When a membrane is bent, it doesn't care. Its only forces are "in the plane," resisting stretching. A combination of the two is called a shell, in which membrane forces mix with bending moments and eighth-order equations appear. In fact the success of eggshells lies exactly in this possibility, to balance a perpendicular applied force by an in-plane internal force. The shell stretches a little, and doesn't break.

The mathematics of beam theory will fit directly into our framework. Therefore we develop it quickly, starting with the differential equation and boundary conditions, and leaving minimum principles for later. We expect A^T to be d^2/dx^2, with a plus sign, since somewhere there will be two integrations by parts. The square of $(d/dx)^T = -d/dx$ should be $(d^2/dx^2)^T = d^2/dx^2$. The second derivative is formally symmetric, like its discrete approximation

$$\frac{d^2}{dx^2} \approx \frac{1}{(\Delta x)^2}\begin{bmatrix} -2 & 1 & & & & \\ 1 & -2 & 1 & & & \\ & \cdot & \cdot & & \cdot & \\ & & & 1 & -2 & 1 \\ & & & & 1 & -2 \end{bmatrix}.$$

This symmetry is destroyed at the endpoints (the corners of the matrix) if A and A^T have different boundary conditions.

The matrix suggests one case in which symmetry is not destroyed. If A comes with $u = 0$ at both ends of the beam, then A^T has $M = 0$. The missing 1's in the first and last rows are chopped off as shown, and symmetry is preserved. In general, the rule for boundary conditions is that the inner product (Au, M) should equal $(u, A^T M)$. When the boundary conditions are homogeneous (zero), we need

$$\int_0^1 \frac{d^2u}{dx^2} M \, dx = \int_0^1 u \frac{d^2M}{dx^2} \, dx. \tag{18}$$

To compare these inner products we integrate both sides by parts. Each side becomes $-\int u' M' dx$, and they differ by the boundary term

$$\left[M \frac{du}{dx} - u \frac{dM}{dx} \right]_{x=0}^{x=1}. \tag{19}$$

There are four important combinations which make this expression zero:

(1)   Simply supported end: $u = 0$ and $M = 0$

(2)   Fixed (or cantilevered) end: $u = 0$ and $\dfrac{du}{dx} = 0$

(3)   Free end: $M = 0$ and $\dfrac{dM}{dx} = 0$

(4)   Sliding clamped end: $\dfrac{du}{dx} = 0$ and $\dfrac{dM}{dx} = 0.$

The conditions on u come from A, and the conditions on M from A^T. The combination $M = du/dx = 0$ overspecifies $M \, du/dx$ and gives no control of $u \, dM/dx$ in (19)—so a combination of M and du/dx is not allowed.

If one of these combinations is present at each end, the integrated term (19) will vanish and we have correct boundary conditions. The example with $u = 0$ and $M = 0$ had both ends simply supported: resting on supports there is no displacement and nothing to produce a bending moment. In other words,

$$Au = \frac{d^2u}{dx^2} \quad \text{with} \quad u(0) = u(1) = 0$$

$$A^T M = \frac{d^2M}{dx^2} \quad \text{with} \quad M(0) = M(1) = 0. \tag{20}$$

This is a symmetric case. In an unsymmetric case, for example with one end fixed and the other free, we have

$$Au = \frac{d^2u}{dx^2} \quad \text{with} \quad u(0) = \frac{du}{dx}(0) = 0$$

$$A^T M = \frac{d^2M}{dx^2} \quad \text{with} \quad M(1) = \frac{dM}{dx}(1) = 0. \tag{21}$$

This corresponds to $u(0) = 0$ and $w(1) = 0$ for an elastic rod. The exercises ask for the deflection of such a beam under a uniform load $f = 1$, and it should be a polynomial since d^4u/dx^4 is constant.

Now we leave the boundary conditions and turn to the differential equation. It can be written as a single equation $A^T C A u = f$, or as a pair of equations

$$M = c\frac{d^2u}{dx^2} \quad \text{and} \quad \frac{d^2M}{dx^2} = f.$$

The first is $M = CAu$ and the second is $A^T M = f$. Of course it is easy to eliminate M and arrive at the fundamental equation of linearized beam theory:

3C If a load $f(x)$ is applied to a beam with bending stiffness $c(x)$, the resulting deflection satisfies

$$\frac{d^2}{dx^2}\left(c\frac{d^2u}{dx^2}\right) = f(x). \tag{22}$$

This fourth-order equation has **two conditions at each end**. If they are homogeneous they produce $Mu' = uM'$ at the boundaries.

Interpolation: Displacements and Slopes

After writing down the four types of end conditions, I realized that the list might be misleading. It suggests that the conditions must force u or its derivatives to be *zero* at each end. That is not at all true. The boundary conditions can equally well be inhomogeneous, as in

$$u(0) = a, \frac{du}{dx}(0) = b, u(1) = c, \frac{du}{dx}(1) = d. \tag{23}$$

The ends of this beam are still fixed, but they are not necessarily lined up. In fact we can have zero force, $f = 0$, and the beam will bend in order to satisfy these boundary conditions. It is curved by the forces at its ends.

The deflection is still governed by the differential equation

$$\frac{d^2}{dx^2}\left(c\frac{d^2u}{dx^2}\right) = f, \quad \text{or} \quad \frac{d^4u}{dx^4} = 0,$$

assuming that c is constant. Any third degree polynomial like $x^3 + 1$ is a solution to this equation; the fourth derivative of a cubic is automatically zero. But there is only one cubic polynomial whose four coefficients match all four conditions in (23). It is the **Hermite cubic**, and we can write it out in full:

$$u = a(x-1)^2(2x+1) + b(x-1)^2x + cx^2(3-2x) + dx^2(x-1). \tag{24}$$

This is certainly a third-degree polynomial, so $d^4u/dx^4 = 0$. Each term is chosen to match one of the boundary conditions in (23). The coefficient of d is $x^3 - x^2$, and it is zero at both ends. Its slope $3x^2 - 2x$ equals zero at $x = 0$ and one at $x = 1$. The graph is drawn upside down in Fig. 3.5b, showing the shape of a beam when one end is held horizontal and the other is fixed at $-45°$. The left figure also shows the coefficient of a—the shape of a beam when one end is raised but still cantilevered.

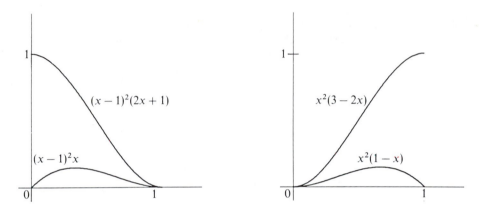

Fig. 3.5. Cubics for the four boundary conditions: coefficients of a, b, c, and $-d$.

These cubics also appear in problems that have nothing to do with beams. They are used for *interpolation*—constructing a curve that goes through a given set of points. It is a different process from least squares fitting, which only approximates the given values. The least squares fit comes as close as possible with a small number of degrees of freedom, whereas *interpolation is exact*. The curve will go right through the points—it must have enough degrees of freedom to do so—and the question is what kind of curve to use.

Suppose we are given the height and also the slope of the unknown curve at the points $x = 0, 1, \ldots, n$. That is $2n + 2$ pieces of information, and they can be matched with a single polynomial of high degree. The degree would be $2n + 1$, producing a polynomial with $2n + 2$ coefficients. However the result would be absolutely terrible. Even if the values and slopes are taken from a smooth curve, the interpolating polynomial is almost certain to oscillate. Away from the interpolation points it is far from the right curve. A much better alternative is to fit the data by a *piecewise polynomial*, which has low degree in each interval between interpolation points.

When heights and slopes are both given, the natural choice is a piecewise cubic. It is exactly the one constructed above in (24). The four coefficients of the cubic satisfy four conditions—two at each end of every interval. These "Hermite cubic" pieces fit together with matching height and slope at the interpolation points. Of course the second and third derivatives at those points can be expected to jump, as in the following example:

$$\text{Heights} \quad u_0 = 0, u_1 = 1, u_2 = 0 \quad \text{at} \quad x = 0, 1, 2$$

$$\text{Slopes} \quad s_0 = 1, s_1 = 0, s_2 = 0 \quad \text{at} \quad x = 0, 1, 2.$$

The cubic between 0 and 1 comes directly from (24), with $a = d = 0$ and $b = c = 1$ (since $s_0 = u_1 = 1$):

$$u(x) = (x - 1)^2 x + x^2(3 - 2x) = -x^3 + x^2 + x.$$

The second piece goes from $x = 1$ to $x = 2$. It also comes from (24), but now the height $u_1 = 1$ is at the left end of the interval. Thus $a = 1$ and $b = c = d = 0$. At the same time x in (24) is changed to $x - 1$, to shift the interval to $1 \leq x \leq 2$, so that

$$u(x) = (x - 2)^2(2x - 1) = 2x^3 - 9x^2 + 12x - 4.$$

At the junction $x = 1$, these cubics $u(x)$ have the same height $u = 1$ and the same slope $u' = 0$. The second derivative jumps from -4 to -6, and the third derivative jumps from -6 to 12. It is hardly possible to see that effect, and we have exchanged the total smoothness of a sensitive high degree polynomial for the stability of a low degree piecewise polynomial.

Cubic Splines

Now comes the more usual interpolation, when only the heights are given. We know the values $u_0, u_1, \ldots, u_n$ at the points $x = 0, 1, \ldots, n$, and we want to fit them with a curve. Again a high degree polynomial is unstable, and a low degree piecewise polynomial is better. The easiest method of all is to connect the points by straight lines—which is piecewise linear interpolation. But then the slopes change suddenly and are unreliable. A much better choice is to use **cubic splines**.†

A cubic spline is a piecewise cubic in which not only the function u and the slope du/dx but also **the second derivative d^2u/dx^2 is continuous**. Physically, it comes from bending a long thin beam to give it the correct heights $u_0, \ldots, u_n$ at the interpolation points. I think of rings at those points, and the beam going through the rings. At all other points the force is zero, the beam is free to choose its own shape, and the solution to the beam equation $d^4u/dx^4 = 0$ is an ordinary cubic. But something must happen at an interpolation point like $x = 0$, where the ring imparts a concentrated load of unknown magnitude f_0 and

$$\frac{d^4u}{dx^4} = f_0 \delta(x). \tag{25}$$

---

† The name and the idea come from naval architects.

Remember that the delta function is a "spike" of infinite height and infinitesimal width, and it has unit area:

$$\int \delta(x)dx = 1 \quad \text{and} \quad \int f(x)\delta(x)dx = f(0). \tag{26}$$

It is not a genuine function, but it has been legalized as a *distribution*—which means that $\delta(x)$ is known by its effect on smooth functions $f(x)$.

Integrating both sides of (25), we discover that the third derivative of u jumps by f_0:

$$\int \frac{d^4u}{dx^4}\, dx = \int f_0 \delta(x)dx \quad \text{or} \quad \left[\frac{d^3u}{dx^3}\right]_{x=0-}^{x=0+} = f_0.$$

In other words d^3u/dx^3 is a *step function*. It is constant on each side of $x=0$, but those constants differ by f_0. At $x=1$ there is another step of height f_1. When we integrate the step function d^3u/dx^3, we conclude that the second derivative d^2u/dx^2 is continuous.

Now the exact shape of the spline can be determined. If slopes $s_0, \ldots, s_n$ were known at the interpolation points, the problem would already be solved. Between those points it is a cubic, and earlier in the section we constructed the only cubic— the Hermite cubic—with the required heights $u_0, \ldots, u_n$ and the required slopes. At present the slopes are unknown, but we have $n+1$ new conditions; d^2u/dx^2 is continuous at each point $x=0, \ldots, n$. This will give $n+1$ equations for the slopes, as follows.

In the infinite interval to the left of $x=0$, the beam is straight (but not necessarily horizontal). There is nothing to produce curvature, so that $d^2u/dx^2 = 0$ at $x=0$. To make the second derivative also zero coming from the right of $x=0$, (24) requires

$$2s_0 + s_1 = 3u_1 - 3u_0. \tag{27}$$

This is the first of the $n+1$ equations. At the next node $x=1$, the continuity of d^2u/dx^2 connects the cubic on the left to the cubic on the right. With some patience this calculation gives

$$(2s_1 + s_2) + (2s_1 + s_0) = (3u_2 - 3u_1) + (3u_1 - 3u_0). \tag{28}$$

There is a similar equation at each of the points $x=1, \ldots, n-1$ and an equation like (27) at $x=n$. Altogether these $n+1$ equations for the unknown slopes go into matrix form as

$$\begin{bmatrix} 2 & 1 & & & \\ 1 & 4 & 1 & & \\ & \cdot & \cdot & \cdot & \\ & & 1 & 4 & 1 \\ & & & 1 & 2 \end{bmatrix} \begin{bmatrix} s_0 \\ s_1 \\ \cdot \\ s_{n-1} \\ s_n \end{bmatrix} = 3 \begin{bmatrix} u_1 - u_0 \\ u_2 - u_0 \\ \cdot \\ u_n - u_{n-2} \\ u_n - u_{n-1} \end{bmatrix}. \tag{29}$$

The matrix is symmetric and positive definite, and the slopes are found from this equation. The piecewise cubic which has these slopes, and which takes on the given values $u_0, \ldots, u_n$, is the **interpolating spline**.

3D With heights and slopes both given, there is one possible piecewise cubic. It is determined in every interval by two conditions at each end, as in (24), and its first derivative is continuous. If only the heights $u_0, \ldots, u_n$ are given, the slopes can be chosen according to (29) and then the piecewise cubic becomes a spline; its second derivative is also continuous.

EXAMPLE $\quad u_0 = 0,\, u_1 = 1,\, u_2 = 4,\, u_3 = 1,\, u_4 = 0$

The equation (29) for the slopes is

$$
\begin{bmatrix}
2 & 1 & & & \\
1 & 4 & 1 & & \\
 & 1 & 4 & 1 & \\
 & & 1 & 4 & 1 \\
 & & & 1 & 2
\end{bmatrix}
\begin{bmatrix}
s_0 \\ s_1 \\ s_2 \\ s_3 \\ s_4
\end{bmatrix}
= 3
\begin{bmatrix}
1 \\ 4 \\ 0 \\ -4 \\ -1
\end{bmatrix}.
\tag{30}
$$

The solution has $s_1 = 3$ and $s_3 = -3$, with the other slopes $s_0 = s_2 = s_4 = 0$. This is a very special spline, drawn in Fig. 3.6. It starts with everything zero except the third derivative; $u = x^3$ over the first interval. After four intervals it returns to zero—again with $u = du/dx = d^2u/dx^2 = 0$. If it is extended to the additional values $u_5 = 0, u_6 = 0, \ldots$, the spline will be identically zero on these extra intervals.

This example is the spline that "dies" the fastest. It begins with $u = 0$ to the left of $x = 0$, and returns to $u = 0$ in only four intervals. It is called a basic spline, or **B-spline**—and all others are combinations of these special four-interval splines.

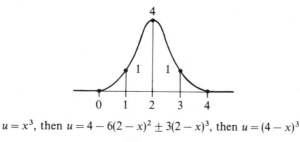

$u = x^3$, then $u = 4 - 6(2 - x)^2 \pm 3(2 - x)^3$, then $u = (4 - x)^3$

Fig. 3.6. The B-spline: a cubic with continuous second derivative.

There is also a minimum principle. The spline has the smallest bending energy

$$P(u) = \int_0^n \left(\frac{d^2 u}{dx^2}\right)^2 dx \tag{31}$$

among all functions with the correct heights $u_0, \ldots, u_n$ at the points $x = 0, \ldots, n$ (where the rings are). That gives the Euler equation $d^4 u/dx^4 = 0$ away from the rings. In every interval u is a cubic, and the spline bends as little as possible. It is an excellent way to pass a curve through the prescribed points.

EXERCISES

3.2.1   For $P(u) = \frac{1}{2} \int_0^1 \left(\frac{d^2 u}{dx^2}\right)^2 dx$, find the first variation $\delta P/\delta u$ from the linear term in $P(u + v)$.

3.2.2   What function $u(x)$ with $u(0) = 0$ and $u(1) = 0$ minimizes

$$P(u) = \int_0^1 \left[\frac{1}{2}\left(\frac{du}{dx}\right)^2 + x\, u(x)\right] dx?$$

3.2.3   What function $w(x)$ with $dw/dx = x$ (and unknown integration constant) minimizes

$$Q(w) = \int_0^1 \frac{w^2}{2} dx?$$

With no boundary condition on w this is dual to Ex. 3.2.2.

3.2.4   What functions u and w minimize P and Q with $dw/dx = x$ and $u(0) = w(1) = 0$? Verify the strong duality $-P = Q$.

3.2.5   With two conditions $w(0) = w(1) = 0$ show that no function satisfies $dw/dx = x$. With no conditions on u show that P has no minimum (except $-\infty$). This is the unstable case, unsupported and allowing rigid motions $u = $ constant.

3.2.6   From the differential equation $-d/dx(c\, du/dx) = f$, derive the weak form (4) by multiplying by test functions v and integrating.

3.2.7   Show that $P(u) + Q(w) \geq 0$ for any admissible u and w:

$$\int_0^1 \left[\frac{c}{2}\left(\frac{du}{dx}\right)^2 - fu + \frac{1}{2c}w^2\right] dx \geq 0 \quad \text{when} \quad f = -\frac{dw}{dx} \quad \text{and} \quad u(0) = w(1) = 0.$$

3.2.8   Show that $P(u) + Q(w) = 0$ for the optimal u and w, which satisfy $w = cu'$.

3.2.9   If $u(0) = 0$ is changed to $u(0) = a$, this essential condition for $P(u)$ must again be a natural condition for $Q(w)$. That requires a change in Q:

$$Q(w) = \int_0^1 \frac{1}{2c} w^2 dx + aw(0) \quad \text{subject to} \quad -\frac{dw}{dx} = f \quad \text{and} \quad w(1) = 0.$$

Introduce the multiplier u for the constraint and show that the vanishing of the first variation $\delta L/\delta w$ does lead to $u(0) = a$. (The new term $aw(0)$ is the work done at the boundary; when w is perturbed to $w + V$, the difference $V(0)$ appears in $\delta L/\delta w$.)

3.2.10 If the ends of a beam are fixed (zero boundary conditions) and the force is $f = 1$ with $c = 1$, solve $d^4u/dx^4 = 1$ and then find M. Why does it have to be done in that order?

3.2.11 With simply supported ends, the boundary conditions make it possible to solve $M'' = f$ directly for M without going first to u. This is a statically determinate beam. Find M and u if $f = 1$ and $c = x$.

3.2.12 What is the shape of a uniform beam under zero force, $f = 0$ and $c = 1$, if $u(0) = u(1) = 0$ at the ends but $du/dx(0) = 1$ and $du/dx(1) = -1$? Sketch this shape.

3.2.13 The *step function* $s(x)$ is zero for $x \le 0$ and jumps to one for $x > 0$; its derivative is the delta function with $\delta(x) = 0$ for every $x \ne 0$ but $\int \delta(x)dx = 1$. To make this reasonable, consider the step s as a limit of functions s_n which have slope n between $x = 0$ and $x = 1/n$; elsewhere $s_n' = 0$. Compute $\int s_n' dx$ and $\int (s_n')^2 dx$, and as $n \to \infty$ verify formally that $\int \delta = 1$ and $\int \delta^2 = \infty$.

3.2.14 What is the maximum height of the Hermite cubic $(x - 1)^2 x$ in Figure 3.5?

3.2.15 Prove that the coefficient matrix for spline interpolation in equation (29) is positive definite. One method of proof is to rewrite the quadratic

$$x^T A x = 2x_0^2 + 2x_0 x_1 + 4x_1^2 + 2x_1 x_2 + 4x_2^2 + \cdots$$

as a sum of many squares.

3.2.16 Which cubic spline has the values $u = 0,4,8,12$ at the nodes $x_0 = 0$, $x_1 = 1$, $x_2 = 2$, $x_3 = 3$? Show that the governing equation (29) is satisfied.

3.2.17 For the B-spline in Fig. 3.6, find the cubic in the second interval from its height and slope at $x = 1$ and $x = 2$.

3.2.18 (a) Find the cubic that replaces (24) if the right end of the interval is moved from $x = 1$ to $x = h$.
  (b) Find the new form of (27) for this case, giving $d^2u/dx^2 = 0$ at $x = 0$.
Note: The new form of (28) at $x_1 = h$, if the second interval goes to $x_2 = h + H$, is a typical spline condition for unequally spaced points:

$$\frac{2s_1 + s_2}{H} + \frac{2s_1 + s_0}{h} = \frac{3}{H^2}(u_2 - u_1) + \frac{3}{h^2}(u_1 - u_0).$$

3.2.19 There are 16 coefficients in the four cubics between the points x_0, x_1, x_2, x_3, x_4. Describe in words the 16 conditions on a cubic spline; five of them are the prescribed heights u_0, u_1, u_2, u_3, u_4.

3.3 ■ LAPLACE'S EQUATION AND POTENTIAL FLOW

There is one more rich source of examples leading to the triple product A^TCA. As in the last two sections, the examples involve derivatives; the applications are continuous rather than discrete. The difference is that the equilibrium equation $A^TCAu = f$ will be a *partial differential equation*. Its solution $u(x,y)$ is defined in a region of the $x - y$ plane. The "nodal variable" u gives the steady state distribution of potential or pressure or temperature, and the "edge variable" $w = CAu$ gives the flow.

Perhaps we can reveal in advance the most important equation of this section. It is *Laplace's equation*, and it fits exactly into the framework of A^TCA (in fact with $C = I$). We write it in its usual form, with plus signs, and also in a form that displays A^TA:

$$\frac{\partial^2 u}{\partial x^2} + \frac{\partial^2 u}{\partial y^2} = 0 \quad \text{or} \quad \begin{bmatrix} -\dfrac{\partial}{\partial x} & -\dfrac{\partial}{\partial y} \end{bmatrix} \begin{bmatrix} \dfrac{\partial}{\partial x} \\ \dfrac{\partial}{\partial y} \end{bmatrix} u = A^TAu = 0.$$

It is easy to recognize which part is A. But it remains to see why this is the fundamental example of a steady state partial differential equation, and how it corresponds so completely to a network equation. It is equivalent to Kirchhoff's laws (in their continuous form). It comes from a principle of minimum energy, as the equation of equilibrium. And vector calculus will explain the minus signs in A^T.

It also remains to *solve* Laplace's equation, but that can hardly be done within one section. The solution is approached analytically in the next chapter by Fourier series and complex variables and integral equations, and then numerically in Chapter 5.

We begin with an *ideal fluid*, which has no viscosity to dissipate energy. It is in steady two-dimensional flow, with a velocity vector v that gives the direction and magnitude of the flow. The horizontal and vertical components are $v_1(x,y)$ and $v_2(x,y)$; the velocity vector is $v = (v_1, v_2)$. Since the flow is steady, it must satisfy a continuous analogue of Kirchhoff's laws:

Voltage law: The velocity should come from a *potential* $u(x,y)$
Current law: There should be *conservation of fluid* (provided there are no sources or sinks).

To express these laws as differential equations is a perfect application of vector calculus. It uses the two fundamental theorems—Stokes' theorem and the divergence theorem. Those may already be familiar, or you may meet them here for the first time. In their discrete form they are almost self-evident (Exercises 3 and 4). In their continuous form they decide which flows are irrotational and which are conservative. Those properties are connected by a third application of vector calculus, linking A (the gradient) to A^T (the divergence, with a minus sign). To see that one is the transpose of the other, so that the inner products

$$\int \int \text{grad } u \cdot w \; dx \; dy \quad \text{and} \quad -\int \int u \text{ div } w \; dx \; dy$$

differ only by boundary terms, is a direct application of "*Green's formula.*" This is the analogue of $(Au)^T w = u^T (A^T w)$, and it is integration by parts in two dimensions.

I think the best plan is to construct first a table of analogies. That will make the goal clear. The rest of the section explains how the laws in the table come directly from physical principles, or indirectly from minimum principles, with examples. We mention that there is disagreement between different fields about plus and minus signs. In the distribution of temperature or voltage, it is customary for v to be the *negative* of the potential gradient. The operator A becomes $A = -\text{grad}$, so that the flow is away from a higher temperature or potential to a lower one. This convention holds also for electrostatics and magnetostatics (which fit into the same framework) and in all those cases $A^T = +\text{div}$. For fluid mechanics and solid mechanics the signs are reversed. The final equation $A^T C A u = f$ will be ***Poisson's equation***—Laplace's equation with a nonzero right hand side.

There are no surprises in the table of analogies, except for the notations ∇ and $\nabla \cdot$ and ∇^2 (or Δ) for the gradient and divergence and Laplacian. The gradient $\nabla = (\partial/\partial x, \partial/\partial y)$ is called "del"—I think it is a nickname for delta—and the Laplacian is "del squared."

($v = Au$)   The velocity is the gradient of the potential,

$$v = \text{grad } u = \nabla u \quad \text{or} \quad (v_1, v_2) = \left(\frac{\partial u}{\partial x}, \frac{\partial u}{\partial y} \right).$$

($w = Cv$)   The velocity times the mass density c (often written ρ) is the momentum density or flow rate $w = cv$.

($A^T w = f$)   Conservation of mass is

$$\text{div } w = \nabla \cdot w = 0 \quad \text{or} \quad \frac{\partial w_1}{\partial x} + \frac{\partial w_2}{\partial y} = 0.$$

This becomes $-\text{div } w = f$ if there are sinks ($f > 0$) or sources ($f < 0$).

($A^T C A u = f$)   The three previous steps combine to give the equilibrium equation

$$-\text{div}(c \text{ grad } u) = -\frac{\partial}{\partial x}\left(c \frac{\partial u}{\partial x} \right) - \frac{\partial}{\partial y}\left(c \frac{\partial u}{\partial y} \right) = f. \qquad (1)$$

With constant density c this is Poisson's equation

$$-\frac{\partial^2 u}{\partial x^2} - \frac{\partial^2 u}{\partial y^2} = \frac{f}{c}. \qquad (2)$$

For incompressible fluids without sources or sinks ($f = 0$) it is Laplace's equation

$$\text{div grad } u = \nabla^2 u = \Delta u = \frac{\partial^2 u}{\partial x^2} + \frac{\partial^2 u}{\partial y^2} = 0. \tag{3}$$

Boundary conditions: On each part of the boundary either
   (1) the potential $u = u_0$ is given, or
   (2) the flow rate through the boundary is prescribed by

$$w \cdot n = (c \text{ grad } u) \cdot n = c \frac{\partial u}{\partial n} = F.$$

In the last line, the normal vector $n = (n_1, n_2)$ is perpendicular to the boundary and points outward with unit length: $n_1^2 + n_2^2 = 1$. The derivative of u in the direction of n is the *normal derivative* $\partial u / \partial n$.

This section concentrates on steady irrotational incompressible flow and Laplace's equation. That is reasonable for fluids, or at least for liquids. For the flight of an airplane it is not reasonable; air is not incompressible (and the flight may not be so steady). However one case that is of crucial importance to aircraft designers can be reduced to Laplace's equation. It is the linearized problem

$$(1 - M^2) \frac{\partial^2 u}{\partial x^2} + \frac{\partial^2 u}{\partial y^2} = 0, \tag{4}$$

where M is the **Mach number**—the ratio of the plane's velocity to the speed of sound. If $M > 1$ the flow is supersonic and we cannot reach Laplace's equation; the coefficient $1 - M^2$ is negative. If $M < 1$ the flow is subsonic, and by simply rescaling x—the distance in the flight direction—the constant coefficient $1 - M^2$ disappears to leave Laplace's equation.

It is almost true that the battle between the Boeing 767 and the Airbus depends on the solution to this equation.† Those planes settle down in flight to steady flow. Their velocity and altitude stay virtually constant, and the designer can compute from (4) the ratio of lift to drag. He chooses the shape to increase that ratio. This is going on in all aircraft companies, and the competition of 767 versus Airbus is not decided at this writing. (The margin will be less than 5%.) There is so much at stake that designers rely on experiments in a wind tunnel at the same time as experiments on a computer. The cost of both is enormous; even the design of the 747 used 20,000 wind tunnel hours. The numerical solution of (4) is much cheaper than an accurate wind tunnel, but when these approximations break down—as they do—the equation becomes more complicated. It may be replaced by a nonlinear potential equation, or by Euler's equation (still inviscid), or eventually by the Navier-Stokes

---

† It also depends on politics.

equations for unsteady viscous flow. At present a three-dimensional Navier-Stokes solution is out of range, even with a supercomputer.

Kirchhoff's Laws: Velocity Potential and Continuity Equation

For electrical circuits there were two statements of the voltage law: (1) the sum of voltage drops around every loop was zero, and (2) there was a potential at the nodes. The voltage drops were potential differences, so the sum around a loop was automatically zero. For fluids, this corresponds to *irrotational flow*. A vortex is not allowed; you cannot mix a drink or pour cream into coffee. Nevertheless some important flows are left, and the continuous case has three ways to express the absence of rotation. The first two are directly analogous to the discrete case:

(1) The velocity has *zero circulation*:

$$\int v_1 dx + v_2 dy = 0 \quad \text{around any closed path}$$

(2) The velocity is the *gradient* of a potential function:

$$v = \text{grad } u \quad \text{or} \quad \begin{bmatrix} v_1 \\ v_2 \end{bmatrix} = \begin{bmatrix} \partial u/\partial x \\ \partial u/\partial y \end{bmatrix}. \tag{5}$$

This second one is like du/dx in Section 3.1; now there are partial derivatives $\partial u/\partial x$ and $\partial u/\partial y$. They are the components of $v = Au$; taking the gradient corresponds to computing differences in potential. Every point in the region is a node, and the "voltage drops" between neighboring points are described by the two partial derivatives. This means that the flow variable has two components—the velocity is a vector—while the potential has one. In some rough way there are still more edges than nodes, $m > n$, even though m and n are now infinite. The velocity points in the direction in which u changes most rapidly, and its magnitude is

$$|v| = (v_1^2 + v_2^2)^{1/2} = \left(\left(\frac{\partial u}{\partial x} \right)^2 + \left(\frac{\partial u}{\partial y} \right)^2 \right)^{1/2} = |\text{grad } u|. \tag{6}$$

EXAMPLE If the potential is $u = xy$ then the velocity components are $v_1 = \partial u/\partial x = y$ and $v_2 = \partial u/\partial y = x$. The magnitude of the velocity is $|v| = (y^2 + x^2)^{1/2}$, equal to the distance from the origin. The curves $u = $ constant (the *equipotential curves*) are the dotted hyperbolas in Fig. 3.7. The velocity vector is everywhere perpendicular to these curves of constant potential, since it points in the direction of most rapid change. The fluid is carried along the solid lines (the *streamlines*) which cross the x-axis and go out to infinity. It is irrotational flow past a 45° wedge.

In this example the streamlines also happen to be hyperbolas, of the form $x^2 - y^2 = $ constant. They follow the velocity vector, and they are perpendicular to the other family $u = xy = $ constant. At the origin we have $v = (y,x) = (0,0)$ and the fluid is motionless.

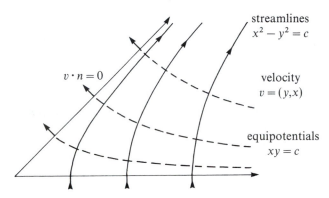

Fig. 3.7. Streamlines and equipotentials: irrotational flow.

The continuous case offers a third form of Kirchhoff's voltage law. It is a major improvement on condition 1, which asks us to check every loop for zero circulation. Instead it is enough to test the **vorticity** around every point. The flow is irrotational if

(3) $\boxed{\text{vorticity} = \text{curl } v = \dfrac{\partial v_2}{\partial x} - \dfrac{\partial v_1}{\partial y} \text{ is everywhere zero.}}$ (7)

The velocity in the example was $(v_1,v_2) = (y,x)$. It came from a potential, and its vorticity was zero: curl $v = 1 - 1 = 0$. We connect potentials to zero vorticity after mentioning two points:

(i)   The curl is really a vector and $\partial v_2/\partial x - \partial v_1/\partial y$ is only one of its components. For two-dimensional flow the other components are zero, so we call this number the curl until reaching three dimensions.

(ii)   The relation between zero circulation and zero vorticity comes from **Stokes' theorem**: The circulation of any vector field is the integral of the curl over the region inside the loop:

$$\int_C v_1 \, dx + v_2 \, dy = \iint_S \text{curl } v \, dx \, dy.$$ (8)

All loops have zero circulation when curl $v = 0$, and vice versa.

That was the first application of vector calculus; conditions 1 and 3 are equivalent. The other link, between velocities that are gradients of potentials and velocities with no curl, is absolutely direct:

$$\text{if} \quad v = \left(\frac{\partial u}{\partial x}, \frac{\partial u}{\partial y} \right) \quad \text{then} \quad \text{curl } v = \frac{\partial}{\partial x}\left(\frac{\partial u}{\partial y} \right) - \frac{\partial}{\partial y}\left(\frac{\partial u}{\partial x} \right) = 0.$$ (9)

The curl of any gradient is automatically zero; $\partial^2 u/\partial y \partial x$ agrees with $\partial^2 u/\partial x \partial y$. If v comes from a potential, it has no vorticity. The opposite is also true: if curl $v = 0$ and the region S is simply connected, then v is the gradient of a potential. We can say it more briefly: in a region without holes, *irrotational flow* and *potential flow* are the same thing.

This completes the first step in our framework. The relation between velocity v and potential u identifies A as the gradient:

$$A = \text{grad} = \begin{bmatrix} \partial/\partial x \\ \partial/\partial y \end{bmatrix} \quad \text{and} \quad v = Au = \begin{bmatrix} \partial u/\partial x \\ \partial u/\partial y \end{bmatrix}.$$

The letter v has replaced e for fluid flow, and b is zero. The next step is to find C.

The analogue of Ohm's law is especially simple. It connects the velocity to the **flow rate**: $w = Cv$. In the case of fluids C measures the *density*.† The velocity is multiplied by the density to find the flow rate w (which has a direction, since w is a vector like v). For most liquids and for many flows in gases, the density is virtually constant in space and in time. There is no significant expansion or compression of the fluid, and C is just a multiple of the identity. We can often suppose that the density equals 1, by choosing the right units, and then $w = v = \text{grad } u$.

Now comes the third and last step, to find A^T. It can be taken either mathematically or physically, by formally "transposing" the gradient A or by establishing the continuous form of Kirchhoff's current law. As a mathematician I have to admit that the second way is better. We are considering the continuity equation $A^T y = 0$, which guarantees **conservation of mass**: fluid is not created or destroyed. Like the voltage law, there is more than one way to express this condition of continuity:

> (1′)   The net flow through any closed curve must be zero
> (2′)   The **divergence** of w at any point must be zero:
>
> $$\text{div } w = \frac{\partial w_1}{\partial x} + \frac{\partial w_2}{\partial y} = 0. \tag{10}$$

Roughly speaking, the divergence measures the rate of flow leaving the point. When the density is constant, in the incompressible case $w = cv$, (10) means that div $v = 0$ and the velocity is also "divergence-free."

Condition 1′ is global—it applies to all regions bounded by curves. In contrast, 2′ can be tested at every point. The link between them is our second application of vector calculus. It is contained in the **divergence theorem**:

$$\iint_S \text{div } w \, dx \, dy = \int_C w \cdot n \, ds. \tag{11}$$

---

† The usual notation for density is ρ. In this section we stick to c, so it can also represent conductivity of heat or of current.

The unit vector n points out of the region, perpendicular to the curve C. Therefore $w \cdot n$ (the same as $w^T n = w_1 n_1 + w_2 n_2$) is the rate of flow that goes *through* C. The length s measures the arc it passes through, so integrating $w \cdot n\, ds$ gives the total outward flow—the flow out minus the flow in. The divergence theorem matches this outward flow to the sources and sinks inside S.

Suppose the divergence is zero; there are no sources or sinks. By equation (11), the net flow through C is zero. Thus condition 1' holds if 2' holds. The two conditions are actually equivalent, because if the right side of (11) is zero around every curve then div $w = 0$ on the left side. The global statement of no net flow and the local statement of no divergence both express conservation of mass.

This completes the vector calculus that turns Kirchhoff's laws into differential equations, and we stop to summarize the result.

3E   The continuous form of the **voltage law** is the absence of vorticity: curl $v = 0$ for irrotational flow. Then the velocity is the gradient of a potential $u(x,y)$:

$$v = \text{grad } u = \begin{bmatrix} \partial u/\partial x \\ \partial u/\partial y \end{bmatrix}. \tag{12}$$

The continuous analogue of the **current law** is conservation of mass, with no sources or sinks. Then $w = cv$ has zero divergence:

$$\text{div } w = \frac{\partial w_1}{\partial x} + \frac{\partial w_2}{\partial y} = 0. \tag{13}$$

When the density c is constant these laws combine to give **Laplace's equation**:

$$\text{div grad } u = \frac{\partial}{\partial x}\left(\frac{\partial u}{\partial x}\right) + \frac{\partial}{\partial y}\left(\frac{\partial u}{\partial y}\right) = 0. \tag{14}$$

Laplace's equation is the equation of equilibrium. A good part of the book is devoted to solving it, so we pause to emphasize again where it comes from: When $c = 1$ we have

$$
\begin{array}{ll}
\text{KVL:} & (v_1, v_2) = (u_x, u_y) \\
\text{KCL:} & \partial v_1/\partial x + \partial v_2/\partial y = 0
\end{array}
\quad \Rightarrow u_{xx} + u_{yy} = 0.
$$

This is the equation $A^T A u = 0$, or div grad $u = 0$, or $\nabla^2 u = 0$, a special but absolutely essential example of $A^T C A u = f$.

The gradient takes the place of the matrix A—it gives derivatives instead of differences. The divergence takes the place of A^T. We see below that the transpose of $A = $ grad is $A^T = -$div, just as the transpose of $\partial/\partial x$ was $-\partial/\partial x$. The minus sign is needed to make $(Au)^Tw = u(A^Tw)$, and it means that

$$A^TA = [-\partial/\partial x \quad -\partial/\partial y] \begin{bmatrix} \partial/\partial x \\ \partial/\partial y \end{bmatrix} = -\frac{\partial^2}{\partial x^2} - \frac{\partial^2}{\partial y^2}.$$

This is symmetric and positive semidefinite. It is made positive definite by a boundary condition like $u = 0$, just as the discrete $A_0^TA_0$ was made positive definite by grounding a node.

EXAMPLE 1 In the hyperbolas of Fig. 3.7 the potential was $u = xy$ and the velocity had components $v_1 = \partial u/\partial x = y$ and $v_2 = \partial u/\partial y = x$. Since it comes from a potential, this velocity must be curl-free: $\partial v_1/\partial y - \partial v_2/\partial x = 0$. Now we check that it is divergence-free:

$$\text{div } v = \frac{\partial v_1}{\partial x} + \frac{\partial v_2}{\partial y} = 0 + 0 = 0.$$

Thus we have not only curl grad $u = 0$, which is automatically true, but also div grad $u = 0$, which assures conservation of mass. In other words, this example satisfies Laplace's equation:

$$\text{div grad } u = 0 \quad \text{or} \quad \frac{\partial^2 u}{\partial x^2} + \frac{\partial^2 u}{\partial y^2} = 0 \quad \text{if} \quad u = xy.$$

Fluid can travel with constant density, in steady flow, along those hyperbolas.

EXAMPLE 2 Laplace's equation is satisfied by $u = xy$ but it is not satisfied by all potentials. The voltage law could be satisfied without the current law. For the potential $u = x^2 + y^2$, the velocity vector is $v = (\partial u/\partial x, \partial u/\partial y) = (2x, 2y)$; the flow goes out radially from the origin. It is irrotational because it comes from a potential. However it is not divergence-free:

$$\text{div } v = \frac{\partial}{\partial x}(2x) + \frac{\partial}{\partial y}(2y) = 4 \quad \text{or} \quad \frac{\partial^2 u}{\partial x^2} + \frac{\partial^2 u}{\partial y^2} = 4.$$

We have to create fluid to maintain this flow. Note that fluid has to be created everywhere (at a uniform rate 4) and not just at the origin. It is interesting to apply the divergence theorem in this case, when C is a circle of radius R:

$$\iint_S \text{div } v \, dx \, dy = \int_C v \cdot n \, ds \quad \text{or} \quad \iint 4 \, dx \, dy = \int 2R \, ds.$$

Both sides equal $4\pi R^2$.

There is one sign convention still to be decided, whether $f > 0$ means creation or destruction of fluid. We show below that the latter choice is natural; f should measure the strength of the *sink*. The continuity equation will be $A^T w = f$, or

$$-\text{div } w = f. \tag{15}$$

The example with $u = x^2 + y^2$ has a source rather than a sink, and f is uniformly negative (it equals $-4/c$).

Boundary Conditions and Green's Formula

We go back to an important point. After discovering that $Au = \text{grad } u$, it was mentioned that A^T could be found either mathematically or physically. By requiring conservation of mass, which is the analogue of Kirchhoff's current law, $A^T w$ was identified as minus the divergence of w. To make that connection complete, with boundary conditions accounted for, we turn to the mathematical approach. Formally it is straightforward:

$$\text{if} \quad A = \begin{bmatrix} A_1 \\ A_2 \end{bmatrix} = \begin{bmatrix} \dfrac{\partial}{\partial x} \\ \dfrac{\partial}{\partial y} \end{bmatrix} \quad \text{then} \quad A^T = [A_1^T \ A_2^T] = \begin{bmatrix} -\dfrac{\partial}{\partial x} & -\dfrac{\partial}{\partial y} \end{bmatrix}. \tag{16}$$

The minus signs came in Section 3.1 from integration by parts; here it will be the same. The goal is to achieve $(Au, w) = (u, A^T w)$, or $\iint w \cdot \text{grad } u \, dx \, dy = \iint u(A^T w) \, dx \, dy$. The inner product of two functions is again an integral, now over a region in the $x - y$ plane. Integration by parts will determine the right form for $A^T w$. In one dimension it was $\int w \, \partial u/\partial x = -\int u \, \partial w/\partial x$, and now it is *Green's formula*:

$$\iint_S \left(w_1 \frac{\partial u}{\partial x} + w_2 \frac{\partial u}{\partial y} \right) dx \, dy = \iint_S u \left(-\frac{\partial w_1}{\partial x} - \frac{\partial w_2}{\partial y} \right) dx \, dy + \int_C u \, w \cdot n \, ds. \tag{17}$$

This is our third application of vector calculus. The left side is the inner product of w with Au. Whatever multiplies u on the right side has to be $A^T w$. It is the *divergence* of w, with the anticipated minus sign:

$$\boxed{A^T w = -\frac{\partial w_1}{\partial x} - \frac{\partial w_2}{\partial y} = -\text{div } w = -\nabla \cdot w.} \tag{18}$$

That completes the theory leading to Laplace's equation, except for the boundary integral that is still waiting on the right side of (17).

To go with $\operatorname{grad}^T = -\operatorname{div}$, there are boundary conditions. When they are homogeneous (zero) they must guarantee that the boundary term, the integral around C in (17), will be zero. Since the integrand is u times $w \cdot n$, this goal is achieved if *either* $u = 0$ *or* $w \cdot n = 0$ at all points of the boundary. Therefore there are boundary conditions of two types:

(1) The potential u is prescribed.
(2) The outflow $w \cdot n$ is prescribed.

The first type comes with A and the second comes with A^T. One is a "Dirichlet condition" and the other is a "Neumann condition." The Neumann condition corresponds to the network matrix A_0, with no nodes grounded and therefore no condition on u. The Dirichlet condition means that all "boundary nodes" are grounded. In general we can separate the boundary into two parts with no overlap, and require $u = 0$ in one part and $w \cdot n = 0$ in the other.†

More generally, we can prescribe $u = u_0(x,y)$ or $w \cdot n = F(x,y)$; the boundary conditions can be nonzero. We met the same question for electrical networks, when a voltage was applied between two nodes; if $x_4 = 0$ and $x_3 = V$, then columns 3 and 4 are both removed from A_0 to give A. The nonzero value V is shifted to the right side of the equation, where it multiplies column 3 and contributes to b. Thus u_0 is like a voltage source and F is like a current source—applied at the boundary.

There is only one restriction on the boundary condition $w \cdot n = F$. If the outflow is prescribed everywhere (a pure Neumann problem) the divergence theorem gives

$$\iint \operatorname{div} w \, dx \, dy = \int w \cdot n \, ds \quad \text{or} \quad \iint (-f) \, dx \, dy = \int F \, ds. \tag{19}$$

In this case the prescribed flow F through the boundary must balance the sources inside. If there are no sources or sinks, then $f = 0$ and the outflow must satisfy $\int F \, ds = 0$. This is the same restriction we met for discrete current sources, and also for forces applied to an unsupported truss; the system as a whole must be in equilibrium.

With the boundary conditions set, we now require equilibrium at every individual point. That gives the differential equation. In the best case, with constant density $c = 1$ and no sources or sinks, the three separate equations are $v = \operatorname{grad} u$, $w = v$, and $\operatorname{div} w = 0$. They combine to give Laplace's equation

$$\boxed{\operatorname{div} \operatorname{grad} u = \frac{\partial^2 u}{\partial x^2} + \frac{\partial^2 u}{\partial y^2} = 0.}$$

Its solutions are called *harmonic functions*. Of course it is the boundary conditions $u = u_0$ or $w \cdot n = F$ that make something happen, and determine the flow.

---

† It would be too much to specify *both* u and $w \cdot n$, or both components of w. *A second-order equilibrium equation has one boundary condition at every boundary point. A fourth-order equation has two.*

EXAMPLE 1   Certainly $u = 1$ and $u = x$ and $u = y$ and $u = xy$ satisfy Laplace's equation. The second derivatives in the equation produce zeros. Therefore if some combination $c_1 + c_2 x + c_3 y + c_4 xy$ happens to satisfy the boundary conditions, that is the solution we want. If $u_0(x,y) = 4y$ on the boundary, then $u = 4y$ in the whole region. The corresponding velocity is $v = \text{grad } u = (0,4)$, and there is uniform flow in the y-direction.

The same solution appears if the correct values of $w \cdot n$ are prescribed on the boundary. Suppose the region is the unit square and the density is $c = 1$ (so that $v = w$). On top of the square the normal vector $n = (0,1)$ points upwards, and to match $w = (0,4)$ we require $w \cdot n = 4$. On the bottom n is $(0, -1)$, pointing out of the square, and $w \cdot n = -4$. These numbers agree with $\partial u / \partial n$, which is $\pm \partial u / \partial y$ on those two sides. On the vertical sides n is $(1,0)$ or $(-1,0)$, in the x-direction, and $w \cdot n = 0$. The integral of $w \cdot n$ around the square is $\int F \, ds = 4 + 0 - 4 + 0 = 0$, as derived in (19) from the divergence theorem. There are no sources, and flow out equals flow in. The solution to Laplace's equation with these boundary conditions on $\partial u / \partial n$ is again $u = 4y$—or more generally it is $u = 4y + C$, since the potential can be raised or lowered by a fixed constant without affecting the flow.

We cannot independently prescribe u and $\partial u / \partial n$ on the boundary. Either one is sufficient to determine the flow.

EXAMPLE 2   The probability of applying Example 1 is extremely small. Most boundary values $u_0(x,y)$ cannot be matched by a combination of those four simple solutions $1, x, y, xy$. Therefore we look for more polynomials that satisfy Laplace's equation, and there is a neat trick that finds them all. (We could check first that x^2 fails and y^2 fails, but $u = x^2 - y^2$ succeeds: $u_{xx} + u_{yy} = 0$.) The trick is to notice that $u = (x + iy)^n$ always satisfies the equation, because

$$\frac{\partial^2 u}{\partial x^2} = n(n-1)(x + iy)^{n-2} \quad \text{and} \quad \frac{\partial^2 u}{\partial y^2} = i^2 n(n-1)(x + iy)^{n-2}.$$

Since $i^2 = -1$, these have opposite signs and add to zero; we have a solution of Laplace's equation. Unfortunately, $(x + iy)^n$ is not real. But it can give *two* polynomial solutions, by taking its real and imaginary parts. For $n = 1$ this produces $\text{Re}(x + iy) = x$ and $\text{Im}(x + iy) = y$; the solutions $u = x$ and $u = y$ are already known. For $n = 2$ it yields

$$\text{Re}(x + iy)^2 = \text{Re}(x^2 + 2ixy - y^2) = x^2 - y^2 \quad \text{and} \quad \text{Im}(x + iy)^2 = 2xy,$$

again known. For $n = 3$ we find the new possibilities

$$\text{Re}(x + iy)^3 = x^3 - 3xy^2 \quad \text{and} \quad \text{Im}(x + iy)^3 = 3x^2 y - y^3.$$

With an infinite sequence of polynomial solutions we can hope to match (in the next chapter) the boundary values $u_0(x,y)$ or $F(x,y)$.

Poisson's Equation

So far we have studied the best case $c = 1$ and $f = 0$. The fluid was incompressible, there were no sources or sinks, and the result was Laplace's equation $u_{xx} + u_{yy} = 0$. With no effort we can fit that equation into the general framework of this chapter, which speaks for itself in Fig. 3.8.

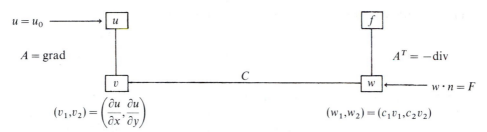

Fig. 3.8. The framework for $-\dfrac{\partial}{\partial x}\left(c_1\dfrac{\partial u}{\partial x}\right) - \dfrac{\partial}{\partial y}\left(c_2\dfrac{\partial u}{\partial y}\right) = f$.

The framework allows the possibility that $c_1 \neq c_2$. The fluid can be anisotropic, distinguishing the x-direction from the y-direction. In addition, the coefficients c can vary with x and y; such a material is inhomogeneous. It is even possible to have different materials in different regions, with a jump in $c(x,y)$ along the line between them. In Section 3.1 w was continuous across such an interface. Here w is a vector, and the same reasoning shows that $w \cdot n$—the component *perpendicular to the interface*—is continuous. We may have different flows parallel to the interface, but across it the flow which leaves one region must enter the other.

If none of these things happen, and the material is homogeneous and isotropic with $c = 1$, there is still the possibility of a source term f. In that case we have *Poisson's equation*

$$-u_{xx} - u_{yy} = f(x,y). \tag{20}$$

Either u or $w \cdot n$ is still prescribed at every point of the boundary, and we want to give examples other than fluid flow.

1. Temperature distribution. Suppose there is a uniform heat source of strength $f = 2$ under a square plate. The temperature is fixed at $u_0 = 0$ on the top and bottom edges of the square, and the side edges are insulated: $w \cdot n = 0$ and no heat can flow out. The problem is to find the steady state distribution $u(x,y)$ of temperature in the square.

2. Deflection of a membrane. Suppose a square membrane is subject to a uniform load $f = 2$. The deflection is fixed at $u_0 = 0$ on the top and bottom edges, and the side edges are free of force: $w \cdot n = 0$. The problem is to find the deflection $u(x,y)$ throughout the square.

For heat flow c is the conductivity; in the membrane problem c is the tension. In both cases a constant value $c = 1$ leads to Poisson's equation

$$-\frac{\partial^2 u}{\partial x^2} - \frac{\partial^2 u}{\partial y^2} = 2 \quad \text{in the square} \quad -1 \le x,y \le 1. \tag{21}$$

On the boundary there is a mixture of Dirichlet and Neumann conditions, or static and kinematic conditions, or displacement and force conditions:

$$u = 0 \quad \text{on} \quad y = \pm 1 \quad \text{and} \quad w \cdot n = \frac{\partial u}{\partial n} = 0 \quad \text{on} \quad x = \pm 1. \tag{22}$$

In mathematical terms, this is an "elliptic boundary-value problem."

The solution is $u = 1 - y^2$. This function satisfies the differential equation, it vanishes for $y = \pm 1$, and it is independent of x. On the left and right sides

$$w = c \operatorname{grad} u = \begin{bmatrix} 0 \\ -2y \end{bmatrix} \quad \text{and} \quad n = \begin{bmatrix} \pm 1 \\ 0 \end{bmatrix} \quad \text{and} \quad w \cdot n = 0.$$

The membrane is shaped like a parabola in the y-direction, weighed down by the load. In the x-direction it is completely flat.

Minimum Principles

The minimum principle that gives Laplace's equation is one of the most famous in mathematics. It comes directly from the quadratic $P = \frac{1}{2}(Au)^T C(Au)$, in which A is the gradient and C is the identity:

$$P(u) = \frac{1}{2} \iint |\operatorname{grad} u|^2 \, dx \, dy = \frac{1}{2} \iint \left[\left(\frac{\partial u}{\partial x} \right)^2 + \left(\frac{\partial u}{\partial y} \right)^2 \right] dx \, dy. \tag{23}$$

The function that minimizes P while satisfying the boundary condition $u = u_0(x,y)$ is the solution to the **Dirichlet problem**—the combination of Laplace's equation in the interior and $u = u_0$ on the boundary. It is the continuous analogue of a square mesh of resistors.† The discrete problem on such a network is the basic finite difference approximation to $u_{xx} + u_{yy} = 0$.

There is a new feature of the dual. As always it minimizes $Q = \frac{1}{2}w^T C^{-1} w - w^T b$ subject to $A^T w = f$, and in our case f is zero; we have Laplace's equation and not Poisson's. The question is, where do the boundary values u_0 enter this problem?

---

† In mechanical terms, the continuous membrane is analogous to a network of springs, as on old bunk beds. Section 3.6 verifies that Laplace's equation appears when the calculus of variations is applied to $P(u)$.

They are decisive in determining u, and they must affect w. Apparently this happens through b, although we have been imagining that b is zero. It is like a network with one voltage fixed at $x_3 = V$; the potential x_3 disappears from the list of unknowns, but the voltage source reappears in b. In the present case, for differential equations, the quadratic is

$$Q(w) = \tfrac{1}{2}w^T C^{-1} w - w^T b = \tfrac{1}{2} \int\!\!\int (w_1^2 + w_2^2)\, dx\, dy - \int_C u_0 w \cdot n\, ds. \qquad (24)$$

The boundary value u_0 enters a boundary integral in Q. Then the minimum subject to $A^T w = f$ (which is div $w = 0$) is achieved by $w = Cv = CAu = \text{grad } u$. The minimization of P gives the potential, and the minimization of Q gives the flow.

If we had started with Q, u would become the *Lagrange multiplier* for the continuity constraint div $w = 0$. For differential equations, the Lagrange multiplier is no longer a number or a vector. It becomes a function $u(x,y)$, because there is a constraint at every point. In the opposite case, when the outflow $w \cdot n = F$ is prescribed instead of the potential u_0, the boundary term disappears from Q and turns up in P:

$$P(u) = \tfrac{1}{2} \int\!\!\int \left[\left(\frac{\partial u}{\partial x} \right)^2 + \left(\frac{\partial u}{\partial y} \right)^2 \right] dx\, dy - \int u\, F\, ds. \qquad (25)$$

You see the give and take between P and Q. When $u = u_0$ is an *essential* boundary condition, there is no condition on w. Instead, w feels the effect of u_0 in a *natural* way, by minimizing Q in (24). When $w \cdot n = F$ is an essential condition imposed on w, there is no condition on u; the minimizing u in (25) naturally satisfies $F = \text{grad } u \cdot n$ at the boundary. And if one part of the boundary has $u = u_0$, the other part needs $w \cdot n = F$. We close with two remarks:

1. Every solution to Laplace's equation satisfies a *maximum principle*. The largest and smallest values of u, in any region, are attained on the boundary and not in the interior (unless $u = $ constant). This is the continuous analogue of the fact that $(A^T A)^{-1}$ had only positive entries when A was an incidence matrix. In the discrete case each value of u is an average of neighboring values. In the continuous case (Section 4.5) $u(x,y)$ is the average of u over every circle centered at x,y.

2. For *computer vision* a crucial problem is to recognize edges. In one dimension, the intensity $u(x)$ is discontinuous at a sharp edge. In reality the corners in u are rounded off, and the edge is recognized by an inflection point $u'' = 0$ at the middle of the jump. In two dimensions the Laplacian enters; we look for curves along which $u_{xx} + u_{yy} = 0$. We are not solving Laplace's equation, but testing with a sensor to find an edge.

Without smoothing most pictures have edges everywhere. Therefore the first step is a convolution with the normal probability distribution, to reduce the noise:

$$u \rightarrow \int\!\!\int u(x - s,\, y - t) e^{-(x^2 + y^2)/2\sigma^2}\, dx\, dy.$$

That removes irrelevant edges. A large σ leaves only the big picture; a small σ allows a closer look. Pattern recognition is an ***inverse problem***—to recover the coloring book from the finished picture—like recovering the coefficients of a differential equation from its solutions.

EXERCISES

3.3.1 (a) Show that $u = x^3 - 3xy^2$ satisfies Laplace's equation.
(b) Do the same for $s = 4x^3y - 4xy^3$, and explain where this comes in the list of polynomial solutions.
(c) Substitute $x = \cos\theta$ and $y = \sin\theta$ in s and simplify to an expression involving 4θ.

3.3.2 Verify that $u = e^x \cos y$ and $s = e^x \sin y$ both satisfy Laplace's equation, and sketch the equipotentials $u = $ constant and the streamlines $s = $ constant.

3.3.3 *Discrete divergence theorem*: Why is the flow across the "cut" in the figure equal to the sum of the flows from the individual nodes A,B,C,D? *Note*: This is true even if flows like $d_1 - d_6$ from nodes like A are nonzero. If the current law holds and each node has zero net flow, then the exercise says that the flow across every cut is zero.

3.3.4 *Discrete Stokes theorem*: Why is the voltage drop around the large triangle equal to the sum of the drops around the small triangles? *Note*: This is true even if voltage drops like $d_1 + d_7 + d_6$ around triangles like ABC are nonzero. If the voltage law holds and the drop around each small triangle is zero, then the exercise says that $d_1 + d_2 + d_3 + d_4 + d_5 + d_6 = 0$.

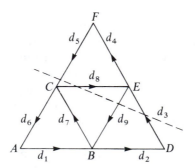

3.3.5 On a graph the analogue of the gradient is the edge-node incidence matrix A_0. The analogue of the curl is the loop-edge matrix R with a row for each independent loop and a column for each edge. Draw a graph with four nodes and six directed edges, write down A_0 and R, and confirm that $RA_0 = 0$ in analogy with curl grad $= 0$.

3.3.6 Why does the flow rate $w = (\partial s/\partial y, -\partial s/\partial x)$ satisfy div $w = 0$ for any "stream function" $s(x,y)$?

3.3.7 If the density is $c = 1$ then

$$w = \begin{bmatrix} \dfrac{\partial s}{\partial y} \\[2mm] -\dfrac{\partial s}{\partial x} \end{bmatrix} \text{ is equal to } v = \begin{bmatrix} \dfrac{\partial u}{\partial x} \\[2mm] \dfrac{\partial u}{\partial y} \end{bmatrix}.$$

Show from these Cauchy-Riemann equations $\partial u/\partial x = \partial s/\partial y$ and $\partial u/\partial y = -\partial s/\partial x$ that both u and s satisfy Laplace's equation.

3.3.8 The curves $u(x,y) = $ constant are orthogonal to the family $s(x,y) = $ constant if grad u is perpendicular to grad s. These gradient vectors are at right angles to the curves, which can be equipotentials and streamlines. Construct a suitable $s(x,y)$ from the geometry and verify

$$(\text{grad } u)^T (\text{grad } s) = \frac{\partial u}{\partial x}\frac{\partial s}{\partial x} + \frac{\partial u}{\partial y}\frac{\partial s}{\partial y} = 0 \text{ if}$$

(a) $u(x,y) = y$: equipotentials are parallel horizontal lines
(b) $u(x,y) = x - y$: equipotentials are parallel 45° lines
(c) $u(x,y) = \log(x^2 + y^2)^{1/2}$: equipotentials are concentric circles.

3.3.9 A differential equation like $dy/dx = f(x,y)$ gives a family of curves depending on the initial value $y(0)$, and $dy/dx = -1/f(x,y)$ gives the orthogonal curves. (The product of the slopes is -1, the usual condition for a right angle; the gradients are in the orthogonal directions $(1,f)$ and $(1,-1/f)$.) Solve $y' = -1/f$ for the second family if the first family is
(a) $y = e^x + $ constant, from $dy/dx = e^x = f$
(b) $y = \frac{1}{2}x^2 + $ constant, from $dy/dx = x = f$
(c) $xy = $ constant, from $dy/dx = -y/x = f$.

3.3.10 In Stokes' law (8), let $v_1 = -y$ and $v_2 = 0$ to show that the area of S equals the line integral $-\int_C y \, dx$. Find the area of an ellipse ($x = a \cos t$, $y = b \sin t$, $x^2/a^2 + y^2/b^2 = 1$, $0 \le t \le 2\pi$).

3.3.11 By computing curl v, show that $v = (y^2, x^2)$ is not the gradient of any function u but that $v = (y^2, 2xy)$ is such a gradient—and find u.

3.3.12 By computing div w, show that $w = (x^2, y^2)$ does not have the form $(\partial s/\partial y, -\partial s/\partial x)$ for any function s. Show that $w = (y^2, x^2)$ does have that form, and find the "stream function" s.

3.3.13 If $u = x^2$ in the square $S = \{-1 < x,y < 1\}$, verify the divergence theorem (11) when $w = $ grad u:

$$\iint_S \text{div grad } u \, dx \, dy = \int_C n \cdot \text{grad } u \, ds.$$

If a different u satisfies Laplace's equation in S, what is the net flow through C?

3.3.14 What potential has the gradient $v = (u_x, u_y) = (2xy, x^2 - y^2)$? Sketch the equipotentials and streamlines for flow into a 30° wedge (Fig. 3.7 was 45°), and show that $v \cdot n = 0$ on the upper boundary $y = x/\sqrt{3}$. The streamlines have $s = xy^2 - \frac{1}{3}x^3 = $ constant.

3.3.15 Solve Poisson's equation $u_{xx} + u_{yy} = 4$ by trial and error if $u = 0$ on the circle $x^2 + y^2 = 1$.

3.3.16 Find a quadratic solution to Laplace's equation if $u = 0$ on the axes $x = 0$ and $y = 0$ and $u = 3$ on the curve $xy = 1$.

3.3.17 Laplace's equation in polar coordinates is

$$\frac{\partial^2 u}{\partial r^2} + \frac{1}{r} \frac{\partial u}{\partial r} + \frac{1}{r^2} \frac{\partial^2 u}{\partial \theta^2} = 0.$$

Show that $u = r \cos \theta + r^{-1} \cos \theta$ is a solution, and express it in terms of x and y. Find $v = (u_x, u_y)$ and verify that $v \cdot n = 0$ on the circle $x^2 + y^2 = 1$. This is the velocity of flow past a circle.

3.3.18 Show that $u = \log r$ satisfies Laplace's equation except at $r = 0$.

3.3.19 Suppose $\delta P / \delta u = \int\int \left[\dfrac{\partial u}{\partial x} \dfrac{\partial v}{\partial x} + \dfrac{\partial u}{\partial y} \dfrac{\partial v}{\partial y} - fv \right] dx \, dy$. Use Green's formula, changing the u in (17) to v and changing w to grad u, to write

$$\frac{\delta P}{\delta u} = \int\int_S v \, [?] \, dx \, dy + \int_C v \, [??] \, ds.$$

If this is zero for all v, find the differential equation and the natural boundary condition satisfied by u.

VECTOR CALCULUS IN THREE DIMENSIONS ■ 3.4

Three dimensions are important too. The velocity vector becomes $v = (v_1, v_2, v_3)$, with three components. The third component is the velocity in the z-direction, and it leads to simple changes in the gradient and divergence:

$$v = \text{grad } u \quad \text{means} \quad v_1 = \frac{\partial u}{\partial x}, v_2 = \frac{\partial u}{\partial y}, v_3 = \frac{\partial u}{\partial z}$$

$$\text{div } v = 0 \quad \text{means} \quad \frac{\partial v_1}{\partial x} + \frac{\partial v_2}{\partial y} + \frac{\partial v_3}{\partial z} = 0$$

$$\text{div grad } u = 0 \quad \text{means} \quad \frac{\partial^2 u}{\partial x^2} + \frac{\partial^2 u}{\partial y^2} + \frac{\partial^2 u}{\partial z^2} = 0$$

Thus Laplace's equation looks for a potential function like $u = xy + z$, for which the sum of all three second derivatives is zero. This particular function corresponds to the velocity field $v = \text{grad } u = (y, x, 1)$. The new component contributes a constant upward velocity $v_3 = 1$, in addition to the components $v_1 = y$ and $v_2 = x$ which carried the flow along the hyperbolas of Fig. 3.7. The fluid now climbs as it travels around those hyperbolas. Its divergence is zero, so that no fluid is created or destroyed and we have steady flow—but it is no longer in a plane.

Whenever there is a **potential function**, the problems of engineering and science are enormously simplified. In our present case it is a velocity potential, and it leads to Laplace's equation. Soon it will be a pressure or a displacement. In discrete problems it was a nodal potential, and equilibrium led to $A^T C A u = 0$—a discrete form of Laplace's equation. (With f on the right side it is a discrete form of Poisson's equation.) In economics the potential becomes the "price," and money flows like water. It is the same in physics and biology and chemistry; if there is a potential then the flow is governed by its derivative or its gradient or its nodal differences. Equilibrium comes from the drive to equalize potentials—in other words to minimize the potential energy.

The natural question is: When does a potential exist? For two-dimensional flows that was decided by the requirement curl $v = 0$. Our goal is to find a similar test in three dimensions. But before that we want to see in one more way the simplification brought by potentials.

Mathematically it is a problem of **line integrals**—the meaning and calculation of

$$W = \int_P^Q F \cdot dr = \int_P^Q F_1 \, dx + F_2 \, dy + F_3 \, dz. \tag{1}$$

Physically it is a problem of **work**. F is the force vector (F_1, F_2, F_3) and r is the position vector (x,y,z). The infinitesimal displacement $dr = (dx,dy,dz)$ is also a vector—the change in position. The work during that change is $F \cdot dr$, the *inner product* of the force vector and the movement vector.

We emphasize that the work can be zero, even when a force is acting. If a ball is tied to a string and going in a circle, the string applies a force to the ball. However the force and the movement are perpendicular. The force is radial, the movement is tangential, and their inner product $F \cdot dr$ (or $F^T dr$) is *zero*. No work is done, and the energy in the ball is not increased.

That is different from hitting a tennis ball. As it hits your racket, the force is opposite to the motion of the ball. The inner product $F \cdot dr$ is negative, and the ball loses kinetic energy fast. Its velocity drops to zero, but the force of the racket is still acting. Therefore the motion dr goes back toward the opponent, $F \cdot dr$ is positive, and the ball picks up speed. To complicate things the ball and racket are elastic, so kinetic and potential energy are both present.

The integral (1) goes along a space curve from P to Q. When that curve is a straight line, dx, dy, and dz are proportional. In general they are not. One way to describe the curve is to give $x(t)$, $y(t)$, $z(t)$—the position at every time t.

EXAMPLE 1 $\qquad\qquad (x, y, z) = (t, 2t, -4t);\ F = (0, 0, -mg)$

This is the straight line case: $dx = dt$, $dy = 2dt$, $dz = -4dt$. The path goes down at a slant, starting from $P = (0,0,0)$ and reaching $Q = (2,4,-8)$ at time $t = 2$. The force is gravity, acting down, and the work it does is

$$W = \int_P^Q F \cdot dr = \int_{t=0}^2 4mg\, dt = 8mg.$$

The fall in height was 8; gravity ignored movement in x and y.

EXAMPLE 2 $\qquad\qquad (x, y, z) = (\cos t, \sin t, t);\ F = (0, x, 0)$

In the $x - y$ plane this is a circle because $x = \cos t$ and $y = \sin t$, but the third coordinate $z = t$ makes the path climb. The curve is a helix, and the work is $\int F \cdot dr = \int x\, dy = \int \cos^2 t\, dt$. It is positive, and furthermore the work depends on the path. A helix with $z = \frac{1}{2}t$ would need two turns to reach the same point (at $t = 4\pi$); the work would be doubled. We see below that this force cannot come from a potential.

There are four main topics—the **tangent** to a curve, the **speed** along it, its **arc length**, and the **potential** (if there is one).

(1) $\quad \dfrac{dr}{dt} = \left(\dfrac{dx}{dt}, \dfrac{dy}{dt}, \dfrac{dz}{dt} \right)$ is a tangent vector along the curve

(2) $\quad \left| \dfrac{dr}{dt} \right| = \sqrt{ \left(\dfrac{dx}{dt} \right)^2 + \left(\dfrac{dy}{dt} \right)^2 + \left(\dfrac{dz}{dt} \right)^2 }$ is the speed

(3) $\displaystyle\int ds = \int \left|\frac{dr}{dt}\right| dt$ is the arc length

Since s is traditional for arc length, speed is not s but $ds/dt = |dr/dt|$. The example has $r = (t, 2t, -4t)$. The tangent is $dr/dt = (1, 2, -4)$. The speed is $\sqrt{1+4+16}$ $= \sqrt{21}$. The arc length is $s = \int \sqrt{21}\, dt = \sqrt{21}\, t$.

An especially nice case is to move on the curve with speed $ds/dt = 1$. Then the arc length agrees exactly with the elapsed time, $s = t$. The example would have to slow down by the factor $\sqrt{21}$, so that the position vector is

$$r = \left(\frac{t}{\sqrt{21}}, \frac{2t}{\sqrt{21}}, \frac{-4t}{\sqrt{21}}\right) \quad\text{and}\quad \frac{dr}{dt} = \left(\frac{1}{\sqrt{21}}, \frac{2}{\sqrt{21}}, \frac{-4}{\sqrt{21}}\right).$$

Now dr/dt is the unit tangent vector, with length 1 because the speed is 1. In that case the length of the second derivative gives the curvature $|d^2r/dt^2|$. Here the second derivative is zero and the curvature is zero, because the path is straight.

Arc length depends only on the shape of the path, not on how fast it is traveled. Using s says *where*; using t also says *when*. If you travel with speed 1, so that $s = t$, then the two coincide. In formulas that involve time, like the position or velocity or acceleration of Halley's comet, t is needed. In formulas that have nothing to do with time, like the length of the path or its curvature or the work $\int F \cdot dr$ done by the sun on the comet, s is more natural. It is intrinsic to the path, where the position $r = (x,y,z)$ as a function of t depends on the traveler.

We come back to the **potential**. Some forces have it; others do not. When there is a potential, $\int F \cdot dr$ is not only independent of speed along the path—it is **independent of the path itself**. The integral from P to Q depends only on those endpoints, not on the curve between them. Every path from one to the other involves the same amount of work. In particular, a closed path that returns to its starting point (so that $P = Q$) involves no work at all. A force associated with such a potential is called *conservative*.

In one dimension that always happens. The value of an integral is decided at its endpoints; that is the fundamental theorem of calculus:

$$\int_P^Q F\, dx = u(Q) - u(P).$$

The potential u is the integral, and the force $F = du/dx$ is its derivative. In more dimensions the existence of an integral like u is not automatic. Suppose the force components are $(F_1, F_2) = (y, 0)$. The force is in the x-direction but its strength is y, so $F \cdot dr$ is $y\, dx$. If we take two paths $(x,y) = (t, t^2)$ and $(x,y) = (t, t^3)$, both starting at $(0,0)$ and ending at $(1,1)$, the work along those paths is different:

$$W_1 = \int y\, dx = \int t^2 dt \quad\text{and}\quad W_2 = \int y\, dx = \int t^3\, dt.$$

The force $F = (y,0)$ is not conservative, and to go along one of those paths and back along the other requires work.

When is there a potential? It requires that every component of F comes from the same u. In three dimensions that means

$$F_1 = \frac{\partial u}{\partial x}, \quad F_2 = \frac{\partial u}{\partial y}, \quad F_3 = \frac{\partial u}{\partial z}. \tag{2}$$

Then F is the gradient of u, and the work is

$$\boxed{\int_P^Q F_1 \, dx + F_2 \, dy + F_3 \, dz = \int_P^Q du = u(Q) - u(P).} \tag{3}$$

In other words, $F \cdot dr$ has to be the exact differential of u. It is always possible to start from u and go backwards; $u = -mgz$ is the potential for $F = (0,0,-mg)$, since $\partial u/\partial x = 0$ and $\partial u/\partial y = 0$. That is the force of gravity. It comes from a potential, it is conservative, and in our example from $(0,0,0)$ to $(2,4,-8)$ it did not matter that the path was a straight line. That simplified the integration along the path, but we now have a better way: Find the difference in potential at the endpoints.

For the force $F = (y,0)$ this method failed; F is not a gradient. The two paths above gave different values for the work. More than that, you can see directly that there is no potential:

$$y = \frac{\partial u}{\partial x} \quad \text{and} \quad 0 = \frac{\partial u}{\partial y} \quad \text{have no solution } u.$$

The latter requires u to be independent of y and then the former is impossible. That argument is easy, but for forces more complicated than $(y,0)$ a new idea is needed. The requirements $F_1 = \partial u/\partial x$, $F_2 = \partial u/\partial y$, $F_3 = \partial u/\partial z$ give three equations in one unknown; for most forces they cannot be solved. Fortunately, it is easy to discover the test that F must pass:

$$\text{if} \quad F_1 = \frac{\partial u}{\partial x} \quad \text{and} \quad F_2 = \frac{\partial u}{\partial y} \quad \text{then} \quad \frac{\partial F_1}{\partial y} = \frac{\partial F_2}{\partial x}. \tag{4}$$

The cross-derivatives $\partial^2 u/\partial x \partial y$ and $\partial^2 u/\partial y \partial x$ must agree. Similarly

$$\frac{\partial F_1}{\partial z} = \frac{\partial F_3}{\partial x} \quad \text{and} \quad \frac{\partial F_2}{\partial z} = \frac{\partial F_3}{\partial y}. \tag{5}$$

In two dimensions, the test (4) is curl $F = 0$. It is the continuous analogue of the voltage law, leading to a potential. In three dimensions the curl comes into its own!

The tests (4) and (5) agree precisely with curl $F = 0$, and we concentrate on the three-dimensional curl after applying these tests to several examples.

(1) $F = (y,0)$ $\dfrac{\partial F_1}{\partial y} = 1$ and $\dfrac{\partial F_2}{\partial x} = 0$ (no potential)

(2) $F = (y,x)$ $\dfrac{\partial F_1}{\partial y} = 1$ and $\dfrac{\partial F_2}{\partial x} = 1$ (potential $u = xy$)

(3) $F = (y^2,x^2)$ $\dfrac{\partial F_1}{\partial y} = 2y$ and $\dfrac{\partial F_2}{\partial x} = 2x$ (no potential)

(4) $F = (x,y)$ $\dfrac{\partial F_1}{\partial y} = 0$ and $\dfrac{\partial F_2}{\partial x} = 0$ (potential $u = \frac{1}{2}x^2 + \frac{1}{2}y^2$)

(5) $F = (yz,xz,xy)$ curl $F = 0$ (potential $u = xyz$)

Gradient, Divergence, and Curl

The gradient and divergence are pretty straightforward. The gradient is the natural extension of the derivative. When a function $u(x,y,z)$ depends on several variables, and those variables move a little, the value of u will move:

$$du = \frac{\partial u}{\partial x}\, dx + \frac{\partial u}{\partial y}\, dy + \frac{\partial u}{\partial z}\, dz$$

$$= (\text{grad } u) \cdot (dr). \tag{6}$$

The vector grad u contains the three partial derivatives, when u depends on three variables. From (6) you can see what is coming: If the force is $F = \text{grad } u$, then the work is $\int F \cdot dr = \int du = $ change in u. Work equals change in potential, and it does not depend on the path—*provided there is a potential.*

The divergence goes the opposite way, from a vector to a scalar. The divergence of (v_1,v_2,v_3) is the sum $\partial v_1/\partial x + \partial v_2/\partial y + \partial v_3/\partial z$. In applications that vector is not a force F but a flow rate w. We are concerned not with the voltage law but with the current law. What enters is not the component of F along the path, but the component of w *perpendicular* to the path. In three dimensions it is the component $w \cdot n$ perpendicular to a surface. Integrating that component gives the net flow outward—across the path or through the surface.

As always, there is special interest in closed paths and closed surfaces, and in vectors w for which the net flow is zero. There will be a test for that, but the test is not to ask whether w is a gradient. Instead we ask whether w is a curl.

That brings us to the third operation, and probably the one that is hardest to visualize. The curl takes a vector to a vector. It is like a matrix, but it is a matrix of derivatives:

$$\text{curl } v = \begin{bmatrix} 0 & -\dfrac{\partial}{\partial z} & \dfrac{\partial}{\partial y} \\[2mm] \dfrac{\partial}{\partial z} & 0 & -\dfrac{\partial}{\partial x} \\[2mm] -\dfrac{\partial}{\partial y} & \dfrac{\partial}{\partial x} & 0 \end{bmatrix} \begin{bmatrix} v_1 \\[2mm] v_2 \\[2mm] v_3 \end{bmatrix}. \tag{7}$$

This gives the **rotation** of the vector field v.

EXAMPLE 1   Suppose v is a pure rotation. At the point $r = (x, y, z)$, the vector is $v = (-y, x, 0)$. It is aimed perpendicular to r ($v^T r = 0$) and if you follow v it carries you in a circle. The z-coordinate will not change since $v_3 = 0$. The curl of v is

$$\begin{bmatrix} 0 & -\dfrac{\partial}{\partial z} & \dfrac{\partial}{\partial y} \\[2mm] \dfrac{\partial}{\partial z} & 0 & -\dfrac{\partial}{\partial x} \\[2mm] -\dfrac{\partial}{\partial y} & \dfrac{\partial}{\partial x} & 0 \end{bmatrix} \begin{bmatrix} -y \\[2mm] x \\[2mm] 0 \end{bmatrix} = \begin{bmatrix} 0 \\[2mm] 0 \\[2mm] 2 \end{bmatrix}. \tag{8}$$

The curl points straight up, along the z-axis and perpendicular to the planes of rotation. The factor 2 is unfortunate but unavoidable.

EXAMPLE 2   If v is rotation around another axis, then curl v should point along that axis. Suppose the axis is in the direction $(\omega_1, \omega_2, \omega_3)$ instead of $(0,0,1)$. How do we describe the rotation? At every point $r = (x, y, z)$, the vector v should be perpendicular to r and also perpendicular to ω. The way to produce such a vector is to take the **cross product** of r and ω:

$$v = \omega \times r = \begin{bmatrix} 0 & -\omega_3 & \omega_2 \\[2mm] \omega_3 & 0 & -\omega_1 \\[2mm] -\omega_2 & \omega_1 & 0 \end{bmatrix} \begin{bmatrix} x \\[2mm] y \\[2mm] z \end{bmatrix} = \begin{bmatrix} \omega_2 z - \omega_3 y \\[2mm] \omega_3 x - \omega_1 z \\[2mm] \omega_1 y - \omega_2 x \end{bmatrix}. \tag{9}$$

That is a pure rotation, whose curl does point along ω.

Remark   There is a more compact and frequently used form of the cross product and curl. They are "determinants" of strange-looking 3 by 3 matrices:

$$u \times v = \begin{vmatrix} i & j & k \\ u_1 & u_2 & u_3 \\ v_1 & v_2 & v_3 \end{vmatrix} \quad \text{and} \quad \text{curl } v = \begin{vmatrix} i & j & k \\ \partial/\partial x & \partial/\partial y & \partial/\partial z \\ v_1 & v_2 & v_3 \end{vmatrix}. \tag{10}$$

The first gives the cross product of two vectors (i, j, k are unit vectors in the coordinate directions). Taking determinants in the usual way, the cross product is

$$u \times v = i(u_2 v_3 - u_3 v_2) + j(u_3 v_1 - u_1 v_3) + k(u_1 v_2 - u_2 v_1).$$

It is perpendicular to u. When the components in parentheses are multiplied by u_1, u_2, u_3 and added, the result is zero. Similarly it is perpendicular to v. The cross product $v \times u$ is the negative of $u \times v$, just as determinants change sign when two rows are reversed. Geometrically the cross product gives the third direction in a "right-handed" orientation of the axes u, v, $u \times v$.

The curl is a cross product with the operator "del":

$$\text{curl } v = \nabla \times v = i \left(\frac{\partial v_3}{\partial y} - \frac{\partial v_2}{\partial z} \right) + j \left(\frac{\partial v_1}{\partial z} - \frac{\partial v_3}{\partial x} \right) + k \left(\frac{\partial v_2}{\partial x} - \frac{\partial v_1}{\partial y} \right). \tag{11}$$

That agrees with the matrix product in (7). We see in these three components exactly the expressions that had to be zero in order for a force to come from a potential. That is our main point.

3F   For any function $u(x,y,z)$ the curl of the gradient is zero:

$$\text{curl grad } u = 0. \tag{12}$$

A force is the gradient of a potential, $F = \text{grad } u$, when curl $F = 0$. A velocity is the gradient of a velocity potential, $v = \text{grad } u$, when curl $v = 0$. In a simply-connected region the velocity is irrotational and the force is conservative:

$$\int v \cdot dr = 0 \quad \text{and} \quad \int F \cdot dr = 0 \text{ around closed paths.} \tag{13}$$

The identity (12) is simply an identity of cross derivatives:

$$\text{curl grad } u = \left(\frac{\partial}{\partial y} \frac{\partial u}{\partial z} - \frac{\partial}{\partial z} \frac{\partial u}{\partial y}, \frac{\partial}{\partial z} \frac{\partial u}{\partial x} - \frac{\partial}{\partial x} \frac{\partial u}{\partial z}, \frac{\partial}{\partial x} \frac{\partial u}{\partial y} - \frac{\partial}{\partial y} \frac{\partial u}{\partial x} \right) = 0. \tag{14}$$

Each pair like $\partial^2 u/\partial y \partial z$ and $\partial^2 u/\partial z \partial y$ is identical. The construction of u from F asks us to solve three equations $F_1 = \partial u/\partial x$, $F_2 = \partial u/\partial y$, $F_3 = \partial u/\partial z$—but the condition curl $F = 0$ makes it possible. A pure rotation $v = \omega \times r$ is not a gradient, because curl $v \neq 0$.

EXAMPLE   The normalized gravitational force decreases with $r^2 = x^2 + y^2 + z^2$:

$$F = \left(\frac{x}{r^3}, \frac{y}{r^3}, \frac{z}{r^3} \right), \text{ curl } F = 0, \text{ and } u = -\frac{1}{r}.$$

We have to mention one more point. There is a second identity, in addition to curl grad $u = 0$. It enters on the other side of our framework, where the flow obeys the current law: div $w = 0$. There we want vector fields that are *divergence-free* instead of curl-free, and fortunately they are easy to find. Let $S = (S_1, S_2, S_3)$ be any vector; its three components are functions of x, y, z. Then if w is chosen to be curl S, copying (11), the divergence of w will be zero:

$$\text{div } w = \frac{\partial}{\partial x}\left(\frac{\partial S_3}{\partial y} - \frac{\partial S_2}{\partial z} \right) + \frac{\partial}{\partial y}\left(\frac{\partial S_1}{\partial z} - \frac{\partial S_3}{\partial x} \right) + \frac{\partial}{\partial z}\left(\frac{\partial S_2}{\partial x} - \frac{\partial S_1}{\partial y} \right) = 0. \quad (15)$$

The equality of mixed derivatives now gives div curl $S = 0$.

3G   For any vector $S(x,y,z)$ the divergence of the curl is zero:

$$\text{div curl } S = 0. \quad (16)$$

A flow rate is the curl of a *vector potential*, $w = \text{curl } S$, when div $w = 0$. Such a flow conserves mass. For a closed path around a two-dimensional region or a closed surface around a three-dimensional region, div $w = 0$ yields the conservation law

$$\int w \cdot n \, ds = 0 \quad \text{or} \quad \iint w \cdot n \, dA = 0. \quad (17)$$

The boundary integral of $w \cdot n$ equals the interior integral of div w—that is the divergence theorem—and if div $w = 0$ then mass is conserved.

The simplest and most important case is in two dimensions (a plane flow). There S is $(0,0,s)$, w is $(w_1, w_2, 0)$, and

$$
\begin{array}{ll}
\text{div } w = 0 \quad \text{means} & \dfrac{\partial w_1}{\partial x} + \dfrac{\partial w_2}{\partial y} = 0 \\[3mm]
w = \text{curl } S \quad \text{means} & w_1 = \dfrac{\partial s}{\partial y}, \ w_2 = -\dfrac{\partial s}{\partial x}.
\end{array} \quad (18)
$$

If w has one of those properties, it also has the other. Every vector $(\partial s/\partial y, -\partial s/\partial x)$ has zero divergence, and every vector with zero divergence comes from a *stream function* s. The curves $s(x,y) = $ constant are the *streamlines* along which the fluid travels.

EXAMPLE 1 $w = (x, -y, 0)$, div $w = 0$, $w = $ curl S, $S = (0, 0, xy)$

EXAMPLE 2 $w = (x, -2y, z)$, div $w = 0$, $w = $ curl S, $S = (-yz, 0, xy)$.

You might guess that the two identities curl grad $= 0$ and div curl $= 0$ are not totally independent. In fact, I have finally realized that *they are transposes of one another.*† The curl is its own transpose, and if we call it R then the identities curl grad $= 0$ and div curl $= 0$ are exactly

$$RA = 0 \quad \text{and} \quad A^T R^T = 0. \tag{19}$$

A is still the gradient, $-A^T$ is still the divergence, and $A^T A u = 0$ is still Laplace's equation—in any number of dimensions.

We will meet the stream function again for real fluids, which have viscosity. And we had better not close without emphasizing that div grad $u = 0$ *is not an identity!* It is Laplace's equation, and it has to be solved.

Electricity and Magnetism

The neatest application of these vector identities comes for *Maxwell's equations*, which describe electric fields and magnetic fields exactly in the language of $A^T C A$. At least that framework is correct in the static case, where the two fields can be studied separately. We start there.

The electric field comes from a scalar potential, $E = -$grad u. Thus $A = -$grad, and then C is multiplication by the dielectric constant ε: $D = \varepsilon E$. The third equation is Gauss' law div $D = \rho = $ charge density. Altogether we have a three-dimensional Poisson equation $A^T C A u = $ div $(-\varepsilon$ grad $u) = \rho$. With constant ε this is Laplace's equation in the absence of charge.

The magnetic induction B is different from E. It is *divergence-free* instead of curl-free. Magnetic fields have rotation but no divergence, because magnetic charges do not exist (at least for Maxwell). B can still come from a potential, but now it is a *vector potential*. We depend on the identity div curl $S = 0$, and take $B = $ curl S. Every divergence-free B is the curl of a vector potential S.‡ Then at the next step, the material constant is the permeability μ. Its reciprocal appears in the step $H = \mu^{-1} B$ (just as conductance is the reciprocal of resistance). The final equation involving A^T is Ampere's law curl $H = j = $ current density. In units other than meter-kilogram-second-coulomb, the speed of light will appear in these equations.

These two systems, electric and magnetic, appear side by side in Fig. 3.9. Supplemented by boundary conditions they complete the static case. However two

---

† Which gives a way of remembering these damn identities.

‡ The vector potential is usually denoted by A, but that letter already appears in $A^T C A$. For magnetostatics $A = $ curl and also $A^T = $ curl. I apologize to physicists for the temporary innovations u and S.

other terms also appear, hidden in boxes but all too visible. They are the corrections necessary for *electrodynamics*, when the fields are changing with time. In the static case we had curl $E = 0$, so that E was the gradient of a potential. In the dynamic case that is false. The potentials are time-varying and *the electric and magnetic fields are coupled*. In a conductor, the static picture allows no current; all charges rearrange themselves to make the field zero. In the dynamic picture this starts to happen with the speed of light, but still it takes time. There are currents in the conductor. The system is set into motion, it never actually reaches equilibrium, and we will see that it is governed by the wave equation rather than Laplace's equation.

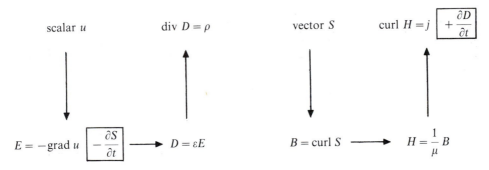

Fig. 3.9. The framework for electrodynamics: Maxwell's equations.

One point about this framework. The author hoped, and he hopes the reader will hope, that the two side-by-side systems fit into one. The time derivative has to join with the space derivatives. By some miracle this does happen, starting with the potentials u, S_1, S_2, S_3. In the space-time analogue of curl, the derivatives of those potentials give the six field components—which are E and B. Then the material properties lead to $D = \varepsilon E$ and $H = B/\mu$, and the transpose of A should yield equilibrium. The proposed A and A^T appear in

$$\begin{bmatrix} E \\ B \end{bmatrix} = \begin{bmatrix} -\mathrm{grad} & -\partial/\partial t \\ 0 & \mathrm{curl} \end{bmatrix} \begin{bmatrix} u \\ S \end{bmatrix} \quad \text{and} \quad \begin{bmatrix} \rho \\ j \end{bmatrix} = \begin{bmatrix} \mathrm{div} & 0 \\ -\partial/\partial t & \mathrm{curl} \end{bmatrix} \begin{bmatrix} D \\ H \end{bmatrix}. \quad (20)$$

I am sorry to say that if you look carefully you will notice one catch. The transpose of $-\partial/\partial t$ should be $+\partial/\partial t$. The symmetry of the dynamic framework requires a change to $i\partial/\partial t$, which is not at all unreasonable. The four dimensions of space-time are x, y, z, and it, when distance is measured by $x^2 + y^2 + z^2 - t^2$.

That miracle could be expressed in other forms, some more elegant than (20). We mention that the symmetry between A and A^T is recovered when A is the "exterior derivative"—a true generalization of the curl. But perhaps there are no absolute miracles. The difficulty with signs has to appear somewhere, and in that framework ε changes to $-\varepsilon$. Thus positive definiteness is lost—which is unavoidable because the wave equation is indefinite.

To reach the simplest wave equation, let ε and μ be constant; the medium is homogeneous. Then for the fields in Fig. 3.9, Maxwell's equations in MKSQ units†
are

$$\text{div } E = \frac{\rho}{\varepsilon} \qquad\qquad \text{div } B = 0$$

$$\text{curl } E = -\frac{\partial B}{\partial t} \qquad \text{curl } B = \mu j + \mu\varepsilon\,\frac{\partial E}{\partial t}. \qquad (21)$$

Suppose the source terms ρ and j are absent, and use the identities

$$\text{curl curl } E = \text{grad div } E - \nabla^2 E, \ \text{curl curl } B = \text{grad div } B - \nabla^2 B.$$

Both divergences are zero, and (21) leads directly to

$$\mu\varepsilon\,\frac{\partial^2 E}{\partial t^2} = \nabla^2 E \quad\text{and}\quad \mu\varepsilon\,\frac{\partial^2 B}{\partial t^2} = \nabla^2 B. \qquad (22)$$

Both fields satisfy the same wave equation. The wave velocity is $1/\sqrt{\mu\varepsilon}$. Purely static measurements of ε and μ match this velocity of electromagnetic waves with the measured speed of light. A decisive step for physics was to identify those waves *as* light.

Remark   The potential contains as always an extra degree of freedom. The scalar u is determined only up to a constant. For fields that constant can be decided at infinity, rather than by grounding a node, but in every case it disappears in grad u. The vector potential has much more freedom. Since we take the curl of S, any term that has zero curl can be added without changing the physics. The vectors with zero curl are exactly the vectors that are gradients: curl grad $= 0$. Thus an arbitrary gradient can be added to S (which will change its divergence). The choice can be based on mathematical convenience, but it must be made.

In the static case, the favorite choice by far is to have div $S = 0$. In two dimensions that was automatically achieved by the stream function—the divergence of $S = (0,0,s(x,y))$ is certainly zero. In three dimensions it has the great advantage of changing $A^T A = $ curl curl into the Laplacian. By the vector identity above, curl curl S equals $-\nabla^2 S$ because grad div $S = 0$. The interesting decision is in the dynamic case, where the right choice (in these units) is to make

$$\text{div } S = -\mu\varepsilon\,\frac{\partial u}{\partial t}. \qquad (23)$$

---

† Also called rationalized *mks* units. At one time so many units were defined by independent experiments that by official Act of Congress, Ohm's law was repealed.

It is easy to say why. One more time derivative gives

$$\mu\varepsilon\frac{\partial^2 u}{\partial t^2} = -\operatorname{div}\frac{\partial S}{\partial t} = \operatorname{div}(\operatorname{grad} u + E) = \nabla^2 u + \frac{\rho}{\varepsilon}. \tag{24}$$

The scalar potential also satisfies the wave equation! So does the vector potential S. Then taking the curl of both sides recovers the wave equations for E and B.

Until now E and B have dominated electromagnetic theory; the potentials were regarded as secondary and Maxwell's equations were primary. However that supremacy is not so certain. It is possible that in the end the potentials will give the more fundamental description. That will be decided by quantum electrodynamics. Meanwhile we return to the integration theorems on which the whole section depends.

Vector Calculus

This book does not devote a hundred pages to vector analysis and the evaluation of double and triple integrals. I hope that is acceptable. It seems more important to emphasize the application that really does matter—the equivalence between an integral form and a differential form of the same conservation law. If an integral (like total mass) is conserved, then there is a differential equation that will do it. The equivalence has to connect volume integrals to surface integrals and surface integrals to line integrals. That is the central idea of multi-dimensional calculus, and we try not to lose sight of it.

This section uses only x-y-z coordinates. Immediately afterward the gradient and divergence and curl will be expressed in other coordinates—especially cylindrical and spherical. The volume integrals remain the same, except that $r\,dr\,d\theta\,dz$ and $r^2\sin\varphi\,dr\,d\theta\,d\varphi$ replace $dx\,dy\,dz$.

The problem in its simplest form is to compute a double integral like $\iint \partial U/\partial x \, dx \, dy$. $U(x,y)$ is a given function; the two-dimensional region of integration is S. If it happens that $U = x$, then $\partial U/\partial x = 1$ and we are computing the area of S. If $U = y$, then $\partial U/\partial x = 0$ and the computation must give zero. The idea is to copy the fundamental theorem of calculus:

$$\int_a^b \frac{dU}{dx}\,dx = U(b) - U(a).$$

That converts a one-dimensional integral to a "zero-dimensional integral." The points b and a have dimension zero, and they are the boundary of the one-dimensional interval. The difference $U(b) - U(a)$ is an integral over that boundary. The plus and minus signs show whether the vector pointing out at the boundary is in the direction of $+x$ or $-x$.

The extension to two dimensions follows the same principle:

$$\iint_S \frac{\partial U}{\partial x}\, dx\, dy = \int_C U\, dy = \int_C U\, n_x ds. \tag{25}$$

The integral in the x-direction gives U, as it did in one dimension. It should leave an integral over y, and it does. The agreement with $n_x ds$ is seen in Fig. 3.10, where n is the unit vector normal to the boundary. Its first component is the fraction dy/ds, the vertical step divided by the arc length. (The first component of the unit vector *tangent* to the boundary would be $t_x = dx/ds$.) One advantage of n and s is that they are not dependent on the coordinate system—their definitions are geometrical and "coordinate-free." They come again in the integral of $\partial V/\partial y$, which uses strips in the y-direction:

$$\iint_S \frac{\partial V}{\partial y}\, dx\, dy = -\int_C V\, dx = \int_C V\, n_y ds. \tag{26}$$

You may ask, why the minus sign in $-\int V\, dx$? The integral of $\partial V/\partial y\, dx\, dy$ should be $+V\, dx$. You are right, but the direction of the curve C causes the minus sign. At the small triangle in the figure, the curve is moving to the left and dx along the curve is negative.

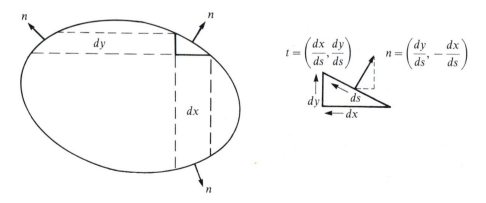

Fig. 3.10. Integration over strips of width $dy = n_x ds$ and $dx = -n_y ds$.

With the right choice of U and V, these formulas yield the key identities of two-dimensional vector calculus. Imagine a potential u or a velocity $v = (v_1, v_2)$ or a flow rate $w = (w_1, w_2)$—and for the moment let them be completely arbitrary, not subject to any of Kirchhoff's laws:

(1) $U = v_2$ and $V = -v_1$ give **Stokes' theorem** for $\iint$ curl v:

$$\iint_S \left(\frac{\partial v_2}{\partial x} - \frac{\partial v_1}{\partial y} \right) dx\, dy = \int_C v_2 dy + v_1 dx = \int_C v \cdot t\, ds$$

This is *circulation*, so curl = circulation per unit area.

(2) $U = w_1$ and $V = w_2$ give the **divergence theorem** for $\iint$ div w:

$$\iint_S \left(\frac{\partial w_1}{\partial x} + \frac{\partial w_2}{\partial y} \right) dx\, dy = \int_C w_1 dy - w_2 dx = \int_C w \cdot n\, ds$$

This is *flux*, so divergence = flux per unit area.

(3) $U = uw_1$ and $V = uw_2$ give **Green's formula** for integration by parts:

$$\iint_S \left(\frac{\partial u}{\partial x} w_1 + u \frac{\partial w_1}{\partial x} + \frac{\partial u}{\partial y} w_2 + u \frac{\partial w_2}{\partial y} \right) dx\, dy = \int_C u\, w \cdot n\, ds.$$

The left side contains (grad $u) \cdot w + u$(div w) and illustrates again why grad^T $= -$div.

A host of consequences now come for velocities and flow rates that satisfy Kirchhoff's laws. Some appeared earlier, others are new, and we give only a short list:

if   curl $v = 0$ then the circulation is zero

if   div $w = 0$ then the flux is zero

if $u_{xx} + u_{yy} = 0$ then $\int$(grad $u \cdot n)ds = 0$ (take $w =$ grad u)

if $u_{xx} + u_{yy} = 0$ and $u = 0$ on C then $u = 0$ everywhere in S

The last says that Laplace's equation with zero boundary conditions has only the solution $u = 0$. It comes from choosing $w = (\partial u/\partial x, \partial u/\partial y)$ in Green's formula. Since $u_{xx} + u_{yy} = 0$ and $u = 0$ on C, all that is left is

$$\iint_S \left[\left(\frac{\partial u}{\partial x} \right)^2 + \left(\frac{\partial u}{\partial y} \right)^2 \right] dx\, dy = 0.$$

Because of the squares, both derivatives are identically zero. The solution is constant and has to be zero. In other words, *there is no nullspace*. For the same reason, Poisson's equation $u_{xx} + u_{yy} = f$ with $u = u_0$ on the boundary cannot have two solutions. Their difference $u^1 - u^2$ would satisfy Laplace's equation, vanish on the boundary, and be zero everywhere.

Each formula extends in its own way to three dimensions:

(1)   On a curved surface Stokes' theorem becomes $\iint n \cdot \text{curl } v \, dS = \int v_1 dx + v_2 dy + v_3 dz$. When the curl is zero and $v = \text{grad } u$, the line integral is zero on closed paths.

(2)   The three-dimensional divergence theorem is

$$\iiint \text{div } w \, dV = \iint w \cdot n \, dS = \text{net flux.}$$

(3)   Green's integration by parts is $\iiint (\nabla u \cdot w + u \nabla \cdot w) dV = \iint u \, w \cdot n \, dS$. With $u = 1$ it changes back to the divergence theorem.

The underlying principle is expressed beautifully but almost too concisely by $\int_S d\omega = \int_{dS} \omega$, in the calculus of "differential forms."

Orthogonal Coordinate Systems

For some problems the x-y-z coordinate system is not appropriate. It is easy to visualize, its axes are not only orthogonal but straight, and Laplace's equation looks as convenient as possible—but on a circle or cylinder or sphere the boundary conditions become miserable. It is worth changing coordinates to replace $x = \sqrt{1 - y^2 - z^2}$ by $r = 1$. It is certainly worth applying symmetry around an axis (which is extremely common) to remove θ and reduce to a simpler equation. For that we need the gradient, divergence, and curl in the new coordinates. They are messy but unavoidable.

The two most important systems are **cylindrical** and **spherical**. They follow the rules for any orthogonal system, in which the ordinary x-y-z coordinates are expressed in terms of new u-v-w coordinates: for example

$$x = u \cos v, \, y = u \sin v, \, z = w. \tag{27}$$

Those are cylindrical coordinates, very thinly disguised: u is r and v is θ. They are polar coordinates in the $x - y$ plane, with the z-direction unchanged (but renamed w). If you will allow me to keep this disguise, the formulas can be derived for any orthogonal system while maintaining an example that can be checked at every step.

The first step is to find the "*scale factors*" a,b,c. If u is changed, the point with coordinates u,v,w will move. When the change is to $u + du$, the distance moved is

$$\sqrt{dx^2 + dy^2 + dz^2} = \sqrt{\left(\frac{\partial x}{\partial u}\right)^2 + \left(\frac{\partial y}{\partial u}\right)^2 + \left(\frac{\partial z}{\partial u}\right)^2} \, du = a \, du.$$

In the example these derivatives are $\cos v, \sin v, 0$. Therefore the scale factor for du is $a = 1$—because $\cos^2 v + \sin^2 v = 1$. There is also a scale factor b to measure the effect of a change in v:

$$b = \sqrt{\left(\frac{\partial x}{\partial v}\right)^2 + \left(\frac{\partial y}{\partial v}\right)^2 + \left(\frac{\partial z}{\partial v}\right)^2} = \sqrt{u^2 \sin^2 v + u^2 \cos^2 v + 0} = u.$$

This is the r in $r\,d\theta$—the movement due to a change in θ. The third scale factor is $c = 1$, because z is really the same as w (that will change in spherical coordinates).

These factors also give the scale for surface area and volume. In general the volume $dx\,dy\,dz$ of a small cube changes to $|J|\,du\,dv\,dw$, where $|J|$ is the determinant of the first-derivative matrix (or Jacobian matrix):

$$|J| = \det \begin{bmatrix} \partial x/\partial u & \partial x/\partial v & \partial x/\partial w \\ \partial y/\partial u & \partial y/\partial v & \partial y/\partial w \\ \partial z/\partial u & \partial z/\partial v & \partial z/\partial w \end{bmatrix} = \det \begin{bmatrix} \cos v & -u\sin v & 0 \\ \sin v & u\cos v & 0 \\ 0 & 0 & 1 \end{bmatrix}.$$

This equals u, which is the r in the volume element $r\,dr\,d\theta\,dz$.

That determinant is simple when J has orthogonal columns, as in this example. The inner products between columns are zero. Then **the system is orthogonal and the volume scale is** $|J| = abc$. The columns have lengths a,b,c and the determinant is their product. The small cube has turned into a small box, with sides $a\,du$, $b\,dv$, and $c\,dw$. In practice, only orthogonal systems are used! That also simplifies surface area—a rectangle on the surface $w = $ constant has area $(a\,du)(b\,dv)$, and in the example that is the familiar $r\,dr\,d\theta$. In some books a is h_1, b is h_2, c is h_3.

Now we turn to the great triad of gradient, divergence, and curl. Each has a geometrical meaning that is independent of coordinates (they are tensors). The **gradient** is defined so that $(\mathrm{grad}\,f) \cdot (\text{change in position}) \approx (\text{change in } f)$, whatever the coordinates. To express $\mathrm{grad}\,f$ in terms of u, v, w we need unit vectors i_1, i_2, i_3 in the coordinate directions—which are radial, circumferential, and axial in Fig. 3.11. The change in position involves the scale factors a, b, c so the gradient must have those in the denominator:

$$\nabla f = \frac{1}{a}\frac{\partial f}{\partial u}i_1 + \frac{1}{b}\frac{\partial f}{\partial v}i_2 + \frac{1}{c}\frac{\partial f}{\partial w}i_3$$

$$= \frac{\partial f}{\partial r}i_r + \frac{1}{r}\frac{\partial f}{\partial \theta}i_\theta + \frac{\partial f}{\partial z}i_z. \tag{28}$$

The change in position is $a\,du\,i_1 + b\,dv\,i_2 + c\,dw\,i_3$, and the gradient correctly gives the change in f:

$$(\nabla f) \cdot (\text{change in position}) = \frac{\partial f}{\partial u}du + \frac{\partial f}{\partial v}dv + \frac{\partial f}{\partial w}dw = df. \tag{29}$$

Note that we write f for a scalar function (earlier it was u) and F for a vector.

The **divergence** goes the other way. It starts with a vector and produces a scalar.

If the vector is $F = F_1 i_1 + F_2 i_2 + F_3 i_3$, its divergence is

$$\nabla \cdot F = \frac{1}{abc} \left[\frac{\partial}{\partial u}(bcF_1) + \frac{\partial}{\partial v}(acF_2) + \frac{\partial}{\partial w}(abF_3) \right]$$

$$= \frac{1}{r} \left[\frac{\partial}{\partial r}(rF_r) + \frac{\partial}{\partial \theta}(F_\theta) + \frac{\partial}{\partial z}(rF_z) \right]. \tag{30}$$

Exercise 3.4.26 checks that div is still $-\text{grad}^T$, as it was in x-y-z coordinates. Finally there is the **curl** of F, which is a vector:

$$\nabla \times F = \frac{1}{abc} \begin{vmatrix} ai_1 & bi_2 & ci_3 \\ \partial/\partial u & \partial/\partial v & \partial/\partial w \\ aF_1 & bF_2 & cF_3 \end{vmatrix}. \tag{31}$$

We emphasize that the key identities curl grad $f = 0$ and div curl $F = 0$ are true in every coordinate system.

EXAMPLE 1   The unit vector $F = i_\theta$ points in the θ-direction and the flow rotates around the z-axis. With $a = c = 1$ and $b = r$, and with $F_1 = F_3 = 0$ and $F_2 = 1$, its curl is

$$\nabla \times i_\theta = \frac{1}{r} \begin{vmatrix} i_r & ri_\theta & i_z \\ \partial/\partial r & \partial/\partial \theta & \partial/\partial z \\ 0 & r & 0 \end{vmatrix} = \frac{1}{r} i_z \frac{\partial r}{\partial r} = \frac{1}{r} i_z.$$

The curl of this F points perpendicular to F, along the axis of rotation.

EXAMPLE 2   The **Laplacian** is still div grad, or $-A^T A$:

$$\nabla^2 f = \frac{1}{r} \left[\frac{\partial}{\partial r}\left(r \frac{\partial f}{\partial r}\right) + \frac{\partial}{\partial \theta}\left(\frac{1}{r} \frac{\partial f}{\partial \theta}\right) + \frac{\partial}{\partial z}\left(r \frac{\partial f}{\partial z}\right) \right] \tag{32}$$

This must agree with the ordinary Laplacian, say for $f = r^2 = x^2 + y^2$:

$$f_{xx} + f_{yy} + f_{zz} = 2 + 2 + 0 = 4 \quad \text{and} \quad \frac{1}{r} \frac{\partial}{\partial r}\left(r \frac{\partial f}{\partial r}\right) = \frac{1}{r} \frac{\partial}{\partial r}(2r^2) = 4.$$

In **spherical coordinates** r changes to the full distance $\sqrt{x^2 + y^2 + z^2}$. The longitude θ is still measured around the z-axis, and the co-latitude φ is the angle

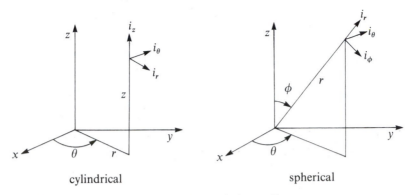

cylindrical   spherical

Fig. 3.11. Unit vectors in cylindrical and spherical coordinates.

from the pole in Fig. 3.11.† These are u, v, and w:

$$x = r \cos \theta \sin \varphi, \ y = r \sin \theta \sin \varphi, \ z = r \cos \varphi.$$

The scale factors are $a = 1$, $b = r \sin \varphi$, and $c = r$. Therefore

$$\text{grad } f = \ \nabla f \ = \frac{\partial f}{\partial r} i_r + \frac{1}{r \sin \varphi} \frac{\partial f}{\partial \theta} i_\theta + \frac{1}{r} \frac{\partial f}{\partial \varphi} i_\varphi$$

$$\text{div } F = \nabla \cdot F \ = \frac{1}{r^2} \frac{\partial}{\partial r} (r^2 F_r) + \frac{1}{r \sin \varphi} \frac{\partial}{\partial \theta} (F_\theta) + \frac{1}{r \sin \varphi} \frac{\partial}{\partial \varphi} (F_\varphi \sin \varphi)$$

$$\nabla^2 f = \nabla \cdot \nabla f = \frac{1}{r^2} \frac{\partial}{\partial r} \left(r^2 \frac{\partial f}{\partial r} \right) + \frac{1}{r^2 \sin^2 \varphi} \frac{\partial^2 f}{\partial \theta^2} + \frac{1}{r^2 \sin \varphi} \frac{\partial}{\partial \varphi} \left(\sin \varphi \frac{\partial f}{\partial \varphi} \right)$$

For some reason the Laplacian of $f = r^2$ is now $\frac{1}{r^2} \frac{\partial}{\partial r} (2r^3) = 6.$

EXERCISES

3.4.1   Find the velocity vector dr/dt, the speed $ds/dt = $ length of dr/dt, and the arc length s up to time t, for the paths $r = (0, \sin 2t, \cos 2t)$ and $r = (t^2, t^2, -t^2)$. What are the shapes of these paths?

3.4.2   If $r = (x, y, z) = (1, t, t^2)$ find the velocity vector dr/dt and the speed $|dr/dt|$. Compute $\int F \cdot dr$ between $t = 0$ and $t = 1$ for the forces $F = (z, y, x)$ and $F = (z, z, x + y)$.

3.4.3   From the tests (4) and (5) decide whether $F = (z, y, x)$ or $F = (z, z, x + y)$ has zero curl and is the gradient of a potential $u(x, y, z)$. When possible find u.

---

† Half the world reverses θ and φ. The situation is unbelievable.

3.4.4 What third components of $F=(yz^2,xz^2,\ )$ make this force conservative—satisfying (5), equal to a gradient, and having curl $F=0$? Find the potential and the integral $\int F \cdot dr$ along any path from $P=(0,0,0)$ to $Q=(1,2,0)$.

3.4.5 Find potentials for $F=(x^2y^3,x^3y^2,0)$ and $F=(2xyz,x^2z,x^2y)$.

3.4.6. Compute the velocity $v=dr/dt$ and the acceleration $a=d^2r/dt^2$ for the path $r=(t\cos t, t\sin t, 0)$. Which part of a is parallel to the path and which is the "Coriolis part" perpendicular to the path? Find the speed $ds/dt=$ length of v.

3.4.7 Velocity is $v=dr/dt=(dr/ds)(ds/dt)$ and acceleration is

$$a=\frac{dv}{dt}=\frac{d^2r}{ds^2}\left(\frac{ds}{dt}\right)^2+\frac{dr}{ds}\frac{d^2s}{dt^2}.$$

Which of these two terms is parallel to v and which is perpendicular? Find v and its length ds/dt and the acceleration a for the speeding-up circular motion $r=(\cos t^2,\sin t^2,0)$.

3.4.8 In pure rotation the velocity is the cross-product

$$v=(\omega_1,\omega_2,\omega_3)\times(x,y,z)=(\omega_2 z-\omega_3 y,\omega_3 x-\omega_1 z,\omega_1 y-\omega_2 x).$$

Show that curl $v=2\omega$ and div $v=0$ and v is perpendicular to ω. Is v the gradient of a potential u? What choice of ω gives rotation in the $x-y$ plane?

3.4.9 Illustrate by example that $\nabla\times v$ need not be perpendicular to v.

3.4.10 Using $\nabla=(\partial/\partial x,\partial/\partial y,\partial/\partial z)$ and its two identities $\nabla\times\nabla u=$ curl grad $u=0$ and $\nabla\cdot(\nabla\times v)=$ div curl $v=0$, write out and verify
  (a) $\nabla\cdot(uv)=\nabla u\cdot v+u\nabla\cdot v$
  (b) $\nabla\times(uv)=\nabla u\times v+u\nabla\times v$
  (c) $\nabla\cdot(v\times w)=w\cdot\nabla\times v-v\cdot\nabla\times w$
  (d) $\nabla\times(v\times w)=(w\cdot\nabla)v-w(\nabla\cdot v)-(v\cdot\nabla)w+v(\nabla\cdot w)$
  (e) $\nabla^2v=\nabla(\nabla\cdot v)-\nabla\times(\nabla\times v)$
Which of (a)–(e) produce vectors and which produce scalars?

3.4.11 Using the matrix form (7) for the curl and the column vector $[\partial/\partial x\ \partial/\partial y\ \partial/\partial z]^T$ for the gradient, verify that curl grad $=0$.

3.4.12 Find vector potentials $S_1=(0,0,s)$ and S_2 such that curl $S_1=(y,x^2,0)$ and curl $S_2=(2x,3y,-5z)$. Check first for zero divergence.

3.4.13 At least formally, the curl is symmetric:

$$\text{curl } v=\begin{bmatrix} 0 & -\dfrac{\partial}{\partial z} & \dfrac{\partial}{\partial y} \\[2mm] \dfrac{\partial}{\partial z} & 0 & -\dfrac{\partial}{\partial x} \\[2mm] -\dfrac{\partial}{\partial y} & \dfrac{\partial}{\partial x} & 0 \end{bmatrix}\begin{bmatrix} v_1 \\[2mm] v_2 \\[2mm] v_3 \end{bmatrix}=\begin{bmatrix} \dfrac{\partial v_3}{\partial y}-\dfrac{\partial v_2}{\partial z} \\[2mm] \dfrac{\partial v_1}{\partial z}-\dfrac{\partial v_3}{\partial x} \\[2mm] \dfrac{\partial v_2}{\partial x}-\dfrac{\partial v_1}{\partial y} \end{bmatrix}.$$

The transpose of this matrix R is R itself, since transposing a derivative changes its sign. Confirm that curl $=$ curl^T using integration by parts: with vectors v and S that are zero on the sides of a cube, write out the terms in

$$\iiint S \cdot \operatorname{curl} v \, dx \, dy \, dz = \iiint v \cdot \operatorname{curl} S \, dx \, dy \, dz.$$

3.4.14 In three dimensions why is div curl grad u automatically zero (in two ways)?

3.4.15 (i) Show that the velocity $v = (x, -y)$ has div $v = 0$, and find a stream function $s(x, y)$ such that $v = \operatorname{curl} s = (\partial s/\partial y, -\partial s/\partial x)$.
  (ii) Show that also $v = (x/(x^2 + y^2), y/(x^2 + y^2))$ is divergence-free and $s = \theta$ (the polar angle, otherwise written $\tan^{-1}(y/x)$) gives $v = \operatorname{curl} s$. *Note*: $s = \theta$ is not a genuine stream function since it changes by 2π as you go around the origin (where v is infinite and the region has a hole).

3.4.16 Let A_0 be the gradient without boundary conditions, and let $Au = \operatorname{grad} u$ restricted by the boundary condition $u = 0$.
  (a) What functions are in the nullspaces of A_0 and A? They satisfy grad $u = 0$ and the associated boundary conditions.
  (b) What are the boundary conditions (on w) and the nullspaces of A_0^T and A^T?

3.4.17 (a) If curl $v = 0$ and div $w = 0$ in a three-dimensional volume V, with $w \cdot n = 0$ on the boundary, show that v and w are orthogonal: $\iiint v^T w \, dV = 0$.
  (b) How could an arbitrary vector field $f(x, y, z)$ be split into $v + w$?

3.4.18 Combine Maxwell's equations (21) and the choice (23) with the vector identity for curl curl S to find the wave equation satisfied by the vector potential S.

3.4.19 From Maxwell's equations deduce conservation of charge: $\partial \rho / \partial t + \operatorname{div} j = 0$.

3.4.20 If S is changed to $S^* = S + \nabla G$, show that altering the other potential to $u^* = u - \partial G/\partial t$ leaves E and B unchanged. This is a *gauge transformation* of the potentials.

3.4.21 Why is $\int x \, dx = 0$ around a closed curve?

3.4.22 Why does $\int x \, dy$ around a closed curve give the area inside?

3.4.23 For $F = (xy, -xy)$ find $\int F \cdot t \, ds$ and $\int F \cdot n \, ds$ around the unit square $0 \le x, y \le 1$, using Stokes' theorem and the divergence theorem.

3.4.24 Prove that $\iint (\varphi_{xx} + \varphi_{yy}) \, dx \, dy = \int \partial \varphi / \partial n \, ds$ by the right choice of u and w in Green's formula.

3.4.25 If $w = (x, y, z)$ what is the flux $\iint w \cdot n \, dS$ out of a unit cube and a unit sphere? Compute both sides in the divergence theorem.

3.4.26 (a) With formula (29) for the gradient verify that (28) equals df.
  (b) Integrate by parts to show from (30) that the divergence is still the transpose of the gradient (with a minus sign):

$$(Af)^T F = f^T A^T F \quad \text{is} \quad \iiint \nabla f \cdot F \, abc \, du \, dv \, dw = - \iiint f \nabla \cdot F \, abc \, du \, dv \, dw.$$

3.4.27 In cylindrical coordinates compute
(a) grad θ (c) div i_r (e) div grad $\theta = \nabla^2\theta$
(b) grad r^4 (d) curl grad θ (f) div curl u_θ

3.4.28 Parabolic cylinder coordinates have $x = uv$, $y = \frac{1}{2}(u^2 - v^2)$, $z = w$. Find the scale factors a, b, c. What is Laplace's equation $\nabla^2 f = 0$ in those coordinates?

3.4.29 (a) Check the scale factors in spherical coordinates and find the gradient of $1/r$.
(b) Compute grad φ, div grad φ, and curl grad φ (φ = polar angle).
(c) Why is the Laplacian of r^2 equal to 6 and not 4?

3.4.30 Show that $u = r \cos \varphi + \frac{1}{2} r^{-2} \cos \varphi$ satisfies $\nabla^2 u = 0$, and also $\partial u/\partial r = 0$ on the unit sphere. Find the velocity field $v = $ grad u for flow past a sphere.

3.4.31 From (32) find Laplace's equation in polar coordinates:

$$\nabla^2 f = \frac{\partial^2 f}{\partial r^2} + \frac{1}{r}\frac{\partial f}{\partial r} + \frac{1}{r^2}\frac{\partial^2 f}{\partial \theta^2} = 0.$$

3.4.32 If the points on a curve are $r = (x, y, z)$, the *line element* in $\int f\,ds$ is

$ds = (1 + y_x^2)^{1/2}dx$ if the curve is $y = y(x)$ in the plane $z = 0$

$ds = |r_t|dt = (x_t^2 + y_t^2 + z_t^2)^{1/2}dt$ if $x = x(t)$, $y = y(t)$, $z = z(t)$

$ds = (a^2\,du^2 + b^2\,dv^2 + c^2\,dw^2)^{1/2}$ if $x = x(u, v, w)$, $y = \ldots$ in orthogonal coordinates

$ds = (r_u \cdot r_u du^2 + 2r_u \cdot r_v du\,dv + r_v \cdot r_v dv^2)^{1/2}$ if $x = x(u, v) \ldots$ in any coordinates.

Compute the distance $\int ds$ around the circle of radius 3 in three ways, from (a) $y^2 = 9 - x^2$, (b) $x = 3 \cos t$, $y = 3 \sin t$, (c) $u = 3$, $v = \theta$, $w = 0$ in cylindrical coordinates.

3.4.33 With $w = $ grad v in Green's formula show that

$$\iint (u\nabla^2 v + \nabla u \cdot \nabla v)\, dx\, dy = \int u\frac{\partial v}{\partial n}\, ds$$

$$\iint (u\nabla^2 v - v\nabla^2 u)dx\, dy = \int \left(u\frac{\partial v}{\partial n} - v\frac{\partial u}{\partial n}\right) ds$$

3.4.34 If A is the gradient and C^{-1} is the curl, the matrix in our framework includes all three of the key operators in vector calculus:

$$M = \begin{bmatrix} C^{-1} & A \\ A^T & 0 \end{bmatrix} = \begin{bmatrix} \text{curl} & \text{grad} \\ -\text{div} & 0 \end{bmatrix} = \begin{bmatrix} 0 & -\partial/\partial z & \partial/\partial y & \partial/\partial x \\ \partial/\partial z & 0 & -\partial/\partial x & \partial/\partial y \\ -\partial/\partial y & \partial/\partial x & 0 & \partial/\partial z \\ -\partial/\partial x & -\partial/\partial y & -\partial/\partial z & 0 \end{bmatrix}$$

Show that M^2 is diagonal! It is $-\Delta^2 I$, and the multiplication verifies the useful identity curl curl $-$ grad div $= -\Delta^2$ in 3.4.10 (e)—as well as curl grad $= 0$ and div curl $= 0$.

3.5 ■ EQUILIBRIUM OF FLUIDS AND SOLIDS

This section goes from elastic bars and elastic beams to elastic solids. The problems become three-dimensional, but the framework remains the same. That makes it possible to understand the principles of continuum mechanics without seeing every detail. We do not and cannot avoid the change in dimension, which means an important change in the unknowns. The displacement u becomes a **vector** with three components, the strain e becomes a **matrix** involving nine derivatives of u, and the elastic coefficient c changes most of all. It becomes a **tensor** that multiplies the strain to give the stress. Then equilibrium is a balance between internal forces and external forces (like $A^T y = f$) and there we stop.

Actually the strain and stress are also tensors. They are second-order and they look indistinguishable from matrices. At the end we discuss the invariance under coordinate transformations—the intrinsic geometrical meaning—that makes a tensor a tensor.

For fluids we go further. There are new terms not seen in Laplace's equation for potential flow. The viscosity term is linear and symmetric, and fits directly into the framework. The pressure term also fits when the fluid is incompressible; pressure is then the Lagrange multiplier for the continuity equation div $v = 0$. However the "advection term" is nonlinear and more or less spoils everything. It enters because the fluid is moving and not just passing forces along. The underlying laws are still conservation of mass and (what is new) conservation of momentum. Each term is important for some flows and unimportant for others, depending on the Reynolds number and the compressibility and the vorticity and the stationarity. We hope to organize these possibilities in a coherent way.

Fluids can be studied without solids, and solids without fluids. It is absolutely reasonable to go forward without either one, and to use this section as a convenient reference to both.

We begin with **solids**. For small rotations and small strains the equations are straightforward—because they are linear. The complications come from writing out all the terms, when vectors are replaced by matrices and matrices by fourth-order (also called fourth-rank) tensors. In fact Einstein invented the "summation convention" to avoid repeating the symbol $\sum$ every time an index was repeated and a sum was required.† Nevertheless a tensor can make a subject look bad; a linear transformation on 3 by 3 matrices has 81 degrees of freedom. It needs a 4-dimensional array with 3^4 entries c_{ijkl}. Fortunately the stress and strain are *symmetric* 3 by 3 matrices, and the transformation is itself symmetric, which after some calculation reduces 81 to 21. Even that number is virtually never met in practice. The most common materials are taken to be *isotropic*; their behavior is invariant under rotation, the same in all directions. Because the strain is a symmetric matrix it has

(a)   three real eigenvalues (the **principal strains**)

---

†Einstein would write $Ax = b$ as $a_{ij}x_j = b_i$, without the Σ.

(b)   three perpendicular eigenvectors (the **principal directions**).

The same is true for the stress. Furthermore, for an isotropic material the principal directions are the same! This leaves only *two* degrees of freedom for the material law (Hooke's law) connecting strain to stress. The law can multiply the strain by a constant, normally written 2μ. It can also add a multiple of the identity matrix, which does not change the eigenvectors—the principal directions. That multiple must be linear and must maintain rotational invariance; it turns out to be a constant λ times the trace—the sum of the eigenvalues. Thus Hooke's law becomes:

$$\text{stress} = 2\mu \text{ (strain)} + \lambda \text{ (trace of strain) } I.$$

The two constants appear in several equivalent forms:

(1)   as the Lamé constants μ and λ
(2)   as shear modulus μ and "bulk modulus" $K = \lambda + \frac{2}{3}\mu$
(3)   as Young's constant E and Poisson's ratio v

The last have the most direct physical meaning. Young's modulus enters when a rod of length L and area A is stretched by a force F. The tensile stress is F/A, the force per unit area. The strain is the non-dimensional ratio $\Delta L/L$, and Young's constant is stress/strain. It is found by a one-dimensional test. The other constant v accounts for the contraction that occurs in directions perpendicular to the force. Poisson's ratio is that contraction, divided by the extension in the force direction. The ratio is typically near .3. When it reaches .5 the material is *incompressible*; its volume does not change. At that point the second Lamé constant λ becomes infinite and new equations are required. Incompressibility is very commonly assumed for fluids—we consider the compressible case only once—but for solids the standard stress-strain law (the constitutive law corresponding to C in $A^T C A$) includes changes in volume.

Strain and Displacement

We turn next to A. For a discrete network, an incidence matrix gave differences in potential. For a discrete truss A had to compute differences in two directions, horizontal and vertical; $\cos\theta$ and $\sin\theta$ entered the matrix. For a continuum we expect derivatives of all three displacements u_1, u_2, u_3 in all three directions x_1, x_2, x_3 (or x, y, z). However the strain is not just the first-derivative matrix (the Jacobian matrix J) with entries $\partial u_i / \partial x_j$. Those are abbreviated to $u_{i,j}$, with a comma to indicate the derivative, but that would have no reason for symmetry: $u_{i,j}$ can be entirely different from $u_{j,i} = \partial u_j / \partial x_i$. **The strain is the symmetric Jacobian $\frac{1}{2}(J + J^T)$:**

$$e_{ij} = \tfrac{1}{2}\left(\frac{\partial u_i}{\partial x_j} + \frac{\partial u_j}{\partial x_i}\right) = \begin{bmatrix} u_{1,1} & \tfrac{1}{2}(u_{1,2} + u_{2,1}) & \tfrac{1}{2}(u_{1,3} + u_{3,1}) \\ - & u_{2,2} & \tfrac{1}{2}(u_{2,3} + u_{3,2}) \\ - & - & u_{3,3} \end{bmatrix}.$$

The strain matrix is symmetric because rotations, which enter the antisymmetric part $\frac{1}{2}(J - J^T)$, involve no stretching. The stress is also symmetric, because those rotations involve no internal forces. Its symmetry is the absence of torque.

In the strain matrix, the diagonal entries $\partial u_i/\partial x_i$ give the **extensions** and the off-diagonal entries give the **shear**. That is something new; it did not arise for one-dimensional bars. The strain matrix that has $e_{12} = e_{21} = 1$ and all other entries zero is an instance of shear, while a diagonal matrix is a pure extension. We pick out four special displacements and the corresponding strains:

$$\text{stretch in one direction:} \quad u = \begin{bmatrix} x_1 \\ 0 \\ 0 \end{bmatrix} \quad \text{and} \quad e = \begin{bmatrix} 1 & 0 & 0 \\ 0 & 0 & 0 \\ 0 & 0 & 0 \end{bmatrix}$$

$$\text{dilation in all directions:} \quad u = \begin{bmatrix} x_1 \\ x_2 \\ x_3 \end{bmatrix} \quad \text{and} \quad e = \begin{bmatrix} 1 & 0 & 0 \\ 0 & 1 & 0 \\ 0 & 0 & 1 \end{bmatrix}$$

$$\text{simple shear:} \quad u = \begin{bmatrix} 2x_2 \\ 0 \\ 0 \end{bmatrix} \quad \text{and} \quad e = \begin{bmatrix} 0 & 1 & 0 \\ 1 & 0 & 0 \\ 0 & 0 & 0 \end{bmatrix}$$

$$\text{pure shear:} \quad u = \begin{bmatrix} x_2 \\ x_1 \\ 0 \end{bmatrix} \quad \text{and} \quad e = \begin{bmatrix} 0 & 1 & 0 \\ 1 & 0 & 0 \\ 0 & 0 & 0 \end{bmatrix}.$$

The first comes from pulling in one direction (with Poisson's ratio $v = 0$ so the other directions don't contract). The second comes from uniform pressure in all directions. The simple shear is like cars passing each other in straight lanes—the farther out the lane, the faster the traffic. Pure shear has the same strain matrix as simple shear, and only the boundary conditions on u can tell the difference. In fact all these *constant strains* involve no external forces within the body, and are decided by the boundary conditions—in the same way that $u = a + bx + cy$ satisfies Laplace's equation. In each case the stress is also constant, its derivatives in $A^T\sigma$ are zero, and the force is $f = 0$.

Pure shear deserves a closer look. It turns a square like $0 \le x_1, x_2 \le 1$ into a parallelogram, stretched in the $45°$ direction. The displacement $u = (x_2, x_1, 0)$ is actually far too large to be considered small; the point $x = (1,0,0)$ is displaced by $u = (0,1,0)$ and moves to $(1,1,0)$. The other corner $x = (0,1,0)$ is displaced by $u = (1,0,0)$ and moves to the same point. The square is stretched not only into a parallelogram but onto a straight line! All u and e above should be multiplied by .01, to be more realistic. In fact e is the infinitesimal strain matrix.

What are the principal strains for pure shear? The eigenvalues of the last matrix are 1, -1, and 0. The eigenvectors are $X_1 = (1,1,0)$, $X_2 = (1,-1,0)$, $X_3 = (0,0,1)$. The first is the stretched $45°$ direction, the second is the contracted $-45°$ direction, the third is the untouched x_3 direction. *A shear is just a stretch in its three principal directions*. This shear has $\Delta L/L = 1$ in the 45° direction, so lengths are doubled. It has $\Delta L/L = -1$ in the second direction, so lengths go to zero (the parallelogram was squeezed to a line). It has $\Delta L = 0$ in the third direction.

Every symmetric matrix is like this shear. It can be diagonalized by an orthogonal matrix (a rotation of coordinates). Then the principal directions become the axes, the shears disappear, and we recognize the pure extensions or contractions that were always present:

3H Every strain matrix is a three-way stretch in the directions of its eigenvectors. Every nonsymmetric matrix J is a three-way stretch followed (or preceded) by a rotation.

Both statements come directly from linear algebra. The first is $Q^T A Q = \Lambda$, the *diagonalization* of a symmetric strain matrix A (or e). The second is $J = QB$, the *polar decomposition* of an arbitrary matrix. It is an orthogonal Q times a symmetric B. Section 1.6 also provided the singular value decomposition $J = Q_1 \Sigma Q_2$, which expresses J as a pure stretch in the three *coordinate* directions, followed *and* preceded by rotations. The matrix Σ is diagonal where B is only symmetric—B stretches in its principal directions, while Σ acts in the coordinate directions.

We need to justify taking the symmetric part $e = \frac{1}{2}(J + J^T)$. It is J that gives the first-order movement of each point, apart from a rigid motion whose derivatives are zero. But it is the entries in e that give the first-order stretching. If two points start out separated by the vector Δ, then after displacement they are separated by $\Delta + J\Delta$. The distance between them is the length of this vector, and the cross terms combine J with J^T to arrive at e:

$$|\Delta + J\Delta| = (|\Delta|^2 + \Delta^T(J + J^T)\Delta + |J\Delta|^2)^{1/2} = |\Delta| + \frac{\Delta^T e \Delta}{|\Delta|} + \cdots$$

The strain-displacement relation is $e = Au$. It connects the kinematic variables, in contrast to the static equation that comes next—the balance of forces.

Stress and Force

We turn to the stresses and the equilibrium equation. They complete the cycle from u to $e = Au$ to $\sigma = Ce$ to $f = A^T\sigma$. The notation σ for the stress tensor replaces w in our standard framework.† Otherwise the framework is unchanged. The only question is to identify A^T.

In the previous section that question had two answers, both leading to $\text{grad}^T = -\text{div}$. One was mathematical, through Green's formula for integration by

---

† The strain might have been ε, but that is too painful for a mathematician to write—ε is always a small number headed for zero.

parts. The other was physical, through conservation of mass. Here there is a third approach to A^T, through the principle of virtual work: At equilibrium, the work of the external forces on any virtual displacement equals the work of the internal stresses. It is a physical statement but the key step is still integration by parts.

Green's formula will have a boundary term which we explain in advance. For fluids it contained $w \cdot n$, the flow rate through the boundary. For solids it contains $\sigma^T n$, the product of the stress matrix σ and the unit normal vector n. That product is a vector, called the **surface traction**. It gives the force from inside the body on each piece of the surface. It is again a mixture of shear, pushing the boundary to the side, and extension or compression, pushing the boundary out or in. The extensional part is $n^T \sigma^T n$, the "normal component of the normal component" of stress. But all components of $\sigma^T n$ must be in balance across every interface, including the boundary surface S between the body and the outside world.

The boundary conditions specify either this traction $\sigma^T n = F$ or the displacement $u = u_0$. When u_0 is given, the traction comes from solving the equations. It is the force required to maintain the displacement—the reaction force at the supports, in the case of a truss. When F is given the situation is reversed. The boundary displacement comes from solving the equations, and it is the Lagrange multiplier for the constraint $\sigma^T n = F$. You will see how $u^T \sigma^T n$ replaces $uw \cdot n$ when Green's formula moves up to matrices—and this boundary term is zero if either $u = 0$ or $\sigma^T n = 0$.

The integration by parts formula is

$$\iiint e^T \sigma \, dV = -\iiint u^T \, \mathrm{div}\, \sigma \, dV + \iint u^T \sigma^T n \, ds. \tag{1}$$

It holds for any u and any σ. The strain e comes from u in the way we already know: $e = Au$ is $\frac{1}{2}(J + J^T)$. The matrix inner product $e^T \sigma$ is the sum of all nine terms $e_{ij}\sigma_{ij}$. Apart from boundary conditions, the left side of (1) is the inner product of Au with σ. Therefore the volume integral on the right must be the inner product of u with $A^T\sigma$, and this identifies the transpose of A:

$$A^T\sigma = -\mathrm{div}\, \sigma = - \begin{bmatrix} \dfrac{\partial \sigma_{11}}{\partial x_1} + \dfrac{\partial \sigma_{12}}{\partial x_2} + \dfrac{\partial \sigma_{13}}{\partial x_3} \\[2ex] \dfrac{\partial \sigma_{21}}{\partial x_1} + \dfrac{\partial \sigma_{22}}{\partial x_2} + \dfrac{\partial \sigma_{23}}{\partial x_3} \\[2ex] \dfrac{\partial \sigma_{31}}{\partial x_1} + \dfrac{\partial \sigma_{32}}{\partial x_2} + \dfrac{\partial \sigma_{33}}{\partial x_3} \end{bmatrix}. \tag{2}$$

Then the equation of equilibrium—which is mathematically $A^T\sigma = f$ and physically a balance of forces—is

$$\mathrm{div}\, \sigma + f = 0. \tag{3}$$

That represents three equations. The divergence in (2) has three components—each balanced by a component of the body force f. When that force is zero, which is common, equilibrium reduces to div $\sigma = 0$. For dynamic problems the right side of (3) would become ρu_{tt}—mass density times acceleration equals force density—which is Newton's law for a body in motion. In equilibrium the three equations are written out below, and with great self-restraint we forego the minimum principles.

3I The strain-displacement equation $e = Au$, the stress-strain law $\sigma = Ce$, and the force balance $f = A^T \sigma$ are

$$e_{ij} = \tfrac{1}{2}\left(\frac{\partial u_i}{\partial x_j} + \frac{\partial u_j}{\partial x_i}\right) \tag{4}$$

$$\sigma_{ij} = 2\mu e_{ij} + \lambda(e_{11} + e_{22} + e_{33})\delta_{ij} \tag{5}$$

$$f_i = -\frac{\partial \sigma_{i1}}{\partial x_1} - \frac{\partial \sigma_{i2}}{\partial x_2} - \frac{\partial \sigma_{i3}}{\partial x_3} \tag{6}$$

The boundary conditions specify each component either of $u = u_0$ or of $\sigma^T n = F$.

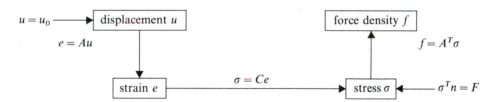

Fig. 3.12. The framework for continuum mechanics.

The Torsion of a Rod

This example is good because most of the strain and stress components are zero. It starts with a vertical rod, a cylinder whose cross-section need not be a circle. By putting your hands on the top and bottom, and twisting the top while holding the bottom, all those cross-sections will turn. The boundary conditions are $\sigma^T n = 0$ on the side of the rod, where there is no force, and $\sigma_{33} = 0$ on the top and bottom—where there is no *vertical* force. The other conditions are $u_1 = u_2 = 0$ on the bottom and $u_1 = -\theta x_2 h$, $u_2 = \theta x_1 h$ on the top—where h is the height (the x_3 component) and θ determines the angle of twist. Subject to these conditions we look for u and e and σ; the internal force f is zero.

It is the best to see the main features physically. The turning of cross-sections will increase linearly with their height x_3, so we expect

$$u_1 = -\theta x_2 x_3, \quad u_2 = \theta x_1 x_3, \quad u_3 = w(x_1, x_2).$$

The *warping function w* is identical for all cross-sections, which start flat but become warped as they turn. Their movement out of the plane is w, still to be determined. The strains and stresses are symmetric matrices:

$$e = Au = \begin{bmatrix} 0 & 0 & \frac{1}{2}(\partial w/\partial x_1 - \theta x_2) \\ 0 & 0 & \frac{1}{2}(\partial w/\partial x_2 + \theta x_1) \\ - & - & 0 \end{bmatrix}$$

$$\sigma = Ce = \begin{bmatrix} 0 & 0 & \mu(\partial w/\partial x_1 - \theta x_2) \\ 0 & 0 & \mu(\partial w/\partial x_2 + \theta x_1) \\ - & - & 0 \end{bmatrix}$$

Certainly $\sigma_{33} = 0$ on the top and bottom—it is zero everywhere. On the sides, where $n = (n_1, n_2, 0)$ points outwards, multiplying by σ^T gives zero automatically in two components. The only serious boundary condition is

$$(\sigma^T n)_3 = \mu \left(\frac{\partial w}{\partial x_1} - \theta x_2 \right) n_1 + \mu \left(\frac{\partial w}{\partial x_2} + \theta x_1 \right) n_2 = 0. \tag{7}$$

The equilibrium equation div $\sigma = 0$ is similar. Its first two components are automatic and the only serious equation is

$$\frac{\partial}{\partial x_1} \left(\frac{\partial w}{\partial x_1} - \theta x_2 \right) + \frac{\partial}{\partial x_2} \left(\frac{\partial w}{\partial x_2} + \theta x_1 \right) + \frac{\partial}{\partial x_3} 0 = 0. \tag{8}$$

That is *Laplace's equation* for w, after discarding the zeros.

To be honorable we have to show it has a solution. If $w = w_0$ were given there would be no doubt, but (7) gives the normal derivative $\nabla w \cdot n = \theta(x_2 n_1 - x_1 n_2) = F$. This is the Neumann problem, which can only be solved when the total flux through the boundary is zero. If div grad $w = 0$ then $\int \nabla w \cdot n \, ds = \int F \, ds$ must be zero. To show that F is all right we apply the divergence theorem to $v = (\theta x_2, -\theta x_1)$, whose divergence is zero:

$$\int F \, ds = \int v \cdot n \, ds = \iint \text{div } v \, dx \, dy = 0.$$

Thus a warping function can be found to complete the displacement.

Final remark   We have referred to tensors without saying what that word means. The strain e and the stress σ are tensors. Properly speaking, u and f are tensors. And more important, *A and C and A^T are tensors*. They are linear transformations between u and e, between e and σ, and between σ and f. The stress is also a linear transformation; if you give it a direction vector n it produces the force $\sigma^T n$ on the plane perpendicular to n. What is special about these transformations, and earns them the name "*tensor*," is that they have an intrinsic geometrical definition—not dependent on the coordinate system.

The statement $u = x_2$ has no meaning until the $x_1 - x_2 - x_3$ coordinates are known; x_2 is not a tensor. On the other hand the formula $e = \text{grad } u$ does have a meaning; the gradient is a tensor. It is true that to *compute* the gradient we need coordinates—if they are rectangular, as you and I immediately assumed they were for $u = x_2$, then grad $u = (0,1,0)$. But if they were cylindrical coordinates, so that u is actually θ, the gradient can go forward in that system—and it involves $1/r$. The gradient and divergence look different for different systems; but in themselves, they do not change. The laws of mechanics are coordinate-free (or we are absolutely lost).

Those laws were written in rectangular coordinates, but a tensor analyst is prepared for other systems. A rotation of axes gives the easy test that a tensor must pass, to verify its **invariance under coordinate transformations**. For the theory of relativity, which includes invariance under moving coordinate changes, Einstein had to go further. He needed curvilinear coordinates and the Schwarz-Christoffel symbols that come with derivatives. That would carry us pretty far, and not with the speed of light, but we do mention three tensors in addition to A and C and A^T. One is the curl; the second is the Laplacian div grad; the third computes the acceleration from the velocity. It involves $v \cdot \nabla$, which is $\Sigma \, v_i \, \partial/\partial x_i$ in rectangular coordinates, and it will soon be seen as basic for fluids.

Fluid Mechanics

For a fluid, equilibrium is not quite the right word. The flow can be *steady*, so that nothing depends on time. Nevertheless there is flow—a movement of fluid— just as there was movement of electrons within a network in "equilibrium." A steady flow satisfies a conservation law like Kirchhoff's current law, but its velocity does not have to come from a potential. The analogue of the voltage law was irrotational flow, with zero vorticity, but in a moment that law will be suspended. It is replaced by conservation of momentum.

First we consider **conservation of mass**. For a steady flow div $w = 0$ as before. That is the differential form of mass conservation; the integral form is "flow rate in equals flow rate out," when the region contains no sources. The flow rate and velocity are related by $w = \rho v$.† If the density is constant then also div $v = 0$. But there is a distinction between div $w = 0$ and div $v = 0$ which needs to be discussed. It reflects the distinction between an **Eulerian** description and a **Lagrangian** description of the flow.

The Eulerian description gives the velocity at each point. The Lagrangian description gives the velocity of each particle. The fluid is flowing past Euler, who

---

†We surrender to the right notation ρ for density. Velocity is still v, although it is u or q in many books—where the potential becomes φ.

sits at a point and watches Lagrange go by. To Euler the continuity equation is

$$\frac{\partial \rho}{\partial t} + \text{div}\,(\rho v) = 0. \tag{9}$$

To Lagrange the same equation is

$$\frac{D\rho}{Dt} + \rho\,\text{div}\,v = 0. \tag{10}$$

Both take a crucial step beyond the static conservation laws seen up to now. Mass is still conserved, but it need not be stationary. The flow out can differ from the flow in, which means that the density can change. For Euler that change produces $\partial \rho / \partial t$; when it is absent we are back to div $w = 0$.

The novelty is in the **material derivative** or **substantial derivative** D/Dt, the rate of change as it would look to Lagrange. To see the difference we compare div (ρv) to ρ div v:

$$\text{div}\,(\rho v) = \frac{\partial}{\partial x_1}(\rho v_1) + \frac{\partial}{\partial x_2}(\rho v_2) + \frac{\partial}{\partial x_3}(\rho v_3) = \rho\,\text{div}\,v + v\cdot\text{grad}\,\rho. \tag{11}$$

The last term has to be part of $D\rho/Dt$, if (9) and (10) are the same. Even in steady flow, when $\partial \rho / \partial t = 0$, the material derivative $D\rho/Dt$ need not be zero—if Lagrange is carried by the flow to points of different density. Those are the new ingredients in $D\rho/Dt$, the velocity v that carries him and the density change grad ρ. A corresponding term enters Df/Dt for any function, not only the density function $f = \rho$, and we emphasize that **advection term** $v\cdot$ grad f:

3J   For any function $f(t,x)$, the ordinary derivative $\partial/\partial t$ at points fixed in space (Euler) is related to the material derivative D/Dt at points moving with the fluid velocity $v(t,x)$ (Lagrange) by the **transport rule**:

$$\frac{Df}{Dt} = \frac{\partial f}{\partial t} + v\cdot\text{grad}\,f. \tag{12}$$

The equation div $v = 0$ comes from the continuity equation (9) when ρ is constant, but it also comes from (10) when the fluid is **incompressible**:

if $D\rho/Dt = 0$ (incompressibility) then   div $v = 0$.

The density need not be constant and the flow need not be steady to have div $v = 0$ and incompressibility.

We test the transport rule (12) on an example in one space dimension. Suppose $f = tx$. After a small time, t becomes $t + dt$. To Lagrange, x also changes. He travels with velocity v and moves to the point $x + v\,dt$. To him f changes from tx to $(t + dt)(x + v\,dt)$. Therefore

$$Df = x\,dt + tv\,dt \quad \text{or} \quad \frac{Df}{Dt} = \frac{\partial f}{\partial t} + v\frac{\partial f}{\partial x}.$$

In general $f(t,x,y,z)$ changes to $f(t + dt,\ x + v_1\,dt,\ y + v_2\,dt,\ z + v_3\,dt)$ as f is transported by the fluid. The time derivative contains $\partial f/\partial t$ and also $v \cdot \nabla f$, verifying (12).

Note   A more fundamental approach to the continuity equation starts with one of the *integral* forms of mass conservation, again without sources or sinks:

$$\frac{\partial}{\partial t}\int_{V_E} \rho\,dV = -\int_{S_E} \rho v \cdot n\,dS \quad \text{or} \quad \frac{D}{Dt}\int_{V_L} \rho\,dV = 0. \tag{13}$$

The first volume V_E is fixed in space; the rate of change to Euler is flow in minus flow out. The second volume V_L moves with the fluid; the rate of change to Lagrange is zero. The two are connected by the transport theorem—the integral form of the transport rule. The proof rests on a change of variables from Eulerian to Lagrangian coordinates, and on the formula $D|J|/Dt = |J|\,\text{div}\ v$ for the change in the Jacobian determinant.

Now we meet Df/Dt when f is v itself.

Acceleration and the Balance of Momentum

We want to express Newton's law $F = ma$. In fluids as elsewhere, force equals mass times acceleration. That is the balance of momentum—its rate of change equals the net force. In statics both sides were zero; there was no acceleration and no unbalanced force. The difference F between internal and external forces was zero (which gave the equilibrium equation). Now there is something to be said on both sides of the equation. We begin with the acceleration.

The important point is that the acceleration is not $\partial v/\partial t$. It is the change of momentum of a *fluid particle* to which Newton's law applies, not the change at a point being watched by Euler. The particle is moving with the fluid velocity v. To find the vector Dv/Dt we apply the transport rule (12) to each component: in the x-direction the acceleration is

$$\frac{Dv_1}{Dt} = \frac{\partial v_1}{\partial t} + v \cdot \text{grad}\ v_1.$$

The other two components of acceleration are similar. The three scalar equations combine into a vector equation

$$\frac{Dv}{Dt} = \frac{\partial v}{\partial t} + (v \cdot \text{grad})v = \frac{\partial v}{\partial t} + \sum v_i \frac{\partial}{\partial x_i} v. \tag{14}$$

The parenthesis in grad)v is a reminder to take the gradient of each velocity component. Then multiplication by the mass density ρ gives one side of Newton's law, and the force density goes on the other side.

The force is partly external and partly internal. For the external part, like gravity, it is usual to write f as force per unit mass. Then ρf appears in Newton's law; in many problems it can be neglected. It is the internal stress T that is crucial, and here the two main categories of fluids part company:

> A **perfect fluid** allows no tangential stress: $T = -pI$
> A **viscous fluid** has internal friction: $T = -pI + \sigma$

In the first case all the stress comes from pressure. The force on every surface is perpendicular to that surface. When pushed the fluid pushes straight back; a perfect fluid cares nothing for shear. In the viscous case there is a stress matrix σ like the one for solids, which has entries off the diagonal. The difference is that the fluid is moving! The displacement changes to the displacement *rate*—in other words the velocity. The strain changes to the strain rate D. The stress is T and the framework ends with the equation of motion:

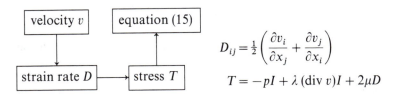

$$D_{ij} = \tfrac{1}{2}\left(\frac{\partial v_i}{\partial x_j} + \frac{\partial v_j}{\partial x_i}\right)$$

$$T = -pI + \lambda\,(\text{div } v)I + 2\mu D$$

The stress-strain relation is still linear; it is the material law for a Newtonian fluid. It includes the pressure. That requires no motion, and it points exactly to the distinction between a solid and a fluid: The stress in a motionless fluid must be a multiple $-pI$ of the identity. But the pressure in a moving fluid is not just hydrostatic pressure, based on depth below the surface.

Now come the great equations of fluid mechanics. Conservation of mass is already set; that is the continuity equation (9–10). Conservation of momentum is Newton's law $ma = F$, and the two sides of that momentum balance are

$$\rho \frac{Dv}{Dt} = \rho f + \text{div } T. \tag{15}$$

That is the **equation of motion** for a fluid. The divergence of the stress gives the internal force, as it did for solids, and what remains is to compute div T for perfect fluids and viscous fluids.

$$\text{Perfect fluids:} \quad T = \begin{bmatrix} -p & & \\ & -p & \\ & & -p \end{bmatrix} \quad \text{and div } T = \begin{bmatrix} -\partial p/\partial x_1 \\ -\partial p/\partial x_2 \\ -\partial p/\partial x_3 \end{bmatrix} = -\text{grad } p .$$

Viscous incompressible fluids: div $T = -\text{grad } p + \mu \nabla^2 v$.

That last calculation is more complicated; it is developed in the exercises. For compressible flow, the pressure p is a function of density and temperature; there is an equation of state. For incompressible flow p is the Lagrange multiplier for the continuity equation div $v = 0$. In practice this is the most important case, and we have reached the two most important equations:

3K The motion of a perfect fluid obeys **Euler's equations**:

$$\rho \frac{Dv}{Dt} = \rho f - \text{grad } p. \tag{16}$$

A viscous incompressible fluid obeys the **Navier-Stokes equations**:

$$\rho \frac{Dv}{Dt} = \rho f - \text{grad } p + \mu \nabla^2 v. \tag{17}$$

Incompressibility and continuity are expressed by $D\rho/Dt = 0$ and div $v = 0$.

By comparison with div $v = 0$, the Euler and Navier-Stokes equations look formidable. Unfortunately they are. They are made nonlinear by the advection term in the acceleration, which as in (14) is $Dv/Dt = \partial v/\partial t + (v \cdot \text{grad})v$. Even for steady flow that nonlinear term is present. No one is sure, for three-dimensional flows, whether these equations always have solutions. Probably Euler is even more at risk than Navier-Stokes, which has viscosity to dissipate energy. However turbulence still occurs when the ratio $v = \mu/\rho$ is sufficiently small. Then advection dominates dissipation, and a smooth laminar flow becomes unstable.

We take up the two equations in turn.

Euler and Bernoulli Equations

In many cases Euler's equations can be integrated once. What is needed is for every term to be a gradient. The trick is to rewrite the advection by using the vector identity

$$(v \cdot \text{grad})v = \tfrac{1}{2} \text{grad}(v_1^2 + v_2^2 + v_3^2) - v \times \text{curl } v. \tag{18}$$

Suppose the flow is irrotational, so that curl $v = 0$ and the last term is removed. Suppose also that f is a conservative force like gravity: it is the gradient of a force potential, $f = -\text{grad } G$. Then Euler's equation divided by ρ is

$$\frac{\partial v}{\partial t} + \tfrac{1}{2}\text{grad}(v_1^2 + v_2^2 + v_3^2) = -\frac{1}{\rho}\text{grad } p - \text{grad } G. \tag{19}$$

If $\partial v/\partial t = 0$ and $\rho = \text{constant}$, this contains only gradients. It says that the gradient of $\tfrac{1}{2}(v_1^2 + v_2^2 + v_3^2) + p/\rho + G$ is zero. Therefore that function must be constant— which yields the most directly useful equation in nonlinear fluid mechanics.

3L   For steady irrotational flow at constant density, **_Bernoulli's equation_** is

$$\tfrac{1}{2}(v_1^2 + v_2^2 + v_3^2) + \frac{p}{\rho} + G = \text{constant.} \tag{20}$$

Higher velocity means lower pressure.

This partly accounts for the possibility of throwing a curveball. When the ball has overspin, the air below it moves faster and lowers the pressure—causing the ball to sink. Bernoulli does not account for a knuckleball, which spins so slowly that it can drift the other way. In fact we would need a different Bernoulli equation—a different first integral of the equation of motion—to permit rotation.

EXAMPLE 1 $v = (y,x,0)$ with   curl $v = 0$ and   div $v = 0$

This is plane potential flow, as in Section 3.3. The fluid travels along hyperbolas and v is the gradient of a velocity potential $u = xy$. In that section pressure was not mentioned; Kirchhoff's laws were enough. Here the pressure comes immediately from Bernoulli's equation:

$$\frac{p}{\rho} = -\tfrac{1}{2}(y^2 + x^2).$$

EXAMPLE 2   A tank with a hole

Suppose the height of the fluid above the hole is h. How fast does the fluid come out? It obeys Bernoulli's law, with a force potential gz from gravity. Therefore

$$\tfrac{1}{2}(v_1^2 + v_2^2 + v_3^2) + \frac{p}{\rho} + gz = \text{constant.}$$

At the top of the fluid every term is zero. At the hole, the pressure is also zero—because the hole is open. The z-coordinate there is $-h$. Multiplying by 2 and taking square roots, the speed is $\sqrt{2gh}$—which is curiously the same as if particles were in free fall down to the hole.

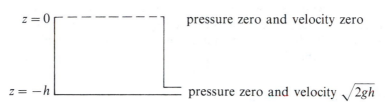

$z = 0$ pressure zero and velocity zero

$z = -h$ pressure zero and velocity $\sqrt{2gh}$

Bernoulli's law is tremendously valuable; one more case will be mentioned. Irrotational flow has a velocity potential—curl $v = 0$ means $v = \mathrm{grad}\, u$—so the *unsteady* equations also have gradients everywhere. But now all functions can depend on time: (19) is the gradient of

$$\frac{\partial u}{\partial t} + \tfrac{1}{2}(v_1^2 + v_2^2 + v_3^2) + \int \frac{dp}{\rho} = \text{function of } t.$$

To go further than Bernoulli, we need to distinguish 2-dimensional flow from 3-dimensional flow. You will see the difference after taking the curl of both sides of Euler's equation (16). On the right side the curl of a gradient is zero, and we assume $f = 0$. On the left side we meet a miserable identity but the result is extremely satisfactory: if $\omega = \mathrm{curl}\, v$ then

$$\boxed{\frac{D\omega}{Dt} = (\omega \cdot \mathrm{grad})v.} \tag{21}$$

In two-dimensional flow that right hand side is zero! The velocity is $(v_1, v_2, 0)$, the *"vorticity"* is $\omega = (0, 0, \omega_3)$, and nothing depends on x_3. Therefore $D\omega/Dt = 0$ and there is a new conservation law: *the vorticity ω is conserved* along every streamline. In fact there is a stream function in two dimensions, where the continuity equation is

$$\mathrm{div}\, v = \frac{\partial v_1}{\partial x} + \frac{\partial v_2}{\partial y} = 0 \quad \text{with solution } (v_1, v_2) = \left(\frac{\partial s}{\partial y}, \ -\frac{\partial s}{\partial x} \right).$$

The curl of this velocity vector has

$$\omega_3 = -\frac{\partial}{\partial x}\left(\frac{\partial s}{\partial x} \right) - \frac{\partial}{\partial y}\left(\frac{\partial s}{\partial y} \right) = -\nabla^2 s, \tag{22}$$

This is the **vorticity-stream function** formulation: a nonlinear equation $D\omega_3/Dt = 0$

for the vorticity, and a linear equation $\omega_3 = -\nabla^2 s$ for the stream function. In this approach the "primitive variables" of velocity and pressure are found later.

In three dimensions vortices are stretched and the flow is much more complex. However it is still true that vortex lines and vortex sheets move with the fluid. Equation (21) has become the basis for a very powerful numerical method—the *vortex method*—which follows discrete vortices through violent motion: turbulent combustion, boundary layers, instability at high Reynolds numbers, and general breakdown. It is at the heart of computational fluid dynamics, and competes with the primitive variables for supremacy.

EXAMPLE   The "ballerina effect"

The velocity $v = (x, -y, 0)$ follows the hyperbolas of Fig. 3.13. It has no curl and no divergence, so there is a potential and a stream function:

$$v = \operatorname{grad} u = \operatorname{grad} \tfrac{1}{2}(x^2 - y^2) \quad \text{and} \quad v = \operatorname{curl} s = \operatorname{curl} xy.$$

This is potential flow and u satisfies Laplace's equation. The movement of a fluid particle is given by $dr/dt = v$:

$$\frac{dx}{dt} = x \quad \text{so} \quad x = x_0 e^t, \quad \frac{dy}{dt} = -y \quad \text{so} \quad y = y_0 e^{-t}.$$

The x-coordinate increases as the y-coordinate decreases. Fluid inside a circle at time zero moves into an ellipse at time t:

$$\text{the circle} \quad x_0^2 + y_0^2 = 1 \quad \text{is stretched to} \quad e^{-2t} x^2 + e^{2t} y^2 = 1.$$

It is like an ice-skater with her arms extended—but so far there is no rotation.

That comes from a shear flow: $w = (0, 0, y)$ has div $w = 0$ but curl $w = (1, 0, 0)$. It has a stream function $s = -\tfrac{1}{2} y^2$ but no potential. The fluid moves in the z-direction (Fig. 3.13) but some particles go faster than others. There is rotation around the x-axis (no particle actually goes *around* the axis!) since that is the direction of curl w.

Now we combine the two flows. The velocities cannot be added, since $v + w$ does not satisfy the (nonlinear) equations of motion. The mixture of potential flow and shear flow gives the *unsteady* velocity vector

$$v = (x, -y, e^t y) \quad \text{with div } v = 0 \quad \text{and curl } v = (e^t, 0, 0).$$

It is three-dimensional, with particles going up or down while their projection moves along the hyperbolas. The vorticity $\omega = \operatorname{curl} v = (e^t, 0, 0)$ satisfies equation (21). The circle that stretched into an ellipse is now spinning around the x-axis because of ω.

This is the "***ballerina effect***." She becomes tall and thin, like the ellipse, and she spins faster and faster like an ice-skater. I am a little sorry she is spinning around the x-axis.

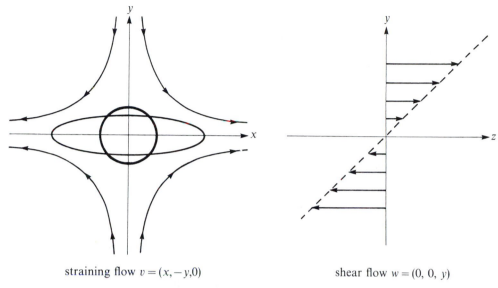

straining flow $v = (x, -y, 0)$ shear flow $w = (0, 0, y)$

Fig. 3.13. Stretching of vortices and spinning of ballerinas.

The Navier-Stokes Equations

Those equations we do not intend to solve. But one valuable thing can be done, which is important throughout applied mathematics. It is to recognize how two different problems can be "*similar*". After a change of units they are identical. The key is to find the right *dimensionless number*, and here it will be the **Reynolds number**. If it is the same for two flows, and if the regions and boundary conditions are also similar, then to find one flow is to find the other.

The basic idea is simple. The Navier-Stokes equation is

$$\frac{\partial v}{\partial t} + \sum v_i \frac{\partial v}{\partial x_i} = -\frac{1}{\rho} \operatorname{grad} p + v\nabla^2 v. \tag{23}$$

The kinematic viscosity coefficient is $v = \mu/\rho$ and the force f was set to zero. The size of the region is determined by a length L, and the velocity is based on a mainstream velocity V. The units of viscosity, length, and velocity are

$$v = \mathrm{cm}^2/\mathrm{sec}, \quad L = \mathrm{cm}, \quad V = \mathrm{cm}/\mathrm{sec}.$$

Therefore ***the ratio*** $\mathrm{Re} = VL/v$ ***is dimensionless***.

That is the Reynolds number. It measures the relative importance of convection and viscosity. It is the only parameter in the Navier-Stokes equations, if we rescale to the dimensionless variables

$$\bar{v} = \frac{v}{V}, \quad \bar{x} = \frac{x}{L}, \quad \bar{t} = \frac{tV}{L}, \quad \frac{\bar{p}}{\bar{\rho}} = \frac{p}{\rho V^2}.$$

Substituting into (13), and remembering that the length L appears in the gradient and L^2 appears in ∇^2, every term in (23) has the factor V^2/L—whose removal leaves a dimensionless equation

$$\frac{\partial \bar{v}}{\partial \bar{t}} + \sum \bar{v}_i \frac{\partial \bar{v}}{\partial \bar{x}_i} = -\frac{1}{\rho} \text{ grad } \bar{p} + \frac{\nabla^2 \bar{v}}{\text{Re}}. \tag{24}$$

That leads to the **law of similarity**.

3M   All similar flows, sharing the same Reynolds number Re, can be found by rescaling the solution of (24). Flows at different speeds past different spheres and with different viscosities depend on the single number VL/v.

Note that v is the ratio of μ to ρ (nothing to do with Poisson's ratio for solids). Surprisingly, it is larger for air than for water—which is more viscous but much more dense.

It remains to comment on boundary conditions. Here perfect fluids and viscous fluids are different. That difference is present in the higher derivative $\nabla^2 v$, and in the correspondingly higher number of boundary conditions. At a solid surface the viscous fluid has **no slip**, whereas without viscosity it can flow *along* the boundary (but not through it):

$$\boxed{\text{viscous flow: } v = 0 \quad \text{inviscid flow: } v \cdot n = 0}$$

EXAMPLE   Flow between parallel planes

The fluid is between the planes $z = h$ and $z = -h$. It flows in the x-direction with velocity $v = (v_1, 0, 0)$, and the continuity equation is div $v = \partial v_1/\partial x = 0$. Therefore the velocity is independent of x—and everything is independent of y. The problem is to find the function $v_1(z)$—the velocity at different heights. The nonlinear term $(v \cdot \text{grad})v$ is zero, which allows a simple solution to Euler and Navier-Stokes.

Without viscosity, Euler is $0 = -\text{grad } p$. The pressure is constant and any velocity profile $v_1(z)$ satisfies the boundary condition $v \cdot n = (v_1, 0, 0) \cdot (0, 0, 1) = 0$. No fluid goes into the planes.

The Navier-Stokes boundary condition is $v = 0$, and that is *not* automatic. Viscosity prevents flow along as well as into the planes. Since Dv/Dt is zero, the

Navier-Stokes equation becomes grad $p = \mu \nabla^2 v$:

$$\frac{\partial p}{\partial x} = \mu \left(\frac{\partial^2 v_1}{\partial x^2} + \frac{\partial^2 v_1}{\partial y^2} + \frac{\partial^2 v_1}{\partial z^2} \right) = \mu \frac{\partial^2 v_1}{\partial z^2}, \quad \frac{\partial p}{\partial y} = 0, \quad \frac{\partial p}{\partial z} = 0.$$

From the last two equations, the pressure depends only on x. So does $\partial p / \partial x$, while $\mu \partial^2 v_1 / \partial z^2$ depends only on z. They must equal the same constant c. Thus the pressure is linear, $p = cx + p_0$. *The velocity profile is quadratic:*

$$\mu \frac{\partial^2 v_1}{\partial z^2} = c \quad \text{or} \quad v_1 = \frac{c}{2\mu} (z^2 - h^2). \tag{25}$$

The flow is fastest along the centerline $z = 0$, and $v_1 = 0$ at the boundaries $z = \pm h$. This is called **plane Poiseuille flow**.

The Stokes Equations

Limiting cases are always important. Starting from Navier-Stokes, one extreme is Euler's equation; the viscosity disappears and the acceleration Dv/Dt takes over. The Reynolds number goes to infinity, and the flow can be extremely wild. At the other extreme, when the motion is very steady and very slow, the acceleration can be neglected and we have a linear equation. The Reynolds number has gone to zero. These are exactly the assumptions in **Stokes flow**.

The only remaining terms are the internal forces due to pressure and viscosity, balancing an external force such as gravity. The internal forces come from pressure gradients and from the Laplacian: in two dimensions

$$\text{grad } p = \begin{bmatrix} \dfrac{\partial p}{\partial x} \\[2mm] \dfrac{\partial p}{\partial y} \end{bmatrix} \quad \text{and} \quad \mu \nabla^2 v = \mu \begin{bmatrix} \dfrac{\partial^2 v_1}{\partial x^2} + \dfrac{\partial^2 v_1}{\partial y^2} \\[2mm] \dfrac{\partial^2 v_2}{\partial x^2} + \dfrac{\partial^2 v_2}{\partial y^2} \end{bmatrix}.$$

The pressure takes the place of a potential, and μ measures the viscosity. Then the Stokes equations (without Navier) are

$$\boxed{\begin{aligned} -\mu \nabla^2 v + \text{grad } p &= f \\ \text{div } v &= 0. \end{aligned}} \tag{26}$$

They give the simplest model for real (and very viscous) fluids. They also appear in the numerical solution of the Navier-Stokes equation, when it is split into a Stokes step and a nonlinear step.

Our brief study has two goals. The first is to fit the Stokes model into our favorite framework

$$\begin{bmatrix} C^{-1} & A \\ A^T & 0 \end{bmatrix} \begin{bmatrix} v \\ p \end{bmatrix} = \begin{bmatrix} f \\ 0 \end{bmatrix}.$$

That is easy to do. A represents the gradient, exactly as it did in ideal flow. The flow is driven by pressure differences grad p rather than potential differences grad u. A^T is again the divergence; the minus sign is no problem in div $v = 0$. The big change comes in C, which is more complicated than we have ever seen it. C^{-1} is a multiple of Laplace's operator, $C^{-1}v = -\mu \nabla^2 v$, so the underlying principle

$$\text{Minimize} \quad Q(v) = \tfrac{1}{2} v^T C^{-1} v \quad \text{subject to} \quad A^T v = 0$$

becomes

$$\text{Minimize} \quad Q(v) = \frac{\mu}{2} \int \int |\text{grad } v|^2 \, dx \, dy \quad \text{subject to} \quad \text{div } v = 0.$$

The pressure is the Lagrange multiplier for the constraint div $v = 0$.

The second goal is to discover a new equation (new to this book). It comes from solving div $v = 0$. In the discrete case the currents that satisfied Kirchhoff's law were combinations of "loop currents." Now we have done the same for the continuous case, by using vector calculus: v is a curl. *The loop currents are replaced by a stream function $s(x,y)$:*

$$\text{if} \quad \text{div } v = 0 \quad \text{then} \quad (v_1, v_2) = \left(\frac{\partial s}{\partial y}, -\frac{\partial s}{\partial x} \right) \quad \text{for some } s.$$

Substituting for v in equation (26), with no forces f, leaves

$$-\mu \left(\frac{\partial^2}{\partial x^2} + \frac{\partial^2}{\partial y^2} \right) \frac{\partial s}{\partial y} + \frac{\partial p}{\partial x} = 0$$

$$\mu \left(\frac{\partial^2}{\partial x^2} + \frac{\partial^2}{\partial y^2} \right) \frac{\partial s}{\partial x} + \frac{\partial p}{\partial y} = 0.$$

Now differentiate the first with respect to y and the second with respect to x, subtract to cancel p, and divide by μ:

$$\frac{\partial^4 s}{\partial x^4} + 2 \frac{\partial^4 s}{\partial x^2 \partial y^2} + \frac{\partial^4 s}{\partial y^4} = 0.$$

We have arrived at a fourth order equation for s, and it is the result of applying the Laplace operator twice:

$$\left(\frac{\partial^2}{\partial x^2} + \frac{\partial^2}{\partial y^2} \right)^2 s = 0. \tag{27}$$

This is the **biharmonic equation**. We could use it in fluid mechanics, for steady motion (in 2D) of a very viscous fluid. It is fundamental also to solid mechanics, where $d^4u/dx^4 = 0$ gave the displacement of a one-dimensional beam. Equation (27) describes the displacement of an unloaded *plate*. But we can hardly expect to write down the solution of a fourth order partial differential equation with general boundary conditions. Therefore we leave it unsolved, as the extreme example of a study that began with equilibrium on a network and led us here.

In that study the analogies are as important as the analysis. If you have seen how Kirchhoff's current law corresponds to zero divergence, and how his voltage law leads to potential differences and derivatives and gradients, you have understood the most essential point. In any problem of equilibrium that is the pattern to look for. Together with those laws you should find a third one, describing the properties of the material. Then the framework is complete, and the equilibrium can be computed.

EXERCISES

3.5.1   A rigid motion is a combination of a constant displacement U (a translation of the whole body) and a pure rotation: $u = U + \omega \times r$. Find the strain Au.

3.5.2   From the laws (4–6) find the strain and stress and external force for each of the following displacements:
   (a) $u = (\lambda_1 x_1, \lambda_2 x_2, \lambda_3 x_3)$ (stretching)
   (b) $u = (-x_2 v(x_1), v(x_1), 0)$ (bending of a beam)
   (c) $u = (\partial\varphi/\partial x_1, \partial\varphi/\partial x_2, \partial\varphi/\partial x_3)$ (displacement potential)
   (d) $u = (0, 0, u_3(x, y))$ (antiplane shear)

3.5.3   The principle of virtual work is

$$\iiint v^T f \, dV + \iint v^T F \, dS = \iiint (Av)^T \sigma \, dV$$

for any "virtual displacement" v. It is the weak form of the equilibrium equation; the perturbation v is required to be zero only where $u = u_0$ is prescribed. Use Green's formula (1) for integration by parts to reach the strong form div $\sigma + f = 0$ in V, $\sigma^T n = F$ on S.

3.5.4   In two-dimensional flow the continuity equation div $w = 0$ was solved by means of a stream function: $w = (\partial s/\partial y, -\partial s/\partial x)$ is divergence-free for any s. In two-dimensional elasticity show that div $\sigma = 0$ is solved by an Airy stress function:

$$\sigma = \begin{bmatrix} \dfrac{\partial^2 A}{\partial y^2} & -\dfrac{\partial^2 A}{\partial x \partial y} & 0 \\[2mm] -\dfrac{\partial^2 A}{\partial x \partial y} & \dfrac{\partial^2 A}{\partial x^2} & 0 \\[2mm] 0 & 0 & 0 \end{bmatrix} \qquad \text{has div } \sigma = 0 \quad \text{for any } A.$$

3.5.5   Find Young's modulus and Poisson's ratio from the Lamé constants μ and λ, by inverting the relations

$$2\mu = \frac{E}{1+v} \quad \text{and} \quad \lambda = \frac{vE}{(1+v)(1-2v)}.$$

3.5.6   Verify that $A^T C A u = f$ can be rewritten as

$$\mu \text{ curl curl } u - (\lambda + 2\mu) \text{ grad div } u = f.$$

3.5.7   (More important than the above) Invert the isotropic stress-strain relation $\sigma_{ij} = 2\mu e_{ij} + \lambda(\text{tr } e)\delta_{ij}$ to find the strain from the stress: $e = C^{-1}\sigma$. Show first that tr σ is $(2\mu + 3\lambda)$tr e, and substitute for tr e to find

$$e_{ij} = \frac{1}{2\mu} \sigma_{ij} - \frac{\lambda}{2\mu + 3\lambda} (\text{tr } \sigma)\delta_{ij}.$$

3.5.8   Show that the strain energy $\frac{1}{2} e^T C e$ is $\mu \Sigma\Sigma e_{ij}^2 + \frac{1}{2} \lambda(\text{tr } e)^2$. From the previous exercise find the complementary energy $\frac{1}{2}\sigma^T C^{-1}\sigma$.

3.5.9   What are P and Q and the minimum principles that govern equilibrium continuum mechanics?

3.5.10   In the torsion of a *circular* rod show that $\nabla w \cdot n = \theta(x_2 n_1 - x_1 n_2)$ is zero around the side—where n points radially outward. Then $w = 0$ everywhere and circular cross-sections are not warped.

3.5.11   For a rod with the square cross-section $-1 \le x_1, x_2 \le 1$ find $F = \theta(x_2 n_1 - x_1 n_2)$ on each of the four sides and verify that $\int F \, ds = 0$.

3.5.12   For an infinitely long rod with fixed boundary suppose all internal forces and displacements are in the x_3-direction, but are independent of the distance x_3 along that axis: $u = (0,0,u_3(x_1,x_2))$. Find the strains, the stresses, the equilibrium equation, and the boundary condition, all in terms of u_3.

3.5.13   There are six independent entries in a 3 by 3 symmetric matrix. Looked at differently, it has three eigenvalues; show how the eigenvectors provide the other three degrees of freedom.

3.5.14   For the displacement $u = (x,xy,xyz)$, find the matrix J with entries $\partial u_i / \partial x_j$ and split it into a symmetric strain part e and a skew-symmetric rotational part r.

3.5.15   Find the material derivatives $D\rho/Dt$ and Dv/Dt when

$$\rho = x^2 + y^2, \quad v = (y,x,0) \quad \text{and} \quad \rho = ze^t, \quad v = (x,0,-z).$$

Which satisfies the continuity equation (10) and which is incompressible?

3.5.16   Apply the divergence theorem to Euler's conservation of mass (13), and then use Green's formula $\text{div}^T = -\text{grad}$ to reach his form (9) of the continuity equation.

3.5.17   Beer in a rotating barrel has the force potential $G = -gz - \frac{1}{2}\omega^2(x^2 + y^2)$, from gravity and centrifugal force. Its velocity is zero with respect to the barrel. Why does the surface of the beer have the parabolic shape $z = -\omega^2(x^2 + y^2)/2g$?

3.5.18   For the same liquid in the same barrel, the velocity with respect to a resting observer is $v = (-\omega y, \omega x, 0)$. Verify the equation of continuity, and reduce Euler's equation to $\rho\omega^2 x = \partial p/\partial x$, $\rho\omega^2 y = \partial p/\partial y$, $-\rho g = \partial p/\partial z$. Find p.

3.5.19   An ideal fluid is flowing at pressure p and velocity v along a pipe of area A. If the area shrinks to $\frac{1}{2}A$, what is the new velocity (by conservation of fluid) and what is the new pressure (by Bernoulli)?

3.5.20   If the punctured tank also has holes at depths $d = \frac{1}{4}h$, $\frac{1}{2}h$, and $\frac{3}{4}h$, at what speeds does the fluid come out? In the time it takes to fall to the base of the tank (which is $\sqrt{2(h-d)g}$) which of the three jets goes out the farthest?

3.5.21   Why does Bernoulli's equation not hold for the Poiseuille flow in (25)?

3.5.22   (a)   Verify the key vector identity (18).
(b)   Take the curl of both sides to reach $-(\text{curl } v \cdot \text{grad})v$ when div $v = 0$ (the "miserable identity" of equation (21) for the vorticity).

3.5.23   In a viscous incompressible fluid, with div $v = 0$ and $\sigma_{ij} = \mu(\partial v_i/\partial x_j + \partial v_j/\partial x_i)$, show that the force div σ from that stress is $\mu\nabla^2 v$—which enters the Navier-Stokes equation.

3.5.24   Suppose the gravitational acceleration g is added to the Navier-Stokes equation (23), with the dimension cm/sec^2. Show that the *Froude number* $F = V^2/Lg$ is dimensionless. Two gravity flows are similar if they share both Re and F.

3.5.25   The *compressible* "gas dynamics equation" comes from

$$\rho(v \cdot \text{grad})v = -\text{grad } p = -c^2 \text{ grad } \rho \quad \left(c^2 = \frac{dp}{d\rho} = \text{sound speed}\right).$$

Continuity is $v \cdot \text{grad } \rho + \rho \text{ div } v = 0$. Taking inner products above with v, substitute ρ div v on the right and cancel ρ. For irrotational flow let $v = (u_x, u_y)$ and deduce that

$$u_x^2 u_{xx} + 2u_x u_y u_{xy} + u_y^2 u_{yy} = c^2(u_{xx} + u_{yy}).$$

This can be linearized by the "hodograph transformation" from (x, y) to (u_x, u_y)—changing independent variables by a mapping onto the velocity plane.

3.5.26   Show that $s = y^3$ satisfies the biharmonic equation (27), and find the corresponding velocity and pressure: $v = \text{curl } s$ and $\mu\nabla^2 v = \text{grad } p$. Along what line is $v = 0$? Sketch the flow in the half-plane $y > 0$.

3.5.27   A viscous fluid in a horizontal pipe of radius R has velocity $v_1 = c(y^2 + z^2 - R^2)/4\mu$ and pressure $p = cx + p_0$.
(a)   Verify that the Navier-Stokes equation and the boundary condition $v = 0$ are satisfied as in (25) (Poiseuille flow).
(b)   By integrating v_1 over the circle find $-\pi c R^4/8\mu$ as the net flow rate—which gives the classical experiment to determine μ.

3.6 ■ CALCULUS OF VARIATIONS

One theme of this book is the relation of equations to minimum principles. *To minimize P is to solve $P' = 0$.* There may be more to it, but that is the main point. When P was a quadratic $\frac{1}{2}x^T A x - x^T b$, or more generally when the unknown was a vector $x_1, \ldots, x_n$, there was no difficulty in reaching $P' = 0$. Solving it may be difficult, but the equation itself comes directly from $\partial P / \partial x_j = 0$. For quadratic P the equation is $Ax = b$, and it is symmetric positive definite at a minimum.

In a continuous problem the form of $P' = 0$ is not so easy to find. The unknown is a function, and P is usually an integral. The derivative P' is the *first variation*; the equation is $\delta P / \delta u = 0$. It has a weak form and a strong form. It is only for an elastic bar, when P was quadratic and the equation was linear and the problem was one-dimensional, that we already know both forms:

$$P = \int \left[\frac{c}{2} (u')^2 - fu \right] dx \qquad \int c u' v' dx = \int f v \, dx \qquad -(cu')' = f.$$

Our goal is to get beyond this first example of $\delta P / \delta u$.

The basic idea should be simple and it is: *Perturb u by a test function v.* Comparing $P(u)$ with $P(u + v)$, the linear term in the difference is the first variation. That is $\delta P / \delta u$, and it is zero for every admissible v. To carry out this program is to go from ordinary calculus to the calculus of variations. We do it in several steps:

(1) One-dimensional problems $P = \int F(u, u') dx$, not necessarily quadratic
(2) Constraints, not necessarily linear
(3) Two-dimensional problems $P = \int\int F(u, u_x, u_y) dx \, dy$
(4) Time-dependent equations in which $u' = du/dt$.

At each step the examples will be as familiar (and famous) as possible. In two dimensions that means Laplace's equation, and minimal surfaces in the nonlinear case. In time-dependent problems it means $F = ma$, and relativity in the nonlinear case. I hope you reach that part. In one dimension we rediscover the circle.

This section is also the opening to *control theory*—the modern form of the calculus of variations. Its constraints are differential equations, and Pontryagin's maximum principle yields solutions. But something has to be left for the next book.

Remark To go backwards from the strong form to the weak, *multiply by v and integrate.* For matrices the strong form is "$A^T C A u = f$." The weak form is "$v^T A^T C A u = v^T f$ for every v." Compare that with $\int v' c u' dx = \int v f dx$ above.

One-dimensional Problems

The basic problem is to minimize P with two boundary conditions:

$$P(u) = \int_0^1 F(u, u') dx \quad \text{with} \quad u(0) = a \quad \text{and} \quad u(1) = b.$$

The best u defeats every other candidate $u + v$ that satisfies these boundary conditions; thus $v(0) = v(1) = 0$. When v and v' are small the correction terms come from the first derivatives:

$$F(u + v, u' + v') = F(u, u') + v \frac{\partial F}{\partial u} + v' \frac{\partial F}{\partial u'} + \cdots \quad \text{(at each point)}$$

$$P(u + v) = P(u) + \int_0^1 \left(v \frac{\partial F}{\partial u} + v' \frac{\partial F}{\partial u'} \right) dx + \cdots \quad \text{(after integrating)}.$$

The integrated term is the first variation and we have already reached the **weak form**:

$$\frac{\partial P}{\delta u} = \int_0^1 \left(v \frac{\partial F}{\partial u} + v' \frac{\partial F}{\partial u'} \right) dx = 0 \quad \text{for every } v. \tag{1}$$

This is the equation for u. It is like a "directional derivative" for an ordinary function; the derivative in each direction v must be zero. The strong form looks for a single derivative which—if it is zero—makes all these directional derivatives zero. It comes from integrating by parts in equation (1):

$$\int_0^1 \left(v \frac{\partial F}{\partial u} - v \frac{d}{dx} \left(\frac{\partial F}{\partial u'} \right) \right) dx + \left[v \frac{\partial F}{\partial u'} \right]_0^1 = 0.$$

The boundary term vanishes because $v(0) = v(1) = 0$. Inside the integral v is arbitrary, so to guarantee zero we need the **strong form**:

$$\boxed{\frac{\partial F}{\partial u} - \frac{d}{dx} \left(\frac{\partial F}{\partial u'} \right) = 0.} \tag{2}$$

This is the **Euler equation** for a one-dimensional problem in the calculus of variations.

EXAMPLE The shortest path between two points

The points are $(0,a)$ and $(1,b)$; F is $\sqrt{1 + (u')^2}$. Calculus gives $P = \int F\, dx$ as the length of the arc between the points. F depends only on u' and the weak form involves only $\partial F / \partial u'$ (this derivative brings the square root into the denominator):

$$\int_0^1 v' \frac{u'}{\sqrt{1 + (u')^2}} dx = 0 \quad \text{for every } v. \tag{3}$$

We emphasize that the quantity multiplying v' does not have to be zero. If it is a constant then (3) is still satisfied; the integral of v' is certain to be zero (because

$v(0) = v(1) = 0$). Thus the strong form allows $\partial F / \partial u'$ to be constant, and actually *forces* it to be constant: Euler's equation (2) is

$$-\frac{d}{dx}\frac{u'}{\sqrt{1+(u')^2}} = 0 \quad \text{or} \quad \frac{u'}{\sqrt{1+(u')^2}} = c.$$

This integration is always possible when F is independent of u. It leaves a first-order equation $\partial F / \partial u' = c$. Squaring both sides, u' is also constant:

$$(u')^2 = c^2(1+(u')^2), \quad \text{or} \quad u' = \frac{c}{\sqrt{1-c^2}}, \quad \text{or} \quad u = \frac{c}{\sqrt{1-c^2}}x + d.$$

The optimal u is linear (*no surprise*). The constants c and d are chosen to match $u(0) = a$ and $u(1) = b$, and the shortest curve connecting two points is a straight line. It is a minimum, not a maximum or a saddle point, because the second derivative of F is positive; $\partial^2 F / \partial u'^2 = (1+(u')^2)^{-3/2} > 0$.

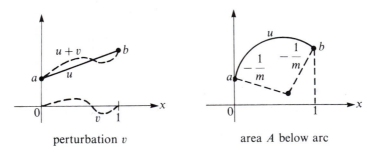

perturbation v area A below arc

Fig. 3.14. Shortest paths: straight line (unconstrained) and circle (constrained).

Constrained Problems

Suppose only certain paths between the points are allowed. We cannot go in a straight line because it violates a constraint. We take the constraint to be $\int u \, dx = A$, and look for ***the shortest curve which has area A below it***:

$$\text{Minimize} \quad P(u) = \int F(u')dx \quad \text{with} \quad u(0) = a, \quad u(1) = b, \quad \int_0^1 u \, dx = A.$$

To solve this problem the constraint should be built into P. There will be a Lagrange multiplier—here called m. It is a number and not a function, because there is one constraint rather than a constraint at every point. The Lagrangian with built-in constraint is

$$L = P + (\text{multiplier})(\text{constraint}) = \int (F + mu)dx - mA.$$

It is useful to go once more through the steps of optimization. By the calculus of variations u must satisfy the Euler equation:

$$\frac{\partial(F + mu)}{\partial u} - \frac{d}{dx}\left[\frac{\partial(F + mu)}{\partial u'}\right] = m - \frac{d}{dx}\frac{u'}{\sqrt{1 + (u')^2}} = 0.$$

Again this is favorable enough to be integrated:

$$mx - \frac{u'}{\sqrt{1 + (u')^2}} = c \quad \text{or} \quad u' = \frac{mx - c}{\sqrt{1 - (mx - c)^2}}.$$

After one more integration we reach the equation of a circle:

$$u = \frac{-1}{m}\sqrt{1 - (mx - c)^2} + d \quad \text{or} \quad (mx - c)^2 + (mu - d)^2 = 1. \tag{4}$$

The shortest path is a circular arc. It goes high enough to enclose area A. Dividing equation (4) by m^2, the radius of the circle is $-1/m$ and its center is at $(c/m, d/m)$. Those three numbers are determined by the three conditions $u(0) = a$, $u(1) = b$, $\int u\, dx = A$. If A has exactly the right value, m can be zero (the constraint does nothing) and the solution is a circle of infinite radius—which brings back a straight line.

We summarize the one-dimensional case, allowing F to depend also on u''— which introduces v'' into the weak form and needs two integrations by parts to reach the Euler equation. When F changes with x the Euler equation does not change, because it is u and not x that is perturbed.

3N The first variation of $P = \int_0^1 F(u, u', u'', x)\, dx$ is

$$\frac{\delta P}{\delta u} = \int_0^1 \left(v\frac{\partial F}{\partial u} + v'\frac{\partial F}{\partial u'} + v''\frac{\partial F}{\partial u''}\right) dx = 0 \quad \text{at a minimum.}$$

The Euler equation is

$$\frac{\partial F}{\partial u} - \frac{d}{dx}\left(\frac{\partial F}{\partial u'}\right) + \frac{d^2}{dx^2}\left(\frac{\partial F}{\partial u''}\right) = 0.$$

If there are constraints they bring with them Lagrange multipliers.

Applications are everywhere, and we mention one (of many) in sports. What angle is optimal in shooting a basketball? The force of the shot depends on the launch angle—for line drives or for sky hooks it is greater than for a soft shot in between. It is minimized at 45° if the ball leaves your hand ten feet up; for shorter people the angle is about 50°. What is interesting is that the same angle solves a second optimization problem: to have the largest margin of error and still score. The condition is $P' = 0$ in basketball and $\delta P/\delta u = 0$ in track—where the strategy to minimize the time $P(u)$ has been analyzed in tremendous detail.

Two-dimensional Problems

In two dimensions the principle is the same. The starting point is a quadratic P, without constraints, representing the potential energy over a plane region S:

$$P(u) = \iint_S \left[\frac{c}{2}\left(\frac{\partial u}{\partial x}\right)^2 + \frac{c}{2}\left(\frac{\partial u}{\partial y}\right)^2 - f(x,y)u(x,y) \right] dx\, dy.$$

If this energy is minimized at u, then $P(u+v) \geq P(u)$ for every v. Therefore we mentally substitute $u + v$ in place of u, and look for the term that is linear in v. That term is the first variation, which must be zero:

$$\frac{\delta P}{\delta u} = \iint_S \left[c\frac{\partial u}{\partial x}\frac{\partial v}{\partial x} + c\frac{\partial u}{\partial y}\frac{\partial v}{\partial y} - fv \right] dx\, dy = 0. \tag{5}$$

This is the equation of virtual work. It holds for every admissible v, and it is the weak form of the Euler equation. The strong form requires as always an integration by parts, in which the boundary conditions take care of the boundary terms. Inside S, that integration leaves

$$\iint_S \left[-\frac{\partial}{\partial x}\left(c\frac{\partial u}{\partial x}\right) - \frac{\partial}{\partial y}\left(c\frac{\partial u}{\partial y}\right) - f \right] v\, dx\, dy = 0. \tag{6}$$

Now the strong form appears. The integral is zero for every v, and by the "fundamental lemma" of the calculus of variations, the term in brackets is forced to be zero everywhere:

$$-\frac{\partial}{\partial x}\left(c\frac{\partial u}{\partial x}\right) - \frac{\partial}{\partial y}\left(c\frac{\partial u}{\partial y}\right) = f(x,y) \quad \text{throughout} \quad S. \tag{7}$$

This is the Euler equation $A^T C A = f$, or $-\nabla \cdot c\nabla u = f$. If the y variable is removed, we are back to a one-dimensional rod.

With no extra effort we can find P for any linear equation

$$\boxed{au_{xx} + 2bu_{xy} + cu_{yy} = 0.} \tag{8}$$

For convenience let a, b, and c be constant. The corresponding quadratic is

$$P(u) = \tfrac{1}{2} \int\!\!\int \left[a \left(\frac{\partial u}{\partial x} \right)^2 + 2b \left(\frac{\partial u}{\partial x} \right) \left(\frac{\partial u}{\partial y} \right) + c \left(\frac{\partial u}{\partial y} \right)^2 \right] dx \, dy.$$

If we minimize P we expect to reach (8) as its Euler equation. But there is more to it than that. To *minimize* P it should be *positive definite*. Inside the integral is an ordinary 2 by 2 quadratic $ax^2 + 2bxy + cy^2$. It is true that partial derivatives have replaced x and y, but at any point they are just numbers. Therefore the test for positive-definiteness is still $ac > b^2$, as it was in Chapter 1. (We can make $a > 0$ in advance.) That test decides whether or not equation (8) can be solved with arbitrary boundary values $u = u_0$. In this positive definite case the equation is called "elliptic;" minimization is justified. That is the first of the three fundamental classes of partial differential equations:

30   Equation (8) is ***elliptic, parabolic,*** or ***hyperbolic*** according to the matrix $\begin{bmatrix} a & b \\ b & c \end{bmatrix}$:

  $ac > b^2$:   elliptic boundary-value problems (equilibrium equations)

  $ac = b^2$:   parabolic initial-value problems (diffusion equations)

  $ac < b^2$:   hyperbolic initial-value problems (wave equations)

The principal examples are $u_{xx} + u_{yy} = 0$, $u_{xx} - u_t = 0$, and $u_{xx} - u_{tt} = 0$.

Laplace's equation is elliptic, with $a = c = 1$ in the identity matrix. The heat equation is parabolic, with $b = c = 0$; the matrix is singular and its determinant is zero. That is the borderline between elliptic and hyperbolic.† The wave equation has $a = 1$ and $c = -1$, an indefinite matrix corresponding to a saddle point. It asks for initial values instead of boundary values—two conditions on part of the boundary and no conditions on another part, instead of one condition everywhere.

   Chapter 6 takes up initial-value problems; here we stay with elliptic equations and consider their boundary conditions. ***They can specify u, or they can specify the normal component $w \cdot n$.*** Those alternatives appear immediately in comparing (5) with (6), to find the boundary term. It comes from Green's formula in Section 3.3:

$$\int\!\!\int_S c\nabla u \cdot \nabla v \, dx \, dy = - \int\!\!\int_S (\nabla \cdot (c\nabla u))v \, dx \, dy + \int_C (c\nabla u \cdot n)v \, ds. \qquad (9)$$

---

   † This marginal parabolic case needs help from lower-order terms: $u_{xx} = 0$ is not parabolic but $u_{xx} = u_y$ is.

Both sides equal $\int\int fv$; that is the weak form. The first term on the right yields the strong form $-\nabla \cdot (c\nabla u) = f$ of equation (7). Therefore the boundary integral in (9) must be zero, and that must be achieved by the boundary conditions. Strictly speaking, they are part of the strong form.

There are two ways to make this boundary integral zero. If boundary values $u = u_0$ are given, and $u + v$ shares those values, then $v = 0$ on the boundary. That kills the last integral in (9). On the other hand, if u is not given and v is free, we must impose $c\nabla u \cdot n = 0$. This is the natural boundary condition $w \cdot n = 0$, which also kills the integral. It goes with A^T, where the essential condition $u = u_0$ goes with A. Notice that the whole vector $w = c\,\text{grad}\,u$ is not zero, since that would force $u = \text{constant}$ and the problem would be overdetermined. It is exactly the product $(c\nabla u \cdot n)v$ that needs to be zero on the boundary—one factor $w \cdot n$ or the other factor v, and nothing more.

The Minimal Surface Problem

Now we are ready for a nonlinear partial differential equation, almost the first in this book. The corresponding energy will not be a quadratic $P(u)$. It will be the exact energy $E(u)$, from which P originally came as an approximation. If there is a thin membrane covering the set S—the usual description is a soap bubble over the set—then stretching this membrane requires energy. The energy is proportional to the **surface area of the soap bubble**. In the right units the material constant will be $c = 1$, and the problem is to minimize the surface area

$$E(u) = \int\int_S \left[1 + \left(\frac{\partial u}{\partial x}\right)^2 + \left(\frac{\partial u}{\partial y}\right)^2 \right]^{1/2} dx\,dy. \tag{10}$$

When $u = 0$, the bubble is flat and the expression in brackets reduces to 1. That is the minimum if $u = 0$ is admissible. But suppose the bubble is created on a piece of wire that goes around S at a varying height $u_0(x,y)$. Then the trivial solution $u = 0$ is not allowed; the bubble has to stick to the wire. The bent wire imposes a boundary condition $u = u_0(x,y)$ at the edge of S, and the **minimal surface problem** is to find the smallest area $E(u)$ subject to this boundary condition. Surface tension minimizes the area.

Before finding the equation for u, we compare E with P:

$$\left[1 + \left(\frac{\partial u}{\partial x}\right)^2 + \left(\frac{\partial u}{\partial y}\right)^2 \right]^{1/2} = \left[1 + \frac{1}{2}\left(\frac{\partial u}{\partial x}\right)^2 + \frac{1}{2}\left(\frac{\partial u}{\partial y}\right)^2 + \cdots \right] \quad \text{(at each point)}$$

$$E(u) = \int\int 1\,dx\,dy + P(u) + \cdots. \quad \text{(after integrating)}$$

The square root of $1 + a$ has been changed to $1 + \frac{1}{2}a + \cdots$. The constant term is $\int\int 1\,dx\,dy = $ area of S; that is fixed. The next term is important; it is exactly the quadratic $P(u)$. When only that part is kept, the equation is linear. It gives a good approximation for a wire that is nearly flat. This is the fundamental linearizing assumption in a large part of applied mathematics—that terms like a^2 are comparatively small, and can be ignored.

When those terms are not ignored, and we stay with $E(u)$, the test for a minimum is still $E(u) \leq E(u + v)$. It is harder to compute the part $\delta E/\delta u$ that is linear in v, but it is still possible: a change inside the square root gives

$$(1 + a + b)^{1/2} = (1 + a)^{1/2} + \frac{1}{2}\frac{b}{(1 + a)^{1/2}} + O(b^2). \tag{11}$$

The part from u is still $1 + a$, and the correction from v is b:

$$1 + a = 1 + \left(\frac{\partial u}{\partial x}\right)^2 + \left(\frac{\partial u}{\partial y}\right)^2, \quad b = 2\frac{\partial u}{\partial x}\frac{\partial v}{\partial x} + 2\frac{\partial u}{\partial y}\frac{\partial v}{\partial y} + \cdots.$$

When we integrate equation (11) over the set S, the result is

$$E(u + v) = E(u) + \iint\limits_{S} \frac{1}{(1 + a)^{1/2}}\left(\frac{\partial u}{\partial x}\frac{\partial v}{\partial x} + \frac{\partial u}{\partial y}\frac{\partial v}{\partial y}\right) dx\, dy + \cdots. \tag{12}$$

Now $\delta E/\delta u$ is exposed. It is the integral in (12) and it is zero; that is the weak form of the Euler equation. Because of the square root of $1 + a$, it is not linear in u. (It is always linear in v; that was the whole point of the first variation!) Integrating by parts to remove the derivatives from v and then requiring zero for all v produces the Euler equation in its strong form:

$$-\frac{\partial}{\partial x}\left(\frac{1}{(1 + a)^{1/2}}\frac{\partial u}{\partial x}\right) - \frac{\partial}{\partial y}\left(\frac{1}{(1 + a)^{1/2}}\frac{\partial u}{\partial y}\right) = 0. \tag{13}$$

This is the minimal surface equation. It is not easy to solve, because of the square root in the denominator. We show below how it got there, by differentiating the square root in E itself.

Perhaps it is only natural that the most important nonlinear equation in geometry should reduce to the most important linear equation, but still it is beautiful. The linearization of (13) approximates the square root by 1; and the result is Laplace's equation.

Nonlinear Equations

Those examples are typical of the general case, which we describe quickly. The variational problem starts with an integral $E = \iint F\, dx\, dy$ that depends on u and its derivatives:

$$F = F(x, y, u, D_1 u, D_2 u, \ldots).$$

For an elastic bar there was only $D_1 u = \partial u / \partial x$. For a soap bubble there is also $D_2 u = \partial u / \partial y$. Higher derivatives are allowed, and we can think of u itself as $D_0 u$. Then the comparison of $E(u)$ with $E(u + v)$ leads to the weak form. That comparison uses an ordinary expansion like $F(u + v) = F(u) + F'(u)v + O(v^2)$, but when F depends on several derivatives there are more terms:

$$F(u + v) = F(u) + \sum \frac{\partial F}{\partial D_i u} D_i v + \cdots. \tag{14}$$

We take **the derivatives of F with respect to u and u_x and u_y** and any other $D_i u$ on which F depends.

The weak form deals with the terms that are linear in v. The strong form integrates those terms by parts, to lift the derivatives from v and put them onto the part involving u:

$$\iint \left(\frac{\partial F}{\partial D_i u} \right) (D_i v) dx dy \rightarrow \iint \left[D_i^T \left(\frac{\partial F}{\partial D_i u} \right) \right] v \, dx \, dy.$$

The transpose is $D_i^T = -D_i$ for derivatives of odd order (with an odd number of integrations by parts and minus signs). For even derivatives it is $D_i^T = +D_i$.

Here, buried inside the calculus of variations, is the real source of $A^T C A$! The derivative A^T is D_i^T and Au is $D_i u$. In between, C can be nonlinear. Normally it *is* nonlinear. But when F is the quadratic $\frac{1}{2} c (Du)^2$, then $D^T \partial F / \partial Du$ becomes $D^T (cDu)$—which is exactly the linear $A^T C A u$.

We summarize the general case before testing it on examples.

3P   Each problem in the calculus of variations can be stated in three equivalent forms:

Variational form: Minimize $E(u) = \iint\limits_S F(u) \, dx \, dy$

Weak form: $\dfrac{\partial E}{\partial u} = \iint\limits_S \left(\sum \dfrac{\partial F}{\partial D_i u} \right) (D_i v) dx \, dy = 0$ for all v

Euler equation: $\sum D_i^T \left(\dfrac{\partial F}{\partial D_i u} \right) = 0.$

When F is a quadratic function of u and its derivatives, the expressions $\partial F / \partial D_i u$ are linear and so is the Euler equation.

The examples will show that $\partial F / \partial D_i u$, the derivative of F with respect to a derivative of u, is much easier to find than it looks.

EXAMPLE 1 $$F = u^2 + u_x^2 + u_y^2 + u_{xx}^2 + u_{xy}^2 + u_{yy}^2$$

The derivatives of F are $2u$, $2u_x$, $2u_y$, ..., $2u_{yy}$. They are derivatives with respect to u and u_x and the other $D_i u$, *not* with respect to x! The weak form is

$$2 \iint [uv + u_x v_x + u_y v_y + u_{xx} v_{xx} + u_{xy} v_{xy} + u_{yy} v_{yy}] \, dx \, dy = 0.$$

The Euler equation in its strong form integrates every term by parts:

$$2[u - u_{xx} - u_{yy} + u_{xxxx} + u_{xyxy} + u_{yyyy}] = 0.$$

It is linear because F is quadratic; the minus signs come with odd derivatives.

EXAMPLE 2 $$F = (1 + u_x^2)^{1/2} \quad \text{or} \quad F = (1 + u_x^2 + u_y^2)^{1/2}$$

The derivatives with respect to u_x and u_y bring the square root into the denominator. The shortest path equation and the minimal surface equation are

$$-\frac{d}{dx} \frac{u_x}{(1 + u_x^2)^{1/2}} = 0 \quad \text{and} \quad -\frac{\partial}{\partial x}\left(\frac{u_x}{F}\right) - \frac{\partial}{\partial y}\left(\frac{u_y}{F}\right) = 0.$$

Every term in these equations fits into the pattern of $A^T C A$, and for the first time in the book *the framework becomes nonlinear*:

$$u \xrightarrow{} e = Au \xrightarrow[\text{nonlinear}]{C} w = C(e) = \frac{\partial F}{\partial e} \xrightarrow{} A^T w = A^T C(Au) = f.$$

Nonlinear Constitutive Laws and Nonquadratic Energies

That diagram is worth a chapter of words. It is the shortest path to nonlinear equilibrium equations. In a linear spring, the force $w = ce$ was proportional to e. The internal strain energy was $F = \int ce\,de = \frac{1}{2}ce^2$. *In a nonlinear spring the constitutive law is $w = C(e)$*. The relation of force to stretching, or stress to strain, or current to voltage, or flow to pressure, will no longer be given by a straight line. The energy density is still $F = \int C(e)\,de$, and minimizing the total potential energy still gives the equilibrium equation:

$$P(u) = \int [F(Au) - fu] \, dx \text{ is minimized when } A^T C(Au) = f.$$

The first variation of P leads to $\int [C(Au)(Av) - fv] \, dx = 0$ for every v (the weak form). Then $A^T C(Au) = f$ is the Euler equation (the strong form). For the nonlinear equivalent of positive definiteness, *the function $C(e)$ should be increasing*. The line

$w = ce$ had a constant slope $c > 0$, and now the slope of w is changing—but it is still positive. That makes the law monotonic and the energy convex and the Euler equation elliptic.

An example is the *power law* $w = C(e) = e^{p-1}$ with $p > 1$. The energy in a spring is its integral $F = e^p/p$. The stretching is $Au = du/dx$ and the equilibrium equation is $A^T C(Au) = (-d/dx)(du/dx)^{p-1} = f$.

We have only one thing to add. It has to do with complementary energy, which is a function of w instead of e. It starts with the *inverse constitutive law* $e = C^{-1}(w)$—for example $e = w^{1/(p-1)}$. The strain comes from the stress or the voltage from the current; the arrow in our framework is reversed. Graphically, we are looking at Fig. 3.15a from the side. **The area under that curve is the complementary energy** $F^*(w)$ $= \int C^{-1}(w)dw$. It is chosen exactly so as to give the twin equations

$$w = C(e) = \frac{\partial F}{\partial e} \quad \text{and} \quad e = C^{-1}(w) = \frac{\partial F^*}{\partial w}. \tag{15}$$

The symmetry is perfect and the dual minimum principle still holds:

$$Q(w) = \int F^*(w)dx \text{ is a minimum subject to } A^T w = f(x).$$

A Lagrange multiplier u takes Q to $L = \int [F^*(w) - uA^T w + uf]dx$. Then at the minimum we recover the two fundamental equations of equilibrium, now nonlinear:

$$\partial L/\partial w = 0 \quad \text{is} \quad C^{-1}(w) - Au = 0$$
$$\partial L/\partial u = 0 \quad \text{is} \quad A^T w \quad\quad = f.$$

A sign has changed from the electrical case $e = -Au$, but nothing else is different. The first equation gives $w = C(Au)$ and then the second is $A^T C(Au) = f$.

Since these nonlinear things are in front of us, why not take the last step? It is never seen in advanced calculus, but there is nothing so incredibly difficult. It is the

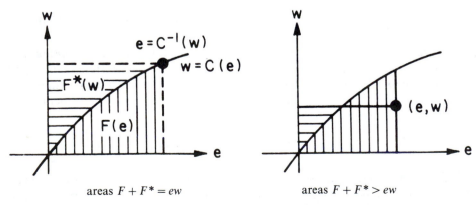

areas $F + F^* = ew$ areas $F + F^* > ew$

Fig. 3.15. The graphs of $w = C(e)$ and $e = C^{-1}(w)$, and the areas F^* and F.

direct link between F and F^*, known as the **_Legendre-Fenchel transform_**:

$$F^*(w) = \max_{e} \, [ew - F(e)] \quad \text{and} \quad F(e) = \max_{w} \, [ew - F^*(w)]. \qquad (16)$$

For the first maximum, differentiate with respect to e. That brings back $w = \partial F/\partial e$, which is the correct $C(e)$. The maximum itself is $e\,\partial F/\partial e - F$, which agrees with the area F^*. (The figures show graphically that the areas satisfy $ew - F = F^*$ on the curve and $ew - F < F^*$ off the curve, so the maximum is F^* as desired.) Similarly, the second maximum in (16) leads to $e = \partial F^*/\partial w$. That is the constitutive law in the other direction, $e = C^{-1}(w)$. The whole nonlinear theory is there, provided the material laws are conservative—the energy in the system should be constant. This conservation law seems to be destroyed by dissipation, or more spectacularly by fission, but in some ultimate picture of the universe it must remain true.

In the next paragraphs kinetic energy appears for the first time. The problems are dynamic instead of static. F will be the Lagrangian and F^* is the Hamiltonian. Later in the book there are equations $mu_{tt} + cu_t + ku = 0$ and $u_{xx} = u_t$ with friction and damping and diffusion; we hesitate to stretch our framework that far. At the very end of the book the Legendre transform reappears at full strength, in constrained optimization. There F and F^* are more general convex functions (with nonnegative second derivatives) and **_we recognize that F^{**} is F_**. Here we compute F^* for the power law and verify that it agrees with $\int C^{-1}(w)\,dw$:

EXAMPLE The power law $F(e) = e^p/p$ (we look at $e > 0$ and $w > 0$)

Differentiating $ew - F(e)$ gives $w - e^{p-1} = 0$, so the conjugate in (16) is

$$F^* = ew - F(e) = w^{1/(p-1)}w - \frac{1}{p}w^{p/(p-1)} = \frac{p-1}{p}\,w^{p/(p-1)} = \frac{1}{q}w^q.$$

F^* is also a power law, with _dual exponent_ $q = p/(p-1)$. This matches the area under $C^{-1}(w) = w^{1/(p-1)}$, because integration will increase that exponent to $1 + 1/(p-1) = q$. _A more symmetric relation is $p^{-1} + q^{-1} = 1$._

Dynamics and Least Action

Fortunately or unfortunately, the world is not in equilibrium. The energy stored in springs and beams and nuclei and people is waiting to be released. When the external forces change, the equilibrium is destroyed. Potential energy is converted to kinetic energy, the system becomes dynamic, and it may or may not find a new steady state. If damping penalizes the kinetic energy, then the transients decay and a new equilibrium is reached. If there is instability (negative damping) the energy grows. In between, when the system is conservative, the transients are neutrally stable and they neither grow nor decay. The energy can change from potential to kinetic to potential to kinetic, but the total energy $K + P$ remains constant. It is like the earth around the sun or a child on a frictionless swing. The force dP/du is no longer zero, the static balance is upset, and the system oscillates.

To describe that motion we need either an equation or a variational principle. Chapter 6 works with equations (Newton's laws). This section derives those laws from the *principle of least action*:

> If the initial state is $u(t_0)$ and the final state is $u(t_1)$, the path between those states minimizes the action integral
>
> $$A = \int_{t_0}^{t_1} (\text{kinetic energy} - \text{potential energy})dt.$$

It is probably better to claim only that $\delta A/\delta u = 0$—the path is always a stationary point but not in every circumstance a minimum. We have a difference of energies and positive definiteness can be lost. The action may be quadratic and the equation for u may be linear, but it could describe a saddle point. Laplace's equation will be overtaken by the wave equation. First come three examples to show how the *global law* of least action (the variational principle) produces the *local law* $F = ma$.

EXAMPLE 1   A ball of mass m attracted by gravity

The only degree of freedom is the ball's height u. The energies are

$$K = \text{kinetic energy} = \frac{1}{2} m \left(\frac{du}{dt}\right)^2; \ P = \text{potential energy} = mgu.$$

The action is $A = \int (\frac{1}{2}m(u')^2 - mgu)dt$, and $\delta A/\delta u$ follows from the rules of this section—with the difference that t replaces the space variable x. When the true path u is compared to its neighbors $u + v$,

$$\frac{\delta A}{\delta u} = \int_{t_0}^{t_1} (mu'v' - mgv)dt = 0 \quad \text{for every } v.$$

That is the weak form of Newton's law. You recognize the momentum mu' as the derivative of $\frac{1}{2}m(u')^2$ with respect to the velocity u'. This is customary in the calculus of variations, the derivative with respect to a function. The strong form is the Euler equation

$$-\frac{d}{dt}\left(m \frac{du}{dt}\right) - mg = 0. \tag{18}$$

That is Newton's law $mu'' = -mg$, or $ma = F$. *The action integral is minimized by the path that follows Newton's law.*

The framework is not changed. The place of A is taken by d/dt, its transpose is $-d/dt$, the material constant is m, and the external force is $f = mg$. The balance between $A^T C A u$ and f is Newton's law—a balance of **inertial forces** instead of mechanical forces. Figure 3.16 identifies the edge variables as the velocity and momentum, and the mass (or the density) replaces c.

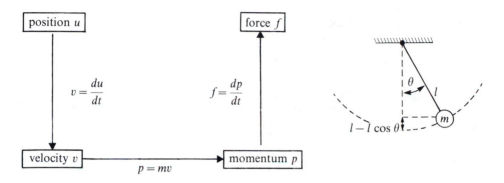

Fig. 3.16. Newton's law for a ball; sketch of a pendulum.

EXAMPLE 2   A simple pendulum with mass m

The pendulum is a weightless rod of length l with a mass at the end. Its position is given by its angle from the vertical. The state variable u is the angle θ rather than the height of the mass—which is $l - l \cos \theta$, equal to zero when the rod hangs straight down and $\theta = 0$. The height still yields the potential energy, and the velocity of the mass is l times $d\theta/dt$:

$$K = \text{kinetic energy} = \frac{1}{2} ml^2 \left(\frac{d\theta}{dt} \right)^2 ; P = \text{potential energy} = mg(l - l \cos \theta)$$

In this problem the equation will not be linear. K is quadratic but P involves $\cos \theta$. The Euler equation follows the usual rule for an integral $\int F(\theta, \theta') dt$, with $F = K - P$:

$$\frac{\partial F}{\partial \theta} - \frac{d}{dt} \left(\frac{\partial F}{\partial \theta'} \right) = 0 \quad \text{or} \quad -mgl \sin \theta - \frac{d}{dt} \left(ml^2 \frac{d\theta}{dt} \right) = 0.$$

This is the equation of a simple pendulum:

$$\frac{d^2\theta}{dt^2} + \frac{g}{l} \sin \theta = 0. \tag{19}$$

When the angle is small and $\sin \theta$ is approximated by θ, the equation becomes linear. A linear clock keeps time, but not the right time.

EXAMPLE 3   A vertical line of springs and masses

The potential energy in the springs is $P = \frac{1}{2}u^T A^T C A u$. The kinetic energy has $\frac{1}{2}m_i(du_i/dt)^2$ from each mass. The difference $K - P$ goes into the action integral, and there is an Euler equation $\delta A/\delta u_i = 0$ for each mass. In matrix form this is the basic equation of undamped motion:

$$\boxed{Mu'' + A^T C A u = 0.}$$

M is the diagonal mass matrix, $A^T C A$ is the positive definite stiffness matrix, and u gives the displacements measured from equilibrium. The system oscillates around that equilibrium and the total energy $K + P$ is constant.

EXAMPLE 4   Waves in an elastic bar

This is just a *continuum of masses and springs*. The action integral is completely analogous to the discrete case,

$$A = \int_{t_0}^{t_1} \int_{x=0}^{1} \left[\frac{1}{2} m \left(\frac{du}{dt} \right)^2 - \frac{1}{2} c \left(\frac{du}{dx} \right)^2 \right] dx\, dt,$$

except that we integrate over x instead of adding the energies in a finite number of springs. Similarly the mass is distributed with density $m(x)$; again we integrate. The action is a double integral, involving x as well as t, but the rules for the Euler equation cover that case:

$$-\frac{\partial}{\partial t} \left(m \frac{\partial u}{\partial t} \right) - \frac{\partial}{\partial x} \left(-c \frac{\partial u}{\partial x} \right) = 0.$$

That is the **wave equation** $mu_{tt} = cu_{xx}$. With constant density m and elastic constant c, the wave speed is $\sqrt{c/m}$ — faster when the bar is stiffer and slower when the bar is heavier.

We are anticipating Chapter 6 and will stop. Staying with the calculus of variations, there are two important comments:

(1)   When u is independent of time, the kinetic energy is $K = 0$ and the least action principle reduces to $\delta P/\delta u = 0$. The dynamic problem goes back to the static problem.

(2)   The dual variable, from all the evidence of preceding chapters, should be in the right place in Fig. 3.16. It should be and is the **momentum**. It is $p = mv = m\, du/dt$, and the kinetic energy is

$$K = \frac{1}{2} mv^2 = \frac{1}{2m} (mv)^2 = \frac{1}{2m} p^2.$$

This is the "complementary" kinetic energy, expressed in terms of p instead of v. Note that m moves to the denominator, just as c did in the elastic energy $w^2/2c$. There is always a choice between primal and dual, and Hamilton found the best combination. His energy function depends on the momentum p and the displacement u; he eliminated the velocity. For a ball it is

$$H = K + P = \frac{1}{2m} p^2 + mgu. \tag{20}$$

The **Hamiltonian** is the total energy in the system, and it takes us directly to the equations of motion:

3Q   The energy $H = K + P$ is a function of displacement u and momentum p, and the equations of motion are **Hamilton's equations**

$$\frac{\partial H}{\partial p} = \frac{du}{dt} \quad \text{and} \quad \frac{\partial H}{\partial u} = -\frac{dp}{dt}. \tag{21}$$

The derivative dH/dt is zero and the total energy is constant:

$$\frac{dH}{dt} = \frac{\partial H}{\partial p}\frac{dp}{dt} + \frac{\partial H}{du}\frac{du}{dt} = u'p' - p'u' = 0. \tag{22}$$

This is the essence of classical mechanics. It is tied to Hamilton and not to Newton. For that reason it survived the revolution brought by Einstein. We will see that H has a relativistic form and even a quantum mechanical form. Before that we go back to a ball and a spring:

$$\text{(ball)} \quad H = \frac{1}{2m} p^2 + mgu \quad \text{(spring)} \quad H = \frac{1}{2m} p^2 + \frac{1}{2} cu^2$$

For the ball, Hamilton's equations $\partial H/\partial p = u'$ and $\partial H/\partial u = -p'$ are

$$\frac{p}{m} = u' \quad \text{and} \quad mg = -p', \quad \text{or} \quad mu'' = -mg.$$

This force $-mg$ never changes. For the spring it changes all the time:

$$\frac{p}{m} = u' \quad \text{and} \quad cu = -p', \quad \text{or} \quad mu'' + cu = 0.$$

That is the fundamental equation of oscillation. The spring force cu acts against the

displacement. Sooner or later the mass changes direction and passes through
equilibrium at top speed, when the force reverses to bring it back. In the physical
plane the mass goes up and down. In the $u - p$ plane (the *phase plane*) the picture is
different. The motion stays on the energy surface $H =$ constant, which is the ellipse
in Fig. 3.17. Each oscillation of the spring is a trip around the ellipse. The four
extreme points are marked on the axes and sketched on the right.

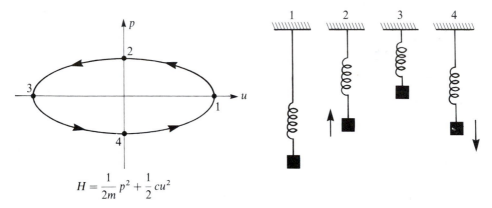

$$H = \frac{1}{2m}p^2 + \frac{1}{2}cu^2$$

Fig. 3.17. (1) tension (2) motion (3) compression (4) motion: constant energy $H = K + P$.

With more springs this figure moves into higher dimensions. There are $2n$ axes
$u_1, \ldots, u_n, p_1, \ldots, p_n$ and the ellipse becomes an ellipsoid. Hamilton's equations are
$\partial H/\partial p_i = du_i/dt$ and $\partial H/\partial u_i = -dp_i/dt$. They lead again to $Mu'' + Ku = 0$, with
mass and stiffness matrices—which is completely analogous to the scalar case. So
is the wave equation, in which $K = A^T C A$ becomes a second derivative (with minus
sign) in space:

$$mu_{tt} - cu_{xx} = 0 \quad \text{or} \quad mu_{tt} - \text{div}(c \text{ grad } u) = 0.$$

For the nonlinear pendulum we return to the phase plane in Section 6.2.

Relativity and Quantum Mechanics

These few paragraphs are an attempt, by a total amateur, to correct the action
integral by the rules of relativity. The famous lectures by Feynman propose the
term $-mc^2\sqrt{1 - (v/c)^2}$ in the Lagrangian $K - P$. At $v = 0$ this is Einstein's formula
$e = mc^2$ for the energy in a rest mass m. It becomes part of the potential energy P.
As the velocity increases from zero there is also a part corresponding to K. For
small x the square root of $1 - x$ is approximately $1 - \frac{1}{2}x$, which linearizes the
problem and brings back Newton—just as linearizing the minimal surface

equation brought back Laplace. The energy is

$$F(v) = -mc^2\sqrt{1-(v/c)^2} \approx -mc^2\left(1 - \frac{1}{2}\frac{v^2}{c^2}\right).$$

The new term $\frac{1}{2}mv^2$ is the familiar kinetic energy.†

Now, trusting in duality, we look for the corresponding F^* in the Hamiltonian. It will be a function of p, not v. In fact the first step is to find $p = \partial F/\partial v$. Before relativity F was $\frac{1}{2}mv^2$ and its derivative was $p = mv$. Now the momentum is

$$p = \frac{\partial F}{\partial v} = \frac{mv}{\sqrt{1-(v/c)^2}}, \tag{23}$$

and it becomes infinite as v approaches the speed of light. Then the new term F^*, according to (16), is

$$F^*(p) = \max_v[pv - F(v)] = mc^2\sqrt{1 + (p/mc)^2}. \tag{24}$$

The maximum of $pv - F(v)$ occurs at (23)—which we solved for v and substituted into $pv - F(v)$ to find (24).

The new term F^* is again Einstein's energy mc^2 when the system is at rest. As p increases from zero, the next contribution from the square root gives

$$F^* \approx mc^2\left(1 + \frac{1}{2}\frac{p^2}{m^2c^2}\right).$$

That correction is the old kinetic energy $\dfrac{1}{2m}p^2$. Roughly speaking, Newton found the low-order term in the energy that Einstein computed exactly. Perhaps the universe is like a minimal surface in space-time, and to Laplace and Newton it looked nearly flat.

It is risky to add anything about quantum mechanics, which works with probabilities instead of definite events. It is a mixture of differential equations (Schrödinger) and matrix equations (Heisenberg). Somehow the event at which $\delta A/\delta u = 0$ must be extremely probable, since it almost always occurs. Feynman's explanation is that each possible trajectory of the system has a phase factor $e^{iA/h}$ multiplying its probability amplitude. The small number h (Planck's constant) means that a slight change in A completely alters the phase. There are strong cancelling effects from nearby paths unless the phase is stationary. In other words $\delta A/\delta u = 0$ at the most probable path. This prediction of stationary phase applies equally to light rays and particles. Optics follows the same principles as mechanics,

---

† Note how kinetic and potential energy are mixed together by relativity.

and light travels by the fastest route; action becomes time. If Planck's constant could go to zero, the deterministic principles of least action and least time would appear and the path would be not only probable but certain.

EXERCISES

3.6.1   What are the weak form and the strong form of the linear beam equation—the Euler equation for $P = \int [\frac{1}{2} c(u'')^2 - fu] dx$?

3.6.2   Minimizing $P = \int (u')^2 dx$ with $u(0) = a$ and $u(1) = b$ also leads to the straight line through these two points. Write down the weak form and the strong form.

3.6.3   Find the Euler equations (strong form) for

(a) $\displaystyle\int [(u')^2 + e^u] dx$ (b) $\displaystyle\int uu' \, dx$ (c) $\displaystyle\int x^2(u')^2 dx$

3.6.4   If $F(u,u')$ is independent of x, as in almost all our examples, show from the Euler equation and the chain rule that $H = u' \partial F / \partial u' - F$ is constant. This is dual to the fact that $\partial F / \partial u'$ is constant when F is independent of u.

3.6.5   If the speed is x the travel time is

$$T = \int_0^1 \frac{1}{x} \sqrt{1 + (u')^2} \, dx \quad \text{with} \quad u(0) = 0 \quad \text{and} \quad u(1) = 1.$$

(a)   From the Euler equation what quantity is constant (Snell's law)?
(b)   Can you integrate once more to find the optimal path $u(x)$?

3.6.6   With the constraints $u(0) = u(1) = 0$ and $\int u \, dx = A$, show that the minimum value of $P = \int (u')^2 dx$ is $12A^2$. Introduce a multiplier m, solve the Euler equation for u, and verify that $A = -m/24$. Then the derivative $dP/dA = 24A$ is $-m$ as the theory predicts.

3.6.7   For the shortest path constrained by $\int u \, dx = A$, what is unusual about the solution in Fig. 3.14 as A becomes large?

3.6.8   Suppose the constraint is $\int u \, dx \geq A$, with inequality allowed. Why does the solution remain a straight line as A becomes small? Where does the multiplier m remain?
Note: This is typical of inequality constraints: either the Euler equation is satisfied or the multiplier is zero.

3.6.9   Suppose the constrained problem is reversed, and we *maximize* the area $P = \int u \, dx$ subject to fixed length $l = \int \sqrt{1 + (u')^2} \, dx$, with $u(0) = a$ and $u(1) = b$.
(a)   Form the Lagrangian and solve its Euler equation for u
(b)   How is the multiplier M related to m in the text?
(c)   When do the constraints eliminate all functions u?

3.6.10   Find by ordinary calculus the shortest broken-line path between $(0,1)$ and $(1,1)$ that goes first to the horizontal axis $y = 0$. Show that the best path treats this axis like a mirror: angle of incidence = angle of reflection.

3.6.11 The principle of maximum entropy selects the probability distribution that maximizes $H = -\int u \log u \, dx$. Introduce Lagrange multipliers for the constraints $\int u \, dx = 1$ and $\int xu \, dx = 1/a$, and find by differentiation an equation for u. On the interval $0 < x < \infty$ show that the most likely distribution is $u = ae^{-ax}$.

3.6.12 If the second moment $\int x^2 u \, dx$ is also known show that Gauss wins again: the maximizing u is the exponential of a quadratic. If only $\int u \, dx = 1$ is known, the most likely distribution is $u = $ constant. The *least* information comes when only one outcome is possible, say $u(6) = 1$, since $u \log u$ is then identically zero.

3.6.13 A path that climbs around a cylinder has $x = \cos \theta$, $y = \sin \theta$, $z = u(\theta)$:

$$\text{its length is } L = \int \sqrt{dx^2 + dy^2 + dz^2} = \int \sqrt{1 + (u')^2} \, d\theta.$$

Show that $u' = $ constant satisfies Euler's equation. What kind of path is $(x,y,z) = (\cos \theta, \sin \theta, c\theta)$?

3.6.14 Starting with the nonlinear equation $-u'' + \sin u = 0$, multiply by v and integrate the first term by parts to find the weak form. What integral P is minimized by u?

3.6.15 Find the Euler equations (strong form) for

(a) $P(u) = \dfrac{1}{2} \int \int \left[\left(\dfrac{\partial^2 u}{\partial x^2} \right)^2 + 2 \left(\dfrac{\partial^2 u}{\partial x \partial y} \right)^2 + \left(\dfrac{\partial^2 u}{\partial y^2} \right)^2 \right] dx \, dy$

(b) $P(u) = \dfrac{1}{2} \int \int (yu_x^2 + u_y^2) \, dx \, dy$ (c) $E(u) = \int u\sqrt{1 + (u')^2} \, dx$

(d) $P(u) = \dfrac{1}{2} \int \int (u_x^2 + u_y^2) \, dx \, dy$ with $\int \int u^2 dx \, dy = 1$.

3.6.16 Show that the Euler equations for

$$\int \int \frac{\partial^2 u}{\partial x^2} \frac{\partial^2 u}{\partial y^2} \, dx \, dy \quad \text{and} \quad \int \int \left(\frac{\partial^2 u}{\partial x \partial y} \right)^2 dx \, dy$$

are the same. (Presumably the two integrals are equal if the boundary conditions are zero.)

3.6.17 Sketch the graph of $p^2/2m + mgu = $ constant in the $u - p$ plane. It is an ellipse, parabola, or hyperbola? Mark the point where the ball reaches maximum height and begins to fall.

3.6.18 Draw a second spring and mass hanging from the first. If the masses are m_1, m_2 and the spring constants are c_1, c_2, the energy is

$$H = K + P = \frac{1}{2m_1} p_1^2 + \frac{1}{2m_2} p_2^2 + \frac{1}{2} c_1 u_1^2 + \frac{1}{2} c_2 (u_2 - u_1)^2.$$

Find the four Hamilton's equations $\partial H/\partial p_i = du_i/dt, \partial H/\partial u_i = -dp_i/dt$, and the matrix equation $Mu'' + Ku = 0$.

3.6.19 The Hamiltonian for a pendulum (with $u = \theta$) is $H = p^2/2m + mgl(1 - \cos u)$. Write out Hamilton's equations (21) and eliminate p to find the equation of a pendulum.

3.6.20   Verify that the energy $\frac{1}{2}e^T Ce$ and the complementary energy $\frac{1}{2}w^T C^{-1}w$ are conjugate. As in equation (16), this means that $\frac{1}{2}w^T C^{-1}w = \max[e^T w - \frac{1}{2}e^T Ce]$.

CHAPTER 3 IN OUTLINE: EQUILIBRIUM IN THE CONTINUOUS CASE

3.1   One-dimensional Problems—analogies between discrete and continuous
   A to A^T: Integration by Parts—the rule is $(Au)^T w = u^T(A^T w)$
   Sturm-Liouville Problems—the solution to $-cu'' + qu = f$
   Singular Perturbations—boundary layer as $c \to 0$

3.2   Differential Equations of Equilibrium—the Euler equation $\delta P/\delta u = 0$
   Minimum Principles—essential and natural boundary conditions
   Complementary Minimum Principle for w—minimize $Q = \int w^2/2c \, dx$
   Fourth-order Equations—the beam equation $(cu'')'' = f$ has $A = (d/dx)^2$
   Interpolation: Displacements and Slopes—four conditions on a cubic
   Cubic Splines—continuous second derivatives at the nodes

3.3   Laplace's Equation and Potential Flow—$A^T Au = -\text{div grad } u = -u_{xx} - u_{yy}$
   Boundary Conditions and Green's Formula—$(\text{grad})^T = -\text{div}$
   Poisson's Equation—$\text{div} (c \text{ grad } u) = f$
   Minimum Principles—Laplace minimizes $P = \frac{1}{2} \int \int (u_x^2 + u_y^2) \, dx \, dy$

3.4   Vector Calculus in Three Dimensions—potentials and work $\int F \cdot dr$
   Gradient, Divergence, and Curl—curl grad $u = 0$ and div curl $S = 0$
   Electricity and Magnetism—Maxwell's equations, static and dynamic
   Vector Calculus—interior integrals equal boundary integrals
   Orthogonal Coordinate Systems—cylindrical and spherical scale factors

3.5   Equilibrium of Fluids and Solids—stress-strain law
   Strain and Displacement—$e = \frac{1}{2}(J + J^T)$, examples of shear
   Stress and Force—equilibrium div $\sigma + f = 0$
   The Torsion of a Rod—warping functions and tensors
   Fluid Mechanics—continuity equation and transport rule
   Acceleration and Momentum Balance—$\rho Dv/Dt = \text{div } T$: perfect and viscous
   Euler and Bernoulli Equations—$\frac{1}{2}v^2 + p/\rho = c$; vorticity and stream function
   The Navier-Stokes Equations—similarity and the Reynolds number
   The Stokes Equations—grad $p = \mu\nabla^2 v$ without acceleration

3.6   Calculus of Variations—the first variation $\delta P/\delta u = 0$: $\partial F/\partial u = (\partial F/\partial u')'$
   Constrained Problems—Lagrange multiplier for $\int u \, dx = A$
   Two-dimensional Problems—elliptic, parabolic, and hyperbolic
   The Minimal Surface Problem—minimize $\int \int (1 + u_x^2 + u_y^2)^{1/2} \, dx \, dy$
   Nonlinear Equations—variational form, weak form, strong form
   The Energies F and F^*—Legendre transform yields complementary energy
   Dynamics and Least Action—$Mu'' + Ku = 0$ and Hamilton's equations
   Relativity and Quantum Mechanics—Einstein's energy $F = -mc(c^2 - v^2)^{1/2}$

4

ANALYTICAL METHODS

FOURIER SERIES AND ORTHOGONAL EXPANSIONS ■ 4.1

How do we introduce a subject as important as Fourier series? The cowardly way is tremendously tempting—to choose a typical function and expand it in a sum of sines and cosines. We could show how to find each term in the sum. Then we could convert to a different but equivalent sum, using complex exponentials e^{ikx} and e^{-ikx} to replace $\sin kx$ and $\cos kx$. And we could try to show that these infinite series (most functions contain infinitely many frequencies k) reproduce the chosen function.

Those are necessary things to do, but not first. We are dealing with the most valuable transformation in mathematical analysis. It represents an arbitrary function as a combination of pure harmonic functions, and we need to explain how it succeeds. At one level the reason can be found in the properties of pure harmonics:

(1) They are **orthogonal.** The inner product of any two different harmonics is zero:

$$\int_0^{2\pi} e^{ikx} e^{-ilx}\, dx = 0 \quad \text{if} \quad k \neq l. \tag{1}$$

The same is true of $\cos kx \cos lx$ and $\cos kx \sin lx$ and $\sin kx \sin lx$—their integrals over an interval of length 2π are also zero. In (1), the integral is $e^{i(k-l)x}/i(k-l)$ and it has the same value at $x = 2\pi$ and $x = 0$. Over a complete period, the area under a sinusoid is zero.

(2) They are **complete.** No function can be orthogonal to all harmonics. If we

find every Fourier component of a function—the projection of f in the direction of every harmonic—then those components add up to f.

(3) They are **convenient**. Other families are orthogonal and complete, but none can match the simplicity of Fourier series. The coefficients in $a_k \cos kx$ or $b_k \sin kx$ or $c_k e^{ikx}$ can be calculated efficiently. We will do it by hand for square waves and delta functions; for human speech the results are in Fig. 4.1.

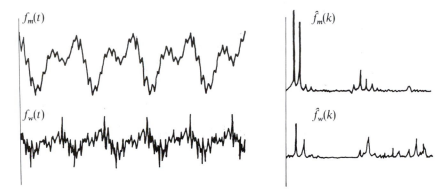

Fig. 4.1. Man and woman saying ee: Waveforms and harmonics.

These three properties are not the whole story. At a deeper level they are the effects, not the causes, of the real reason for success. Fourier analysis goes beyond the mechanical process of splitting a function into simpler pieces and recombining them. It solves differential equations; it tests the stability of difference equations; it describes the response of every system that is linear and stationary (or time-invariant). The Fourier and Laplace transforms work in the frequency domain to give information in the state domain. They apply to equilibrium problems (here), to signal processing and convolution (in 4.2), to sampling problems (in 4.3), and to dynamic problems (in 5.5 and 6.3), for one reason:

> *Every harmonic e^{ikx} is an eigenfunction of every*
> *derivative and every finite difference.*

Thus a Fourier series is an expansion into eigenvectors (in the discrete case) or eigenfunctions (in the continuous case). The harmonics are orthogonal because the eigenvectors of a symmetric or skew-symmetric problem are always orthogonal. The harmonics are complete because the eigenvectors of such a problem are always complete. Symmetric and skew-symmetric problems can be diagonalized by $A = Q \Lambda Q^T$, and the columns of Q are exactly the harmonics. The simplicity and importance of Fourier analysis come from the simplicity and importance of linear stationary processes—and especially difference and differential equations with constant coefficients.

Derivatives like d/dx and d^2/dx^2 have eigenvalues ik and $(ik)^2$:

$$\frac{d}{dx}e^{ikx} = ike^{ikx} \quad \text{and} \quad \frac{d^2}{dx^2}e^{ikx} = -k^2 e^{ikx}.$$

The Laplacian has eigenvalues

$$\left(\frac{\partial^2}{\partial x^2} + \frac{\partial^2}{\partial y^2}\right)e^{i(k_1 x + k_2 y)} = -(k_1^2 + k_2^2)e^{i(k_1 x + k_2 y)}.$$

One-sided and centered differences have eigenvalues

$$\frac{e^{ik(x+h)} - e^{ikx}}{h} = \left(\frac{e^{ikh} - 1}{h}\right)e^{ikx} \quad \text{and} \quad \frac{e^{ik(x+h)} - e^{ik(x-h)}}{2h} = \frac{i \sin kh}{h}e^{ikx}.$$

If you prefer sines and cosines you may be restricted to even derivatives and centered differences. They take sines to sines and cosines to cosines, while the odd derivatives and one-sided differences mix them up. But the severest limits on Fourier analysis are the restrictions on coefficients and boundary conditions:

(1)   The coefficients must be **constant**.
(2)   Harmonics must be capable of satisfying the **boundary conditions**.

The first eliminates all but the most special equations. The second eliminates all but the most regular regions. Fourier series can solve Laplace's equation on a square or a circle, but not on an ellipse. We will be satisfied with a circle.

We begin by expanding $f(x)$ into a sum of harmonics. The frequencies k are integers and the functions $\cos kx$ and $\sin kx$ and e^{ikx} are **periodic**: they are not changed when x is increased by 2π. This is the normal meaning of the words "Fourier series," in which the function is given on an interval of length 2π. It becomes periodic by reproducing itself on each succeeding interval. We want to emphasize here, before the analysis starts, that this periodic case is not the only possibility—although it does come first:

(i) $f(x) = \displaystyle\sum_{k=-\infty}^{\infty} c_k e^{ikx}$ (**Fourier series**: $0 \le x < 2\pi$)

The expansion into $a_0 + a_1 \cos x + b_1 \sin x + a_2 \cos 2x + \cdots$ is completely equivalent.

A second possibility is to allow every frequency k, integer or not, changing the sum over whole numbers into an integral over all k:

(ii) $f(x) = \displaystyle\int_{-\infty}^{\infty} c(k)e^{ikx}\, dk$ (**Fourier integral**: $-\infty < x < \infty$)

The function no longer has to be periodic as x goes from $-\infty$ to $+\infty$.

The third possibility goes in the opposite direction. The function is given at only

n points and it needs only n frequencies $k = 0, \ldots, n - 1$:

(iii) $f(x) = c_0 + c_1 e^{ix} + \cdots + c_{n-1} e^{i(n-1)x}$ (***discrete Fourier series***).

This equation holds at n equally spaced points $x = 0, 2\pi/n, 4\pi/n, \ldots$ below 2π.

The n values of f form one vector and the n coefficients c_k form another vector. Discrete Fourier analysis goes back and forth between them and the Fast Fourier Transform does it quickly.

   There is a fourth possibility (and a fifth ...) but it is time to compute ordinary Fourier coefficients c_k. Separating f into harmonics is the analysis, and recombining the harmonics into f is the synthesis. This section develops the classical applications of Fourier series and orthogonal functions, and the next section (which I hope you will read) is more modern.

The Fourier Coefficients

   Suppose $f(x)$ is a function with period 2π. Its graph from 2π to 4π is a repetition of its graph from 0 to 2π, and the graph continues to repeat itself along the whole line. Thus $f(x + 2\pi) = f(x)$ for every x; that is the statement of periodicity. The functions $\cos x$ and $\sin x$ are 2π-periodic (and so is a constant function). The functions $\cos 2x$ and $\sin 2x$ oscillate twice as fast; they have period 2π and also period π. The idea of Fourier series is to obtain every periodic function f, even one that has jumps (repeated at intervals of length 2π!) as a combination of these pure harmonics:

$$f(x) = a_0 + a_1 \cos x + b_1 \sin x + a_2 \cos 2x + b_2 \sin 2x + \cdots \tag{2}$$

All terms on the right are 2π-periodic, and our formulas could use any interval of length 2π. The usual choices are $-\pi$ to π (which we pick) or 0 to 2π. For a moment we skip the slightly special coefficient a_0, and compute the others.

   To find a_k, the coefficient of a typical $\cos kx$, here is the trick. Multiply both sides of (2) by the function $\cos kx$, and integrate from $-\pi$ to π. The left side becomes $\int f(x) \cos kx\, dx$. On the right side *every term except one is zero*, and the only surviving term is the interesting one:

$$\int_{-\pi}^{\pi} f(x) \cos kx\, dx = \int_{-\pi}^{\pi} a_k \cos kx \cos kx\, dx = a_k \pi. \tag{3}$$

The average value of the cosine squared is $\frac{1}{2}$, so its integral is π. It is equally likely to be near one or near zero; a direct proof uses $\cos^2 kx = \frac{1}{2} + \frac{1}{2} \cos 2kx$, which oscillates around $\frac{1}{2}$. Dividing both sides of equation (3) by π gives the Fourier coefficient we want:

$$a_k = \frac{1}{\pi} \int_{-\pi}^{\pi} f(x) \cos kx\, dx. \tag{4}$$

This is Euler's formula. You see how strongly it depended on the orthogonality of the harmonics. Each number a_k can be found separately, because all the others disappear when you multiply by $\cos kx$ and integrate. Every cosine is orthogonal to every other cosine and to every sine:

$$\int_{-\pi}^{\pi} \cos kx \cos lx \, dx = 0 \quad \text{and} \quad \int_{-\pi}^{\pi} \cos kx \sin lx \, dx = 0. \tag{5}$$

These are easy to check. The last integrand is a sum of two sines, $\frac{1}{2} \sin(k+l)x + \frac{1}{2} \sin(k-l)x$, and the area under any sine wave is zero.

The same method produces b_k, the typical sine coefficient: Return to (2), multiply both sides by $\sin kx$, and integrate. The right side is again left with one term (this time it is $b_k \pi$) because of identities like (5) for the sines. Therefore $b_k \pi$ matches the left hand side, and the coefficient is

$$b_k = \frac{1}{\pi} \int_{-\pi}^{\pi} f(x) \sin kx \, dx. \tag{6}$$

Finally we go back for the constant term a_0. It is the coefficient of $\cos 0x$ (which is 1). To be systematic both sides of equation (2) should be multiplied by 1—consider it done. Integrating from $-\pi$ to π leaves $2\pi a_0$ alone on the right side, and the only novelty is in its extra factor 2:

$$a_0 = \frac{1}{2\pi} \int_{-\pi}^{\pi} f(x) \, dx. \tag{7}$$

In other words, the constant a_0 is the average of the function f. The other harmonics oscillate around zero and combine to give $f - a_0$.

We collect the basic steps of Fourier analysis and then show by example how the series reproduces the original function f.

4A The Fourier series of a 2π-periodic function is

$$f(x) = a_0 + \sum_{k=1}^{\infty} (a_k \cos kx + b_k \sin kx)$$

with a_0, a_k, b_k as above. Equivalently it is

$$f(x) = \sum_{k=-\infty}^{\infty} c_k e^{ikx} \quad \text{with} \quad c_k = \frac{1}{2\pi} \int_{-\pi}^{\pi} f(x) e^{-ikx} \, dx. \tag{8}$$

Each term in these series is the projection of f onto one of the pure harmonics $\cos kx$ or $\sin kx$ or e^{ikx}.

Note that the coefficients c_k come from a single unified formula (8). That is an important advantage. The series is

$$f(x) = c_0 + c_1 e^{ix} + c_{-1} e^{-ix} + c_2 e^{2ix} + c_{-2} e^{-2ix} + \cdots$$

and the function can be real even if the separate terms are complex. The coefficients are again found by orthogonality: Multiply both sides of this series by e^{-ikx} (not e^{+ikx}!) and integrate over a period. The only terms remaining after the integration are

$$\int_{-\pi}^{\pi} f(x) e^{-ikx}\, dx = \int_{-\pi}^{\pi} c_k e^{ikx} e^{-ikx}\, dx = 2\pi c_k.$$

This is formula (8) for the complex coefficient c_k.

For Fourier series you can choose the real form or the complex form; both are in use. The sines and cosines keep everything real, but they involve two calculations where $\int f e^{-ikx}\, dx$ is only one. For Fourier integrals over the whole line and for discrete Fourier transforms over a finite set of points, it seems pretty certain that the complex form has won.

Remark 1   The relation of a_k and b_k to c_k comes directly from the link between the cosine and sine and complex exponential:

$$e^{ikx} = \cos kx + i \sin kx \quad \text{and} \quad e^{-ikx} = \cos kx - i \sin kx. \tag{9}$$

Multiplying by $f(x)$ and integrating over a period,

$$\int f e^{-ikx}\, dx = \int f \cos kx\, dx - i \int f \sin kx\, dx.$$

With $1/2\pi$ on the left and $1/\pi$ on the right, this is

$$c_k = \tfrac{1}{2} a_k - \tfrac{1}{2} i b_k \quad \text{and} \quad c_{-k} = \tfrac{1}{2} a_k + \tfrac{1}{2} i b_k.$$

In the opposite direction the sine and cosine can be recovered from the exponentials of frequency k and $-k$:

$$\cos kx = \tfrac{1}{2}(e^{ikx} + e^{-ikx}) \quad \text{and} \quad \sin kx = \frac{1}{2i}(e^{ikx} - e^{-ikx}).$$

Therefore multiplication by $f(x)$ and integration leads to

$$a_k = c_k + c_{-k} \quad \text{and} \quad b_k = i(c_k - c_{-k}).$$

Remark 2   The Fourier coefficients need and use the function f over the whole interval. That contrasts sharply with the Taylor coefficients, which are concentrated at one point. A Taylor expansion $f(0) + xf'(0) + x^2f''(0)/2 + \cdots$ uses derivatives; Fourier uses integrals. They are both possible for very smooth functions, like analytic functions, but Fourier is much more general.

Remark 3   The inner product of complex vectors is not $u^T v$ but $\bar{u}^T v$. The same is true for complex functions; the sign of the imaginary part of u is reversed.† Therefore integrating e^{ikx} times e^{-ikx} gave the inner product of a function with itself (the square of its length):

$$\text{if}\quad u = \begin{bmatrix} 1 \\ i \end{bmatrix}\quad \text{then}\quad \|u\|^2 = \begin{bmatrix} 1 & -i \end{bmatrix} \begin{bmatrix} 1 \\ i \end{bmatrix} = 2$$

$$\text{if}\quad u = e^{ikx}\quad \text{then}\quad \|u\|^2 = \int e^{-ikx} e^{ikx}\, dx = 2\pi$$

$$\text{if}\quad u = e^{ikx}\quad \text{and}\quad v = e^{ilx}\quad \text{then}\quad \bar{u}^T v = 0.$$

This is the orthogonality we needed.

Examples of Fourier Series

EXAMPLE 1   (very simple)

$$f = \cos^2 x$$

The identity $\cos^2 x = \frac{1}{2}(1 + \cos 2x)$ means that a_0 and a_2 must be $\frac{1}{2}$. All other integrals like $\int \cos^2 x \cos 5x\, dx$ must be zero.

EXAMPLE 2   (very important)

$$f = \delta(x)$$

More correctly f is a **row of delta functions**, one at every point $x = 0, 2\pi, -2\pi, 4\pi, -4\pi, \ldots$ Then it is periodic, with one impulse in each period. Since integrating any product $\delta(x)g(x)$ picks out the value of g at $x = 0$, the Fourier coefficients of a delta function are

$$a_0 = \frac{1}{2\pi} \int \delta(x)\, dx = \frac{1}{2\pi}$$

$$a_k = \frac{1}{\pi} \int \delta(x) \cos kx\, dx = \frac{1}{\pi} \qquad (\text{since } \cos 0 = 1)$$

$$b_k = \frac{1}{\pi} \int \delta(x) \sin kx\, dx = 0 \qquad (\text{since } \sin 0 = 0)$$

$$c_k = \frac{1}{2\pi} \int \delta(x) e^{-ikx}\, dx = \frac{1}{2\pi} \qquad (\text{since } e^0 = 1).$$

---

† Otherwise the vector $(1, i)$ has length $1^2 + i^2 = 0$.

How can a combination of harmonics approach a delta function? For the cosine series (sines are absent since $b_k = 0$) the first terms are

$$\frac{1}{2\pi}[1 + 2 \cos x + 2 \cos 2x + \cdots + 2 \cos Nx]. \tag{10}$$

For the complex series every c_k is $1/2\pi$ and the corresponding terms are

$$\frac{1}{2\pi}[1 + e^{ix} + e^{-ix} + e^{2ix} + e^{-2ix} + \cdots + e^{iNx} + e^{-iNx}]. \tag{10'}$$

These are **partial sums** of the Fourier series for the delta function. The complex terms from frequencies k and $-k$ should match the real terms of the same frequency, and they do:

$$e^{ikx} + e^{-ikx} = 2 \cos kx.$$

What is more striking is that these partial sums (call them P_N) come out neatly. The sum of cosines in (10) needs a trigonometric identity but the complex case is simpler: (10') contains the geometric progression $1 + e^{ix} + e^{2ix} + \cdots + e^{2Nix}$, all multiplied by e^{-iNx}. It can be simplified to

$$P_N = \sum_{-N}^{N} e^{ikx} = \frac{e^{i(N+1/2)x} - e^{-i(N+1/2)x}}{e^{i(1/2)x} - e^{-i(1/2)x}} = \frac{\sin (N + \frac{1}{2})x}{\sin \frac{1}{2}x}. \tag{11}$$

Divided by 2π, this should approach a delta function. Figure 4.2 indicates how it does. At $x = 0$ every term in the sum is 1, so the height is $P_N = 2N + 1$. The "spike" is getting proportionately thinner, but P_N does not go to zero outside that spike! It oscillates faster and faster, with amplitude $1/\sin \frac{1}{2}x$. On every interval away from $x = 0$, the area under the curve goes to zero. The oscillations cancel themselves out as N gets large, except at $x = 0$. However **the area under the whole curve does not change:**

$$\int_{-\pi}^{\pi} P_N \, dx = \int_{-\pi}^{\pi} [1 + 2 \cos x + \ldots + 2 \cos Nx] \, dx = \int_{-\pi}^{\pi} 1 \, dx = 2\pi.$$

The integral of f times P_N is similar. It goes to zero on intervals away from $x = 0$. It is concentrated near the origin, where $f(x) \approx f(0)$. Then $\int f(x)P_N(x)dx \to 2\pi f(0)$, and in this integrated sense, $P_N/2\pi \to \delta$.

EXAMPLE 3   (odd but typical)

$$f(x) = x.$$

Since f must be 2π-periodic, it cannot remain equal to x outside the basic interval from $-\pi$ to π. It has to repeat itself, with a jump discontinuity at the end of each

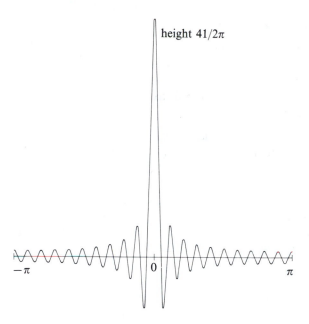

Fig. 4.2. The Fourier series for the delta function: $\dfrac{1}{2\pi} \displaystyle\sum_{-20}^{20} e^{ikx}$.

period (Fig. 4.3). The cosine terms disappear from the Fourier series because

$$\int_{-\pi}^{\pi} x \cos kx \, dx = 0 \quad \text{for all} \quad k.$$

The cosines are **even** functions whereas $f = x$ is an **odd** function. In one case $\cos(-kx) = \cos kx$; in the other, f reverses sign when x does. Therefore the product $x \cos kx$ reverses sign, and the integral from $-\pi$ to 0 cancels the integral from 0 to π.

However the $\sin kx$ terms (which are odd like f) are all present:

$$b_k = \frac{1}{\pi} \int_{-\pi}^{\pi} x \sin kx \, dx = \frac{1}{\pi} \left[\frac{-x \cos kx}{k} + \frac{\sin kx}{k^2} \right]_{-\pi}^{\pi} = \frac{2}{k} \quad \text{or} \quad \frac{-2}{k}.$$

The k^2 term is zero because $\sin k\pi = 0$, and in the other term $\cos k\pi$ alternates between -1 and 1—producing the sine series

$$x = 2 \left(\frac{\sin x}{1} - \frac{\sin 2x}{2} + \frac{\sin 3x}{3} - \cdots \right). \tag{12}$$

Figure 4.3 shows this series after 1, 2, 3, 4, 5, and 10 terms. Convergence is clear but not especially fast. The interesting point is $x = \pi$, where the function jumps and there are two things to notice:

(1)   The series gives the answer zero at $x = \pi$. This answer is halfway between the values of f on opposite sides of the discontinuity.

(2)   Near the jump there is an overshoot of about 18%. This is the **Gibbs phenomenon**, which does not disappear as more Fourier terms are kept—it moves closer to the jump.

Both 1 and 2 are typical of Fourier series at a discontinuity. Since the step function at a jump is the integral of a delta function, the overshoot and oscillations in Fig. 4.3 must be the integral of Fig. 4.2. The area under the central arch in Fig. 4.2 is nearly 1.18.†

Another typical and important feature of jumps is the slow decay of the Fourier coefficients. As in (12), the kth coefficient is of order $1/k$; that reappears for the square wave in the exercises. For its integral the hat function, a_k and b_k decay like $1/k^2$. For its derivative the delta function, a_k and b_k are constant.

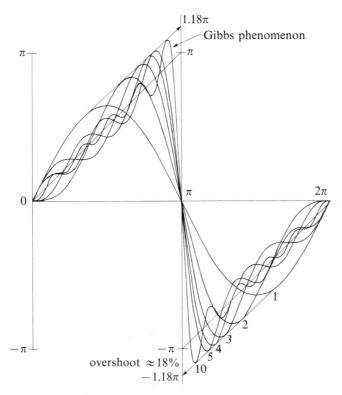

Fig. 4.3. Partial sums of the Fourier series for $f = x$.

---

† This number is not the same in all books.

Sine Series and Cosine Series

The first two examples produced only cosines; the third function $f(x) = x$ produced only sines. There will be no mystery if we look again at even and odd functions:

$$f \text{ is } \textbf{\textit{even}} \text{ if } \quad f(-x) = f(x) \qquad \text{for all } x$$

$$f \text{ is } \textbf{\textit{odd}} \text{ if } \quad f(-x) = -f(x) \quad \text{for all } x$$

In either case it is enough to know f over the *half*-period $0 \le x < \pi$, since in the other half its graph is either a mirror image (the even case) or its negative.

Cosines are even and sines are odd. The function x^k is even if k is even, and odd if k is odd. Exponentials e^{ikx} and most other functions are neither. But every function has an even part and an odd part,

$$f = f_e + f_o, \quad \text{with} \quad f_e(x) = \frac{f(x) + f(-x)}{2} \quad \text{and} \quad f_o(x) = \frac{f(x) - f(-x)}{2}$$

The even part yields the cosine terms and the odd part yields the sines.

4B If f is odd then all a_k are zero and all b_k can be computed from the half-period between 0 and π:

$$b_k = \frac{1}{\pi} \int_{-\pi}^{\pi} f(x) \sin kx \, dx = \frac{2}{\pi} \int_{0}^{\pi} f(x) \sin kx \, dx. \tag{13}$$

Similarly b_k is zero if f is even, and a_k comes from a half-period.

That idea can be turned around. Suppose f is known over the half-period $0 < x < \pi$. Then the b_k from (13) are the coefficients in its **Fourier sine series**

$$f(x) = b_1 \sin x + b_2 \sin 2x + b_3 \sin 3x + \cdots \tag{14}$$

This series will produce f on the other half-period, where it was originally not known, provided f is extended in the right way: Since the sines are odd, f should extend past $x = 0$ as an odd function. Then the Fourier series for f has no cosine terms and (14) is correct over the whole period.

That is an odd periodic extension of f. The opposite case keeps only the cosine coefficients

$$a_0 = \frac{1}{\pi} \int_{0}^{\pi} f(x) \, dx \quad \text{and} \quad a_k = \frac{2}{\pi} \int_{0}^{\pi} f(x) \cos kx \, dx.$$

These give the **Fourier cosine series**, and it matches f over a whole period provided f is extended to be even: $f(-x) = f(x)$.

EXAMPLE $f(x) = 1$ on the half-period $0 < x < \pi$.

If f is odd then it equals -1 on the other half-period; the function is a *square wave*. Its Fourier sine series is in the exercises, and cannot be given away here. If f is even then $f = 1$ everywhere, and its Fourier cosine series is simply $a_0 = 1$ (there are no other terms).

Properties of Fourier Series

The mathematical literature on Fourier series is enormous, and from it we select three main points:

1.   Each Fourier coefficient a_k or b_k or c_k is the best possible choice in the mean-square sense. In other words, the error

$$E = \int_{-\pi}^{\pi} \left[f(x) - \sum_{0}^{N} (A_k \cos kx + B_k \sin kx) \right]^2 dx$$

is a minimum when A_k is Fourier's a_k and B_k is b_k. For proof we check a typical derivative $\partial E / \partial B_k$, which should be zero at the minimum:

$$-\frac{1}{2} \frac{\partial E}{\partial B_k} = \int_{-\pi}^{\pi} \sin kx \left[f(x) - \sum_{0}^{N} (A_k \cos kx + B_k \sin kx) \right] dx$$

$$= \int_{-\pi}^{\pi} \sin kx [f(x) - B_k \sin kx] \, dx \tag{15}$$

since orthogonality removes all other terms. (15) is zero and E is minimized if

$$B_k = \frac{\int f(x) \sin kx \, dx}{\int \sin^2 kx \, dx} = \frac{1}{\pi} \int f(x) \sin kx \, dx.$$

This is exactly the Fourier coefficient, so the best choice of B_k is b_k.

The interpretation is this: The sines and cosines are a perpendicular set of axes in "function space." Fourier analysis projects f onto each of these axes. The 1-dimensional projections like $a_3 \cos 3x$ are the components of f in these orthogonal directions. It is orthogonality that allows us to find each term separately, and it is completeness that allows the sines and cosines (or the exponentials e^{ikx}) to reproduce f.

2.   Since projections are never larger than the original, no piece of the Fourier series can be larger than f:

$$\int_{-\pi}^{\pi} (a_0 + a_1 \cos x + \cdots + b_N \sin Nx)^2 \, dx \le \int_{-\pi}^{\pi} (f(x))^2 \, dx. \tag{16}$$

This is Bessel's inequality (Exercise 4.1.13). In the limit as $N \to \infty$, when the Fourier series on the left reconstructs the function f on the right, the inequality becomes an equality. Integrating the left side (all cross terms vanish by orthogonality) it is *Parseval's formula*:

$$2\pi a_0^2 + \pi(a_1^2 + b_1^2 + a_2^2 + b_2^2 + \cdots) = \int_{-\pi}^{\pi} (f(x))^2 \, dx. \tag{17}$$

That is a beautiful formula.† It is also the key to the most famous infinite-dimensional space, called *Hilbert space*, which can be described in a paragraph.

In one form Hilbert space contains functions; in another form it contains vectors with infinitely many components. Not all functions and not all vectors are allowed. The functions must have "finite length"—the integral of $(f(x))^2$ must be finite, which excludes a delta function and also excludes $f = 1/x$. It includes square waves and $\sin x$ and even some unbounded functions like $\log |\sin x|$; the area under f^2 is finite. For vectors the test is similar—*the sum of squares must be finite*—which excludes $(1, 1, 1, \ldots)$ but includes $(1, \frac{1}{2}, \frac{1}{3}, \ldots)$. The connection between functions and vectors is in Parseval's formula (17). Hilbert space contains the function f exactly when its vector form contains the Fourier coefficients of f. The length is the same in the function space L^2 and the infinite vector space l^2, when adjusted by the factors π and 2π in Parseval's formula. These spaces allow the number of dimensions to become infinite without losing the geometry of lengths and angles and inner products.

3. Starting from f we found its Fourier series, and we have been assuming that this infinite series converges back to f. That is assuming a lot! The convergence is most natural in the mean-square sense, in other words in Hilbert space. There the distance between f and its Fourier series does go to zero (Parseval's formula) as more terms are included in the series. The distance is measured by squaring and integrating, but this mean-square convergence does not guarantee convergence at every particular point x.

The best point to look at is $x = 0$. It is actually typical although it looks special. Since $\sin 0 = 0$ and $\cos 0 = 1$, the Fourier series at that point is simply $a_0 + a_1 + \cdots$. The formulas for those coefficients give

$$a_0 + a_1 + a_2 + \cdots = \frac{1}{2\pi} \int_{-\pi}^{\pi} f(x)[1 + 2\cos x + 2\cos 2x + \cdots] \, dx \tag{18}$$

and the question is whether this approaches the value of f at $x = 0$.

That is not at all obvious, but fortunately we already hold the key. The expression that now multiplies f is *the Fourier series for the delta function* which appeared earlier in (10). The right side of (18) should converge, if there is any justice,

---

† With the exponentials $c_k e^{ikx}$ it is even better; the left side is $2\pi \sum |c_k|^2$.

to $\int f(x)\delta(x)\,dx = f(0)$. This proof succeeds if f is smooth enough to have a derivative. It might fail at points where f is less smooth, but to the surprise of most mathematicians convergence was recently proved at "almost every" point.†

Solution of Laplace's Equation

Our first application of Fourier series is to differential equations. In this chapter that means equilibrium problems, and we concentrate on Laplace's equation

$$\frac{\partial^2 u}{\partial x^2} + \frac{\partial^2 u}{\partial y^2} = 0. \tag{19}$$

The idea is to construct u as an infinite series, choosing its coefficients to match $u_0(x, y)$ along the boundary. The boundary values u_0 are given and the problem is to find $u(x, y)$ inside the region. Everything depends on the shape of the boundary, and we take the most important case—a circle of radius 1.

To begin we need enough solutions to Laplace's equation. For a circle that is not difficult. The solutions can be written as polynomials in x and y, and in terms of r and θ they are even simpler: their two forms are

$$\text{Re}(x + iy)^n \quad \text{and} \quad \text{Im}(x + iy)^n: \quad 1, x, y, x^2 - y^2, 2xy, \ldots$$

$$\text{Re } z^n \quad \text{and} \quad \text{Im } z^n: \quad 1, r\cos\theta, r\sin\theta, r^2\cos 2\theta, r^2\sin 2\theta, \ldots$$

They satisfy Laplace's equation for many reasons. They come from $z = x + iy$, two y derivatives of $(x + iy)^n$ produce $i^2 = -1$ and cancel the effect of two x derivatives, and (most convincing) everything disappears when $r^n \cos n\theta$ and $r^n \sin n\theta$ are substituted into

$$\frac{1}{r}\frac{\partial}{\partial r}\left(r\frac{\partial u}{\partial r}\right) + \frac{1}{r^2}\frac{\partial^2 u}{\partial \theta^2} = 0. \tag{20}$$

That is Laplace's equation in polar coordinates.

Combinations of these special solutions give the general solution

$$u = a_0 + a_1 r\cos\theta + b_1 r\sin\theta + a_2 r^2\cos 2\theta + b_2 r^2\sin 2\theta + \cdots \tag{21}$$

All that remains is to choose the constants a_k and b_k to make $u = u_0$ on the boundary. The radius is $r = 1$, so we must have

$$u_0(\theta) = a_0 + a_1\cos\theta + b_1\sin\theta + a_2\cos 2\theta + b_2\sin 2\theta + \cdots \tag{22}$$

---

† When f is in the Hilbert space L^2, the points where convergence fails can be covered by intervals whose total length is arbitrarily small.

This is exactly the Fourier series for u_0—which is periodic, since θ and $\theta + 2\pi$ give the same point on the circle. The constants a_k and b_k must be the Fourier coefficients. Thus the problem is completely solved—if an infinite series is acceptable as the solution.†

4C The solution to Laplace's equation on a unit circle with $u = u_0(\theta)$ on the boundary is (21), if a_k and b_k are chosen to be the Fourier coefficients of u_0. The exponential form of this solution uses the complex coefficients c_k:

$$u(r, \theta) = \sum_{k=-\infty}^{\infty} c_k r^{|k|} e^{ik\theta}.$$

It is interesting that Poisson managed to sum this infinite series; his formula gives u directly from u_0. At any point inside the circle the solution to Laplace's equation is

$$u(r, \theta) = \frac{1}{2\pi} \int_{-\pi}^{\pi} u_0(\varphi) \frac{1 - r^2}{1 + r^2 - 2r \cos(\theta - \varphi)} \, d\varphi. \tag{23}$$

At $r = 0$ the fraction disappears and u is the average value $\int u_0(\varphi) \, d\varphi / 2\pi$—where φ is the variable of integration around the boundary.

Physically that is correct; the steady state temperature at the center is the average temperature around the circle. If we leave the center and go toward the boundary (say $r \to 1$ along the ray at angle θ) the fraction in (23) somehow approaches a delta function. It becomes concentrated at the point where $\varphi = \theta$. At all other points the fraction goes to zero as r goes to 1, because of $1 - r^2$, and the result is exactly what we want: the integral picks out u_0 at the point $\varphi = \theta$ and

$$u(r, \theta) \to u_0(\theta) \quad \text{as} \quad r \to 1.$$

In other words, the boundary values of u are u_0.

Poisson's formula gives the response at r, θ to the boundary values u_0. To check it we could verify that the fraction in (23) satisfies Laplace's equation, or else connect it to Fourier series by the identity

$$\frac{1 - r^2}{2(1 + r^2 - 2r \cos \theta)} = \frac{1}{2} + r \cos \theta + r^2 \cos 2\theta + \cdots \tag{24}$$

In this section the Fourier approach is natural. First we replace θ by $\theta - \varphi$, changing each $\cos k\theta$ in equation (24) to $\cos k\theta \cos k\varphi + \sin k\theta \sin k\varphi$. Then

---

† If not we could change to n terms that match u_0 at n equally spaced points around the circle, using the Fast Fourier Transform.

multiplying by u_0/π and integrating, the left side becomes Poisson's formula and the right side produces all the right Fourier coefficients:

$$\frac{1}{2\pi} \int_{-\pi}^{\pi} u_0(\varphi) \, d\varphi + \frac{1}{\pi} \sum_{k=1}^{\infty} r^k \int_{-\pi}^{\pi} u_0(\varphi)(\cos k\theta \cos k\varphi + \sin k\theta \sin k\varphi) \, d\varphi$$

$$= a_0 + a_1 r \cos \theta + b_1 r \sin \theta + \cdots$$

This is $u(r, \theta)$, from the Fourier series in (21), and Poisson's formula is confirmed.

EXAMPLE 1 $u_0 = $ delta function at the boundary point (1, 0).

The response to an impulse at the point $\varphi = 0$ on the circle is

$$u(r, \theta) = \frac{1}{2\pi} \frac{1 - r^2}{1 + r^2 - 2r \cos \theta}.$$

At the center, where $r = 0$, the average of the delta function is $u = 1/2\pi$.

EXAMPLE 2 $u_0(\theta) = \theta, \ -\pi < \theta < \pi.$

The Fourier series for u_0 was calculated in equation (12):

$$\theta = 2\left(\frac{\sin \theta}{1} - \frac{\sin 2\theta}{2} + \frac{\sin 3\theta}{3} - \cdots \right).$$

Therefore the Fourier series (21) for u is

$$u(r, \theta) = 2\left(\frac{r \sin \theta}{1} - \frac{r^2 \sin 2\theta}{2} + \frac{r^3 \sin 3\theta}{3} - \cdots \right).$$

On the circle u has a sudden jump at $\theta = \pi$, but inside the circle that jump is gone. The powers r^k damp the high frequencies and *Laplace's equation always has smooth solutions*.

Orthogonal Functions

We cannot leave the impression that the only orthogonal functions are the sines and cosines. They are undoubtedly the most important, but they are not unique. There are plenty of other possibilities, and there are at least five ways to find them:

(1) We could start with the powers $1, x, x^2, \ldots$ and force them to be orthogonal—by subtracting from each x^k its projection onto the (already ortho-

gonalized) polynomials of lower degree. This is the Gram-Schmidt idea, and it produces the **Legendre polynomials** $P_0 = 1$, $P_1 = x$, $P_2 = x^2 - \frac{1}{3}$, ...

(2) We could invent examples like the **Walsh functions** in Fig. 4.4:

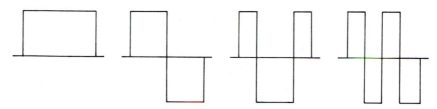

Fig. 4.4. The first four orthogonal Walsh functions.

They look like digital sines and cosines or vectors of 1's and -1's.

(3) We could combine $\sin kx$ with $\sin ly$ into a **double** Fourier sine series like

$$f(x, y) = b_{11} \sin x \sin y + b_{21} \sin 2x \sin y + b_{12} \sin x \sin 2y + \cdots. \qquad (25)$$

This applies to a function f that is odd and periodic in both x and y. The interval from $-\pi$ to π has changed into a "period square," and if we know f in one square $-\pi \leq x, y < \pi$ then we know it everywhere. In this case the orthogonal functions are the products $\sin kx \sin ly$. They are orthogonal over the square instead of the interval. They have two indices k and l but they are really just an orthogonal sequence and (25) is an orthogonal expansion.

There is a similar sequence $\cos kx \cos ly$ for even functions (including $k = 0$ and $l = 0$). There will be a larger sequence that mixes sines and cosines, for functions that are neither even or odd. And there will be a corresponding sequence of exponentials—*orthogonal*, *periodic*, and *complete*:

$$f(x, y) = \sum_{-\infty}^{\infty} \sum_{-\infty}^{\infty} c_{kl} e^{ikx} e^{ily}. \qquad (26)$$

These "double Fourier series" can solve Poisson's equation in a square (Ex. 4.1.26).

(4) We could write each of the usual cosines as a polynomial in $\cos \theta$:

$$\cos 2\theta = 2 \cos^2 \theta - 1, \cos 3\theta = 4 \cos^3 \theta - 3 \cos \theta, \ldots$$

Changing variables to $x = \cos \theta$ gives the **Chebyshev polynomials** $T_0 = 1$, $T_1 = x$, $T_2 = 2x^2 - 1$, $T_3 = 4x^3 - 3x$, ... They are orthogonal because the cosines are orthogonal. However there is something new from the change of variables:

$$\int_{-\pi}^{\pi} \cos k\theta \cos l\theta \, d\theta = 0 \text{ becomes } \int_{-1}^{1} T_k(x) T_l(x) \frac{dx}{\sqrt{1 - x^2}} = 0. \qquad (27)$$

This is weighted orthogonality, and the *weight function* is $w = 1/\sqrt{1-x^2}$. It comes from $dx = -\sin\theta\, d\theta = -\sqrt{1-x^2}\, d\theta$. We will see that weight functions present no problem.

(5)   We could look for the **eigenfunctions** of a symmetric differential equation. Like the eigenvectors of a symmetric matrix, these eigenfunctions are guaranteed to be orthogonal. If the equation is $d^2u/dx^2 = \lambda u$ we are back to sines and cosines and pure exponentials. But there is no shortage of equations that have non-constant coefficients, and every symmetric equation has its own orthogonal functions.

With the possible exception of the Walsh functions, this fifth method *contains all the others*. There is a differential equation $d/dx[(1-x^2)du/dx] = \lambda u$ for the Legendre polynomials. There is a differential equation $\partial^2 u/\partial x^2 + \partial^2 u/\partial y^2 = \lambda u$ for double Fourier series. There is a weighted differential equation $d/dx(w^{-1}\, du/dx) = \lambda wu$ for the Chebyshev polynomials, including their weight $w = 1/\sqrt{1-x^2}$. I am reluctant to tell you too many more examples—Laguerre with weight e^{-x}, or Hermite with weight $e^{-x^2/2}$, or hypergeometric functions, or Jacobi functions, or ultraspherical polynomials which are often named after Gegenbauer.† But there is one that cannot be missed.

We describe that one (Bessel functions) after showing that all orthogonal expansions work in the same way. Suppose a function $f(x)$ is given; it is to be expanded into a sum of orthogonal functions. If those functions are $T_0, T_1, T_2, \ldots$, then we want

$$f(x) = c_0 T_0 + c_1 T_1 + c_2 T_2 + \cdots \tag{28}$$

The problem is to find each coefficient c_k. The method is to multiply the equation by $T_k(x)$ *and by the weight function* $w(x)$. Then integrate over the interval on which the functions T_k are orthogonal:

$$\int fT_k w\, dx = c_0 \int T_0 T_k w\, dx + \cdots + c_k \int T_k^2 w\, dx + \cdots \tag{29}$$

On the right side only one term survives, as in Fourier series. There T_k was $\cos kx$, the weight was $w = 1$, and the integral from $-\pi$ to π was equal to π. Here we have a more general case but the same formula:

4D   The coefficients in a weighted orthogonal expansion like (28) are

$$c_k = \frac{\int fT_k w\, dx}{\int T_k^2 w\, dx}. \tag{30}$$

---

† Even the well-known Krawtchouk had polynomials.

The weight is like a "mass matrix" M in the eigenvalue problem. Instead of $Ax = \lambda x$ that becomes $Kx = \lambda Mx$, just as the right side of the differential equation contains λwu. The eigenvectors are not orthogonal in the usual sense, but where our functions are w-orthogonal the eigenvectors are M-orthogonal: $x_i^T Mx_j = 0$. The matrix M should be positive definite just as the weight w should be positive.

Now we are ready for the most important example of weighted orthogonality.

Bessel Functions

A double Fourier series is perfect for a square. Virtually any function is a combination of the exponentials $e^{ikx}e^{ily}$, in other words a combination of sines and cosines of kx and ly. On a circle those functions are not so satisfactory. If f depends only on the angle θ, periodic functions are still all right. But if f depends also on r, or if it depends *only* on r (which is the quite common case of radial symmetry) then new functions are needed.

Physically the best example is a circular drum. If you strike it, it oscillates. Its motion contains a mixture of oscillations that go at their own frequencies. I cannot call them harmonics, because the frequencies are no longer in an arithmetical progression. The goal is to discover those frequencies and to find the corresponding shapes (the eigenfunctions) of the drumhead.†

The problem is governed by Laplace's equation in polar coordinates:

$$\frac{\partial^2 u}{\partial r^2} + \frac{1}{r}\frac{\partial u}{\partial r} + \frac{1}{r^2}\frac{\partial^2 u}{\partial \theta^2} = -\lambda u. \tag{31}$$

The boundary condition is $u = 0$ at $r = 1$; the circumference of the drum is fastened down. It is the coefficients $1/r$ and $1/r^2$ that destroy the Fourier series in r, although we can expect the usual series in θ. A change of variables has altered the constant coefficients in $u_{xx} + u_{yy} = -\lambda u$, just as surely as a change to a variable material. There was no choice, when the region became a circle.

The natural idea is to separate r from θ, and to look for eigenfunctions of the special form $u = A(\theta)B(r)$. This is called **separation of variables**. It succeeded for squares, where $A(x)$ was e^{ikx} and $B(y)$ was e^{ily}. It needs an exceptional geometry, but the circle is exceptional. We will reach an *ordinary* differential equation for A and another one for B; each depends only on a single variable, r or θ. In fact the θ equation will have constant coefficients. Its solutions will be $A = e^{in\theta}$, and to shortcut the rest of the paragraph you can put $u = e^{in\theta}B(r)$ into Laplace's equation. Writing for posterity we go more slowly and substitute $u = A(\theta)B(r)$:

$$AB'' + \frac{1}{r}AB' + \frac{1}{r^2}A''B = -\lambda AB. \tag{32}$$

---

† For oscillations of a sphere instead of a circle, Legendre replaces Bessel (Ex. 4.1.32).

Now multiply by r^2, divide by AB, and separate r from θ:

$$\frac{r^2 B'' + rB' + \lambda r^2 B}{B} = -\frac{A''}{A}. \tag{33}$$

The left side depends only on r. The right side depends only on θ. Therefore both must be constant. If the constant is n^2, then on the right $A'' = -n^2 A$. This gives the dependence on θ as $e^{in\theta}$ or $e^{-in\theta}$ (or sine and cosine). It also establishes that n is an integer, since we must have the same value at $\theta = 0$ and $\theta = 2\pi$.

That leaves an ordinary differential equation for B:

$$\boxed{r^2 B'' + rB' + \lambda r^2 B = n^2 B.} \tag{34}$$

This is one form of **Bessel's equation**. We are interested (for each n) in solutions that are zero on the boundary of the circle. Then the eigenfunctions $u = AB$ of the Laplacian will be $\cos n\theta\, B$ and $\sin n\theta\, B$. The eigenvalues λ, which affect B, remain to be discovered.

To solve Bessel's equation is not easy; to find a solution using a few familiar functions is impossible. The direct approach is to assume an infinite series $B = \Sigma\, c_m r^m$ and determine its coefficients c_m. That technique fills whole chapters in textbooks on applied mathematics, which I frankly think is unreasonable. It is not a technique to be used in a new situation, partly because for numerical purposes it is so involved. Therefore we do it only once, in the radially symmetric case $n = 0$ with no dependence on θ, to see what a Bessel function looks like. The technique asks us to substitute $B = \Sigma\, c_m r^m$ into $r^2 B'' + rB' + \lambda r^2 B = 0$:

$$\Sigma\, c_m m(m-1) r^m + \Sigma\, c_m m r^m + \lambda \Sigma\, c_m r^{m+2} = 0.$$

The third sum is unlike the others. It starts at r^2 with the coefficient λc_0. It multiplies each power r^m by λc_{m-2}, so that comparing coefficients of r^m in all three sums leads to

$$c_m m(m-1) + c_m m + \lambda c_{m-2} = 0. \tag{35}$$

In other words $\lambda c_{m-2} = -m^2 c_m$. Suppose c_0 is chosen, say $c_0 = 1$. Then for $m = 2$ this recursion gives $c_2 = -\lambda/2^2$. Similarly c_4 is $-\lambda/4^2$ times c_2, and c_6 is determined from c_4. Each step gives one more coefficient in the power series expansion of B:

$$B(r) = 1 - \frac{\lambda r^2}{2^2} + \frac{\lambda^2 r^4}{2^2 4^2} - \frac{\lambda^3 r^6}{2^2 4^2 6^2} + \cdots. \tag{36}$$

This is a **Bessel function** of order zero. If λ were 1 the standard notation would be $J_0(r)$. With λ present B is $J_0(\sqrt{\lambda} r)$. The subscript 0 means that the frequency n in the θ direction is zero.

It remains to determine the eigenvalues λ. They come from the boundary condition $B = 0$ at $r = 1$. The eigenfunction is $J_0(\sqrt{\lambda}r)$ and the requirement is $J_0(\sqrt{\lambda}) = 0$. The best way to appreciate these functions is by comparison with the cosine, whose series we know and whose behavior we know:

$$\cos(\sqrt{\lambda}r) = 1 - \frac{\lambda r^2}{2!} + \frac{\lambda^2 r^4}{4!} - \frac{\lambda^3 r^6}{6!} + \cdots. \tag{37}$$

Again we set $r = 1$ and pick out the values at which $\cos\sqrt{\lambda} = 0$. The zeros of the cosine, although you couldn't tell it from the series, are at $\sqrt{\lambda} = \pi/2, 3\pi/2, 5\pi/2, \ldots$. They occur at regular intervals with constant spacing π. The zeros of the Bessel function are almost that regular (fortunately for our ears). They occur at

$$\sqrt{\lambda} \approx 2.4, 5.5, 8.65, 11.8, 14.9, \ldots$$

and their spacing converges rapidly to π. In fact the Bessel function $J_0(r)$ approaches $\sqrt{2/\pi r}\cos(r - \pi/4)$, which looks like the cosine with its graph shifted by $\pi/4$ and its amplitude slowly decreasing. Figure 4.5 shows this function up to its third zero, at $\sqrt{\lambda_3}$. To get that far requires the r^{22} term in the series (36); the results of stopping at earlier terms are displayed.

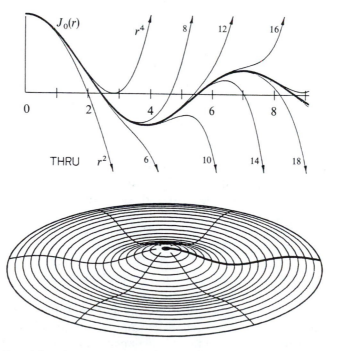

Fig. 4.5. The Bessel function $J_0(r)$ and a drum at frequency λ_3.

You must see the analogy between the Bessel function and the cosine. $B(r)$ comes from the oscillations of a circular drum; $C(x) = \cos(k - \frac{1}{2})\pi x$ comes from the oscillations of a square. In one case θ has been removed and the drum is oscillating radially. In the other case y has been removed and the square is flapping in the x-direction. The center of the drum and the left side of the square are free; the slopes there are zero. The boundary of the drum and the right side of the square are fixed; the functions are zero. That determines the frequencies $\sqrt{\lambda}$ for the drum and $\pi/2$, $3\pi/2$, $5\pi/2$, ... for the square. B and C are eigenfunctions of Laplace's equation (without θ and y):

$$\frac{1}{r}\frac{d}{dr}\left(r\frac{dB}{dr}\right) = -\lambda B \quad \text{and} \quad \frac{d^2C}{dx^2} = -\left(k - \frac{1}{2}\right)^2\pi^2 C.$$

The first eigenfunction B drops from 1 to 0, like the cosine. It is the first arch in the figure, stretched to cross zero at $r = 1$. The second eigenfunction crosses zero and comes up again. The kth eigenfunction, like the kth cosine, has k arches. These are the modes of pure oscillation, each with its own frequency, and Fig. 4.5 shows mode $k = 3$—reaching zero for the third time at the boundary.

Both of these sequences are eigenfunctions of a symmetric problem (Laplace's equation). Therefore **there must be orthogonality**. The Bessel functions $B_k = J_0(\sqrt{\lambda_k}\, r)$ are orthogonal over a circle,

$$\int_0^{2\pi}\int_0^1 B_k(r)\, B_l(r)\, r\, dr\, d\theta = 0 \quad \text{if } k \neq l. \tag{38}$$

The cosines are orthogonal over a square,

$$\int_0^1\int_0^1 \cos(k - \tfrac{1}{2})\pi x\, \cos(l - \tfrac{1}{2})\pi x\, dx\, dy = 0 \quad \text{if } k \neq l. \tag{39}$$

The θ integral and the y integral make no difference and can be ignored. Thus the Bessel functions are orthogonal over the interval from 0 to 1, and so are the cosines. Their boundary conditions are identical, zero slope at the left endpoint and zero value at the right endpoint.

Notice the one thing that is different! There is a weighting factor r in (38) and no weighting factor in (39). The Bessel functions are orthogonal with respect to this weight $w = r$; for the cosines w is 1. The weight does not spoil the over-riding advantage of orthogonality, that the coefficients in an expansion like $f = \Sigma\, c_k B(\sqrt{\lambda_k}\, r)$ can be computed directly from formula (30).

In closing we describe the other oscillations of a drum. If they depend on θ then more Bessel functions will appear. The separation of variables into $A(\theta)B(r)$ gave possibilities other than $A = 1$ (the radially symmetric case considered so far). The equation was $A'' + n^2A = 0$ and it allowed $A = \cos n\theta$ or $A = \sin n\theta$ for every integer n. Then the equation for B was (34):

$$rB'' + B' + \left(\lambda r - \frac{n^2}{r} \right) B = 0. \tag{40}$$

For $\lambda = 1$ this has a solution $J_n(r)$ which is finite at $r = 0$. That is the *Bessel function of order n.* (All other solutions blow up at $r = 0$; they involve Bessel functions of the second kind.) For every positive λ the solution is just rescaled to $J_n(\sqrt{\lambda}\, r)$. At $r = 1$ the boundary condition requires $J_n(\sqrt{\lambda}) = 0$; that picks out the eigenvalues. The products $A(\theta)B(r) = \cos n\theta\, J_n(\sqrt{\lambda_k}\, r)$ and $\sin n\theta\, J_n(\sqrt{\lambda_k}\, r)$ are the eigenfunctions. They give the shape of the drum in its pure oscillations, and Fig. 4.6 indicates roughly what they look like.

The simplest guide is the nodal lines along which the drum does not move. They are like the zeros of the sine function, where a violin string is still. For the drum we are in two dimensions and the eigenfunctions are $A(\theta)B(r)$. There is a nodal line from the center whenever $A = 0$ and a nodal circle whenever $B = 0$. For different values of n (the frequency in $\cos n\theta$) and k (the oscillation number in the r direction), the figure shows where the drumhead is motionless. The oscillations themselves are functions of time—they are solutions $A(\theta)B(r)e^{i\sqrt{\lambda}\,t}$ of the wave equation in a circle.

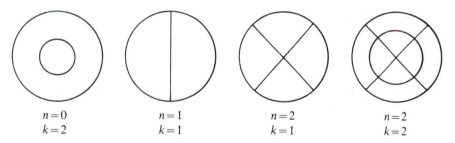

$$
\begin{array}{cccc}
n=0 & n=1 & n=2 & n=2 \\
k=2 & k=1 & k=1 & k=2
\end{array}
$$

Fig. 4.6. Nodal lines of drum = zero lines of $A(\theta)B(r)$.

Finally we mention a problem that is unsolved as of Christmas 1984. *Can you hear the shape of a drum?* If you know the eigenvalues λ, does that determine the boundary of the drumhead? I think the eigenvalues above, for a circle, do not occur for any other shape. But whether two different drums could sound the same, no one knows.

EXERCISES

4.1.1   Find the Fourier series on $-\pi < x < \pi$ for
  (a) $f(x) = \sin^3 x$, an odd function
  (b) $f(x) = |\sin x|$, an even function
  (c) $f(x) = x^2$, integrating either $x^2 \cos kx$ or the sine series for $f = x$
  (d) $f(x) = e^x$, using the complex form of the series.
What are the even and odd parts of $f(x) = e^x$ and $f(x) = e^{ix}$?

4.1.2 A square wave has $f(x) = -1$ on the left side $-\pi < x < 0$ and $f(x) = +1$ on the right side $0 < x < \pi$.

(1) Why are all the cosine coefficients $a_k = 0$?

(2) Find the sine series $\Sigma\, b_k \sin kx$ from equation (6).

4.1.3 Find this sine series for the square wave f in another way, by showing that

(a) $df/dx = 2\delta(x) - 2\delta(x + \pi)$ extended periodically

(b) $2\delta(x) - 2\delta(x + \pi) = \dfrac{4}{\pi}(\cos x + \cos 3x + \cdots)$ from (10)

Integrate each term to find the square wave f.

4.1.4 At $x = \pi/2$ the square wave equals 1. From the Fourier series at this point find the alternating sum that equals π:

$$\pi = 4(1 - \tfrac{1}{3} + \tfrac{1}{5} - \tfrac{1}{7} \cdots).$$

4.1.5 From Parseval's formula the square wave sine coefficients satisfy

$$\pi(b_1^2 + b_2^2 + \cdots) = \int_{-\pi}^{\pi} |f(x)|^2 \, dx = \int_{-\pi}^{\pi} 1 \, dx = 2\pi.$$

Derive another remarkable sum $\pi^2 = 8(1 + \tfrac{1}{9} + \tfrac{1}{25} + \cdots)$.

4.1.6 Around the unit circle suppose u is a square wave

$$u_0 = \begin{cases} +1 & \text{on the upper semicircle} & 0 < \theta < \pi \\ -1 & \text{on the lower semicircle} & -\pi < \theta < 0 \end{cases}$$

From the Fourier series for the square wave write down the Fourier series for u (the solution (21) to Laplace's equation). What is the value of u at the origin?

4.1.7 If a square pulse is centered at $x = 0$ to give

$$f(x) = 1 \quad \text{for} \quad |x| < \frac{\pi}{2}, f(x) = 0 \quad \text{for} \quad \frac{\pi}{2} < |x| < \pi,$$

draw its graph and find its Fourier coefficients a_k and b_k.

4.1.8 Suppose f has period T instead of 2π, so that $f(x) = f(x + T)$. Its graph from $-T/2$ to $T/2$ is repeated on each successive interval and its real and complex Fourier series are

$$f(x) = a_0 + a_1 \cos\frac{2\pi x}{T} + b_1 \sin\frac{2\pi x}{T} + \cdots = \sum_{-\infty}^{\infty} c_j e^{ij2\pi x/T}.$$

Multiplying by the right functions and integrating from $-T/2$ to $T/2$, find a_k, b_k, and c_k.

4.1.9 Establish by integration of x or otherwise the odd Fourier series

$$x(\pi - |x|) = \frac{8}{\pi}\left(\frac{\sin x}{1} + \frac{\sin 3x}{27} + \frac{\sin 5x}{125} + \cdots\right), 0 < x < \pi.$$

4.1.10 What constant function is closest in the least square sense to $f = \cos^2 x$? What multiple of $\cos x$ is closest to $f = \cos^3 x$?

4.1.11 Sketch the graph and find the Fourier series of the even function $f = 1 - |x|/\pi$ (extended periodically) in either of two ways: integrate the square wave or compute (with $a_0 = \frac{1}{2}$)

$$a_k = \frac{1}{\pi} \int_{-\pi}^{\pi} f(x) \cos kx \, dx = \frac{2}{\pi} \int_0^{\pi} \left(1 - \frac{x}{\pi}\right) \cos kx \, dx.$$

4.1.12 Sketch the 2π-periodic half wave with $f(x) = \sin x$ for $0 < x < \pi$ and $f(x) = 0$ for $-\pi < x < 0$. Find its Fourier series.

4.1.13 Integrate the left side of (16) to find Bessel's inequality for the squares of the Fourier coefficients a_k and b_k.

4.1.14 (a) Find the lengths of the vectors $u = (1, \frac{1}{2}, \frac{1}{4}, \frac{1}{8}, \ldots)$ and $v = (1, \frac{1}{3}, \frac{1}{9}, \ldots)$ in Hilbert space and test the Schwarz inequality $|u^T v|^2 \le (u^T u)(v^T v)$.
   (b) For the functions $f = 1 + \frac{1}{2}e^{ix} + \frac{1}{4}e^{2ix} + \cdots$ and $g = 1 + \frac{1}{3}e^{ix} + \frac{1}{9}e^{2ix} + \cdots$ use part (a) to find the numerical value of each term in

$$\left| \int_{-\pi}^{\pi} \bar{f}(x)g(x)dx \right|^2 \le \int_{-\pi}^{\pi} |f(x)|^2 \, dx \int_{-\pi}^{\pi} |g(x)|^2 \, dx.$$

Substitute for f and g and use orthogonality (or Parseval).

4.1.15 In the solution to Laplace's equation with $u_0 = \theta$ on the boundary, (26) is the imaginary part of $2(z - z^2/2 + z^3/3 \cdots) = 2 \log (1 + z)$. Confirm that on the unit circle $z = e^{i\theta}$, the imaginary part of $2 \log (1 + z)$ agrees with θ.

4.1.16 If the boundary condition for Laplace's equation is $u_0 = 1$ for $0 < \theta < \pi$ and $u_0 = 0$ for $-\pi < \theta < 0$, find the Fourier series solution $u(r, \theta)$ inside the unit circle. What is u at the origin?

4.1.17 With boundary values $u_0(\theta) = 1 + \frac{1}{2}e^{i\theta} + \frac{1}{4}e^{2i\theta} + \cdots$, what is the Fourier series solution to Laplace's equation in the circle? Sum the series.

4.1.18 (a) Verify that the fraction in Poisson's formula satisfies Laplace's equation for each φ.
   (b) What is the response $u(r, \theta)$ to an impulse at the point $(0, 1)$ on the circle at the angle $\varphi = \pi/2$?
   (c) If $u_0(\varphi) = 1$ in the quarter-circle $0 < \varphi < \pi/2$ and $u_0 = 0$ elsewhere, show that at points on the horizontal axis (and especially at the origin)

$$u(r, 0) = \frac{1}{2} + \frac{1}{2\pi} \tan^{-1} \left(\frac{1 - r^2}{-2r}\right) \quad \text{by using}$$

$$\int \frac{d\varphi}{b + c \cos \varphi} = \frac{1}{\sqrt{b^2 - c^2}} \tan^{-1} \left(\frac{\sqrt{b^2 - c^2} \sin \varphi}{c + b \cos \varphi}\right).$$

4.1.19 A plucked string goes linearly from $f(0) = 0$ to $f(p) = 1$ and back to $f(\pi) = 0$. The linear part $f = x/p$ reaches to $x = p$, followed by $f = (\pi - x)/(\pi - p)$ to $x = \pi$. Sketch f as an odd function and find a plucking point p for which the second harmonic $\sin 2x$ will not be heard ($b_2 = 0$).

4.1.20 Show that $P_2 = x^2 - \frac{1}{3}$ is orthogonal to $P_0 = 1$ and $P_1 = x$ over the interval $-1 \le x \le 1$. Can you find the next Legendre polynomial by choosing c to make $x^3 - cx$ orthogonal to P_0, P_1, and P_2?

4.1.21 Using formula (30) with $f = |x|$, find the first 3 coefficients in the Legendre expansion $|x| = c_0 P_0 + c_1 P_1 + c_2 P_2 + \cdots$. Sketch $|x|$ and $c_0 P_0 + c_1 P_1 + c_2 P_2$ on the same graph for $-1 \le x \le 1$. To what functions is the difference of those graphs orthogonal?

4.1.22 If all orthogonal functions T_k are multiplied by 10 what happens to their coefficients c_k in (30)?

4.1.23 A function $f(x, y)$ of two variables in the square $-\pi < x, y < \pi$ can have the *double* Fourier series

$$f(x, y) = \sum\sum c_{jk} e^{ijx} e^{iky} \quad \text{(complex form with } -\infty < j, k < \infty)$$

$$f(x, y) = a_{00} + a_{10} \cos x + a_{01} \cos y + a_{11} \cos x \cos y + \dots \text{ (even)}$$

$$f(x, y) = b_{11} \sin x \sin y + b_{21} \sin 2x \sin y + b_{12} \sin x \sin 2y + \dots \text{ (odd)}$$

By multiplying by the right functions and integrating over the square, give formulas for c_{jk} and b_{jk}. (If f is neither even or odd its real series will also include all products $\cos kx \sin lx$ and $\sin kx \cos lx$.)

4.1.24 Find the double Fourier coefficients c_{jk} if f in the square is

(a) a two-dimensional impulse $\delta(x, y)$: for any g, $\iint g\delta \, dxdy = g(0, 0)$
(b) a line of one-dimensional impulses $\delta(x)$: $\iint g\delta(x) \, dxdy = \int g(0, y) \, dy$
(c) $\cos^2 x \cos^2 y$

4.1.25 From the sine series for x in equation (12) and a similar series for y find the coefficient b_{kl} in the double sine series for $f = xy$.

4.1.26 If f has the double sine series $\sum\sum b_{kl} \sin kx \sin ly$, show that Poisson's equation $-u_{xx} - u_{yy} = f$ is solved by the double sine series $u = \sum\sum b_{kl} \sin kx \sin ly/(k^2 + l^2)$. This is the solution with $u = 0$ on the boundary of the square $-\pi < x, y < \pi$.

4.1.27 Find from $\partial E/\partial C_k = 0$ the coefficients C_k that minimize the error

$$E = \int [f - C_1 T_1 - C_2 T_2 \cdots]^2 \, w \, dx,$$

assuming that T_1, T_2, ... are orthogonal with weight w over the interval of integration. Compare with the coefficients c_k in equation (30).

4.1.28 Rodrigues' formula for the Legendre polynomials is $P_n = (2^n n!)^{-1} \, d^n(x^2 - 1)^n/dx^n$. Show that this gives $P_1 = x$ and $P_2 = (3x^2 - 1)/2$, and prove orthogonality by integrating $\int_{-1}^{1} P_2 P_1 \, dx$ by parts. Why does the formula always produce a polynomial of degree n?

4.1.29 The polynomials $1, x, y, x^2 - y^2, 2xy, \dots$ solved Laplace's equation in two dimensions. Find five independent combinations of $x^2, y^2, z^2, xy, xz, yz$ that satisfy $u_{xx} + u_{yy} + u_{zz} = 0$. With spherical polynomials of all degrees we can match $u = u_0$ on the surface of a sphere.

4.1.30 Show that two eigenfunctions u_1 and u_2 of a Sturm-Liouville problem $(pu')' + qu + \lambda wu = 0$ are orthogonal with weight w. Multiply the equation for u_1 (with $\lambda = \lambda_1$) by u_2; multiply the equation for u_2 (with $\lambda_2 \neq \lambda_1$) by u_1; subtract and integrate over the interval. With zero boundary conditions integrate $u_2(pu_1')'$ and $u_1(pu_2')'$ by parts to show that $\int u_1 u_2 \, wdx = 0$.

4.1.31 Fit the Bessel equation (40) into the framework of a Sturm-Liouville equation $(pu')' + qu + \lambda wu = 0$. What are p, q, and w? What are they for the Legendre equation $(1 - x^2)P'' - 2xP' + \lambda P = 0$?

4.1.32 Show that the first Legendre polynomials $P_0 = 1$, $P_1 = \cos \varphi$, $P_2 = \cos^2 \varphi - \frac{1}{3}$ are eigenfunctions of Laplace's equation $(wu_\varphi)_\varphi + w^{-1}u_{\theta\theta} = -\lambda wu$ with $w = \sin \varphi$ on the surface of a sphere. Find the eigenvalues λ of these *spherical harmonics*. The Legendre polynomials $P_n(\cos \varphi)$ are the eigenfunctions that are independent of the longitude θ.

4.1.33 Compare the $n!$ beneath r^n in the cosine series with $2^2 4^2 \cdots n^2$ in the Bessel series (36). Write the latter as $2^n[(n/2)!]^2$ and use Stirling's formula $n! \approx \sqrt{2\pi n} \, n^n e^{-n}$ to show that the ratio of these coefficients approaches $\sqrt{\pi n/2}$. They have the same alternating signs and the two series are very similar.

4.1.34 Substitute $B = \Sigma \, c_m r^m$ into Bessel's equation (40) and show from the analogue of (35) that λc_{m-2} must equal $(n^2 - m^2)c_m$. This recursion starts from $c_n = 1$ and successively finds $c_{n+2} = \lambda/(n^2 - (n-2)^2)$, c_{n+4}, ... as the coefficients in a "Bessel function of order n:"

$$B_n(r) = r^n \left[1 + \frac{\lambda r^2}{n^2 - (n+2)^2} + \frac{\lambda^2 r^4}{(n^2 - (n+2)^2)(n^2 - (n+4)^2)} + \cdots \right]$$

$$= \frac{n!}{2^n} \sum_{k=0}^{\infty} \frac{(-1)^k (\sqrt{\lambda}/2)^{2k+n}}{k!(k+n)!}.$$

4.1.35 Where are the drum's nodal lines in Fig. 4.6 if $n = 1$, $k = 2$ or $n = 2$, $k = 3$?

4.1.36 Explain why the third Bessel eigenfunction $B = J_0(\sqrt{\lambda_3} \, r)$ is zero at $r = (\lambda_1/\lambda_3)^{1/2}$, $r = (\lambda_2/\lambda_3)^{1/2}$, and $r = 1$.

4.2 ■ DISCRETE FOURIER SERIES AND CONVOLUTION

We come now to reality. The truth is that the digital computer has totally defeated the analog computer. The input is a sequence of numbers and not a continuous function. The output is another sequence of numbers, whether it comes from a digital filter or a finite element stress analysis or an image processor. The question is whether the special ideas of Fourier analysis still have a part to play, and the answer is absolutely *yes*. The discrete Fourier transform takes n numbers $f_0, ..., f_{n-1}$ into n coefficients $c_0, ..., c_{n-1}$, and the patterns that are hidden in f stand out more distinctly in c.

If the transformation is·linear, then f and c must be connected by a matrix. Certainly Fourier analysis is linear. We want to propose a typical problem and study the matrix that governs it:

Find the coefficients c_0, c_1, c_2, c_3 such that

$$c_0 + c_1 + c_2 + c_3 = 2$$
$$c_0 + ic_1 - c_2 - ic_3 = 4$$
$$c_0 - c_1 + c_2 - c_3 = 6 \tag{1}$$
$$c_0 - ic_1 - c_2 + ic_3 = 8.$$

The input sequence is $f = 2, 4, 6, 8$. The equations look for a 4-term Fourier series that matches f at four equally spaced points. Instead of reproducing f for all x, the discrete transform only satisfies the four conditions

$$c_0 + c_1 e^{ix} + c_2 e^{2ix} + c_3 e^{3ix} = \begin{matrix} 2 \\ 4 \\ 6 \\ 8 \end{matrix} \quad \text{at} \quad \begin{matrix} x = 0 \\ x = \pi/2 \\ x = \pi \\ x = 3\pi/2 \end{matrix}$$

Those are the 4 equations in (1), and the output sequence is c_0, c_1, c_2, c_3. It is the discrete Fourier transform, which reproduces the 4 values of f by a finite Fourier series—ending after 4 terms. At $x = 2\pi$ the series returns to $f_0 = 2$ and continues periodically.

The problem is to solve equation (1). We could start by adding the four separate equations. There are tremendous cancellations on the left side and they leave $4c_0 = 20$. Then the constant term $c_0 = 5$ is the average of $f = 2, 4, 6, 8$, just as in the continuous case $c_0 = (2\pi)^{-1} \int f(x)dx$.

There is a similar way to find c_1. Multiply the four equations by $1, -i, -1, i$, and add. Now everything cancels on the left except $4c_1$. The right side is $2 - 4i - 6 + 8i = -4 + 4i$. After dividing by 4 we have $c_1 = -1 + i$, and there must be a good method for c_2 and c_3.

To see it more clearly we write the coefficient matrix as

$$A = \begin{bmatrix} 1 & 1 & 1 & 1 \\ 1 & i & i^2 & i^3 \\ 1 & i^2 & i^4 & i^6 \\ 1 & i^3 & i^6 & i^9 \end{bmatrix}.$$

This is identical to the matrix in (1); that equation is $Ax = f$. The last coefficient i^9 was simplified to i, and i^6 became -1. The other entries also agree. In the present form it will be easier to recognize A^{-1}, and infinitely easier to see how the pattern extends beyond $n = 4$.

The matrix A is symmetric but not Hermitian; a_{kj} equals a_{jk} but not $\bar{a}_{jk}$. When we take complex conjugates, changing every i to $-i$, the matrix becomes

$$\bar{A} = \begin{bmatrix} 1 & 1 & 1 & 1 \\ 1 & (-i) & (-i)^2 & (-i)^3 \\ 1 & (-i)^2 & (-i)^4 & (-i)^6 \\ 1 & (-i)^3 & (-i)^6 & (-i)^9 \end{bmatrix}.$$

Then the whole trick behind the discrete Fourier transform lies in the relation between $\bar{A}$ and A and A^{-1}. The product of $\bar{A}$ and A gives 4's on the diagonal and 0's everywhere else:

$$\boxed{\bar{A}A = 4I \quad \text{so the inverse is} \quad A^{-1} = \tfrac{1}{4}\bar{A}.}$$

The solution to $Ac = f$ is the set of Fourier coefficients $c = \tfrac{1}{4}\bar{A}f$. Taken row by row this produces c_0, c_1, c_2, c_3. The first two rows give

$$c_0 = \tfrac{1}{4}(f_0 + f_1 + f_2 + f_3) \quad \text{and} \quad c_1 = \tfrac{1}{4}(f_0 - if_1 + (-i)^2 f_2 + (-i)^3 f_3),$$

which already led to $c_0 = 5$ and $c_1 = -1 + i$. The other two are

$$c_2 = \tfrac{1}{4}(f_0 - f_1 + f_2 - f_3) = -1 \quad \text{and}$$
$$c_3 = \tfrac{1}{4}(f_0 + (-i)^3 f_1 + (-i)^6 f_2 + (-i)^9 f_3) = -1 - i.$$

Elimination is not needed to solve equation (1), since the inverse matrix is explicitly known.

The Discrete Transform for Arbitrary n

The pattern found above is not limited to $n = 4$. For every n the matrix connecting f to c can be written down and inverted. It represents n equations, each one requiring the finite Fourier series $\Sigma\, c_k e^{ikx}$ (with n terms) to agree with f. There is agreement at n equally spaced points between 0 and 2π; their spacing is $2\pi/n$. The first point is at $x = 0$:

$$c_0 + c_1 + \cdots + c_{n-1} = f_0.$$

The next point at $x = 2\pi/n$ involves the complex number $w = e^{2\pi i/n}$:

$$c_0 + c_1 w + c_2 w^2 + \cdots + c_{n-1} w^{n-1} = f_1.$$

The third point $x = 4\pi/n$ involves $e^{4\pi i/n}$, which is w^2:

$$c_0 + c_1 w^2 + c_2 w^4 + \cdots + c_{n-1} w^{2(n-1)} = f_2.$$

The remaining equations bring higher powers of w, and the full problem is

$$\begin{bmatrix} 1 & 1 & 1 & \cdot & 1 \\ 1 & w & w^2 & \cdot & w^{n-1} \\ 1 & w^2 & w^4 & \cdot & w^{2(n-1)} \\ \cdot & \cdot & \cdot & \cdot & \cdot \\ 1 & w^{n-1} & w^{2(n-1)} & \cdot & w^{(n-1)^2} \end{bmatrix} \begin{bmatrix} c_0 \\ c_1 \\ c_2 \\ \cdot \\ c_{n-1} \end{bmatrix} = \begin{bmatrix} f_0 \\ f_1 \\ f_2 \\ \cdot \\ f_{n-1} \end{bmatrix} \tag{2}$$

For $n = 4$ the key number w was i, the fourth root of unity: $i^4 = 1$. It was equal to $e^{2\pi i/4}$, the coefficient matrix was A, and its last entry was $w^9 = i^9$. In general $w = e^{2\pi i/n}$ lies on the unit circle $|w| = 1$ in the complex plane, at an angle $2\pi/n$ from the horizontal. Its square is twice as far around the circle, and its nth power is all the way around. In other words $w^n = 1$ and w is an nth **root of unity**:

$$\boxed{w = e^{2\pi i/n} \quad \text{and} \quad w^n = e^{2\pi i} = 1.} \tag{3}$$

Its conjugate $\bar{w} = e^{-2\pi i/n}$ is another nth root, lying at the angle $-2\pi/n$ below the horizontal (Fig. 4.7). Notice from the figure that $1 + w + w^2 + \cdots + w^{n-1} = 0$.

The matrix in equation (2) consists entirely of powers of w. Its entries are w^{jk}, if the rows j and the columns k are numbered from 0 to $n - 1$. *We denote this matrix by F, for Fourier*. It is n by n, and when the order n needs to be remembered it will be added as a subscript. Thus F_4 was 4 by 4 and w_4 is i. The important thing is that F has a convenient inverse:

4E   For every n the matrices F and $\bar{F}$ formed from w and $\bar{w}$ satisfy

$$F\bar{F} = \bar{F}F = nI \quad \text{or} \quad F^{-1} = \frac{1}{n}\bar{F}. \tag{4}$$

The discrete Fourier transform of a sequence f is $c = \frac{1}{n}\bar{F}f$.

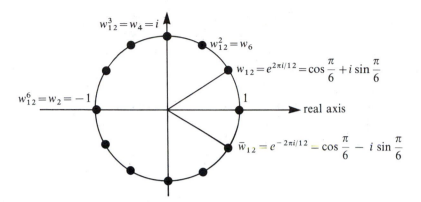

Fig. 4.7. The $n = 12$ roots of unity, including w_n and $\bar{w}_n$.

In the cases $n = 2$ and $n = 3$ this means that

$$F_2 = \begin{bmatrix} 1 & 1 \\ 1 & -1 \end{bmatrix} \quad \text{has} \quad F_2^{-1} = \frac{1}{2}\begin{bmatrix} 1 & 1 \\ 1 & -1 \end{bmatrix}$$

$$F_3 = \begin{bmatrix} 1 & 1 & 1 \\ 1 & e^{2\pi i/3} & e^{4\pi i/3} \\ 1 & e^{4\pi i/3} & e^{8\pi i/3} \end{bmatrix} \quad \text{has} \quad F_3^{-1} = \frac{1}{3}\begin{bmatrix} 1 & 1 & 1 \\ 1 & e^{-2\pi i/3} & e^{-4\pi i/3} \\ 1 & e^{-4\pi i/3} & e^{-8\pi i/3} \end{bmatrix}.$$

In the n by n case row j of $\bar{F}$ and column k of F combine to give

$$1 \cdot 1 + \bar{w}^j w^k + \bar{w}^{2j} w^{2k} + \cdots + \bar{w}^{(n-1)j} w^{(n-1)k} = \begin{cases} 0 & \text{if } j \neq k \\ n & \text{if } j = k \end{cases} \tag{5}$$

With $j = k$ all terms are 1 and the sum is n. With j different from k, this orthogonality between vectors corresponds to the orthogonality $\int e^{-ijx} e^{ikx} dx = 0$ between functions. There we knew the integral. Here we have n terms of a geometric series and we know the sum:

$$1 + r + r^2 + \cdots + r^{n-1} = \frac{1 - r^n}{1 - r} \quad \text{with} \quad r = \bar{w}^j w^k.$$

But $r^n = 1$ follows from $\bar{w}^n = 1$ and $w^n = 1$, so this sum is zero—which proves (5). Thus F^{-1} is the same as $\bar{F}/n$, and the formula

$$c_k = \frac{1}{n} \sum_{j=0}^{n-1} f_j \bar{w}^{jk} = \frac{1}{n} \sum f_j e^{-ijk2\pi/n} \quad \text{is parallel to} \quad \frac{1}{2\pi} \int_0^{2\pi} f(x) e^{-ikx} dx. \tag{6}$$

In fact the first is an accurate way to compute the second, when the integral is not known. Approximating the integral by the sum is called the "trapezoidal rule." For smooth periodic functions (and large enough n) the two are extremely close.

Notice that the columns of $F/\sqrt{n}$ are orthogonal unit vectors in complex n-dimensional space. It is like an orthogonal matrix Q, except that it is complex. In the language of matrices, $U = F/\sqrt{n}$ is a **unitary matrix**: its inverse is the same as $\bar{U}^T$, the transpose of its conjugate. These factors $\sqrt{n}$ (and $\sqrt{2\pi}$ in the continuous case) make the formulas more elegant, but in the end it seems better to keep n with F^{-1} and away from F.

Discrete Convolutions

It is remarkable enough that the Fourier matrix F is so easy to invert. To transform a sequence we just multiply by $\bar{F}/n$. To transform back we multiply by F. If that were all (and up to 1965 it was all) the discrete transform would have an important place. Now there is more. The multiplications by $\bar{F}$ and F, which are matrices containing powers of $\bar{w}$ and w, can be done in an extremely fast and ingenious way. Instead of n^2 separate multiplications they require less than $n \log_2 n$. This rearrangement of $\bar{F}$ and F is called the **Fast Fourier Transform**.

Section 5.5 will describe the basic idea of the FFT. The matrix F_2 is related to F_4, and F_4 to F_8, and F_8 to F_{16}, so the transform of order 16 comes from two transforms of order 8. Those come from four transforms of order 4 and eight of order 2. That accounts for the factor n in the cost, and the steps that combine these smaller transforms account for $\log_2 n$. Newer and more subtle approaches have been discovered, but this fast transform of Cooley and Tukey is fundamental.

We want to say why F is so important. There is one calculation that appears constantly in applications, and there are two ways to do it. The direct method is contained in the definition. The indirect method uses F and F^{-1}. It looks like the longer way, but because of these matrices it is faster. The calculation is called **convolution**, and the rule that allows it to be done through F and F^{-1} is the **convolution rule**.

The definition comes first. Suppose we "multiply" two vectors of length n; what is their product? It could be taken component by component, so that

$$c = (c_0, \ldots, c_{n-1}) \text{ times } d = (d_0, \ldots, d_{n-1}) \text{ equals } cd = (c_0 d_0, \ldots, c_{n-1} d_{n-1}).$$

That is easy and fast, but normally it is not the product we want. The output needed in applications is the *convolution product*. It will look more complicated, but it also starts with two vectors $f = (f_0, \ldots, f_{n-1})$ and $g = (g_0, \ldots, g_{n-1})$, and it produces another vector of length n. That vector is denoted by $f * g$:

$$\boxed{\begin{aligned} f * g = (f_0 g_0 + f_1 g_{n-1} + f_2 g_{n-2} + \cdots + f_{n-1} g_1, \ldots, \\ f_0 g_{n-1} + f_1 g_{n-2} + \cdots + f_{n-1} g_0). \end{aligned}}$$

Each component of this convolution is the sum of n terms. The first component contains all products $f_j g_k$ with $j + k = 0$ or $j + k = n$. The later components are $(f * g)_l$ and they contain all products with $j + k = l$ or $j + k = l + n$. Notice that *we do not distinguish between indices that differ by n.*† The convolution of $(1, 2, 3)$ with $(4, 5, 6)$ is $(1 \cdot 4 + 2 \cdot 6 + 3 \cdot 5, \ 1 \cdot 5 + 2 \cdot 4 + 3 \cdot 6, \ 1 \cdot 6 + 2 \cdot 5 + 3 \cdot 4) = (31, 31, 28)$.

This product $f * g$ is chosen to match a multiplication like

$$(1 + w + w^2)(1 + w + w^2) = 1 + 2w + 3w^2 + 2w^3 + w^4.$$

To find the coefficient of w^2 in the answer we combine the three ways it can be reached: 1 times w^2, w times w, and w^2 times 1. This corresponds to $f_0 g_2 + f_1 g_1 + f_2 g_0$ in the convolution product; that is the key point. However the correspondence is not yet perfect. We started with vectors $f = (1, 1, 1)$ and $g = (1, 1, 1)$ coming from $1 + w + w^2$. The product $f * g$ should have *three* components, not five. To fold back $2w^3$ and w^4 we use the fundamental property of w (for $n = 3$): w^3 is the same as 1, and w^4 is the same as w. The correct convolution product $(1, 1, 1) * (1, 1, 1)$ is $(3, 3, 3)$. The corresponding product of sums is

$$(1 + w + w^2)(1 + w + w^2) = 3 + 3w + 3w^2. \tag{7}$$

This is absolutely correct if $w^3 = 1$.

We give more examples below, and the applications to partial differential equations come in Section 5.5. Those will explain why convolutions are important. First we want to explain why convolutions are fast. That is a separate question, and the answer lies in the **convolution rule**:

4F The convolution of f and g corresponds to ordinary multiplication of their transforms. The vector $f * g$ can be computed by transforming separately to $c = F^{-1} f$ and $d = F^{-1} g$, multiplying c and d component by component, and then transforming back with an extra factor n:

$$f * g \text{ is } n \text{ times } F \text{ times } cd.$$

Thus the easy multiplication cd has its place. The opposite statement is also true: Multiplication of f and g corresponds to convolution of their transforms. The same rules will extend to Fourier series and Fourier integrals, but before the examples and proofs we emphasize the conclusion that is significant in practice.

The convolution rule combines with the Fast Fourier Transform into a quick way to compute $f * g$. It uses the fast transform three times:

> Compute c from f, d from g, and $f * g$ from ncd.

In between are the n steps in the multiplication cd. The result is an algorithm that is

---

† That is called *aliasing*. We will see it again.

indirect but fast; it takes $O(n \log n)$ steps instead of the n^2 multiplications in the definition of $f * g$.

In the example given earlier the transforms of $f = g = (1, 1, 1)$ would have been $c = d = (1, 0, 0)$. Then the product cd is also $(1, 0, 0)$. Transforming back (with multiplication by $n = 3$) gives the convolution $f * g = (3, 3, 3)$ in agreement with equation (7). The savings are not clear for such a small problem—in fact there are no savings—but for higher degree polynomials this is the right algorithm. Our first example shows that it applies to ordinary polynomials in x as well as to polynomials in the special number $w = e^{2\pi i/n}$ that produced periodicity.

EXAMPLE 1   *Multiplication of polynomials f and g*

Multiplying $f_0 + f_1 x + f_2 x^2$ by $g_0 + g_1 x + g_2 x^2$ is exactly like convolution, with one essential difference. The product is a polynomial of degree 4. It has five coefficients, where f and g have three. Since convolution produces an output sequence of the same length as the inputs, we add two zeros to the input:

$$f = (f_0, f_1, f_2, 0, 0) \quad \text{and} \quad g = (g_0, g_1, g_2, 0, 0).$$

Their convolution (with $n = 5$) is

$$f * g = (f_0 g_0, f_0 g_1 + f_1 g_0, f_0 g_2 + f_1 g_1 + f_2 g_0, f_1 g_2 + f_2 g_1, f_2 g_2).$$

You recognize these as the coefficients of the product fg. The "periodic" aspect of discrete convolution has been suppressed by adding the two zeros. Normally with $n = 5$ the products $f_1 g_4$ and $f_2 g_3$ and $f_3 g_2$ and $f_4 g_1$ all contribute to $(f * g)_0$, because $1 + 4 = 2 + 3 = 5$ and 5 is not distinguished from 0. In discrete convolution every index is taken "*modulo n.*" The zeros in f and g ensure that all these aliased terms will be zero.

EXAMPLE 2   *Multiplication of numbers f and g*

Multiplying 303 by 222 is also a convolution, with one new difficulty. As before we add zeros to avoid problems with periodicity. The sequence $f * g$ is

$$(3, 0, 3, 0, 0) * (2, 2, 2, 0, 0) = (6, 6, 12, 6, 6).$$

The difficulty is that $f * g$ can have components larger than 9. Those will not be normal decimal digits even though they are correct:

$$303 \text{ times } 222 \text{ equals } 67266.$$

A final step is needed to change 12 in the 100's place to $2(100) + 1(1000)$. However all components like 12 involve far fewer digits than a direct multiplication of two-n-digit numbers. Thus the convolution rule and the FFT allow fast multiplication of decimals—or of numbers to any base b. For binary multiplication ($b = 2$) the computer has its own method wired in, but for large bases b—possibly useful in cryptography—convolution is a reasonable alternative.

Now we come to the ***matrix form of convolution***.

EXAMPLE 3 *Multiplication by circulant matrices*

A circulant is a constant-diagonal matrix with a special form:

$$
C = \begin{bmatrix}
f_0 & f_{n-1} & \cdot & \cdot & f_1 \\
f_1 & f_0 & f_{n-1} & \cdot & f_2 \\
\cdot & f_1 & f_0 & \cdot & \cdot \\
\cdot & \cdot & \cdot & \cdot & f_{n-1} \\
f_{n-1} & f_{n-2} & \cdot & f_1 & f_0
\end{bmatrix}.
$$

It is "periodic," since the lower diagonals fold around to appear again as upper diagonals. C is n by n and has n diagonals. An example of multiplication with $n = 4$ is

$$
Cg = \begin{bmatrix}
2 & 8 & 6 & 4 \\
4 & 2 & 8 & 6 \\
6 & 4 & 2 & 8 \\
8 & 6 & 4 & 2
\end{bmatrix}
\begin{bmatrix}
g_0 \\ g_1 \\ g_2 \\ g_3
\end{bmatrix}
=
\begin{bmatrix}
2g_0 + 8g_1 + 6g_2 + 4g_3 \\
4g_0 + 2g_1 + 8g_2 + 6g_3 \\
6g_0 + 4g_1 + 2g_2 + 8g_3 \\
8g_0 + 6g_1 + 4g_2 + 2g_3
\end{bmatrix}.
\tag{8}
$$

What you see on the right is precisely a discrete convolution! Those are the four components of $f * g$. The circulant matrix is exactly arranged so that multiplication by C corresponds to convolution by $f = (2, 4, 6, 8)$. The numbers 2, 4, 6, 8 go into the first column of C, and by coming around periodically they result in a pure convolution.

For any other matrix we would look for the eigenvalues and eigenvectors. We do that here too, expecting them to be special because C is so special. In fact the eigenvectors can be identified first, because ***all 4 by 4 circulant matrices have the same eigenvectors***. One eigenvector is the vector of 1's:

$$
\begin{bmatrix}
2 & 8 & 6 & 4 \\
4 & 2 & 8 & 6 \\
6 & 4 & 2 & 8 \\
8 & 6 & 4 & 2
\end{bmatrix}
\begin{bmatrix}
1 \\ 1 \\ 1 \\ 1
\end{bmatrix}
= 20
\begin{bmatrix}
1 \\ 1 \\ 1 \\ 1
\end{bmatrix}.
$$

Another eigenvector contains the powers of w: $Cx = \lambda x$ is

$$
\begin{bmatrix} 2 & 8 & 6 & 4 \\ 4 & 2 & 8 & 6 \\ 6 & 4 & 2 & 8 \\ 8 & 6 & 4 & 2 \end{bmatrix} \begin{bmatrix} 1 \\ w \\ w^2 \\ w^3 \end{bmatrix} = (2 + 8w + 6w^2 + 4w^3) \begin{bmatrix} 1 \\ w \\ w^2 \\ w^3 \end{bmatrix}.
$$

The first row is certainly correct. The second row is identical except multiplied by w: $4 + 2w + 8w^2 + 6w^3$ agrees with $(2 + 8w + 6w^2 + 4w^3)w$. The terms 4 and $4w^4$ are the same because $w^4 = 1$. In this example $n = 4$ and $w = i$; the eigenvalue is $2 + 8i + 6i^2 + 4i^3 = -4 + 4i$. We recognize this as 4 times the Fourier coefficient $c_1 = -1 + i$, just as 20 was 4 times the Fourier coefficient $c_0 = 5$.

This pattern holds for all eigenvectors and all circulant matrices: **The eigenvectors of C are the columns of the Fourier matrix F.** In other words, the matrix F diagonalizes every circulant. That lies at the foundation of discrete Fourier analysis, and for this specific example it means

$$
C \begin{bmatrix} 1 & 1 & 1 & 1 \\ 1 & w & w^2 & w^3 \\ 1 & w^2 & w^4 & w^6 \\ 1 & w^3 & w^6 & w^9 \end{bmatrix} = \begin{bmatrix} 1 & 1 & 1 & 1 \\ 1 & w & w^2 & w^3 \\ 1 & w^2 & w^4 & w^6 \\ 1 & w^3 & w^6 & w^9 \end{bmatrix} \begin{bmatrix} 20 & & & \\ & -4+4i & & \\ & & -4 & \\ & & & -4-4i \end{bmatrix}. \tag{9}
$$

That matrix equation is remarkable. It is $CF = F\Lambda$; if we multiply by F^{-1} it becomes $C = F\Lambda F^{-1}$. It is the usual $S\Lambda S^{-1}$ factorization of Chapter 1; the eigenvector matrix S is the Fourier matrix F. The diagonal matrix Λ contains the eigenvalues, and *they are 4 times the Fourier coefficients of $f = (2, 4, 6, 8)$.* The eigenvalues are the components of $4c$. This factor 4 is the n that enters the convolution rule.

I have finally realized that $C = F\Lambda F^{-1}$ is a disguised form of the convolution rule. In fact it *is* the rule, written in matrix notation. The multiplication Cg was recognized in (8) as $f * g$, and now it is separated into three steps—each coming from one of the factors in $F\Lambda F^{-1}$:

1. $F^{-1}g$ gives the discrete transform of g, called d
2. Λ multiplies by nc to give ncd
3. F transforms back and the result is $f * g$.

In the applications these steps are fast by the FFT. From the point of view of theory this has become more than an example; it is a matrix proof of the convolution rule. It will be restated more clearly, together with the rule in the opposite direction:

4G Suppose the sequences f and g have discrete transforms $c = F^{-1}f$ and $d = F^{-1}g$. Then the transform of $f * g$ is n times the component-by-component product cd:

$$\text{transform of } f * g = n \text{ (transform of } f)(\text{transform of } g). \tag{10}$$

$$\text{transform of } fg = (\text{transform of } f) * (\text{transform of } g). \tag{11}$$

To verify the factor n we take an example with $n = 2$ and $w = -1$:

$$F = \begin{bmatrix} 1 & 1 \\ 1 & -1 \end{bmatrix} \text{ and } F^{-1} = \frac{1}{n}\bar{F} = \frac{1}{2}\begin{bmatrix} 1 & 1 \\ 1 & -1 \end{bmatrix}$$

$$f = (4, 2) \text{ has transform } c = F^{-1}f = (3, 1)$$

$$g = (8, 2) \text{ has transform } d = F^{-1}g = (5, 3)$$

$$f * g = (4 \cdot 8 + 2 \cdot 2, \ 4 \cdot 2 + 2 \cdot 8) = (36, 24) \text{ has transform } 2cd = (30, 6)$$

$$fg = (32, 4) \text{ has transform } c * d = (3 \cdot 5 + 1 \cdot 3, \ 3 \cdot 3 + 1 \cdot 5) = (18, 14).$$

This convolution rule will appear again in the next section for functions on the whole line $-\infty < x < \infty$. Their transforms contain all frequencies $-\infty < k < \infty$. The sums become integrals and we give a different proof. Here, in the typical case $n = 3$, the exercises ask you to show explicitly that the rule is correct.

Signal Processing and Constant-diagonal Matrices

Our first applications of Fourier series were extremely classical. They used the ability of infinite series to represent functions and to solve constant-coefficient equations, especially Laplace's equation. But somehow they did not use the idea of *frequencies*. That idea becomes much more important for signals, when an input at time t produces a response at other times. Sooner or later time had to enter this chapter, and by measuring the response b we want to determine the signal u.

Suppose the signals are sent and received at discrete times $t = 0, \pm 1, \pm 2, \ldots$. The amplitude of the signal at time $t = j$ is u_j, not yet known. We do know the response to each signal—and especially to the standard unit signal $u_0 = 1$. **The response at time j to that signal at time zero is a_j.** There is an immediate response of a_0, the later responses are $a_1, a_2, \ldots$, and the earlier responses (if any) are $a_{-1}, a_{-2}, \ldots$. A new signal at a different time will produce its own responses, and we make two fundamental assumptions:

Linearity: The responses to separate signals are added
                The responses to an amplified signal cu are c times the responses to u

Stationarity: A unit signal at time k produces the same responses as at time zero, but the response a_j is shifted to time $t = k + j$.

Linearity combines superposition and scaling. It has been so fundamental in this book that we have hardly mentioned it. Stationarity is something different; it allows the whole system to be shifted in time without changing its characteristics. The system is said to be *time-invariant*. Traditionally these systems have appeared in engineering textbooks and not in applied mathematics—but that is not right.

There is a third property which might or might not be present:

> ***Causality:***   The response never comes before the signal: $a_j = 0$ for $j < 0$.

This chapter *does not* make that assumption. Causality is naturally present when the problem has a definite starting time and a short duration. In control problems the effect follows the cause; the output is never earlier than the input. On the other hand causality is not so convenient when the problem extends over a long time; and when it deals with very long strings of inputs and outputs. That is the case in data processing and in communications. The signals last so long that it is better to idealize them as extending indefinitely from $t = -\infty$ and $t = +\infty$, and to use Fourier methods.

We can make that distinction more clearly. Causal problems go from $t = 0$ to $t = \infty$ and they use the Laplace transform (Section 6.3). They are "one-sided". They correspond to lower triangular matrices, while our present problems might easily produce symmetric matrices. The response of an airport runway to the landing of a plane is essentially symmetric: a load at $x = 0$ produces deflections at $x = 0, \pm 1, \pm 2, \ldots$ and the deflections at x and $-x$ are approximately equal. This example suggests that x could be the variable rather than t; time has a distinguished direction but space is more neutral. However the beauty of the mathematics is to compare the Laplace transform with the Fourier transform (or in the discrete case the z-transform with Fourier series) and the best variable is t.

Our basic problem is simple. Given the responses $\ldots, b_{-1}, b_0, b_1, \ldots$ we want to know the signals $\ldots, u_{-1}, u_0, u_1, \ldots$ that produced them.† The response vector b depends linearly on the unknown signal u, and the problem is to invert that relationship—*to find u from b*. The first step is to see how b comes from the signal through multiplication by an infinite matrix A:

$$
\begin{bmatrix}
\cdot & \cdot & \cdot & \cdot & & \cdot \\
a_1 & a_0 & a_{-1} & a_{-2} & \cdot & \\
a_2 & a_1 & a_0 & a_{-1} & a_{-2} & \\
\cdot & a_2 & a_1 & a_0 & a_{-1} & \\
\cdot & \cdot & \cdot & \cdot & \cdot &
\end{bmatrix}
\begin{bmatrix}
\cdot \\
u_{-1} \\
u_0 \\
u_1 \\
\cdot
\end{bmatrix}
=
\begin{bmatrix}
\cdot \\
b_{-1} \\
b_0 \\
b_1 \\
\cdot
\end{bmatrix}
. \qquad (12)
$$

You must look at that equation $Au = b$. The signal $u_0 = 1$ multiplies the "middle column" to give the outputs $b_j = a_j$. Those are the correct responses to the standard

---

† The dots emphasize that these sequences go back to $-\infty$ as well as forward to $+\infty$. The sequences are "doubly infinite" and so are the matrices that multiply them.

unit signal. For other signals the requirement of linearity is certainly met (it is automatic in every matrix multiplication). The requirement of stationarity appears in the fact that all columns are the same—except that they are shifted in time. This produces a **constant-diagonal matrix**, also known as a **Toeplitz matrix** or a **convolution matrix**. The entries down each diagonal are constant, and the i, j entry of A depends only on the difference between i and j: it is a_{i-j}. The main diagonal contains only $a_{i-i} = a_0$. These matrices could have been designed with Fourier analysis in mind.

EXAMPLE 1   The **shift matrix** S with $a_1 = 1$.

Suppose each signal is received one time step later, without distortion. The output is a perfect copy of the input, shifted one step forward in time:

$$Su = \begin{bmatrix} 0 & 0 & . & . & . \\ 1 & 0 & 0 & . & . \\ 0 & 1 & 0 & 0 & . \\ . & 0 & 1 & 0 & 0 \\ . & . & 0 & 1 & 0 \end{bmatrix} \begin{bmatrix} u_{-2} \\ u_{-1} \\ u_0 \\ u_1 \\ . \end{bmatrix} = \begin{bmatrix} . \\ u_{-2} \\ u_{-1} \\ u_0 \\ u_1 \end{bmatrix}. \tag{13}$$

This is the shift matrix, and it would be easy to recover the input from the output— just shift one step back. The inverse of this doubly-infinite matrix is its transpose, with 1's above the diagonal.

Notice that the shift does not behave like an ordinary lower triangular matrix (whose inverse should also be lower triangular). In the finite case it would be in trouble; if it is chopped off to be n by n then it no longer has an inverse. There will be zeros on and above the diagonal, and the determinant is zero. If it is infinite in only *one* direction, toward $t = +\infty$, it is in half that trouble. S^T is a left-inverse, since $S^T S = I$, but SS^T will start with a row of zeros. We can shift forward and then back, but not back and then forward—if there is a cutoff at $t = 0$. When S is infinite in both directions it does have an inverse S^T.

EXAMPLE 2   Decaying responses $a_0 = \frac{1}{2}, a_1 = \frac{1}{4}, a_2 = \frac{1}{8}, \dots$.

If half the signal is received immediately, then a quarter is received, then an eighth, and so on, the equation $Au = b$ is

$$\begin{bmatrix} \frac{1}{2} & 0 & 0 & . & . \\ \frac{1}{4} & \frac{1}{2} & 0 & 0 & . \\ \frac{1}{8} & \frac{1}{4} & \frac{1}{2} & 0 & 0 \\ . & \frac{1}{8} & \frac{1}{4} & \frac{1}{2} & 0 \\ . & . & \frac{1}{8} & \frac{1}{4} & \frac{1}{2} \end{bmatrix} \begin{bmatrix} . \\ u_{-1} \\ u_0 \\ u_1 \\ . \end{bmatrix} = \begin{bmatrix} . \\ b_{-1} \\ b_0 \\ b_1 \\ . \end{bmatrix}. \tag{14}$$

Now it is a more serious problem to solve for u. Fourier analysis can find the inverse of a doubly infinite matrix, and this time A^{-1} is also lower triangular. It will again have constant diagonals.

Here is the central idea. Let the signals u_k be the Fourier coefficients of a *signal function*

$$u(t) = \cdots + u_0 + u_1 e^{it} + u_2 e^{2it} + \cdots = \sum_{k=-\infty}^{\infty} u_k e^{ikt}.$$

Let the responses b_k be the Fourier coefficients of the *response function*

$$b(t) = \cdots + b_0 + b_1 e^{it} + b_2 e^{2it} + \cdots = \sum_{k=-\infty}^{\infty} b_k e^{ikt}.$$

The problem of finding the vector u from the vector b is transformed into: *Find the function $u(t)$ from the function $b(t)$.*

For the special signal $u_0 = 1$, the response is known. The signal function is $u(t) = 1$ and the response function is

$$a(t) = \cdots + a_0 + a_1 e^{it} + a_2 e^{2it} + \cdots = \sum_{k=-\infty}^{\infty} a_k e^{ikt}.$$

We now show that this same $a(t)$ connects *every* signal to its response, transforming the matrix equation $Au = b$ into a multiplication of functions:

4H   The response $b(t)$ comes from the signal $u(t)$ through **multiplication by** $a(t)$:

$$a(t)\,u(t) = b(t). \tag{15}$$

This is the Fourier equivalent of $Au = b$. The inverse matrix A^{-1} also has constant diagonals and its entries are the Fourier coefficients of $1/a(t)$:

$$\frac{1}{a(t)} = g(t) = \cdots + g_0 + g_1 e^{it} + g_2 e^{2it} + \cdots \tag{16}$$

The j, k entry of A^{-1} is g_{j-k} and the solution $u = A^{-1}b$ is

$$u_j = \sum g_{j-k} b_k \quad \text{or} \quad u(t) = g(t)\,b(t) = \frac{b(t)}{a(t)}. \tag{17}$$

This is Fourier analysis at its best.

In the shift example, with response $a_1 = 1$ below the diagonal of S, $a(t)$ is e^{it}. Its reciprocal is $g = e^{-it}$. Therefore $g_{-1} = 1$ and S^{-1} is the matrix with 1's above the diagonal. The inverse reverses the shift, and algebraically it comes from the term with $j - k = -1$ (or $k = j + 1$):

$$u_j = \sum g_{j-k} b_k = b_{j+1} \quad \text{for the shift.}$$

Before solving a harder example we need to see why multiplication by A transforms into multiplication by a.

The equation $a(t) u(t) = b(t)$ is the key to everything.† It comes from comparing multiplication of matrices with multiplication of series:

> the jth component of Au is $\Sigma\, a_{j-k} u_k$
>
> the coefficient of e^{ijt} in $a(t)u(t)$ is also $\Sigma\, a_{j-k} u_k$.

The first is ordinary matrix multiplication with the special entries $a_{jk} = a_{j-k}$. The second looks for e^{ijt} in the product

$$(\cdots + a_0 + a_1 e^{it} + a_2 e^{2it} + \cdots)(\cdots + u_0 + u_1 e^{it} + u_2 e^{2it} + \cdots).$$

For e^{2it} there are contributions from $a_0 u_2$ and $a_1 u_1$ and $a_2 u_0$—and from all other pairs $a_{2-k} u_k$ whose indices add to 2. This case $j = 2$ is completely typical. Those are the same pairs that go into multiplication by a constant-diagonal matrix, and the components of $b = Au$ agree with the coefficients of $b(t) = a(t)u(t)$.

This sum $\Sigma\, a_{j-k} u_k$ is the **convolution** of the sequence of a's with the sequence of u's. It produces a third sequence, the coefficients in $a(t)u(t)$. It is exactly like discrete convolution—except that periodicity is gone. The sums go from $k = -\infty$ to $k = +\infty$, and the jth component of the answer contains all pairs $a_{j-k} u_k$ whose indices add to j.

Here is the key point: The matrix A is inverted by finding the Fourier coefficients of $g = 1/a$. The product $A^{-1}A = I$ agrees with $ga = 1$. As evidence of the power of this idea we invert the lower triangular matrix in (14). Its entries down each column are $\frac{1}{2}, \frac{1}{4}, \frac{1}{8}, \ldots$ and

$$a(t) = \tfrac{1}{2}(1 + \tfrac{1}{2}e^{it} + \tfrac{1}{4}e^{2it} + \cdots) = \frac{1}{2}\frac{1}{1 - \tfrac{1}{2}e^{it}} = \frac{1}{2 - e^{it}}.$$

The reciprocal is simply $g = 2 - e^{it}$, so the inverse matrix must be

$$A^{-1} = \begin{bmatrix} -1 & 2 & & \\ & -1 & 2 & \\ & & -1 & 2 \\ & & & \end{bmatrix}$$

Multiplying this matrix into (14) gives $A^{-1}A = I$ as required, and $u = A^{-1}b$ is

$$u_j = 2b_j - b_{j-1}.$$

---

† It is the convolution rule! There will be a similar transformation in Section 6.3 from a differential equation $u'' + u = f$ to a multiplication $(s^2 + 1)U = F$. There f is the *input*, u is the *output*, and $1/(s^2 + 1)$ is the **transfer function**.

This is the signal that produces b_j; the problem is solved. The responses $b_0 = \frac{1}{2}$, $b_1 = \frac{1}{4}$, ... correspond to the special signal $u_0 = 1$.

EXAMPLE 3 The symmetric case $a_1 = a_{-1} = \frac{1}{2}$

For this matrix $a(t) = \frac{1}{2}e^{-it} + \frac{1}{2}e^{it} = \cos t$. Its reciprocal is $g = 1/\cos t$, and there is a problem: The cosine is zero at $t = \pi/2$ (and $t = -\pi/2$). At those points g is infinite. A normal and reasonable inverse fails to exist because there is a solution to $Au = 0$— a sequence of signals that alternates between 1 and -1 at every other step:

$$Au = \begin{bmatrix} \cdot & \frac{1}{2} & 0 & \cdot & \cdot \\ \frac{1}{2} & 0 & \frac{1}{2} & 0 & \cdot \\ 0 & \frac{1}{2} & 0 & \frac{1}{2} & 0 \\ \cdot & 0 & \frac{1}{2} & 0 & \frac{1}{2} \\ \cdot & \cdot & 0 & \frac{1}{2} & \cdot \end{bmatrix} \begin{bmatrix} 0 \\ 1 \\ 0 \\ -1 \\ 0 \end{bmatrix} = 0.$$

This signal produces no output. The response to every $+1$ is canceled by the response to -1 on each side. In fact there are two signals of this kind, since a shift through one step gives a different vector u', again with no output. The nullspace of A is 2-dimensional, containing u and u'.

Of course not all symmetric cases go wrong. If there is a response like $a_0 = \frac{5}{4}$ as well as $a_{-1} = a_1 = \frac{1}{2}$, then $a = \frac{5}{4} + \cos t$. That function is never zero. Its reciprocal $g = 1/a$ has a normal Fourier series that could be computed by factoring a:

$$a = \frac{1}{2}e^{-it} + \frac{5}{4} + \frac{1}{2}e^{it} = (1 + \frac{1}{2}e^{-it})(1 + \frac{1}{2}e^{it})$$

$$g = \frac{1}{1 + \frac{1}{2}e^{-it}} \frac{1}{1 + \frac{1}{2}e^{it}} = (1 - \frac{1}{2}e^{-it} + \frac{1}{4}e^{-2it} \cdots)(1 - \frac{1}{2}e^{it} + \frac{1}{4}e^{2it} \cdots)$$

(18)

You might guess that the first line corresponds to an LDL^T factorization of A (with $D = I$), and the second line is a factorization of A^{-1}.

We take note of the test for invertibility:

4I   The constant-diagonal matrix A is invertible if $a(t)$ is never zero.

This test was failed by $a = \cos t$ and passed by $a = \cos t + \frac{5}{4}$. Strictly speaking it is necessary and sufficient for invertibility when A is a band matrix. Then $a(t)$ is a polynomial and its zeros are clear. If $a(t)$ is a more general function we need to specify exactly which matrices are acceptable as inverses, but still the essential condition will be $a(t) \neq 0$.

This section should say why the Fourier method applies so well to constant-diagonal matrices A. The real reason is found in the eigenvectors and eigenvalues (as always). The equation $Ax = \lambda x$ for these infinite matrices is

$$
\begin{bmatrix}
\cdot & \cdot & \cdot & & \cdot & & \cdot \\
a_1 & a_0 & a_{-1} & a_{-2} & a_{-3} \\
a_2 & a_1 & a_0 & a_{-1} & a_{-2} \\
a_3 & a_2 & a_1 & a_0 & a_{-1} \\
\cdot & \cdot & \cdot & & \cdot & & \cdot
\end{bmatrix}
\begin{bmatrix}
e^{2it} \\
e^{it} \\
1 \\
e^{-it} \\
\cdot
\end{bmatrix}
= a(t)
\begin{bmatrix}
e^{2it} \\
e^{it} \\
1 \\
e^{-it} \\
\cdot
\end{bmatrix}. \tag{19}
$$

The function $a(t)$ is the eigenvalue! Furthermore there is one of these "eigenvalues" for every t between $-\pi$ and π. The powers $1, w, \ldots, w^{n-1}$, which were equally spaced around the circle, are replaced by the powers e^{it} which *fill* the circle. The discrete series has become an infinite series, and the sequence f at n points has become a continuous function.

All the information in A is reflected in a. When A is symmetric, $a(t)$ is real. When A is positive definite, $a(t)$ is positive. When A is invertible, $a(t)$ is not zero. The eigenvectors are the same for all constant-diagonal matrices A, so every A is diagonalized by the same Fourier matrix F—which is now doubly infinite:

$$
F^{-1}AF = \Lambda = \text{multiplication by } a(t). \tag{20}
$$

A convolution goes into a multiplication when you take Fourier transforms. That is the effect of F and F^{-1}. To interpret (20) in infinite dimensions would take patience, so we leave that for the finite Fourier transform and emphasize only one main point: When A becomes multiplication by $a(t)$ and B becomes multiplication by $b(t)$, then

$$AB \text{ becomes multiplication by } ab$$

$$A^{-1} \text{ becomes multiplication by } 1/a$$

$$A^T A \text{ becomes multiplication by } |a|^2.$$

EXERCISES

4.2.1   Find the discrete transforms $c = F^{-1}f$ of the following sequences: $f = (1, 1, 1, 1)$, $f = (1, 0, 1, 0)$, $f = (1, -1)$.

4.2.2   Find the inverse transforms Fc of $c = (1, 1, 1, 1)$, $c = (0, 0, 1, 0)$, and $c = (2, 4, 6, 8)$.

4.2.3   Write down the 4 by 4 unitary matrix $U = F_4/\sqrt{4}$ and verify that its second and third columns are orthogonal. What is U^{-1}?

4.2.4   What is the discrete (periodic) convolution of $(1, 2, 3)$ and $(3, 2, 1)$?

4.2.5 (a)   Find the convolution of $f = (1, 1, 1, 1)$ and $g = (1, 0, 1, 0)$.
  (b)   From Exercise 1 find their transforms c and d, the component-by-component product cd, and its inverse transform $F(4cd)$.

4.2.6   The numbers $5, -1 + i, -1, -1 - i$ that gave complex interpolation of $2, 4, 6, 8$ also give real interpolation by

$$g = 5 + (-1 + i)e^{ix} - \cos 2x + (-1 - i)e^{-ix} = 5 - 2 \cos x - 2 \sin x - \cos 2x.$$

Show that $g = 2, 4, 6, 8$ at the points $x = 0, \pi/2, \pi, 3\pi/2$ and graph $g(x)$ from 0 to 2π. *Note*: At these four points e^{3ix} equals e^{-ix} and e^{2ix} equals $\cos 2x$, allowing $c_0 + c_1 e^{ix} + c_2 e^{2ix} + c_3 e^{3ix}$ to change to g.

4.2.7   Compute $c = F^{-1} f$ for $f = (4, 2, 0, 2)$ and find $f * f$ from the convolution rule. Check by direct (periodic) convolution of f with f.

4.2.8   Verify the convolution rule (10) for $n = 3$, comparing $\frac{1}{3}(f_0 + f_1 + f_2)(g_0 + g_1 + g_2)$ on the right side with the first component of

$$F^{-1}(f * g) = \frac{1}{3} \begin{bmatrix} 1 & 1 & 1 \\ 1 & \bar{w} & \bar{w}^2 \\ 1 & \bar{w}^2 & \bar{w}^4 \end{bmatrix} \begin{bmatrix} f_0 g_0 + f_1 g_2 + f_2 g_1 \\ f_0 g_1 + f_1 g_0 + f_2 g_2 \\ f_0 g_2 + f_1 g_1 + f_2 g_0 \end{bmatrix}.$$

4.2.9   Verify the opposite convolution rule (11) for $n = 3$. The first components of $F^{-1}(fg)$ and $(F^{-1}f) * (F^{-1}g)$ are $\frac{1}{3}(f_0 g_0 + f_1 g_1 + f_2 g_2)$ and

$$\frac{1}{9}[(f_0 + f_1 + f_2)(g_0 + g_1 + g_2) + (f_0 + \bar{w}f_1 + \bar{w}^2 f_2)(g_0 + \bar{w}^2 g_1 + \bar{w}^4 g_2)$$
$$+ (f_0 + \bar{w}^2 f_1 + \bar{w}^4 f_2)(g_0 + \bar{w}g_1 + \bar{w}^2 g_2)].$$

Why are these two expressions equal if $w = w_3 = e^{2\pi i/3}$?

4.2.10 (a)   Multiply the numbers $f = 1111$ and $g = 3333$ by convolution of two sequences of length $n = 7$.
  (b)   Explain why the components of $f * g$ can never exceed $81n$ if f and g are decimals with n digits.

4.2.11   Let C be the "second-difference" circulant matrix

$$C = \begin{bmatrix} 2 & -1 & 0 & -1 \\ -1 & 2 & -1 & 0 \\ 0 & -1 & 2 & -1 \\ -1 & 0 & -1 & 2 \end{bmatrix}.$$

What are its eigenvalues and eigenvectors? How can you tell it is not invertible? Connect its eigenvalues to the function $2 - e^{ix} - e^{-ix}$ at $x = 0, \pi/2, \pi, 3\pi/2$.

4.2.12   Since the preceding matrix C is symmetric it has real eigenvalues and it should have real eigenvectors. Find two real eigenvectors for $\lambda = 2$. What is their relation to $(1, i, i^2, i^3)$ and $(1, -i, (-i)^2, (-i)^3)$?

4.2.13 Write down the eigenvectors and eigenvalues (in terms of $w = w_3$) of

$$C = \begin{bmatrix} 4 & 1 & 1 \\ 1 & 4 & 1 \\ 1 & 1 & 4 \end{bmatrix} \quad \text{and} \quad C_f = \begin{bmatrix} f_0 & f_2 & f_1 \\ f_1 & f_0 & f_2 \\ f_2 & f_1 & f_0 \end{bmatrix}.$$

Under what condition is the matrix C_f invertible?

4.2.14 How do you know from $a = 4 + e^{ix} + e^{-ix}$ that the matrix above is symmetric positive definite? Compare a^2 with the entries in C^2.

4.2.15 The previous function a equals 6, 3, 3 at the points $x = 0, 2\pi/3, 4\pi/3$. Therefore $g = 1/a = g_0 + g_1 e^{ix} + g_2 e^{-ix}$ equals $1/6, 1/3, 1/3$ at those points. Find g_0, g_1, g_2 and show that they are the entries of C^{-1}.

4.2.16 If a unit signal at time zero produces only the weaker response $a_1 = \frac{1}{2}$, what is the infinite matrix in $Au = b$ that connects input u to output b? What is $a(t)$? What is A^{-1}?

4.2.17 What sequences $b_j = \Sigma\, a_{j-k} u_k$ come from the convolution of

(a) $a_0 = 1, a_1 = 2, u_0 = 3, u_1 = 4$, all other $u_k = 0 = a_k$.
(b) $a = (\dots, 0, 1, 0, \dots)$ with $a_0 = 1$ and any $u(\dots, u_{-1}, u_0, u_1, \dots)$.

Write both parts also as an infinite matrix-infinite vector multiplication.

4.2.18 (a) For the lower triangular matrix L with $l_{ii} = 1$ and $l_{ii-1} = \frac{1}{2}$ corresponding to $l(t) = 1 + \frac{1}{2} e^{it}$, what is $A = LL^T$?
(b) L^T corresponds to the conjugate function $\bar{l} = 1 + \frac{1}{2} e^{-it}$; verify that $a(t) = l(t)\bar{l}(t)$.
(c) What is L^{-1}?

4.2.19 (a) To invert the infinite symmetric matrix with diagonals $a_{-1} = a_1 = \frac{1}{2}$ and $a_0 = \frac{5}{4}$, show that the reciprocal of $a(t)$ in (18) splits into the "partial fractions"

$$g = \frac{1}{(1 + \frac{1}{2} e^{-it})(1 + \frac{1}{2} e^{it})} = \frac{-2e^{-it}/3}{1 + \frac{1}{2} e^{-it}} + \frac{4/3}{1 + \frac{1}{2} e^{it}}.$$

(b) Expanding the last fraction into $\frac{4}{3}(1 - \frac{1}{2} e^{it} + \frac{1}{4} e^{2it} \cdots)$ gives $g_0 = 4/3$ on the main diagonal of A^{-1} and $g_1 = -2/3, g_2 = 1/3, \dots$ on the lower diagonals. The other fraction gives the symmetric entries above the diagonal. Multiply this A^{-1} by A and verify that the diagonal entries are 1's.

4.2.20 For the *two-sided shift* S in (13), the matrix $A = S - \lambda I$ corresponds to the function $a(t) = e^{it} - \lambda$. Show from $4I$ that A is invertible if $|\lambda| \neq 1$, and find the "eigenvector" that satisfies $Sx = e^{it}x$. Note: The *spectrum* of S contains all λ for which $S - \lambda I$ is not invertible—even if, as here, there is no eigenvector of finite length.

4.2.21 Show that $a * u$, whose jth component is $\Sigma_{k=-\infty}^{\infty}\, a_{j-k} u_k$, is the same as $u * a$, whose jth component is $\Sigma_{l=-\infty}^{\infty}\, u_{j-l} a_l$. Similarly $f * g = g * f$ for discrete convolution with finite sums and periodicity.

4.2.22   The symmetric doubly-infinite matrix A with 1 on the main diagonal and $\frac{1}{2}, \frac{1}{4}, \frac{1}{8}, \dots$ on the neighboring diagonals has

$$a(t) = \sum_{-\infty}^{\infty} \frac{1}{2^{|j|}} e^{ijt} = \frac{1}{1 - \frac{1}{2}e^{it}} + \frac{1}{1 - \frac{1}{2}e^{-it}} - 1.$$

Simplify a and $g = 1/a$, and find the matrix A^{-1}.

4.2.23   Show that multiplication of the *singly*-infinite matrices

$$U = \begin{bmatrix} 1 & \frac{1}{2} & 0 & \cdot \\ 0 & 1 & \frac{1}{2} & \cdot \\ 0 & 0 & 1 & \cdot \\ \cdot & \cdot & \cdot & \cdot \end{bmatrix} \quad \text{times } L = \begin{bmatrix} 1 & 0 & 0 & \cdot \\ \frac{1}{2} & 1 & 0 & \cdot \\ 0 & \frac{1}{2} & 1 & \cdot \\ \cdot & \cdot & \cdot & \cdot \end{bmatrix} \quad \text{equals } A = \begin{bmatrix} \frac{5}{4} & \frac{1}{2} & 0 & \cdot \\ \frac{1}{2} & \frac{5}{4} & \frac{1}{2} & \cdot \\ 0 & \frac{1}{2} & \frac{5}{4} & \cdot \\ \cdot & \cdot & \cdot & \cdot \end{bmatrix}$$

but in the opposite order LU does *not* give A. For indices that start at 0 instead of $-\infty$ the Gauss factorization $A = LU$ does not correspond to multiplication of transfer functions. In the other order, the Wiener-Hopf factorization UL maintains constant diagonals.

4.2.24   If $f = \Sigma \, a_k e^{ikx}$ and $g = \Sigma \, u_k e^{ikx}$ what are the coefficients in the corresponding sum for fg?

4.2.25   Find the eigenvalues and eigenvectors of the circular shift matrix

$$A = \begin{bmatrix} 0 & 1 & 0 & 0 \\ 0 & 0 & 1 & 0 \\ 0 & 0 & 0 & 1 \\ 1 & 0 & 0 & 0 \end{bmatrix}.$$

What is the discrete transform of $f = (0, 0, 0, 1)$? What is its convolution with $g = (g_0, g_1, g_2, g_3)$?

4.2.26   (*The first step in the Fast Fourier Transform*) For $n = 4$,

$$Fx = \begin{bmatrix} 1 & 1 & 1 & 1 \\ 1 & i & i^2 & i^3 \\ 1 & i^2 & i^4 & i^6 \\ 1 & i^3 & i^6 & i^9 \end{bmatrix} \begin{bmatrix} x_0 \\ x_1 \\ x_2 \\ x_3 \end{bmatrix} = \begin{bmatrix} x_0 + x_1 + x_2 + x_3 \\ x_0 + ix_1 + i^2x_2 + i^3x_3 \\ x_0 + i^2x_1 + i^4x_2 + i^6x_3 \\ x_0 + i^3x_1 + i^6x_2 + i^9x_3 \end{bmatrix}.$$

The FFT starts with $x_0 \pm x_2$ and $x_1 \pm x_3$ and it quickly computes

$$(x_0 + x_2) + (x_1 + x_3), (x_0 - x_2) + i(x_1 - x_3), (x_0 + x_2) - (x_1 + x_3), (x_0 - x_2) - i(x_1 - x_3).$$

Verify that those are exactly the four components of Fx.

FOURIER INTEGRALS ■ 4.3

The Fourier series was perfect for periodic functions. The only frequencies needed were the integers—or the multiples of $2\pi/T$ when the period was T. For other functions, which are not periodic, all frequencies have to be allowed and the series has to be replaced by an integral. The coefficients a_k and b_k and c_k that gave the weight of each harmonic will turn into the Fourier transform $\hat{f}(k)$—which is defined on the whole line $-\infty < k < \infty$. It still measures the density of e^{ikx} in the function f. In making this change, you will see how the important things survive.

The first step is to find a formula for $\hat{f}(k)$. Roughly speaking, we start with the known formula for the Fourier coefficients and let the period T approach infinity. The function $f_T(x)$ will be chosen to agree with $f(x)$ over one interval, from $-T/2$ to $T/2$, and then continue periodically. As $T \to \infty$ the Fourier coefficients of f_T should approach (with the right scaling) the Fourier transform of f.

For f_T there is a choice between cosines and sines with coefficients a_k and b_k, and exponentials e^{ikx} with coefficients c_k. The exponentials are better. In one sense they always have been; there was a single unified formula for $c_k = \frac{1}{2}(a_k - ib_k)$. Unlike a_0, the coefficient c_0 needs no special treatment. The Fourier integral is already complicated enough, and in everyday practice (as well as theory) the exponentials are used.

When the period is T the Fourier coefficients are

$$c_k = \frac{1}{T} \int_{x=-T/2}^{T/2} f_T(x)\, e^{-iKx}\, dx, \quad \text{with } K = k\frac{2\pi}{T}. \tag{1}$$

With those weights the harmonics of period T combine to reproduce f_T:

$$f_T(x) = \sum_{k=-\infty}^{\infty} c_k e^{iKx} = \sum_{k=-\infty}^{\infty} \frac{1}{T} e^{iKx} \left[\int_{x=-T/2}^{T/2} f_T(x) e^{-iKx}\, dx \right]. \tag{2}$$

As T gets larger two things happen. The function f_T agrees with f over a longer interval, and *the sum approaches an integral*. Each step in the sum changes k by 1, so K changes by $2\pi/T$; that is ΔK. When T goes to infinity the factor $1/T = \Delta K/2\pi$ goes to zero. Then the sum in (2) yields an integral with respect to K:

$$f(x) = \int_{K=-\infty}^{\infty} \frac{dK}{2\pi} e^{iKx} \left[\int_{x=-\infty}^{\infty} f(x) e^{-iKx}\, dx \right]. \tag{3}$$

This identity is the key to the Fourier integral. The right side starts with f, transforms it to a function of K, and then transforms back.

To appreciate formula (3) you have to take it apart. It contains two "dummy variables" of integration, K and x; those letters can be changed. We are free to write k instead of K, and we will. We could also alter x inside the brackets to another symbol, to avoid confusion with x outside, but we won't. The formula separates into two steps, first inside the brackets and then outside. One step computed the Fourier transform and the second step inverts it—like computing the coefficients c_k in the periodic case and then adding all $c_k e^{ikx}$ to recover f.

These are the key formulas for a nonperiodic function.

4J   The *Fourier transform* of f is a function $\hat{f}$ depending on frequency:

$$\hat{f}(k) = \int_{x=-\infty}^{\infty} f(x)\, e^{-ikx}\, dx. \tag{4}$$

The *inverse Fourier transform* of $\hat{f}$ brings back the original function f:

$$f(x) = \frac{1}{2\pi} \int_{k=-\infty}^{\infty} \hat{f}(k)\, e^{ikx}\, dk. \tag{5}$$

The first step analyzes f into its spectrum of harmonics—the transform $\hat{f}$ is the "spectral density"—and the second step reconstructs f. With infinite integrals we need restrictions on the functions (and a more rigorous proof) to be certain that f reappears. The difficulty is minimal if f and $\hat{f}$ drop quickly toward zero, but unless they are identically zero one thing is impossible: *They cannot both vanish everywhere outside a finite band.* This is related to Heisenberg's uncertainty principle, proved below. First we give the examples on which the whole section is based.

A List of Essential Transforms

EXAMPLE 1 $f(x) = $ delta function

$$\hat{f}(k) = \int e^{-ikx}\, \delta(x)dx = 1 \quad \text{for all frequencies } k$$

EXAMPLE 2 $f(x) = $ square pulse $= \begin{cases} 1 & -L \leq x \leq L \\ 0 & |x| > L \end{cases}$

$$\hat{f}(k) = \int_{-L}^{L} e^{-ikx}\, dx = \frac{e^{-ikL} - e^{ikL}}{-ik} = \frac{2 \sin kL}{k}$$

When f is an even function of x, as in these examples, $\hat{f}$ is an even function of k. The next example is neither even or odd:

EXAMPLE 3 $f(x) = $ decaying pulse $= \begin{cases} e^{-ax} & x > 0 \\ 0 & x < 0 \end{cases}$

$$\hat{f}(k) = \int_0^{\infty} e^{-ax} e^{-ikx}\, dx = \left[\frac{e^{-(a+ik)x}}{-(a+ik)} \right]_0^{\infty} = \frac{1}{a+ik}.$$

That pulse goes in one direction only. There is also an even pulse that goes in both directions, and an odd pulse that is negative when x is negative:

EXAMPLE 4 $f(x) = $ even decaying pulse $= e^{-a|x|}$

$$\hat{f}(k) = \text{even part} = \frac{1}{a+ik} + \frac{1}{a-ik} = \frac{2a}{a^2+k^2}.$$

EXAMPLE 5 $f(x) = $ odd decaying pulse $= \begin{cases} e^{-ax} & x>0 \\ -e^{ax} & x<0 \end{cases}$

$$\hat{f}(k) = \text{odd part} = \frac{1}{a+ik} - \frac{1}{a-ik} = \frac{-2ik}{a^2+k^2}$$

The figure shows f and $\hat{f}$ for this odd case. Notice the jump at $x=0$.

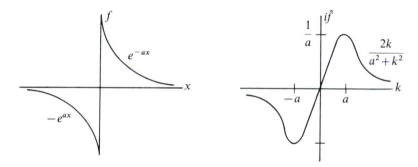

Fig. 4.8. The odd decaying pulse and its transform.

These examples are safe when a is positive. The integrals are finite because e^{-ax} drops quickly to zero. But they are infinitely more interesting (and more dangerous) at the limiting value $a=0$. The exponential becomes $e^0 = 1$—it no longer disappears at infinity—and the functions that emerge in the limit are the most basic transform pairs in engineering.

We take the three decaying pulses in reverse order, first odd then even then one-sided. Of course with $a=0$ they no longer decay.

EXAMPLE 5' $f(x) = $ **sign function** $= \begin{cases} 1 & x>0 \\ -1 & x<0 \end{cases}$

$$\hat{f}(k) = \text{limit of } \frac{-2ik}{a^2+k^2} = \frac{2}{ik}$$

This is infinite at $k=0$. The second graph in Fig. 4.8 approaches a hyperbola as a goes to zero, with an infinite jump at the center.

EXAMPLE 4′ $f(x) = \textbf{\textit{constant function}} = 1$ for all x

$$\hat{f}(k) = \text{limit of } \frac{2a}{a^2 + k^2} = \begin{cases} ? & \text{if } k = 0 \\ 0 & \text{if } k \neq 0 \end{cases}$$

EXAMPLE 3′ $f(x) = \textbf{\textit{step function}} = \begin{cases} 1 & x > 0 \\ 0 & x < 0 \end{cases}$

$$\hat{f}(k) = \tfrac{1}{2}\,(\text{example } 4' + \text{example } 5')$$

The constant function has a new problem. At $k = 0$ Example 4 has $\hat{f} = 2/a$ and it blows up as a approaches zero (see the next figure). That is more serious than the infinite jump; it is a spike. We suspect a delta function in the limit, or a multiple of a delta function, and that is confirmed by two pieces of evidence:

(a)   The integral of $\hat{f}$ in 4′ is $\int_{-\infty}^{\infty} \dfrac{2a\,dk}{a^2 + k^2} = 2\pi$ for every a, even with $\hat{f}$ approaching zero. This should remain true at $a = 0$. For that we need $\hat{f} = 2\pi\delta$ in the limit.

(b)   Since the transform of δ was 1, a further transform should return from 1 to $2\pi\delta$. Then division by 2π in (5) reproduces δ.

We conclude that $\hat{f} = 2\pi\delta$ **is the transform of** $f = 1$.

The step function is $\tfrac{1}{2}$ (sign function + constant function). Therefore its transform is also the average of 4′ and 5′:

$$\hat{f}(k) = \pi\delta(k) + \frac{1}{ik}. \tag{6}$$

That would not have been easy to guess.

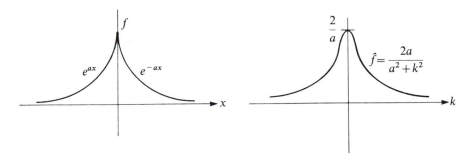

Fig. 4.9. The even decaying pulse and its transform.

All these examples have gone in one direction, from the function $f(x)$ to its transform $\hat{f}(k)$. We have taken Fourier's word that the inversion formula (5) will reconstruct f. It is sometimes easier in that backward direction and sometimes harder. A particular case is the return from $\hat{f} = 2\pi\delta$. The inversion formula gives

$$f(x) = \frac{1}{2\pi} \int_{-\infty}^{\infty} 2\pi\delta(k)e^{ikx}\, dk = e^{i0x} = 1 \text{ for all x.}$$

That is correct. The transform of $f = 1$ was just discovered to be $\hat{f} = 2\pi\delta$. Therefore the inverse transform should and does bring back $f = 1$. This may be the right way to discover that $\hat{f} = 2\pi\delta$ in the first place.

It is harder to go from $\hat{f} = 1/(a + ik)$ back to the decaying pulse:

$$\frac{1}{2\pi} \int_{-\infty}^{\infty} e^{ikx} \frac{dk}{a + ik} = \begin{array}{ll} e^{-ax} & \text{if} \quad x > 0 \\ 0 & \text{if} \quad x < 0 \end{array} \tag{7}$$

That is not in a standard table of integrals. However it is a perfect example of complex integration; there is a way to compute integrals from $-\infty$ to ∞ without knowing what they are between other points a and b. This integration and the proof of (7) will come at the very end of the chapter.

Energy and the Uncertainty Principle

There was a formula for Fourier series that connected the energy in f to the energy in its Fourier coefficients. One side contained the length of f in the Hilbert space L^2; the other side contained the length of the sequence c_k in the Hilbert space l^2. The equation was Parseval's:

$$\int_{-\pi}^{\pi} |f(x)|^2\, dx = 2\pi \sum_{-\infty}^{\infty} |c_k|^2. \tag{8}$$

It was proved by multiplying $\Sigma\, c_k e^{ikx}$ by $\Sigma\, \bar{c}_k e^{-ikx}$ and integrating. Now we want a similar equation for the Fourier transform.

4K The function f and its transform $\hat{f}$ satisfy Plancherel's formula

$$2\pi \int_{-\infty}^{\infty} |f(x)|^2\, dx = \int_{-\infty}^{\infty} |\hat{f}(k)|^2\, dk. \tag{9}$$

More generally the inner product of any functions f and g satisfies

$$2\pi \int_{-\infty}^{\infty} f(x)\overline{g}(x)dx = \int_{-\infty}^{\infty} \hat{f}(k)\overline{\hat{g}}(k)dk. \tag{9'}$$

I suspect that Parseval and Plancherel were the same person, or almost.

EXAMPLE   The decaying pulse $f = e^{-ax}$ has energy

$$\int_{-\infty}^{\infty} |f(x)|^2 \, dx = \int_{0}^{\infty} e^{-2ax} \, dx = \frac{1}{2a}.$$

Its transform has

$$\int_{-\infty}^{\infty} \left| \frac{1}{a+ik} \right|^2 dk = \int_{-\infty}^{\infty} \frac{dk}{a^2 + k^2} = \frac{\pi}{a}, \text{ confirming (9)}.$$

From the energy equation (9) we can prove an uncertainty principle. The classical case was discovered by Heisenberg; it deals with position and momentum. As one is measured more exactly the other becomes less exact. If light goes through a hole, then the narrower the hole the less we know about the angle at which the photons pass through. A screen on the other side will eventually show a bigger image from a smaller hole (?) There is a similar uncertainty for the phase and amplitude of an oscillation—and for many other pairs. Here the principle involves f and $\hat{f}$, and the mathematical basis is the same. If one is concentrated in a narrow band the other fills a wide band. An impulse $f = \delta$ of zero width has a transform $\hat{f} = 1$ of infinite width. For most functions the bandwidth (really the half-width W) is not so easy to measure, but probability suggests the right way. It corresponds to the standard deviation, where we sample points at random under a curve like $e^{-x^2/2}$. W^2 is the expected value (average) of x^2 in the sample. Imitating that idea and normalizing for the energy in f, the half-widths W_x in space and W_k in frequency are

$$W_x^2 = \frac{\int x^2 |f(x)|^2 \, dx}{\int |f(x)|^2 \, dx} \quad \text{and} \quad W_k^2 = \frac{\int k^2 |\hat{f}(k)|^2 \, dk}{\int |\hat{f}(k)|^2 \, dk}.$$

All integrals go from $-\infty$ to ∞, and the uncertainty principle is:

> *Every function has $W_x W_k \geq \frac{1}{2}$.*

The proof uses the Schwarz inequality. The cosine of the angle between $f'(x)$ and $xf(x)$ is at most one, even in Hilbert space: $|a^T b|^2 \leq (a^T a)(b^T b)$ becomes

$$\left| \int xf(x)f'(x) \, dx \right|^2 \leq \left(\int |xf(x)|^2 \, dx \right) \left(\int |f'(x)|^2 \, dx \right). \tag{10}$$

Integrating the left side by parts gives a new left side:

$$\int xf(x)f'(x) \, dx = \left[x \frac{f(x)^2}{2} \right]_{-\infty}^{\infty} - \int \frac{f(x)^2}{2} \, dx. \tag{11}$$

The integrated term is zero at $\pm\infty$ whenever the bandwidths are finite.

The last term in (10) is the energy in the derivative f'. Anticipating (13) below, its Fourier transform is $ik\hat{f}(k)$. Then Plancherel's energy equation allows us to switch between $\int |f'|^2\, dx$ and $\int |k\hat{f}(k)|^2\, dk$, remembering the factor 2π in (9). Similarly we switch between $\int |f|^2/2$ and $\int |\hat{f}|^2/2$ in (11), with another factor 2π to cancel the first. After these substitutions (10) becomes

$$\left(\int \frac{f(x)^2}{2}\, dx\right)\left(\int \frac{|\hat{f}(k)|^2}{2}\, dk\right) \le \left(\int |xf(x)|^2\, dx\right)\left(\int |k\hat{f}(k)|^2\, dk\right).$$

Taking square roots, this is the uncertainty principle $W_x W_k \ge \frac{1}{2}$.

Remark 1. Quantum mechanics associates position with the multiplication $xf(x)$ and momentum with the differentiation df/dx (in other words with $ik\hat{f}(k)$). These operations do not commute:

$$\frac{d}{dx}(xf) - x\frac{d}{dx}f = f, \quad \text{or} \quad BA - AB = I.$$

That fits our case into the general form of Heisenberg's uncertainty principle, easily proved by the Schwarz inequality:

$$\| Af \| \, \| Bf \| \ge \tfrac{1}{2}|f^T(AB - BA)f|.$$

Remark 2. It is possible to remove the factor 2π from (9) by a slight change in $\hat{f}$. Suppose the present $\hat{f}$ is divided by $\sqrt{2\pi}$; call the new transform F. After squaring, the 2π disappears and F has the same length as f. The transform from f to F is the continuous analogue of an orthogonal matrix Q, which preserves the length of every vector:

$$(Qx)^T(Qx) = x^TQ^TQx = x^Tx.$$

More correctly, it is the analogue of a complex matrix U that preserves the length of every *complex* vector:

$$(\overline{Ux})^T(Ux) = \bar{x}^T\bar{U}^TUx = \bar{x}^Tx.$$

A matrix with this property $\bar{U}^TU = I$ is called *unitary*. Real orthogonal matrices are included. In the continuous case Plancherel's theorem says that **the Fourier transform is unitary**. We need only remember that the factor $\sqrt{2\pi}$ divided from the transform must be multiplied back in the inverse transform, so that

$$F(k) = \frac{1}{\sqrt{2\pi}}\int e^{-ikx} f(x)dx \quad \text{can reconstruct}$$

$$f(x) = \frac{1}{\sqrt{2\pi}}\int e^{ikx} F(k)dk. \tag{12}$$

This is the most symmetric form of the Fourier integral.

Derivatives, Integrals, and Shifts

Suppose we know the Fourier transform of f. Then there is a simple rule for the transform of df/dx. For Fourier series

$$f = \sum_{-\infty}^{\infty} c_k e^{ikx} \quad \text{leads to} \quad \frac{df}{dx} = \sum_{-\infty}^{\infty} ikc_k e^{ikx},$$

so the coefficient c_k in f becomes ikc_k in the derivative. For Fourier transforms the rule is the same:

$$f = \frac{1}{2\pi} \int_{-\infty}^{\infty} \hat{f}(k) e^{ikx} dk \quad \text{leads to} \quad \frac{df}{dx} = \frac{1}{2\pi} \int_{-\infty}^{\infty} ik\hat{f}(k) e^{ikx} dk.$$

Therefore differentiation takes $\hat{f}(k)$ to $ik\hat{f}(k)$.

You see the underlying reason: the exponential e^{ikx} is an eigenfunction with eigenvalue ik,

$$\frac{d}{dx} e^{ikx} = ik \, e^{ikx}. \tag{13}$$

Fourier's formulas simply express f as a combination of these eigenfunctions.†

The rule for indefinite integrals is the opposite. Since integration is the inverse of differentiation, we *divide* by ik instead of multiplying:

$$f = \sum_{-\infty}^{\infty} c_k e^{ikx} \quad \text{leads to} \quad \int f = \sum_{-\infty}^{\infty} \frac{c_k}{ik} e^{ikx} + \text{constant}. \tag{14}$$

Again the exponentials are eigenfunctions, but there is one exception: The case $k = 0$ is ruled out. The derivative has a zero eigenvalue (d/dx is not invertible) and the constant functions are the eigenvectors. The derivative of c_0 is zero, and in the opposite direction that produces the constant of integration—anything in the nullspace of d/dx can be added to the integral.

Integration also divides the Fourier transform by ik, and at $k = 0$ it is undecided. That uncertainty already appeared for the delta function and step function and sign function:

$$\text{if} \quad f = \delta \quad \text{then} \quad \hat{f} = 1$$

$$\text{if} \quad f = \int \delta = \text{step function} \quad \text{then} \quad \hat{f} = \frac{1}{ik} + \pi\delta(k).$$

---

† Fourier series can be regarded as a special case of Fourier transforms. When f is periodic, $\hat{f}$ is a line of delta functions of strengths c_k.

This step goes from 0 to 1. Shifting it down by $\frac{1}{2}$ changes it to half of the sign function, whose transform was $2/ik$:

$$\text{if } f = \int \delta = \tfrac{1}{2} \text{ (sign function) then } \hat{f} = \frac{1}{ik}.$$

This time the constant of integration is zero, and $\pi\delta(k)$ disappears.

Note that the step function has derivative δ. Multiplying its transform by ik brings back $\hat{\delta} = 1$.

There is a third operation on f that yields a simple change in $\hat{f}$. It is a shift, or *translation*, of the graph of f. Suppose we move the graph a distance d to the right and call the resulting function F. Thus $F(x) = f(x - d)$. The Fourier transform of F is

$$\hat{F}(k) = \int_{-\infty}^{\infty} e^{-ikx} F(x)dx = \int_{-\infty}^{\infty} e^{-ikx} f(x - d)\, dx$$

$$= \int_{-\infty}^{\infty} e^{-ik(y+d)} f(y)dy = e^{-ikd} \hat{f}(k). \tag{15}$$

Because of the shift, the transform is multiplied by e^{-ikd}. This is especially clear for the delta function (Fig. 4.10) where an impulse at $x = d$ picks out the number

$$\int e^{-ikx} \delta(x - d)dx = e^{-ikd}.$$

The same rule holds for periodic functions and Fourier series. Shifting f through d multiplies each c_k by e^{-ikd}. Shifting it back multiplies by e^{+ikd}. And multiplying the function by an exponential will shift its transform. To summarize:

4L Suppose the transform of $f(x)$ is $\hat{f}(k)$. Then

(1) the transform of $\dfrac{df}{dx}$ is $ik\hat{f}(k)$

(2) the transform of $\displaystyle\int_a^x f(x)dx$ is $\dfrac{\hat{f}(k)}{ik} + c\delta(k)$

(3) the transform of $F(x) = f(x - d)$ is $e^{-ikd}\hat{f}(k)$

(4) the transform of $g(x) = e^{ixd} f(x)$ is $\hat{f}(k - d)$.

One consequence is that a smooth f gives a rapidly decreasing $\hat{f}$. When all derivatives of f are nice functions, all transforms $(ik)^n \hat{f}(k)$ must be nice—so $\hat{f}$ decreases rapidly for large k. Reversing the roles of f and $\hat{f}$, a rapidly decreasing f corresponds to a smooth $\hat{f}$. The decaying pulse provides an example: It is not smooth but e^{-ax} goes quickly to zero. Its transform $1/a + ik$ is smooth but decreases slowly. The bell-shaped Gaussian $f = e^{-x^2/2}$ and its transform $\hat{f} = \sqrt{2\pi}\, e^{-k^2/2}$ illustrate how functions can be both smooth and rapidly decreasing.

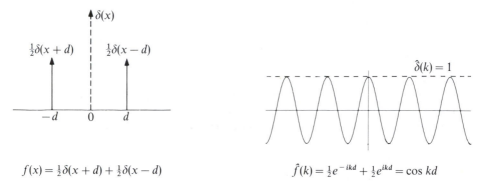

$$f(x) = \tfrac{1}{2}\delta(x+d) + \tfrac{1}{2}\delta(x-d) \qquad\qquad \hat{f}(k) = \tfrac{1}{2}e^{-ikd} + \tfrac{1}{2}e^{ikd} = \cos kd$$

Fig. 4.10. Shifts by $\pm d$ multiply the transform by $e^{\mp ikd}$

Convolution and Green's Functions

With these rules for derivatives we can solve differential equations (provided they have constant coefficients and no problems with boundaries). Consider the example

$$-\frac{d^2u}{dx^2} + a^2u = h(x), \quad -\infty < x < \infty. \tag{16}$$

Taking the Fourier transform of each term gives

$$-(ik)^2 \hat{u}(k) + a^2 \hat{u}(k) = \hat{h}(k).$$

Then the transform of the solution is

$$\hat{u}(k) = \frac{\hat{h}(k)}{a^2 + k^2}. \tag{17}$$

By inverting $\hat{u}$ we reach the solution $u(x)$.

The most important case has a delta function on the right side. Its transform is $\hat{h} = 1$. Then $\hat{u} = 1/(a^2 + k^2)$ is also a familiar transform. It comes from the even decaying pulse after division by $2a$:

$$G(x) = \frac{1}{2a}\, e^{-a|x|}.$$

This special solution $u = G$, when the right side is $h = \delta$, is the **Green's function** of the differential equation.† Away from $x = 0$ it satisfies $-G'' + a^2 G = 0$. At $x = 0$ its slope is $\frac{1}{2}$ from the left and $-\frac{1}{2}$ from the right. Therefore $-G''$ is a delta function as required.

Knowing G, we can solve the differential equation with any right side h. The transform of the solution was found above:

$$\hat{u}(k) = \frac{\hat{h}(k)}{a^2 + k^2} = \hat{G}(k)\,\hat{h}(k).$$

Therefore the question is: What function has the transform $\hat{G}\hat{h}$? The answer is not $u = Gh$. The solution u is not the product of the Green's function and the right hand side but the **convolution** of G and h.

4M   The convolution of G and h is the function

$$u(x) = \int_{y=-\infty}^{\infty} G(x-y)\,h(y)dy. \tag{18}$$

It is written $u = G * h$ and its Fourier transform is $\hat{u} = \hat{G}\hat{h}$ (the convolution rule).

This integral of $G(x-y)\,h(y)$ corresponds exactly to the sum $\Sigma\, G_{j-k}h_k$ in a discrete convolution. In Section 4.2 we multiplied Fourier coefficients; here we multiply Fourier transforms. A direct proof of the convolution rule computes $\hat{u}$ and produces $\hat{G}\hat{h}$:

$$\hat{u}(k) = \int_{x=-\infty}^{\infty} e^{-ikx}\, u(x)dx$$

$$= \int_{x=-\infty}^{\infty} \int_{y=-\infty}^{\infty} e^{-ik(x-y)} e^{-iky} G(x-y)h(y)dy dx.$$

On the right, integrate first with respect to x. The terms e^{-iky} and $h(y)$ can move outside since they are independent of x. Then changing $x - y$ to z separates the two integrals:

$$\hat{u}(k) = \int_{y=-\infty}^{\infty} e^{-iky}h(y) \int_{z=-\infty}^{\infty} e^{-ikz}\, G(z)dzdy = \hat{h}(k)\hat{G}(k). \tag{19}$$

This proves the convolution rule for any two functions. When G is the fundamental solution and h is the right hand side of a differential equation, it solves the equation.

---

† In mathematics it is the *fundamental solution*; in engineering it is the *impulse response*.

That looks more magical than it is. Any function is a superposition of delta functions:

$$h(x) = \int_{-\infty}^{\infty} \delta(x - y)\, h(y)dy \quad \text{or} \quad h = \delta * h.$$

The solution for δ is G, so the solution for h is a superposition of G's:

$$u(x) = \int_{-\infty}^{\infty} G(x - y)\, h(y)dy \quad \text{or} \quad u = G * h.$$

EXAMPLE $h(x) = \delta(x - d) = $ delta function moved to $x = d$

The transform is $\hat{h} = e^{-ikd}$. The transform of u is $e^{-ikd}/(a^2 + k^2)$. We can find u itself in three ways:

(1)   If $\hat{u}$ is multiplied by e^{-ikd} then u is shifted through d. Thus $u(x) = G(x - d)$.

(2) $u(x) = G * h = \int G(x - y)\, \delta(y - d)\, dy = G(x - d)$.

(3)   When the right side h is moved through d so is u.

This is simple, and very different from Laplace's equation in a circle. There Green's function had to change as the impulse h moved toward the boundary. Here there is no boundary and the entire problem shifts by d. It is more like Laplace's equation in free space, where the Green's function is $1/4\pi r$—and r is the distance from wherever the impulse is located.

Integral Equations

We turn to equations in which the unknown function u occurs *inside an integral*. For some (probably good) reason they are not as popular as differential equations. On T-shirts with Maxwell's equations, the integral form is printed on the back or not at all. Nevertheless integral equations are completely parallel to matrix equations, and they were responsible for the early development of infinite-dimensional space (Hilbert space). If you allow me to write the matrix equation as $Ku - \lambda u = f$ instead of $Ax = b$, the symmetric case can be quickly described by a table of analogies:

Matrix equation	*Integral equation*
$Ku - \lambda u = f$	$\displaystyle\int_a^b K(x, y)u(y)dy - \lambda u(x) = f(x)$
symmetric matrix $K_{ij} = K_{ji}$	symmetric kernel $K(x, y) = K(y, x)$
eigenvalues $\lambda_1, \ldots, \lambda_n$	eigenvalues $\lambda_1, \lambda_2, \ldots \to 0$
orthonormal eigenvectors	orthonormal eigenfunctions
$\quad u_1, \ldots, u_n$	$\quad u_1(x), u_2(x), \ldots$

solution $u = \sum \dfrac{u_i^T f}{\lambda_i - \lambda} u_i$ solution $u(x) = \sum \dfrac{\int u_i f dx}{\lambda_i - \lambda} u_i(x)$

solvable whenever $\lambda \neq \lambda_1, \dots \lambda_n$ solvable whenever $\lambda \neq \lambda_1, \lambda_2, \dots$
  and $\lambda \neq 0$

solvable for $\lambda = \lambda_i$ only if   solvable for $\lambda = \lambda_i$ only if
$u_i^T f = 0$ $\int u_i f dx = 0$

constant-diagonal matrix: convolution equation:
$K_{ij} = k_{i-j}$ $K(x, y) = k(x - y)$

positive definite if   positive definite if
$a = \sum k_j e^{-ij\theta} > 0$ $\hat{k}(\omega) = \int k(x)e^{-i\omega x} dx > 0$

The matrix case is mostly old but partly new. $K - \lambda I$ is singular when λ is an eigenvalue; the eigenvectors give the nullspace. In the symmetric case they also give the left nullspace, to which every f in the column space must be perpendicular. Forgive this descent into linear algebra; it is the last two lines of the table that deserve attention. Infinite matrices with constant diagonals were solved by Fourier series in Section 4.2. Integral equations with convolution kernels $k(x - y)$ will be solved below, provided the finite limits a, b are changed to $-\infty, \infty$. Equations with finite a and b will not be solved, because they are the hardest of all. The points a and b are like boundaries and they make difficulties for transforms. Some of the convolution ideas can be saved—Levinson's algorithm solves constant-diagonal problems faster than elimination, and Wiener-Hopf solved integral equations with limits $0, \infty$—but the Fourier transform strongly prefers the whole line.

The last entries in the table are significant. If the problem is positive definite we are glad to know it, because that guarantees a solution for every negative λ. The case $\lambda = -1$ gives the standard form of a "Fredholm integral equation of the second kind." Our example will fit into that form; it is the best. The limiting case $\lambda = 0$ removes λu from the integral equation and leaves an equation of the "first kind." It looks easier, with only an integral on the left side, but it is much less certain to have a solution. It is somewhere between being positive definite and positive semidefinite. In the matrix case, either $\lambda = 0$ is an eigenvalue or it is not. For integral equations $\lambda = 0$ is the limit of an infinite sequence of eigenvalues, and we are safer with equations of the second kind.

The test for positive definiteness needs an explanation. In the matrix case, K is a finite piece of a doubly infinite matrix A. A is a convolution matrix, with constant diagonals, and its transfer function is $a = \sum k_j e^{ij\theta}$. At the end of Section 4.2 we remarked that A is positive definite when this function a is positive. It follows that every finite section K of the infinite matrix is also positive definite.

For positive definiteness in the integral equation we again start with the whole line: $a = -\infty$ and $b = +\infty$. Then the integral is a convolution $k*u$ and the eigenfunctions are $u = e^{i\omega x}$:

$$k*u = \int_{-\infty}^{\infty} k(x - y)e^{i\omega y} dy = \int_{-\infty}^{\infty} k(z)e^{i\omega(x-z)} dz = \hat{k}(\omega)e^{i\omega x}. \tag{20}$$

The eigenvalue is the Fourier transform $\hat{k}(\omega)$.† When $\hat{k}$ is positive for all frequencies the problem is positive definite. It remains so when the limits a and b become finite. However $e^{i\omega x}$ does not remain an eigenfunction over a finite interval and explicit solutions become hard to find.

I have to admit that even with limits $-\infty, \infty$ the eigenfunctions $e^{i\omega x}$ are not completely normal. There are too many. We do not expect a continuum of eigenfunctions, one for every ω, but that is what appears. Any function $f(x)$ can be written as a combination of these eigenfunctions—that is the Fourier transform!— and the integral equation can be solved explicitly. We give the general formula and a specific example.

4N   After Fourier transform, $\int_{-\infty}^{\infty} k(x - y)u(y)dy - \lambda u(x) = f(x)$ becomes

$$(\hat{k}(\omega) - \lambda)\hat{u}(\omega) = \hat{f}(\omega). \tag{21}$$

Therefore $\hat{u} = (\hat{k} - \lambda)^{-1}\hat{f}$ and the integral equation is solved by the inversion

$$u(x) = \frac{1}{2\pi} \int_{-\infty}^{\infty} \frac{\hat{f}(\omega)}{\hat{k}(\omega) - \lambda} e^{i\omega x} d\omega. \tag{22}$$

This is the exact analogue of the solution formula in the table. There u was a sum of eigenfunctions with $\lambda_i - \lambda$ in the denominator; here it is an integral over eigenfunctions $e^{i\omega x}$ divided by $\hat{k} - \lambda$. The solution is secure provided λ is not an eigenvalue. That is guaranteed if λ is negative and the transform $\hat{k}$ is positive. Otherwise f can resonate with the eigenfunctions and the solution u is out of control.

EXAMPLE 1   A case with $\lambda = 0$: $\int_{-\infty}^{\infty} e^{-|x - y|}u(y)dy = f(x)$

The kernel $e^{-|x-y|}$ is the two-sided decaying pulse. Its transform is $\hat{k} = 2/(1 + \omega^2)$. It is real because the pulse is an even function; the equation is symmetric. Since $\hat{k}$ is positive the integral equation is positive definite. To solve it we apply the Fourier transform to both sides:

$$\frac{2}{1 + \omega^2} \hat{u}(\omega) = \hat{f}(\omega) \quad \text{or} \quad u(x) = \frac{1}{2\pi} \int_{-\infty}^{\infty} \left(\frac{1}{2} + \frac{\omega^2}{2}\right) \hat{f}(\omega)e^{i\omega x}d\omega. \tag{23}$$

This can be greatly simplified. The term with factor $\frac{1}{2}$ is nothing but $\frac{1}{2}f(x)$; we are just recovering f from its transform. The term with $\frac{1}{2}\omega^2$ is $-\frac{1}{2}f''$. (Two derivatives

---

† The frequency is ω instead of k because here $k(x - y)$ is the kernel.

multiply the transform by $(i\omega)^2$.) Therefore the solution to the integral equation is

$$u(x) = \tfrac{1}{2}f(x) - \tfrac{1}{2}f''(x). \tag{24}$$

This example touched the dangerous value $\lambda = 0$ and escaped with nothing worse than two derivatives of the forcing function f. If f is smooth then u is satisfactory; if f is a step function then the solution is more singular (f' is δ and f'' is the doublet δ').

Strangely enough this example seems to be the reverse of Green's functions. There we started with a differential equation like (24), and its solution was a convolution integral. Here we start with the integral equation and its solution is (24).

EXAMPLE 2 $\tfrac{1}{2}\int_{-\infty}^{x} u(y)dy - \tfrac{1}{2}\int_{x}^{\infty} u(y)dy + u(x) = f(x)$

The kernel $k(x - y)$ is $+\tfrac{1}{2}$ when y is less than x and $-\tfrac{1}{2}$ when y exceeds x. In other words

$$k(x) = \frac{1}{2} \text{ (sign function)} = \frac{1}{2}\frac{x}{|x|} \quad \text{and} \quad \hat{k}(\omega) = \frac{1}{i\omega}.$$

This is not a symmetric kernel. In fact it is *skew-symmetric*; the transform $\hat{k}$ of the sign function is pure imaginary. The corresponding matrix equation would have $+\tfrac{1}{2}$ everywhere below the diagonal and $-\tfrac{1}{2}$ above. The Fourier transforms $\hat{k}(\omega)\hat{u}(\omega) + \hat{u}(\omega) = \hat{f}(\omega)$ of both sides of the integral equation are

$$\left(\frac{1}{i\omega} + 1\right)\hat{u}(\omega) = \hat{f}(\omega) \quad \text{or} \quad \hat{u}(\omega) = \frac{i\omega}{1 + i\omega}\hat{f}(\omega).$$

Then the solution is the inverse transform (Exercise 4.3.22):

$$u(x) = \frac{1}{2\pi}\int_{-\infty}^{\infty} \frac{i\omega}{1 + i\omega}\hat{f}(\omega)e^{i\omega x}d\omega = \int_{-\infty}^{x} e^{y-x}f'(y)dy. \tag{25}$$

It is easy for integral equations to become complicated. There is only one more type we should mention—the **Volterra equation**. Its upper limit is not b or ∞ but x. Instead of definite integrals Volterra studied equations containing indefinite integrals:

$$\int_{a}^{x} K(x, y)u(y)dy + cu(x) = f(x). \tag{26}$$

If $K(x, y)$ is again a convolution kernel $k(x - y)$, the difference is that the convolution now stops halfway. The integral ends at $y = x$ and the kernel drops to zero. Fredholm equations corresponded to a symmetric matrix or a skew-symmetric matrix, where Volterra equations correspond to a lower triangular matrix. The method of solution is still a transform, but now it is the **Laplace transform**.

In other words, a Volterra equation is the integral form of an initial-value problem. It has a starting time $a = 0$ or $a = -\infty$, but no final time b. The past enters but not the future. It is a causal problem in time rather than a boundary-value problem in space, and therefore it is left to Chapter 6.

The most valuable thing is to tabulate the eigenfunctions and eigenvalues for convolutions of all kinds. Section 4.2 introduced discrete convolutions with finite sequences, and also with infinite sequences. In one case the matrix was an N by N circulant, in the other it was doubly infinite. This section introduced continuous convolutions from $-\infty$ to ∞ (Fredholm) or from $-\infty$ to x (Volterra). In every convolution **the eigenfunctions are exponentials and the eigenvalues are the transforms**.

Convolution	*Eigenfunctions*	*Frequencies*	*Eigenvalues*
$\displaystyle\sum_{j=0}^{N-1} a_{i-j} u_j$	$u_j = w^{jk}$	$k = 0, \ldots, N-1$	$\displaystyle\sum_{0}^{N-1} a_j \bar{w}^{jk}$
$\displaystyle\sum_{j=-\infty}^{\infty} a_{i-j} u_j$	$u_j = e^{ijt}$	$0 \le t < 2\pi$	$\displaystyle\sum_{-\infty}^{\infty} a_j e^{-ijt}$
$\displaystyle\int_{y=-\infty}^{\infty} k(x-y)u(y)dy$	$u(x) = e^{i\omega x}$	$-\infty < \omega < \infty$	$\displaystyle\int_{-\infty}^{\infty} k(x)e^{-i\omega x}dx$
$\displaystyle\int_{y=-\infty}^{x} k(x-y)u(y)dy$	$u(x) = e^{izx}$	$\text{Im } z \le 0$	$\displaystyle\int_{0}^{\infty} k(x)e^{-izx}dx$

Each line follows the same pattern, with different names. In line 1 the eigenvalues are the discrete Fourier transform. In line 2 the last column is the transfer function. In line 3, for a continuous convolution, it is the Fourier transform $\hat{k}(\omega)$. In line 4, when the convolution is lower triangular, $k(x)$ starts at $x = 0$ and the transform is one-sided. In every case the test for invertibility is the same:

> **The unknowns u can be recovered from these convolutions**
> **provided the eigenvalues are not zero.**

The recovery process is a deconvolution, in other words an inverse transform. The forward transform has separated the frequencies; the eigenfunctions are uncoupled. Then the inverse divides by the eigenvalues—which explains why they cannot be zero.

The table is a highly condensed summary of equations that can be solved directly by Fourier methods. They are convolution equations, or constant-diagonal equations. They are symmetric positive definite when the eigenvalues are real and positive. They are certainly special—there was not space for a complete theory of integral equations—but they are indisputably the simplest, and also the most important in applications.

We end the section with a remarkable case in which the function is recovered from samples of itself, through knowledge of its Fourier transform.

The Sampling Theorem

A function is **band-limited** if all its frequencies lie in a band $-W < k < W$. For the higher frequencies, the transform $\hat{f}(k)$ is zero. Intuitively, such a function cannot oscillate as rapidly as e^{iWx}. What is amazing is that the values of f at a discrete set of points (the multiples $x = n\pi/W$) completely determine the function. From the samples at intervals of π/W we can reconstruct f on the whole line $-\infty < x < \infty$. This rate of sampling is the **Nyquist rate**—two samples for every period of e^{iWx}.

Note that if the oscillations e^{iWx} and e^{-iWx} are allowed, the sampling can fail. The function might be $f(x) = \sin Wx$, which is zero at all the sampling points $x = n\pi/W$. Those samples cannot distinguish between $f = \sin Wx$ and $f = 0$. Similarly if the sampling rate is slower than the Nyquist rate it can fail. At the proper rate, however, f can be perfectly reproduced by an explicit formula.

40 If $\hat{f}(k) = 0$ outside the interval $-W < k < W$, then the values of f at the sampling points $x = n\pi/W$ can be interpolated to reproduce f:

$$f(x) = \sum_{n=-\infty}^{\infty} f\left(\frac{n\pi}{W}\right) \frac{\sin(\pi x - nW)}{(\pi x - nW)}. \tag{27}$$

Proof Suppose $W = \pi$, and use the inverse transform of $\hat{f}$ to bring back f:

$$f(x) = \frac{1}{2\pi} \int_{-\infty}^{\infty} \hat{f}(k) e^{ikx} dk = \frac{1}{2\pi} \int_{-\pi}^{\pi} \hat{f}(k) e^{ikx} dk. \tag{28}$$

We reduced the interval to $-\pi < k < \pi$ since $\hat{f} = 0$ elsewhere. Now look at the particular point $x = -n$; that is the key. The integral of $\hat{f}(k) e^{-ikn}$ on the right is the same as the nth Fourier coefficient of the function $\hat{f}(x)$—provided we temporarily think of $\hat{f}$ as 2π-periodic instead of zero outside the basic interval from $-\pi$ to π. With these coefficients $c_n = f(-n)$ the Fourier series gives $\hat{f}$ in the basic interval:

$$\hat{f}(k) = \sum_{n=-\infty}^{\infty} f(-n) e^{ink} \quad \text{for} \quad -\pi < k < \pi.$$

When this new form of $\hat{f}$ is put into the inversion formula (28) and $e^{ik(x+n)}$ is integrated, it gives the sampling theorem:

$$f(x) = \frac{1}{2\pi} \int_{-\pi}^{\pi} \left(\sum_{n=-\infty}^{\infty} f(-n)\, e^{ink} \right) e^{ikx}\, dk$$

$$= \sum_{n=-\infty}^{\infty} f(-n) \left[\frac{e^{ik(x+n)}}{2\pi i(x+n)} \right]_{-\pi}^{\pi}$$

$$= \sum_{n=-\infty}^{\infty} f(-n)\, \frac{\sin \pi(x+n)}{\pi(x+n)} = \sum_{n=-\infty}^{\infty} f(n)\, \frac{\sin \pi(x-n)}{\pi(x-n)}.$$

Reversing the sign of n at the last step has no effect on a sum from $-\infty$ to ∞. And if W is different from π, the same argument applies to a $2W$-periodic function—or we can rescale the x variable by π/W and the k variable by W/π, to complete the proof. Realistically we would sample 5 or 10 times in each period, and not just twice, to avoid being drowned by noise.

Band-limited functions are exactly what "band-pass filters" are designed to achieve. They multiply the transform $\hat{f}$ of the input signal by a function that is nearly $\hat{a} = 1$ for the frequencies to be kept and $\hat{a} = 0$ for the frequencies to be destroyed. Of course the filter does that by convolving the function. The convolution of f with $a = (\sin Wx)/\pi x$ multiplies $\hat{f}$ by $\hat{a}$ and leaves it limited to the band $-W < k < W$.

EXERCISES

4.3.1   Find the transform $\hat{g}$ of the one-sided ascending pulse

$$g(x) = e^{ax} \quad \text{for } x < 0, \quad g(x) = 0 \quad \text{for } x > 0.$$

4.3.2   Find the Fourier transforms (with $f = 0$ outside the ranges given) of

(a) $f(x) = 1$ for $0 < x < L$
(b) $f(x) = 1$ for $x < 0$
(c) $f(x) = \int_0^1 e^{ikx}\, dk$
(d) the finite wave train $f(x) = \sin x$ for $0 < x < 10\pi$

4.3.3   Find the inverse transforms of

(a) $\hat{f}(k) = \delta(k)$ (b) $\hat{f}(k) = e^{-|k|}$ (separate $k < 0$ from $k > 0$).

4.3.4   Apply Plancherel's formula $2\pi \int |f|^2 dx = \int |\hat{f}|^2 dk$ to

(1) the square pulse $f = 1$ for $-1 < x < 1$, to find $\displaystyle \int_{-\infty}^{\infty} \frac{\sin^2 t}{t^2}\, dt$

(2) the even decaying pulse, to find $\displaystyle \int_{-\infty}^{\infty} \frac{dt}{(a^2 + t^2)^2}.$

Note The next three exercises involve $f = e^{-x^2/2}$ and its transform $\hat{f} = \sqrt{2\pi}\, e^{-k^2/2}$.

4.3.5 Verify Plancherel's energy equation for $f = \delta$ and $f = e^{-x^2/2}$. Infinite energy is allowed.

4.3.6 What are the half-widths W_x and W_k of the bell-shaped function $f = e^{-x^2/2}$ and its transform? Show that equality holds in the uncertainty principle.

4.3.7 What is the transform of $xe^{-x^2/2}$? What about $x^2 e^{-x^2/2}$, using **4L**?

4.3.8 Show that the odd pulse (Example 5) is $-1/a$ times the derivative of the even pulse (Example 4). Therefore the transform of the odd pulse should be what factor times the transform of the even pulse?

4.3.9 The decaying pulse e^{-ax} has derivative $-ae^{-ax}$ (and 0 for $x < 0$), so that differentiation seems to multiply its Fourier transform by $-a$ instead of ik. How can this be?

4.3.10 Solve the differential equation

$$\frac{du}{dx} + au = \delta(x)$$

by taking Fourier transforms to find $\hat{u}(k)$. What is the solution u (the Green's function for this equation)?

4.3.11 Take Fourier transforms of the unusual equation

$$\text{(integral of } u) - (\text{derivative of } u) = \delta$$

to find $\hat{u}$ (using **4L**). Do you recognize u?

4.3.12 The convolution $C = f * g$ of the decaying pulse and ascending pulse (Ex. 1) is

$$C(x) = \int_{-\infty}^{\infty} f(x-y)g(y)\,dy \quad \text{with transform } \hat{C} = \hat{f}\hat{g} = \frac{1}{a+ik}\frac{1}{a-ik} = \frac{1}{a^2+k^2}.$$

Find C by recognizing this transform and also by explicitly computing the integral.

4.3.13 The square pulse with $f = 1$ for $-\frac{1}{2} < x < \frac{1}{2}$ has transform $\hat{f} = (2/k)\sin k/2$. Graph the "hat function" $h = f * f$ whose transform is $\hat{f}^2$. (The cubic B-spline is $h * h = f * f * f * f$ and its transform is $\hat{f}^4$.)

4.3.14 Show that the Fourier transform of gh is the convolution $\hat{g} * \hat{h}/2\pi$ by repeating the proof of the convolution rule—but with e^{+ikx} to produce the inverse transform.

4.3.15 The derivative of the delta function is the *doublet* δ'. It is a "distribution" concentrated at $x = 0$ and from integration by parts it picks out not $f(0)$ but $-f'(0)$:

$$\int f(x)\,\delta'(x)\,dx = -\int f'(x)\,\delta(x)\,dx = -f'(0).$$

(a) Why should the Fourier transform of δ' be ik?
(b) What does the inverse formula (5) give for $\int ke^{ikx}\,dk$?
(c) Exchanging k and x, what is the Fourier transform of $f(x) = x$?

4.3.16   If $f(x)$ is an even function then the integrals for $x > 0$ and $x < 0$ combine into

$$\hat{f}(k) = \int_{-\infty}^{\infty} f(x)e^{-ikx}dx = 2\int_{0}^{\infty} f(x)\cos kx\, dx$$

$$f(x) = \frac{1}{2\pi}\int_{-\infty}^{\infty} \hat{f}(k)e^{ikx}dk = \frac{1}{\pi}\int_{0}^{\infty} \hat{f}(k)\cos kx\, dx$$

Find $\hat{f}$ in this way for the even decaying pulse $e^{-a|x|}$. What are the corresponding formulas for sine transforms when f is odd?

4.3.17   If f is a line of delta functions explain why $\hat{f}$ is too:

the transform of $f = \sum_{n=-\infty}^{\infty} \delta(x - 2\pi n)$ is $\hat{f} = \sum_{n=-\infty}^{\infty} \delta(k - n)$.

The footnote after equation (13) may be useful.

4.3.18   (a)   Why is $F(x) = \sum_{n=-\infty}^{\infty} f(x + 2\pi n)$ a 2π-periodic function?

(b)   Show that its Fourier coefficient $c_k = \dfrac{1}{2\pi}\int_{-\pi}^{\pi} Fe^{-ikx}dx$ equals $\hat{f}(k)/2\pi$.

(c)   From $F(x) = \sum c_k e^{ikx}$ at $x = 0$ find *Poisson's summation formula*:

$$\sum_{n=-\infty}^{\infty} f(2\pi n) = \sum_{k=-\infty}^{\infty} \hat{f}(k)/2\pi$$

4.3.19   If $u(x) = 1$ then it is an eigenfunction for convolution: $k * 1$ is a multiple of 1. Prove this directly and show that $k(0)$ is the multiple. The same argument for $u = e^{i\omega x}$ gave the eigenvalue $\hat{k}(\omega)$ in equation (20).

4.3.20   Another proof of positive definiteness when $\hat{k}(\omega) > 0$ is to show that the quadratic form $u^T K u$ is positive for every u. If K is a convolution then

$$u^T K u = \int_{-\infty}^{\infty}\int_{-\infty}^{\infty} k(x - y)u(y)u(x)dydx = \frac{1}{2\pi}\int_{\infty} \hat{k}(\omega)|\hat{u}(\omega)|^2 d\omega > 0.$$

Use the convolution rule on the y-integral and Plancherel's formula (9') on the x-integral to establish this identity.

4.3.21   Apply Fourier transforms to $\int_{-\infty}^{\infty} e^{-|x-y|}u(y)dy - 2u(x) = f(x)$ to show that the solution is $u = -\frac{1}{2}f + \frac{1}{2}g$, where g comes from integrating f twice. (Its transform is $\hat{g} = \hat{f}/(i\omega)^2$.) If $f = e^{-|x|}$ find u and verify that it solves the integral equation.

4.3.22   (a)   If $f(x) = e^{i\omega x}$ confirm that the solution $u(x)$ given by (25) is $i\omega e^{i\omega x}/(1 + i\omega)$ and that it solves the integral equation of Example 2.

(b)   In the first integral in (25) identify the functions whose transforms are $1/(1 + i\omega)$ and $i\omega\hat{f}(\omega)$. Then the second form of (25) comes from the convolution rule.

4.3.23   (a)   Take Fourier transforms to find $\hat{u}(\omega)$ if

$$4\int_{y=-\infty}^{\infty} e^{-|x-y|}u(y)dy + u(x) = f(x), \quad -\infty < x < \infty.$$

(b) Express $\left[\dfrac{8}{1+\omega^2}+1\right]^{-1}$ as $1-\dfrac{8}{\omega^2+9}$ and find its inverse transform g.

(c) Write u as a convolution $f*g$ by the convolution rule.

4.3.24 Add two more types of convolution to the table of eigenfunctions, frequencies, and eigenvalues:

(i) finite continuous: $\int_0^{2\pi} a(x-y)u(y)dy$ where a and u are 2π-periodic

(ii) one-sided discrete: $\sum\limits_{j=-\infty}^{i} a_{i-j}u_j$.

4.3.25 Why does the sampling formula $\sum f(n)\sin\pi(x-n)/\pi(x-n)$ give the correct value $f(0)$ at $x=0$?

4.3.26 Suppose the Fourier transform of f is $\hat{f}(k)=1$ for $-\pi<k<\pi$, $\hat{f}(k)=0$ elsewhere. Check that the sampling theorem is correct.

4.3.27 Take Fourier transforms in the equation $d^4G/dx^4-2a^2d^2G/dx^2+a^4G=\delta$ to find the transform $\hat{G}$ of the fundamental solution. How would it be possible to find G?

4.3.28 What is $\delta*\delta$?

4.3.29 Suppose g is the mirror image of f, $g(x)=f(-x)$. Show from (4) that $\hat{g}(k)=\hat{f}(-k)$. If f is an even function (equal to its own mirror image, so that $f=g$) then so is $\hat{f}$.

4.3.30 Suppose g is a stretched version of f, $g(x)=f(ax)$. Show that $\hat{g}(k)=a^{-1}\hat{f}(k/a)$ and illustrate with the even pulse $f=e^{-|x|}$.

4.3.31 If $f=e^{-x^2/2}$ has transform $\hat{f}=\sqrt{2\pi}\,e^{-k^2/2}$, use the previous exercise to find the transform of $g=e^{-a^2x^2/2}$. Then show that $e^{-x^2/2}*e^{-x^2/2}=\sqrt{\pi}\,e^{-x^2/4}$, transforming the left side by the convolution rule (18) and the right side by the choice $a^2=\frac{1}{2}$.

Note on the transform $\hat{f}=\sqrt{2\pi}\,e^{-k^2/2}$: This is calculated in Exercise 6.4.4 and it comes also from the identity

$$\hat{f}(k)=\int_{-\infty}^{\infty}e^{-x^2/2}e^{-ikx}dx=e^{-k^2/2}\int_{-\infty}^{\infty}e^{-(x+ik)^2/2}dx.$$

The last integral is $\sqrt{2\pi}$ when $k=0$, and the change from $x+ik$ to x is justified by Cauchy's theorem in Section 4.5.

4.3.32 What is $\hat{f}$ if $f(x)=e^{5x}$ for $x\le0$, $f(x)=e^{-3x}$ for $x\ge0$?

4.3.33 Propose a definition of the two-dimensional Fourier transform. Given $f(x,y)$ what is $\hat{f}(k_1,k_2)$? Given $\hat{f}(k_1,k_2)$, what integral like (5) will invert the transform and recover $f(x,y)$?

4.3.34 Find the function $f(x)$ whose Fourier transform is $\hat{f}(k)=e^{-|k|}$.

4.4 ■ COMPLEX VARIABLES AND CONFORMAL MAPPING

It is no surprise that real equations can have complex solutions. The real equation $x^2 + 1 = 0$ led to the invention of i (and also $-i$!) in the first place. That was declared to be the solution and the case was closed. No doubt someone asked about $x^2 + i = 0$, but fortunately there was an answer: The square roots of complex numbers are again complex numbers. You must allow combinations $x + iy$, with a real part and an imaginary part, but beyond that it is not necessary to go. The complex numbers are algebraically closed, and every polynomial with complex coefficients has a full set of complex roots. Of course the word complex allows the possibility that $y = 0$ and the number is actually real.

This "fundamental theorem of algebra" was applied in Chapter 1 to $Ax = \lambda x$. That is another equation whose solutions can be complex when A is real. The eigenvalues of a real skew-symmetric matrix are automatically imaginary, and the eigenvalues of an orthogonal matrix can be any numbers $e^{i\theta}$ on the unit circle. In this section we solve one more real equation by complex methods—it is a partial differential equation, namely Laplace's equation—but first comes a brief review of concepts like the unit circle and the complex conjugate.

The best way to imagine complex numbers is to put them into a plane. The number $z = -\sqrt{2} + \sqrt{2}\,i$ has real part $x = -\sqrt{2}$ and imaginary part $y = \sqrt{2}$. Note that the imaginary part is a real number; it is not $\sqrt{2}\,i$. The distance from the origin is $r = \sqrt{x^2 + y^2} = 2$; it is the *absolute value* $r = |z|$. The angle has $\tan\theta = y/x = -1$, and it is $135°$ or $3\pi/4$. This point is shown at the left of Fig. 4.11.

Its reciprocal $1/z$ is shown at the right. The use of two planes allows us to visualize a function like $f(z) = z^{-1}$ by plotting the image point in the second plane. In this case we need $w = 1/(-\sqrt{2} + \sqrt{2}\,i)$, and there is an established way to divide complex numbers:

$$w = \frac{1}{x + iy} = \frac{1}{x + iy}\frac{x - iy}{x - iy} = \frac{x - iy}{x^2 + y^2}.$$

Numerator and denominator are multiplied by the **conjugate** of $z = x + iy$, which is $\bar{z} = x - iy$. The denominator becomes the real number $z\bar{z} = r^2 = x^2 + y^2$. Thus the real part of $1/z$ is x/r^2 and the imaginary part is $-y/r^2$.

This point $w = 1/z$ is at a distance $1/r$ from the origin and at an angle whose tangent is $-y/x$. In other words its angle reverses to $-\theta$. This trick for division is not necessary in polar coordinates, where we go immediately to $1/r$ and $-\theta$:

$$\text{if} \quad z = re^{i\theta} \quad \text{then} \quad \frac{1}{z} = \frac{1}{re^{i\theta}} = \frac{1}{r}e^{-i\theta}.$$

A circle of radius r in the z-plane is mapped to a circle of radius $1/r$ in the w-plane. Moving counterclockwise around one circle takes us clockwise around the other.

The point in the left figure is $z = 2\cos\theta + 2i\sin\theta$, and the point $1/z$ in the right figure is $w = \frac{1}{2}\cos\theta - \frac{1}{2}i\sin\theta$. It is an amazing fact that every circle, whether its center is the origin or not, goes into another circle when $w = 1/z$.

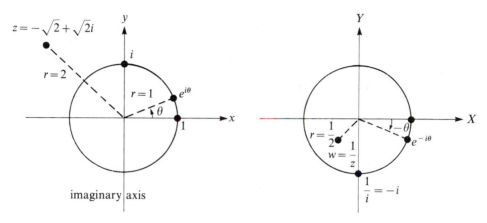

Fig. 4.11. The complex z-plane and w-plane with $w = 1/z$.

The circle of radius $r = 1$ is special. It is the **unit circle** in the z-plane, and it transforms to the unit circle in the w-plane. Its points have the form

$$z = e^{i\theta} = \cos\theta + i\sin\theta \quad \text{with} \quad r^2 = \cos^2\theta + \sin^2\theta = 1.$$

Thus $|z| = 1$. The point $w = 1/z$ is at the angle $-\theta$; it is the conjugate $\bar{z} = \cos\theta - i\sin\theta$. And we might notice that whenever a complex number z is divided by its conjugate $\bar{z}$, that gives a point on the unit circle:

$$\frac{z}{\bar{z}} = \frac{re^{i\theta}}{re^{-i\theta}} = e^{2i\theta} \quad \text{or} \quad \frac{x+iy}{x-iy} = \frac{x+iy}{x-iy}\frac{x+iy}{x+iy} = \frac{(x^2-y^2)+2ixy}{x^2+y^2}.$$

This shows again the advantage of polar coordinates for multiplication or division. (For addition they can't beat $z + \bar{z} = 2x$.) In the displayed equation on the right we would have to check the real part squared plus the imaginary part squared, while on the left we know immediately that the absolute value is 1; $e^{2i\theta}$ is on the unit circle. The general rule in multiplying $re^{i\theta}$ by $Re^{i\varphi}$ is to multiply radii and add angles, so the product is $rRe^{i(\theta+\varphi)}$.

The applications below use functions like z^2 and e^z and $\log z$, and the basic properties of complex numbers are reviewed in the first exercises.

Analytic Functions and Laplace's Equation

Laplace's equation $u_{xx} + u_{yy} = 0$ is the basic partial differential equation of equilibrium. It has to be solved, and help comes from a very unlikely source:

complex numbers. The solutions will be real functions $u(x, y)$, depending on real numbers x and y, but the key idea is to consider the complex combination $z = x + iy$. This will be useless in three dimensions—there is no way to invent the kind of super-complex numbers that would be needed—but the idea is fantastically successful in the plane.

The reason it succeeds is simple. Any decent function $f(z) = f(x + iy)$ is automatically a solution of Laplace's equation. In verifying this fact we will reach two real solutions of the equation and find the connection between them.

The key is that the derivatives $\partial f / \partial x$ and $\partial f / \partial y$ are identical except for one extra factor i from the chain rule. When f depends on the combination $z = x + iy$, its derivatives are

$$\frac{\partial f}{\partial x} = \frac{\partial f}{\partial z}\frac{\partial z}{\partial x} = \frac{\partial f}{\partial z} \quad \text{and} \quad \frac{\partial f}{\partial y} = \frac{\partial f}{\partial z}\frac{\partial z}{\partial y} = i\frac{\partial f}{\partial z}. \tag{1}$$

Multiplying the first equation by i makes the right sides the same, and the left sides are

$$i\frac{\partial f}{\partial x} = \frac{\partial f}{\partial y}. \tag{2}$$

That is *the* fundamental equation satisfied by $f(x + iy)$. We come back to it below. An example is the function $f = (x + iy)^2$, whose x derivative is $2(x + iy)$ and whose y derivative is $2i(x + iy)$. As a *function of a complex variable* it is $f = z^2$ and its derivative is $2z$.

The remaining step is easy. We differentiate again in equation (1) to produce another factor i from the y derivative:

$$\frac{\partial^2 f}{\partial x^2} = \frac{\partial^2 f}{\partial z^2} \quad \text{and} \quad \frac{\partial^2 f}{\partial y^2} = i^2\frac{\partial^2 f}{\partial z^2} = -\frac{\partial^2 f}{\partial z^2}. \tag{3}$$

The minus sign is the key. It immediately gives Laplace's equation:

$$\frac{\partial^2 f}{\partial x^2} + \frac{\partial^2 f}{\partial y^2} = 0. \tag{4}$$

Every function $f = (x + iy)^n$ is a solution to Laplace's equation. That was noticed in an earlier section; it gave the polynomials $1, x, y, x^2 - y^2, \dots$. Now there is no longer any restriction to polynomials; a function like $e^z = e^{x+iy}$ is typical of the new solutions $f(x + iy)$. It solves the differential equations written above, including $i\,\partial f / \partial x = \partial f / \partial y$. It is a combination of the special solutions $(x + iy)^n$, since e^z can be expanded in a "power series" with infinitely many terms:

$$e^{x+iy} = 1 + (x + iy) + \frac{(x + iy)^2}{2!} + \cdots = \sum_{n=0}^{\infty} \frac{(x + iy)^n}{n!}. \tag{5}$$

In any region where such a series converges—this one converges for every x and y because $n!$ grows so quickly—the sum is called an **analytic function**.

A second example, with coefficients that do not include $n!$, is

$$f = \frac{1}{1-z} = 1 + z + z^2 + \cdots = \sum_{n=0}^{\infty} (x + iy)^n.$$

This series also converges to an analytic function, but the convergence is only inside the unit circle $x^2 + y^2 = 1$. On that circle every term in the series has magnitude $|z^n| = 1$ and convergence fails. The difficulty can be traced to the particular point $z = 1$ (which means $x = 1$, $y = 0$) where the function $1/1 - z$ is in trouble; the denominator is zero and f has a singularity. The power series degenerates into $1 + 1 + 1 + \cdots$ and $1/1 - z$ blows up. At every other point $z \neq 1$, inside the unit circle or not, the function is analytic. For proof we can move the center of the power series (Exercise 4.4.11) to the point $z = a$ and test the convergence of

$$\frac{1}{1-z} = c_0 + c_1(z - a) + c_2(z - a)^2 + \cdots.$$

This series converges in a circle $|z - a| < 1 - a$, so the function is certainly analytic at the center. The circle reaches to the nearest (and only) singularity at $z = 1$.

Now come the real solutions of Laplace's equation. The special functions $(x + iy)^n$ gave two solutions, by taking their real and imaginary parts. The same will be true for every analytic function $f(x + iy)$. We write the two parts as u and s:

$$\boxed{f(x + iy) = u(x, y) + is(x, y).} \tag{6}$$

For the example e^{x+iy} it is easy to find the real part u and the imaginary part s; multiplying $e^{iy} = \cos y + i \sin y$ by e^x leads to

$$u = e^x \cos y \quad \text{and} \quad s = e^x \sin y.$$

When $f = u + is$ is substituted in (4), this u and this s should both satisfy Laplace's equation—and they do. That gives a remarkable source of solutions, two at a time, as the real and imaginary parts of any $f(x + iy)$.

You may be surprised that we got two solutions from one idea. The reason is that $x - iy$ would have been as good a combination as $x + iy$. The second derivatives of $f(x - iy)$ lead to $(-i)^2$, which produces the same minus sign that was decisive in equation (3). Then the right mixtures of $f(x + iy)$ and $f(x - iy)$ give the real solutions u and s. But $f(x - iy)$ is *not* analytic, since it violates the crucial equation $i \, \partial f / \partial x = \partial f / \partial y$. Even the simplest case $f = x - iy$ fails that test—and is to be avoided.

The last and most important step is to find the link between u and s. For that we substitute $f = u + is$ into $i \, \partial f / \partial x = \partial f / \partial y$:

$$i \left(\frac{\partial u}{\partial x} + i \frac{\partial s}{\partial x} \right) = \left(\frac{\partial u}{\partial y} + i \frac{\partial s}{\partial y} \right).$$

Separating the real parts from the imaginary parts, there are two requirements:

$$\frac{\partial u}{\partial x} = \frac{\partial s}{\partial y} \quad \text{and} \quad \frac{\partial u}{\partial y} = - \frac{\partial s}{\partial x}. \tag{7}$$

These are the **Cauchy-Riemann equations**. They give a conclusive test for an analytic function. If they are satisfied then $u + is$ is a combination of the powers $(x + iy)^n$. The equations determine s from u, or u from s, apart from a constant of integration.

We intend to show that when u is the potential, s is the stream function. First the steps above need to be summarized:

4P   A function $f(z)$ is **analytic** at $z = a$ if in a neighborhood of that point

(1)   it depends on the combination $z = x + iy$ and satisfies $i \partial f / \partial x = \partial f / \partial y$
(2)   its real and imaginary parts are connected by the Cauchy-Riemann equations $u_x = s_y$ and $u_y = - s_x$
(3)   it is the sum of a convergent power series $c_0 + c_1(z - a) + c_2(z - a)^2 + \cdots$

If these conditions are satisfied then the real functions u and s satisfy Laplace's equation.

It follows from the Cauchy-Riemann equations that $u_{xx} = s_{yx} = - u_{yy}$, leading again to Laplace's equation $u_{xx} + u_{yy} = 0$. Our goal is to find the solution u that matches given boundary conditions, and the first step is to improve the shape of the boundary.

Conformal Mapping

There are two difficulties in solving Laplace's equation with boundary conditions:

$$\frac{\partial^2 u}{\partial x^2} + \frac{\partial^2 u}{\partial y^2} = 0 \quad \text{in the region } S$$

$$u = u_0 \quad \text{on the boundary } C. \tag{8}$$

One difficulty is $u_0(x, y)$; it may not be easy to think of a function $f(z)$ whose real part has this boundary value. The other difficulty is the boundary itself. If the region S is special, we have a chance. We can find simple functions $f(z)$ with simple boundary values, and combine them to match u_0. If S is not special, then the only hope is to improve it by changing variables. It is remarkable that complex numbers can contribute to that step too.

In principle we could try any new variables X and Y. They are given in terms of the old variables by some functions $X(x, y)$ and $Y(x, y)$, and they bring changes in Laplace's equation. The second derivative $\partial^2/\partial x^2$ becomes a combination of $\partial/\partial X$ and $\partial^2/\partial X^2$ and $\partial/\partial Y$ and $\partial^2/\partial Y^2$. Similarly $\partial^2/\partial y^2$ becomes more complicated; the equation gets worse as the boundary gets better. But if X and Y are the real and imaginary parts of a function $F(z) = F(x + iy)$, something wonderful happens. The complications all cancel and the equation for $U(X, Y)$, in the new variables, is still Laplace's equation!

The change from x and y to X and Y is called a **conformal mapping**, when it is given by an analytic function $F(z)$. If $F = z^2$, for example, then $X = x^2 - y^2$ and $Y = 2xy$. The point whose coordinates used to be $(x, y) = (1, 1)$ now has coordinates $(X, Y) = (0, 2)$. The function which used to be $u = xy(x^2 - y^2)$ is now $U = XY/2$. It satisfied Laplace's equation $u_{xx} + u_{yy} = 0$, and now it satisfies $U_{XX} + U_{YY} = 0$. This is a general principle, and it can be understood without the torture of the chain rule.

4Q Suppose the combination $X + iY$ is a function F of the combination $x + iy$. Then if $U(X, Y)$ satisfies Laplace's equation in the X, Y variables, the corresponding $u(x, y)$ satisfies Laplace's equation in the x, y variables. This function u is $U(X(x, y), Y(x, y))$.

The proof is almost simpler than the statement. U is the real part of some analytic function $f(X + iY)$. But $X + iY$ is $F(x + iy)$. Thus in the x, y variables u is the real part of $f(F(z))$. It comes from a function that depends only on $z = x + iy$—and real parts of functions of z automatically satisfy Laplace's equation.

In other words, when the combination $X + iY$ depends on the combination $x + iy$, we never lose touch with Laplace's equation.

EXAMPLE $f = e^{X+iY}$ and $X + iY = x^2 - y^2 + 2ixy$.

$U = e^X \cos Y$ satisfies Laplace's equation and so does $u = e^{x^2-y^2} \cos 2xy$. The function f is e^w and w is z^2 so the composite function is e^{z^2} and its real part is u.

There is one danger in this change of variables: The derivative $F' = dF/dz$ should not be zero. When $F' \neq 0$, the mapping in the opposite direction is also conformal. The derivative of the inverse map is $1/F'$, and it is in trouble if $F' = 0$. In the examples F was z^2, and its derivative $2z$ vanishes at the origin. Therefore the point $(0, 0)$ must not be inside the region S for that change of variables.

It is not forbidden that $(0, 0)$ is on the boundary of S. Suppose S is the wedge between the x-axis and the line $y = x$ (Fig. 4.12). The angle is $45°$ and the vertex is at $(0, 0)$. We can improve the shape of that boundary by increasing the angle to $90°$; that is achieved by $F(z) = z^2$. Squaring any complex number has the effect of doubling its angle θ, and at the same time squaring its distance r from the origin. Thus θ goes to 2θ and r goes to r^2. If we write z in polar form $re^{i\theta}$, instead of rectangular form $x + iy$, then $z^2 = r^2 e^{2i\theta}$.

The points in the $45°$ wedge, when they are squared, fill up the $90°$ wedge. The points on the boundary line $y = x$, when they are squared, move to the imaginary axis. The point $1 + i$ in the figure goes to $(1 + i)^2 = 2i$.

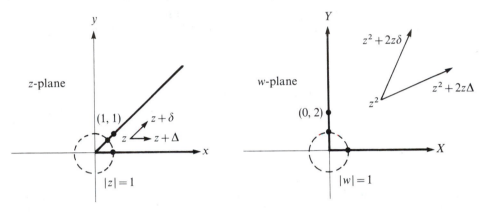

Fig. 4.12. The conformal mapping from z to $z^2 = w = X + iY = (x + iy)^2$.

The geometry of this mapping is interesting. Rays from the origin are rotated onto other rays, with the angle doubled. Circles around the origin are taken onto other circles, with the radius squared. The circle that stays invariant is the unit circle, with radius $r = 1$. It contains the points $z = e^{i\theta} = \cos\theta + i\sin\theta$. Of course the individual points move around the circle—their angles are doubled. Only the point $z = 1$ and the origin $z = 0$ are fixed by the conformal mapping, and satisfy $z^2 = z$.

Question   Why not use the mapping that just doubles θ without squaring r? That takes the $45°$ wedge to the $90°$ wedge, and it takes $re^{i\theta}$ to $re^{2i\theta}$. It looks simple but it has a terrible flaw: It is *not conformal*. It has not preserved the special combinations $x + iy$ and $re^{i\theta}$. Therefore this map will not preserve Laplace's equation.

Conformal mappings have an extra property, which seems in this case unbelievable. They also *preserve angles*. Somehow, while doubling every θ, the angle between lines is not changed! The figure shows $z + \Delta$ going to $z^2 + 2z\Delta$ and $z + \delta$ going to $z^2 + 2z\delta$. The small segments Δ and δ, whose squares we ignore, are multiplied by the same number $2z$—which rotates both segments in the same way. The little triangle is amplified in the second figure but its angle is not changed. The straight rays going out from z should be curved in the w-plane, because of the neglected terms Δ^2 and δ^2, but near z this is negligible and the rays meet at the same angle in the z and w planes. The amplification factor is $|2z|$, from the derivative of z^2.

At the origin you will not believe that angles are preserved; and you are right. The 45° angle is undeniably doubled to 90°. The mapping is not conformal at that point, where the derivative is $F'(z) = 2z = 0$. It is this vanishing of F' that is forbidden inside the region S.

The map from z to z^2 simplifies Laplace's equation in the wedge. Suppose there are mixed boundary conditions: $u = 0$ on the x-axis and its derivative $\partial u / \partial n = 0$ on the 45° line. Here n is in the direction perpendicular to the boundary. Since the x-axis goes to the X-axis when z is squared, the first condition looks the same: $U = 0$ on the line $Y = 0$. Since points on the 45° line $y = x$ go to points on the Y-axis, and right angles remain right angles (because angles are preserved), the condition on that axis is $\partial U / \partial n = 0$. In the new coordinates of Fig. 4.13 the solution is easy to find. The normal n is now in the X direction and $\partial U / \partial X$ must be zero. U can be a function of Y alone. It satisfies Laplace's equation $U_{XX} + U_{YY} = 0$, which reduces to $U_{YY} = 0$ and produces $U = cY + d$. It also satisfies $U = 0$ when $Y = 0$, so the second constant is $d = 0$. The other constant in $U = cY$ is not determined by the problem—or rather, it is still to be determined by conditions at infinity. Figure 4.13 shows the solution $U = cY$, with straight lines in the w-plane and hyperbolas in the z-plane.

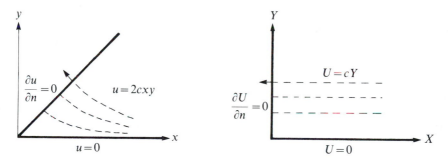

Fig. 4.13. The solutions $u = 2cxy$ and $U = cY$ to Laplace's equation.

Returning to the $x - y$ plane, $U = cY$ becomes $u = 2cxy$. This satisfies Laplace's equation, it vanishes on the horizontal line $y = 0$, and its derivative on the 45° line is zero in the direction of the normal $n = (-1, 1)/\sqrt{2}$:

$$\frac{\partial u}{\partial n} = n \cdot \text{grad } u = n_1 \frac{\partial u}{\partial x} + n_2 \frac{\partial u}{\partial y}$$

$$= \left(\frac{-1}{\sqrt{2}}\right) 2cy + \frac{1}{\sqrt{2}} 2cx = 0 \quad \text{on } y = x.$$

The potential u is the same as it was in Fig. 3.7 and its level curves—the equipotentials $2cxy = \text{constant}$—are still hyperbolas. The function $s = c(y^2 - x^2)$ will be the stream function.

Remark   Conformal mapping separates difficulties with the geometry from difficulties with the boundary conditions. You look first for a mapping F that simplifies the geometry. Then you find a function $U(X, Y)$ that satisfies the new boundary conditions. If $u = u_0$ is given at a boundary point x, y, it becomes $U = u_0$ at the corresponding point X, Y—which is moved by the mapping.

According to Riemann's mapping theorem, every region without holes (except the entire $x - y$ plane) can be mapped conformally onto every other such region. In principle, we can always make the boundary into a circle or a straight line. The difficulty is to do it in practice. Mappings are most successful when the original boundary is composed of circular arcs or line segments, and the "Schwarz-Christoffel maps" can now be done numerically. They take a polygon, for example a triangle, onto the whole line $Y = 0$. For our wedge that is achieved by $z \to z^4$, and in general an angle θ is straightened by the power $z^{\pi/\theta}$. We want to describe three important mappings and use the last one to study flow past an airfoil.

Important Conformal Mappings

1. $w = e^z = e^{x+iy}$.

This map compresses the whole real axis, where $y = 0$, into half the real axis; $w = e^x$ is always positive. It has a similar effect on the line $y = \pi$ but this time w is always negative; the extra factor is $e^{iy} = e^{i\pi} = -1$. The mapping can turn the two parallel boundaries of an infinite horizontal strip into a *single line*. Points inside the strip, lying between $y = 0$ and $y = \pi$, go above this line in the w-plane.

Thus a strip can change into a half-plane. That is made possible by taking all points at the left end of the strip, where $x = -\infty$, into the same point $e^{x+iy} = 0$.

2. $w = (az + b)/(cz + d)$.

This mapping takes every circle into another circle! It is called a *bilinear* transformation, or often a *linear* transformation with the extra word "fractional" understood. The constants a, b, c, d can be real or complex numbers. If we decide where three points should go then a, b, c, d are determined. The inverse map from w back to z is also linear, since

$$w = \frac{az + b}{cz + d} \quad \text{leads to} \quad z = \frac{-dw + b}{cw - a}.$$

Note that *a straight line is a special case of a circle*. The radius is infinite and the center is at infinity, but it is still a circle. With linear mappings some circles can go to straight lines and vice versa. We find them for particular choices of a, b, c, d:

(i) $w = ax + b$: The whole plane is shifted by b and expanded or contracted by a. Lines remain lines and $c = 0, d = 1$.

(ii) $w = 1/z$: The plane is "inverted". The circle of radius r around 0 becomes a circle of radius $1/r$; angles are reversed as in Fig. 4.11. The point $z = re^{i\theta}$ goes to $w = r^{-1}e^{-i\theta}$. The exterior of the unit circle $|z| = 1$ goes to the interior of $|w| = 1$. Since a large z means a small w, and $z = 1$ goes to $w = 1$, the vertical line $x = 1$ must go to a circle that contains the points $w = 0$ and $w = 1$. It is the circle $|w - \frac{1}{2}| = \frac{1}{2}$ in Fig. 4.14.

(iii) $w = (z - z_0)/(\bar{z}_0 z - 1)$: Here $z_0 = x_0 + iy_0$ is any point inside the unit circle and $\bar{z}_0 = x_0 - iy_0$ is its conjugate. The choice $z = z_0$ certainly produces $w = 0$. Furthermore *the circle* $|z| = 1$ *goes to the circle* $|w| = 1$, as we prove by multiplying by $\bar{z}$:

$$\bar{z}w = \frac{\bar{z}(z - z_0)}{\bar{z}_0 z - 1} = -\frac{z_0\bar{z} - 1}{\bar{z}_0 z - 1} \quad \text{because } \bar{z}z = |z|^2 = 1.$$

The numerator and denominator on the right are complex conjugates, so their ratio has magnitude 1. On the left side $|\bar{z}| = 1$. Therefore $|w| = 1$.

This mapping can retain a circular boundary while moving any chosen point to the center.

EXAMPLE If $z_0 = \frac{1}{2}$ then $w = \dfrac{2z - 1}{z - 2}$. The points $z = 1$ and $z = -1$ are exchanged and a point like $z = i$ stays on the unit circle. The numerator has absolute value $|2i - 1| = \sqrt{5}$ and the denominator has $|i - 2| = \sqrt{5}$. The point at $z = 0$ goes to $w = \frac{1}{2}$ and $z = \frac{1}{2}$ goes to $w = 0$ (Fig. 4.14). The point at $z = 2$ goes to $w = \infty$ and $z = \infty$ goes to $w = 2$.

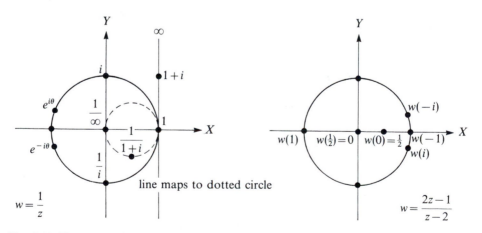

line maps to dotted circle

Fig. 4.14. Two maps that preserve the unit circle.

(iv) $w = (z - i)/(z + i)$: The real axis $z = x$ is taken to the unit circle $|w| = 1$. If z is real, then the numerator and denominator are conjugates: $|x - i| = |x + i|$ and

$|w| = 1$. The halfplane above the real axis goes inside the unit circle, and the point $z = i$ goes to the origin. With this map we can always exchange halfplanes and circles. Those have the simplest boundaries and the real problem of conformal mapping is to reach either a circle or a halfplane.

3. $w = \dfrac{1}{2}\left(z + \dfrac{1}{z}\right).$

This mapping has a rather remarkable property. If $z = e^{i\theta}$ lies on the unit circle, then the corresponding point w is

$$w = \tfrac{1}{2}(e^{i\theta} + e^{-i\theta}) = \tfrac{1}{2}(\cos\theta + i\sin\theta + \cos\theta - i\sin\theta) = \cos\theta.$$

Since the cosine is real, and stays between -1 and 1, the whole unit circle goes to a *part of a line*: if $|z| = 1$ then $-1 \le w \le 1$. The points outside the unit circle fill the rest of the w-plane. So do the points inside the circle. Each w comes from the two values $z = w \pm (w^2 - 1)^{1/2}$. (We solved the quadratic $z^2 - 2wz + 1 = 0$.) A multiple-valued inverse will be extremely awkward for most regions, but it gives no trouble if the original region S is exactly the exterior of the unit circle. Then S goes to the whole w-plane, the circle shrinks to a line segment, and other circles $|z| = R$ are stretched to ellipses (Fig. 4.15).

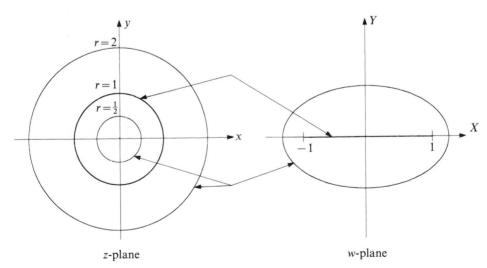

z-plane   w-plane

Fig. 4.15. The map from z to $w = \tfrac{1}{2}(z + z^{-1})$.

Best of all are the curved lines in the z-plane that become straight in the w-plane. Those curves are the **lines of flow around a circle** (Fig. 4.16). If we add a third dimension coming out of the page, they are the **streamlines** around a circular cylinder. One boundary condition requires constant velocity at infinity; far from the circle the flow is uniform and horizontal. The other boundary condition $\partial u / \partial n = 0$ prevents flow into the circle, which is a solid obstacle. In the w-plane these conditions become so simple you have to smile. The second gives $\partial U/\partial Y = 0$ on the line segment from -1 to 1, since the normal direction to the circle has changed into the normal direction (the Y direction) to the straight line. Therefore the potential can be $U = cX + d$. As in our earlier example, any linear function—and only a linear function—will be independent of Y and also satisfy Laplace's equation $U_{XX} = 0$.

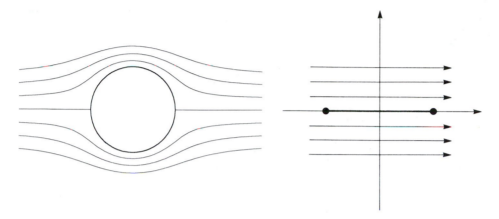

Fig. 4.16. Streamlines $Y = y - \dfrac{y}{x^2 + y^2} = \text{constant}$ around a circle.

As always the potential includes a constant d that can be set to zero. The constant c is fixed by the velocity at infinity. To go back to the z-plane we look at the conformal mapping:

$$X + iY = \frac{1}{2}\left(z + \frac{1}{z}\right) = \frac{1}{2}\left(x + iy + \frac{1}{x + iy}\frac{x - iy}{x - iy}\right)$$

has real and imaginary parts

$$X = \frac{1}{2}\left(x + \frac{x}{x^2 + y^2}\right) \quad \text{and} \quad Y = \frac{1}{2}\left(y - \frac{y}{x^2 + y^2}\right). \tag{9}$$

Thus the streamlines in Fig. 4.16 are the curves $Y = \text{constant}$.

Of course most airfoils are not circles. A true airfoil has a cross-section like the one in the next figure, with a sharp trailing edge at $w = 1$. What is really amazing is that this shape comes from a circle by the same conformal mapping. The circle must go through $z = 1$ to produce the singularity at $w = 1$, but the center is moved away from the origin. Then the streamlines around the circle become streamlines around the airfoil. Note that to map the airfoil onto the line segment from -1 to 1 we actually use three mappings: from the airfoil to the circle in Fig. 4.17, then from that circle to the unit circle, and finally from the unit circle to the line in Fig. 4.16.

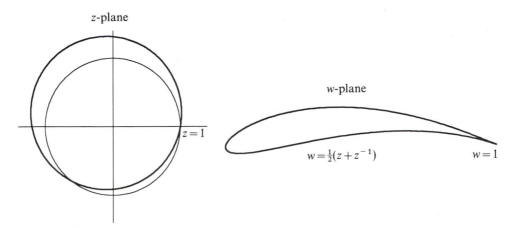

Fig. 4.17. The mapping from an offset circle to an airfoil.

I am afraid there is one more practical point. As it is, the plane will not fly. It needs *circulation* around the airfoil to create lift. The potential needs an extra term $k\theta$, where $\theta(x, y)$ is the usual angle to the point (x, y). It is surprising that we can add that term to our previous solution and find a new solution of the same boundary value problem. That has to be checked:

1. θ is the imaginary part of $\log z = \log re^{i\theta}$, so it satisfies Laplace's equation
2. θ is constant on rays perpendicular to the unit circle, so $\partial\theta/\partial n = 0$ on that boundary
3.   At infinity $\partial\theta/\partial x = \partial\theta/\partial y = 0$, so the extra $k\theta$ leaves the flow uniform at infinity.

Thus the potential is not unique! It is not even single-valued, since θ increases by 2π as we travel around the origin. That is the source of the circulation, and the conformal mapping has a new term:

$$F(z) = \frac{1}{2}\left(z + \frac{1}{z}\right) - ik \log z. \tag{10}$$

This adds the extra term $k\theta$ to the real part u, and gives the new streamlines in Fig. 4.18. When it is applied to the airfoil, it presents a new question: Which value of k does the actual solution choose? Nature must select one potential from this family of mathematical solutions, and Kutta and Joukowsky guessed which one. Their hypothesis was that the circulation adjusts itself to make the velocity at the trailing edge finite. This is confirmed by an overwhelming number of experiments. It gives the flow in the next figure, and makes it possible to compute the lifting force $-2\pi\rho Vk$—which depends on the density ρ, the velocity V at infinity, and the circulation k. That is our ultimate example of a conformal mapping.

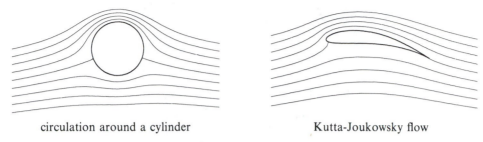

circulation around a cylinder   Kutta-Joukowsky flow

Fig. 4.18. Flows with circulation.

Two-dimensional Fluid Flows

We go back to the basic properties of $f(z) = f(x + iy)$. Its real part $u(x, y)$ satisfies Laplace's equation. That was the equation for a potential in Section 3.3, governing the flow of an ideal fluid. Therefore every analytic function $f(z)$ leads to a potential and a flow pattern. The horizontal and vertical velocities are given by the gradient:

$$v = \begin{bmatrix} v_1 \\ v_2 \end{bmatrix} = \begin{bmatrix} \dfrac{\partial u}{\partial x} \\ \dfrac{\partial u}{\partial y} \end{bmatrix}. \tag{11}$$

This velocity is divergence-free, since the equation div $v = 0$ is

$$\frac{\partial v_1}{\partial x} + \frac{\partial v_2}{\partial y} = 0 \quad \text{or} \quad \frac{\partial^2 u}{\partial x^2} + \frac{\partial^2 u}{\partial y^2} = 0, \tag{12}$$

which is satisfied by u. An incompressible fluid can maintain this velocity without sources or sinks or rotation; it is pure potential flow, independent of time.

The imaginary part of $f = u + is$ also has a part to play. It is the **stream function**. Section 3.3 introduced such a function with the following observation. For any $s(x, y)$ the vector $v = (\partial s/\partial y, -\partial s/\partial x)$ is automatically divergence-free:

$$\text{div } v = \frac{\partial}{\partial x}\left(\frac{\partial s}{\partial y}\right) + \frac{\partial}{\partial y}\left(-\frac{\partial s}{\partial x}\right) = 0.$$

The divergence is zero because the x and y derivatives can be taken in either order. Thus a velocity field can come *either* from a potential function u *or* from a stream function s.

What is the relation between u and s, if they give the same velocity? To answer that we look at the two ways to construct v:

$$v = \begin{bmatrix} \dfrac{\partial u}{\partial x} \\[2mm] \dfrac{\partial u}{\partial y} \end{bmatrix} \quad \text{and} \quad v = \begin{bmatrix} \dfrac{\partial s}{\partial y} \\[2mm] -\dfrac{\partial s}{\partial x} \end{bmatrix}. \tag{13}$$

If these agree then u and s satisfy the Cauchy-Riemann equations! Therefore u and s are the real and imaginary parts of a single analytic function $f(x + iy)$. They are called **conjugate functions**. The curves $u = $ constant are the equipotentials, and the orthogonal curves $s = $ constant are the streamlines. The flow travels in the direction of the velocity v, which is the gradient of u. It is perpendicular to the curves $u = $ constant. In other words, the velocity points along the streamlines. If you drop a leaf on the fluid it follows one of those lines. We give several examples below.

The boundary condition at a solid obstacle is $\partial u/\partial n = 0$. This prevents flow into the obstacle: $n \cdot v = n \cdot \text{grad } u = \partial u/\partial n = 0$. Therefore the obstacle boundary is a streamline. In each application we hope to find a conformal mapping that makes this boundary straight. However the solution in the new variables is not always a linear function $U = cX$ or $U = cY$. It may depend on other boundary conditions, including flows that are prescribed across open parts of the boundary—away from the obstacle.

The stream function has its own physical interpretation. The difference $s(P) - s(Q)$ measures the mass flowing across a curve between P and Q (if the density of fluid is 1). The flow is the same across every curve connecting these points. If the points are on the same streamline then $s(P) = s(Q)$; nothing flows across a streamline. We summarize and give examples:

4R   Each analytic function $f(x + iy) = u + is$ corresponds to a steady flow in which u is the potential, s is the stream function, and the velocity comes from either one by (13).

EXAMPLE 1 $f(z) = z$ with potential $u = x$ and stream function $s = y$

This is a uniform flow from left to right. The velocity is $v = (\partial u/\partial x, \partial u/\partial y) = (1, 0)$, a unit vector in the x-direction. The streamlines are horizontal. Across the curve between P and Q in Fig. 4.19a, the flow depends only on the y-components of P and Q. The rate of mass flow across the curve is $s(P) - s(Q)$, if the fluid has unit density. The speed of the flow is $|v| = |df/dz| = 1$.

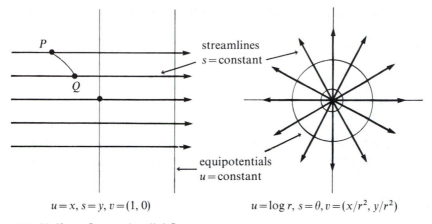

$u = x, s = y, v = (1, 0)$ $u = \log r, s = \theta, v = (x/r^2, y/r^2)$

Fig. 4.19. Uniform flow and radial flow.

EXAMPLE 2 $f(z) = \log z = \log re^{i\theta}$ *with* $u = \log r$ *and* $s = \theta$

This is a radial flow leaving the origin, and its velocity components are the derivatives of $\log r$:

$$v = \begin{bmatrix} \dfrac{1}{r}\dfrac{\partial r}{\partial x} \\[2mm] \dfrac{1}{r}\dfrac{\partial r}{\partial y} \end{bmatrix} = \begin{bmatrix} \dfrac{x}{r^2} \\[2mm] \dfrac{y}{r^2} \end{bmatrix}.$$

The magnitude of v is $1/r$, with infinite speed at the origin. The streamlines are the rays $\theta = $ constant, and the equipotentials are the circles $r = $ constant. For $f(z) = i \log z$ these are reversed and the flow travels around the circles.

The function $\log z$ is not analytic at the origin. The potential $u = \log r$ will violate Laplace's equation at that one point, and in fact

$$\frac{1}{2\pi}\left(\frac{\partial^2 u}{\partial x^2} + \frac{\partial^2 u}{\partial y^2}\right) = \text{``delta function at the origin.''} \tag{14}$$

To check the factor 2π, integrate both sides of (14) inside a circle of radius r. The left side is the divergence of $v/2\pi$, by (12). From the divergence theorem its integral can be computed from $v \cdot n$ on the circle. Since v points in the same outward direction as the unit vector n, we have $v \cdot n = |v| = 1/r$. Then

$$\iint_S \text{div } v \; dxdy = \int_C v \cdot n \; ds = \int_C \frac{1}{r} r d\theta = 2\pi.$$

The strength of the source is 2π. That is the rate at which mass leaves the origin and flows across each circle. It is also the difference between s at a point P and s at the *same* point $Q = P$, if we travel from P to P on a circle around the origin. The increase in $s = \theta$ is 2π, which agrees with the flow across the circle.

Green's Functions and Electric Fields

That example $u = \log r$ is important. It allows us to go from Laplace's equation to Poisson's; there can be sources or sinks on the right hand side. We can solve

$$\frac{\partial^2 u}{\partial x^2} + \frac{\partial^2 u}{\partial y^2} = h(x, y) \tag{15}$$

with $u = 0$ on the circular boundary $r = 1$.

This equation is solved in (14) if h is a delta function at the origin. The solution is $(\log r)/2\pi$; note that $\log 1 = 0$ and the boundary condition is satisfied. Now we allow a delta function at a different point (x_0, y_0) inside the circle. If that point can be moved to the origin while the unit circle stays where it is, the problem is solved. To do that conformally, preserving Laplace's equation, we return to the fractional linear transformations given earlier. The one that preserved the unit circle and took $z_0 = x_0 + iy_0$ to the origin was

$$w = \frac{z - z_0}{\bar{z}_0 z - 1} \tag{16}$$

In the new variables the delta function is at the origin: $z = z_0$ corresponds to $w = 0$, and Laplace's equation holds elsewhere. This is the problem solved by $(\log r)/2\pi$. Since the absolute value r in the new variables is $|w|$, we have $u(x, y) = \frac{1}{2\pi}\log|w|$.

To repeat, this solves (15) when h is the delta function $\delta(x - x_0, y - y_0)$. The source is concentrated at (x_0, y_0), where $w = 0$ and u is infinite. The potential is zero on the unit circle.

There is a special virtue to delta functions. Any other function h can be written as a *superposition* of delta functions. At all points inside the circle

$$h(x, y) = \int\int_S h(x_0, y_0)\delta(x - x_0, y - y_0)dx_0\,dy_0. \tag{17}$$

On the right side everything is zero except at $(x_0, y_0) = (x, y)$. There the delta function has a unit impulse, and multiplied by the factor h that reproduces the left side.

Since Poisson's equation is linear, superposition can be applied there too. If its right side is a sum $h_1 + h_2$, and those two parts correspond to solutions u_1 and u_2, then the complete solution is $u = u_1 + u_2$. In the present case h is not a sum of two terms but an integral; the superposition (17) has a delta function for every point. Since integration is the continuous form of summation, this superposition of right-hand sides is matched by a superposition of solutions:

$$u(x, y) = \iint_S h(x_0, y_0) \frac{1}{2\pi} \log |w|\, dx_0 dy_0. \tag{18}$$

This is a general formula for the solution of Poisson's equation. It combines the right side h of the particular problem with a Green's function $(\log |w|)/2\pi$ that is known once and for all.

We emphasize that the Green's function involves both $z = x + iy$ and $z_0 = x_0 + iy_0$. G is the *response* at (x, y) from an impulse at (x_0, y_0). It is the influence function or the "fundamental solution"—which depends on the geometry and boundary conditions because it obeys those conditions. For a circle the computations were made easy by knowing the conformal mapping (16). For every equation and every boundary condition the Green's function gives a kind of *inverse*, as we see below.

4S The Green's function G for a linear differential equation $Ku = h$ is the solution when h is a delta function at x_0, y_0: $KG = \delta(x - x_0, y - y_0)$. For Poisson's equation the Green's function with $u = 0$ on the unit circle is

$$G(x, y; x_0, y_0) = \frac{1}{2\pi} \log \left| \frac{z - z_0}{\bar{z}_0 z - 1} \right| \quad \text{and} \quad \frac{\partial^2 G}{\partial x^2} + \frac{\partial^2 G}{\partial y^2} = \delta. \tag{19}$$

For bounded h the solution to $Ku = h$ is a superposition of these fundamental solutions:

$$u(x, y) = \iint_S h(x_0, y_0) G(x, y; x_0, y_0) dx_0\, dy_0. \tag{20}$$

To see that (20) is correct, apply K to both sides. On the right side KG is δ and (20) turns into the identity (17). Thus $Ku = h$.

It is $KG = \delta$ that reveals G as the inverse of K. That is clearest if compared with the finite-dimensional case, where K is a matrix. When K multiplies a column of K^{-1} it produces a column of I. A column of I is like a delta function; the nonzero part is concentrated at a single point. Writing $h(x, y)$ as an integral of delta functions is like writing a vector h as $\sum h_j \delta_j$; δ_j is the jth column of the identity matrix. In other words, (17) corresponds to the identity $h = Ih$.

In the continuous case there is a "column" for every point x_0, y_0. In the identity the column is δ, concentrated at the point. In K^{-1} the column is G, the response to δ. Then (20) is exactly the formula $u = K^{-1}h$, with an integral instead of a sum over all columns. The response to a distributed source h is the sum or integral of responses to point sources δ.

If K is symmetric then so is its inverse G. The response at x_0, y_0 to an impulse at x, y is the same as the response at x, y to an impulse at x_0, y_0:

$$G(x, y; x_0, y_0) = G(x_0, y_0; x, y) \quad \text{corresponds to } G_{ij} = G_{ji}.$$

This is true of (20); exchanging z and z_0 has no effect on $\log|w|$. It will be true throughout our whole framework $K = A^T C A$. We emphasize that equations other than $u_{xx} + u_{yy} = h$ have Green's functions. G was found by conformal mapping in this special case; for $d^2u/dx^2 = h$ it is in the exercises, and in three dimensions it changes from $(\log r)/2\pi$ to $1/4\pi r$.

EXAMPLE   The electric field intensity comes from the potential by

$$E = -\operatorname{grad} u = - \begin{bmatrix} \partial u/\partial x \\ \partial u/\partial y \\ \partial u/\partial z \end{bmatrix}.$$

The minus sign is normal in electrostatics but otherwise this agrees with $v = \operatorname{grad} u$ for fluids. In a region free of charge the divergence of E is zero and u satisfies Laplace's equation. If the charge density is ρ then $\operatorname{div} E = \rho$ (Maxwell's equation). A point charge corresponds to a delta function on the right side of the equation, and the potential is $u = 1/4\pi r$. A line of charges along the z-axis has potential $u = (\log r)/2\pi$.

The theory of electrostatics copies the theory of ideal fluids with one significant exception. On a solid surface in the fluid the boundary condition was $\partial u/\partial n = 0$; there is no flow into the obstacle. On a conducting surface in the field the boundary condition is $u = \text{constant}$; there is no potential difference (or charge would flow along the surface to remove it). It is the *tangential* component of E that vanishes, where before it was the *normal* component of v. The equipotentials in the fluid correspond to lines of force in the electric field. The stream function s corresponds to the negative of the electric potential. They are both constant along the boundary. Complex variables will hardly notice the difference, since this reversal of $u + is$ requires only multiplication by i.

EXERCISES

4.4.1   For the complex numbers $z = 1 + i$ and $w = 3 - 4i$,

    (a)   find their sum and product
    (b)   find their positions in the complex plane
    (c)   find the positions of their conjugates $\bar{z} = 1 - i$ and $\bar{w} = 3 + 4i$
    (d)   find their absolute values $|z|$ and $|w|$
    (e)   write z and $\bar{z}$ in polar form (z is $|z|e^{i\theta}$) by finding θ.

4.4.2 Find the real and imaginary parts of

(a) $z = e^{-2i\theta}$

(b) $z = \dfrac{1}{1+i}$ $\left(\text{multiply by } \dfrac{1-i}{1-i}\right)$

(c) $\log z = \log(re^{i\theta})$

(d) $i \log i \log \log i$

4.4.3 What can you say about

(a) the sum of a complex number z and its conjugate $\bar{z}$?

(b) the conjugate of a number $z = e^{i\theta}$ on the unit circle?

(c) the product of two numbers on the unit circle?

(d) the sum of two numbers on the unit circle?

(e) the suspicious formula $e^{2\pi i a} = (e^{2\pi i})^a = 1^a = 1$?

4.4.4 Find the absolute value (or modulus) $|z|$ if

(a) $z = e^i$

(b) $z = \dfrac{1}{3 - 4i}$

(c) $z = \dfrac{3+i}{3-i}$

(d) $z = (3 + 4i)^2$

(e) $z = e^{3+4i}$

4.4.5 Find the real and imaginary parts of the analytic functions

(a) $f = 1 + i(x + iy)$

(b) $f = e^{(x+iy)^2}$

(c) $\cos(x + iy) = \frac{1}{2}(e^{i(x+iy)} + e^{-i(x+iy)})$.

Verify that u and s satisfy the Cauchy-Riemann equations.

4.4.6 The derivative df/dz of an analytic function is also analytic; it still depends on the combination $z = x + iy$. Find df/dz if $f = 1 + z + z^2 + \cdots$ or $f = z^{1/2}$ (away from $z = 0$).

4.4.7 Are the following functions analytic?

(a) $f = |z|^2 = x^2 + y^2$

(b) $f = \text{Re } z = x$

(c) $f = \sin z = \sin x \cosh y + i \cos x \sinh y$.

Can a function satisfy Laplace's equation without being analytic?

4.4.8 If $u(x, y) = x + 4y$, find its conjugate function $s(x, y)$ from the Cauchy-Riemann equations. If $s = (1 + x)y$, find u. If $u = x^2$, why does no s satisfy those equations?

4.4.9 Decompose $f = 1/z$ into $u + is$ by making the denominator real:

$$f = \frac{1}{x + iy} = \frac{1}{x + iy}\frac{x - iy}{x - iy} = \frac{x - iy}{x^2 + y^2}.$$

Verify that u and s satisfy the Cauchy-Riemann equations. Are the curves $u = $ constant and $s = $ constant hyperbolas or ellipses (or neither)?

4.4.10   The Cauchy-Riemann equations in polar coordinates, where $z = re^{i\theta}$, must still come from the chain rule:

$$\frac{\partial f}{\partial r} = \frac{\partial f}{\partial z}\frac{\partial z}{\partial r} = \frac{\partial f}{\partial z}e^{i\theta} \quad \text{and} \quad \frac{\partial f}{\partial \theta} = \frac{\partial f}{\partial z}\frac{\partial z}{\partial \theta} = \frac{\partial f}{\partial z}ire^{i\theta}.$$

(a)   Multiply the first by ir to find the relation between $\partial f/\partial r$ and $\partial f/\partial \theta$
(b)   Substituting $f = u(r, \theta) + is(r, \theta)$ into that relation, find the Cauchy-Riemann equations connecting u and s
(c)   Show that these equations are satisfied by the powers $f = z^n = r^n e^{in\theta}$, for which $u = r^n \cos n\theta$ and $s = r^n \sin n\theta$, and also by $u = \log r$ and $s = \theta$ (from $f = \log z$)
(d)   Combine the Cauchy-Riemann equations in (b) into the polar coordinate form of Laplace's equation:

$$\frac{\partial}{\partial r}\left(r\frac{\partial u}{\partial r}\right) + \frac{1}{r}\frac{\partial^2 u}{\partial \theta^2} = 0.$$

4.4.11   The function $1/(1 - z)$ has a singularity at $z = 1$, but around any other point a it admits the power series

$$\frac{1}{1 - z} = \frac{1}{(1 - a) - (z - a)} = \frac{1}{1 - a}\left(1 + \frac{z - a}{1 - a} + \left(\frac{z - a}{1 - a}\right)^2 + \cdots\right).$$

This geometric series converges when the repeated factor $r = (z - a)/(1 - a)$ has magnitude below 1. Sketch the regions in the complex plane given by $|r| < 1$ for the three cases $a = 0$, $a = 2$, $a = i$.

4.4.12   The following series are convergent for any $|z| < 1$:

$$-\log(1 - z) = z + \frac{z^2}{2} + \frac{z^3}{3} + \cdots \quad \text{and} \quad \frac{1}{(1 - z)^2} = 1 + 2z + 3z^2 + \cdots$$

Identify the term by term derivative of the first, and the term by term integral of the second. Where is the singularity that prevents convergence in a larger region like $|z| < 2$?

4.4.13   For the exponential mapping $w = e^z = e^{x+iy}$, show that

(1)   each horizontal line $y = b$ is changed into a ray from the origin (at what angle with the horizontal?)
(2)   each vertical line $x = c$ is changed into a circle around the origin (of what radius?)

4.4.14   If $z = 1 + iy$ show that $w = 1/z$ is on the circle $|w - \frac{1}{2}| = \frac{1}{2}$, in agreement with Fig. 4.14.

4.4.15   The equation of a circle $|z - z_0|^2 = R^2$ can be rewritten as

$$p z\bar{z} + qz + \bar{q}\bar{z} + r = 0 \quad (p, r \text{ real}).$$

Substitute $w = 1/z$ and show that this gives a similar equation for a circle in the w-plane. Thus inversion maps circles to circles.

4.4.16 Show that the linear transformation $w = (az + b)/(cz + d)$ is the result of the three simple transformations

$$z \rightarrow z_1 = cz + d, \;\; z_1 \rightarrow z_2 = 1/z_1, \;\; z_2 \rightarrow w = \frac{a}{c} + \frac{bc - ad}{c} z_2.$$

Since all three take circles into circles (not necessarily centered at the origin) so does any linear transformation.

4.4.17 For the map $w = \frac{1}{2}(z + z^{-1})$ in Fig. 4.15, what happens to points $z = x > 1$ on the real axis? What happens to points $0 < x < 1$? What happens to the imaginary axis $z = iy$?

4.4.18 (a) For the same mapping let $z = re^{i\theta}$ and show that $w = X + iY$ has

$$X = \tfrac{1}{2}(r + r^{-1})\cos\theta \quad \text{and} \quad Y = \tfrac{1}{2}(r - r^{-1})\sin\theta.$$

(b) Using $\cos^2\theta + \sin^2\theta = 1$ find the equation of the ellipse in the $X - Y$ plane that comes from the circle $r = 2$ (and also from $r = \frac{1}{2}$) in the $x - y$ plane.
(c) Using $(r + r^{-1})^2 - (r - r^{-1})^2 = 4$ find the equation of the hyperbola in the $X - Y$ plane that comes from the ray $\theta = \pi/4$. Sketch it into Fig. 4.15; the hyperbola should be perpendicular to the ellipse since the ray was perpendicular to the circle in the $x - y$ plane.

4.4.19 Derive grad $u \cdot$ grad $s = 0$ from the Cauchy-Riemann equations. Since the equipotential curves are perpendicular to grad u and the streamlines are perpendicular to grad s, the equation grad $u \cdot$ grad $s = 0$ confirms that these curves are perpendicular.

4.4.20 Given $f = u + is$, suppose we take $s(x, y)$ as the potential instead of $u(x, y)$. For the flow with this potential, what is the stream function? What function $F(z)$ will produce this flow?

4.4.21 (a) With a delta function at $x = 0$ solve

$$\frac{d^2 u}{dx^2} = \delta(x) \quad \text{for } -1 \le x \le 1, \text{ with } u(1) = u(-1) = 0.$$

Away from $x = 0$ the equation is $d^2u/dx^2 = 0$. Thus u is $ax + b$ on one side and $cx + d$ on the other; it is continuous at $x = 0$ but u' jumps by 1. Find a, b, c, d to obtain this one-dimensional Green's function (not log r as in 2D).
(b) With the delta function moved to $x = x_0$ solve the same problem. The solution is now the Green's function $G(x, x_0)$ for a source at $x = x_0$.
(c) Compute $u(x) = \int_{-1}^{1} G(x, x_0)dx_0$ and verify that it is the correct solution to $d^2u/dx^2 = 1$ with $u(1) = u(-1) = 0$.
(d) What would be the solution to $d^2u/dx^2 = h(x)$?

4.4.22 For the mapping $w = \sin z$ show how real points in the w-plane correspond to boundary points of a "blocked channel" above the interval from $z = -\pi/2$ to $z = \pi/2$. Sketch the streamlines in the z-plane that correspond to horizontal lines in the w-plane.

4.4.23 Solve Laplace's equation in the 45° wedge if the boundary condition is $u = 0$ on both sides $y = 0$ and $y = x$.

(a) Where does $F(z) = z^4$ map the wedge?
(b) Find a solution with zero boundary conditions other than $u \equiv 0$.

4.5 ■ COMPLEX INTEGRATION

There is another reward for working with functions of $x + iy$. Integration around a closed curve becomes exceptionally simple. It is no longer necessary to integrate in the usual sense, by finding a function whose derivative is f. Instead one looks inside the curve for points where f is not analytic. In the example $f = 1/z$ this occurs at the origin, and the behavior at $z = 0$ determines the integral around a curve anywhere in the plane.

If the curve encloses no singularities of f then **the integral is zero**. That is the fundamental theorem of complex integration. The integral starts as

$$\int_C f(z)dz = \int_C (u + is)(dx + idy),$$ (1)

and it can be split into the real and imaginary parts

$$\int_C (u\,dx - s\,dy) + i \int_C (u\,dy + s\,dx).$$ (2)

Cauchy's theorem says that both parts are zero if f is an analytic function in a region containing C. The only qualification is that the region should be simply connected—in other words, without holes.

4T   *Cauchy's Theorem*: If f is analytic in a simply connected region containing the closed curve C, then

$$\int_C f(z)\,dz = 0.$$ (3)

Holes are excluded because otherwise the theorem would fail. The function $1/z$ is analytic in the region between $|z| = 1$ and $|z| = 3$, but its integral around the closed circle $|z| = 2$ is not zero. The correct value is $2\pi i$, because of the singularity at $z = 0$, and we cannot permit that point to be hidden inside a hole.

To prove Cauchy's theorem we need to use the hypothesis that f is a smooth and *single-valued* function of the combination $z = x + iy$. This excludes not only $f = 1/z$, $f = 1/z^2$, ..., which are not smooth, but also all fractional powers $f = z^\alpha$ (if the curve encloses $z = 0$). The square root of z changes sign as z circles the origin, and z^α goes from r^α to $r^\alpha e^{2\pi i\alpha}$. If α is not an integer then $e^{2\pi i\alpha} \neq 1$ and f is not single-valued. We mentioned earlier that a convergent power series produced an analytic f, and vice versa, but the most convenient property is even simpler: if f is analytic then u and s satisfy the Cauchy-Riemann equations. From these equations we now show that the integral (3) is zero.

One proof uses Stokes' theorem, that the integral around C equals an integral over the set inside it: for any v_1 and v_2,

$$\int_C v_1 \, dx + v_2 \, dy = \iint_S \text{curl } v \, dx \, dy = \iint_S \left(\frac{\partial v_2}{\partial x} - \frac{\partial v_1}{\partial y} \right) dx \, dy. \tag{4}$$

Applied to $v_2 = u$ and $v_1 = s$ the last integral is zero, because

$$\frac{\partial u}{\partial x} - \frac{\partial s}{\partial y} = 0 \quad \text{(by Cauchy-Riemann).}$$

Applied to $v_1 = u$ and $v_2 = -s$ the other part of (2) is also zero, because

$$-\frac{\partial s}{\partial x} - \frac{\partial u}{\partial y} = 0 \quad \text{(again by Cauchy-Riemann).}$$

Thus the real and imaginary parts both vanish in (2), proving that $\int f(z)dz = 0$. This is Cauchy's theorem.†

The important point is that the integral of $f(z)$ from P to Q does not depend on the path. It depends only on P and Q. If we come back from Q to P, that cancels the part from P to Q and the integral around the closed loop is zero. The condition for a line integral to depend only on its endpoints is simple and very useful: *the integrand should be an exact differential*:

$$v_1 = \frac{\partial U}{\partial x} \quad \text{and } v_2 = \frac{\partial U}{\partial y} \quad \text{for some function } U(x, y). \tag{5}$$

In that case $v_1 \, dx + v_2 \, dy$ equals dU, the derivative of U. Its integral is $U(Q) - U(P)$ and obviously depends only on the endpoints.

The test for (5) is again $\partial v_1/\partial y = \partial v_2/\partial x$. (The cross derivatives U_{xy} and U_{yx} must be equal.) Therefore the integral of $f(z)dz = f(z)dx + if(z)dy$ depends only on the endpoints, provided this test is passed by $v_1 = f$ and $v_2 = if$:

$$\frac{\partial f}{\partial y} = i \frac{\partial f}{\partial x}. \tag{6}$$

But this is the fundamental equation for analytic functions of $x + iy$. The integral of $f(z)dz$ depends only on the endpoints P and Q, and around a closed curve ($P = Q$) it is zero.

---

†We should have mentioned earlier the pronunciation Koshi (French not Japanese!).

To tell the truth, this whole discussion is really about Kirchhoff's voltage law. It had three forms in the continuous case (Section 3.3) and we have used them all. Either the integral around every closed loop is zero, or the curl $\partial v_2/\partial x - \partial v_1/\partial y$ is zero, or the flow comes from a potential U.

Cauchy's Integral Formula

Now we allow a function $f = 1/(z - a)$ that is analytic at most points but not all. Imagine a circle around the point $z = a$ (where f blows up). If the circle has radius r then its typical point is $z = a + re^{i\theta}$, and the integral around the circle can be computed explicitly:

$$\int \frac{dz}{z-a} = \int_0^{2\pi} \frac{ire^{i\theta}d\theta}{re^{i\theta}} = \int_0^{2\pi} id\theta = 2\pi i. \tag{7}$$

The answer is $2\pi i$, regardless of the radius r. In fact the answer is $2\pi i$, **regardless of the shape of the path.** For any simple closed curve C around a, circular or not, the integral is

$$\boxed{\int_c \frac{dz}{z-a} = 2\pi i.} \tag{8}$$

Proof Choose a circle around a that crosses the path C. The integral around the circle is certainly $2\pi i$. Since the integral forward around the path and backward around the circle is zero, (8) is proved. Note that Cauchy's theorem $\int f(z)dz = 0$ applies to this region between the path and the circle, as long as the singular point $z = a$ is somewhere else (Fig. 4.20).

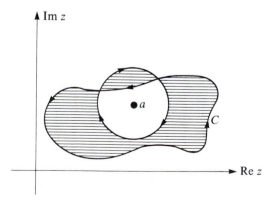

Fig. 4.20. The integral of $\dfrac{1}{z-a}$ is $2\pi i$.

This is a point to emphasize: *Paths can be moved as long as they avoid singularities.* A circle around a can be changed to a larger circle—but it cannot change to a circle that does not enclose $z = a$. The former integral gives $2\pi i$, the latter gives zero, and to change one circle into the other it would have to pass through a.

Probably (8) is the most important integral in complex analysis. It applies not only to the particular function $1/(z-a)$, but to any analytic function f divided by $z - a$. We lose analyticity by this division, but we recover it again by changing to

$$F(z) = \frac{f(z) - f(a)}{z - a}.$$

The numerator is also zero at $z = a$, and the ratio F is completely regular. Therefore we can integrate this analytic F and obtain zero. The two parts of F give the same integral, and since $f(a)$ is constant the integral is easy:

$$\int \frac{f(z)dz}{z - a} = \int \frac{f(a)dz}{z - a} = 2\pi i f(a). \tag{9}$$

That is **Cauchy's integral formula** for the value of f at a point:

4U   If a is any point inside the simple closed curve C, and f is analytic inside and on C, then

$$f(a) = \frac{1}{2\pi i} \int_C \frac{f(z)dz}{z - a}. \tag{10}$$

To know f on C is to know f everywhere inside; there is not much freedom for analytic functions. Equation (8) was the special case $f = 1$. This formula for $f(a)$ leads directly to a **maximum principle**:

If f is analytic then the maximum of $|f(z)|$ is reached on the boundary of S.

For proof let $z = a + re^{i\theta}$ travel on a small circle around any point a. The value of f at the center is the average of f around the circle, since (10) gives

$$f(a) = \frac{1}{2\pi} \int_0^{2\pi} f(z)d\theta. \tag{11}$$

Therefore $|f(a)|$ cannot be larger than all the values $|f(z)|$ on the circle. The maximum cannot occur at an interior point unless f is constant and the maximum occurs everywhere. The same is true for any solution $u(x, y)$ of Laplace's equation; it is the real part of some f and we may take the real part of both sides in (11). Then $u(a)$ is the average of u around the circle. The maximum of u must be found at a boundary point.

Cauchy's formula (10) is like the earlier representation of $h(x, y)$ as a combination of delta functions. Here f is a combination of the functions $1/(z - a)$. We can regard a as variable, since the formula holds for every a, and then the derivative of (10) is

$$f'(a) = \frac{1}{2\pi i} \int_c \frac{f(z)dz}{(z - a)^2}. \tag{12}$$

Higher derivatives are also possible, but this already shows that very few analytic functions stay bounded.

If f is analytic and bounded in the whole plane, it must be a constant.

Proof   Let C be a circle around a of radius R. Its length is $2\pi R$ and $1/(z - a)^2$ has magnitude $1/R^2$. Thus the integral in (12) is less than $\max |f|/R$ for any R. Therefore the integral must be zero, so $f' = 0$ from (12) and f is a constant.

This is Liouville's theorem. Analytic functions cannot do what they want. On the real line a function like $1/(1 + x^2)$ can stay between 0 and 1. In the complex plane that cannot happen. The function $f(z) = 1/(1 + z^2)$ must have singularities somewhere, and they are on the imaginary axis at $z = \pm i$.

Our last application of complex analysis will put these singularities to use.

Singularities and Residues

Integrating an analytic function around a closed path gives $\int f(z)dz = 0$. If f has a singularity inside the loop, this integral is no longer expected to vanish. But the loop can move close to the singularity without changing the integral. The first figure shows a path being moved from $|z| = 1$ to a small circle $|z| = r$, when the only singularity is at the origin. The integral around this path C is zero (no poles are inside). The integrals up and down the connectors will cancel (when they are moved close together). Therefore going forward around the big circle and backward around the small circle gives zero; the radius of the circle does not matter.

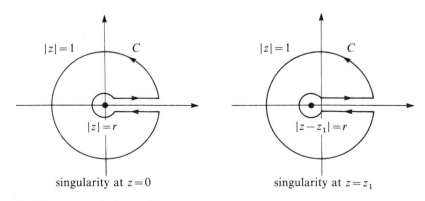

Fig. 4.21. Paths around singularities.

The purpose is this. We want to compute definite integrals like

$$I = \int_0^{2\pi} \cos^2\theta \, d\theta \quad \text{and} \quad J = \int_0^{2\pi} \frac{d\theta}{20 + 2\cos\theta}.$$

One is easy, the other less so. But both can be done without knowing the functions whose derivatives are $\cos^2\theta$ and $(20 + 2\cos\theta)^{-1}$. These are real integrals, but a change to $z = e^{i\theta}$ sends them around the unit circle. That means a conversion from θ to z:

$$\cos\theta = \frac{1}{2}(e^{i\theta} + e^{-i\theta}) = \frac{1}{2}\left(z + \frac{1}{z}\right), \quad dz = ie^{i\theta}d\theta, \quad d\theta = \frac{dz}{iz}.$$

Then the two integrals are

$$I = \int \frac{1}{4}\left(z + \frac{1}{z}\right)^2 \frac{dz}{iz} \quad \text{and} \quad J = \int \frac{dz}{iz(20 + z + z^{-1})}.$$

The key idea is to **look only at their singularities**. The first has a singularity at $z = 0$ and nowhere else:

$$I = \frac{1}{4i}\int\left(z + \frac{2}{z} + \frac{1}{z^3}\right) dz. \tag{13}$$

The integral of z is automatically zero, because z is analytic. The integral of $1/z^3$ is accidentally zero, in spite of the singularity. The integral of $1/z$ is not zero; it is $2\pi i$. Therefore the whole result depends only on the number $2/4i$ that multiplies $1/z$:

$$I = \frac{2}{4i}2\pi i = \pi.$$

In the language of "residues," our function has a residue of $2/4i$ at $z = 0$, and the integral is $2\pi i$ times the residue.

The other integral J will make this point more clearly. It does not have a singularity at $z = 0$, because z^{-1} cancels z. However the denominator does vanish when

$$20z + z^2 + 1 = 0, \quad \text{or} \quad z = -10 \pm \sqrt{99}.$$

These are the two singularities, z_1 and z_2, at which the integrand blows up. The one at $z_2 = -10 - \sqrt{99}$ is outside the unit circle and does not affect the integral. The one at $z_1 = -10 + \sqrt{99}$ is inside, and the path can go around z_1 instead of the origin (Fig. 4.21). Integrating $1/(z - z_1)$ will lead to the usual result $2\pi i$, and our whole problem is *to find the residue*—the number that multiplies this singular part $1/(z - z_1)$.

For that we write the quadratic as

$$\frac{1}{i(20z + z^2 + 1)} = \frac{1}{i(z - z_1)(z - z_2)}.$$

When z is close to z_1, the factor $z - z_2$ is close to $z_1 - z_2 = 2\sqrt{99}$. That is the reason for moving the path of integration; we can concentrate on z near z_1. The residue we want is $1/i2\sqrt{99}$, and the integral itself is

$$J = \frac{1}{i2\sqrt{99}} 2\pi i = \frac{\pi}{\sqrt{99}}.$$

From these two examples we now draw the general rule.

In both cases the singularity was a **pole**. For J it was a simple pole (a pole of order 1) and for I it was a pole of order 3. Approaching a pole from any direction makes the function blow up, as it did for $1/z$ or $1/z^3$ or $1/(z - z_1)$. With a pole at $z = a$ the function looks like

$$f(z) = f_0(z) + \frac{c_1}{z - a} + \frac{c_2}{(z - a)^2} + \cdots + \frac{c_N}{(z - a)^N} \tag{14}$$

where f_0 is analytic at a. The last power N is the order of the pole. However it is the coefficient c_1 that decides the integral—since all higher powers give $\int dz/(z - a)^k = 0$. The integral of f around a reduces to $2\pi i$ times c_1:

$$\boxed{\int f(z)dz = \int \frac{c_1}{z - a} = 2\pi i c_1.}$$

This coefficient c_1 is the residue. It is a real or complex number, and in the examples it was $2/4i$ and $1/i2\sqrt{99}$.

4V   If the path C encloses one or more poles of f then the integral is $2\pi i$ times the sum of the residues at those poles:

$$\int_C f(z)dz = 2\pi i[\text{Res}(a_1) + \cdots + \text{Res}(a_r)]. \tag{15}$$

At a simple pole the residue is given by

$$c_1 = \text{Res}(a) = \lim_{z \to a} (z - a)f(z). \tag{16}$$

This is the residue theorem. It is proved by changing C to a path that circles close to the poles. They key formula (16) for the residue comes from (14); with $N = 1$ (a simple pole) we multiply through by $z - a$ and then c_1 appears as z approaches a.

EXAMPLE 1 The residue of $\dfrac{3z}{z^2 + 1}$ at the pole $z = i$ is

$$\lim_{z \to i} \frac{3z(z - i)}{(z - i)(z + i)} = \frac{3}{2}.$$

EXAMPLE 2 The residues of $\dfrac{z + 5}{\sin z}$ at the poles $z = 0$ and $z = \pi$ are

$$\lim_{z \to 0} \frac{(z + 5)z}{\sin z} = 5 \quad \text{and} \quad \lim_{z \to \pi} \frac{(z + 5)(z - \pi)}{\sin z} = -(\pi + 5).$$

These limits can be found by l'Hôpital's rule: When numerator and denominator go to zero, take the ratio of their derivatives. As $z \to \pi$ in Example 2, the ratio $(2z + 5 - \pi)/\cos z$ approaches $-(\pi + 5)$.

EXAMPLE 3 The residue of $\dfrac{e^z}{z^2}$ is not found from (16) because $z = 0$ is a double pole. Instead we can look directly for the coefficient of $1/z$:

$$\frac{e^z}{z^2} = \frac{1}{z^2}\left(1 + z + \frac{z^2}{2} + \cdots\right) \quad \text{has residue } c_1 = 1.$$

Alternatively we can find a formula for the residue at a pole of order N. The pole is removed when we multiply (14) by $(z - a)^N$, but now the crucial number c_1 is combined with $(z - a)^{N-1}$. It takes $N - 1$ derivatives to separate it out:

$$c_1 = \text{Res}(a) = \lim_{z \to a} \frac{1}{(N - 1)!} \frac{d^{N-1}}{dz^{N-1}} [(z - a)^N f(z)]. \tag{17}$$

For $N = 1$ this is the same formula as (16). We emphasize that these formulas apply only to *poles*, and not to an **essential singularity**—like the one at $z = 0$ for

$$e^{1/z} = 1 + \frac{1}{z} + \frac{1}{2z^2} + \frac{1}{6z^3} + \cdots.$$

The residue is 1, which multiplies $1/z$— but where the expansion (14) stopped at a finite power $-N$, this one goes forever. (It is a *Laurent expansion*, with all positive and negative powers allowed.) For z near zero, the function $e^{1/z}$ is extremely erratic. Approached from the positive side, $e^{1/z}$ goes to $+\infty$; from the negative side it goes to zero; in the imaginary direction it is $e^{1/iy}$ and goes faster and faster around the unit circle. The only value that is never attained is $e^{1/z} = 0$.

A third type of singularity occurs when a function is not single-valued. The example $f = \sqrt{z}$ has a **branch point** at $z = 0$. As you circle that point the square root $e^{i\theta/2}$ changes from $e^{i0} = 1$ to $e^{i\pi} = -1$. Riemann would say that you have gone to another branch—which would not happen on a small circle around a point like $z = 4$. That path stays on one branch and the square root stays near 2 (or near -2). Somehow the two branches are joined at $z = 0$ in such a way that each trip around the origin starts on one branch and ends on the other. There is a "two-sheeted Riemann surface," like a parking garage that brings you back to the ground after two loops.

The function $(z - i)^{1/3}$ has three sheets joined at the branch point $z = i$. The function $z^{1/2}(1 - z)^{1/3}$ needs six sheets, joined in some unbelievable way at 0 and 1. On the other hand $z^{1/2}(1 - z)^{1/2}$ has only two possible values and two sheets; a wide circle around both branch points leaves you on the same sheet. The function $\log z$ has a branch point at $z = 0$, since $\log r + i\theta$ increases by $2\pi i$ as θ goes from 0 to 2π, but now it needs *infinitely many sheets*. There are infinitely many values for the logarithm; $\log 1$ can be 0 or $2\pi i$ or any multiple of $2\pi i$, since $e^{n2\pi i} = 1$. Similarly the functions z^π and z^i have infinitely many branches, because $e^{\pi \log z}$ and $e^{i \log z}$ have infinitely many values. From these garages the only way out is to turn around.

The residue theorem does not apply at branch points. However the path can still be moved near them, through a region where f is analytic. The novelty is that the integrals along the connectors no longer cancel since f jumps to a new value after circling the branch point. As an example we compute

$$K = \int_0^\infty \frac{\sqrt{x}}{x^2 + 1}\, dx$$

by using the odd-shaped path in Fig. 4.22 below. The integral we want comes from the straight part 1 as R goes to infinity. The integral along part 3 is *another copy* of K; it does not cancel because $\sqrt{z}$ changed sign when z went around 0. The integral on part 2 approaches zero, since the path has length $2\pi R$ and $\sqrt{z}/(z^2 + 1)$ has magnitude approximately $\sqrt{R}/R^2$. (The product with $2\pi R$ is small.) Finally part 4 of the path introduces the residues at the two poles. Factoring $z^2 + 1$ into $(z - i)(z + i)$, (16) gives

$$2\pi i\,\mathrm{Res}(i) = \lim_{z \to i} \frac{2\pi i \sqrt{z}}{z + i} = \pi\sqrt{i} \quad \text{and} \quad 2\pi i\,\mathrm{Res}(-i) = \lim_{z \to -i} \frac{2\pi i \sqrt{z}}{z - i} = -\pi\sqrt{-i}.$$

The whole integral is zero because $\sqrt{z}/(z^2 + 1)$ is analytic inside the loop:

$$0 = \int_1 + \int_2 + \int_3 + \int_4 \to K + 0 + K - \pi\sqrt{i} + \pi\sqrt{-i}.$$

Since the square root of $i = e^{\pi i/2}$ is $e^{\pi i/4}$, and similarly for $-i = e^{3\pi i/2}$, we have

$$2K = \pi e^{\pi i/4} - \pi e^{3\pi i/4} \quad \text{or} \quad K = \pi \cos\frac{\pi}{4} = \frac{\pi}{\sqrt{2}}.$$

I was very relieved that K turned out to be real.

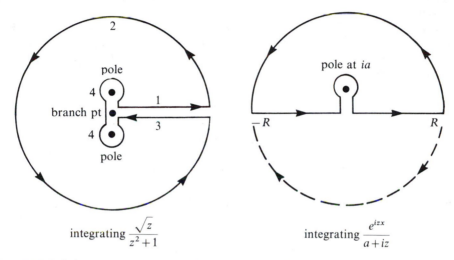

integrating $\dfrac{\sqrt{z}}{z^2 + 1}$ integrating $\dfrac{e^{izx}}{a + iz}$

Fig. 4.22. Infinite contours as $R \to \infty$: branch point and poles.

Fourier Integrals and Laplace Integrals

An important application of complex integration is to compute (or to invert) a transform. The section on Fourier integrals was typical; it was seldom that we could test the inversion formula by going backwards. It is even more typical to have no way of going forward either; most infinite integrals $\int f(x)e^{-ikx}\,dx$ are impossible. But it is exactly the fact that the limits *are* infinite—the endpoints are $-\infty$ and $+\infty$—that gives complex residue methods a chance. We illustrate with the inverse transform of $\hat{f} = 1/(a + ik)$:

$$f(x) = \frac{1}{2\pi} \int_{k=-\infty}^{\infty} \frac{e^{ikx}}{a + ik}\,dk. \tag{18}$$

This combines harmonics of all frequencies and should give back a decaying pulse.

The interesting point is that this does not seem to be in a standard table of integrals. Nevertheless it can be done. The idea is to change k to z and go around the closed semicircle in Fig. 4.22b. The integral on the straight part from $-R$ to R

approaches (18). If the integral on the circular part approaches zero, then the answer comes from the residue at the pole $z = ai$:

$$f(x) = \frac{2\pi i}{2\pi}\,\text{Res}(ai) = i\lim_{z \to ai}\frac{e^{izx}(z - ai)}{a + iz} = e^{iaix} = e^{-ax}.$$

That is our exponential pulse. However it is correct only for positive x. For negative x we should recover $f(x) = 0$, since the pulse was one-sided. What is different about the integral when x is negative?

In a word, the semicircle above the horizontal axis is the wrong path. The function e^{izx} was small on that semicircle for positive x, but for negative x it is enormous. At the point $z = iR$ on the circle, e^{izx} equals e^{-Rx}. For $x < 0$ this is exponentially large and the complex variable method falls apart—unless we change to the semicircle *below* the horizontal. On that path the function is small for negative x. Furthermore there is no pole inside that semicircle! The pole is still at $z = ia$, above the horizontal. The integral of an analytic function (with no poles) is zero. As $R \to \infty$ in the dotted semicircle we finally obtain 0 in (18):

$$f(x) = e^{-ax} \text{ for } x > 0 \quad \text{and} \quad f(x) = 0 \text{ for } x < 0.$$

The Fourier inversion formula is confirmed. We mention that the semicircular integrals approach zero because of the term $|e^{izx}| = e^{-xR\sin\theta}$. The path length πR nullifies $1/|a + iz| = O(1/R)$, leaving $\int e^{-xR\sin\theta}\,d\theta$ to approach zero. Without this help from the exponential, $\int dz/(a + iz)$ would diverge.

Notice something remarkable as $a \to 0$. The pole approaches the origin, and in the limit it falls right onto the path. That is unacceptable. Therefore the path makes a small semicircle around 0, and because it is a semicircle we take only *half* of the residue.† The integral for $a = 0$ is $+\frac{1}{2}$ for positive x and $-\frac{1}{2}$ for negative x, and we recover $f(x) = \frac{1}{2}$ (sign function).

We turn now to *Laplace transforms*. Their purpose is to solve initial-value problems, so a proper introduction will come in Section 6.3. Two remarks are sufficient: First, they are very close to the Fourier transform, and second, they are not easy to invert. It is the second point which brings them into this chapter. Their inversion formula is a rather mystical integral, the only one you ever see in which the path of integration is *the imaginary axis*. The integral goes from $z = -i\infty$ to $z = +i\infty$. It travels up the complex plane instead of across. Often it moves onto a line that is parallel to the imaginary axis, from $a - i\infty$ to $a + i\infty$, in order to keep all poles to the left. We want to show how such a path can appear, and how the residues at the poles can yield the integral.

First we define the Laplace transform $F(s)$. It starts with a function $f(t)$, coming from an initial-value problem. For negative t this function is *identically zero*. For positive t it is often a "transient" that decays exponentially fast; it might be e^{-at}.

---

† We could go back to (7) and get πi instead of $2\pi i$.

Then the Laplace transform of f is a function of s:

$$F(s) = \int_0^\infty f(t)e^{-st}\, dt. \tag{19}$$

It is the Fourier transform except that the imaginary number ik has been replaced by the real number s. In other words, when the frequency k is chosen to be $-is$ *the Fourier transform $\hat{f}(-is)$ equals the Laplace transform $F(s)$.*

The Laplace transform in the case $f = e^{-at}$ is the integral of $e^{-(a+s)t}$, which means division by $a + s$:

$$F(s) = \int_0^\infty e^{-(a+s)t}\, dt = \left[\frac{e^{-(a+s)t}}{-(a+s)} \right]_{t=0}^\infty = \frac{1}{a+s}.$$

At $t = 0$ the exponential equals 1. At $t = \infty$ it has dropped to zero. You should notice that this function f is our *one-sided pulse*. It jumps to 1 at $t = 0$ and then decays. Its Fourier transform—the example on which Section 4.3 was based—was $1/(a + ik)$. Its Laplace transform must therefore be $1/(a + s)$, since s has replaced ik.

The Laplace inversion formula copies the Fourier formula, with a 90° twist. To reconstruct f from its harmonics Fourier used (with t instead of x)

$$f(t) = \frac{1}{2\pi} \int_{-\infty}^\infty \hat{f}(k)e^{ikt}\, dk.$$

In the Laplace transform ik becomes s. Therefore s goes up the imaginary axis from $-i\infty$ to $+i\infty$, and $ds = idk$. With that simple but inscrutable change of variables we reach the Laplace inversion formula:

$$f(t) = \frac{1}{2\pi i} \int_{-i\infty}^{i\infty} F(s)e^{st}\, ds. \tag{20}$$

Our purpose is to study typical examples. Think of s as z.

EXAMPLE 1 $F(s) = \dfrac{1}{a+s}$ and $f(t) = \dfrac{1}{2\pi i} \displaystyle\int_{-i\infty}^{i\infty} \dfrac{e^{st}}{a+s}\, ds$

The inverse transform f should be the one-sided decaying pulse. Complex methods require the path up the imaginary axis to be closed by a semicircle, with radius $R \to \infty$. We choose the semicircle to the *left* of the imaginary axis, where the real part of s is negative and e^{st} is small. Then the integral is given by the residue at the pole $s = -a$:

$$\text{Res}(-a) = \lim_{s \to -a} \frac{s+a}{s+a}e^{st} = e^{-at}.$$

The factor $2\pi i$ that multiplies residues will cancel the factor $2\pi i$ in the inversion formula, and we recover the pulse $f = e^{-at}$.

EXAMPLE 2 $F(s) = \dfrac{1}{(a+s)^2}$ and $f(t) = \dfrac{1}{2\pi i} \displaystyle\int_{-i\infty}^{i\infty} \dfrac{e^{st}}{(a+s)^2}\, ds$

The path is closed by the same semicircle and the only novelty is the double pole. In principle the residue is the coefficient of $1/(a+s)$, which might seem to be zero for $e^{st}/(a+s)^2$. But remember that everything—including the exponential—must be expanded in powers of $a + s$:

$$\frac{e^{st}}{(a+s)^2} = \frac{e^{-at}e^{(a+s)t}}{(a+s)^2} = \frac{e^{-at}}{(a+s)^2}\left[1 + t(a+s) + \frac{t^2(a+s)^2}{2!} + \cdots\right].$$

The coefficient of $1/(a+s)$ is the residue te^{-at}, and this is $f(t)$. It also comes from the double pole formula (17), as the derivative of e^{zt} at $z = -a$.

The special case $a = 0$ is important. In the first example $F = 1/s$ corresponds to $f = 1$. In the second example $F = 1/s^2$ corresponds to $f = t$. In general $F = 1/s^{n+1}$ is the Laplace transform of $f = t^n/n!$ Those have poles at $s = 0$, on the path of integration. However a shift to the vertical line from $a - i\infty$ to $a + i\infty$, with the real number a chosen large enough, leaves all poles to the left. The semicircle will contain the poles, and the residues add to $f(t)$. That completes the inversion of F.

In Chapter 6 the Laplace transform solves initial-value problems in the same way that the Fourier transform solves boundary-value problems. All functions are decomposed into their frequency components; then the differential equation falls apart. You have to accept e^{-st} as if it were a harmonic (it is, but the frequency is imaginary) and superposition gives the answer. The final solution is a transient from Laplace plus a steady state from Fourier.

EXERCISES

4.5.1   If the line integral along C from P to Q depends only on P and Q, why does the case $Q = P$ (a closed loop) give

$$\int_C = \int_{-C} \quad \text{and} \quad \int_C = -\int_{-C} \quad \text{so} \quad \int_C = 0?$$

4.5.2   Compute the following integrals:

(a) $\int dz/z$ from 1 to i, the short way and long way on the circle $z = e^{i\theta}$
(b) $\int x\, dz$ around the unit circle, where $x = \cos\theta$ and $z = e^{i\theta}$—or alternatively where $x = \frac{1}{2}(z + z^{-1})$

(c) $\int dz/z$ around the circle $|z - 2i| = 1$

(d) $\int y^2 \, dx + 2xy \, dy$ from $P = (0, 0)$ to $Q = (1, 1)$, noticing the exact differential (of what function U?)

(e) $\int dz/z$ for a path that winds three times around $z = 0$. (A simple closed curve winds only once.)

4.5.3 (a) Compute $\int dz/z^2$ around the circle $z = re^{i\theta}$, by substituting for z and dz and integrating directly.

(b) Despite the pole at $z = 0$ this integral is zero. What is the residue of $1/z^2$ at the pole?

(c) Why is $\int dz/z^2$ also zero around circles that are not centered at the origin?

4.5.4 Draw two circular disks in the complex plane, one not containing the origin and the other one centered at $z = 0$. In the first disk, mark the points where the absolute values $|z^2|$ and $|1/z^2|$ attain a maximum. In the second, mark the points where $|z^2|$ and the real part $x^2 - y^2$ and the imaginary part $2xy$ attain a maximum. Where does $|1/z^2|$ attain a maximum in the second disk?

4.5.5 If $f(z) = z^2$ on the circle $z = a + re^{i\theta}$ around the point a, substitute directly into Cauchy's integral formula (10) and show that it correctly gives $f(a) = a^2$.

4.5.6 Show that Cauchy's integral formula (10) for $f(a)$ reduces to the average value (11) at the center of a circle. What is the average value of e^z around the unit circle?

4.5.7 Find the location of the poles, and the residues, for

(a) $\dfrac{1}{z^2 - 4}$ (b) $\dfrac{z + 3}{z - 3}$ (c) $\dfrac{1}{(z^2 - 1)^2}$

(d) $\dfrac{e^z}{z^3}$ (e) $\dfrac{1}{1 - e^z}$ (f) $\dfrac{1}{\sin z}$

4.5.8 Evaluate the following integrals around the unit circle:

(a) $\displaystyle\int \dfrac{dz}{z^2 - 2z}$ (b) $\displaystyle\int \dfrac{e^z \, dz}{z^2}$ (c) $\displaystyle\int \dfrac{dz}{\sin z}$

4.5.9 By complex integration compute the real integrals

(a) $\displaystyle\int_0^{2\pi} \cos^4\theta \, d\theta$ (b) $\displaystyle\int_0^{2\pi} \dfrac{d\theta}{a + \cos \theta}, \quad a > 1$ (c) $\displaystyle\int_0^{2\pi} \cos^3\theta \, d\theta$

4.5.10 Find the poles above the real axis and evaluate

(a) $\displaystyle\int_{-\infty}^{\infty} \dfrac{dx}{(1 + x^2)^2}$ (b) $\displaystyle\int_{-\infty}^{\infty} \dfrac{dx}{4 + x^2}$ (c) $\displaystyle\int_{-\infty}^{\infty} \dfrac{dx}{x^2 - 2x + 3}$

4.5.11 Find all the poles, branch points, and essential singularities of

(a) $\dfrac{1}{z^4 - 1}$ (b) $\dfrac{1}{\sin^2 z}$ (c) $\dfrac{1}{e^z - 1}$ (d) $\log(1 - z)$

(e) $\sqrt{4 - z^2}$ (f) $z \log z$ (g) $e^{2/z}$ (h) $\dfrac{e^z}{z^e}$

The point $z = \infty$ can be included by setting $w = 1/z$ and studying $w = 0$. Thus $z^3 = 1/w^3$ has a triple pole at $z = \infty$, $e^z = e^{1/w}$ has an essential singularity, and $\log z = -\log w$ has a branch point.

4.5.12 Find residues at $z = \pm i$ and use Fig. 4.22 to show that

$$\int_{-\infty}^{\infty} \frac{dx}{1 + x^2} = \pi \quad \text{and} \quad \int_{-\infty}^{\infty} \frac{e^{ikx} dx}{1 + x^2} = \pi e^{-|k|}.$$

4.5.13 Invert the following Laplace transforms to find $f(t)$ for $t \geq 0$:

(a) $F(s) = \dfrac{1}{s^2 + 1}$ (b) $F(s) = \dfrac{s}{s^2 + 1}$ (c) $F(s) = \dfrac{1}{(a + s)^3}$ (a triple pole)

CHAPTER 4 IN OUTLINE: ANALYTICAL METHODS

4.1 Fourier Series and Orthogonal Expansions—e^{ikx} as an eigenfunction of d/dx
The Fourier Coefficients—formulas for a_k, b_k, and c_k
Examples of Fourier Series—the coefficients for $f = \delta$ and $f = x$
Sine Series and Cosine Series—odd and even functions from 0 to π
Properties of Fourier Series—least squares approximations in Hilbert space
Solution of Laplace's Equation—$\sum a_k r^k \cos k\theta$ and Poisson's formula
Orthogonal Functions—the expansion $f = c_0 T_0 + c_1 T_1 + \cdots$
Bessel Functions—oscillations of a circular drum

4.2 Discrete Fourier Series and Convolution—the Fourier matrix F
The Discrete Transform for Arbitrary n—F^{-1} is $\bar{F}/n$
Discrete Convolutions—convolution rule and circulants $C = F\Lambda F^{-1}$
Signal Processing—convolution matrices $A_{ij} = a_{i-j}$

4.3 Fourier Integrals—the transform from f to $\hat{f}$ and its inverse
A List of Essential Transforms—delta functions, pulses, step functions
Energy and the Uncertainty Principle—energies $\int |f|^2 = 2\pi \int |\hat{f}|^2$
Derivatives, Integrals and Shifts—transforms of f', $\int f$, $f(x - d)$, $e^{ixd} f$
Convolution and Green's Functions—$\hat{u} = \hat{G}\hat{h}$ and fundamental solutions
Integral Equations—convolution kernels $K(x - y)$, solution by transform
The Sampling Theorem—band-limited f sampled at the Nyquist rate

4.4 Complex Variables and Conformal Mapping—the z-plane and w-plane
Analytic Functions and Laplace's Equation—Cauchy-Riemann equations
Conformal Mapping—boundary change preserving Laplace's equation
Important Conformal Mappings—e^z, $(az + b)/(cz + d)$, $\frac{1}{2}(z + z^{-1})$
Two-dimensional Fluid Flows—$f = u + is$: potential and stream function
Green's Functions and Electric Fields—superposition of point sources

4.5 Complex Integration—Cauchy's theorem $\int f(z)dz = 0$
Cauchy's Integral Formula—maximum principle, $2\pi i f(a) = \int f(z)dz/(z - a)$
Singularities and Residues—real integrals by complex methods
Fourier Integrals and Laplace Integrals—transforms computed from poles

5

NUMERICAL METHODS

LINEAR AND NONLINEAR EQUATIONS ■ 5.1

From its title, this chapter should study computational methods for at least the four great problems of numerical analysis:

1. equilibrium equations
2. eigenvalue problems
3. initial-value problems
4. optimization with or without constraints.

But we are only halfway in the book! And, to tell the truth, the first two problems provide more than enough to do. Therefore the initial-value problem will be studied numerically as well as analytically in Chapter 6, and algorithms for optimization will accompany the theory in Chapters 7 and 8. This leaves some numerical problems untouched, but it is sufficient for a very wide range of applications.

We emphasize the classical problems $Ax = b$ and $Ax = \lambda x$, governed by symmetric matrices. They are so important, and they come in so many forms, that the first question is one of organization. To make this chapter coherent we distinguish three types of algorithms: *direct*, *semi-direct*, and *iterative*. The first work directly with the matrix A, hammering it into shape. The second require only multiplications Av, for different vectors v; semi-direct methods aim to avoid storing and manipulating all of A. If the matrix is sparse then Av can be formed very quickly. The correct answer is eventually reached, but even when semi-direct methods are slower they have an important advantage: They give a good and frequently sufficient answer before the end. It is useless to stop halfway in Gaussian elimination, but in the middle of the conjugate gradient method you have something useful.

The methods in the third class, the iterative methods, try to avoid A as much as possible. They may take only its main diagonal D (not to be confused with the pivots) and solve $Dx_{k+1} = (D - A)x_k + b$. Starting from any initial guess x_0, this provides a sequence of approximations $x_1, x_2, \ldots$, none of which will satisfy $Ax = b$. But the limit of that sequence (if it has a limit!) does satisfy the original equation: $Dx_\infty = (D - A)x_\infty + b$ is the same as $Ax_\infty = b$. Thus the limit of an iterative method is the correct solution. If some other matrix M replaces D (as it will) then $Mx_{n+1} = (M - A)x_n + b$ is solved at every step. M tries to make this step easy and at the same time to be near A.

For iterative methods the test for convergence applies to the eigenvalues of $M^{-1}A$. They must lie in a unit circle around 1. For semi-direct methods a preconditioner M is also important and $M^{-1}A$ should again be near the identity, but the word "near" will have a different meaning. The essential thing is that M can give tremendous acceleration, and at this moment the search is on—every numerical analyst has a favorite preconditioner and you have a perfect chance to find a better one.

For the eigenvalue problem it must be admitted that no method can be totally direct: The eigenvalues cannot be found in a finite number of steps. Nevertheless the matrix can be made tridiagonal and the methods are closely parallel to those for $Ax = b$. We classify them below in a very rough table. The QR method is superior for most ordinary problems—it is astonishingly good—but for large sparse matrices other algorithms may win.

	$Ax = b$	$Ax = \lambda x$
direct	elimination: square A (or $A^T C A$) orthogonalization: rectangular A	tridiagonalization followed by shifted QR
semi-direct	conjugate gradients (preconditioned)	Lanczos algorithm
iterative	overrelaxation, multigrid, ...	power methods

Most of these methods succeed by constructing vectors that are orthogonal. Therefore Section 5.2 begins with this most desirable case. We write Q instead of A when the columns are orthonormal, and we solve $Qx = b$. The solution is exact when Q is square and it is a projection when Q is rectangular; in both cases it is $x = Q^T b$. Then the real problem is to achieve orthogonality when it is not present at the beginning. We construct Q from A in three ways—by Gram-Schmidt and Householder in 5.2 and by Lanczos in 5.3. Those lead directly to the methods in the table. In the applications to equilibrium and oscillation, the fundamental matrix is not A but $A^T C A$.

The table does not mention special direct algorithms for special matrices. The most important is the *Fast Fourier Transform*, to be explained in Section 5.5. It leads to fast approximations to partial differential equations (spectral methods) provided the geometry allows Fourier analysis to operate. The key is to multiply Fx or $F^{-1}b$ as quickly as possible, using the properties of the Fourier matrix F.

Normally this needs n^2 separate multiplications but the FFT uses only $\frac{1}{2}n \log n$—an improvement that totally revived problems which seemed to be computationally hopeless. The applications go far beyond Laplace's equation, to the digital processing of images and speech and music and radar.

Finally we come to the numerical solution of differential equations. This is at the center of scientific computing. It starts from a continuous problem and must end with a discrete algorithm. The crucial step is the choice of approximation, and there are two leading contenders—*finite difference methods* and *finite element methods*. For equilibrium problems we give first place to finite elements; for initial-value problems we emphasize finite differences. Section 5.4 describes the solution by finite elements of Poisson's equation, and it overcomes the geometrical limitations of complex variables and Fourier series. The same principle, using trial functions and test functions that are polynomials within each element, extends to all other differential equations. This principle has carried the finite element method, which 25 years ago was nowhere, into a central position in computational mechanics. At the end it leads (as difference equations also do) to very large and very sparse systems, which have been attacked by every method that numerical analysts could think of—and still refuse to give up.

Direct Methods for $Ax = b$

Of course the most important direct method is Gaussian elimination. Orthogonality is not involved. The n by n system $Ax = b$ is solved with a fixed number of multiplications and additions—$n^3/3$ on the left side of the equation and n^2 on the right side. Symmetry reduces the former to $n^3/6$, since half the matrix contains all the information. If A is also positive definite then the factors L and D are properly scaled, and there is no need to look down each column for the largest pivot and a possible row exchange—as we should do when A is unsymmetric or indefinite.

These operation counts apply to full matrices. If A contains a large number of zeros, and their location is known, the algorithm can be much faster. Roughly speaking, the best place for zeros is at the beginning of a row—before any nonzero entries. Those leading zeros survive throughout elimination, and every one saves a whole row operation; the multiplier is $l_{ij} = 0$ and no elimination step is required. A zero in the middle of a row is likely to become nonzero before it is the target for elimination, and nothing is saved. That destruction of zeros, known as "fill-in," is the chief difficulty for large sparse systems.

We single out two cases that are favorable. The first is a *band matrix*, when A is zero outside a band of width w on each side of the main diagonal (Fig. 5.1). The matrix keeps that form during elimination. The row containing the pivot has only $w + 1$ nonzeros, and there are only w nonzeros below the pivot. In other words, this stage of elimination needs $w^2 + w$ multiplications. With n columns the algorithm is completed in less than $(w^2 + w)n$ steps on the left side and $(2w + 1)n$ on the right, which can be far below $n^3/3$ and n^2. The same is true for a "variable band" matrix, in which w varies from row to row but stays much smaller than n. The figure shows the zeros in L and U.

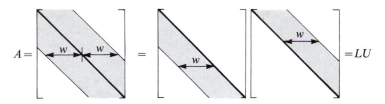

Fig. 5.1. A band matrix and its factors.

For a one-dimensional problem like $-d^2u/dx^2 = f$ the band is very narrow. In fact A is tridiagonal, if we make the natural 3-point approximation to the second derivative:

$$\frac{d^2u}{dx^2} \approx \frac{u(x+h) - 2u(x) + u(x-h)}{h^2}.$$

The right side approaches the left side as h goes to zero, but we stop at a finite meshsize h. Then the differential equation $-d^2u/dx^2 = f$ is replaced by a *finite difference* approximation for the values of u at the meshpoints $x = h, 2h, \ldots, Nh$:

$$Au = \begin{bmatrix} 2 & -1 & & & \\ -1 & 2 & -1 & & \\ & -1 & \cdot & & \\ & & & \cdot & -1 \\ & & & -1 & 2 \end{bmatrix} \begin{bmatrix} u_1 \\ u_2 \\ \cdot \\ \cdot \\ u_N \end{bmatrix} = h^2 \begin{bmatrix} f(h) \\ f(2h) \\ \cdot \\ \cdot \\ f(Nh) \end{bmatrix}.$$

Here $u(0) = u(1) = 0$ are the boundary conditions, and $(N+1)h = 1$. This matrix can also come from linear finite elements, or from $A^T C A$ with C as the identity matrix and A as an incidence matrix—with -1 on the main diagonal followed by $+1$ on the next diagonal. The essential point is the speed at which a tridiagonal system can be solved. It is much better to know A than to know the full matrix A^{-1}, even though the desired solution is $h^2 A^{-1}f$.

In two dimensions we meet reality. There the unknowns correspond to the points of a grid. The derivatives $\partial^2 u/\partial x^2$ and $\partial^2 u/\partial y^2$ in Laplace's equation or Poisson's equation both have 3-point approximations—one in the x-direction and the other in the y-direction. Combined, they give the 5-point molecule in the figure; the matrix has $+4$ along the main diagonal and four -1's on a typical row. (At the edge of the grid the boundary condition $u = 0$ removes some -1's from the diagonals in Fig. 5.2.) Unfortunately, the bandwidth is greatly increased in two dimensions. If the unknowns are ordered a row at a time, then every point has to wait through a whole row for the neighbor above it to appear, and **the "5-point matrix" has width** $w = N$. There is no ordering of unknowns that can make this bandwidth smaller; it is the price of forcing a two-dimensional array onto a line!

Since the matrix A is of order $n = N^2$—this is the number of unknowns—the operation count $w^2 n$ equals N^4. That is far below the exorbitant cost $n^3 = N^6$ of a

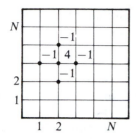
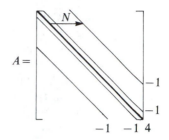

Fig. 5.2. Finite difference model of Laplace's equation: 5-point scheme.

full matrix, but it is still very large. Many of the zeros inside the band will be lost to fill-in. Therefore we try for a better ordering, in which the band can be squeezed more tightly in some places at the price of widening in other places. The result will be a varying band, called a *profile* or an *envelope*. Its shape decides the cost of elimination. In the following example it is shaped like an arrow.

Each ordering of the unknowns corresponds to a permutation matrix P. The rows are renumbered together with the columns, which produces the new matrix PAP^T. It is still symmetric and positive definite; P has just reordered the equations and the unknowns. We mention one choice that will make the discussion specific. The idea is to **dissect** the square into rectangles that are separated by thin strips S. With two rectangles (Fig. 5.3) the rows are cut in half and the bandwidth is reduced to $N/2$. Therefore the cost $w^2 n$ is divided by 4. The last row of the matrix comes from the strip S and is not banded at all—that is where we pay the price, but for one row it is not expensive. If we are bolder and use $N^{1/2}$ rectangles instead of two, then their separators S produce many more rows—but with some care the operation count drops from N^4 to $N^{7/2}$. However that is not the last word. The rectangles can be dissected again, in the y-direction, and that can continue. In the end we have a process of *nested dissection* in which the count drops to $O(N^3)$—but at an increased cost for overhead, which means that dissection is not the cure for all difficulties. It is a **domain decomposition**; the separator produces the arrowhead.

There is a third component of cost which is often the most critical. It is the cost of **storage**. Clearly we will never set aside N^4 registers for the entries of A, most of which are zero. But there could be N^3 nonzeros in its factors L and L^T, and that number needs to be controlled. The nonzero entries of a sparse matrix can be

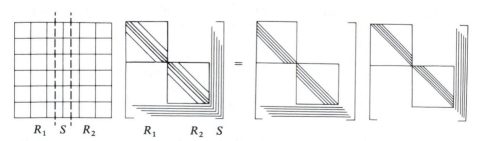

Fig. 5.3. Dissection: An "arrow matrix" factored into $A = LU$.

placed in a data structure that allows for deletions and insertions, as elimination creates and destroys zeros. A code like SPARSPAK even begins with a symbolic factorization, ignoring the actual entries and distinguishing only between zeros and nonzeros. It will order the rows automatically, by selecting at each step the row with fewest nonzeros in the matrix waiting to be factored. This is the **minimum degree algorithm**. For a matrix from the five-point scheme on a square, dissection is better and the fast Poisson solvers in Section 5.5 are better still.† But in other problems the symbolic step first analyzes the nonzero structure of LDL^T, and later the actual entries of the matrix give the numbers that go into L and D.

There is one more point. The entries can be computed when they are needed in elimination, and not before. This corresponds to the frontal method for finite elements, in which the stiffness matrix K is assembled from submatrices (Section 5.4) and the assembly is done at the last possible moment. With a good ordering we can work with small full submatrices—which is especially attractive on vectorized or parallel hardware.

Those last few words open up a part of applied mathematics—the computer science part—that is changing everything else. Perhaps we can speak briefly about **scientific computing**: the right combination of mathematical models and numerical analysis and good software. That has become central to the whole international effort of engineering and science. It takes place on an enormous scale, but it does not make the effort any less personal; there is a tremendous individual challenge in mastering those tools. My hope is that this book will help you with the model and the algorithm. At the same time we do not forget the third part, the computer itself—even if codes are not given for A^TCA on the Cray or Fujitsu. The book can propose improvements in the model or the algorithm, but a new generation of computers is on its way and the greatest success will come in fitting the three parts together.

I believe that improving the algorithm is often the easiest. Therefore the immediate goal is to introduce the principles on which efficient and stable numerical methods can be based. It is a delicate judgment, to know how much to say. On the one hand, the methods that survive all seem to use the most basic ideas of applied mathematics. That is an empirical fact. They are not based on temporary ideas or insignificant tricks, and to tell you about finite elements or Householder matrices or Fourier transforms is a pleasure. Those subjects would be worth teaching if computers didn't exist. On the other hand, an important algorithm will receive (or already has received) the attention of a professional. He intends to write software that can be used without knowing its contents and without taking this course. Codes have been assembled into a series of widely available packages, starting with EISPACK for the eigenvalue problem and LINPACK for $Ax = b$ (square or rectangular). The series extends to MINPACK, FUNPACK, GRAPHPAK, ELLPACK, LLSQ, LP/PROTRAN, and B-spline—and this list is surely not complete. From most of those names you can guess their purpose, and

---

† There was an Olympic competition for this special problem, won by West Germany. Multigrid methods took the silver. Fast Poisson solvers are available in FISHPACK.

two important sources of software are

International Mathematical and
    Statistical Libraries
7500 Bellaire Boulevard
Houston, TX 77036
(800)222-IMSL

Numerical Algorithms Group
256 Banbury Rd., Oxford, UK
1101 31st St., Downers Grove,
IL 60515
(865)511245 or (312)971-2337

The appendix surveys the software available for scientific computing, including the new electronic mail source called "netlib."

Thus our goal is to explain the underlying principles of numerical analysis, some to be applied now and others to be here for reference. The first step turns away from $Ax = b$ to remedy a grievous omission.

Nonlinear Equations and Newton's Method

If there is one topic about which Chapters 1–4 leave the author feeling guilty, it is nonlinear equations. Virtually everything has been linear. That cannot continue, and the next chapters will discuss nonlinear differential equations and nonlinear optimization. Here we want to catch up on the basic problem of n nonlinear equations in n unknowns—and to show (ironically enough) how they are solved by linearization.

A single nonlinear equation has the form $g(x) = 0$. The graph of g is a curve, and we want to compute the point x^* where it crosses the horizontal axis. There are many methods for this one-dimensional problem, but we shall emphasize those that remain competitive when there are n functions $g_1, \ldots, g_n$ and n unknowns $x_1, \ldots, x_n$. In that case the equations are still $g(x) = 0$, in vector form, or written out more fully they are

$$g_1(x_1, \ldots, x_n) = 0$$
$$\ldots$$
$$g_n(x_1, \ldots, x_n) = 0.$$

In a linear problem g would become $Ax - b$. The inhomogeneous part b is included in $g(x)$. We expect to solve $g(x) = 0$ by starting from an initial guess and improving it. The current approximation will be x^k and the question is how to compute the next vector x^{k+1}.

Everything depends on the information available at x^k. Normally we can calculate g at that point, to see how close $g^k = g(x^k)$ is to zero. If it is possible, we also want the derivative g' at x^k. That will lead to *Newton's method*, as follows:

Keep the first derivative terms in the approximation

$$g(x^{k+1}) \approx g(x^k) + g'(x^k)(x^{k+1} - x^k). \tag{1}$$

To aim for $g(x^{k+1}) = 0$, set the right side to zero:

$$g'(x^k)(x^{k+1} - x^k) = -g(x^k). \tag{2}$$

This is a linear equation for the next guess x^{k+1}. It is simplest for the case $n = 1$, with one equation and one unknown. In that case, Newton's method finds the *crossing point of the tangent line* (Fig. 5.4). Of course that point x^{k+1} is not the same as the crossing point x^* of the curve—unless g is linear and the curve is straight, when the method converges in one step. However, x^{k+1} should be a major improvement on x^k.

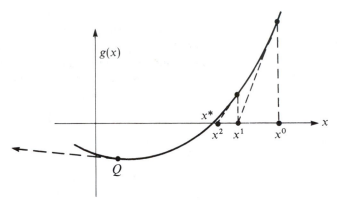

Fig. 5.4. Newton's method: the curve and the tangent line.

For vectors the picture is similar but harder to draw. There are n surfaces $g_i(x)$ each with a tangent plane at x^k. The surfaces intersect in a curve that crosses the horizontal at x^*. The planes intersect in a tangent line that crosses at x^{k+1}. The method succeeds if x^{k+1} is close to x^*. Figure 5.4 appears again in higher dimensions, with all n of the x-axes in the horizontal plane.

Fortunately the algebra stays virtually unchanged for all n. Instead of a single number dg/dx, the derivative g' becomes a matrix (it is the *Jacobian matrix J*). It contains the first derivatives $\partial g_i/\partial x_j$ of the n functions g_i with respect to the n variables x_j. The multiplication $J(x^{k+1} - x^k)$ still gives the first-order correction to g in equation (1). Thus Newton's equation (2) is the linear system

$$\boxed{J^k(x^{k+1} - x^k) = - g^k.} \qquad (3)$$

Unless n is small we recommend elimination rather than matrix inversion.

One curious point: The n-dimensional case combines the central ideas of calculus and linear algebra. Calculus gives the partial derivatives in J and matrix multiplication adds them correctly. The tangent planes come from differentiation; their intersection comes from Gaussian elimination. In one dimension we need only calculus and for $g = Ax - b$ we need only algebra. In that case the derivative matrix J is A. Newton's equation becomes $A(x^{k+1} - x^k) = b - Ax^k$, or $Ax^{k+1} = b$, and the first step gives the exact answer.

Normally Newton's method† converges quickly. If x^k is near the solution x^*, and

---

† Also known as Newton-Raphson. We take the unjustified liberty of burying Raphson.

if in that neighborhood J is invertible, then the error $x^{k+1} - x^*$ will have magnitude proportional to the **square** of $x^k - x^*$. That is *quadratic convergence*; the linear terms in (1) were exact. Unavoidably, an initial guess x^0 that is far from x^* may lead to difficulty. Convergence can fail if the derivative changes sign and the tangent goes the wrong way (as it would from point Q in Fig. 5.4). This is the result of depending on local information at x^k. When x^0 is close to x^* or when J is everywhere positive definite, success is secure. At points where J is not positive definite, a multiple of the identity is often added to make it so.

EXAMPLE 1 $g(x) = x^2 - b$ and $g'(x) = 2x$ and $x^* = \sqrt{b}$

Newton's method computes $\sqrt{b}$ by averaging x and b/x: $g'\Delta x = -g$ is

$$2x^k(x^{k+1} - x^k) = b - (x^k)^2 \quad \text{or} \quad x^{k+1} = \frac{1}{2}\left(x^k + \frac{b}{x^k}\right).$$

It always converges and the squaring of the error is impressive. We can see it by subtracting the exact solution $\sqrt{b}$ from both sides:

$$x^{k+1} - \sqrt{b} = \frac{1}{2}\left(x^k + \frac{b}{x^k}\right) - \sqrt{b} = \frac{1}{2x^k}(x^k - \sqrt{b})^2.$$

At every step the difference $x^k - \sqrt{b}$ is squared.

The penalty for quadratic convergence is the cost of each step. If every partial derivative $\partial g_i / \partial x_j$ is simple then that cost is acceptable. In many n-dimensional problems it is hard enough to find all the derivatives once, at x^0, and it is natural to reuse the same derivative matrix J^0 at every step. That method is called *modified Newton*:

$$\boxed{J^0(x^{k+1} - x^k) = -g(x^k).} \tag{4}$$

Its convergence is slower—the lines in Fig. 5.5 stay parallel to each other instead of tangent to the curve—but each step is faster. With J^0 factored into LU it requires only a forward-back substitution to find each x^{k+1}. The real cost lies in evaluating g at x^k and in the larger number of steps. Modified Newton succeeds if J is not changing too rapidly.

EXAMPLE 2 $g(x) = x^2 - b$ and $J^0 = g'(x^0) = 2x^0$

The modified Newton method is $2x^0(x^{k+1} - x^k) = b - (x^k)^2$. If x^0 is close to the answer then the error is multiplied by a factor less than one—but it is not squared!

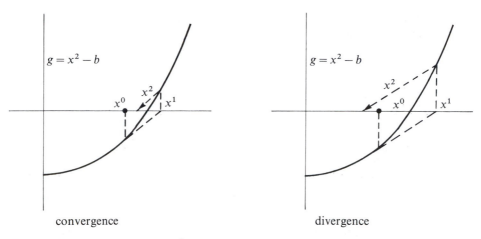

convergence   divergence

Fig. 5.5. Modified Newton for $x^2 - b = 0$: parallel lines replace tangents.

After some manipulation we capture the multiplying factor in brackets:

$$(x^{k+1} - \sqrt{b}) = \left[1 - \frac{x^k + \sqrt{b}}{2x^0} \right](x^k - \sqrt{b}).$$

If x_0 is too small this factor is below -1, and the error grows at every step (oscillating from side to side). Modified Newton can quite easily diverge.

Between Newton and modified Newton is a third possibility known as **quasi-Newton**. It changes J at every step, not by recomputing the derivatives $\partial g_i / \partial x_j$ but by using information from the step itself. After the first step we know the column vectors $\Delta x = x^1 - x^0$ and $\Delta g = g(x^1) - g(x^0)$. This indicates the derivatives of the g_i in the direction of Δx. Then the next $J = J^1$ is adjusted to satisfy $J^1 \Delta x = \Delta g$, for example by the rank-one update

$$J^1 = J^0 + \frac{(\Delta g - J^0 \Delta x)(\Delta x)^T}{(\Delta x)^T (\Delta x)}. \tag{5}$$

That is hardly more expensive than modified Newton, which keeps J fixed at J^0. The convergence is faster than linear and there is a more highly recommended "BFGS formula" to maintain symmetry if J^0 is symmetric. The full theory would carry us too far.

What if the derivatives of g are simply impossible and J cannot be found even once? In desperation we can take J to be the identity matrix, or a multiple I/α of the identity—which makes the iteration step even simpler:

$$\boxed{x^{k+1} - x^k = -\alpha g(x^k).} \tag{6}$$

This is less impulsive than it looks. It is the method of ***steepest descent***, and it is a natural idea if we want the minimum of a function $P(x_1, \ldots, x_n)$:

> Suppose the graph of P is a bowl-shaped surface with its lowest point at x^*. At that point the first derivatives are zero:
>
> $$g_i = \frac{\partial P}{\partial x_i} = 0 \quad \text{at} \quad x = x^*.$$
>
> Those are the n equations in n unknowns. The vector g is the gradient of P and it is zero at the minimum. To move from a height $P(x^k)$ on the surface toward the lowest point $P(x^*)$, the steepest descent method
>
>   (i)   goes in the tangent direction $-g(x^k)$ in which the surface is steepest and P decreases most rapidly
>   (ii) stops at the point x^{k+1} in this direction at which P is as small as possible.

It is like water descending a mountain. At each point it moves in the direction of steepest descent. However there is a crucial difference: water changes direction continuously but the algorithm must take finite steps. Each iteration can *start* in the steepest direction, but only in exceptional cases will it hit x^*. It is like a skier who can't turn; he goes forward until his path begins to climb. At that point, which is x^{k+1}, he chooses a new direction of steepest descent and makes another run. His path is illustrated in Fig. 5.6a, where the dotted lines are ***contours*** (level lines) of the function $P(x)$. On each contour P is constant; perpendicular to the contour is the direction of steepest descent. Thus each new step starts at right angles to the level line.

Note how P adds a new generation to the picture. Imagining the functions g_i as the "parents" of $\partial g_i / \partial x_j$, we now have a grandparent; it is P. With the linear case written separately on the right, the three generations contain a function, then a vector, and finally a matrix:

> Grandparent $P(x_1, \ldots, x_n)$ $[= \tfrac{1}{2} x^T A x - x^T b]$
>
> Parents $g_i = \partial P / \partial x_i$ $[= \text{components of } Ax - b]$
>
> Grandchildren $J_{ij} = \dfrac{\partial g_i}{\partial x_j} = \dfrac{\partial^2 P}{\partial x_j \partial x_i}$ $[= \text{entries of } A].$

All the later generations would be tensors.

A set of functions g_i may or may not have a common ancestor like P. When there is a grandparent, the matrix in the third generation is symmetric: $J = J^T$ because $\partial^2 P / \partial x_j \partial x_i$ equals $\partial^2 P / \partial x_i \partial x_j$. If P is minimized at x^* then J is symmetric positive definite at that minimum.

There are serious difficulties with the method of steepest descent. When the graph of P drops into a long narrow valley, this method tends to travel back and

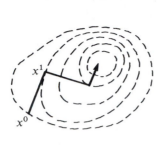

descending the contours of P

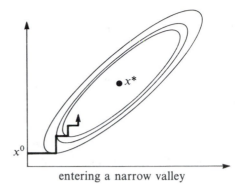

entering a narrow valley

Fig. 5.6. The directions of steepest descent.

forth across the valley—in very short steps. It creeps slowly toward the minimum, as in the figure on the right. The rate of convergence for a linear $g = Ax - b$ is governed by the ratio $(\lambda_{max} - \lambda_{min})/(\lambda_{max} + \lambda_{min})$, which can be very close to 1. In practice the method often stops far from the true solution x^* because of round-off error.

EXAMPLE 3 $\qquad P(x) = 2x_1^2 - 2x_1 x_2 + x_2^2 + 2x_1 - 2x_2$

The derivatives of P are $g_1 = 4x_1 - 2x_2 + 2$ and $g_2 = -2x_1 + 2x_2 - 2$. If the method starts at the origin, the steepest direction points toward $-g(0, 0) = (-2, 2)$. A step of size α brings us to $P(-2\alpha, 2\alpha) = 20\alpha^2 - 8\alpha$. Differentiating, the choice $\alpha = 1/5$ minimizes P in that direction and the first step ends at $x^1 = (-2/5, 2/5)$. At that point we recompute the steepest direction and take a second step:

$$g(x^1) = \left(-\frac{2}{5}, -\frac{2}{5}\right), \quad P(x^1 - \alpha g(x^1)) = \frac{4\alpha^2 - 8\alpha - 20}{25},$$

$$\alpha = 1, \quad x^2 = x^1 - 1 \cdot g(x^1) = \left(0, \frac{4}{5}\right).$$

It turns out that each pair of steps brings us to $x^{2k} = (0, 1 - (1/5^k))$. Even on this problem with quadratic P and linear $g = Ax - b$, we do not reach the minimum x in a finite time. What is worse, we do not get close to it quickly.

Something has to be done. Most good methods use a *quadratic model*—they will find the exact answer if P is exactly quadratic, and they regard any function $P(x)$ as approximately quadratic near each point. In that model we can choose directions d^k that are much better than the steepest direction $-g^k$. A great favorite is the *conjugate gradient method*:

$$\boxed{x^{k+1} - x^k = \alpha d^k \quad \text{with direction} \quad d^k = -g^k + \beta d^{k-1}.} \qquad (7)$$

You see the change; there is a correction term βd^{k-1} to what would otherwise be steepest descent. The conjugate gradient choice is $\beta = g_k^T(g_k - g_{k-1})/g_{k-1}^T g_{k-1}$, because of its remarkable property in the linear case $g = Ax - b$: The new direction d^k is A-orthogonal† not only to d^{k-1} but to **all previous directions** (see Section 5.3). In that case the algorithm reaches x^* in n steps, and comes close to it sooner, while steepest descent is still crossing the valley.

This completes our selection of nonlinear methods: *Newton, modified Newton, quasi-Newton, steepest descent, and conjugate gradients.* The second and fourth use less information than Newton; modified Newton uses only J^0 and steepest descent uses only g. Those take the most iterations, and quasi-Newton and conjugate gradients are major improvements at moderate cost.

EXAMPLE 4   The first and second derivatives of $P = \frac{1}{4}(x_1^4 - 4x_1x_2 + x_2^4)$ are

$$g = \begin{bmatrix} x_1^3 - x_2 \\ x_2^3 - x_1 \end{bmatrix} \quad \text{and} \quad J = \begin{bmatrix} 3x_1^2 & -1 \\ -1 & 3x_2^2 \end{bmatrix}.$$

There are three real solutions to $g(x) = 0$: the vector (x_1, x_2) can be $(0, 0)$, or $(1, 1)$ or $(-1, -1)$. At these points P is 0, or $-\frac{1}{2}$, or $-\frac{1}{2}$. The Jacobian J is symmetric because it has the ancestor P, and it is positive definite at $(1, 1)$ and $(-1, -1)$ but not at the saddle-point $(0, 0)$:

$$J = \begin{bmatrix} 3 & -1 \\ -1 & 3 \end{bmatrix} \quad \text{or} \quad J = \begin{bmatrix} 0 & -1 \\ -1 & 0 \end{bmatrix}.$$

Newton's method $J^k(x^{k+1} - x^k) = -g^k$ has quadratic convergence, as a computer will verify. If it starts near the unstable point at $(0, 0)$, it will slowly accelerate toward a minimum (I don't know which one). Modified Newton stays with J^0 and has linear convergence—or fails, depending on x^0. Quasi-Newton is more likely to succeed; it is fast by computer but not by hand. Steepest descent chooses $x^{k+1} = x^k - \alpha g^k$, and substituting into P leaves a one-dimensional problem for α. (It could be solved by Newton's method!) This example has enough symmetry to give steepest descent a fair chance, or perhaps an unfair chance, against conjugate gradients. I would be grateful to know what you find.

EXAMPLE 5   Equilibrium of a nonlinear material frequently has the form

$$g(x) = A^T C(Ax) - f = 0 \quad \text{with nonlinear } C.$$

The stress $C(Ax)$ is a nonlinear function of the strain Ax. When f is a large force, it may be hard to choose a good x^0—and the success of all methods depends heavily on that initial guess. Therefore it is common to apply f in small increments. We mention the "Riks method" as a way to choose force levels that lead to

---

† A-orthogonal means that $(d^k)^T A d^{k-1} = 0$. The inner product is weighted by A.

approximately equal changes in x—and thus allow the algorithm to move cautiously until the full force f is reached.

The important thing is to recognize the derivative matrix $J = g'$ when g has the form $A^T C(Ax) - f$. It is an old friend, slightly altered by nonlinearity: $J = A^T C' A$. Here C' is the slope of the stress-strain curve, or the current-voltage curve, *at the point Ax^k*, where J is computed. In the linear case those curves were straight lines; by Hooke's law and Ohm's law the slope C' was the same at all points. For a nonlinear material the slope varies. Each step of Newton's method brings a new matrix $A^T C' A$ (unless we use modified Newton) but our basic framework is still dominant.

EXERCISES

5.1.1   Which zeros are "filled in" when elimination reduces A to triangular form,

$$A = \begin{bmatrix} 4 & 1 & 0 & 0 & 1 \\ 1 & 4 & 1 & 0 & 0 \\ 0 & 1 & 4 & 1 & 0 \\ 0 & 0 & 1 & 4 & 1 \\ 1 & 0 & 0 & 1 & 4 \end{bmatrix} = LDL^T?$$

Mark nonzeros by x's in the 5 by 5 upper triangular $U = DL^T$ or the lower triangular L (this is the usual elimination of Sec. 1.3 but without computing the numbers).

5.1.2   Why do elimination and back substitution with a banded matrix use about $2wn$ multiplications on the right side of $Ax = b$? How many multiplications are involved in the ordinary product $A^{-1}b$?

5.1.3   What are the pivots for

$$A = \begin{bmatrix} .001 & 1 \\ 1 & 1 \end{bmatrix}$$

and what are the pivots if the rows are exchanged?

5.1.4   What are the coefficients in a 7-point scheme for Poisson's equation $-u_{xx} - u_{yy} - u_{zz} = f$ in three dimensions? Find the bandwidth w for a natural ordering if there are N meshpoints in each direction. What is $w^2 n$ for $N = 10$?

5.1.5   The "9-point scheme" gives a more accurate approximation to $-u_{xx} - u_{yy} = f$, with a molecule

When the unknowns are ordered by rows, why is the bandwidth $w = N + 1$? Can you prove that no ordering achieves a smaller bandwidth?
Hint: In any ordering, look at the meshpoints numbered 1 and N^2. If they are adjacent, row 1 of K has width $w = N^2 - 1$. If they are two jumps apart, w is at least $\frac{1}{2}(N^2 - 1)$. How many jumps apart can they be when the 9-point scheme allows diagonal neighbors? For the 5-point scheme, the proof of $w \geq N$ is more complicated and may be unpublished.

5.1.6 (a) Write down Newton's method for the solution of

$$g_1 = x_1 + \sin x_2 = 0 \quad \text{and} \quad g_2 = x_1 \cos x_2 + x_2 = 0.$$

(b) Is the Jacobian matrix J symmetric? If so, find P such that $g_1 = \partial P/\partial x_1$ and $g_2 = \partial P/\partial x_2$.

5.1.7 Sketch the graph of $g(x) = xe^{-x}$ and draw the path of Newton's method from $x^0 = \frac{1}{2}$ and from $x^0 = 1$.

5.1.8 For a single nonlinear equation $g(x) = x^4 - 16 = 0$, show that Newton's method goes from x^k to $x^{k+1} = 3x^k/4 + 4/(x^k)^3$. If x^k is nearly correct, say $x^k = 2/(1 - E)$, find the leading coefficients in $x^{k+1} = 2 + cE + dE^2 + \cdots$

5.1.9 Draw a curve for which the modified Newton method has x^k converging to x^* from one side, and another curve for which x^k jumps from one side of x^* to the other. How is this decided by g at x^*?

5.1.10 Suppose $g(x) = 0$ is rewritten as $x = f(x)$, by defining $f(x) = x - g(x)$. The method of *successive substitution* solves $x = f(x)$ by the iteration $x^{k+1} = f(x^k)$.
(a) By returning to g compare with steepest descent: What is the steplength α?
(b) If $g(x) = ax - b$ then $x^{k+1} = (1 - a)x^k + b$. Starting from $x^0 = 1$ find x^1, x^2, and the typical x^k.
(c) For which values of a does x^k converge to the correct $x^* = b/a$?

5.1.11 If successive substitution converges to x^* then the change from step to step is

$$x^{k+1} - x^k = f(x^k) - f(x^{k-1}) \approx f'(x^*)(x^k - x^{k-1}).$$

Why do we need $|f'(x^*)| < 1$ to have convergence?

5.1.12 For $P(x) = x_1^2 + 2x_2^2 - 2x_1 - 8x_2$ compute $g_1 = \partial P/\partial x_1$ and $g_2 = \partial P/\partial x_2$ and find the minimizing x^*. Take one steepest descent step starting from the origin $x^0 = (0,0)$; what is the direction $-g$ and where does the step end? Find the new steepest direction $-g(x^1)$ and check that it is perpendicular to the first direction.

5.1.13 What graph would have an incidence matrix A such that $A^T A$ is the standard matrix (Fig. 5.2) for the 5-point scheme?

5.1.14 With alternating "red" and "black" meshpoints, the center of each 5-point molecule has opposite color from its four neighbors. Display this numbering with $3 \times 3 = 9$ interior meshpoints and write down the 5-point matrix K. After elimination, where will the upper triangular matrix have zeros?

5.1.15 (*recommended*) What does Newton's method do, experimentally, for $x_1^3 - x_2 = 0$, $x_2^3 - x_1 = 0$?

5.2 ■ ORTHOGONALIZATION AND EIGENVALUE PROBLEMS

This chapter is really about the advantages of orthogonality. There are many, and they appear everywhere in scientific computation. Some of them go back to Pythagoras, who discovered that the square of the length of a vector b is

$$\|b\|^2 = b^T b = b_1^2 + b_2^2 + \cdots + b_n^2. \tag{1}$$

At least he would have discovered it if Euclid had asked him. Pythagoras certainly knew it to be true in two dimensions, when b_1 and b_2 come from a right triangle; the length of the hypotenuse is $\|b\|$. The formula is remarkable, partly because it remains so simple in spaces of any dimension. It is not easy to visualize 6-dimensional space, but the one thing you can hope to imagine is the presence of six orthogonal axes. The vector b has a component in each of those directions. If Pythagoras had found the length by starting with the first two components, and then admitting one new component at a time (producing a right triangle at every step, by orthogonality), he would have arrived at (1).

Our goal in this chapter is to see how computations are improved by orthogonality. We begin with linear systems $Ax = b$—aiming to find the exact solution or the least squares solution. The key is to go from general matrices A to orthogonal matrices Q. Then the solution is immediate. It is the same in the continuous case, where a least square fit by $1, x, \ldots, x^{10}$ is numerically impossible but after orthogonalization it is easy. For eigenvalues a series of orthogonalizations is the basis for the **QR method**, which is the best algorithm for matrices of reasonable size. It has become an axiom that every step in a numerical method is improved by orthogonality—of vectors or functions or descent directions or polynomials.

Underlying the whole chapter is the perpendicularity of the axes; without that formula (1) would fail. It depended on having two (or six) orthogonal directions, but those directions are not unique. By rotating and reflecting them we find others. To turn the geometry into algebra we think of one "standard" set of perpendicular axes, and the components $b_1, \ldots, b_6$ of every vector are given with respect to those axes. The axes are a coordinate system, and $b_1, \ldots, b_6$ are the coordinates. Two vectors in six-dimensional space are perpendicular if Pythagoras' law is correct on the triangle they form; a is orthogonal to b if

$$\|a + b\|^2 = \|a\|^2 + \|b\|^2.$$

Expanding the left side into $(a_1 + b_1)^2 + \cdots + (a_6 + b_6)^2$, it is clear what must happen. The sum of the six terms $2a_i b_i$ must disappear. Therefore a **is orthogonal to** b **when their inner product is zero**:

$$a^T b = a_1 b_1 + \cdots + a_6 b_6 = 0. \tag{2}$$

The vectors in the six coordinate directions, starting with $e_1 = (1,0,0,0,0,0)$, do satisfy this condition of perpendicularity—or we would be hopelessly inconsistent.

They are the columns of the six by six identity matrix, and they are mutually orthogonal.

Orthonormal Vectors

Now comes the step that puts orthogonality to work. Suppose we have six mutually orthogonal unit vectors; call them $q_1, ..., q_6$. The inner product of q_i with q_j is zero, except when $i = j$. The inner product of q_i with itself is one, since it is a unit vector (its length is $\|q_i\| = 1$). Such a set of vectors is called *orthonormal*:

$$q_i^T q_j = \begin{cases} 0 & \text{if} \quad i \neq j, \text{ giving orthogonality} \\ 1 & \text{if} \quad i = j, \text{ normalizing the lengths.} \end{cases} \tag{3}$$

These vectors go in six perpendicular directions, and it must be possible to write any vector b as a combination of $q_1, ..., q_6$. The only problem is to discover the right combination

$$c_1 q_1 + c_2 q_2 + \cdots + c_6 q_6 = b. \tag{4}$$

The numbers c_i come from an easy trick. It works because the vectors are orthogonal, and it uses the same idea that lies behind Fourier series (finite or infinite). The idea is to take the inner product of both sides of (4) with q_i—or in other words, to *multiply* (4) *by* q_i^T:

$$q_i^T (c_1 q_1 + \cdots + c_i q_i + \cdots + c_6 q_6) = q_i^T b. \tag{5}$$

The left side collapses because of orthogonality to only one term, the one given by $c_i q_i^T q_i$. From the normalization $q_i^T q_i = 1$ this equals c_i, and we have the simplest formula in scientific computation:

5A The components of b with respect to orthonormal vectors q_i are

$$c_i = q_i^T b. \tag{6}$$

The vector $c_i q_i$, or $q_i q_i^T b$, is the projection of b in the direction of q_i, and by adding these projections we recover b:

$$c_1 q_1 + \cdots + c_6 q_6 = q_1 q_1^T b + \cdots + q_6 q_6^T b = b. \tag{7}$$

EXAMPLE In three-dimensional space, suppose the orthonormal vectors are

$$q_1 = \begin{bmatrix} \frac{2}{3} \\ \frac{2}{3} \\ -\frac{1}{3} \end{bmatrix}, \quad q_2 = \begin{bmatrix} -\frac{1}{3} \\ \frac{2}{3} \\ \frac{2}{3} \end{bmatrix}, \quad q_3 = \begin{bmatrix} \frac{2}{3} \\ -\frac{1}{3} \\ \frac{2}{3} \end{bmatrix} \quad \text{with } b = \begin{bmatrix} 0 \\ 0 \\ 1 \end{bmatrix}.$$

Then the numbers $c_i = q_i^T b$ are $-\frac{1}{3}, \frac{2}{3}, \frac{2}{3}$, and $-\frac{1}{3} q_1 + \frac{2}{3} q_2 + \frac{2}{3} q_3 = b$.

The second step is to put this into the language of matrices. Suppose the six vectors q_i are the columns of Q. Then the 36 inner products $q_i^T q_j$ appear when Q^T multiplies Q; row i of Q^T meets column j of Q. Because the vectors are orthogonal, $q_i^T q_j = 0$ off the diagonal of $Q^T Q$; on the diagonal we find $q_i^T q_i = 1$. Therefore $Q^T Q$ reduces to the 6 by 6 identity matrix. Q is an **orthogonal matrix**—it should really be called orthonormal, but it's too late to change—and **its transpose is equal to its inverse**.

5B   A square orthogonal matrix has all the following properties:

$$Q^T Q = I, \quad \text{and} \quad QQ^T = I, \quad \text{and} \quad Q^T = Q^{-1}. \tag{8}$$

Its columns are orthonormal, and multiplication by Q preserves lengths:

$$\| Qx \| = \| x \| \quad \text{for every vector } x. \tag{9}$$

The last property is immediate from $\| Qx \|^2 = (Qx)^T (Qx) = x^T Q^T Q x = x^T x = \| x \|^2$. The product $Q^T Q$ is replaced by I, and in every application of orthogonality this fact must be used. The whole of statement (3), that the vectors q_i are orthonormal, can be condensed into the matrix formula $Q^T Q = I$.

Note   Not only does Q have orthonormal columns, it also has orthonormal rows. This follows from $QQ^T = I$. Since Q^T is a left inverse of Q and Q is square, Q^T is also a right inverse—which sounds obvious but is quite remarkable. The rows point in completely different directions from the columns, but they are orthonormal whenever the columns are.

We return to put $c_1 q_1 + \cdots + c_6 q_6 = b$ into a neater form. The left side is a combination Qc of the columns of Q. Therefore the equation is just $Qc = b$. Multiplying both sides by Q^{-1}, which equals Q^T, **the solution becomes** $c = Q^T b$. Of course this must agree with the earlier formula $c_i = q_i^T b$, and it does:

$$c = Q^T b = \begin{bmatrix} -q_1^T- \\ \vdots \\ -q_6^T- \end{bmatrix} \begin{bmatrix} | \\ b \\ | \end{bmatrix} = \begin{bmatrix} q_1^T b \\ \vdots \\ q_6^T b \end{bmatrix}.$$

The coefficients are given, all six at once, by $c = Q^T b$.

Note that $c_1^2 + \cdots + c_6^2 = b_1^2 + \cdots + b_6^2$, or $\| c \| = \| Qc \| = \| b \|$. This comes directly from (9); multiplication by Q preserves lengths. It is really another statement of Pythagoras' law, that in any orthonormal system the distance squared is a sum of six squares. In the standard system the coordinates are the b_i; in the new system the coordinates are the c_i. The distance to a point is independent of the axes and it is given equally well by $\| b \|$ or $\| c \|$.

Least Squares

What happens with only two orthonormal vectors in six-dimensional space? The vectors q_1 and q_2 can still be the columns of Q, but this matrix is no longer square; it is a 6 by 2 matrix like

$$Q = \begin{bmatrix} .7 & .7 \\ .1 & .1 \\ .3 & -.3 \\ .4 & -.4 \\ .5 & -.5 \\ 0 & 0 \end{bmatrix}.$$

The columns are orthonormal but they are not a basis; most vectors b are not combinations of these two columns. In other words $Qc = b$ will usually have *no solution*. Q is not invertible, its rows are not orthonormal, and $QQ^T \neq I$. Nevertheless something is saved, because $Q^T Q$ is still the 2 by 2 identity matrix. The equation $Q^T Q = I$ does not lead to an exact solution to $Qc = b$—in general that is impossible—but it does make the least squares solution easy.

One method is to use the normal equations from Section 1.4. For that we simply multiply by the transposed matrix, which gives $Q^T Qc = Q^T b$ or $c = Q^T b$. *The solution is still $Q^T b$*, and the best coefficients c_i are still the inner products $q_i^T b$. The sum $c_1 q_1 + c_2 q_2$ does not reproduce b itself, since that is impossible. Instead it becomes the ***projection*** of b onto the column space of Q. This is the vector p, the combination of the columns that is closest to b.

EXAMPLE Suppose Q has only one column (call it q). The one-dimensional projection onto the line through q is the multiple $p = cq$ that is closest to b. From our formula, $c = q^T b$ and $p = cq = qq^T b$. The vector $b - p$ is left over (Fig. 5.7). It is the "error," the part of b that is not in the direction of q. ***This error is perpendicular to*** q:

$$q^T (b - p) = q^T b - cq^T q = 0 \quad \text{since } q^T q = 1.$$

To project $b = [1 \; 2 \; 3 \; 4]$ onto the line through $a = [1 \; 1 \; 1 \; 1]$, we make a into a unit vector $q = a/\|a\| = [\frac{1}{2} \frac{1}{2} \frac{1}{2} \frac{1}{2}]$ and then compute

$$c = q^T b = 5 \quad \text{and} \quad p = cq = [\tfrac{5}{2} \tfrac{5}{2} \tfrac{5}{2} \tfrac{5}{2}].$$

Projections are easy to visualize, especially a two-dimensional projection in three-dimensional space. The orthonormal columns q_1 and q_2 span a plane and most points b are not on that plane. The point closest to b is $p = c_1 q_1 + c_2 q_2$, a combination of the two one-dimensional projections. The error $b - p$ is perpendicular and it is $c_3 q_3$—if we imagine a third orthogonal direction. The

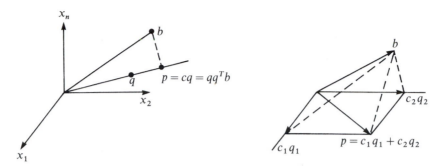

Fig. 5.7. Projections onto a line and a plane.

coefficients are the same $q_i^T b$ as before. For n columns within m-dimensional space $(n < m)$, the projection is $p = QQ^T b$:

5C If the columns of an m by n matrix Q are orthonormal, so that $Q^T Q = I$, then the numbers $c_i = q_i^T b$ give
   (1)   The best least squares solution to $Qc = b$
   (2)   The projection $p = c_1 q_1 + \dots + c_n q_n$ of b onto the column space of Q
   (3)   The coefficients that minimize the distance $\| c_1 q_1 + \dots + c_n q_n - b \|$ between b and any combination of the columns.

These all describe a projection onto the subspace with basis $q_1, \dots, q_n$.

Non-orthogonal Columns

The equation $Qc = b$ was easy to solve. Its solution is $c = Q^T b$, provided the columns of Q are orthonormal. When that fails—most columns are linearly independent without being orthonormal—there has to be a change in the formula. The solution is not $x = A^T b$, because the columns of A are coupled by the matrix $A^T A$, and this is no longer the identity. We have come to the typical problem of least squares.

The overdetermined system is $Ax = b$, with m equations and $n < m$ unknowns. The direct approach is through the normal equations, which multiply by A^T:

$$A^T A x = A^T b, \quad \text{or} \quad x = (A^T A)^{-1} A^T b.$$

In practice the inverse of $A^T A$ is virtually never computed. It appears in the formula but not in the computation. Instead $A^T A x = A^T b$ is solved by elimination and back substitution. The real question is whether to compute $A^T A$.

In many cases the normal equations are satisfactory, but there is a second approach which avoids them. This is especially desirable when the columns of A are barely independent; $A^T A$ is then 'ill-conditioned" and roundoff errors will be amplified. That can never happen with orthogonal matrices. Therefore we prefer to work with orthonormal combinations $q_1, \ldots, q_n$ of the original columns $a_1, \ldots, a_n$. We are **orthogonalizing the columns of** A. Of course there are many ways to choose orthogonal vectors in the column space. But one of those ways is exceptional, because it has the following properties:

  (1) q_1 is a multiple of a_1
  (2) q_1 and q_2 are orthonormal in the plane of a_1 and a_2
  (3) q_1, q_2, q_3 are orthonormal in the space spanned by a_1, a_2, a_3—and
so on.

Each q_k begins with the corresponding column a_k. It subtracts the projection of a_k onto each vector $q_1, \ldots, q_{k-1}$ that is already established. That leaves a vector v in a new direction, orthogonal to all these established directions:

$$q_k \text{ is a multiple of } v = a_k - (q_1^T a_k)q_1 - \cdots - (q_{k-1}^T a_k)q_{k-1}. \tag{10}$$

This is the **Gram-Schmidt** idea. This section describes two important ways of organizing the calculation. First we write down immediately the one important way to organize the result. It is a matrix factorization.

Orthogonalization starts with an m by n matrix A and it ends with an m by n matrix Q. The link between them, if the rules listed above as $(1), (2), (3), \ldots$ are followed, is a square upper triangular matrix R. **The original matrix is factored into** $A = QR$. To make this clearer we display the matrices:

$$A = \begin{bmatrix} & & \\ a_1 & \cdots & a_n \\ & & \end{bmatrix} = \begin{bmatrix} & & \\ q_1 & \cdots & q_n \\ & & \end{bmatrix} \begin{bmatrix} r_{11} & \cdots & r_{1n} \\ & \ddots & \\ & & r_{nn} \end{bmatrix} = QR. \tag{11}$$

Comparing the first columns gives $a_1 = q_1 r_{11}$, so r_{11} is just the length of a_1. Then q_1 is a unit vector as required. The column a_2 is a combination of q_1 and q_2, or q_2 is a combination of a_2 and q_1, exactly as expected in rule (2). The vectors $a_1, \ldots, a_k$ are always combinations of $q_1, \ldots, q_k$—and vice versa, because $AR^{-1} = Q$. The factors Q and R are uniquely determined if all diagonal entries r_{kk} are positive; for the multiple in (10) we divide by the length of v, to make q_k a unit vector.

The factorization $A = QR$ gives immediately the least squares solution to $Ax = b$:

5D   If $A = QR$ with $Q^T Q = I$, then the least squares solution to $Ax = b$ is

$$x = R^{-1} Q^T b. \tag{12}$$

Proof 1: The least squares solution to $Qc = b$ is $c = Q^T b$. Now identify c with Rx and identify $Ax = b$ with $Q(Rx) = b$. Then Rx equals $Q^T b$. This is the equation you solve by back substitution, to find $x = R^{-1} Q^T b$.

Proof 2: The normal equations are

$$A^T A x = A^T b \quad \text{or} \quad R^T Q^T Q R x = R^T Q^T b \quad \text{or} \quad Q^T Q R x = Q^T b \quad \text{or} \quad R x = Q^T b.$$

EXAMPLE
$$a_1 = \begin{bmatrix} 0 \\ 0 \\ 1 \end{bmatrix}, \quad a_2 = \begin{bmatrix} 0 \\ 4 \\ 1 \end{bmatrix}, \quad a_3 = \begin{bmatrix} 5 \\ 1 \\ 1 \end{bmatrix}$$

a_1 is already a unit vector so $q_1 = a_1$ and $r_{11} = 1$. Subtracting from a_2 its projection onto q_1 gives the right direction for q_2:

$$v = a_2 - (q_1^T a_2) q_1 = a_2 - q_1 \quad \text{has} \quad \|v\| = 4 \quad \text{so} \quad q_2 = \tfrac{1}{4}(a_2 - q_1).$$

Thus $q_2 = (0, 1, 0)$. Then subtracting from a_3 its projections onto q_1 and q_2 leaves

$$v = a_3 - (q_1^T a_3) q_1 - (q_2^T a_3) q_2 = a_3 - q_1 - q_2 = (5, 0, 0).$$

Dividing by 5 gives $q_3 = (1, 0, 0)$ and $a_3 = q_1 + q_2 + 5 q_3$. Altogether

$$A = \begin{bmatrix} 0 & 0 & 5 \\ 0 & 4 & 1 \\ 1 & 1 & 1 \end{bmatrix} = \begin{bmatrix} 0 & 0 & 1 \\ 0 & 1 & 0 \\ 1 & 0 & 0 \end{bmatrix} \begin{bmatrix} 1 & 1 & 1 \\ & 4 & 1 \\ & & 5 \end{bmatrix} = QR.$$

Modified Gram-Schmidt

For actual calculations the Gram-Schmidt method has one serious problem. The computed vectors q_k do not turn out to be orthogonal. The idea expressed in (10), of subtracting away the components in each established direction $q_1, \ldots, q_{k-1}$, is numerically unstable. But with a slight rearrangement the algorithm becomes stable and efficient.

The new arrangement is known as "Modified Gram-Schmidt." It is the right way to compute the projection of a_k onto $q_1, \ldots, q_{k-1}$. It starts with $(q_1^T a_k) q_1$. It subtracts that off *immediately*, leaving

$$a_k^1 = a_k - (q_1^T a_k) q_1. \tag{13}$$

Then it projects a_k^1 instead of the original a_k onto q_2, and subtracts that projection:

$$a_k^2 = a_k^1 - (q_2^T a_k^1) q_2. \tag{14}$$

In principle this is identical with Gram-Schmidt, which projects a_k onto both q_1 and q_2 and arrives at v: (13) and (14) give

$$a_k^2 = [a_k - (q_1^T a_k) q_1] - (q_2^T [a_k - (q_1^T a_k) q_1]) q_2$$
$$= a_k - (q_1^T a_k) q_1 - (q_2^T a_k) q_2 = v \quad \text{(since } q_2^T q_1 = 0\text{)}.$$

In practice the repeated one-dimensional projections are much better.

A short code for the algorithm can overwrite, a column at a time, the m by n matrix A by a matrix Q with orthonormal columns:

For $k = 1, \ldots, n$

$$r_{kk} := \left(\sum_{i=1}^{m} a_{ik}^2 \right)^{1/2}$$

For $i = 1, \ldots, m$

$$a_{ik} := a_{ik}/r_{kk}$$

For $j = k + 1, \ldots, n$

$$r_{kj} := \sum_{i=1}^{m} a_{ik} a_{ij}$$

For $i = 1, \ldots, m$

$$a_{ij} := a_{ij} - a_{ik} r_{kj}$$

The last line removes from each later column of A (that is, from columns $k + 1, \ldots, n$) the component in the direction of the newly created unit vector q_k. The entries of this q_k are the numbers a_{ik} on line 4. The inner products $q_k^T a_j$ on line 6 are the multiples r_{kj} of q_k to be removed from the later columns a_j.

The algorithm requires about mn^2 multiplications, half on line 6 and half on line 8. All vectors have length m, each step like (13) or (14) needs $2m$ multiplications, and there are about $n^2/2$ such steps—removing the first column from $n - 1$ later columns, then the second from $n - 2$ later columns, and eventually completing

$$(n - 1) + (n - 2) + \cdots + 1 \approx \frac{n^2}{2} \quad \text{one-dimensional projections.}$$

This number mn^2 is *twice* what is needed to find $A^T A$ for the normal equations ($A^T A$ is symmetric so we do about half of this matrix multiplication). The final steps of least squares—to solve $Rx = Q^T b$ or $A^T A x = A^T b$—are faster than mn^2. Therefore the normal equations win, but at the risk of amplifying the roundoff error when the columns of A are not strongly independent.

The QR Algorithm

This section describes one of the great algorithms of numerical linear algebra. Part of its attraction is that the key idea was so unexpected. The QR factorization orthogonalizes the columns of A, and what has that to do with eigenvalues? After the surprise, the real reasons for its supremacy are

(1) It is simple and very fast
(2) It uses orthogonal matrices and is extremely stable.

It will find all the eigenvalues of a matrix of moderate size, say $n < 100$. The QR method is not so well suited to a large sparse matrix, especially if only a few eigenvalues are required. It normally finds the eigenvalues in decreasing order of magnitude. The eigenvectors can come separately, or we can keep a record of the iteration steps.

The basic step is magically simple. The given square matrix A is factored into QR, and then Q and R are **multiplied in the opposite order**: $A_1 = RQ$. This new matrix is similar to the original one:

$$A_1 = RQ = Q^{-1}QRQ = Q^{-1}AQ. \qquad (15)$$

Therefore A_1 has the same eigenvalues as A: if $A_1 x = \lambda x$ then $AQx = \lambda Qx$. The eigenvectors are changed by Q but the eigenvalues are not changed. The new matrix A_1 is still symmetric, because $Q^{-1} = Q^T$ and the transpose of $Q^T AQ$ is $Q^T AQ$.

The process continues in exactly the same way from A_1 to A_2, and from A_k to A_{k+1}, preserving the eigenvalues at every step:

factor A_k into $Q_k R_k$ and form $A_{k+1} = R_k Q_k$.

This describes the QR algorithm "without shifts." Under fairly general circumstances it converges for both symmetric and unsymmetric matrices. The sequence A_k approaches a diagonal or upper triangular form and the diagonal entries approach the eigenvalues. The following example shows how the off-diagonal entry can be *cubed* at a single step—which is a quite exceptional rate of convergence. We write c for $\cos \theta$ and s for $\sin \theta$.

EXAMPLE
$$A = \begin{bmatrix} c & s \\ s & 0 \end{bmatrix} = \begin{bmatrix} c & s \\ s & -c \end{bmatrix} \begin{bmatrix} 1 & cs \\ 0 & s^2 \end{bmatrix} = QR$$

This QR factorization—by Gram-Schmidt or otherwise—is half of the first step. The columns of A are made orthogonal in Q. Now the QR method reverses the order:

$$RQ = \begin{bmatrix} c + cs^2 & s^3 \\ s^3 & -cs^2 \end{bmatrix}.$$

The off-diagonal entries have gone from s to s^3. We cannot promise that the next step will produce exactly s^9, for a reason that is explained next. But if the algorithm is done correctly, the succeeding steps *will* give entries of order s^9, s^{27}, s^{81}, ... and the matrices become diagonal at terrific speed. Symmetric matrices of order n need three or four QR steps for each new eigenvalue and about $9n^2$ multiplications to find them all. But this requires two crucial improvements in the basic algorithm, including an $O(n^3)$ preprocessing step, as follows:

(1) **Shifts.** If the number b is close to an eigenvalue, the QR factorization converges faster for $A - bI$ than for A. With a shift the step becomes

$$\textbf{\textit{factor }} A_k - b_k I \textbf{\textit{ into }} Q_k R_k \textbf{\textit{ and form }} A_{k+1} = R_k Q_k + b_k I. \tag{16}$$

It is still true that A_{k+1} is similar to A_k and therefore has the same eigenvalues. In practice the last step of adding back $b_k I$ is not necessary, since we can keep a running total of the shifts.

The simplest choice of shift is the last entry on the diagonal; that entry converges first to an eigenvalue. For our 2 by 2 example with $a_{22} = 0$ the shift at the first step was $b = 0$; that explains the appearance of s^3 without a shift. But the second step *will* require a shift to maintain cubic convergence, since the last diagonal entry in A_1 is now $-cs^2$. When the eigenvalue is found, it is removed to leave a smaller matrix and a new series of shifts.

Wilkinson proposed a strategy that is even faster—to choose b_k as an eigenvalue of the last 2 by 2 submatrix in A_k. Of the two eigenvalues he takes the one nearest to the last diagonal entry. Of course for a 2 by 2 matrix he is finished in one step, but the serious applications of the QR algorithm are to larger matrices—and to unsymmetric matrices, where a complex pair $\lambda = a \pm ib$ is indicated by a 2 by 2 block on the diagonal as the QR algorithm proceeds.

(2) **Preprocessing to a tridiagonal A_0.** The factorization of a full matrix into $A = QR$ is expensive, and that is needed at every step. But if A is first reduced to a tridiagonal form A_0, then all later matrices remain tridiagonal. The algorithm is greatly accelerated. Therefore we construct an orthogonal matrix U connecting the original A to a tridiagonal $A_0 = U^{-1}AU$. This is the *direct method* in the eigenvalue problem. There is no direct way to reach a complete diagonalization since there will never be a formula for the roots of $\det(A - \lambda I) = 0$. However a great many zeros can be achieved; we only have to accept one diagonal of nonzeros below the main diagonal. The matrix can be made tridiagonal in a simple sequence of steps before the QR method begins.

We emphasize that the row operations of elimination are not allowed. They change the eigenvalues as they produce zeros in the matrix. It is only by applying one operation to the rows and simultaneously applying the inverse operation to the columns, as in $U^{-1}AU$, that the eigenvalues remain fixed. With this first step from A to A_0, followed by the shifts and QR factorizations that produce $A_1, A_2, \ldots$ and converge to the eigenvalues, the QR algorithm is set. It is the most popular choice in EISPACK, the algorithm package for the eigenvalue problem. It is also available in MATLAB, a useful code for matrix computations on the IBM PC.

To complete this section we want to say more about the actual steps in the calculation. We concentrate on three specific problems:

1. Orthogonalization (from A to Q and R)
2. Tridiagonalization (from A to $A_0 = U^{-1}AU$)
3. Solution of the normal equations $A^T C A x = A^T C b$.

Each of them can be accomplished by calling the right code. Therefore what follows is definitely optional, but it has a purpose. The central achievement of numerical linear algebra is to solve matrix problems from hundreds of different sources, in a way that combines speed and numerical stability. Looking at the algorithms that are most successful, the key seems to be an intelligent use of orthogonal matrices. They are certainly stable; what we need is a fast way to produce them. There is always Gram-Schmidt, but Householder had a better idea.

Householder Matrices

The Gram-Schmidt process started with n independent vectors and ended with n orthonormal vectors. It began with the columns of A (m by n) and it produced the columns of Q (m by n), with $Q^T Q = I$. A and Q were connected by a square upper triangular matrix R, so that $A = QR$.

This seemed natural at the time, but compared to Gaussian elimination there is something bizarre. In elimination we started with A and ended with an upper triangular matrix. The goal was to get ready for back substitution. If that is still true, it should be R that we aim for; Q should be only the matrix that gets us there. The steps of elimination were row operations, recorded in a lower triangular L. By analogy, the steps toward R should be "little orthogonal transformations," and they will combine to give Q.

For elimination there was a choice between producing one zero at a time, or a column of zeros below the pivot. The matrix A could be multiplied by

$$\begin{bmatrix} 1 & & & \\ -l_{21} & 1 & & \\ 0 & 0 & 1 & \\ 0 & 0 & 0 & 1 \end{bmatrix} \quad \text{or} \quad \begin{bmatrix} 1 & & & \\ -l_{21} & 1 & & \\ -l_{31} & 0 & 1 & \\ -l_{41} & 0 & 0 & 1 \end{bmatrix}.$$

The first matrix subtracts a multiple of row 1 from row 2. The second subtracts multiples of row 1 from all other rows. Matrices of the first type multiply together to give the second type. It is their inverses (with plus signs) that take the upper triangular U back to A. The happy accident of elimination is that multiplying these inverses does not mix them together, and the inverse of the whole process is given by L.

For $A = QR$ there is a similar choice. We go from A to R by producing zeros below the diagonal, one at a time or by columns. Lower triangular matrices are no use; we need *orthogonal matrices*. For a single zero, it is sufficient to use a "plane rotation" like

$$\begin{array}{l} \text{row } p \rightarrow \\ \\ \text{row } q \rightarrow \end{array} \begin{bmatrix} c & & -s & \\ & 1 & & \\ s & & c & \\ & & & 1 \end{bmatrix} \quad \text{with} \quad \begin{array}{l} c = \cos\theta \\ \\ s = \sin\theta \end{array}$$

Certainly this is an orthogonal matrix; its columns have length one and they are perpendicular. Its inverse is its transpose. When it multiplies A it alters rows p and q, and the right θ will produce the zero we want. A sequence of plane rotations, chosen in any order that does not spoil zeros already achieved, gives a matrix that goes from A to R. It is still an orthogonal matrix, since it is a product of orthogonal matrices.

Actually there is one other difference from Gram-Schmidt. These rotation matrices are square. Previously Q was rectangular and R was square, but now it is the reverse:

$$A_{m \times n} = Q_{m \times n} R_{n \times n} \text{ is changed to } A_{m \times n} = \bar{Q}_{m \times m} \bar{R}_{m \times n}.$$

The difference is not so great since Q and R have not disappeared:

$$\bar{Q} = \left[\, Q \,\middle|\, \cdots \,\right] \quad \text{and} \quad \bar{R} = \left[\begin{array}{c} R \\ \hline 0 \end{array}\right]. \tag{17}$$

The product $\bar{Q}\bar{R}$ agrees with QR. The zero rows in $\bar{R}$ destroy the final columns of $\bar{Q}$ (which are not zero since they are part of an orthogonal matrix). Of course R itself is triangular; that is the main point. But $\bar{Q}\bar{R}$ does achieve one thing more than QR. It displays not just the n columns of Q but also $m - n$ additional columns (where the dots are) that complete an orthonormal set. The first n columns are combinations of the columns of A, and the last $m - n$ are *perpendicular* to the columns of A. One is an orthonormal basis for the column space, the other for the left nullspace. This extra information in $\bar{Q}$ will help when there is a constraint $A^T y = 0$.

Normally no bars go above these matrices; we remember their shape.

Now come **Householder matrices**. They are square and orthogonal like plane rotations, but they can produce several zeros at once. Fortunately they have a simple form:

$$\boxed{H = I - 2uu^T \text{ for some unit vector } u.}$$

This matrix is obviously symmetric. It is also orthogonal because

$$H^T H = (I - 2uu^T)(I - 2uu^T) = I - 4uu^T + 4uu^Tuu^T = I. \tag{18}$$

We emphasize the difference between the matrix uu^T (column times row) and the number $u^T u = 1$ inside uu^Tuu^T.

EXAMPLES   If $u = \begin{bmatrix} 1 \\ 0 \end{bmatrix}$ then $H = I - 2uu^T = \begin{bmatrix} -1 & 0 \\ 0 & 1 \end{bmatrix}$.

That matrix takes (x, y) to $(-x, y)$, a reflection across the y-axis.

$$\text{If } u = \begin{bmatrix} 1/\sqrt{2} \\ -1/\sqrt{2} \end{bmatrix} \quad \text{then} \quad H = \begin{bmatrix} 0 & 1 \\ 1 & 0 \end{bmatrix}.$$

This matrix takes (x,y) to (y,x), a reflection across the $45°$ line.

$$\text{If } u = \begin{bmatrix} \cos\theta \\ -\sin\theta \end{bmatrix} \quad \text{then} \quad H = \begin{bmatrix} -\cos 2\theta & \sin 2\theta \\ \sin 2\theta & \cos 2\theta \end{bmatrix}.$$

H looks like a plane rotation, but its determinant is -1 (as always) and it must be a reflection across the line at angle θ.

Geometrically, a Householder matrix H takes every vector into its mirror image on the other side of a plane—the plane perpendicular to u. This reflection is sketched in Fig. 5.8, where $Hu = -u$ and $Ha = r$. The matrix $H = I - 2uu^T$ removes *twice* the component in the direction of u, so it carries each vector onto the opposite side of the plane. Only vectors in the plane $u^Tx = 0$ are fixed; they are perpendicular to u, so $Hx = x$.

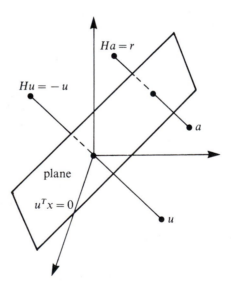

Fig. 5.8. Reflection in a plane by a Householder matrix.

Starting with any nonzero vector v, the normalization $u = v/\|v\|$ gives a unit vector whose Householder matrix is $H = I - 2vv^T/\|v\|^2$. Our goal is to use these matrices to change the columns of A into the columns of R, and we begin with the first column. You can see in the figure that u should be chosen parallel to $a - r$, if H is to take a into r.

5E   The Householder matrix that takes a into r has $v = a - r$:

$$\text{if } a = \begin{bmatrix} a_1 \\ a_2 \\ \vdots \\ a_m \end{bmatrix} \quad \text{and} \quad r = \begin{bmatrix} \|a\| \\ 0 \\ \vdots \\ 0 \end{bmatrix} \quad \text{then } Ha = r. \tag{19}$$

The lengths obey $\|a\| = \|r\|$ since H is an orthogonal matrix.

To prove that $Ha = r$, when $v = a - r$ and $a^T a = r^T r$, we multiply

$$Ha = \left(I - 2\frac{vv^T}{\|v\|^2} \right) a = a - (a - r)\frac{2(a - r)^T a}{(a - r)^T (a - r)}$$

$$= a - (a - r) = r.$$

This identity can be used right away. It gives a matrix (call it H_1) that takes the first column of A to the first column of R. The remaining columns are changed, as in elimination. Then the second step works with the second column of $H_1 A$. This becomes the new vector a, with zero inserted above the pivot. It will turn into $r = (0, r_{22}, 0, \ldots)$ provided H_2 is constructed from this new $v = a - r$. Furthermore the first row and column of $H_1 A$, including the entry r_{12} that is actually present above the pivot, **will not be changed by H_2**—because a and r and v begin with zeros and H_2 begins with 1:

$$H_2(H_1 A) = \left(I - 2\frac{vv^T}{\|v\|^2} \right) \begin{bmatrix} r_{11} & r_{12} & r_{13} \\ 0 & \ulcorner \quad \urcorner & * \\ 0 & | \quad | & * \\ 0 & | \; a \; | & * \\ 0 & \llcorner \quad \lrcorner & * \end{bmatrix} = \begin{bmatrix} r_{11} & r_{12} & r_{13} \\ 0 & r_{22} & r_{23} \\ 0 & 0 & * \\ 0 & 0 & * \\ 0 & 0 & * \end{bmatrix}.$$

In this 5 by 3 case, one more matrix H_3 will complete the orthogonalization $H_3 H_2 H_1 A = R$. Since Householder matrices have $H^{-1} = H^T = H$, this is the same as

$$A = H_1 H_2 H_3 R = QR.$$

These are the $\bar{Q}$ and $\bar{R}$ of (17), with a square matrix $\bar{Q}$ and zeros in the last two rows of $\bar{R}$.

EXAMPLE $A = \begin{bmatrix} 3 & 4 \\ 4 & 0 \end{bmatrix}$ starts with $a = \begin{bmatrix} 3 \\ 4 \end{bmatrix}$, $r = \begin{bmatrix} 5 \\ 0 \end{bmatrix}$, $a - r = \begin{bmatrix} -2 \\ 4 \end{bmatrix}$.

Then the Householder matrix with $v = a - r$ is

$$H = I - 2\frac{vv^T}{v^Tv} = \begin{bmatrix} 1 & 0 \\ 0 & 1 \end{bmatrix} - \frac{2}{20}\begin{bmatrix} 4 & -8 \\ -8 & 16 \end{bmatrix} = \begin{bmatrix} .6 & .8 \\ .8 & -.6 \end{bmatrix}.$$

H is symmetric and orthogonal, and HA is triangular:

$$HA = \begin{bmatrix} .6 & .8 \\ .8 & -.6 \end{bmatrix}\begin{bmatrix} 3 & 4 \\ 4 & 0 \end{bmatrix} = \begin{bmatrix} 5 & 2.4 \\ 0 & 3.2 \end{bmatrix} = R.$$

In this one-step case, H is Q. A larger matrix requires several H_i and their product is the same Q produced by Gram-Schmidt.

One practical point: The matrix product $Q = H_1 \cdots H_n$ is never actually computed. All we store are the vectors v_i that govern each H_i. A multiplication like Hx needs only inner products with the v_i. Therefore the orthogonalization is very fast; to go from A to R takes the same mn^2 operations as Gram-Schmidt. Because of the Householder matrices it is extremely stable. But for sparse matrices, QR fills in more zeros than LDL^T and there is a delicate battle between stability and speed.

Orthogonalization solves the problem of (unweighted) least squares:

5F   The least squares solution to $Ax = b$ is computed by applying Householder transformations $H_1, H_2, \ldots, H_n$ to both sides, leaving

$$\begin{bmatrix} R \\ 0 \end{bmatrix}[x] = \begin{bmatrix} c \\ d \end{bmatrix}. \tag{20}$$

Lengths are unchanged, and the product of the H_i is Q. The solution x is given by back substitution in $Rx = c$, and the error $\|Ax - b\|$ is equal to $\|d\|$.

A second use of Householder is to make a symmetric matrix tridiagonal; A goes to A_0 while preserving its eigenvalues. In that case the matrices H appear *on both sides of* A. The first step takes A to $H_1AH_1^{-1} = H_1AH_1$, acting on columns as well as rows, and we cannot expect zeros in all entries below the pivot. The product H_1A could have those zeros but they would be destroyed by the H_1 that comes on the right. By accepting an extra nonzero below the pivot, and taking

$$a = \begin{bmatrix} a_{11} \\ a_{21} \\ a_{31} \\ \cdot \end{bmatrix} \quad \text{into} \quad r = \begin{bmatrix} a_{11} \\ \alpha \\ 0 \\ \cdot \end{bmatrix},$$

the $n - 2$ zeros in r will remain in the first column of H_1AH_1.

EXAMPLE $A = \begin{bmatrix} 1 & 0 & 1 \\ 0 & 1 & 2 \\ 1 & 2 & 0 \end{bmatrix}$ has $a = \begin{bmatrix} 1 \\ 0 \\ 1 \end{bmatrix}$ and $r = \begin{bmatrix} 1 \\ 1 \\ 0 \end{bmatrix}$

Then $H_1 = I - 2\dfrac{(a-r)(a-r)^T}{(a-r)^T(a-r)} = \begin{bmatrix} 1 & 0 & 0 \\ 0 & 0 & 1 \\ 0 & 1 & 0 \end{bmatrix}$ and $A_0 = H_1 A H_1 = \begin{bmatrix} 1 & 1 & 0 \\ 1 & 0 & 2 \\ 0 & 2 & 1 \end{bmatrix}$.

In this 3 by 3 case the result is already tridiagonal. In the 4 by 4 case a second matrix H_2 is needed for the second column.

Notice that we are claiming zeros *above* the tridiagonal part as well as below. That is because A_0 is automatically symmetric; for $n = 5$ it looks like

$$A_0 = H_3 H_2 H_1 A H_1 H_2 H_3.$$

This is $U^{-1}AU$, with $U = H_1 H_2 H_3$ orthogonal because it is a product of orthogonal matrices. A_0 is tridiagonal, it has the same eigenvalues as A, and it is ready to reveal them—the QR algorithm is ready to begin.

Difficulties with C

The final application of Householder matrices is to the equilibrium equations themselves. Up to now the matrix C has cooperated completely. We pretended it was the identity, in describing the two basic approaches to least squares:

1. Form $A^T A$ and solve $A^T A x = A^T b$
2. Factor A into QR and solve $Rx = Q^T b$.

When C is present, or there is a weighting matrix W (with $C = W^T W$), the least squares problem minimizes the weighted error

$$P(x) = (Ax - b)^T C(Ax - b) = \| W(Ax - b) \|^2.$$

Now we have to account for C.

In method 1, the normal equations become $A^T C A x = A^T C b$. In method 2, the QR factorization works with WA instead of A and Wb instead of b. When C and W are close to diagonal these changes are easy. The same is true of the full equilibrium equations, with $A^T Cb - f$ on the right hand side; that is still simple if C is sparse. The awkward case is the one in which C^{-1} *rather than C is close to diagonal*.

This unhappy possibility occurs in statistics and signal processing. In those applications Gauss announced that C should be V^{-1}—the inverse of the covariance matrix of the measurement errors. When the measurements are independent and V is diagonal, there is no difficulty. If neighboring measurements are correlated then V is still a banded matrix. However V^{-1} is terrible. The zeros disappear and the inverse of a band matrix is a full matrix. In this brief (and optional) section we consider how to avoid working with the full matrix $C = V^{-1}$.

Method 1 was based on $A^T C A$, which will also be full. Therefore we try method 2, beginning with a QR factorization of A alone. This produces a major improvement in both equilibrium equations. If $C^{-1} = V$ in the first equation $C^{-1}y + Ax = b$, and if we multiply by Q^T, then

$$Q^T V y + \begin{bmatrix} R \\ 0 \end{bmatrix} x = \begin{bmatrix} c \\ d \end{bmatrix}. \tag{21}$$

The last two terms are exactly as in (20). We emphasize that this is the square orthogonal matrix Q (sometimes written $\bar{Q}$) whose first n columns come from the columns of A. Its last $m - n$ columns are perpendicular to the columns of A. They appeared at no cost when we used square Householder matrices.

The second equilibrium equation is $A^T y = 0$. It requires y to be perpendicular to the rows of A^T, which are the columns of A. This is exactly the condition satisfied by the final columns of Q! Therefore y should be a combination (with unknown coefficients z_i) of those last $m - n$ columns:

$$y = Q \begin{bmatrix} 0 \\ z \end{bmatrix}. \tag{22}$$

Multiplying by $A^T = R^T Q^T$ does give zero, because $Q^T Q = I$ and

$$A^T y = [R^T | 0^T] \begin{bmatrix} 0 \\ z \end{bmatrix} = 0.$$

Now substitute (22) into the first equilibrium equation (21):

$$Q^T V Q \begin{bmatrix} 0 \\ z \end{bmatrix} + \begin{bmatrix} R \\ 0 \end{bmatrix} x = \begin{bmatrix} c \\ d \end{bmatrix}. \tag{23}$$

This splits into two equations when $Q^T V Q$ is partitioned into blocks:

$$Q^T V Q = \begin{bmatrix} E & F \\ F & G \end{bmatrix}. \tag{24}$$

The multiplications in (23) produce

$$\begin{aligned} Fz + Rx &= c \\ Gz \quad\;\; &= d \end{aligned} \tag{25}$$

and *these* are the equilibrium equations that we solve. The second gives $z = G^{-1}d$ and the first gives $x = R^{-1}(c - FG^{-1}d)$. When V is the identity matrix, F will be zero and we are back to the equation $Rx = c$ of ordinary least squares. If V is sparse we have managed to escape the full matrix V^{-1}, at the cost of some algebra (for which I apologize). At least the conclusion is easy to state:

5G If $A = QR$, then the vector x that minimizes $(Ax - b)^T V^{-1}(Ax - b)$ can be computed from (25) without actually inverting V.

When A is taken to R by Householder matrices or plane rotations, they are applied to both sides of V in (24) and to b in (21). Their product is Q, known implicitly from the factors rather than explicitly. For $n = 10$ and $m = 1000$, we have avoided the millions of multiplications in the full matrix $A^T V^{-1} A$.

There is an enormous least squares project by the Geodetic Survey to readjust the North American Datum—a network of reference points that will eventually be known within a few centimeters. We cannot display $Ax = b$; it is one of the largest computations ever attempted. It uses 6,000,000 observations to find 400,000 unknowns. The equations are actually nonlinear, because of trigonometric identities and distance formulas—but it is their size and sparsity that will be crucial. For the accuracy required I would not form $A^T A$. It is better to find Q and R by Householder matrices or by plane rotations (which have less fill-in and are preferred). A recommended ordering of unknowns comes from the nested dissection of 5.1, known in geodesy as Helmert blocking. And the new feature of such an overwhelming problem is management of the data, which comes from theodolites with one-second accuracy and has to be recalled from storage when the orthogonalization algorithm needs it.

I think the 1983 completion date was very optimistic.

EXERCISES

5.2.1 If Q contains the first n columns of the m by m identity matrix, what is the best least squares solution to $Qc = b$? Here $b = (b_1, \ldots, b_m)$ is any vector in R^m and $c = (c_1, \ldots, c_n)$ is in R^n.

5.2.2 (i) Find $A^T A$ if the columns of A are orthogonal but not orthonormal—their lengths are $l_1 > 0, \ldots, l_n > 0$. (ii) If A has only one column a, the normal equation $A^T Ax = A^T b$ gives $x = a^T b / a^T a$. Find x and the projection p of $b = (1, 1, 1)$ onto the line through $a = (3, 0, 4)$.

5.2.3 The distance (squared) from the point b to the line through a is

$$\| b - p \|^2 = \left\| b - \frac{a^T b}{a^T a} a \right\|^2 = \frac{(b^T b)(a^T a) - (a^T b)^2}{a^T a}$$

Since this cannot be negative, the numerator gives the **Schwarz inequality** $|a^T b| \leq \|a\| \, \|b\|$.

(1) How does this inequality come from the geometric formula

$$a^T b = \|a\| \, \|b\| \cos \theta?$$

(2) By making the right choice of b, show that always

$$(a_1 + a_2 + \cdots + a_m)^2 \leq m(a_1^2 + a_2^2 + \cdots + a_m^2).$$

(3) When does equality hold, in this particular case and in $|a^T b| \leq \|a\| \, \|b\|$?

5.2.4 (a) Show that

$$Q = \frac{1}{2} \begin{bmatrix} 1 & -1 & -1 & -1 \\ -1 & 1 & -1 & -1 \\ -1 & -1 & 1 & -1 \\ -1 & -1 & -1 & 1 \end{bmatrix} \quad \text{is an orthogonal matrix.}$$

(b) Can you find an orthogonal matrix like this in which the first row and column are all positive: $q_{i1} = q_{1j} = 1/2$?

5.2.5 Find the best least squares solution to

$$\frac{1}{2} \begin{bmatrix} 1 & -1 \\ -1 & 1 \\ -1 & -1 \\ -1 & -1 \end{bmatrix} \begin{bmatrix} x_1 \\ x_2 \end{bmatrix} = \begin{bmatrix} 1 \\ 1 \\ 1 \\ 1 \end{bmatrix} = b.$$

What is the projection of b onto the space spanned by these two columns? What is its projection orthogonal to this space?

5.2.6 When can the points x_1 and x_2 on the x-axis be the projections of two orthonormal vectors (x_1, y_1) and (x_2, y_2)? Draw a figure and prove that $x_1^2 + x_2^2 = 1$.

5.2.7 Suppose the three points (x_1, y_1), (x_2, y_2), (x_3, y_3) in a plane are projections of orthonormal vectors $q_1 = (x_1, y_1, z_1)$, $q_2 = (x_2, y_2, z_2)$, $q_3 = (x_3, y_3, z_3)$ in three-dimensional space. From the connection between orthonormal rows and orthonormal columns, show that this is only possible if the vectors (x_1, x_2, x_3) and (y_1, y_2, y_3) are orthonormal.

5.2.8 What is the 3 by 3 projection matrix P onto the line through $a = (1, 2, 2)$?

5.2.9 Write A as QR and compare $x = (A^TA)^{-1}A^Tb$ with $x = R^{-1}Q^Tb$, if

$$A = \begin{bmatrix} 1 & -6 \\ 3 & 6 \\ 4 & 8 \\ 5 & 0 \\ 7 & 8 \end{bmatrix} \quad \text{and} \quad b = \begin{bmatrix} -3 \\ 7 \\ 1 \\ 0 \\ 4 \end{bmatrix}.$$

5.2.10 The projection matrix onto the column space of A is $P = A(A^TA)^{-1}A^T$. If $A = QR$ find a simpler formula for P.

5.2.11 Factor the following matrix into $A = QR$:

$$A = \begin{bmatrix} \frac{2}{3} & 0 \\ \frac{2}{3} & 2 \\ \frac{1}{3} & -1 \end{bmatrix}.$$

5.2.12 What rotation angle θ produces a zero below the diagonal in the matrix product

$$\begin{bmatrix} \cos\theta & -\sin\theta \\ \sin\theta & \cos\theta \end{bmatrix} \begin{bmatrix} a & b \\ c & d \end{bmatrix}?$$

Find the QR factorization of the matrix on the right.

5.2.13 What choice of the rotation angle θ will make A_0 tridiagonal?

$$A_0 = \begin{bmatrix} 1 & & \\ & c & -s \\ & s & c \end{bmatrix} \begin{bmatrix} 2 & 1 & 1 \\ 1 & 2 & 1 \\ 1 & 1 & 3 \end{bmatrix} \begin{bmatrix} 1 & & \\ & c & s \\ & -s & c \end{bmatrix} = U^{-1}AU \quad \text{with} \quad \begin{array}{l} c = \cos\theta \\ s = \sin\theta \end{array}$$

5.2.14 Find the next matrix $A_1 = RQ$ if

$$A = \begin{bmatrix} 3 & 4 \\ 4 & 0 \end{bmatrix} = \begin{bmatrix} .6 & .8 \\ .8 & -.6 \end{bmatrix} \begin{bmatrix} 5 & 2.4 \\ 0 & 3.2 \end{bmatrix} = QR.$$

Confirm that the trace of A equals the trace of A_1.

5.2.15 With $a_{33} = 2$ as the shift, apply one step of shifted QR to

$$A = \begin{bmatrix} 2 & -1 & -1 \\ -1 & 2 & -1 \\ -1 & -1 & 2 \end{bmatrix}.$$

5.2.16 Verify that A_{k+1} is equal to $Q_k^{-1}A_kQ_k$ in (16), so that eigenvalues are preserved at each shifted QR step.

5.2.17 If A_0 is symmetric and tridiagonal show that $A_1 = RQ = Q^{-1}A_0Q$ is also symmetric and tridiagonal. *Hint*: Indicate by x the nonzero entries in a 4 by 4 $A_0 = QR$. Each column of Q comes by Gram-Schmidt from the current column and previous column of the tridiagonal A_0.

5.2.18 If H is a Householder matrix and $Hx = y$, show that also $Hy = x$ (the basic property of reflection). What 4 by 4 matrix gives a reflection across the plane $v^Tx = x_1 + x_2 + x_3 + x_4 = 0$?

5.2.19 $P = I - uu^T$ is like a Householder matrix except that the coefficient 2 is missing. P subtracts from every vector its component in the direction of u; it is a projection *onto* the plane in Fig. 5.8. Show that $P^2 = P$ and compute P for the three vectors

$$u = \begin{bmatrix} 1 \\ 0 \end{bmatrix}, \quad u = \begin{bmatrix} 1/\sqrt{2} \\ -1/\sqrt{2} \end{bmatrix}, \quad u = \begin{bmatrix} \cos\theta \\ -\sin\theta \end{bmatrix}.$$

5.2.20 (a) Find the Householder matrix H that takes $a = (\cos\theta, \sin\theta)$ into $r = (1, 0)$.
(b) Use it to find R in the factorization

$$HA = H \begin{bmatrix} \cos\theta & \sin\theta \\ \sin\theta & 0 \end{bmatrix} = R.$$

(c) Verify that your matrix equals

$$H = \begin{bmatrix} \cos\theta & \sin\theta \\ \sin\theta & -\cos\theta \end{bmatrix}.$$

It is the same as Q in $A = QR$; the columns of A have been made orthonormal.

5.2.21 Find the Householder matrix H_1 that produces zeros below the first entry in

$$A = \begin{bmatrix} 2 & 1 \\ 1 & 0 \\ 2 & 2 \end{bmatrix}.$$

Compute $H_1 A$. Since its (3,2) entry becomes zero H_2 is not needed.

5.2.22 If $H = I - 2uu^T$ show that the unit vector u is an eigenvector and find the eigenvalue. Show also that any x perpendicular to u is an eigenvalue with $\lambda = 1$; in other words $Hx = x$. From the product of eigenvalues find the determinant of H, and from the sum of eigenvalues find the trace.

5.2.23 Construct H to produce zeros below the first entry in HA if

$$A = \begin{bmatrix} 1 & 1 & 1 \\ 1 & 1 & 1 \\ 1 & 1 & 1 \\ 1 & 1 & 1 \end{bmatrix} \quad \text{with} \quad a = \begin{bmatrix} 1 \\ 1 \\ 1 \\ 1 \end{bmatrix} \quad \text{and} \quad r = \begin{bmatrix} \|a\| \\ 0 \\ 0 \\ 0 \end{bmatrix} = \begin{bmatrix} 2 \\ 0 \\ 0 \\ 0 \end{bmatrix}.$$

Compute $HA = R$ and explain this strange behavior.

5.2.24 Find H and the tridiagonal form $A_0 = HAH$ for

$$A = \begin{bmatrix} 1 & 3 & 4 \\ 3 & 2 & 1 \\ 4 & 1 & 0 \end{bmatrix}.$$

5.2.25 Show that V^{-1} is full and compute it from L^{-1} if

$$V = \begin{bmatrix} 1 & & & \\ 1 & 1 & & \\ 0 & 1 & 1 & \\ 0 & 0 & 1 & 1 \end{bmatrix} \begin{bmatrix} 1 & 1 & 0 & 0 \\ & 1 & 1 & 0 \\ & & 1 & 1 \\ & & & 1 \end{bmatrix} = LL^T.$$

SEMI-DIRECT AND ITERATIVE METHODS ■ 5.3

The construction of an iterative method looks pretty relaxed. If the original matrix A is too hard to work with, the numerical analyst replaces it by a simpler matrix M. The difference $A - M$ is moved to the other side of the equation. Starting from $Ax = b$, this splitting produces the completely equivalent system

$$Mx = (M - A)x + b. \tag{1}$$

So far nothing is new; an exact solution will force us back to A. The novelty is to solve (1) *iteratively* by successive substitution. We begin with an initial guess x_0—it may be zero—and then at every step the current guess x_k leads to the next approximation x_{k+1}:

$$\boxed{Mx_{k+1} = (M - A)x_k + b.} \tag{2}$$

In this way the coefficient matrix does change to M, but at a price; the simpler equation (2) has to be solved over and over.

Not every matrix M will work; it is best if it is close to A. On the other hand M has to remain as convenient as possible. The choice $M = A$ gives immediate convergence; $M - A$ is zero and the first and only iteration would be $Ax_1 = b$. At the other extreme is $M = I$, the identity matrix, whose simplicity is hard to beat. But this simplicity is matched, and in some respects even exceeded, by three choices of M that come directly from A:

1. $M =$ diagonal part of A (Jacobi's method)
2. $M =$ triangular part of A (Gauss-Seidel method)
3. $M =$ combination of 1 and 2 (successive overrelaxation)

I admit that those choices do not look especially brilliant. It is surprising how important and effective they are, and our first goal is to understand why.

Remember that the convergence of the vectors $x_0, x_1, x_2, \ldots$ is not at all automatic. We can say this: If they do approach a limit x_∞, then that limit is the correct solution $A^{-1}b$ to the original system. For proof look at (2); in the limit it becomes $Mx_\infty = (M - A)x_\infty + b$, which means that $Ax_\infty = b$. The important question is convergence, and equally important is the *rate* of convergence. If the x_k go quickly toward $A^{-1}b$, then the iterative method becomes an alternative to elimination. It cannot give the exact answer in a finite time, but it may give a good approximation much earlier.

You will see below that convergence is decided by the powers of the matrix $B = M^{-1}(M - A)$. In fact *powers of matrices* are the common element in all the methods of this section. For eigenvalues it will be the powers of A itself, or its inverse, or its shifted inverse $(A - sI)^{-1}$. Semi-direct methods use the vectors $x_0, Ax_0, A^2x_0, \ldots$, again formed by repeatedly multiplying by A. But the actual powers A^2, A^3, ... are never computed in any of these methods! All multiplications are matrices times vectors.

Semi-direct methods do find $A^{-1}b$ in a fixed number of steps, if we exclude roundoff error. When compared for speed alone with direct methods, they lose. They take longer to finish. But in the right circumstances they are terrific as iterative methods. A semi-direct method starts from any initial guess x_0, produces improved approximations x_k, and can be terminated at will—something that Gauss could not do.

The convergence or divergence of iterative methods must depend on M and A. To see how, subtract equation (2) from equation (1). This produces an **error equation** for the differences $e_k = x - x_k$:

$$Me_{k+1} = (M - A)e_k. \tag{3}$$

This is a one-step difference equation, from which the vector b has disappeared. It is even simpler after multiplication by M^{-1}:

$$e_{k+1} = M^{-1}(M - A)e_k = Be_k. \tag{4}$$

Every step of the iteration does the same thing to the error; it is multiplied by $B = I - M^{-1}A$. After k iterations there have been k multiplications, and the current error is related to the original error by $e_k = B^k e_0$.

This formula shows that everything rests on one point. *Do the powers B^k approach zero?* If they do, then the sequence e_k also goes to zero. Therefore the question of convergence is identical to the question of **stability**:

$$e_k \to 0 \quad \text{and} \quad x_k \to x \quad \text{exactly when } B^k \to 0. \tag{5}$$

The test for the powers of B to converge to the zero matrix is absolutely basic.

5H   The powers B^k approach zero if and only if every eigenvalue of B satisfies $|\lambda_i| < 1$. The rate of convergence is governed by the largest of the $|\lambda_i|$, which is called the *spectral radius* of B.

Thus the eigenvalues of B, which may be positive or negative or complex numbers, must lie **inside the unit circle** in the complex plane. If λ is negative then $|\lambda|$ is its absolute value. If $\lambda = a + ib$ then $|\lambda|$ is the square root of $a^2 + b^2$. In every case $|\lambda|$ is the distance r from the origin 0 to the point $\lambda = re^{i\theta}$, and we need $r < 1$.

The importance of 5H is that the eigenvalues are in complete control. In the applications it may not be obvious at a glance whether every $|\lambda(B)| < 1$. This test is passed if the entries of B are small enough; the exercises show that when all absolute row sums are less than one, the same is true of the eigenvalues. However such a condition on the rows (or columns) is not necessary. The key is to recognize that the error $B^k e_0$ splits into n different terms, each one proceeding independently according to its own eigenvalue:

$$B^k e_0 = c_1 \lambda_1^k x_1 + \cdots + c_n \lambda_n^k x_n. \tag{6}$$

Here $x_1, \ldots, x_n$ are the eigenvectors; we assume B has a complete set. Every multiplication by B, at every iteration step, will multiply each eigenvector x_i by one more factor λ_i. Therefore (6) is correct if it is correct at the starting point $k = 0$; the constants c_i come from the initial error by $e_0 = c_1 x_1 + \cdots + c_n x_n$. Equation (6) shows that the error $B^k e_0$ goes to zero when every eigenvalue satisfies $|\lambda_i| < 1$. It is exactly like differential equations, in which the powers λ^k change to exponentials $e^{\lambda t}$ and the stability condition becomes $\text{Re } \lambda < 0$.

You see that the largest $|\lambda_i|$ is dominant in (6). It governs the rate at which the errors approach zero. There are some slight exceptions: If a coefficient c_i happens to be zero then the eigenvector x_i is not activated, at least in exact arithmetic. But convergence is required from any error e_0, including those (the great majority) that contain all x_i. Also an eigenvalue may happen to be repeated, without contributing its full share of eigenvectors. In that case the Jordan form of B supplies the missing terms; it is J^k rather than the diagonal matrix Λ^k that governs the powers of B. The powers of J involve $k\lambda^k$ if one eigenvector is missing, and higher polynomials $p(k)\lambda^k$ if several eigenvectors are missing—but as k increases these terms are still controlled by λ. Thus the spectral radius, the largest $|\lambda_i|$, is decisive.

5H can be summarized in a sentence. The powers of B are

$$B^k = (S\Lambda S^{-1})(S\Lambda S^{-1}) \cdots (S\Lambda S^{-1}) = S\Lambda^k S^{-1}; \tag{7}$$

their growth or decay is determined by the numbers $\lambda_1^k, \ldots, \lambda_n^k$. Notice in equation (7) why the eigenvalues enter and not the pivots; the powers of LDU or LDL^T give no cancellation but S^{-1} cancels S at each step. The eigenvalues $\lambda = \frac{1}{2}$ are visible in the first example below, whereas $\lambda = 1$ is hidden in the second:

$$B = \begin{bmatrix} 1/2 & 10 \\ 0 & 1/2 \end{bmatrix} \quad \text{looks large but} \quad B^k = \begin{bmatrix} 1/2^k & 20k/2^k \\ 0 & 1/2^k \end{bmatrix} \to 0$$

$$B = \begin{bmatrix} 1/2 & 1/2 \\ 1/2 & 1/2 \end{bmatrix} \quad \text{looks small but} \quad B^k = \begin{bmatrix} 1/2 & 1/2 \\ 1/2 & 1/2 \end{bmatrix} \text{ is not convergent.}$$

The Matrix M

Now comes the key decision—the choice of M. Remember that for a good iterative method there are two conflicting requirements:

1. The equation $Mx_{k+1} = (M - A)x_k + b$ should be easy to solve. Therefore M should be a convenient (and invertible!) matrix. It may be diagonal or triangular.

2. M should be close to A, so that the eigenvalues of $B = M^{-1}(M - A) = I - M^{-1}A$ are as small as possible (and certainly inside the unit circle).

The more the diagonal part of A is dominant, the easier these are.

We want to propose several choices for M. The oldest is the most obvious—M is the *diagonal part* of A. If all a_{ii} are nonzero then that diagonal matrix is extremely easy to invert; we call it D. This choice leads to **Jacobi's method**, and the iteration $Dx_{k+1} = (D - A)x_k + b$ written out in full is

$$a_{11}(x_1)_{k+1} = (-a_{12}x_2 - a_{13}x_3 - \cdots - a_{1n}x_n)_k + b_1$$
$$\vdots$$ (8)
$$a_{nn}(x_n)_{k+1} = (-a_{n1}x_1 - a_{n2}x_2 - \cdots - a_{nn-1}x_{n-1})_k + b_n$$

When A is sparse, the terms on the right are mostly zero and this step from x_k to x_{k+1} is easy. The important question is whether the iteration converges, and how quickly.

EXAMPLE 1 $A = \begin{bmatrix} 2 & -1 \\ -1 & 2 \end{bmatrix}, \quad M = \begin{bmatrix} 2 & \\ & 2 \end{bmatrix}, \quad B = I - M^{-1}A = \begin{bmatrix} 0 & \frac{1}{2} \\ \frac{1}{2} & 0 \end{bmatrix}.$
(JACOBI)

If the components of x are v and w, the Jacobi step is

$$\begin{matrix} 2v_{k+1} = w_k + b_1 \\ 2w_{k+1} = v_k + b_2 \end{matrix} \quad \text{or} \quad x_{k+1} = \begin{bmatrix} 0 & \frac{1}{2} \\ \frac{1}{2} & 0 \end{bmatrix} x_k + \tfrac{1}{2}b.$$ (9)

The decisive matrix B has eigenvalues $\pm\frac{1}{2}$, which means that the error is cut in half (one more bit becomes correct) at every step. In this example, which is much too small to be typical, the convergence is fast.

If we try to imagine a larger matrix A, there is an immediate difficulty with the Jacobi iteration (8). *It requires us to keep all the components of x_k until the calculation of x_{k+1} is complete.* A much more natural idea, which requires only half as much storage, is to start using each component of the new x_{k+1} as soon as it is computed. Then x_{k+1} takes the place of x_k a component at a time, and the old vector x_k can be destroyed as fast as x_{k+1} is created. The first equation finds the first component as in (8), and the next equation operates immediately with that new first component:

$$a_{22}(x_2)_{k+1} = -a_{21}(x_1)_{k+1} + (-a_{23}x_3 - \cdots - a_{2n}x_n)_k + b_2.$$ (10)

The last equation will use the new values exclusively:

$$a_{nn}(x_n)_{k+1} = (-a_{n1}x_1 - a_{n2}x_2 - \cdots - a_{nn-1}x_{n-1})_{k+1} + b_n.$$ (11)

This is called the **Gauss-Seidel method**, even though Gauss didn't know about it and Seidel didn't recommend it. Nevertheless it is a good method. M becomes the lower triangular part of A, when all terms in x_{k+1} are moved to the left side. On the right side, $M - A$ is strictly upper triangular.

EXAMPLE 2
(GAUSS-SEIDEL)
$$A = \begin{bmatrix} 2 & -1 \\ -1 & 2 \end{bmatrix}, \quad M = \begin{bmatrix} 2 & 0 \\ -1 & 2 \end{bmatrix}, \quad B = I - M^{-1}A = \begin{bmatrix} 0 & \frac{1}{2} \\ 0 & \frac{1}{4} \end{bmatrix}$$

A single Gauss-Seidel step takes the components v_k and w_k into

$$\begin{aligned} 2v_{k+1} &= w_k + b_1 \\ 2w_{k+1} &= v_{k+1} + b_2 \end{aligned} \quad \text{or} \quad \begin{bmatrix} 2 & 0 \\ -1 & 2 \end{bmatrix} x_{k+1} = \begin{bmatrix} 0 & 1 \\ 0 & 0 \end{bmatrix} x_k + b.$$

The eigenvalues of B are again decisive, and of course they are easy to find: They are $\frac{1}{4}$ and 0. The error is divided by 4 every time, so *a single Gauss-Seidel step is worth two Jacobi steps*. Since both methods require the same number of operations—we just use the new values instead of the old, and actually save on storage—the Gauss-Seidel method is better. (For parallel machines Jacobi may win!)

There is a way to make it better still. It was discovered during the years of hand computation (probably by accident) that convergence is faster if we go beyond the Gauss-Seidel correction $x_{k+1} - x_k$. Roughly speaking, the ordinary method converges monotonically; the approximations x_k stay on the same side of the solution x. Therefore it is natural to try introducing an *overrelaxation factor* to move closer to the solution. This factor is always called "omega;" $\omega = 1$ recovers Gauss-Seidel. With $\omega > 1$ the method is *successive overrelaxation*, and it is known by the initials SOR. The optimal choice of ω depends on the problem, but it never exceeds 2. It is often near 1.9.

Overrelaxation stays with the lower triangular M of Gauss-Seidel except that it divides the diagonal entries by ω. For the same 2 by 2 matrix each SOR step has M on the left side and $M - A$ on the right:

EXAMPLE 3
(SOR)
$$\begin{bmatrix} 2/\omega & 0 \\ -1 & 2/\omega \end{bmatrix} x_{k+1} = \begin{bmatrix} -2 + 2/\omega & 1 \\ 0 & -2 + 2/\omega \end{bmatrix} x_k + b.$$

The crucial matrix $I - M^{-1}A$, whose eigenvalues govern the convergence, becomes

$$B = \begin{bmatrix} 1 & 0 \\ 0 & 1 \end{bmatrix} - \begin{bmatrix} \frac{1}{2}\omega & 0 \\ \frac{1}{4}\omega^2 & \frac{1}{2}\omega \end{bmatrix} \begin{bmatrix} 2 & -1 \\ -1 & 2 \end{bmatrix} = \begin{bmatrix} 1 - \omega & \frac{1}{2}\omega \\ \frac{1}{2}\omega(1 - \omega) & 1 - \omega + \frac{1}{4}\omega^2 \end{bmatrix}.$$

The optimal choice of ω makes the largest eigenvalue of B as small as possible. The whole point of overrelaxation is to discover this optimal ω—either in advance, as we now do here, or during the computations. To start, the product of the eigenvalues equals the determinant:

$$\lambda_1 \lambda_2 = \det B = (1 - \omega)^2. \tag{12}$$

This is a general rule; in the n by n case the determinant of B is $(1 - \omega)^n$. That already explains why we never go as far as $\omega = 2$, since the product of the eigenvalues would be too large to allow all $|\lambda_i| < 1$; the iteration could not converge. The 2 by 2 case also holds the key to the behavior of the eigenvalues, which is this: At $\omega = 1$ the Gauss-Seidel eigenvalues are 0 and $\frac{1}{4}$, and as ω increases these eigenvalues approach one another. The largest one gets smaller, and *at the optimal ω the two eigenvalues are equal*. At that moment they both equal $\omega - 1$, to give the right determinant.† This value of ω is easy to compute, because the sum $2(\omega - 1)$ of the eigenvalues equals the trace—the sum along the diagonal of B:

$$\lambda_1 + \lambda_2 = 2(\omega - 1) = 2 - 2\omega + \tfrac{1}{4}\omega^2. \tag{13}$$

This quadratic equation yields the optimal value $\omega_{opt} = 4(2 - \sqrt{3}) \approx 1.07$. At that point the two eigenvalues $\lambda_1 = \lambda_2 = \omega - 1$ are approximately 0.07, which is a major reduction from the Gauss-Seidel value $\lambda = \frac{1}{4}$. In this 2 by 2 example the right choice of ω has again doubled the rate of convergence, because $(\frac{1}{4})^2 \approx 0.07$.

The discovery that such an improvement could be produced so easily, almost as if by magic, was the starting point for 20 years of enormous activity in numerical analysis. The first problem was to develop the theory of overrelaxation, and Young's thesis in 1950 contained the solution—a simple formula for the optimal ω. The key step was to find a connection between the eigenvalues λ of $I - M^{-1}A$ and the eigenvalues μ of the original Jacobi matrix $I - D^{-1}A$, when $M = D$ is the diagonal of A. That connection is expressed by

$$(\lambda + \omega - 1)^2 = \lambda\omega^2\mu^2. \tag{14}$$

It is valid for a wide class of finite difference matrices, and if we take $\omega = 1$ (Gauss-Seidel) it yields $\lambda^2 = \lambda\mu^2$. In that case $\lambda = 0$ and $\lambda = \mu^2$. This is confirmed by Examples 1 and 2, in which $\mu = \pm\frac{1}{2}$ and $\lambda = 0$, $\lambda = \frac{1}{4}$. It is completely typical of the relation between Jacobi and Gauss-Seidel: The matrices in Young's class have eigenvalues μ that occur in plus-minus pairs, and the corresponding λ are 0 and μ^2. By using the latest approximations to x, we double the rate of convergence.

The important problem is to do better still; we want to choose ω so that the largest λ will be minimized. Fortunately, this problem is already solved. Young's equation (14) is nothing but the characteristic equation for our 2 by 2 example, and the best ω was the one that made the two roots equal to $\omega - 1$. Exactly as in (13), where $\mu^2 = \frac{1}{4}$, this leads to

$$2(\omega - 1) = 2 - 2\omega + \mu^2\omega^2, \quad \text{or} \quad \omega = \frac{2(1 - \sqrt{1 - \mu^2})}{\mu^2}. \tag{15}$$

The only difference is that for a large matrix this pattern will be repeated for a number of different pairs $\pm\mu_i$—and we can only make a single choice of ω. The

---

† If ω is further increased, the eigenvalues become a complex conjugate pair—both have $|\lambda| = \omega - 1$, so their product is still $(\omega - 1)^2$ and their modulus is now increasing with ω.

largest of these pairs gives the Jacobi eigenvalue $\mu_{\max}$, and it also gives the largest value of ω and of $\lambda = \omega - 1$. Since our goal is to make $\lambda_{\max}$ as small as possible, that extremal pair $\pm\mu_{\max}$ specifies the best choice of ω:

$$\omega_{\mathrm{opt}} = \frac{2(1 - \sqrt{1 - \mu_{\max}^2})}{\mu_{\max}^2} \quad \text{and} \quad \lambda_{\max} = \omega_{\mathrm{opt}} - 1. \tag{16}$$

This is Young's formula for the optimal overrelaxation factor.

The whole theory applies directly to finite difference equations, which are a constant source of large sparse matrices. For Laplace's equation in a square and its 5-point difference approximation, we can compute the improvement brought by ω. Section 5.1 gave the form of the matrix A: It has $+4$ along the main diagonal and -1 on the four diagonals from neighboring mesh points. If there are N^2 unknowns, coming from N vertical lines and N horizontal lines inside the square, then the width of the mesh is $h = 1/(N+1)$. According to Exercise 5.3.8, the largest eigenvalue of the Jacobi iteration (when M is the diagonal matrix $4I$) is $\mu = \cos \pi h$. This is less than one, so the iteration converges. The other iterative methods converge more quickly:

Jacobi: $\quad \mu = \cos \pi h \approx 1 - \frac{1}{2}\pi^2 h^2$

Gauss-Seidel: $\quad \lambda = \mu^2 = \cos^2\pi h \approx 1 - \pi^2 h^2$

Successive overrelaxation: $\quad \lambda = \dfrac{1 - \sin \pi h}{1 + \sin \pi h} \approx 1 - 2\pi h.$

Note the crucial point! There is an order of magnitude improvement, from h^2 to h, in the distance between λ and 1. The eigenvalues of B are moved further away from the unit circle. This can only be appreciated by an example.

Suppose $N = 21$, which is moderate. The matrices are of order $N^2 = 441$. The meshwidth is $h = 1/22$, so that $\cos \pi h = .99$ and the Jacobi method is slow. Gauss-Seidel is twice as fast, with largest eigenvalue $\cos^2\pi h = .98$. But since $\sin \pi h = \sqrt{.02} = .14$, the convergence factor with overrelaxation will be

$$\lambda_{\max} = \frac{1 - .14}{1 + .14} = .75.$$

The error is reduced by 25% at every step. Therefore a *single* SOR step is the equivalent of 29 Jacobi steps: $(.99)^{29} \approx .75$.

That is a striking result from such a simple idea. It was the first major improvement on the iterations of the nineteenth century, and it has been followed by others—although SOR itself is still alive. In our problem it would change to *symmetric overrelaxation*, sweeping alternately forward and backward through the mesh. In practice it might be done by blocks—solving for a row of difference approximations at a time.

We have not sufficiently emphasized the use of **blocks**. It is natural to put a row of meshpoints together. If coupled by a tridiagonal matrix there is no difficulty— and most block algorithms are slightly faster and more stable than point algorithms. Of course there must be stability to start with, as the following example shows: Place the horizontal differences $-1, 2, -1$ in M and leave the vertical differences (in the y-direction) in $A - M$. With $N^2 = 9$ unknowns the iteration will look like

$$\begin{bmatrix} H & & \\ & H & \\ & & H \end{bmatrix} u_{k+1} = \begin{bmatrix} -2I & I & 0 \\ I & -2I & I \\ 0 & I & -2I \end{bmatrix} u_k + b \quad \text{with} \quad H = \begin{bmatrix} 2 & -1 & 0 \\ -1 & 2 & -1 \\ 0 & -1 & 2 \end{bmatrix}.$$

The new vector u_{k+1} can be found quickly, but is it any good? The answer is *no*. This is a completely unstable iteration; errors can be amplified. For $N = 2$ the error equation $Me_{k+1} = (M - A)e_k$ is

$$\begin{bmatrix} 2 & -1 & & \\ -1 & 2 & & \\ & & 2 & -1 \\ & & -1 & 2 \end{bmatrix} e_{k+1} = \begin{bmatrix} -2 & 0 & 1 & 0 \\ 0 & -2 & 0 & 1 \\ 1 & 0 & -2 & 0 \\ 0 & 1 & 0 & -2 \end{bmatrix} e_k, \quad \text{say with } e_0 = \begin{bmatrix} 1 \\ 1 \\ -1 \\ -1 \end{bmatrix}.$$

Since e_1 is $-3e_0$ and $e_k = (-3)^k e_0$, the iteration fails. It can be saved by taking *two half-steps* each iteration—an M_h in the horizontal direction, as above but with an increased diagonal, and then a corresponding M_v in the vertical direction:

$$M_h u_{k+1/2} = (M_h - A)u_k + b, \quad M_v u_{k+1} = (M_v - A)u_{k+1/2} + b.$$

This is the *alternating direction method*, which reduces the error vector at each step by $B = (I - M_v^{-1}A)(I - M_h^{-1}A)$.

Finally we mention a promising new idea. It is a direct attack on the real difficulty with elimination, that the factors in $A = LU$ or $A = LDL^T$ fill too wide a band around the diagonal. The triangular L is not nearly as sparse as the original A. For the purpose of iteration, we throw away this fill-in and use **incomplete LU** factors—say L_0 and U_0. The *ILU* preconditioner is then $M = L_0 U_0$, which makes $Mx_{k+1} = (A - M)x_k + b$ easy to solve. In practice L_0 can be the lower triangular part of A, with its main diagonal adjusted to give zero row sums in $A - M$. This has been successful on systems of order greater than a million.

Multigrid Methods

The classical iterations have a serious drawback. They hammer away with an unchanging matrix B. Convergence is controlled by the largest eigenvalue, which usually means the errors of lowest frequency. The smooth component of the error holds back convergence, while high frequency errors are rapidly killed. The multigrid philosophy is to change gears and make every step count, rather than suffering this effect of different scales.

The idea is to *change the grid*. Errors that are smooth on a grid of width h can be attacked on a coarser grid. Errors that are invisible on the coarse grid of width $2h$ (because their frequency is too high to resolve) are no problem on the fine grid. Starting from an approximation x_h on the fine grid, there is a transfer down and back:

1. Compute the residual error $r_h = b - A_h x_h$
2. Transfer this vector to a coarse grid: $r_{2h} = Cr_h$
3. Solve the coarse system $A_{2h} y_{2h} = r_{2h}$
4. Transfer the correction to the fine grid: $y_h = Fy_{2h}$
5. Make the correction to x: $x_h^{\text{new}} = x_h + y_h$

Before and after, the fine grid iteration with $B = M^{-1}(M - A_h)$ is applied as usual. Thus a 2-grid cycle multiplies the error by

$$B^* = B[I - FA_{2h}^{-1}CA_h]B.$$

This process prefers square grids and simple difference matrices A. It achieves by speed what other algorithms do by finesse. In fact the Gauss-Seidel iteration B is better than overrelaxation, which attempts to do what multigrid is already doing faster—reduce the low frequency errors.

The coarsening matrix C combines nine values on the fine mesh to give a single value (at their center) on the coarse mesh. The weights in the figure are multiplied by $\frac{1}{16}$. The refining matrix F does the reverse; each coarse value at a dark circle splits into nine values on the fine mesh. A set of 1's on the fine mesh collapses into 1's on the coarse mesh. Similarly 1's at the dark circles spread into 1's on the fine mesh (because F multiplies the weights by $\frac{1}{4}$). Those are not the only possibilities for C and F, but they are the favorites.

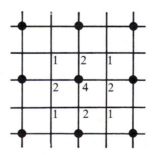

Something is elegant here: C is essentially the transpose of F. The factor 4 in the weights for F and C reflects the factor 4 in the number of mesh points. In a variational approach this elegance is pursued one step further, and the difference operators are related by $A_{2h} = CA_h F$. This is the multigrid analogue of the triple products $A^T C A$ of earlier chapters. We must add that practical multigrid methods involve more special devices than you would care to see in a textbook.

One trick, however, is in the open. It almost suggests itself. The coarse system $A_{2h} y_{2h} = r_{2h}$ need not be solved exactly. It too is subject to iteration and a further coarsening of the grid. This recursive "nesting" makes the method asymptotically optimal: Each digit in the answer costs cn operations. What is remarkable is this linear dependence on n, the number of (fine) meshpoints. The system is solved in a multiple of the time required to write it down! The corresponding power was n^3 for Gaussian elimination of a full matrix. Multigrid is comparable to the value $5n$ for a tridiagonal matrix, except that in two dimensions A is not at all tridiagonal and the observed constant is more like $c = 100$.

One refinement provides a faster start than $x_0 = 0$. The full multigrid method goes recursively upward from an exact solution on a very coarse mesh. The bilinear extrapolation matrices F carry this solution to a reasonable x_0 on the fine mesh. Then the error reduction is as described:

High frequencies are expensive for multigrid (done on fine meshes)
Low frequencies are expensive for single grids (because $\lambda(B) \approx 1$)

Examples and codes are included on MUGTAPE, presently available from the GMD, Postfach 1240, D-5205 West Germany, and from the Weizmann Institute in Israel.

We mention that multigrid methods can be adapted to machines that have a parallel architecture. The hardest problem with parallel machines is communication between processors. For multigrid the *hypercube* array seems to be very well suited: There are 2^d processors connected like the nodes of a cube in d dimensions. (Each node has d neighbors.) Local operations like C and F and M^{-1} and $b - A_h u_h$ can go in parallel on a given grid; it is more delicate to keep all grids going at once. Some algorithms are surviving the shift to parallel computing better than others.

Power Methods for Eigenvalues

We turn to eigenvalue algorithms, where the powers of a matrix are again of central importance. In fact the underlying reason is the same. Any vector u_0 is an unknown combination $c_1 x_1 + \cdots + c_n x_n$ of the unknown eigenvectors, and every time it is multiplied by A there is one more multiplication of x_i by λ_i:

$$u_k = A^k u_0 = c_1 \lambda_1^k x_1 + \cdots + c_n \lambda_n^k x_n. \tag{17}$$

We no longer care whether $|\lambda_i| < 1$. We do care about the dominant eigenvalue, because it comes to the surface in the computations. Suppose the eigenvalues are in increasing order and the largest is all by itself; no other eigenvalue has the same magnitude and λ_n is not repeated. Thus $|\lambda_1| \leq \cdots \leq |\lambda_{n-1}| < |\lambda_n|$, and the power method will compute λ_n. The method starts from u_0 and just keeps multiplying by A:

5I *Ordinary power method.* Suppose A has a single largest eigenvalue. Then if u_0 contains any component $c_n x_n$ of the dominant eigenvector x_n, the sequence $u_k = A^k u_0$ converges (after rescaling) to that eigenvector:

$$\frac{u_k}{c_n \lambda_n^k} = \frac{c_1}{c_n}\left(\frac{\lambda_1}{\lambda_n}\right)^k x_1 + \cdots + \frac{c_{n-1}}{c_n}\left(\frac{\lambda_{n-1}}{\lambda_n}\right)^k x_{n-1} + x_n \quad \text{converges to } x_n.$$

The next largest λ gives the *convergence factor* $r = |\lambda_{n-1}|/|\lambda_n|$.

The scaling will reveal the eigenvalue λ_n. In practice we might keep the first component or the largest component of each u_k equal to one.

This method would be terrific except for a few serious problems:

1. We may not want x_n and λ_n
2. λ_n may not be alone
3. The convergence ratio r may be very close to 1.

All these are very common; you would think that the power method is doomed. But there is a way around each of those limitations, at a fairly reasonable price:

(1) The *block power method* works with several vectors at once, in place of a single u_k. If we start with p orthonormal vectors, multiply them all by A, and apply Gram-Schmidt to orthogonalize them again—that is a single step of the method—then the convergence ratio is reduced to $r' = |\lambda_{n-p}|/|\lambda_n|$. Furthermore we obtain approximations to p eigenvalues and their eigenvectors.

(2) The *inverse power method* operates with A^{-1} instead of A. A single step is $v_{k+1} = A^{-1}v_k$, which means that we solve the linear system $Av_{k+1} = v_k$ (and save L and U for the next step!). Now the theory guarantees convergence to the *smallest* eigenvalue, provided the convergence factor $r'' = |\lambda_1|/|\lambda_2|$ is less than one. Often it is the smallest eigenvalue that is wanted in the applications, and inverse iteration is an automatic choice.

The block inverse power method (known as "subspace iteration" in the finite element literature) works with p orthonormal vectors and finds the p smallest eigenvalues. It is a combination of (1) and (2).

(3) The *shifted inverse power method* is even faster. Suppose that A is replaced by $A - sI$. Then all eigenvalues are shifted by the same amount s, and the convergence factor for the inverse method changes to $r''' = |\lambda_1 - s|/|\lambda_2 - s|$. If s is close to λ_1, r''' will be very small and the convergence is tremendously accelerated. Each step of the method solves $(A - sI)w_{k+1} = w_k$, and this difference equation is satisfied by

$$w_k = \frac{c_1 x_1}{(\lambda_1 - s)^k} + \frac{c_2 x_2}{(\lambda_2 - s)^k} + \cdots + \frac{c_n x_n}{(\lambda_n - s)^k}.$$

For s near λ_1 the first denominator is so near zero that one or two steps make that first term completely dominant. If a good approximation to any eigenvalue has been computed by another algorithm (such as QR), this is a good way to find the eigenvector.†

If λ_1 is not already approximated then the shifted power method has to generate its own choice of s—or, since we can vary the shift at every step if we want to, it must choose the s_k that enters $(A - s_k I)w_{k+1} = w_k$. The simplest possibility is to work with the scaling factors, but in the symmetric case the most accurate choice seems to be the **Rayleigh quotient**

$$s_k = \frac{u_k^T A u_k}{u_k^T u_k}.$$

This quotient has a minimum at the true eigenvector. Its graph is like the bottom of a parabola, and the error $\lambda - s_k$ is roughly the square of the error in the eigenvector. The convergence factors $r''' = |\lambda_1 - s_k|/|\lambda_2 - s_k|$ decrease at every step, and r''' itself is converging to zero. The final result, with these Rayleigh quotient shifts, is cubic convergence of s_k to λ_1.

We test these methods on a small example.

$$A = \begin{bmatrix} 2 & -1 \\ -1 & 2 \end{bmatrix} \text{ has } \lambda_1 = 1 \text{ with } x_1 = \begin{bmatrix} 1 \\ 1 \end{bmatrix}, \lambda_2 = 3 \text{ with } x_2 = \begin{bmatrix} 1 \\ -1 \end{bmatrix}.$$

The ordinary power method gives a sequence (not scaled) like

$$u_0 = \begin{bmatrix} 1 \\ 0 \end{bmatrix}, u_1 = \begin{bmatrix} 2 \\ -1 \end{bmatrix}, u_2 = \begin{bmatrix} 5 \\ -4 \end{bmatrix}, u_3 = \begin{bmatrix} 14 \\ -13 \end{bmatrix}, \ldots, u_\infty = \begin{bmatrix} 1 \\ -1 \end{bmatrix}.$$

The inverse power method gives the sequence (also unscaled)

$$v_0 = \begin{bmatrix} 3 \\ 4 \end{bmatrix}, v_1 = \frac{1}{3}\begin{bmatrix} 10 \\ 11 \end{bmatrix}, v_2 = \frac{1}{9}\begin{bmatrix} 31 \\ 32 \end{bmatrix}, v_3 = \frac{1}{27}\begin{bmatrix} 94 \\ 95 \end{bmatrix}, \ldots, v_\infty = \begin{bmatrix} 1 \\ 1 \end{bmatrix}.$$

The Rayleigh quotient at the first guess is $s = v_0^T A v_0/v_0^T v_0 = 26/25$, and

$$w_0 = \begin{bmatrix} 3 \\ 4 \end{bmatrix} \quad \text{leads to} \quad w_1 = (A - sI)^{-1} w_0 = \frac{25}{49}\begin{bmatrix} 172 \\ 171 \end{bmatrix}.$$

Lanczos Method for Eigenvalues

We come now to **semi-direct methods**. They are also based on a sequence of vectors like u, Au, $A^2 u$, ..., but they keep and use more information than the power method did. In fact the power method kept no information at all; $A^k u$ was written

---

† That is why we concentrate elsewhere on the eigenvalue; the eigenvector is easy to find at the end.

over $A^{k-1}u$. History was forgotten and exact answers were never reached. But after n steps we should know everything about a matrix! If the vectors $u, Au, ..., A^{n-1}u$ are linearly independent, as they usually are, then every vector x is a combination of those basis vectors. Ax is the same combination of $Au, ..., A^n u$. In principle we should not have to go further than that to solve $Ax = \lambda x$ or $Ax = b$.

The difficulty is that the sequence $u, Au, ..., A^{n-1}u$ is not at all convenient. The powers of A push toward the top eigenvector (which justified the power method) and these vectors are not even close to orthogonal. Nevertheless **they can be orthogonalized by a simple recurrence formula**. That produces an orthonormal sequence $q_1, q_2, ...$, and we can stop at any time. The q_j are the columns of an orthogonal matrix Q. If we look at the eigenvalues of $Q^T AQ$, we are using the Lanczos idea. If we look for the combination of the q_j that comes closest to solving $Ax = b$, that is the conjugate gradient idea.

I think it is not necessary to study the full theory before using these algorithms. We will give the steps for the standard methods and then with a preconditioner M. Before that comes the one crucial point of mathematics, the 3-term recurrence formula for the orthonormal q_j. These 3-term formulas appear in the continuous case too; there is one that connects $\cos(n+1)x$ to $\cos nx$ and $\cos(n-1)x$, and there are others for all the classical special functions (Legendre, Bessel, Hermite, ...). For vectors, a 3-term formula signals the presence of a tridiagonal matrix T, and we begin with that matrix.

5J Starting from any unit vector $u = q_1$ and any symmetric matrix A, there is an orthonormal sequence $q_1, q_2, ...$ such that

$$Aq_j = b_{j-1}q_{j-1} + a_j q_j + b_j q_{j+1}. \tag{18}$$

The q_j are the columns of an orthogonal matrix: $Q^T Q = I$ and

$$Q^T AQ = T \quad \text{is tridiagonal.} \tag{19}$$

Both equations (18) and (19) say that $AQ = QT$:

$$
\begin{bmatrix} Aq_1 & \cdots & Aq_n \end{bmatrix} = \begin{bmatrix} q_1 & \cdots & q_n \end{bmatrix} \begin{bmatrix} a_1 & b_1 & & \\ b_1 & a_2 & b_2 & \\ & b_2 & \cdot & \cdot \\ & & \cdot & a_n \end{bmatrix}
$$

The coefficients a_j and b_j in the 3-term recurrence formula are the same as the entries in the symmetric tridiagonal matrix T.

Those entries are easy to find. Multiplying (18) by q_j^T gives $q_j^T Aq_j = a_j$. The other terms disappear by orthogonality. And rewriting (18) as

$$b_j q_{j+1} = Aq_j - a_j q_j - b_{j-1}q_{j-1} \tag{20}$$

gives b_j as the length of the vector r_j on the right hand side (since q_{j+1} is a unit vector). We note the convention $q_0 = 0$ at the start of the recursion. If it should happen that (20) is zero, then $b_j = 0$ and we restart with any q_{j+1} orthogonal to the previous q's. In that exceptional case T splits into two smaller tridiagonal matrices at the row where $b_j = 0$.

Thus the **Lanczos iteration** (pronounced Lanchos) starts with $j = 0$, $q_0 = 0$, $b_0 = 1$, $r_0 = q_1$:

$$q_{j+1} = r_j/b_j$$

$$j := j + 1$$

$$a_j = q_j^T A q_j$$

$$r_j = (A - a_j I)q_j - b_{j-1}q_{j-1}$$

$$b_j = \| r_j \|$$

When the rows of A have c nonzeros, the matrix-vector product Aq_j uses cn multiplications. The whole iteration step can be organized to require only $cn + 4n$. This is where semi-direct methods are most useful—when A is large and sparse. Its reduction to tridiagonal form by Householder transformations would be expensive; plane reflections are somewhat better. The sparsity makes Lanczos very competitive and the size means that we go only part way in constructing the tridiagonal T. Even a small part of T can give excellent approximations to $\lambda_{\max}$ and $A^{-1}b$.

The remarkable thing is that each new vector q_{j+1} is orthogonal to *all earlier* q's. That requires a proof. It is not surprising that q_{j+1} is orthogonal to the two previous vectors q_j and q_{j-1}, because (20) subtracted from Aq_j the right multiples of those two vectors (Exercise 5.3.16). The key point is that Aq_j is already and automatically orthogonal to the other vectors $q_1, \ldots, q_{j-2}$. It is the symmetry of A which makes that true: for $i \leq j - 2$

$$q_i^T(Aq_j) = (Aq_i)^T q_j = (b_{i-1}q_{i-1} + a_i q_i + b_i q_{i+1})^T q_j. \tag{21}$$

All three inner products on the right are zero. (This is a proof by induction; q_j is known to be orthogonal to these previous vectors from the previous step.) Thus $q_i^T(Aq_j)$ is zero, and from (20) $q_i^T q_{j+1}$ must be zero. In other words the new q_{j+1} is orthogonal to all previous q_i, as we wanted to prove. The sequence of q's has two basic properties:

(1)   Each q_{j+1} is a combination of $q_1, Aq_1, \ldots, A^j q_1$
(2)   Each q_{j+1} is orthogonal to all combinations of $q_1, Aq_1, \ldots, A^{j-1}q_1$

The practical problem is to protect this orthogonality from roundoff errors.

EXAMPLE 1 $A = \begin{bmatrix} 1 & 1 & 1 & 1 & 1 \\ 1 & 1 & 1 & 1 & 1 \\ 1 & 1 & 1 & 1 & 1 \\ 1 & 1 & 1 & 1 & 1 \\ 1 & 1 & 1 & 1 & 1 \end{bmatrix}$ with $q_1 = \begin{bmatrix} 1 \\ 0 \\ 0 \\ 0 \\ 0 \end{bmatrix}$ and $Aq_1 = \begin{bmatrix} 1 \\ 1 \\ 1 \\ 1 \\ 1 \end{bmatrix}$

The vector q_1 is set. To make Aq_1 orthogonal to it, we subtract a multiple of q_1. The multiple is $a_1 = q_1^T Aq_1 = 1$ and the subtraction leaves $(0, 1, 1, 1, 1)$. This is the vector $b_1 q_2$ in (20), and its length is $b_1 = \sqrt{4} = 2$. Dividing by 2 the second step starts with the second unit vector:

$$q_2 = \begin{bmatrix} 0 \\ \frac{1}{2} \\ \frac{1}{2} \\ \frac{1}{2} \\ \frac{1}{2} \end{bmatrix} \quad \text{and} \quad Aq_2 = \begin{bmatrix} 2 \\ 2 \\ 2 \\ 2 \\ 2 \end{bmatrix} \quad \text{and} \quad a_2 = q_2^T Aq_2 = 4.$$

At this point we meet something unnatural. The right side of (20) is $Aq_2 - a_2 q_2 - b_1 q_1 = 0$. The sequence ends abruptly with $b_2 = 0$. It can be restarted with a vector like $q_3 = (0, \frac{1}{2}, \frac{1}{2}, -\frac{1}{2}, -\frac{1}{2})$ that is orthogonal to q_1 and q_2. But then Aq_3 is zero, so $a_3 = b_3 = 0$. Again we revive it with $q_4 = (0, \frac{1}{2}, -\frac{1}{2}, \frac{1}{2}, -\frac{1}{2})$. It dies again with $Aq_4 = 0$ and $a_4 = b_4 = 0$. Finally q_5 can be $(0, \frac{1}{2}, -\frac{1}{2}, -\frac{1}{2}, \frac{1}{2})$ and $Aq_5 = 0$ and $a_5 = 0$. Thus the tridiagonal matrix is

$$T = \begin{bmatrix} a_1 & b_1 & & & \\ b_1 & a_2 & b_2 & & \\ & b_2 & a_3 & b_3 & \\ & & b_3 & a_4 & b_4 \\ & & & b_4 & a_5 \end{bmatrix} = \begin{bmatrix} 1 & 2 & & & \\ 2 & 4 & & & \\ & & 0 & & \\ & & & 0 & \\ & & & & 0 \end{bmatrix}.$$

To tell the truth this is exactly what we hope for. The eigenvalues of T are extremely easy to find (not that A was so difficult). They are $\lambda_1 = 5$ and $\lambda_2 = \lambda_3 = \lambda_4 = \lambda_5 = 0$. A and T are matrices of rank one—every row is a multiple of the first row—and they have four zero eigenvalues. We emphasize that A and T always have the same eigenvalues since $Q^{-1}AQ = T$.

The example brings out the main point about semi-direct methods. They can give accurate answers part way. Compare the results of stopping with a 3 by 3 matrix in A and in T. The original matrix A has a submatrix of 1's, with eigenvalues $3, 0, 0$. In contrast the upper left corner of T has the correct eigenvalues $5, 0, 0$. I admit that this is exceptional; the eigenvalues of submatrices are almost never exact. But for a tridiagonal matrix they converge very quickly to the extreme eigenvalues λ_1 and λ_n, and they are easy to compute by the QR algorithm. The cost is in finding the q's, and we come back to the difficulties with that iteration.

EXAMPLE 2 Orthogonalizing the polynomials $1, x, x^2, \ldots$ The matrix A has changed to multiplication by x. It is continuous instead of discrete, but formulas (18)–(20) still apply. They give the *Legendre polynomials* $P_n(x)$, each one orthogonal to all polynomials of lower degree. Their roots are the "Gauss points" that give extra accuracy in numerical integration. At every step xP_n starts as automatically orthogonal to $P_0, P_1, \ldots, P_{n-2}$—and subtracting the right combination of P_{n-1} and P_n is the 3-term recurrence that gives the next polynomial P_{n+1}.

The stumbling block with the Lanczos idea is its numerical instability. In practice, *the vectors q_j are not actually orthogonal.* After a few steps they are barely independent. It may be true in principle that they have no components from the early q_i, but in fact they soon have very large components. The calculation is not just degraded by roundoff error, it is destroyed. Therefore some preventive measure is required, or the Lanczos idea will have to be discarded—as it was in the 1960's. At that time it was seen as a direct method; the whole sequence of q's and the whole triangular T were computed. Now, as a semi-direct method, it only maintains enough orthogonality for part of the calculation.

One possibility is a complete orthogonalization of each q_{j+1} against every preceding q. That is expensive. It is not strictly necessary, because the instability only starts when the method converges! That happened in Example 1, where q_1 and q_2 were ordinary but q_3 could not be determined from (20) because the right side was zero. At that same moment the 2 by 2 submatrix of T contained the correct eigenvalues $\lambda = 0$ and $\lambda = 5$. This was a good thing, but the general rule in numerical analysis is that lucky accidents are not so wonderful. Nearby problems produce a very small number instead of zero.[†] In those problems q_3 will be determined by the algorithms, but badly. The cancellations occurring in (20) will not be exact, and the computed q_{j+1} will have a large component in the direction of any eigenvector that is nearly a combination of $q_1, \ldots, q_j$.

A *selective* orthogonalization is recommended, in which the (few) converged eigenvectors are removed by Gram-Schmidt from each new q_{j+1}. As the calculation continues, each loss of orthogonality becomes a signal that a new eigenvector is available. Its eigenvalue λ is identified from the current submatrix of T—usually it is $\lambda_{\max}$ but sometimes it is at the bottom of the spectrum. The eigenvector is removed from all later q's, or else λ appears over and over as a repeated "ghost" eigenvalue of the computed matrix T.

Suppose this numerical difficulty is under control. We can say in a few words why the Lanczos method is better than the power method, if they both stop after j vectors. The power method has reached $v = A^{j-1}q_1$; its estimate for $\lambda_{\max}$ is the Rayleigh quotient $v^T A v / v^T v$. The Lanczos method has reached the j by j submatrix T_j of T, and *its* estimate for $\lambda_{\max}$ is the largest eigenvalue of T_j. Unlike the power method, *this uses all the vectors* $q_1, \ldots, q_j$. It maximizes the Rayleigh quotient over all combinations of these vectors, instead of accepting the last vector v. The

---

† It is the same in elimination; small pivots are a menace.

Lanczos method is an efficient way to find the best combination and to update it at each step—along with estimates for j other eigenvalues of A.

The Conjugate Gradient Method

It is not only eigenvalues that are simpler when A is tridiagonal. The same is true of linear equations. If we can change $Ax = b$ to $Ty = f$, the computation time drops from n^3 to n. The tridiagonal matrix T, if it is symmetric positive definite, will factor into LDL^T with only two diagonals in L. More important, *there will be a simple recursive formula for L and D*. This is one situation in which we can do part of Gaussian elimination and stop in the middle of the recursion. It is a direct method if it is carried to the end, and it is semi-direct since we are prepared to stop.

How does $Ax = b$ change to $Ty = f$? That uses the orthonormal vectors q_j. If they are the columns of Q, then Q^T is Q^{-1} and $Q^T A Q = T$ is tridiagonal (proved above). Multiplying both sides by Q^T,

$$Ax = b \quad \text{is} \quad (Q^T A Q)(Q^T x) = (Q^T b) \quad \text{which is} \quad Ty = f. \tag{22}$$

A further simplification is available. The Lanczos sequence can start with any unit vector, and we choose $q_1 = b/\|b\|$. All later q's are perpendicular to q_1 (and thus to b). Since those q's are the rows of Q^T, all entries of $f = Q^T b$ are zero after the first. The right side of $Ty = f$ is improved at the same time as the left side, and the new system is

$$Ty = \begin{bmatrix} a_1 & b_1 & & \\ b_1 & a_2 & b_2 & \\ & b_2 & a_3 & \cdot \\ & & \cdot & \cdot \end{bmatrix} \begin{bmatrix} \\ y \\ \\ \end{bmatrix} = \begin{bmatrix} \|b\| \\ 0 \\ 0 \\ \cdot \end{bmatrix} = f. \tag{23}$$

We could solve this in one shot, but that misses the main point. The a's and b's are computed one at a time, and the recursive idea is to *update y at each step* and stop when desired.

The first step uses only a_1, the first entry of T. The second step includes b_1 and a_2 in a 2 by 2 matrix T_2. The third step involves the 3 by 3 matrix T_3 in the corner of equation (23):

$$T_3 y_3 = \begin{bmatrix} a_1 & b_1 & \\ b_1 & a_2 & b_2 \\ & b_2 & a_3 \end{bmatrix} \begin{bmatrix} \\ y_3 \\ \\ \end{bmatrix} = \begin{bmatrix} \|b\| \\ 0 \\ 0 \end{bmatrix}. \tag{24}$$

Each vector y_j comes from chopping $Ty = f$ at component j. It is reasonable to look for a formula that computes y_{j+1} from the previous y_j—which is the basic step of the conjugate gradient method.

We need to say that the formula is not trivial. It does not simply compute one new component of the final vector y. The result $y_1 = \|b\|/a_1$ at step 1 is altered by b_1 and a_2 at step 2, and we have to expect a completely new vector y_{j+1} at each step. That vector comes from the LDL^T factors of T_{j+1}, and those factors change in only two or three entries. In fact the a's and b's never appear in conjugate gradients; it is L and D and the solution which are updated. Before giving the formula we point to its main properties:

1.   The exact solution $y = y_n$ is reached in n steps. Unlike steepest descent, the method terminates.
2.   Each y_j is the best choice among all vectors whose last $n - j$ components are zero. It minimizes the underlying quadratic. In some sense the conjugate gradient method is to steepest descent as the Lanczos method is to the power method—it looks not just at the latest vector but at the best combination of all vectors reached by step j.
3.   The change from y_j to y_{j+1} is "conjugate" to all preceding changes (and all later changes). Here the word conjugate means perpendicular, provided we use the inner product that is natural for the problem. For y that inner product is given by T and for x it is given by A:

$$(y_{i+1} - y_i)^T T(y_{j+1} - y_j) = 0 \quad \text{or} \quad (x_{i+1} - x_i)^T A(x_{j+1} - x_j) = 0. \tag{25}$$

Since $x = Qy$ and $Q^T A Q = T$ these are equivalent. Thus **the direction from x_j to x_{j+1} is "A-orthogonal" to all previous directions**.

To prove (25) we look at the residual after step j. That is the amount $b - Ax_j$ or $f - Ty_j$ by which the equations are not satisfied. For example $f - Ty_3$ is zero except in component 4 (where the asterisk is):

$$Ty_3 = \begin{bmatrix} a_1 & b_1 & & \\ b_1 & a_2 & b_2 & \\ & b_2 & a_3 & b_3 \\ & & b_3 & a_4 \end{bmatrix} \begin{bmatrix} y_3 \\ \hline 0 \end{bmatrix} = \begin{bmatrix} \|b\| \\ 0 \\ 0 \\ \hline * \end{bmatrix}. \tag{26}$$

Since $f - Ty_j$ is nonzero only in component $j + 1$, and $f - Ty_{j+1}$ is nonzero only in component $j + 2$, the difference $T(y_{j+1} - y_j)$ is entirely zero in earlier components. But for $i < j$ that is where $y_{i+1} - y_i$ is *not* zero. Therefore these two vectors are orthogonal as claimed in (25). Converting to $x = Qy$ we have A-orthogonality.

This is more than an idle remark; it suggests a different way of approaching conjugate gradients (and explaining their name). Suppose the vectors $d_1, \ldots, d_n$ are A-orthogonal: $d_i^T A d_j = 0$. Then we ask for the components of x in those directions:

$$x = \alpha_1 d_1 + \alpha_2 d_2 + \cdots + \alpha_n d_n.$$

If the d's were genuinely orthogonal we would multiply by d_j^T to find α_j. All other terms on the right would disappear. When the d's are A-orthogonal those terms still disappear, provided we multiply *not by d_j^T but by $d_j^T A$*:

$$d_j^T A x = \alpha_j d_j^T A d_j. \tag{27}$$

You may think that A makes this more complicated. However in studying $Ax = b$ it has a tremendous advantage: what we know is Ax and not x! Then (27) can give the coefficients $\alpha_1, \alpha_2, \ldots$ exactly because A is present.

The approach to conjugate gradients is this. Start from $x_0 = 0$, where the residual $b - Ax$ is $r_0 = b$. The residual is also the gradient of $P(x) = \frac{1}{2} x^T A x - b^T x$, with a minus sign. It is the direction d_1 of steepest descent, and we go in that direction to a point $x_1 = \alpha_1 d_1$. There we compute the new residual $r = b - Ax_1$ (the negative gradient). This would be the second direction, except that it is not conjugate to the first direction d_1. Therefore we add to it the right multiple of d_1: $d_2 = r_1 + \beta_2 d_1$ will be A-orthogonal to d_1. Then we move in this direction to the new point $x_2 = x_1 + \alpha_2 d_2$. The cycle of calculations is

direction $d_j = r_{j-1} + \beta_j d_{j-1}$;   **new point** $x_j = x_{j-1} + \alpha_j d_j$;   **residual** $r_j = b - Ax_j$.

The number α_j is in (27). After the nth step $x_n = x_{n-1} + \alpha_n d_n$ we have x. The number β_j is chosen to make d_j A-orthogonal to d_{j-1}. Like a and b in the Lanczos method, α and β come from inner products. Several different expressions (and other indices in the cycle) are possible and frequently used. To eliminate a multiplication each new residual is computed recursively and not from $b - Ax$.

The connection to tridiagonal matrices leads to the mathematical identity that makes the method special: Each new direction is automatically A-orthogonal to all previous directions. This corresponds to the ordinary orthogonality proved for Lanczos, of q_{j+1} against all previous q_j. We summarize the links to Lanczos and steepest descent:

(1) The residuals r_j are multiples of the Lanczos vectors q_j

(2) The loss of orthogonality from roundoff that was serious for Lanczos is much less destructive here

(3) The matrix A must now be positive definite

(4) To minimize a non-quadratic $P(x)$ the negative gradient $-g = -(\partial P/\partial x_1, \ldots, \partial P/\partial x_n)$ replaces r in the cycle; the first step is still steepest descent

(5) The difference from steepest descent is that the inexpensive correction $\beta_j d_{j-1}$ makes the directions A-orthogonal.

Steepest descent would succeed in n steps if A were the identity. The cross-sections $P(x) = $ constant would be circles instead of ellipses and the descent directions would be orthogonal. Conjugate gradients *do* succeed in n steps, because they maintain A-orthogonality and *in that inner product the cross-sections are circles*. That explains why the matrix A appears in the following cycle.

5K   The conjugate gradient method starts from $x_0 = 0$ and $r_0 = b$ and computes

$$\beta_j = r_{j-1}^T r_{j-1} / r_{j-2}^T r_{j-2} \quad (\text{except } \beta_1 = 0)$$
$$d_j = r_{j-1} + \beta_j d_{j-1} \quad\quad (\text{except } d_1 = r_0)$$
$$\alpha_j = r_{j-1}^T r_{j-1} / d_j^T A d_j$$
$$x_j = x_{j-1} + \alpha_j d_j$$
$$r_j = r_{j-1} - \alpha_j A d_j.$$

There is only one matrix-vector multiplication in each cycle, and with exact arithmetic x_n is $A^{-1}b$.

The method could start from a better approximation x_0, with $r_0 = b - Ax_0$.

We add one comment on α_j. For nonlinear equations, when P is not a quadratic, α_j is the number at which $P(x_{j-1} + \alpha_j d_j)$ is a minimum. It is the step length, found by a line search starting from x_{j-1} in the direction d_j. When P is quadratic this search finds the α_j given in the cycle above, and each successive step gives one more component $\alpha_j d_j$ of the final solution x.

The choice $\beta_j = g_{j-1}^T (g_{j-1} - g_{j-2}) / g_{j-2}^T g_{j-2}$ is often preferred in the non-quadratic case. If the negative gradient $-g_j$ coincides with the residual r_j this reduces to the formula in the cycle. The conjugate gradient method may take twice as many steps as the quasi-Newton method mentioned in Section 5.1, but with sparse matrices each step is much faster.

EXAMPLE $Ax = b$ is $\begin{bmatrix} 2 & 1 & 1 \\ 1 & 2 & 1 \\ 1 & 1 & 2 \end{bmatrix} \begin{bmatrix} 3 \\ -1 \\ -1 \end{bmatrix} = \begin{bmatrix} 4 \\ 0 \\ 0 \end{bmatrix}.$

From $x_0 = 0$ and $r_0 = d_1 = b$ the first cycle gives $\alpha_1 = \frac{1}{2}$ and $x_1 = \frac{1}{2}b = (2, 0, 0)$. The residual is $r_1 = (0, -2, -2)$. Then the second cycle yields

$$\beta_2 = \frac{8}{16}, \quad d_2 = \begin{bmatrix} 2 \\ -2 \\ -2 \end{bmatrix}, \quad \alpha_2 = \frac{8}{16}, \quad x_2 = \begin{bmatrix} 3 \\ -1 \\ -1 \end{bmatrix}.$$

The correct solution is reached in two steps, where normally it takes three. The reason is that this particular A has only two distinct eigenvalues. The residual r_2 is zero and the cycles stop.

Convergence

The goal of conjugate gradients is to reach $x = A^{-1}b$. However the algorithm never inverts a matrix. Instead of A^{-1} it takes powers of A, increasing the power by one in each cycle. Thus x_1 is a multiple of b; the residual and new direction and next point x_2 involve Ab; in general r_{j-1} and d_j and x_j include $A^{j-1}b$. Each step makes the error as small as possible by adding on the next projection $\alpha_j d_j$. We are minimizing the quadratic P over all vectors x_j that are combinations of $b, Ab, \ldots, A^{j-1}b$. As in Section 2.2, this is the same as minimizing the error $E_j = (x - x_j)^T A(x - x_j)$.

After n steps the algorithm will find x itself. Every vector is a combination of the n vectors $b, \ldots, A^{n-1}b$, and the best combination is the exact solution x. The example computed above was exceptional; it took only two steps to reach x. However that was no accident. When A has only two eigenvalues, A^{-1} is a combination of I and A. Therefore $x = A^{-1}b$ is a combination of b and Ab, and conjugate gradients will guarantee to find that combination in two steps. In general the eigenvalues lie between $a = \lambda_{\min}$ and $b = \lambda_{\max}$, and the vector x_j closest to x gives an error no greater than

$$E_j \leq E_0 \left(\frac{\sqrt{b} - \sqrt{a}}{\sqrt{b} + \sqrt{a}} \right)^{2j}.$$

This bound is pessimistic in practice, depending only on the condition number b/a and not on the position of the eigenvalues between a and b.

In any case these eigenvalues can be changed. That is done by a **preconditioning matrix** M. Instead of solving $Ax = b$ we solve $M^{-1}Ax = M^{-1}b$, and if M is close to A then the coefficient matrix $M^{-1}A$ is close to the identity. Its eigenvalues are near 1 (of course we never compute them) and the new condition number $\lambda_{\max}/\lambda_{\min}$ is smaller than the original b/a. The matrix M must be symmetric positive definite, and easy to invert. The preconditioned method should actually work with the symmetric matrix $S = M^{-1/2}AM^{-1/2}$ in the middle of all inner products, but the cycle can produce this effect without finding a square root of M. The simplest arrangement is to start each cycle with an extra step:

$$\text{Solve } Mz_{j-1} = r_{j-1}$$
$$\beta_j = z_{j-1}^T r_{j-1} / z_{j-2}^T r_{j-2} \quad (\text{except } \beta_1 = 0)$$
$$d_j = z_{j-1} + \beta_j d_{j-1} \quad (\text{except } d_1 = z_0)$$
$$\alpha_j = z_{j-1}^T r_{j-1} / d_j^T A d_j$$
$$x_j = x_{j-1} + \alpha_j d_j$$
$$r_j = r_{j-1} - \alpha_j A d_j$$

With this change the directions are S-orthogonal and the residuals are M-orthogonal (instead of A and I). The matrices that were used in iterative methods $Mx_{n+1} = (M - A)x_n + b$ are natural candidates as preconditioners, provided they are symmetric positive definite. And there is one new twist: convergence no longer requires the eigenvalues of $B = I - M^{-1}A$ to be small. Instead of their size it is their *position* that matters, and if they fall into a few clusters then convergence will be fast. In the extreme case $M^{-1}A$ might have only two distinct eigenvalues and convergence will occur in two cycles.

EXAMPLE A = second difference matrix with $u_0 = u_{n+1} = 0$

M = second difference matrix with $u_0 = u_1$ and $u_{n+1} = 0$.

The matrix A is my favorite but M is easier to invert:

$$
M = \begin{bmatrix} 1 & -1 & & \\ -1 & 2 & -1 & \\ & \cdot & \cdot & \cdot \\ & & -1 & 2 \end{bmatrix} = \begin{bmatrix} 1 & & & \\ -1 & 1 & & \\ & \cdot & \cdot & \\ & & -1 & 1 \end{bmatrix} \begin{bmatrix} 1 & -1 & & \\ & 1 & \cdot & \\ & & \cdot & -1 \\ & & & 1 \end{bmatrix}
$$

$$
A = \begin{bmatrix} 2 & -1 & & \\ -1 & 2 & -1 & \\ & \cdot & \cdot & \cdot \\ & & -1 & 2 \end{bmatrix}.
$$

The difference $M - A$ is completely concentrated in one single entry. Therefore $M^{-1}(M - A)$ is concentrated in the first column and it has $n - 1$ zero eigenvalues. With only one other eigenvalue, conjugate gradients preconditioned by M will end after two cycles. Pursuing this idea might give the fastest possible way to solve $Ax = b$.

This preconditioning applied to the Lanczos method would solve the eigenvalue problem $Ax = \lambda Mx$. It appears that a successful preconditioning of the ordinary eigenvalue problem $Ax = \lambda x$ is still to be found. In every case we are offering to do extra work with M in exchange for faster convergence. It is one of the fundamental problems of numerical analysis to find the right balance.

EXERCISES

5.3.1   The Jacobi iteration for a general 2 by 2 matrix has

$$
A = \begin{bmatrix} a & b \\ c & d \end{bmatrix} \quad \text{and} \quad M = \begin{bmatrix} a & 0 \\ 0 & d \end{bmatrix}.
$$

Find the eigenvalues of $B = M^{-1}(M - A)$. If A is symmetric positive definite, show that the iteration converges.

5.3.2   For Gauss-Seidel the iteration matrices are

$$
M = \begin{bmatrix} a & 0 \\ c & d \end{bmatrix} \quad \text{and} \quad M - A = \begin{bmatrix} 0 & -b \\ 0 & 0 \end{bmatrix}.
$$

Find the eigenvalues of $B = M^{-1}(M - A)$. Give an example of a matrix A for which Gauss-Seidel iteration will not converge.

5.3.3 Compute by direct multiplication the powers of

$$B = \frac{1}{2}\begin{bmatrix} 0 & 1 & 0 & 1 \\ 1 & 0 & 1 & 0 \\ 0 & 1 & 0 & 1 \\ 1 & 0 & 1 & 0 \end{bmatrix}.$$

For what matrix A is this the Jacobi matrix $B = I - D^{-1}A$? Find the eigenvalues of B.

5.3.4 Decide the convergence or divergence of Jacobi and Gauss-Seidel iterations for

$$A = \begin{bmatrix} 1 & 2 & -2 \\ 1 & 1 & 1 \\ 2 & 2 & 1 \end{bmatrix}.$$

Construct M for both methods and find the eigenvalues of $B = I - M^{-1}A$.

5.3.5 In overrelaxation use the triangularity of M and $M - A$, and the relation $m_{ii} = a_{ii}/\omega$ between diagonal entries, to confirm that the determinant of $B = M^{-1}(M - A)$ is $(1 - \omega)^n$. What range of ω, either overrelaxed ($\omega > 1$) or underrelaxed ($\omega < 1$), keeps the determinant and possibly all the $|\lambda_i|$ below 1?

5.3.6 If $\mu = \cos \pi h$ show from (16) that $\lambda = (1 - \sin \pi h)/(1 + \sin \pi h)$. What is the limit of ω_{opt} as $h \to 0$?

5.3.7 If λ is an eigenvalue of C with eigenvector x, show that λ and $-\lambda$ are both eigenvalues of

$$B = \begin{bmatrix} 0 & C \\ C & 0 \end{bmatrix}.$$

What are the two corresponding eigenvectors of B? If the red squares on a chessboard are numbered before the black squares, the 5-point finite difference matrix of order 64 has the same pattern as B except for $+4$ on the diagonal.

5.3.8 For the tridiagonal n by n matrix

$$B = \frac{1}{2}\begin{bmatrix} 0 & 1 & & \\ 1 & 0 & \cdot & \\ & \cdot & \cdot & 1 \\ & & 1 & 0 \end{bmatrix}$$

show that the vector $x = (\sin \pi h, \sin 2\pi h, ..., \sin n\pi h)$ satisfies $Bx = \lambda x$ with eigenvalue $\lambda = \cos \pi h$. Here $h = 1/n + 1$ so $\sin (n + 1)\pi h = 0$.
Note: The other eigenvectors replace π by $2\pi, 3\pi, ..., n\pi$. The other eigenvalues are $\cos 2\pi h, \cos 3\pi h, ..., \cos n\pi h$, all smaller than $\cos \pi h < 1$. B comes from Jacobi iterations for the second-difference matrix with $-1, 2, -1$ on its diagonals.

The following exercises require Gerschgorin's "circle theorem": *Every eigenvalue of B lies in at least one of the circles $C_1, \ldots, C_n$, where C_i has its center at the diagonal entry b_{ii} and its radius $r_i = \Sigma_{j \neq i} |b_{ij}|$ equal to the absolute sum along the rest of the row.*

Proof $Bx = \lambda x$ leads to

$$(\lambda - b_{ii})x_i = \sum_{j \neq i} b_{ij}x_j, \quad \text{or} \quad |\lambda - b_{ii}| \leq \sum_{j \neq i} |b_{ij}| \frac{|x_j|}{|x_i|}.$$

If the largest component of x is x_i, then these last ratios are ≤ 1, and λ lies in the ith circle: $|\lambda - b_{ii}| \leq r_i$. For the matrix in Ex. 5.3.8 all circles are centered at the origin since $b_{ii} = 0$, and all except two have radius $r = \frac{1}{2} + \frac{1}{2} = 1$. By Gerschgorin the eigenvalues satisfy $|\lambda| \leq 1$; the exercise found strict inequality $|\lambda| < 1$ and therefore convergence.

5.3.9 If a matrix B has absolute row sums less than 1,

$$|b_{i1}| + \cdots + |b_{ii}| + \cdots + |b_{in}| < 1 \text{ for each } i,$$

show from Gerschgorin's circles that all eigenvalues satisfy $|\lambda| < 1$.

5.3.10 The matrix

$$A = \begin{bmatrix} 4 & 2 & 1 \\ 1 & 5 & 3 \\ 2 & 4 & 7 \end{bmatrix}$$

is called *diagonally dominant* because each diagonal entry exceeds the sum along the rest of its row. Sketch the three Gerschgorin circles for this matrix. No circle contains $\lambda = 0$ and diagonally dominant matrices are always invertible.

5.3.11 What is the Jacobi matrix $B = I - D^{-1}A$ for this diagonally dominant A? Show that all row sums of B are less than 1 and the Jacobi iteration converges.

5.3.12 Apply five steps of the unscaled power method to $A = \begin{bmatrix} 1 & 1 \\ 1 & 0 \end{bmatrix}$ starting from $u_0 = \begin{bmatrix} 1 \\ 0 \end{bmatrix}$. Do you recognize the series of numbers in the u_k? What is the convergence factor r and to what eigenvector do the u_k converge?

5.3.13 For the same A and u_0, take three inverse power steps without shifts and find the eigenvector to which the sequence converges.

5.3.14 Compute the Rayleigh quotient $u^T Au / u^T u$ for

$$A = \begin{bmatrix} 2 & -1 \\ -1 & 2 \end{bmatrix} \quad \text{and} \quad u = \begin{bmatrix} 1 \\ 1 + c \end{bmatrix}$$

and show that it differs from $\lambda = 1$ by a second-order term (involving c^2).

5.3.15 Show that Aq_i is orthogonal to q_j when $i \leq j - 2$, by using the two Lanczos properties: q_i is a combination of $q_1, \ldots, A^{i-1}q_1$, and q_j is orthogonal to all combinations of $q_1, \ldots, A^{j-2}q_1$.

5.3.16 (a) Show from (18) that q_{j+1} is orthogonal to q_j if $a_j = q_j^T A q_j$.
   (b) Show that q_{j+1} is also orthogonal to q_{j-1}, using (21) for $i = j - 1$.

(c)   Show finally that $Q_j^T A Q_j = T_j$, where the columns of Q_j are $q_1, \ldots, q_j$.

5.3.17   What 3-term recurrence connects $\cos(n+1)\theta$ to $\cos n\theta$ and $\cos(n-1)\theta$? Rewrite it in terms of x for the Chebyshev polynomials $T_n = \cos n\theta$ (with $x = \cos\theta$).

5.3.18   For the matrix

$$A = \begin{bmatrix} 2 & 0 & 1 \\ 0 & 3 & 2 \\ 1 & 2 & 2 \end{bmatrix} \quad \text{and} \quad q_1 = b = \begin{bmatrix} 1 \\ 0 \\ 0 \end{bmatrix}$$

(a)   find the orthogonal vectors q_2 and q_3 in the Lanczos method and the tridiagonal matrix $Q^T A Q = T$
(b)   find the vectors x_j in the conjugate gradient method ending with $x_3 = A^{-1}b = (2, 2, -3)$.

5.3.19   Given vectors d and r, what combination $r + \beta d$ is A-orthogonal to d?

5.3.20   The Cayley-Hamilton theorem says that the product $(A - \lambda_1 I) \cdots (A - \lambda_n I)$ is zero. Multiply by A^{-1} and show that the inverse is a combination of $I, A, \ldots, A^{n-1}$. Therefore $x = A^{-1}b$ is a combination of $b, Ab, \ldots, A^{n-1}b$.

5.3.21   Suppose A has the special form $I + vv^T$ where v is a combination $cq_1 + dw$ of unit vectors.
(a)   If the Lanczos algorithm starts with q_1 show from (20) that its next vector q_2 is w.
(b)   Compute $a_1 = q_1^T A q_1$, $b_1 = q_2^T A q_1$, $a_2 = q_2^T A q_2$ in terms of c and d.
(c)   Show that the 2 by 2 matrix T_2 with these entries has the same eigenvalues 1 and $1 + c^2 + d^2$ as A itself. The algorithm is finished after step 2.

5.3.22   Compare the first conjugate gradient points x_1 with and without preconditioning to the solution $x = (0, 1)$ of $Ax = b$:

$$A = \begin{bmatrix} 9 & 1 \\ 1 & 1 \end{bmatrix}, \quad b = \begin{bmatrix} 1 \\ 1 \end{bmatrix}, \quad M = \begin{bmatrix} 9 & 0 \\ 0 & 1 \end{bmatrix}.$$

5.3.23   If A has only two eigenvalues λ_1 and λ_2 (neither of them zero) what combination $cI + dA$ is equal to A^{-1}?

5.3.24   Solving $Ax = \lambda x$ is equivalent for symmetric matrices to finding stationary points of the *Rayleigh quotient* $R = x^T A x / x^T x$.
(a)   Write out R, $\partial R / \partial x_1$, and $\partial R / \partial x_2$ when $A = \begin{bmatrix} 1 & 5 \\ 5 & 1 \end{bmatrix}$
(b)   Which vectors x minimize and maximize this R?
The gradient of R is $Ax - R(x)x$, so descent methods move from each x_k partly in the direction of Ax_k. Since that direction is provided by the "Krylov sequence" $x_1, Ax_1, A^2x_1, \ldots$, the conjugate gradient method is also an optimal way to minimize and maximize R.

5.3.25   Show that the iteration $Mx_{k+1} = (M - A)x_k + b$ can be done by computing the residual $r = b - Ax_k$ and solving $M(x_{k+1} - x_k) = r$.

5.3.26   Starting from mesh values 1 at four points on a coarse mesh (the corners of a square) what does the multigrid refining matrix F produce on the fine mesh? Conversely, what does the coarsening matrix C do to four 1's on a square of the fine mesh?

5.4 ■ THE FINITE ELEMENT METHOD

How do you go from a continuous problem to a discrete one? That is the *discretization* step, which has to be taken before differential equations can be solved. It produces a finite-dimensional problem and then the algorithms for $Ax = b$ or $Ax = \lambda x$ play their part. This section is about formulating the discrete approximation, which probably takes more insight than the solution step.

There are two basic ways to make a continuous problem discrete. We are thinking here of equilibrium problems, for example Laplace's equation. One way is to choose a finite number of *points*, and to replace derivatives by differences. That we have seen already; it led to the five-point scheme in Section 5.1. The other way is to choose a finite number of *functions*, and to approximate the exact solution by a combination of those trial functions. If the functions are sines and cosines, then we are close to Fourier series—but normally the geometry has to be simple and convenient. If the functions are special to the problem—they could be the eigenfunctions of Laplace's equation—then they may be theoretically perfect but practically useless; they are seldom known in advance. If the functions are *piecewise polynomials*, then the pieces can be chosen to fit the geometry of the problem and the computer can generate the polynomials. This is the idea behind the finite element method. It allows the computer to assemble the discrete problem as well as solve it.

The finite element method has had a remarkable history. It was created for the complicated equations of elasticity and structural mechanics (linear and nonlinear), and there it has essentially superseded finite differences. Now it is everywhere. You cannot do engineering calculations, at least for partial differential equations, without considering it. We don't mean that finite differences or Fourier series are obsolete. The battle is still raging, especially for initial-value problems. With the right geometry sines and cosines can give tremendous accuracy, and fast transforms make it possible to keep hundreds of harmonics; those virtues come together in the spectral method. About finite differences there are two things to say. Standard difference equations are easier to program and faster to run than a full finite element code; for a model problem they are often the best choice. Even when the decision goes to finite elements, the idea of differences is not lost. The choice of piecewise polynomials is successful exactly because it keeps the best features of finite differences. It ends up as an approximation at meshpoints, and the finite element equations end up looking like difference equations. They *are* difference equations, but they are derived on the basis of a more systematic and widely applicable principle—the Rayleigh-Ritz-Galerkin rule, or the variational form, or the method of weighted residuals.

We begin with a one-dimensional example of finite elements.

EXAMPLE $u'' = 2$ with $u(0) = 0$, $u(1) = 1$, and solution $u(x) = x^2$.

The approximation will use two intervals, from 0 to h and from h to 1. There is one

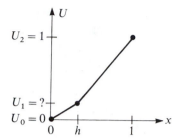

Fig. 5.9. Two linear elements with one unknown U_1.

unknown, the height U_1 at the point $x = h$. The values $U_0 = 0$ and $U_2 = 1$ are known from the boundary conditions, and the trial functions are *piecewise linear*. The true solution $u = x^2$ (a parabola) will be approximated by the broken line in Fig. 5.9, and the finite element principle must decide on the best line.

U_1 is found as follows. Look at the energy

$$P(u) = \int_0^1 \left[\frac{1}{2} \left(\frac{du}{dx} \right)^2 + 2u \right] dx$$

which is minimized by the true solution. Admit only the linear functions $U(x)$ in the figure and compute $P(U)$. Choose the value of U_1 that minimizes $P(U)$.

In the first interval $U = U_1 x/h$ with slope U_1/h:

$$\int_0^h \left[\frac{1}{2} \left(\frac{U_1}{h} \right)^2 + 2 \frac{U_1 x}{h} \right] dx = \frac{U_1^2}{2h} + U_1 h.$$

In the second interval $U = a + bx$ with $b = \dfrac{U_2 - U_1}{1 - h}$ and $a = \dfrac{U_1 - U_2 h}{1 - h}$:

$$\int_h^1 \left[\frac{1}{2} b^2 + 2(a + bx) \right] dx = \frac{(U_2 - U_1)^2}{2(1 - h)} + (U_2 + U_1)(1 - h).$$

Differentiating the sum $P(U)$ from these two intervals,

$$\frac{dP}{dU_1} = \frac{U_1}{h} - \frac{U_2 - U_1}{1 - h} + h + 1 - h = 0.$$

Substituting the known $U_2 = 1$ gives the minimizing value $U_1 = h^2$.

In this case the finite element solution is exact! The true value of $u = x^2$ at the meshpoint is also h^2. Of course the linear U is not equal to the quadratic u inside the intervals, and generally they will not be equal at the nodes. In fact $dP/dU_1 = 0$ was really a difference equation, involving the value of U_2 (which we inserted at the end) and the value $U_0 = 0$ (which was used from the start). It was an ordinary difference equation on an unevenly spaced mesh, and the finite element method gives a systematic way to derive that equation and to approximate u.

We will propose two approaches, one based on minimum principles (Rayleigh-Ritz) and the other on the weak form (Galerkin). The elements are also described in two ways, by trial functions T_i or by polynomials $U = a + bx$ within each element. They are equivalent. After these principles we return to computational examples, and to be realistic they must be in two dimensions. Our main example will be Poisson's equation with zero boundary conditions:

$$-\frac{\partial^2 u}{\partial x^2} - \frac{\partial^2 u}{\partial y^2} = f(x,y) \text{ in the open set } S$$

$$u = 0 \text{ on the boundary } C. \tag{1}$$

This is the strong form of the equation; its derivatives can be approximated directly by differences. It is not the right framework for finite elements. Instead of asking for an equation that holds at each point we need an equation that holds for each function! The right form is the **weak form**—the equation of virtual work. To derive it, multiply (1) by a test function $v(x,y)$ and integrate over S:

$$\iint_S \left[-\frac{\partial^2 u}{\partial x^2} - \frac{\partial^2 u}{\partial y^2} \right] v \, dx \, dy = \iint_S fv \, dx \, dy. \tag{2}$$

This involves functions rather than points. The residual $u_{xx} + u_{yy} + f$ is required to be zero, but only in a "weighted" sense; the weight is provided by v and each choice of v gives one equation. To achieve symmetry between u and v we integrate the left side by parts, requiring v to vanish on the boundary (as u did):

$$\boxed{\iint_S \left[\frac{\partial u}{\partial x} \frac{\partial v}{\partial x} + \frac{\partial u}{\partial y} \frac{\partial v}{\partial y} \right] dx \, dy = \iint_S fv \, dx \, dy.} \tag{3}$$

That is the weak form. It holds for every test function $v(x,y)$. It is the basis for **Galerkin's method**, which approximates (3) in the following way:

1. Choose a finite set of trial functions $T_1(x,y), \ldots, T_n(x,y)$
2. Admit approximations to u of the form $U(x,y) = U_1 T_1 + \cdots + U_n T_n$
3. Determine the n unknown numbers $U_1, \ldots, U_n$ from (3), using n different test functions v to obtain n equations.

To maintain symmetry we choose the test functions to be the same as the trial functions. Then there is a Galerkin equation for each $v = T_i(x,y)$. Substituting into

(3), that equation is

$$\int\int_S \left[\left(U_1\frac{\partial T_1}{\partial x} + \cdots + U_n\frac{\partial T_n}{\partial x}\right)\frac{\partial T_i}{\partial x} + \left(U_1\frac{\partial T_1}{\partial y} + \cdots \right.\right.$$
$$\left.\left. + U_n\frac{\partial T_n}{\partial y}\right)\frac{\partial T_i}{\partial y}\right] dx\, dy = \int\int_S fT_i\, dx\, dy. \qquad (4)$$

This is the key to everything. It will look less complicated in a minute, when written as an ordinary linear equation $KU = F$. The unknown function $u(x,y)$ has been replaced by an unknown vector U. The right side $f(x,y)$ has changed to F. In finite differences there may be no relation between these two discretizations, one on each side of the equation, but for Galerkin they come from the same source—the choice of the trial functions T_i.

You can see from (4) the calculations that are required. On the left side we need the numbers

$$K_{ij} = \int\int_S \left(\frac{\partial T_i}{\partial x}\frac{\partial T_j}{\partial x} + \frac{\partial T_i}{\partial y}\frac{\partial T_j}{\partial y}\right) dx\, dy. \qquad (5)$$

These enter the stiffness matrix K, which is symmetric: $K_{ij} = K_{ji}$. Later we show that K is also positive definite. On the right side we need the n components of the vector F:

$$F_i = \int\int_S fT_i\, dx\, dy. \qquad (6)$$

Then equation (4) is exactly $KU = F$, and its solution gives the Galerkin approximation $U_1 T_1 + \cdots + U_n T_n$.

It is remarkable that the same approximation appears from a different source. Instead of the weak form it begins with the minimum principle. For Poisson's equation that asks us to minimize

$$P(u) = \int\int_S \left[\frac{1}{2}\left(\frac{\partial u}{\partial x}\right)^2 + \frac{1}{2}\left(\frac{\partial u}{\partial y}\right)^2 - fu\right] dx\, dy$$

over all functions that satisfy the essential boundary condition $u = 0$. This is the continuous problem, and it has a finite-dimensional analogue. We narrow the class of admissible functions, and allow only combinations $U = U_1 T_1 + \cdots + U_n T_n$ of the trial functions. That is the **Rayleigh-Ritz method**—to use a finite number of functions and to discretize the minimum principle, as in the computed example above.

To carry out the minimization we substitute U into the energy:

$$P(U) = \int\int_{S} \left[\frac{1}{2}\left(U_1 \frac{\partial T_1}{\partial x} + \cdots + U_n \frac{\partial T_n}{\partial x} \right)^2 + \frac{1}{2}\left(U_1 \frac{\partial T_1}{\partial y} + \cdots + U_n \frac{\partial T_n}{\partial y} \right)^2 \right.$$

$$\left. - f(U_1 T_1 + \cdots + U_n T_n) \right] dx\, dy. \tag{7}$$

This is an ordinary quadratic in the unknowns $U_1, \ldots, U_n$. The coefficients of U_1^2 and U_1 are

$$\frac{1}{2}\int\int \left(\frac{\partial T_1}{\partial x}\frac{\partial T_1}{\partial x} + \frac{\partial T_1}{\partial y}\frac{\partial T_1}{\partial y} \right) dx\, dy \quad \text{and} \quad \int\int f T_1\, dx\, dy.$$

and those numbers are identical with $\frac{1}{2}K_{11}$ and F_1 in Galerkin's equation. The coefficient of $U_i U_j$ in (7) agrees with K_{ij} in (5). Therefore the quantity to be minimized is exactly

$$P(U) = \tfrac{1}{2}U^T K U - U^T F. \tag{8}$$

The minimum of this quadratic, which is unconstrained, occurs at the solution of $KU = F$.

5L For a symmetric problem the Rayleigh-Ritz and Galerkin methods lead to the same discretization $KU = F$. The unknown vector U contains the coefficients in the approximation $U(x,y) = U_1 T_1 + \cdots + U_n T_n$. The Rayleigh-Ritz method minimizes the energy $P(U)$, restricting u to the n degrees of freedom allowed by the trial functions. The Galerkin method starts from the weak form and requires the first variation $\delta P/\delta U$ to be zero for n test functions v.

Galerkin's method is more general; it does not require the problem or its approximation to be symmetric. If we add a first derivative $\partial u/\partial x$ to Laplace's equation the symmetry and the minimum principle are lost. However the change in equation (2) is straightforward. We still reach $KU = F$ but the matrix K is no longer symmetric. It contains the extra skew-symmetric terms

$$K_{ij}^* = \int\int T_i \frac{\partial T_j}{\partial x}\, dx\, dy = - \int\int \frac{\partial T_i}{\partial x} T_j\, dx\, dy = - K_{ji}^*.$$

Galerkin also allows test functions that are different from the trial functions. The extreme case is to use delta functions at n points. When v is a delta function the weak form becomes the strong form (the differential equation). With this choice Galerkin's method reduces to **collocation: The approximation U satisfies the**

differential equation at n points. With v replaced by δ and u by U, (2) becomes

$$-\frac{\partial^2 U}{\partial x^2} - \frac{\partial^2 U}{\partial y^2} = f \text{ at the } n \text{ collocation points.} \tag{9}$$

That gives n equations for the unknown coefficients $U_1, \ldots, U_n$. It involves second derivatives of the T_i where the symmetric form needed only first derivatives. (The smoothness of the trial functions compensates for the roughness of the delta functions.) For piecewise polynomials this extra smoothness can be painful to achieve. To collocate at nodal points we should use cubic splines, where Hermite cubics are simpler—determined by values and slopes but with jumps in second derivatives. For collocation at the two "Gauss points" in each interval, those are also more accurate than splines. We emphasize that collocation can be extremely simple and efficient—the integrals of the weak form are gone and the differential equation (linear or nonlinear) is enforced at n points.

Finite Element Trial Functions

What properties are desirable in the trial functions? That is the crucial point, because the choice of the T_i determines everything else. Ritz used only a few functions ($n = 10$ would have been exceptional) whereas a computer easily accepts $n > 1000$—provided the T_i are simple to work with. Another look at the algorithm reveals what is needed:
1.   The functions T_i must be capable of approximating the true $u(x,y)$
2.   The entries K_{ij} and F_i must be convenient to compute
3. K should be sparse and "well-conditioned."

These requirements are a mixture of engineering mathematics and computer science and numerical analysis. So is the finite element method. It evolved gradually, not always on the foundation set by Ritz and Galerkin. The equation $KU = F$ came first from physical principles and was recognized only later as the approximation developed by Courant from piecewise polynomials. We now go back to the 1960's for the part that was fun—the construction of the trial functions, or "shape functions."

The simplest element divides the region into triangles. Those are the pieces on which the polynomials are defined—a different polynomial in each triangle. Within a triangle the functions will be ***linear***. The approximating U has the form $U = a + bx + cy$; its graph is a plane. Its three coefficients a,b,c are determined by its value *at the vertices of the triangle*. Then if U is known at all vertices, it is known everywhere. In the construction there is a choice between
    (a)   describing each function T_i, and
    (b)   describing U as a polynomial in each piece.

These are equivalent and we are following option (b), which is usually easier. Therefore we now make the connection between pieces: The trial functions are ***continuous across each edge*** between triangles. The reason is simple and beautiful.

> Suppose U is determined in each triangle by its values at the vertices. An edge between triangles connects two vertices that belong to both triangles. Since U is linear along that edge it is known from its values at those shared vertices. Therefore U is the **same** straight-line function along the edge when approached from either triangle.

The graph of U is a collection of flat triangular pieces. They fit together like a roof with no gaps. At boundary vertices we require $U = 0$, which forces U to vanish along each outer edge of the region (which we assume to be a polygon). We denote this finite element by P_1—a polynomial of degree 1 that is continuous between triangles.

What are the trial functions $T_1, \ldots, T_n$? They give the other description of P_1, with a function T_j for every vertex V_j inside S. At that vertex, T_j equals 1; at all other vertices T_j is zero. The graph of T_j is a pyramid; in Fig. 5.10 its base is a pentagon. It rises linearly to 1 in the triangles around V_j, and in all other triangles it is identically zero—since it is zero at all three vertices. The number n of degrees of freedom, which is the dimension of the trial space P_1, equals the number of vertices inside S.

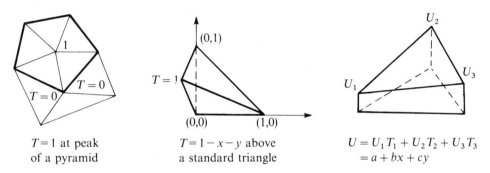

$$T=1 \text{ at peak} \qquad T=1-x-y \text{ above} \qquad U = U_1 T_1 + U_2 T_2 + U_3 T_3$$
$$\text{of a pyramid} \qquad\quad \text{a standard triangle} \qquad\qquad = a + bx + cy$$

Fig. 5.10. Linear functions in triangles: the P_1 element.

We emphasize one attractive feature of the construction. $U_1, \ldots, U_n$ normally just give the coefficients of the T_j in the Galerkin approximation. For finite elements these numbers have a significance of their own: They are the approximations to $u(x, y)$ *at the vertices*. In other words,

$$U_1 T_1 + \cdots + U_n T_n \text{ reduces to } U_i \text{ at the vertex } V_i. \tag{10}$$

The only nonzero term is U_i because all other T_j are zero at that vertex.† This turns $KU = F$ into a difference equation. The approximations U_i are at the nodes and when K is assembled it will be sparse. The equation comes from a variational principle, but it looks like (and is) a difference method. Before computing K for this linear element P_1 we look at four other finite elements:

---

† This is abbreviated by $T_j(V_i) = \delta_{ij}$. The symbol δ_{ij} represents 1 if $i = j$ and 0 otherwise.

1. ***Linear elements in one dimension***: The unknowns U_j are the nodal values, the approximation is $U = a + bx$ in each interval, and a typical "hat function" T is shown in Fig. 5.11.
2. ***Cubic elements in one dimension***: The unknowns are the heights and slopes, $U = a + bx + cx^2 + dx^3$, and the shape functions T_j already appeared in Fig. 3.5. These are the Hermite cubics; splines are not given directly by nodal values and do not qualify as finite elements.
3. ***Bilinear elements on rectangles***: Instead of triangles with 3 vertices the Q_1 element uses rectangles with 4 vertices. The four values of U determine the four coefficients in $U = a + bx + cy + dxy$.
4. ***Quadratic elements on triangles***: Each triangle has six nodes—the three vertices and the midpoints of the three edges—and the P_2 trial function has six coefficients:

$$U = a + bx + cy + dx^2 + exy + fy^2 \text{ in each triangle.} \tag{11}$$

These are all continuous across element boundaries. On an edge between rectangles the functions in Q_1 are determined by two nodal values. They are linear on that edge and give the same straight-line function when approached from either side. The P_2 element is one step more subtle. It is a parabola along each edge; for example it is $a + bx + dx^2$ on the line $y = 0$. The edge contains three nodes, at its midpoint as well as its ends. These three shared nodes produce the same parabola from the triangles on either side.

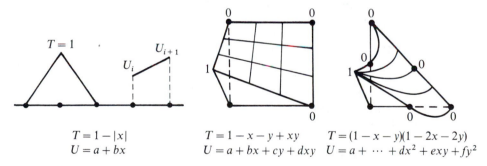

$$T = 1 - |x| \qquad T = 1 - x - y + xy \qquad T = (1 - x - y)(1 - 2x - 2y)$$
$$U = a + bx \qquad U = a + bx + cy + dxy \qquad U = a + \cdots + dx^2 + exy + fy^2$$

Fig. 5.11. Linear on intervals, bilinear on rectangles, quadratic in triangles.

Only the cubics have the same *derivatives* from both sides.

How closely can these trial functions approximate the true solution u? For polynomials that is not difficult to estimate, and the accuracy improves with each reduction in the meshsize h—which is the longest edge of any element. The degree of approximation comes directly from the degree of the trial functions. For linear elements the error is of order h^2. This is the distance between a curve and its straight-line approximation. For the P_2 element that becomes h^3 and for cubics it is h^4. Of course U will not generally equal u at nodal points; nevertheless the error $u - U$ is of order h^{p-1} when the polynomials are complete through degree p. We show below that U is actually the *best* approximation to u—even if it is not exact at the nodes.

The approximation by trial functions is good, and it improves as p increases. The computation of K and F and $K^{-1}F$ is simple, but it gets worse as p increases. In one dimension cubic elements are satisfactory, and in higher dimensions a natural compromise is $p = 2$. One of the best elements is Q_2, biquadratic on rectangles, with 9 polynomial coefficients (up to x^2y^2) and 9 nodes in each rectangle. For complicated problems or for testing, Q_1 and P_1 are less accurate but considerably simpler.

Assembling the Element Matrices

The first goal, after the subdivision into triangles or rectangles, is to compute the matrix K. For Laplace's equation its entries are

$$K_{ij} = \int\int_S \left(\frac{\partial T_i}{\partial x} \frac{\partial T_j}{\partial x} + \frac{\partial T_i}{\partial y} \frac{\partial T_j}{\partial y} \right) dx \, dy. \tag{12}$$

On triangles each function T_j is a pyramid over V_j. If the pyramids do not overlap—if V_i and V_j are not vertices of the same triangle—then K_{ij} will be zero. Thus the stiffness matrix K is sparse; it connects only neighboring vertices. For linear functions the derivatives are constant in each triangle, which simplifies the integrals in (12). Nevertheless that formula is not the best way to find K.

It is better to work with $U = a + bx + cy$ in each triangle, than with a pyramid around every vertex. In each element e the energy is half of

$$\boxed{\int\int_e \left[\left(\frac{\partial U}{\partial x} \right)^2 + \left(\frac{\partial U}{\partial y} \right)^2 \right] dx \, dy = (b^2 + c^2)(\text{area of } e).} \tag{13}$$

This integral contributes an **element stiffness matrix** k_e to the overall matrix K. The problem is to compute the individual matrices k_e and assemble them into K. Each matrix k_e is 3 by 3; its order agrees with the number of coefficients in U and the number of nodes in the element.

On a standard right triangle the calculation is easy. The slope b in the x direction is $U_2 - U_1$. (We look along the baseline.) The slope in the y direction is $c = U_3 - U_1$ and the area is $\frac{1}{2}$. These give the energy integral $\frac{1}{2}(b^2 + c^2)$ in (13), and the element matrix comes from writing that answer in matrix form:

$$\tfrac{1}{2}(U_2 - U_1)^2 + \tfrac{1}{2}(U_3 - U_1)^2 = [U_1 \ U_2 \ U_3] \begin{bmatrix} 1 & -\frac{1}{2} & -\frac{1}{2} \\ -\frac{1}{2} & \frac{1}{2} & 0 \\ -\frac{1}{2} & 0 & \frac{1}{2} \end{bmatrix} \begin{bmatrix} U_1 \\ U_2 \\ U_3 \end{bmatrix}. \tag{14}$$

This matrix is k_e. It is the contribution to K from one triangle. For other triangles it can be computed from the matrix $B^T B$ given below, or from a change of variables that brings back the standard triangle. The result is quite remarkable; the entries of k_e depend only on the tangents of the angles. The integral in (13) turns out to be

$$\frac{1}{2}\left[\frac{(U_2 - U_1)^2}{\tan \theta_3} + \frac{(U_3 - U_1)^2}{\tan \theta_2} + \frac{(U_3 - U_2)^2}{\tan \theta_1}\right], \tag{15}$$

and for the right triangle these tangents were 1,1, and ∞. In that case (15) is the same as (14). In general the element matrix hidden inside (15) is

$$k_e = \begin{bmatrix} c_2 + c_3 & -c_3 & -c_2 \\ -c_3 & c_1 + c_3 & -c_1 \\ -c_2 & -c_1 & c_1 + c_2 \end{bmatrix}, \quad c_i = \frac{1}{2\tan \theta_i}. \tag{16}$$

Then the quadratic (15) is the same as $[U_1 U_2 U_3]k_e[U_1 U_2 U_3]^T$ and assembling these small matrices will give K.

Note that the element matrices are symmetric. They are also positive semidefinite, because they come from integrating $(\partial U/\partial x)^2 + (\partial U/\partial y)^2$. However they are not positive definite; each row adds to zero. In other words the vector $(1,1,1)$ is again in the nullspace. This is unavoidable, because if $U_1 = U_2 = U_3 = 1$ then the function $a + bx + cy$ reduces to $U(x,y) = 1$ and its derivatives are zero. Nevertheless the assembled K is positive definite.

The assembly is easiest for unit intervals in one dimension. There the element matrix is 2 by 2; an element has two nodes. Between nodal values U_1 and U_2 the slope is $U_2 - U_1$, and

$$\int_0^1 \left(\frac{dU}{dx}\right)^2 dx = (U_2 - U_1)^2 = [U_1 \ U_2] \begin{bmatrix} 1 & -1 \\ -1 & 1 \end{bmatrix} \begin{bmatrix} U_1 \\ U_2 \end{bmatrix}. \tag{17}$$

This reveals one element matrix k_e and our task is to combine it with the others. The local numbering U_1, U_2 in the element has to be matched with the global numbering $U_1, \ldots, U_n$ over the whole region. The short vector U_e has only two components from the full vector U. There is a special case in the first interval, if it connects U_1 to the fixed value $U_0 = 0$; its element matrix is only 1 by 1. We accept this as the boundary condition at $x = 0$, and we impose no essential condition at the other end—where $du/dx = 0$ is a natural condition. Then the assembly of K from the k_e is (apart from the factor $1/h$ from the element widths)

$$\begin{bmatrix} 1 & & & \\ & & & \\ & & & \\ & & & \end{bmatrix} + \begin{bmatrix} & & & \\ & 1 & -1 & \\ & -1 & 1 & \\ & & & \end{bmatrix} + \cdots$$

$$+ \begin{bmatrix} & & & \\ & & & \\ & & 1 & -1 \\ & & -1 & 1 \end{bmatrix} = \begin{bmatrix} 2 & -1 & & & \\ -1 & 2 & -1 & & \\ & -1 & \cdot & \cdot & \\ & & \cdot & 2 & -1 \\ & & & -1 & 1 \end{bmatrix} = K. \tag{18}$$

This matrix is not only semidefinite but positive definite. That is the effect of the first element, reflecting the essential boundary condition. The vector $1,1,\dots,1$ is not in the nullspace of the assembled matrix and K is not singular. Its *condition number*—the ratio of largest to smallest eigenvalue—is of order n^2. This exponent is expected when there are second derivatives in the continuous problem, and K is ideal for computations. In fact it is exactly a finite difference matrix.

In two dimensions the assembly of K is interesting. Each 3 by 3 matrix k_e is added into the rows and columns that correspond to the three vertices of its element. The code has to know the numbering system in order to select the right three vertices from the full list $1,\dots,n$. A typical input might contain the information

triangle 1:   nodes 3,1,2
triangle 2:   nodes 3,4,2
...
triangle N: nodes 5,4,n

This local-global numbering will involve boundary vertices also, and later $U = 0$ is enforced at those points. After assembling the three triangles listed above, the matrix will look like

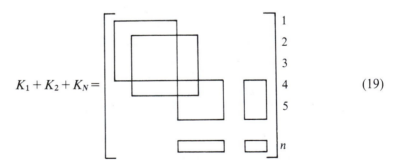

$$K_1 + K_2 + K_N = \qquad\qquad\qquad\qquad\qquad\qquad\qquad\qquad (19)$$

K_i is the result of putting the 3 by 3 matrix k_i into the right place in the large matrix. After assembling the other $N-3$ elements, K will be complete.

For Laplace's equation on a right-angled regular mesh the result is unexpected. K is exactly the matrix for the 5-point scheme! Its coefficients are $4,-1,-1,-1,-1$, even though every point has six neighbors and not four. The two accidental zeros come from (14) and would be nonzero for the irregular triangles in (16). Normally there is an entry off the diagonal for every neighboring node. Nodes on rectangles have 8 neighbors, and for the P_2 element on triangles a vertex has 18. Thus the number of entries in K and its bandwidth climb quickly. In plane elasticity there are horizontal and vertical displacements at every node, as in trusses, and everything is doubled. You can regard finite elements as a systematic way to construct accurate difference equations on an irregular mesh.

That completes a basic description of the finite element method. It combines old ideas of Ritz and Galerkin with the new choice of piecewise polynomials, and it reaches $KU = F$. The principles are clear—but if you are going to use it tomorrow there is more I should say. The following pages bring closer the computations that are actually done with finite elements.

Finite Element Computations

Practically speaking the assembly process is automatic and the work is at the element level. Those integrals have to be done over and over, and in most cases they are not done exactly. Instead of a true integration we use a quadrature formula. The computer evaluates the integrand at a few "Gauss points" inside the element, and averages those values. We illustrate with the three integrals that are needed in the finite element method:

1.   **The force vector** F. For the right side of $KU = F$, the weak form contained $\int \int fv \, dx \, dy$. In finite elements the test functions v are the same as the trial functions T. Thus F_i can be computed from $\int \int f T_i$, one function at a time, but as before there is a better way. The overall integral

$$F_1 U_1 + \cdots + F_n U_n = \int \int f(x,y) U(x,y) \, dx \, dy \tag{20}$$

is computed *an element at a time*. The simplest approximation does everything at a single point, the centroid of the triangle. At that point (call it P) U is the average of its values U_1, U_2, U_3 at the vertices. Therefore

$$\int \int_e f U \, dx \, dy \approx f(P) \frac{U_1 + U_2 + U_3}{3} \quad \text{(area of } e\text{)}. \tag{21}$$

Then the contributions of all triangles are combined—that is another assembly— to determine F. On a regular mesh of right triangles with area $h^2/2$, each vertex belongs to six triangles (see Ex. 5.4.7). The central unknown U_1 appears in six formulas like (21). Its assembled coefficient is

$$F_1 = \frac{h^2}{6} [f(P_1) + \cdots + f(P_6)], \tag{22}$$

an average at the six centroids of the triangles. A difference equation would have selected f *at the vertex*, instead of this average. Under a constant load the result is the same.

For the bilinear element Q_1 on a rectangle the Gauss point is at the center. The answer is exact for combinations of $1, x, y$, and xy. For the biquadratic element Q_2 we average over four Gauss points; they are in the special positions $(\pm 1/\sqrt{3}, \pm 1/\sqrt{3})$ for the standard square $-1 \le x, y \le 1$. The answer is then exact for all bicubic polynomials.† Other quadrilaterals can be mapped to this square by a change of variables $x,y \to X,Y$; the Jacobian $|J| = (\partial x/\partial X)(\partial y/\partial Y) - (\partial y/\partial X)(\partial x/\partial Y)$ becomes an extra factor in the integral. Even a quadrilateral with curved sides is possible. For these "isoparametric elements" and for higher-order quadrature in triangles we refer to our joint book *An Analysis of the Finite Element Method* (Prentice-Hall).

2. **The element matrices k_e.** The actual calculation of k_e fits perfectly into the framework of $A^T C A$. That is how it is organized by structural engineers. The material coefficient in the middle is $c = 1$ for Laplace's equation; in general the elasticity or conductivity can be variable. The important point is the calculation of the derivatives AU, when U is a polynomial in each element. The key is to connect the derivatives of U to its values at the nodes.

On a standard right triangle U is $a + bx + cy$. The link between its two derivatives and its three nodal values is contained in a 2 by 3 matrix B:

$$AU = \begin{bmatrix} \partial U/\partial x \\ \partial U/\partial y \end{bmatrix} = \begin{bmatrix} b \\ c \end{bmatrix} = \begin{bmatrix} U_2 - U_1 \\ U_3 - U_1 \end{bmatrix} = \begin{bmatrix} -1 & 1 & 0 \\ -1 & 0 & 1 \end{bmatrix} \begin{bmatrix} U_1 \\ U_2 \\ U_3 \end{bmatrix} = BU. \quad (23)$$

That says a lot in one line. It expresses AU (the derivatives) in a discrete form BU (the differences). It gives the matrix B that the computer uses to find k_e. In fact k_e is exactly the integral of $B^T C B$ over the element:

$$\iint_e U^T A^T C A U \, dx \, dy \to U^T \left[\iint_e B^T C B \, dx \, dy \right] U = U_e^T k_e U_e.$$

In practice this is often done by one-point integration:

$$k_e = (\text{area of } e)(B^T C B \text{ at the centroid}). \quad (24)$$

The calculation is now in matrix form, and it fits into codes like NASTRAN and ADINA. It is typical of the general rule:

The matrix k_e is a weighted combination of the matrices $B^T C B$ at the integration points. The weights add up to the area of e.

EXAMPLE 1   For Laplace's equation the matrix B in (23) yields

$$B^T B = \begin{bmatrix} 2 & -1 & -1 \\ -1 & 1 & 0 \\ -1 & 0 & 1 \end{bmatrix}.$$

---

† In one dimension N Gauss points give exact answers until $\int x^{2N} \, dx$. A square with N by N points will miss terms involving x^{2N} or y^{2N}.

Multiplied by the area $\frac{1}{2}$ this is the matrix k_e in (14). Multiplied further by the value of c at the centroid it is the element matrix for $-\text{div}(c \text{ grad } u) = f$. The exercises ask for B on an arbitrary triangle and a square.

EXAMPLE 2   For $U = a + bx$ the matrix B is $\begin{bmatrix} -1 & 1 \end{bmatrix}$:

$$\frac{dU}{dx} = b = U_2 - U_1 = \begin{bmatrix} -1 & 1 \end{bmatrix} \begin{bmatrix} U_1 \\ U_2 \end{bmatrix} \quad \text{and} \quad B^T B = \begin{bmatrix} 1 & -1 \\ -1 & 1 \end{bmatrix}.$$

After assembly this gives as before the $-1, 2, -1$ second-difference matrix K.

EXAMPLE 3   For the bilinear Q_1 element B is no longer a constant matrix—and this leads to an important question in finite element computations. Is one integration point enough? The function $U = a + bx + cy + dxy$ has derivatives

$$AU = \begin{bmatrix} dU/dx \\ dU/dy \end{bmatrix} = \begin{bmatrix} b + dy \\ c + dx \end{bmatrix} = B \begin{bmatrix} U_1 \\ U_2 \\ U_3 \\ U_4 \end{bmatrix}.$$

Without calculation we see that B involves x and y. Therefore $B^T C B$ has terms in x^2 and y^2, which will be lost in one-point integration. If the center of the square is $x = y = 0$, those terms are zero but the true integral of x^2 or y^2 is not zero.

Thus more integration points are officially required—and the four Gauss points would be normal for a standard square. However they are four times as expensive as one point. Therefore a major effort is being made to control the damage more cheaply, by adding back a stabilizing term to the nearly singular K. Exercise 5.4.8 identifies the "hourglass" mode, with $U = 1$ or -1 in a checkerboard pattern, which is incorrectly assigned zero energy. It can destroy K^{-1} unless it is controlled.

3.   **The mass matrix M.** The matrix K is not the only one to appear in finite elements. For eigenvalue problems, and later for initial-value problems, there is also a mass matrix M. It comes from the integral of U^2 rather than $(\partial U/\partial x)^2 + (\partial U/\partial y)^2$, so in some respects it is simpler than K. It will enter whenever the differential equation has a term involving only u or $\partial u/\partial t$.

The eigenvalues of a clamped membrane give a good example:

$$-\frac{\partial^2 u}{\partial x^2} - \frac{\partial^2 u}{\partial y^2} = \lambda u \tag{25}$$

with $u = 0$ along the boundary. This is like Laplace's equation, with minus signs to make λ positive, but now λu replaces f on the right. The **weak form of the eigenvalue**

problem is still found from test functions v:

$$\iint \left[\frac{\partial u}{\partial x} \frac{\partial v}{\partial x} + \frac{\partial u}{\partial y} \frac{\partial v}{\partial y} \right] dx\, dy = \lambda \iint uv\, dx\, dy. \tag{26}$$

On the left side nothing is new. Galerkin replaces u by $U = U_1 T_1 + \cdots + U_n T_n$; Ritz uses these trial functions as test functions; the finite element method takes them to be piecewise polynomials. The matrix K is unchanged.

The *mass matrix* appears on the right side, when the trial function T_j multiplies the test function T_i. That produces the entry

$$M_{ij} = \iint T_j T_i\, dx\, dy. \tag{27}$$

The strong form (25) and the weak form (26) change into a matrix problem

$$\boxed{KU = \lambda MU.} \tag{28}$$

This eigenvalue problem is not quite standard. It involves ***two matrices instead of one***. Engineering and physics are usually kind enough to produce symmetric matrices in their eigenvalue problems, but if we multiply by M^{-1} then symmetry is lost. The single matrix in $M^{-1}KU = \lambda U$ is not symmetric. However there is a remedy: Half of M^{-1} on the left side of K and the other half on the right side will leave a symmetric matrix A.

5M   The eigenvalues of $KU = \lambda MU$ are the same as the eigenvalues of $Ax = \lambda x$, if M is factored into LL^T and A is $L^{-1}K(L^T)^{-1}$. If K and M are symmetric positive definite so is A.

The algebra starts from $KU = \lambda MU = \lambda LL^T U$ and multiplies on the left by L^{-1}. Then the change to $x = L^T U$ leaves $L^{-1}K(L^T)^{-1}x = \lambda x$. This is $Ax = \lambda x$. To show that $\lambda > 0$, multiply both sides of $KU = \lambda MU$ by U^T. Since $U^T KU = \lambda U^T MU$, the positive definiteness of K and M makes λ positive.

In theory L can be any matrix for which $M = LL^T$. In practice it comes from elimination. Chapter 1 led to $M = LDL^T$, with ones on the diagonal of L, and Cholesky moved the square roots of the pivots into $\bar{L}$:

$$M = \bar{L}\bar{L}^T \quad \text{with} \quad \bar{L} = LD^{1/2}. \tag{29}$$

This $\bar{L}$, with $d_j^{1/2}$ multiplying column j, is the matrix L in 5M.

The mass matrix M is assembled from element matrices m_e just as K was assembled from k_e. The difference is that we now integrate U^2.

EXAMPLE On a standard interval U equals $U_1 + (U_2 - U_1)x$ and

$$\int_0^1 U^2 \, dx = \tfrac{1}{3}(U_1^2 + U_1 U_2 + U_2^2) = [U_1 \; U_2] \begin{bmatrix} \frac{1}{3} & \frac{1}{6} \\ \frac{1}{6} & \frac{1}{3} \end{bmatrix} \begin{bmatrix} U_1 \\ U_2 \end{bmatrix}.$$

Assembling these element mass matrices as in (18) produces a factor h times

$$\begin{bmatrix} \frac{1}{3} & & & \\ & & & \\ & & & \\ & & & \end{bmatrix} + \begin{bmatrix} \frac{1}{3} & \frac{1}{6} & & \\ \frac{1}{6} & \frac{1}{3} & & \\ & & & \\ & & & \end{bmatrix} + \cdots + \begin{bmatrix} & & & \\ & & & \\ & \frac{1}{3} & \frac{1}{6} \\ & \frac{1}{6} & \frac{1}{3} \end{bmatrix}$$

$$= \frac{1}{6} \begin{bmatrix} 4 & 1 & & & \\ 1 & 4 & 1 & & \\ & 1 & \cdot & \cdot & \\ & & \cdot & 4 & 1 \\ & & & 1 & 2 \end{bmatrix} = M. \qquad (30)$$

Then $KU = \lambda MU$ gives the approximate eigenvalues of a rod. It is fixed at one end by $u(0) = 0$ and free at the other end. If the differential equation is

$$-\frac{d}{dx}\left(c\,\frac{du}{dx}\right) = \lambda \rho u \text{ rather than } -\frac{d^2u}{dx^2} = \lambda u,$$

then the elasticity c multiplies K and the density ρ multiplies M—which explains the name "mass matrix."

To close this discussion of computations we comment on $KU = F$. It is usually solved by elimination, taking advantage of the zeros in K. The unknowns are ordered to maximize this advantage. Iterative methods are not popular (the optimal ω is hard to find) and semi-direct methods are not well known. Elimination is safe and predictable and its cost can often be reduced by Bruce Irons' **frontal method**: Compute and assemble element matrices only *when they are needed*. During the computation there are three types of nodes—those whose elements are all fully assembled, those belonging to some assembled elements and some not yet assembled, and those not yet touched. When a node joins the first group elimination is carried out. The second group is in active storage. A node in the third group becomes active when an element containing that node enters the calculation.

We emphasize the role of the triple products B^TCB. The framework based on A^TCA for the overall problem is copied by the finite element method on each piece. It is a different conception from finite differences.

Accuracy and Convergence of Finite Elements

It remains to show that U is close to u. The approximation $U = U_1 T_1 + \cdots + U_n T_n$ is a combination of the trial functions, but is it the best possible combination? The answer is yes, provided you interpret "best" in the correct way. That property is the key to the accuracy of finite elements.

Unavoidably U is not the ordinary least squares approximation—it is *not* the projection of u onto the subspace of trial functions. Instead it is the best approximation *in the energy*; it is the least squares approximation *weighted by K*. The stiffness matrix enters because it is responsible for the answer $U = K^{-1}F$. This approximation of u by U will be clearest in the finite-dimensional case, using the language of matrices. We keep small letters for the original problem and a capital U for the approximation—which is constrained to lie in the subspace of trial functions.

5N Without constraints, the minimum of $P(u) = \frac{1}{2}u^T k u - u^T f$ occurs at the point $u = k^{-1}f$. With constraints, the minimum of P occurs at the point U which is closest to u in the following sense:

> *The weighted error* $E = (U - u)^T k (U - u)$ *is a minimum.*

The proof comes from writing $P(u)$ as

$$\tfrac{1}{2}u^T k u - u^T f = \tfrac{1}{2}(u - k^{-1}f)^T k (u - k^{-1}f) - \tfrac{1}{2}f^T k^{-1}f.$$

After simplification the right side is identical to the left. The last term on the right is a constant. Therefore if every u is allowed—the unconstrained case—the minimum occurs when the first term on the right is zero, at $u = k^{-1}f$.

When the problem is constrained into a subspace and u becomes U, the minimization is still governed by that first term on the right. It may no longer reach zero. But minimizing P on the left side is the same as minimizing E on the right, which is the point of the proof.

In the continuous case this result remains true. $U(x,y)$ is again the closest approximation to the exact $u(x,y)$, if the error is measured by its "strain energy:"

$$E = (U - u)^T k (U - u) = \iint \left[\left(\frac{\partial}{\partial x}(U - u) \right)^2 + \left(\frac{\partial}{\partial y}(U - u) \right)^2 \right] dx \; dy.$$

The Ritz-Galerkin-finite element method automatically chooses U to be the combination of trial functions that minimizes this error in energy.

The convergence of U to u is now simple. If the trial functions allow it, it will happen! We look at the trial function U^* that matches u at the nodes, and estimate E for that function. Then U will be better than U^* in energy, even if it cannot be better at the nodes. For finite elements these are piecewise polynomials, and the critical question is their degree p. The basic theory can be stated in a sentence: ***The method converges if $p \geqslant 1$, and its error in energy is $E = O(h^{2p})$.*** Linear and bilinear elements have $p = 1$. Quadratics and biquadratics have $p = 2$. The answers are exact for solutions of degree p (this is checked by the patch test) and the errors come from higher terms.

That explains all but the crucial part—how so many engineering problems have been successfully solved. Those problems have complicated geometries and materials and equations. It is the simplicity of finite elements that has brought success. In fact the real difficulty is not to solve the equations, but to set them up. We need a systematic way to approximate physical laws, staying with the underlying mathematics but allowing the computer to do the work. The concept of trial functions was old, but the *choice* of trial functions (as piecewise polynomials) was the right idea at the right time.

EXERCISES

5.4.1   In the linear element P_1 suppose the value of U at every vertex is $U_j = 1$. Describe $U(x, y) = U_1 T_1 + \cdots + U_n T_n$ within the triangles. What shapes should replace triangles in 3-dimensional space, if $U = a + bx + cy + dz$ in each element?

5.4.2   Suppose the square $0 \leq x, y \leq 1$ is cut into two triangles by the $45°$ line $y = x$. If $U = 1$ at the midpoint $(\frac{1}{2}, \frac{1}{2})$ of that diagonal, and $U = 0$ at all other nodes of the quadratic P_2 in Fig. 5.11, find $U(x, y)$ in both triangles.

5.4.3   (a)   Show that a cubic polynomial $p(x, y)$ has ten terms, to match the ten nodes in the triangle below. Why is there continuity across the edges between triangles?
   (b)   Count the terms $1, x, y, \ldots, x^3 y^3$ in a bicubic polynomial, with the exponents of x and y going separately up to 3.

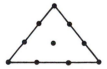

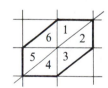

5.4.4   The square drawn above has 8 nodes rather than the usual 9 for biquadratic Q_2. Therefore we remove the $x^2 y^2$ term and keep

$$U = a_1 + a_2 x + a_3 y + a_4 x^2 + a_5 xy + a_6 y^2 + a_7 x^2 y + a_8 xy^2.$$

Find the function which equals 1 at $x = y = 0$ and zero at all other nodes.

5.4.5   Suppose a square has 9 nodes—its corners, edge midpoints, and center. The rows of K will have varying numbers of nonzeros:
   (a)   How many nodes are neighbors of a corner node? They lie in the four squares that meet at the corner.
   (b)   How many nodes are neighbors of a midpoint node?
   (c)   For the node at the center of the square, why is elimination permitted within the element matrix k_e before assembling it with others into K?

5.4.6   If the element stiffness matrix k_e comes from a triangle with node numbers 1,4,2, what 3 by 5 renumbering matrix r_e will put $r_e^T k_e r_e$ into the right positions in a 5 by 5 matrix K?

5.4.7   A regular mesh of horizontal, vertical, and northeast diagonal lines has six triangles adjacent to a typical vertex (third figure above). Check that the element matrices k_e of (15) combine to give the weights 4, $-1, -1, -1, -1$ of the 5-point Laplacian.

5.4.8   (a)   If $U = a + bx + cy + dxy$ find the coefficients a,b,c,d from the four equations $U = U_i$ at the corners $(\pm 1, \pm 1)$ of the standard square.
  (b)   From part (a) write down the 2 by 4 matrix B_0 at the center point $P = (0,0)$ of the square. The derivatives $b + dy$ and $c + dx$ reduce there to b and c:

$$AU = \begin{bmatrix} \partial U/\partial x \\ \partial U/\partial y \end{bmatrix} = \begin{bmatrix} b \\ c \end{bmatrix} = B_0 \begin{bmatrix} U_1 \\ U_2 \\ U_3 \\ U_4 \end{bmatrix}.$$

  (c)   Show that both $(1,1,1,1)$ and $(1,-1,1,-1)$ are in the nullspace of B_0. The first comes from a constant state $U = 1$ and correctly has zero energy; the second comes from an "hourglass" state $U = xy$ and should have positive energy.

5.4.9   From the previous exercise find the approximation $k_e = 4B_0^T B_0$ for the Q_1 bilinear element on the square of area 4. Compare with the correct k_e found by integrating $(b + dy)^2 + (c + dx)^2$ analytically.

5.4.10   For the one-dimensional trial functions in Fig. 5.11 on subintervals of length one, show that

$$\int \left(\frac{dT_j}{dx}\right)^2 dx = 2 \quad \text{and} \quad \int \frac{dT_j}{dx}\frac{dT_{j+1}}{dx} dx = -1.$$

These are the entries K_{jj} and K_{jj+1} in (18). Draw the trial function T_n at the endpoint $x = 1$ and explain why K_{nn} equals 1 instead of 2.

5.4.11   If the boundary condition is $du/dx = 0$ at both ends then

$$K = \begin{bmatrix} 1 & -1 & & \\ -1 & 2 & \cdot & \\ & \cdot & \cdot & -1 \\ & & -1 & 1 \end{bmatrix}.$$

Show that K is singular by finding a vector U with $KU = 0$. Show that the differential problem is also singular, by finding a function $u(x)$ with

$$\frac{d^2u}{dx^2} = 0 \quad \text{and} \quad \frac{du}{dx}(0) = \frac{du}{dx}(1) = 0.$$

5.4.12 If $U = a + bx + cy$ on the triangle below, find b and c from the equations $U = U_1$, $U = U_2$, $U = U_3$ at the nodes. Show that

$$\begin{bmatrix} b \\ c \end{bmatrix} = BU = \begin{bmatrix} -\dfrac{1}{l} & \dfrac{1}{l} & 0 \\ \dfrac{d}{lh} - \dfrac{1}{h} & -\dfrac{d}{lh} & \dfrac{1}{h} \end{bmatrix} \begin{bmatrix} U_1 \\ U_2 \\ U_3 \end{bmatrix}.$$

5.4.13 (a) For the previous matrix B, multiply $B^T B$ by the area $lh/2$ to find the element stiffness matrix k_e for Laplace's equation. For the standard right triangle reduce it to (14).

(b) Show that the off-diagonal entries agree with $c_i = (2 \tan \theta_i)^{-1}$ predicted in (16). The tangents of θ_1 and θ_2 can be read from the figure and c_3 comes from the identity

$$\tan \theta_3 = -\tan(\theta_1 + \theta_2) = \frac{\tan \theta_1 + \tan \theta_2}{\tan \theta_1 \tan \theta_2 - 1}.$$

5.4.14 Show that numerical integration using the centroid in (24) is exactly correct if f is a constant. Is it exact if $f = x$?

5.4.15 From equations (5) and (27) for K_{ij} and M_{ij}, verify that if $U = U_1 T_1 + \cdots + U_n T_n$ then

$$\iint \left[\left(\frac{\partial U}{\partial x} \right)^2 + \left(\frac{\partial U}{\partial y} \right)^2 \right] dx\, dy = U^T K U \quad \text{and} \quad \iint U^2 \, dx\, dy = U^T M U.$$

The Rayleigh quotient $U^T K U / U^T M U$ gives the minimum principle for the smallest eigenvalue of $KU = \lambda MU$. The minimum of the quotient is λ_1.

5.4.16 Find the eigenvalues of $KU = \lambda MU$, if

$$K = \begin{bmatrix} 2 & -1 \\ -1 & 2 \end{bmatrix} \quad \text{and} \quad M = \begin{bmatrix} 1 & 0 \\ 0 & 2 \end{bmatrix}.$$

This gives the oscillation frequencies for two unequal masses in a line of springs.

5.4.17 If the symmetric matrices K and M are not positive definite, $KU = \lambda MU$ may not have real eigenvalues. Construct a 2 by 2 example.

5.4.18 For two linear trial functions on the unit interval, with $U(0) = U(1) = 0$ and stepsize $h = \frac{1}{3}$, show that

$$K = \begin{bmatrix} 6 & -3 \\ -3 & 6 \end{bmatrix} \quad \text{and} \quad M = \frac{1}{18} \begin{bmatrix} 4 & 1 \\ 1 & 4 \end{bmatrix}.$$

Find the lowest eigenvalue of $KU = \lambda MU$ and compare with the true eigenvalue π^2 (the eigenfunction is $u = \sin \pi x$) of $-u'' = \lambda u$.

5.4.19 If the finite element form of $-u'' + u = f$ is $K^* U = F$, explain why $K^* = K + M$—the sum of the earlier stiffness and mass matrices. Find the approximation U_1 at the midpoint $x = \frac{1}{2}$, with linear elements, boundary condition $U_0 = U_2 = 0$, and $f = 1$. What is the exact solution $u(x)$?

5.5 ■ THE FAST FOURIER TRANSFORM

This section begins with one special matrix F. It is an n by n matrix, coming from a discrete Fourier transform on n points. To indicate its size we might have written F_n (and we should have said that there is a matrix for each n). It has a very particular form, and the pattern begins to appear at $n = 3$:

$$F_1 = [1] \quad F_2 = \begin{bmatrix} 1 & 1 \\ 1 & -1 \end{bmatrix} \quad F_3 = \begin{bmatrix} 1 & 1 & 1 \\ 1 & w & w^2 \\ 1 & w^2 & w^4 \end{bmatrix}$$

All entries of F_n are powers of the same number w_n. That number is complex (except for $w_1 = 1$ and $w_2 = -1$) and it is also extremely special; it is an *nth root of* 1. It is found by going around the unit circle through the angle $2\pi/n$:

$$w_n = e^{2\pi i/n} = \cos \frac{2\pi}{n} + i \sin \frac{2\pi}{n}.$$

For $n = 3$ this number is $-\frac{1}{2} + i\frac{\sqrt{3}}{2}$. Its cube is 1. For $n = 4$ it is $w_4 = i$, and the matrix is

$$F_4 = \begin{bmatrix} 1 & 1 & 1 & 1 \\ 1 & i & i^2 & i^3 \\ 1 & i^2 & i^4 & i^6 \\ 1 & i^3 & i^6 & i^9 \end{bmatrix}.$$

The entry in row j and column k—if the numbering is from 0 to $n - 1$, instead of 1 to n—is the number w_n with exponent j times k:

$$\boxed{(F_n)_{jk} = w_n^{jk} = e^{2\pi ijk/n}.}$$

This looks complicated but its nth power is again 1. There are n different nth roots of unity; the whole set $1, w_n, w_n^2, \ldots$ is found in the second row (which is row 1!) of the matrix. They are evenly spaced around the unit circle, at multiples of $2\pi/n$, and the primitive root w_n is the first to be reached after leaving 1.

Other rows contain roots in an interesting order. For example the third row of F_4 is $1, -1, 1, -1$. In a certain way this connects back to F_2 —and the whole inspiration of the Fast Fourier Transform is to find that connection! Similarly F_8 is related to F_4, and F_{80} is related to F_{40}. The speed of the FFT depends on working with highly composite numbers, having many factors; for a prime number like $n = 17$ a completely different algorithm is necessary. A power of 2 will allow us to concentrate on the relation between F_n and $F_{n/2}$, and we make that choice— although most codes allow any factors $n = pq$.

The next paragraphs explain the main idea of the Cooley-Tukey algorithm. It uses a trick (Gauss noticed it earlier) that computes the multiplication $F_n x$ with amazing speed. It is true that you could use the algorithm—codes are readily available—without knowing that trick. But it is neat and you will prefer to know. Then we show how the Fourier matrix F enters in partial differential equations (the heat equation and wave equation as well as Laplace's and Poisson's). For steady-state problems the FFT gives a fast Poisson solver. For initial-value problems it leads to the spectral method. Both are described in this section.

Our interest is in powers like $n = 2^{12}$. There will be $n^2 = 2^{24}$ entries in F_n, and an ordinary matrix-vector product $F_n x$ requires 2^{24} complex multiplications. In itself that is not terrible; it takes a few seconds on a big machine.† But if it is repeated thousands of times, as it is in time series analysis and image processing and elsewhere, the cost of these products $F_n x$ becomes prohibitive. By contrast, the Fast Fourier Transform finds $F_n x$ with only $6 \cdot 2^{12}$ multiplications—it is *more than* 600 *times faster*. It replaces n^2 multiplications by $\frac{1}{2}nl$, when $n = 2^l$. Thus $l = \log_2 n$ and the factor $n = 2^{12}$ is exchanged for $\frac{1}{2}l = 6$. By connecting F_n to $F_{n/2}$, and then to $F_{n/4}$, and eventually to F_1, the usual n^2 steps are reduced to $\frac{1}{2}n \log_2 n$. Practically speaking, we have n instead of n^2.

The fast transform starts with a completely trivial observation: If n equals $2m$ then w_n^2 equals w_m. By squaring w_n its angle $2\pi/n$ is doubled to $2\pi/m$; we go twice as far around the unit circle. The square of $w_4 = i$ is $w_2 = -1$; the $90°$ angle becomes $180°$. In general the square of $e^{2\pi i/n}$ is $e^{4\pi i/n}$, or $e^{2\pi i/m}$, and

$$w_n^2 = w_m \quad \text{if} \quad m = \tfrac{1}{2}n. \tag{1}$$

This identity allows $y = F_n x$ (a vector with n components) to be recovered from two vectors y' and y'' with m components. That is the key idea and it starts by dividing up x itself. The vector $(x_0, x_1, \ldots, x_{n-1})$ is split into two shorter pieces, by separating the even and odd components:

$$x' = (x_0, x_2, \ldots, x_{n-2}) \quad \text{and} \quad x'' = (x_1, x_3, \ldots, x_{n-1}).$$

From these vectors we form $y' = F_m x'$ and $y'' = F_m x''$. Those multiplications involve the half-size matrix F_m, but the results y' and y'' contain enough information to reconstruct y:

50   The first m and the last m components of $y = F_n x$ are

$$
\begin{aligned}
y_j &= y_j' + w_n^j y_j'', \quad j = 0, \ldots, m-1 \\
y_{j+m} &= y_j' - w_n^j y_j'', \quad j = 0, \ldots, m-1.
\end{aligned}
\tag{2}
$$

Thus the three steps are: split x into x' and x'', form $y' = F_m x'$ and $y'' = F_m x''$, and compute y from (2).

---

†I have not tried it on my PC.

The verification of (2) begins by dividing each component of $y = F_n x$ into odd and even parts. The entries of F_n are powers of w_n, and the separation of y_j follows exactly the separation of x:

$$y_j = \sum_{k=0}^{n-1} w_n^{kj} x_k \quad \text{is the same as} \quad \sum_{k=0}^{m-1} w_n^{2kj} x_{2k} + \sum_{k=0}^{m-1} w_n^{(2k+1)j} x_{2k+1}.$$

Each sum on the right has m terms. The first involves the even components x_{2k} and comes from x'. The second contains the components x_{2k+1} of x''. Since $w_n^2 = w_m$, these two parts are the same as

$$y_j = \sum_{k=0}^{m-1} w_m^{kj} x_k' + w_n^j \sum_{k=0}^{m-1} w_m^{kj} x_k''. \tag{3}$$

Now the trick is nearly complete. The two sums are y_j' and y_j'', the components of the half-size transforms $F_m x'$ and $F_m x''$. The first part of identity (2) is exactly the same as (3). For the other part of (2) we need $j + m$ in place of j, and this produces a sign change in (3):

inside the sums, $w_m^{k(j+m)}$ remains w_m^{kj} since $w_m^{km} = 1^k = 1$

outside, $w_n^{j+m} = -w_n^j$ since $w_n^m = e^{2\pi i m/n} = e^{\pi i} = -1$.

The sign change yields the second part of (2) and completes the proof.

Thus F_n uses twice as many multiplications as F_m (to form y' and y'') plus the m extra multiplications by w_n^j in (2). Repeating the same idea, a similar rule connects m to $m/2$ and eventually to $m = 1$, which needs no multiplications. For $n = 2^l$ we claim that the final count is $\frac{1}{2}nl$. This is correct at $l = 0$ and the step from $m = 2^l$ to $n = 2^{l+1}$ follows the rule given above. Twice the count for m, plus m extra multiplications, agrees with the count for n:

$$2(\tfrac{1}{2} 2^l l) + 2^l = \tfrac{1}{2} 2^{l+1}(l+1). \tag{4}$$

The cost is slightly more than linear; the factor $l = \log_2 n$ reflects the multiplications $w_n^j y_j''$ that appear in (2) at every step. However $\frac{1}{2}nl$ is so far below n^2 that discrete Fourier analysis has been completely transformed by formula (2).

EXAMPLE   The steps from $n = 4$ to $m = 2$ are

$$\begin{bmatrix} x_0 \\ x_1 \\ x_2 \\ x_3 \end{bmatrix} \rightarrow \begin{bmatrix} x_0 \\ x_2 \\ x_1 \\ x_3 \end{bmatrix} \rightarrow \begin{bmatrix} F_2 x' \\ \\ F_2 x'' \end{bmatrix} \rightarrow \begin{bmatrix} \\ y \\ \\ \end{bmatrix}.$$

These combine to multiply x by F_4. In fact every step is a matrix multiplication and the product of those matrices must be F_4:

$$
\begin{bmatrix}
1 & 1 & 1 & 1 \\
1 & i & i^2 & i^3 \\
1 & i^2 & i^4 & i^6 \\
1 & i^3 & i^6 & i^9
\end{bmatrix}
=
\begin{bmatrix}
1 & & 1 & \\
& 1 & & i \\
1 & & -1 & \\
& 1 & & -i
\end{bmatrix}
\begin{bmatrix}
1 & 1 & & \\
1 & -1 & & \\
& & 1 & 1 \\
& & 1 & -1
\end{bmatrix}
\begin{bmatrix}
1 & & & \\
& & 1 & \\
& 1 & & \\
& & & 1
\end{bmatrix} . \quad (5)
$$

The vector $x = (2,4,6,8)$ splits into $x' = (2,6)$ and $x'' = (4,8)$. Multiplying by F_2 gives $y' = (8,-4)$ and $y'' = (12,-4)$, and those combine into

$$y_0 = 8 + 12, \quad y_1 = -4 - 4i, \quad y_2 = 8 - 12, \quad y_3 = -4 + 4i.$$

This compares with $(5, -1+i, -1, -1-i)$ computed earlier from $F^{-1} = \bar{F}/4$.

You recognize F_2 appearing twice in the same matrix, in the middle step. At the right is the matrix that separates x into x' and x''. The third matrix carries out formula (2). Starting with F_8 the middle matrix would contain two copies of F_4, and **each of those would be split** as above. Thus the FFT amounts to a giant factorization of the Fourier matrix. The single matrix F with n^2 nonzeros turns into a product of approximately $\log n$ matrices with a total of only $n \log n$ nonzeros.

The Complete FFT and the Butterfly

The first step of the FFT changes multiplication by F_n to two multiplications by $F_m = F_{n/2}$. The even-numbered components $(x_0, x_2, \ldots)$ are transformed separately from $(x_1, x_3, \ldots)$. We give a flow graph (added in the second printing) for $n = 8$:

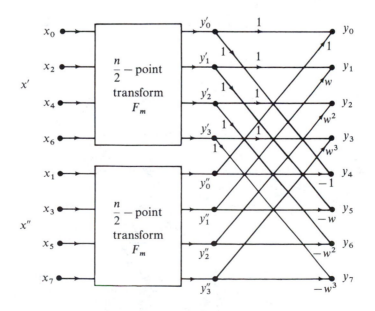

The key idea is **to replace each F_4 box by a similar picture involving two F_2 boxes.** The new factor $w_4 = i$ is the square of the old factor $w = w_8 = e^{2\pi i/8} = (1 + i)/\sqrt{2}$. The top half of the graph changes from F_4 to

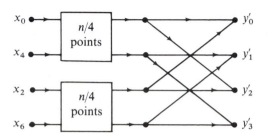

Then each of those boxes for $F_2 = \begin{bmatrix} 1 & 1 \\ 1 & -1 \end{bmatrix}$ is a single butterfly:

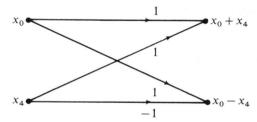

By combining the graphs you can see the whole picture. It shows the order that the n x's enter the FFT and the $\log n$ stages that take them through it—and it also shows the simplicity of the logic:

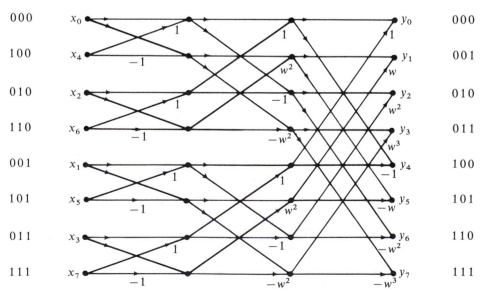

Every stage needs $\frac{1}{2}n$ multiplications so the final count is $\frac{1}{2}n \log n$. There is an amazing rule for the permutation of x's before entering the FFT: Write the subscripts $0, \ldots, 7$ in binary and *reverse the order of their bits*. The subscripts appear in "bit-reversed order" on the left side of the graph. Even numbers come before odd (numbers ending in 0 come before numbers ending in 1) and this is repeated at every stage.

Before the applications, we need to remember what the FFT achieves. It mixes the eight inputs into combinations like $y_0 = x_0 + x_1 + \cdots + x_7$ and $y_1 = x_0 + wx_1 + \cdots + w^7 x_7$. Those two outputs should be traceable on the graph (and $y_2, \ldots, y_7$ are similar). The eight combinations of eight x's are produced without 64 multiplications; 12 will do. The inverse transform F^{-1} just changes w to its conjugate $\bar{w}$—nothing more, except a final division by n. This reorganization has made digital signal processing an overwhelming success. Section 4.2 applied it to convolution; here we apply it to $-u_{xx} - u_{yy} = f$.

Periodic Difference Equations

The Fast Fourier Transform can solve certain finite difference equations with exceptional speed. To operate perfectly it needs a constant-coefficient problem $Ku = f$ on a regular mesh. A specific case is the 5-point scheme for Poisson's equation, in which the rows of K have nonzero entries $4, -1, -1, -1, -1$.

That is a two-dimensional problem and Gaussian elimination is not very impressive. By contrast the FFT is quite remarkable. The natural idea for a square array of unknowns is a ***two-dimensional*** FFT. After transforming each row of the square, we transform each column. There are n one-dimensional transforms of the rows, and n more transforms of the resulting columns. The cost is $n^2 \log n$, and the matrix F now has n^2 rows and n^2 columns. It is a "tensor product" of the one-dimensional transforms. Algebraically, each ordinary entry w^{jk} has been replaced by an n by n block $w^{jk}F$. Numerically, the computation is not hard to organize.

For difference equations there is a slight variation. It is often quicker to use the FFT in one direction only. To connect fast transforms to difference equations we start with the one-dimensional example

$$T_4 = \begin{bmatrix} d & -1 & 0 & -1 \\ -1 & d & -1 & 0 \\ 0 & -1 & d & -1 \\ -1 & 0 & -1 & d \end{bmatrix}. \tag{6}$$

This matrix is "psychologically tridiagonal." It comes from the difference equation

$$-u_{k+1} + du_k - u_{k-1} = f_k. \tag{7}$$

Almost all its entries are zero—or they would be if the matrix were larger. T_{32} would have 32 nonzeros on the main diagonal and 31 on each side. Its only other nonzeros are the two -1's in the corners, which prevent the matrix from being genuinely tridiagonal. They come from ***periodicity*** $u_k = u_{k+32}$ in the difference equation. If the equation asks for u_{-1} we give it u_{31}, and if it asks for u_{32} we give it u_0. That produces the -1's in the corners, and the 32 unknowns are $u_0, u_1, \ldots, u_{31}$.

We want to solve the equation $Tu = f$. This is a one-dimensional difference equation and it does not actually need the FFT. However in two dimensions fast transforms will make all the difference, and we emphasize the main point:

> T is diagonalized by the Fourier matrix F, so that $F^{-1}TF = \Lambda$.

Therefore $Tu = f$ is solved by $u = F\Lambda^{-1}F^{-1}f$. The three steps are (1) multiply by F^{-1} (2) divide by the eigenvalues, which are the numbers $d - w^j - 1/w^j$ on the diagonal of Λ (3) multiply by F. The two multiplications are done by the FFT. For T_4 the four eigenvalues in Λ (all real) are

$$d - 1 - 1, \, d - i - i^{-1} = d, \, d - i^2 - i^{-2} = d + 2, \, d - i^3 - i^{-3} = d. \tag{8}$$

That is typical. The smallest eigenvalue of T_n is always $d - 2$. The matrices are positive definite if $d > 2$, and they are singular at $d = 2$. That choice gives the periodic second difference matrix (the circulant with diagonals $-1, 2, -1$). The vector it takes to zero is our old enemy $(1, 1, \ldots, 1)$.

Now we come to the real problem, in two dimensions. It is *Poisson's equation* in the square $0 < x, y < 1$. The next pages are devoted to constructing a *fast Poisson solver*, and for many readers the details will be optional. The main point is that it can be done. It applies the idea of a two-dimensional FFT to the difference equation

$$(-u_{j+1\,k} + 2u_{jk} - u_{j-1\,k}) + (-u_{jk+1} + 2u_{jk} - u_{jk-1}) = h^2 f_{jk}. \tag{9}$$

This is the "5-point scheme." It is the simplest finite difference approximation to $-u_{xx} - u_{yy} = f(x, y)$, although not the most accurate. The 9-point scheme uses the four additional pairs $(j \pm 1, k \pm 1)$ and the FFT applies to that case too.

The meshpoints lie on a regular grid. Each side of the square is divided into $N + 1$ intervals of length $h = 1/(N + 1)$. There are N^2 meshpoints in the interior of the square and the final form of $Ku = h^2 f$ depends on the boundary conditions. Three cases have special importance:

1. Periodic conditions in both x and y
2. Dirichlet conditions: $u = u_0$ on the boundary
3. Neumann conditions: $\partial u / \partial n = F$ on the boundary.

Here we begin with the *periodic* case, which corresponds to a discrete Fourier series. It is associated with the Fourier matrix F_{N+1}.† Later the Dirichlet condition $u = 0$ will be matched by a sine series and the Neumann condition by cosines.

The unknowns u_{jk} come in a natural order, starting with $u_{00}, u_{10}, \ldots, u_{N0}$ along the bottom of the square. That group will be denoted by U_0. The next row of meshpoints contributes another group U_1. Moving up the square we end with the unknowns along row N. By periodicity the group U_{N+1} at the top of the square is

---

† Thus n in the FFT equals $N + 1$ in the Poisson solver.

the same as U_0 at the bottom. The right side f will be ordered in the same way. Then the matrix for the 5-point scheme becomes

$$K = \begin{bmatrix} T & -I & 0 & \cdot & -I \\ -I & T & -I & \cdot & 0 \\ 0 & -I & T & \cdot & 0 \\ \cdot & \cdot & \cdot & \cdot & -I \\ -I & 0 & 0 & -I & T \end{bmatrix}. \tag{10}$$

K looks very much like T but it actually contains T! It is a **block**-convolution matrix, and its diagonal entries are $d = 4$. Two of the -1's, from the left and right neighbors of each meshpoint, are inside the block T. The -1's from neighbors above and below lead to the blocks $-I$ on each side of the diagonal. K is of order $(N + 1)^2$, T is of order $N + 1$, and every row is connected to its two neighbors:

$$-U_{k+1} + TU_k - U_{k-1} = h^2 f_k. \tag{11}$$

This periodic matrix K has one serious drawback. **It is singular**. Every row adds to zero, and the vector of 1's satisfies $Ku = 0$. The situation is exactly what it was in Chapter 2, where $A_0^T A_0$ was unavoidably singular and a constant could be added to all potentials. In fact this K comes from a graph precisely as it did there. It is the graph with each meshpoint connected to its four neighbors, and with the left and bottom sides of the square connected to the right side and the top. To make it nonsingular we can ground one or more nodes; in fact the Dirichlet condition $u_0 = 0$ will ground *all* boundary nodes. Then K becomes $A^T A$, now invertible.

We are unwilling to give up so soon on the periodic case. Instead the equation will be changed by adding a multiple of the identity. That makes K nonsingular, with all eigenvalues positive. The new term comes from the **three-dimensional** Poisson equation, if the right side is $F = f(x,y)e^{iaz}$ and the solutions are $U = u(x,y)e^{iaz}$. Substituting into $-U_{xx} - U_{yy} - U_{zz} = F$ and cancelling e^{iaz} leaves

$$-u_{xx} - u_{yy} + a^2 u = f. \tag{12}$$

In the difference equation this adds $h^2 a^2$ to the diagonals of T and K.

Now comes the main idea. We know that $F^{-1}TF$ is diagonal; it is Λ. Its entries are $\lambda_j = d - w^j - w^{-j}$, and d is $4 + h^2 a^2$. Our problem is to do something similar to the big matrix K, and the crucial first step is to transform each row separately. These one-dimensional transforms diagonalize every block in the matrix. For $N = 2$, the transforms F^{-1} on the left and F on the right will change K to

$$\begin{bmatrix} F^{-1} & & \\ & F^{-1} & \\ & & F^{-1} \end{bmatrix} \begin{bmatrix} T & -I & -I \\ -I & T & -I \\ -I & -I & T \end{bmatrix} \begin{bmatrix} F & & \\ & F & \\ & & F \end{bmatrix} = \begin{bmatrix} \Lambda & -I & -I \\ -I & \Lambda & -I \\ -I & -I & \Lambda \end{bmatrix}. \tag{13}$$

I hope you see the improvement. Multiplying each row by F^{-1} has changed T to Λ, and the original equation (14) has become (15):

$$-U_{k+1} + TU_k - U_{k-1} = h^2 f_k \tag{14}$$

$$-V_{k+1} + \Lambda V_k - V_{k-1} = h^2 c_k \tag{15}$$

The new unknown V_k is the transform $F^{-1}U_k$ of the old unknown. The new right side c_k is the transform $F^{-1}f_k$ of the old right side. The middle term is $F^{-1}TU_k = F^{-1}TFF^{-1}U_k = \Lambda V_k$. The new equations (15) will fall apart, if we look at the jth component of each V_k—in other words, *we look along column j*:

$$-V_{k+1\,j} + \lambda_j V_{kj} - V_{k-1\,j} = h^2 c_{kj}. \tag{16}$$

That is a problem of size $N + 1$ and not $(N + 1)^2$. It is a *one-dimensional* difference equation, governed by T. There are $N + 1$ of these easy problems, one for each j. Their solutions fill out the vectors V_k, from which we recover the original $U_k = FV_k$.

To express it differently, transforming each row leads us to (15). Then each column gives (16). That could be solved by another transform, completing the two-dimensional FFT. It can also be solved by direct elimination. Either way we have reached a fast Poisson solver that needs only $N^2 \log N$ steps:

5P   The nonsingular system $Ku = f$ is solved in four stages:
   1.   Compute the vector $c_k = F^{-1}f_k$ for each block in f
   2.   Solve $N + 1$ equations of the form (16) with $\lambda_j = 4 + h^2 a^2 - w^j - w^{-j}$
   3.   Reassemble those solutions into the blocks V_k
   4.   Compute $U_k = FV_k$ to find each row of u.
Multiplication by F^{-1} at step 1 and F at step 4 requires $N \log_2 N$ multiplications by the Fast Fourier Transform, and they are done $N + 1$ times—once for each block.

EXAMPLE   Suppose $f = 1$. Then the four stages are:
   1.   Each block $h^2 f_k$ is $(h^2, h^2, \ldots, h^2)$. Its Fourier transform is $h^2 c_k = (h^2, 0, \ldots, 0)$.
   2.   The jth components of these c_k give a vector of zeros on the right side of (16), except in the case $j = 0$ when the right side is $(h^2, h^2, \ldots, h^2)$. The solution in that case with $\lambda_0 = d - 2 = 2 + h^2 a^2$ is $(1/a^2, 1/a^2, \ldots, 1/a^2)$.
   3.   After reassembling these solutions by rows each V_k is $(1/a^2, 0, \ldots, 0)$.
   4.   The inverse transform FV_k is $U_k = (1/a^2, 1/a^2, \ldots, 1/a^2)$.

This agrees with the true solution $u = 1/a^2$ to the equation $-u_{xx} - u_{yy} + a^2 u = 1$.

The operation count $N^2 \log_2 N$ is nearly best possible. For a problem of this size it takes N^2 steps just to write down the answer. A multiple of N^2 would be absolutely best possible, and it can actually be achieved by multigrid.

Fast Poisson Solver

After the periodic case it is easy to fix $u = 0$ along the boundary. This is the Dirichlet problem and a nonzero boundary condition $u = u_0$ is not more difficult. If u_{j0} or u_{0k} is prescribed in the finite difference equation, that term moves to the right side and affects only the vector f. The important changes are in K and T:

$$K = \begin{bmatrix} T & -I & 0 \\ -I & T & -I \\ 0 & -I & T \end{bmatrix} \quad \text{with} \quad T = \begin{bmatrix} 4 & -1 & 0 \\ -1 & 4 & -1 \\ 0 & -1 & 4 \end{bmatrix}.$$

These are now genuinely tridiagonal. K is block tridiagonal of order N^2 and T is tridiagonal of order N. The -1's in the corners, which came from periodicity, have disappeared. The unknowns u_{jk} appear only at the N^2 interior meshpoints, and the boundary values are known. Most rows still have four -1's and the corner meshpoints have only two—since their other neighbors are known from $u = u_0$.

We hope to keep the same algorithm for the new $Ku = f$. However the eigenvectors of T are no longer $(1, w^j, w^{2j}, \ldots)$. The nonperiodic matrix T is not diagonalized by the Fourier matrix F. Fortunately the eigenvectors can still be found, but they are discrete sines instead of discrete exponentials:

$$\begin{bmatrix} 4 & -1 & 0 & \cdot & 0 \\ -1 & 4 & -1 & \cdot & 0 \\ 0 & -1 & 4 & \cdot & 0 \\ \cdot & \cdot & \cdot & \cdot & -1 \\ 0 & 0 & 0 & -1 & 4 \end{bmatrix} \begin{bmatrix} \sin \pi jh \\ \sin 2\pi jh \\ \sin 3\pi jh \\ \cdot \\ \sin N\pi jh \end{bmatrix} = (4 - 2 \cos \pi jh) \begin{bmatrix} \sin \pi jh \\ \sin 2\pi jh \\ \sin 3\pi jh \\ \cdot \\ \sin N\pi jh \end{bmatrix}. \quad (17)$$

The eigenvalues are $\lambda_j = 4 - 2 \cos \pi jh$. The smallest is $\lambda_1 = 4 - 2 \cos \pi h$, which is slightly above 2. The previous $\lambda_0 = 2$ is gone and K is positive definite without the help of $a^2 h^2$. Notice that (17) depended on $\sin 0 = 0$ and $\sin(N + 1)j\pi h = \sin j\pi = 0$, which allowed the eigenvector to satisfy the boundary condition. We can think of the invisible zeroth and $(N + 1)$st components of the eigenvector as being zero without destroying the pattern. In the continuous case there is an eigenfunction $\sin j\pi x$ for every j. The discrete case stops at $j = N$.

Suppose the eigenvectors in (17) are the columns of an N by N "*sine matrix*" S. This matrix diagonalizes T, to give $S^{-1}TS = \Lambda$. The problem is again decoupled within each row, and the four steps of the periodic case are only slightly changed:

1. Compute the N-vector $c_k = S^{-1}f_k$ for each block in f
2. Solve N tridiagonal systems of the form (16)
3. Reassemble those solutions into the N blocks V_k
4. Compute $U_k = SV_k$ to find each row of u.

The sine matrix S takes the place of the Fourier matrix F.

You recognize the key question: Is there a **Fast Sine Transform** to multiply quickly by S and S^{-1}? The answer is *yes*; steps 1 and 4 can be very fast. Since the tridiagonal systems of step 2 are always easy the solution of $Ku = f$ will be extremely efficient. This directly solves the Dirichlet problem on a square and indirectly aids the solution of a host of other problems.

I think it would be excessive to describe the Fast Sine Transform in detail, but the main points are important. They require one more look at the transform

$$f = Fc \quad \text{or} \quad f_j = \sum_{k=0}^{n-1} w^{jk} c_k \quad \text{or} \quad c_k = \frac{1}{n} \sum_{j=0}^{n-1} \bar{w}^{jk} f_j. \tag{18}$$

If f is a real vector, half of those complex multiplications are unnecessary:

   (i)   *if f is real then c_k is the conjugate of c_{n-k}.*
Changing k to $n - k$ will change w^{jk} into its conjugate $w^{j(n-k)} = \bar{w}^{jk}$ since $w^{jn} = 1$. The example $f = (2,4,6,8)$ in an earlier section had $c = (5, -1 + i, -1, -1 - i)$. The components c_0 and $c_{n/2}$ are real, and all components after $c_{n/2}$ are the conjugates of those before. The transform c is determined by n real numbers, not $2n$.

Now consider what happens if f is also odd, so that $f_j = -f_{n-j}$:

   (ii)   *if f is odd then c is odd*
   (iii) *if f is real and odd then c is imaginary and odd.*
(ii) comes from changing j to $n - j$ in equation (18) and using $w^n = 1$. Then (iii) is a combination of (i) and (ii): $\bar{c}_k = c_{n-k}$ and $c_k = -c_{n-k}$ give $\bar{c}_k = -c_k$. In this case $n/2$ real numbers are enough to determine the transform. Each property cuts the effort in half, and with both properties in the sine transform the factor is $1/4$.

We need to say that the actual implementation of these properties takes some ingenuity. Each case has to be related to a complex transform, so the FFT trick can work. The property $\bar{c}_k = c_{n-k}$ for real vectors allows us to transform two vectors f_1 and f_2 at once, by writing $f = f_1 + if_2$, transforming it to c, and recovering c_1 and c_2: $c_{1k} = (c_k + \bar{c}_{n-k})/2$ and $c_{2k} = i(-c_k + \bar{c}_{n-k})/2$. A single real vector of length $2n$ can send its components alternately into f_1 and f_2 and be transformed quickly. There is a further twist for odd and even vectors, and the FST and FCT produce fast solvers with Dirichlet and Neumann boundary conditions.

Separation of Variables

The methods just described for $Ku = f$ share one special property with the differential equations that they come from. Without referring to that property our description could not be complete. It is the fact that the discrete and continuous Poisson equations both permit solutions that have the unusual form

$$\boxed{u(x,y) = A(x)B(y).}$$

This is the idea of *separation of variables*, and the equation and boundary conditions limit the choices of A and B. For the differential equation we look at infinite combinations $\Sigma A_n(x)B_n(y)$ of acceptable choices. For the matrix equation this sum is finite.

We met this idea once before for Bessel functions, and once before that without acknowledging it. The problem was Laplace's equation in a circle, and the separated solutions were $r^n e^{in\theta}$. They were found as powers of $x+iy$. They could also have been found by substituting $A(\theta)B(r)$ into Laplace's equation. The equation is satisfied by $r^n e^{in\theta}$ even for fractional or irrational n, but the exponential has $e^{in0} = e^{in2\pi}$ only if n is an integer. That narrows the choice of A and B. To allow $r = 0$ it is narrowed further; negative powers would produce $r^{-1} = \infty$. We are left with $n \geq 0$, and analytic functions.

We want to apply the same method to Poisson's equation in a square. The choices of A and B are familiar, since with periodic boundary conditions they should both be periodic:

$$A_k(x) = e^{2\pi ikx} \quad \text{and} \quad B_l(y) = e^{2\pi ily}. \tag{19}$$

The products $A_k B_l$ are the eigenfunctions of the Laplacian:

$$\left(-\frac{\partial^2}{\partial x^2} - \frac{\partial^2}{\partial y^2}\right)(A_k B_l) = \lambda(A_k B_l) \quad \text{with} \quad \lambda = (2\pi k)^2 + (2\pi l)^2. \tag{20}$$

If the function $f(x,y)$ is expanded into a double Fourier series $f = \Sigma\Sigma c_{kl} A_k(x)B_l(y)$, separation of variables solves Poisson's equation in a periodic square. The solution is an expansion in eigenfunctions,

$$u = \Sigma\Sigma \frac{c_{kl}}{\lambda_{kl}} e^{2\pi ikx} e^{2\pi ily}.$$

Substituted into $-u_{xx} - u_{yy} = f$, the λ_{kl} disappear and we have equality.

It is exactly this pattern that was copied by the algorithm for $Ku = f$! In each row of the square array, the Fourier matrix F introduced $e^{2\pi ikx}$ and diagonalized T. Then the rearrangement into equation (16) gave $N + 1$ equations going up the square and introduced $e^{2\pi ily}$. The discrete exponentials are fundamental to the matrix K and in fact *they are its eigenvectors*. The analogue of (20) is

$$K(A_k B_l) = \lambda(A_k B_l) \text{ with } \lambda = 4 - e^{2\pi ikh} - e^{-2\pi ikh} - e^{2\pi ilh} - e^{-2\pi ilh}. \tag{21}$$

The components of this eigenvector $A_k B_l$ are the values of $e^{2\pi ikx} e^{2\pi ily}$ at the meshpoints. We remark that $u = K^{-1}f$ could have been written as a finite double Fourier series (an expansion in eigenvectors, similar to the continuous case). The algorithm was more interested in numbers than formulas, so it computed u directly from f—but behind its success stood these eigenvectors of K.

The eigenvalues deserve more comment. For $k = l = 0$, corresponding to an eigenvector of 1's, the eigenvalue is $4 - 1 - 1 - 1 - 1 = 0$. The difference and differential equations both have a zero eigenvalue, and the periodic problem is singular. The next eigenvalues at $k = 1$, $l = 0$ remain close. For larger k and l the discrete and continuous eigenvalues must diverge. The continuous $\lambda = (2\pi k)^2 + (2\pi l)^2$ goes to infinity while the discrete λ never exceeds 8:

$$4 - 2 \cos 2\pi kh - 2 \cos 2\pi lh \le 8.$$

More properly this should be $8/h^2$, which does eventually grow as $h \to 0$.

Now we turn to the nonperiodic case, where separation of variables leads to $A_k B_l = \sin \pi kx \sin \pi ly$. These are the new eigenfunctions, vanishing on the four sides of the square. The factors 2 in $e^{2\pi ikx}$ and $e^{2\pi ily}$ have disappeared. We have odd functions and Fourier sine series from 0 to π, not the full series from 0 to 2π. The eigenvalues also omit the 2:

$$\left(-\frac{\partial^2}{\partial x^2} - \frac{\partial^2}{\partial y^2} \right) (\sin \pi kx \sin \pi ly) = (\pi^2 k^2 + \pi^2 l^2)(\sin \pi kx \sin \pi ly). \tag{22}$$

The continuous solution is a double sine series $u = \Sigma\Sigma \, b_{kl} \sin \pi kx \sin \pi ly$.

This pattern is again copied in the discrete case by K. The sine matrix S diagonalizes each block of the matrix, and the eigenvectors of K come from $A_k B_l = \sin \pi kx \sin \pi ly$ at the meshpoints:

$$K(A_k B_l) = \lambda(A_k B_l) \quad \text{with} \quad \lambda = 4 - 2 \cos \pi kh - 2 \cos \pi lh. \tag{23}$$

The smallest eigenvalue at $k = l = 1$ is now greater than zero:

$$\lambda_{\min} = 4 - 4 \cos \pi h \approx 4 - 4 \left(1 - \frac{\pi^2 h^2}{2!} \right) = 2\pi^2 h^2.$$

Divided by h^2 this agrees with the smallest eigenvalue of the differential equation, also at $k = l = 1$. The largest eigenvalue of K is again near 8.

Remark   The ratio $\lambda_{\max}/\lambda_{\min} \approx 8/2\pi^2 h^2$ is the **condition number** of K. This number measures the sensitivity, or vulnerability, of the equation $Ku = f$. If there is an error in f, from roundoff or from physical measurements, there will be a corresponding correction to u. The question is whether a small error Δf can produce a large change $\Delta u = K^{-1} \Delta f$. That will happen if K is nearly singular. But the computation is independent of scaling, so the right things to compare are *relative changes* $|\Delta u|/|u|$ and $|\Delta f|/|f|$. The condition number of K is the key:

$$\frac{|\Delta u|}{|u|} \le \frac{\lambda_{\max}}{\lambda_{\min}} \frac{|\Delta f|}{|f|}. \tag{24}$$

The extreme case occurs when Δf is in the direction of the first eigenvector, so that $\Delta u = K^{-1} \Delta f = \Delta f / \lambda_{\min}$. At the same time f can be a multiple of the last eigenvector, so that $u = K^{-1}f = f/\lambda_{max}$. With those choices we have equality in (24).† Normally this extreme case does not occur—roundoff has high frequencies rather than low frequencies—but the condition number gives a very valuable rule of thumb: The computer can lose *log c decimal places* to the roundoff errors in elimination.

In Poisson's equation a mesh with $h = 1/100$ would give $c = 4/\pi^2 h^2 \approx 4000$. The matrix is moderately ill-conditioned but losing 3 or 4 decimal places should not be fatal.

Time-dependent Problems

Separation of variables is extremely valuable when there are time derivatives. It is also extremely direct, because the part involving time is only an exponential. For the heat equation it is a decay $e^{-\lambda t}$, and for the wave equation it is an oscillation $e^{i\omega t}$. Those appeared in Chapter 1 for ordinary differential equations, and the idea is the same for partial differential equations. We use it here as a transition to the next chapter.

The key is to find the eigenvectors. They solve the time-dependent problem, by combining with $e^{-\lambda t}$ or $e^{i\omega t}$ into pure exponential solutions. For partial differential equations they are eigen*functions*. That is the step from matrices to derivatives, which takes us directly to the fundamental equations:

> the heat equation is $u_t = u_{xx}$
>
> the wave equation is $u_{tt} = u_{xx}$

On the right side, $\partial^2/\partial x^2$ has negative eigenvalues:

(periodic case)
$$\frac{\partial^2}{\partial x^2} e^{2\pi i k x} = -(2\pi k)^2 e^{2\pi i k x}$$

$$(25)$$

(zero boundary conditions)
$$\frac{\partial^2}{\partial x^2} \sin \pi k x = -(\pi k)^2 \sin \pi k x$$

The separated solutions $A(x)B(t)$ can be written down immediately.

5Q The heat equation has decaying solutions $e^{-\lambda t}v$; the wave equation has oscillating solutions $e^{i\omega t}v$ and $e^{-i\omega t}v$. The eigenvalues $-\lambda = -\omega^2$ appear in (25) with the eigenfunctions v—one for each frequency k. The solutions to the heat and wave equations are combinations of these exponential solutions:

$$u = \sum c_k e^{-\lambda_k t} v_k(x) \quad \text{and} \quad u = \sum (c_k e^{i\omega_k t} + d_k e^{-i\omega_k t}) v_k(x). \qquad (26)$$

---

† For nonsymmetric or indefinite matrices it is the eigenvectors of $K^T K$ that are extreme. The condition number becomes $c = \sigma_{max}/\sigma_{min}$, a ratio of singular values (square roots of eigenvalues of $K^T K$).

In the heat equation, $\partial/\partial t$ produces the factor $-\lambda$ and so does $\partial^2/\partial x^2$. In the wave equation there are two time derivatives, and applied to $e^{i\omega t}$ they also produce $(i\omega)^2 = -\omega^2 = -\lambda$.

The numerical method that corresponds to all these formulas is the **spectral method**, and it is extremely successful for the right equations and boundaries and boundary conditions. It uses fast transforms. On its own ground it can defeat finite elements and finite differences. The spectral method is to time dependence what Poisson solvers are to space dependence, an intelligent way to separate variables. It can be described in a few fairly optional pages.

The Spectral Method

We begin with the periodic heat equation, starting at time 0 from $u_0(x)$:

$$\frac{\partial u}{\partial t} = \frac{\partial^2 u}{\partial x^2} \quad \text{with} \quad u(0,x) = u_0(x) = u_0(x + 2\pi). \tag{27}$$

The exact solution u can be written as an infinite Fourier series. The approximate solution U can be written as a finite series. It is

$$U(t,x) = \sum_{k=0}^{n-1} C_k(t)e^{ikx}, \tag{28}$$

where each $C_k(t)$ is to be determined. The functions that were $e^{2\pi ikx}$ from 0 to 1 have become e^{ikx} from 0 to 2π. That matches the conventions of the discrete transform. The number n will be even, and in most cases a power of 2, to make the Fast Fourier Transform as effective as possible.

Note that U is a function of x; it is not a set of values at meshpoints. But we can have that too. Consider U at the points $x = 0, h, 2h, \ldots$ with uniform spacing $h = 2\pi/n$. That yields a vector $U_0, U_1, \ldots, U_{n-1}$ of mesh values. According to (28), the coefficients $C_0, C_1, \ldots, C_{n-1}$ give the discrete Fourier transform of that vector. The mesh vector U in x-space is the inverse transform of the coefficient vector C in k-space. By computing one we know the other.

We might also compare the spectral method (before actually describing it) to the finite element method. One difference is this: the trial functions T_k were normally 1 at the meshpoint x_k and 0 at the others, whereas e^{ikx} is nonzero everywhere. That is not such an important distinction. We could produce from the exponentials an interpolating function like T_k, which is zero at all meshpoints but one:

$$g_k(x) = \frac{1}{n} \frac{\sin\frac{n}{2}(x - x_k)}{\sin\frac{1}{2}(x - x_k)} \cos\frac{1}{2}(x - x_k). \tag{29}$$

Of course it is not a piecewise polynomial; that distinction is genuine. Each function g_k spreads over the whole interval $0 \leq x \leq 1$, where T_k was zero in all elements not containing x_k. The stiffness matrix K was sparse for finite elements; in the spectral method it is full, like F. Every unknown is connected to every other unknown, but we still have convolution matrices and fast transforms.

There is another difference from finite elements. Those were based on trial functions equal to test functions. To determine the approximation, Galerkin would multiply by e^{-ikx} and integrate. In contrast, the pseudo-spectral method† is closer to collocation. $U(x,0)$ interpolates $u(x,0)$ at the meshpoints. The initial values $C_k(0)$ are coefficients in the *discrete* Fourier series for u_0, where Galerkin chose the ordinary Fourier coefficients.

The discrete method is quicker and easier, but it suffers from *aliasing*. It cannot distinguish $u_0 = e^{inx}$ from $u_0 = 1$. The two are equal at every meshpoint $x = 2\pi k/n$. The spectral method starts from $u_0 = 1$, when the true solution starts from e^{inx}. In the same way, every high frequency k is replaced by its "alias" $k*$ between 0 and $n-1$; $k-k*$ is divisible by n.†† Thus $C_k(0)$ contains not only the ordinary Fourier coefficient c_k of the initial function but also all coefficients that share that same alias:

$$C_k(0) = c_k + c_{k+n} + c_{k-n} + c_{k+2n} + c_{k-2n} + \cdots. \tag{30}$$

This throws extra weight back into the lower frequencies which decay less rapidly. For smooth functions this aliasing effect is compensated for by the rapid decay of the higher Fourier coefficients, so the extra terms in (30) should not overwhelm the coefficient c_k.

Now we come to the method itself. Whether $C_k(0)$ is c_k or the sum (30), the problem is to move forward in time. We need an equation for $\partial U/\partial t$. Unlike finite differences or finite elements, which replace the right side $\partial^2 u/\partial x^2$ by differences at nodes, *the spectral method uses $\partial^2 U/\partial x^2$ exactly*. That is the crucial point.

The Fourier approximation U is a combination of oscillations e^{ikx} up to frequency $n-1$, and we simply differentiate them:

$$\frac{\partial U}{\partial t} = \frac{\partial^2 U}{\partial x^2} \quad \text{becomes} \quad \sum_0^{n-1} \frac{dC_k}{dt} e^{ikx} = \sum_0^{n-1} C_k(t) \frac{d^2}{dx^2} e^{ikx}. \tag{31}$$

Those x-derivatives are $(ik)^2 e^{ikx}$, and frequencies are uncoupled:

$$\frac{dC_k}{dt} = (ik)^2 C_k \quad \text{or} \quad C_k(t) = e^{-k^2 t} C_k(0). \tag{32}$$

---

† This is the most common spectral method. The prefix "pseudo" may gradually disappear.

†† It is aliasing that makes wagon wheels go backwards in old Westerns.

The coefficients decay precisely as they should, because this equation in time can be solved exactly for the C_k. Of course we have only n coefficients, not the infinite sequence needed for the true solution u.

In more general problems, the equation in time will not be solved exactly. It needs a difference method with time step Δt, as Chapter 6 will describe. What is essential about spectral methods is that *there is no* Δx. The derivatives with respect to space variables are computed explicitly and correctly. That can be done in two ways:

1.   Stay with the harmonics e^{ikx} or $\sin kx$ or $\cos kx$, and use the FFT to go between coefficients C_k and mesh values U_k. The coefficients multiply the harmonics, but the mesh values enter the difference equation in time.

2.   Use an expansion $U = \Sigma \, U_k(t)g_k(x)$ that works directly with the values U_k at meshpoints (where $g_k = 1$). There is a "differentiation matrix" D that gives the mesh values of the derivatives, $D_{jk} = g_k'(x_j)$. Then the approximate heat equation becomes $U_t = D^2 U$.

Both methods use the FFT at each time step, to implement the discrete transforms in 1 and to multiply by the constant-diagonal matrix D in 2.

One more remark before summarizing the periodic case. It concerns nonlinear equations, where derivatives are simpler than differences. For a product $U \partial U / \partial x$, we can differentiate by method 1 or 2, transform to mesh values, multiply, and transform back. With the FFT, convolutions have made spectral methods competitive. Now we collect the main points.

5R   For periodic problems the spectral methods use Fourier expansions

$$U(t,x) = \sum_{k=0}^{n-1} C_k(t)e^{ikx} = \sum_{k=0}^{n-1} U_k(t)g_k(x).$$

By computing x-derivatives exactly, partial differential equations in space-time become ordinary differential equations in time. The unknowns are the C_k or the U_k (or both, since they are connected by a discrete Fourier transform). The FFT allows x-derivatives in physical space to translate quickly into multiplications by ik in frequency space.

The fact that x-derivatives are exact makes spectral methods free of *phase error*. Differentiation multiplies e^{ikx} by the right factor ik while differences give the wrong factor iK:

$$\frac{e^{ik(x+h)} - e^{ik(x-h)}}{2h} = iK \, e^{ikx}, \quad \text{with } K = \frac{\sin kh}{h}.$$

For low frequencies, when kh is small and there are enough meshpoints in a wavelength, K is close to k. For higher frequencies K is significantly smaller. In the heat equation this means slower decay: the exponent is $(iK)^2 t$ instead of $(ik)^2 t$. In the wave equation it means a slower wave velocity. The phase is wrong, and the crests of the approximate wave are in the wrong places (Section 6.5). In contrast, the spectral method can follow even the nonlinear wave interactions that lead to turbulence.

Chebyshev Polynomials for Zero Boundary Conditions

When the problem is not periodic, an interpolation at equally spaced points is not successful. To use periodic functions like $\sin \pi k x$ is to ask for trouble at the ends of the interval; especially the derivatives are inaccurate. To use ordinary polynomials (Lagrange interpolation) is to face instability as their degrees increase. One solution was piecewise polynomials—splines for interpolation or finite elements for differential equations. Another solution, much more limited in scope but more accurate when it works, is to redistribute the interpolation points—and to move them toward the boundaries. This is the spectral method as used in practice, since most problems are not periodic.

On the interval from -1 to 1, the favorite choice is

$$x_j = \cos \frac{\pi j}{n}, \quad j = 0, 1, \ldots, n. \tag{33}$$

That choice comes from $n + 1$ points that are equally spaced around a semicircle (Fig. 5.12). The equal spacing in θ becomes an unequal spacing in x, when the points are projected onto the horizontal axis. They are still symmetric about $x = 0$, but they become denser near the boundaries.

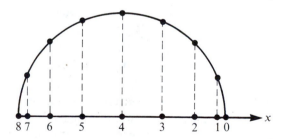

Fig. 5.12. Nine interpolation points for the Chebyshev spectral method.

This change of variables $x = \cos \theta$ is the key to everything. It converts the periodic functions $1, \cos \theta, \cos 2\theta, \cos 3\theta, \ldots$ into ordinary polynomials in x. For example $\cos 2\theta$ is $2 \cos^2 \theta - 1$, which is $2x^2 - 1$. The next is $\cos 3\theta = 4 \cos^3 \theta - 3 \cos \theta$, leading to $4x^3 - 3x$. The functions constructed in this way are the *Chebyshev polynomials*:

$$T_0 = 1, \ T_1 = x, \ T_2 = 2x^2 - 1, \ T_3 = 4x^3 - 3x, \ \ldots, \ T_n(\cos \theta) = \cos n\theta. \tag{34}$$

T_n is an even polynomial when n is even; it is odd when n is odd. None of these polynomials can go above 1 or below -1 in our interval, because T_n is $\cos n\theta$ in disguise.

The striking property of the polynomials is their behavior at the unequally spaced points $\cos \pi j/n$. The rule $T_n(\cos \theta) = \cos n\theta$ gives

$$T_n\left(\cos \frac{\pi j}{n}\right) = \cos n \frac{\pi j}{n} = \begin{array}{ll} 1 & \text{for } j = 0,2,4, \dots \\ -1 & \text{for } j = 1,3,5, \dots \end{array}$$

T_n oscillates at those points just as a cosine oscillates at equal intervals. The Chebyshev method uses these polynomials:

$$U(t,x) = \sum_{k=0}^{n} A_k(t) T_k(x). \tag{35}$$

Any spectral method needs two things to succeed:

1.   A quick way to find U at the meshpoints
2.   A quick formula for exact x-derivatives

They come from the identity $T_k(\cos \theta) = \cos k\theta$. At $x_j = \cos \pi j/n$, U is

$$U_j = \sum A_k T_k(x_j) = \sum_0^n A_k \cos \frac{k\pi j}{n}. \tag{36}$$

Thus a Fast Cosine Transform will link the A_k to the U_j, and you can be sure that it exists. This gives Chebyshev an important advantage over Legendre.

The second requirement (fast derivatives) is met by a special calculation:

$$\sum_0^n A_k \frac{dT_k}{dx} = \sum_0^{n-1} B_k T_k \quad \text{with} \quad \begin{array}{l} B_{n-1} = 2nA_n \\ B_k = B_{k+2} + (2k+2)A_{k+1} \\ B_0 = \tfrac{1}{2}B_2 + A_1 \end{array}$$

Thus the key spectral idea of exact x-derivatives is still possible. It leads to higher accuracy than for finite differences, when the geometry permits. The rate of convergence is "infinite," which means faster than any power $1/n^p$, if the true solution has infinitely many derivatives. Because the mesh is dense near the boundary, the method resolves fast transitions (boundary layers) much more clearly.

However there are disadvantages. The first is the limitation to special geometries. A change of variables might produce a square or a circle, but the flexibility of finite elements will never be reached. The second is the care required with boundary conditions. The spectral method is less robust and forgiving than its competitors; a mistake at the boundary can lead to disaster. The extra conditions needed for finite differences are no longer wanted or allowed. The spectral method has to be understood by the user! And it has other drawbacks, but like the fast Poisson solver it makes separation of variables into a practical algorithm.

Other Applications of the FFT

By emphasizing partial differential equations, we have omitted too many other applications. A few are outlined here. They come from *digital signal processing*, which interprets radar and sonar and digitized speech and telecommunications. It also dominates the search for offshore oil. Water layers produce "ringing" that masks the signals from oil or gas; those primary reflections could not be seen before 1960. Now virtually all exploration data is digital and more than 10^9 systems of equations are solved every year—partly by the FFT and partly by Levinson's algorithm for constant-diagonal (Toeplitz) matrices. The key problem is deconvolution, as we see in three applications:

1. A CT scanner indicates the variation of density in the X-ray area. The input is the density u; the output from the scanner is b. They are related by a convolution that depends on the design of the scanner: $a*u = b$. Thus the problem of analyzing X-rays is exactly the problem of computing u from b. Each designer provides his own *deconvolution rule* $u = g*b$, in which the transfer function g is approximately $1/a$—and the accuracy of this approximation governs the quality of the scanner.

2. The *autocorrelation* of $f(x)$ is its convolution with its mirror image $\bar{f}(-x)$:

$$R(s) = \int_{-\pi}^{\pi} f(x)\bar{f}(x-s)\,dx.$$

This function is crucial to random processes, continuous or discrete. It can reveal hidden periodicities (like sunspot cycles or the alpha rhythm in brain waves). Any contamination of f by unknown noise can be estimated from R. The underlying reason, which is all we can give here, lies in the Fourier coefficients:

$$R = \int_{-\pi}^{\pi} \left(\sum c_k e^{ikx}\right)\left(\sum \bar{c}_k e^{-ik(x-s)}\right) dx = \sum |c_k|^2 \, e^{iks}.$$

The coefficients of R are $|c_k|^2$, never negative, and they give the energy in each harmonic—the "energy spectrum" of f.

3. The FFT is part of the revolution in *digital image processing*. A typical image contains a million values; they are the responses to the original image. Those responses may be compressed for transmission; that is already a fascinating problem, and so is the enhancement of the final image. In between comes the key step of image reconstruction—to reconstruct the picture by recovering the signal from the response.

The problem is 2-dimensional but the central idea is not changed. A point in the original source is like our standard signal $u_0 = 1$. It is an impulse. The response to it is a "point-spread function" $a(x,y)$. When the signal is a collection of sources with density $u(x,y)$, the response is a superposition of these point-spread functions:

$$b(s,t) = \int\int a(s-x,t-y)u(x,y)\, dx\ dy \quad (+\text{noise}).$$

This is a convolution, so the recovery of u is a deconvolution. It uses an approximation to a^{-1}, to return from b to u. It is done discretely, with n sufficiently large to reduce aliasing (which is called *wrap-around* in image processing). It hardly needs saying that all deconvolutions are computed by the convolution rule— transform, multiply, and transform back. With the FFT what else would we do?

EXERCISES

5.5.1   What are F^2 and F^4 for the 4 by 4 Fourier matrix F?

5.5.2   If you form a 4 by 4 submatrix of F_8 by keeping only the entries in its first, third, fifth, and seventh rows and columns, what is that matrix?

5.5.3   Mark all the sixth roots of 1 in the complex plane. What is the complex number w_6? Why is $1 + w + w^2 + w^3 + w^4 + w^5 = 0$?

5.5.4   Write out the analogue of equation (5) that factors the matrix F_6 into two copies of F_3.

5.5.5   Compute $y = F_4 x$ by the three steps of the FFT for the even vector $x' = (2,6,6,6)$ and the odd vector $x'' = (0,-2,0,2)$.

5.5.6   What is $y = F_8 x$ if $x = (2,0,6,-2,6,0,6,2)$?

5.5.7   Since F_n is a symmetric matrix we can transpose the whole FFT:

$$F_n = \begin{bmatrix} I & D \\ I & -D \end{bmatrix}\begin{bmatrix} F_m & \\ & F_m \end{bmatrix}\begin{bmatrix} & P & \end{bmatrix} \text{ becomes } F_n = \begin{bmatrix} & P^T & \end{bmatrix}\begin{bmatrix} F_m & \\ & F_m \end{bmatrix}\begin{bmatrix} I & I \\ D & -D \end{bmatrix}.$$

Draw the new flow graphs for $n = 8$, which are the reverse of the graphs in the text. The vector x enters in the normal order $x_0, \ldots, x_7$, its first half and last half are combined into $x_f + x_l$ and $Dx_f - Dx_l$, and then come the two F_4's. Their output is in the order $\hat{x}_0, \hat{x}_2, \ldots, \hat{x}_5, \hat{x}_7$.
In this sequence the butterflies start at the left of the graph. The bottom one combines $w^3 x_3$ with $-w^3 x_7$, from the last entries of D and $-D$. The F_4 boxes are on the right, and are replaced in the next graph by F_2 boxes.

5.5.8   For $n = 4$ write out the formula for c_0, c_1, c_2, c_3 in (18), and verify
(1) $\bar{c}_1 = c_3$ when f is real, and
(2) $c_1 = -c_3$ and $c_0 = c_2 = 0$ if $f_1 = -f_3$ and $f_0 = f_2 = 0$.

5.5.9   Write down the periodic matrix T_3 with diagonal entries $d = 2$. Show that its eigenvalues are 0,3,3 by using the formula $\lambda_j = d - w^j - w^{-j}$ and also by finding 3 real eigenvectors. Compare your eigenvectors for $\lambda_1 = \lambda_2 = 2$ with the sum and difference of the eigenvectors $(1,w,w^2)$ and $(1,w^2,w^4)$ in equation (7).

5.5.10   Find the six eigenvalues of the periodic matrix T_6 with diagonal d. Verify that their sum equals the trace $6d$.

5.5.11   For the periodic matrix T with -1's in the corners, explain why the whole last column can be nonzero in the upper triangular U that comes from elimination. Why is the operation count still proportional to the order n, as for a genuinely tridiagonal matrix?

5.5.12   Find the incidence matrix A_0 for a graph that has its four edges around a square and verify that $A_0^T A_0$ is the periodic matrix T_4 with diagonal $d = 2$. What vector is in its nullspace? Solve $T_4 u = f$ for $f = (1, -1, 1, -1)$.

5.5.13   For the periodic matrix T_n with diagonal $d = 2$, find the form of T_n^2. To what differential equation does it give a finite difference approximation?

5.5.14   Three consecutive rows of unknowns in the square array satisfy

$$-U_{k-2} + TU_{k-1} - U_k \quad = f_{k-1}$$
$$-U_{k-1} + TU_k \quad - U_{k+1} = f_k$$
$$-U_k \quad + TU_{k+1} - U_{k+2} = f_{k+1}.$$

After multiplying the second by T and adding, what equation links rows $k - 2, k$, and $k + 2$? This is "block-cyclic reduction," separating odd k from even k; it is effective once but unstable if repeated too often.

5.5.15   Verify that the eigenvalue $4 - 2 \cos \pi j h$ is correct in (17).

5.5.16   Suppose $f = e^{2\pi i x} e^{4\pi i y}$. What is the periodic solution to Poisson's equation $-u_{xx} - u_{yy} = f$, using (20)? What is the solution to the discrete equation $Ku = f$, using (21)?

5.5.17   Suppose $f = \sin \pi x \sin 2\pi y$. What is the solution to Poisson's equation with $u = 0$ on the boundary? What is the solution to the finite difference equation $Ku = f$?

5.5.18   Suppose the boundary condition $\partial u / \partial n = 0$ is imitated in the one-dimensional discrete case by $u_0 = u_1$ and $u_N = u_{N+1}$. What is the N by N matrix in $Ku = f$ when the differential equation is $-u'' = f(x)$? Is it singular or nonsingular?

5.5.19   Apply separation of variables to Laplace's equation $u_{rr} + u_r/r + u_{\theta\theta}/r^2 = 0$ in polar coordinates, by substituting $u = A(\theta)B(r)$ and separating into $-A_{\theta\theta}/A = (r^2 B_{rr} + r B_r)/B$. If both sides equal zero, what solutions u do you find? If both sides equal $n^2 \neq 0$, what solutions do you find?

5.5.20   Show from their connections with cosines that the Chebyshev polynomials satisfy a recursion formula and a weighted orthogonality condition:

$$T_{k+1}(x) = 2x T_k(x) - T_{k-1}(x) \quad \text{and} \quad \int_{-1}^{1} \frac{T_j(x) T_k(x) \, dx}{\sqrt{1 - x^2}} = 0 \quad \text{if } j \neq k.$$

5.5.21   With $u = e^{ikx}$ at $t = 0$, solve the heat equation $u_t = u_{xx}$ and its spectral approximation (31). There is aliasing unless $0 \leqslant k < n$.

CHAPTER 5 IN OUTLINE: NUMERICAL METHODS

5.1 Linearity and Nonlinearity—algorithms for $Ax = b$, $Ax = \lambda x$, and $g(x) = 0$
Direct Methods for $Ax = b$—ordering unknowns and ordering LINPACK
Nonlinear Equations and Newton's Method—$J^k(x^{k+1} - x^k) = -g(x^k)$

5.2 Orthogonalization and Eigenvalue Problems—applications of orthogonality
Orthonormal Vectors—the columns of Q with $Q^T Q = I$
Least Squares—the best coefficients $c_i = q_i^T b$
Non-orthogonal Columns—the orthogonalization $A = QR$
Modified Gram-Schmidt—numerically stable orthogonalization
The QR Algorithm—factor $A_k = Q_k R_k$ and form $A_{k+1} = R_k Q_k$
Householder Matrices—reflections $H = I - 2uu^T = H^{-1}$ produce Q and R
Difficulties with C—the solution of $A^T C A x = A^T C b$ for banded C^{-1}

5.3 Semi-direct and Iterative Methods—the error equation $e_{k+1} = Be_k = (I - M^{-1}A)e_k$
The Matrix M—Jacobi, Gauss-Seidel, SOR, ADI, multigrid, incomplete LDL^T
Power Methods for Eigenvalues—blocks, inverses, and shifts of origin
Lanczos Method for Eigenvalues—a 3-term recurrence and a triangular $Q^T A Q$
The Conjugate Gradient Method—recursive solution of $Ax = b$
Convergence and Preconditioning—optimal accuracy for A, improvement for $M^{-1}A$

5.4 The Finite Element Method—Galerkin from weak form, Ritz from minimization
Finite Element Trial Functions—piecewise polynomials P_1, Q_1, P_2 and continuity
Assembling the Element Matrices—the contributions to K from each element
Finite Element Computations—element matrices, mass matrices, eigenvalue problems
Accuracy and Convergence of Finite Elements—energy error h^{2p} is optimal

5.5 The Fast Fourier Transform—$y = F_n x$ in $\frac{1}{2}n \log n$ multiplications
The Complete FFT and the Butterfly—the flow graph with $\log n$ stages
Periodic Difference Equations—solution of 5-point scheme by the FFT
Fast Poisson Solver—sine transform with $u = 0$ on square boundary
Separation of Variables—$u = A(x)B(y)$, exponentials or sines
The Spectral Method—exact derivatives u_{xx} at the meshpoints
Chebyshev Polynomials—approximation $U = \sum A_k(t) T_k(x)$, nodes at $\cos \pi j/n$
Other Applications of the FFT—digital signals and deconvolution

6

INITIAL-VALUE
PROBLEMS

The mathematics in this book springs from two sources: calculus and linear algebra. Those are both at the service of applied mathematics, and they are called on constantly. When the unknown is a vector we need linear algebra. That is sufficient for a discrete problem, provided the system is linear. A problem in which u varies continuously, in space or time or both, needs calculus. If the law involves derivatives du/dx or du/dt, we have passed from matrix equations to differential equations. Algebra organizes our ideas about the dependent variables $u_1, \ldots, u_n$, while calculus deals with the independent variables x or y or t. When both complications are present we have a system of differential equations—possibly partial differential equations and possibly nonlinear. That may not be the ultimate in generality or in difficulty, but within applied mathematics it is pretty close.

This chapter takes the next steps toward that goal. It introduces the time t and velocities du/dt and accelerations d^2u/dt^2. The unknown u, a scalar or a vector, evolves in time from its initial state u_0 at $t = 0$. We start with no space variables, studying ordinary differential equations. Their solution, analytical and later numerical, comes first. Then we admit space along with time, and slopes along with velocities, balancing the rate of change $\partial u/\partial t$ with the gradient $\partial u/\partial x$. There is conservation of mass or momentum or energy. This balance equation allows waves to travel in space-time, and when it is nonlinear those waves can break—producing a shock that travels at supersonic speed and catches the other waves.

Thus we end with nonlinear partial differential equations. Those are more difficult than linear waves and linear responses, where the frequencies and the harmonics are completely dominant. The Laplace transform (in time) joins the Fourier transform (in space) to convert constant-coefficient equations back to linear algebra. Those transforms apply also to finite difference equations, where the crucial problems are stability and accuracy. The goal is always to simplify and to watch the eigenvectors—especially when they are exponentials.

A start was made on ordinary differential equations in Section 1.5. We can review in three lines what was done:

> **1st order equations:** $u' - au = 0$ leads to $u = e^{at}u_0$
> **2nd order equations:** $u'' + pu' + qu = 0$ leads to $u = c_1 e^{\lambda_1 t} + c_2 e^{\lambda_2 t}$
> **1st order systems:** $u' - Au = 0$ leads to $u = e^{At}u_0 = \Sigma\, c_i e^{\lambda_i t} x_i$.

All those problems are linear, with constant coefficients and zero right hand sides. They are the most basic equations, but certainly not the only ones. Therefore we begin to allow time-dependent coefficients $a(t)$ on the left side, forcing terms $f(t)$ on the right side, and nonlinear equations.

A few words about notation. The symbol u' is a shorthand for du/dt. Some authors would write dy/dx, but the whole point is the distinction between time and space (and between initial conditions and boundary conditions). Other authors write $\dot{u}$, but in my classroom the dot looks like ambient dust. The matrix exponential e^{At} is new; it could have appeared earlier but did not. The terms au and Au have moved to the left side of the equation, leaving space on the right side for $f(t)$. You recognize that the constants c_1, c_2 in a second-order equation and $c_1, \ldots, c_n$ in a first-order system are adjusted to match the initial conditions. Higher-order equations and second-order systems are also important but we cannot do everything.

Birth, Death, and Steady State

Suppose the scalar equation with $a = $ constant has a forcing term $f(t)$:

$$\boxed{u' - au = f(t).} \tag{1}$$

This right side will produce something more than $u = e^{at}u_0$. The inhomogeneous term f is the source and we look for its effect on u. One quick method is to find an *integrating factor*, which multiplies both sides and allows us to integrate immediately. For this equation the factor is e^{-at}:

$$e^{-at}(u' - au) = e^{-at}f(t). \tag{2}$$

The differential equation is now "**exact.**" On the left side stands the derivative of $e^{-at}u$; the product rule gives the two terms $e^{-at}u'$ and $-ae^{-at}u$. Therefore we integrate both sides from 0 to t:

$$e^{-at}u(t) - u_0 = \int_0^t e^{-as}f(s)\,ds. \tag{3}$$

Here is another point of notation. It is exasperating to change t to s in equation (2) before integrating, but it is worse not to do it. We do not want $\int_0^t e^{-at} f(t) \, dt$, in which t serves as the variable of integration and also as the endpoint. On the left side of (3) there is no need for s because the integration has been done. It gave $[e^{-as} u(s)]_0^t$ and we evaluated it at both endpoints. Of course at $s = 0$ the exponential was $e^0 = 1$.

To find the solution $u(t)$, move u_0 to the other side of (3) and multiply the equation by e^{at}:

$$u(t) = \int_0^t e^{a(t-s)} f(s) \, ds + e^{at} u_0. \tag{4}$$

There are two parts to this solution. One is the familiar $e^{at} u_0$, which comes from the initial condition. The other is the new part, which adds up all contributions from the source term f. Notice that e^{at} could go inside the integral; it is constant while we integrate with respect to s. The resulting factor $e^{a(t-s)}$ indicates what has happened to the force applied at time s. Each piece of f is swept forward in time by the exponential, just as u_0 is brought all the way from $t = 0$. This is Duhamel's principle, that f contributes like u_0 except that $f(s)$ has only the shorter time $t - s$ in which to evolve. It is linearity that allows these pieces of u to be added together.

This integral is particularly clear when f is an impulse δ acting at time T. Its effect is not felt until it has been applied:

$$\int_0^t e^{a(t-s)} \delta(s - T) \, ds = \begin{cases} 0 & \text{for } t < T \\ e^{a(t-T)} & \text{for } t \geq T \end{cases} \tag{5}$$

Nothing happens until $t = T$, and then it is like introducing $u_0 = 1$—but not as an initial value, because it starts at time T. From then on it is carried by the exponential.

Remark The exponential acts like a *lower triangular matrix L*, which produces no effect before the cause. If f is zero up to time T, so is Lf. The integral in (4) is matrix multiplication with entries $L_{ts} = e^{a(t-s)}$, but it is a continuous matrix and a continuous multiplication; instead of adding we integrate. The multiplication in (5) has $f = \delta$ concentrated at component T, and then $u = Lf$ is column T of the matrix.

EXAMPLE 1 $u' + u = 3$ with $a = -1$ and $u_0 = 5$

Solution $u(t) = \displaystyle\int_0^t e^{-t+s} 3 \, ds + 5e^{-t} = 3 + 2e^{-t}$

Check $u(0) = 5$ and $u' + u = -2e^{-t} + 3 + 2e^{-t} = 3$.

From my experience on this example it is a good idea to check.

The case in which a is negative is extremely important. The solution $e^{at}u_0$ then decays to zero with exponential speed. Such an equation is called **stable**. We see from this example the effect of a constant forcing term $f = 3$. That acts to increase u but the negative exponential is stronger, and the combination carries $u(t)$ toward a **steady state**. The solution $3 + 2e^{-t}$ approaches the constant value $u_\infty = 3$, which is influenced by the forcing term but not by the initial condition. In fact we find the steady state by removing the time derivative u' (which is zero when u is steady) and reducing $u' + u = 3$ to $u = 3$.

The equation models a "birth-death process" in which the population size is u. There are births at a constant rate 3, and there are deaths at a rate proportional to the population. This process is death-dominated, and the initial population of 5 is soon irrelevant. The steady state would be zero except for the 3 people constantly being born.

The opposite equation $u' - u = -3$ is birth-dominated; its solution with $u_0 = 5$ is $2e^t + 3$. It continues to increase despite the deaths (or harvesting) due to $f = -3$. The equation is now **unstable**, and it describes an explosion of population because $a > 0$.

6A The equation $u' = au$ is stable for $a < 0$, unstable for $a > 0$, and neutrally stable for $a = 0$. If $a < 0$ and the forcing term $f(t)$ approaches a constant then so does the solution; the steady limit of $u' - au = f$ is $-au_\infty = f_\infty$.

A second example is a man falling from a plane. If the air resistance is cv, proportional to his velocity, Newton's law is the first-order equation $mv' + cv = mg$. This is $F = ma$, and the man quickly approaches a terminal velocity $v_\infty = mg/c$ (where v' is zero). If he has a parachute then c is large and v_∞ is not fatal.

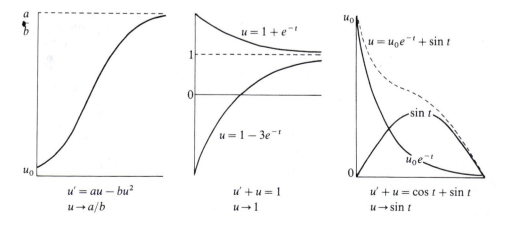

u' = au − bu²
u → a/b

u' + u = 1
u → 1

u' + u = cos t + sin t
u → sin t

Fig. 6.1. Approaches to steady state and steady oscillation.

The other case of tremendous importance is an oscillating source $f = e^{i\omega t}$. In electrical engineering that forcing term is omnipresent; it occurs also in mechanics. It leads not to a steady state but to a **steady oscillation**, as we show by example.

EXAMPLE 2 $u' - au = e^{i\omega t}$ with $a < 0$ and $u_0 = 0$

Solution $u(t) = \displaystyle\int_0^t e^{a(t-s)}e^{i\omega s}\,ds = \dfrac{e^{i\omega t} - e^{at}}{i\omega - a}$

Check $u(0) = 0$ and $u' - au = \dfrac{i\omega e^{i\omega t} - ae^{at} - ae^{i\omega t} + ae^{at}}{i\omega - a} = e^{i\omega t}$.

For large t the decaying term e^{at} can be disregarded and the steady oscillation is $u_\infty(t) = e^{i\omega t}/(i\omega - a)$. The solution takes on the same frequency as the driving force. The amplitude of the force is divided by $|i\omega - a|$, which is dangerous only when $i\omega$ is close to a. In second-order equations and RLC circuits we will see that this resonance can occur.

Variable Coefficients

The next step is to remove the requirement that a is constant:

$$\boxed{u' - a(t)u = 0.} \tag{6}$$

We could look again for an integrating factor and find one. But there is a natural way to solve the equation, by recognizing that it is **separable**. The unknown u can be separated from the time t, and

$$\frac{du}{dt} = a(t)u \quad \text{becomes} \quad \frac{du}{u} = a(t)\,dt.$$

Now each side can be integrated from 0 to t:

$$\log u(t) - \log u_0 = \int_0^t a(s)\,ds. \tag{7}$$

Again the symbol s is used as the variable of integration; we write $h(t)$ for this integral. When $\log u_0$ is moved to the right side with h and we take exponentials, the solution appears:

$$u(t) = e^{h(t)}u_0 \quad \text{where} \quad h(t) = \int_0^t a(s)\,ds.$$

If a is constant, $h(t)$ equals at and we return to $u = e^{at}u_0$. When a is variable, its integral governs the solution. It also enters the integrating factor, which is $e^{-h(t)}$, and that factor allows a source term f at the same time. We go from equation to solution in four undocumented steps:

$$u' - a(t)u = f(t)$$

$$e^{-h(t)}(u' - a(t)u) = e^{-h(t)}f(t)$$

$$\frac{d}{dt}(e^{-h(t)}u(t)) = e^{-h(t)}f(t)$$

$$e^{-h(t)}u(t) - u_0 = \int_0^t e^{-h(s)}f(s)\,ds$$

$$u(t) = \int_0^t e^{h(t)-h(s)}f(s)\,ds + e^{h(t)}u_0.$$

For $h(s) = as$ and $h(t) = at$, the solution specializes to the case for constant coefficients. The integrating factor has one minor advantage over the separation of u and t, in the case $u < 0$ when (7) would need $|u|$ to avoid the logarithm of a negative number. We mention that for systems $u' - A(t)u = f$ the corresponding formula with $h(t) = \int A(s)ds$ would be wrong.

EXAMPLE 3 $u' + 2tu = 2t$ with $u_0 = 0$ and $h(t) = -\int_0^t 2s\,ds = -t^2$

Solution $u(t) = \int_0^t e^{-t^2 + s^2} 2s\,ds = e^{-t^2}[e^{s^2}]_0^t = 1 - e^{-t^2}$

Check $u(0) = 0$ and $u' + 2tu = 2te^{-t^2} + 2t(1 - e^{-t^2}) = 2t.$

It was a complete fluke that this integral was elementary. The forcing term $f = 1$, which looks simpler than $f = 2t$, would have led to $\int e^{s^2}\,ds$ which is not elementary. It is the so-called "error function" and is found only in tables. Most solutions to variable-coefficient problems are worse than that; the equation may be linear but its solution has never been seen. Numerical methods are almost always required, but we give one more exception (which is singular at $t = 0$):

EXAMPLE 4 $u' - \dfrac{1}{t}u = 3t$ with $u_0 = 0$

Attempted solution $h(t) = \int_0^t \dfrac{ds}{s} = [\log s]_0^t = -\infty$

There is **no solution** starting from $t = 0$. However we can start from $t = 1$, in which case $h = \log t$:

Solution
$$u = \int_1^t e^{\log t - \log s} \, 3s \, ds = \int_1^t 3t \, ds = 3t(t-1)$$

Check
$$u(1) = 0 \quad \text{and} \quad u' - \frac{1}{t}u = 6t - 3 - 3t + 3 = 3t.$$

Correction
That solution is OK at $t = 0$! But I think $u(0) = 1$ would be impossible.

Nonlinear Equations

If the world population satisfies $u' = au$ it grows like e^{at}. That exponential growth was the prediction of Malthus. Of course a has to be positive; experimentally it is about 0.02. With this estimate the population doubles every T years, where

$$e^{0.02T} = 2 \quad \text{or} \quad T \approx 35.$$

It is nice to think of exponentially more people reading this book. However they will be chewing on it unless there is a term to slow things down. This competition term is reasonably proportional to u^2, and the **logistic equation** is

$$\boxed{\frac{du}{dt} = au - bu^2.} \tag{8}$$

The coefficient b is small and the growth starts as virtually exponential. However bu^2 becomes important as u gets large. The growth would actually stop if it reached $au - bu^2 = 0$, or $u = a/b$. We will show that this point is approached but never reached.

To solve (8) we recognize it as separable: $dt = du/(au - bu^2)$. The right side can be split (this is the idea of *partial fractions*) into

$$\frac{1}{au - bu^2} = \frac{1}{au} + \frac{b/a}{a - bu}.$$

Starting from $u_0 > 0$ the integration of these terms produces logarithms:

$$\frac{1}{a}(\log u - \log u_0) - \frac{1}{a}(\log(a - bu) - \log(a - bu_0)) = t.$$

Multiplying by a and taking exponentials, that becomes

$$\frac{u}{a - bu} = e^{at}\frac{u_0}{a - bu_0}.$$

From this we observe something remarkable: There is a quantity that has perfect exponential growth but *it is not* u. It is the ratio of u to $a - bu$. The ratio approaches infinity as the denominator goes to zero, at $u = a/b$. This is a rare explicit example in which a nonlinear equation can be made linear.

After clearing denominators the last equation can be solved for u:

$$u = \frac{a}{b + e^{-at}(a - bu_0)/u_0}.$$

As t goes to infinity the population approaches a/b. Its graph is the S-shaped curve in Fig. 6.1, assuming u_0 is below a/b. That curve is experimentally confirmed, for example with protozoa. For people the correction to Malthus is estimated at

$$a = 0.029, \ b = 2.9 \times 10^{-12}, \frac{a}{b} = 10 \text{ billion people.}$$

The world population is on the lower part of the S curve but not far from halfway to u_∞.

Second-Order Equations

If we are serious about applied mathematics, we should concentrate on the equations that arise most often and need to be solved. Second-order equations fit that description. They are frequently linear with constant coefficients, and they come with a driving force $f(t)$. Before introducing f we review the solution of the homogeneous problem

$$u'' + pu' + qu = 0. \tag{9}$$

The key is to look at exponentials $u = e^{\lambda t}$. Substitution gives

$$(\lambda^2 + p\lambda + q)e^{\lambda t} = 0, \tag{10}$$

and two exponents λ_1 and λ_2 solve this equation. They are the roots of $\lambda^2 + p\lambda + q$ and they can be written down immediately:

$$\lambda_1 = \frac{1}{2}(-p + \sqrt{p^2 - 4q}) \quad \text{and} \quad \lambda_2 = \frac{1}{2}(-p - \sqrt{p^2 - 4q}). \tag{11}$$

This presents three different possibilities for λ_1 and λ_2:

(I)   two distinct real roots: $\lambda_1 > \lambda_2$ if $p^2 > 4q$

(II)   two equal real roots: $\lambda_1 = \lambda_2$ if $p^2 = 4q$

(III) two complex conjugate roots: $\overline{\lambda}_1 = \lambda_2$ if $p^2 < 4q$

In the first case the general solution to the equation is

$$u(t) = c_1 e^{\lambda_1 t} + c_2 e^{\lambda_2 t}. \tag{12}$$

In the critical case with $\lambda_1 = \lambda_2$ there appears a new second solution:

$$u(t) = c_1 e^{\lambda_1 t} + c_2 t e^{\lambda_1 t}. \tag{13}$$

The third case returns to (12) but its exponents are now complex. They have the form $\lambda = a \pm ib$, and if we want real solutions they come from the splitting $e^{ibt} = \cos bt + i \sin bt$:

$$u(t) = c_1 e^{(a + ib)t} + c_2 e^{(a - ib)t}$$

$$= c_1 e^{at}(\cos bt + i \sin bt) + c_2 e^{at}(\cos bt - i \sin bt)$$

$$= d_1 e^{at}\cos bt + d_2 e^{at}\sin bt. \tag{14}$$

The solutions grow or decay depending on the real part $a = -p/2$; that decides stability. The case $a = p = 0$ is conservative (neutrally stable) and the solutions have fixed energy. They oscillate with a frequency depending on the imaginary part b. We emphasize that the special form $te^{\lambda t}$ for case II works only in that case. If λ is not a repeated root then this is *not* a solution of (9).

EXAMPLE 1   *Damped spring $mu'' + cu' + ku = 0$ with free oscillations*

This is Newton's law $F = ma$. The unknown u is the displacement of a mass m hanging from the lower end of a spring (Fig. 6.2). The spring stiffness is k. For a mass at rest, the force mg from gravity is balanced by the upward force kx from the spring. Here x is the steady-state displacement, and $kx = mg$ is an infinitely special case of our equilibrium equation $Kx = f$. (The 1 by 1 matrix $K = A^T CA$ is the

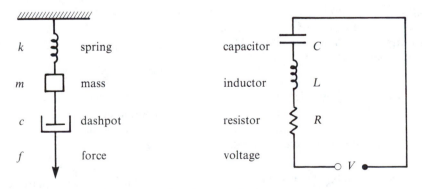

Fig. 6.2. A spring with damping and a circuit with resistance.

number k.) The displacement u in the dynamical problem is measured from this steady-state position. There is free oscillation but it is affected by a new damping term cu', which is proportional to du/dt.

The damping is represented by a "dashpot" in Fig. 6.2. It is like a shock absorber. When it comes from air resistance the effect will be small; if the spring moves in oil or molasses the damping constant c will be larger. Notice that for $c > 0$ the term cu' always acts *against* the direction of movement; it is dissipative and stabilizing. If the velocity $u' = du/dt$ is positive, the mass moves downward (u increases) and the damping force $-cu'$ acts upward. When the mass changes direction so does the damping force. The assumptions of linearity for both the spring force (Hooke's law) and the damping force are accurate for small displacements and velocities.

To fit the equation $mu'' + cu' + ku = 0$ into our model we divide by m. Then p is c/m and q is k/m. The exponents in the solution are

$$\lambda_1 = -\frac{c}{2m} + \frac{1}{2m}\sqrt{c^2 - 4mk} \quad \text{and} \quad \lambda_2 = -\frac{c}{2m} - \frac{1}{2m}\sqrt{c^2 - 4mk}. \tag{15}$$

Thus the comparison of p^2 with $4q$ becomes a comparison of c^2/m^2 with $4k/m$, in other words of c^2 with $4mk$. We expect to find three cases:

(I)   *Overdamping:* $c^2 > 4mk$ with distinct negative roots $\lambda_1 > \lambda_2$

(II)   *Critical damping:* $c^2 = 4mk$ with equal roots $\lambda = -c/2m$

(III)   *Underdamping:* $c^2 < 4mk$ with complex conjugate roots

If c is zero there is no damping at all and we are back to pure imaginary exponents $\lambda = \pm i\sqrt{k/m}$. In this case the spring oscillates forever. Otherwise we should discuss the three cases separately.

(I)   **Overdamping.** The displacement is $u = c_1 e^{\lambda_1 t} + c_2 e^{\lambda_2 t}$. With both components negative it approaches zero; the spring comes to rest at equilibrium. There are no imaginary parts and no oscillations. I think that $u = e^{-t} - 2e^{-8t}$ would start out negative, become positive by $t = 1$, and then fall back to equilibrium. Those constants $c_1 = 1$ and $c_2 = -2$ would come from an initial position $u(0) = -1$ and initial velocity $u'(0) = 15$.

(II)   **Critical damping.** Here $u = (c_1 + c_2 t)e^{\lambda t}$, with the extra factor t because of the repeated root. Again the spring might pass once through equilibrium, at the time when $c_1 + c_2 t = 0$, but it returns to rest. If c_1 and c_2 have the same sign the spring goes straight to equilibrium (and gets there at $t = \infty$).

(III)   **Underdamping.** Now λ_1 and λ_2 have imaginary parts from the square root of $c^2 - 4mk$, and there are oscillations. They go on indefinitely (Fig. 6.3b). But there is

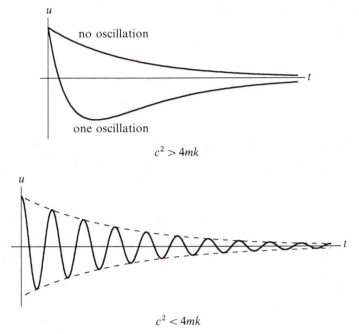

Fig. 6.3. Spring with overdamping, critical damping, and underdamping

also a real part $-k/2m$ in both exponents, so the amplitude is falling. I think this is the behavior of a typical spring:

$$u = e^{at}(d_1 \cos bt + d_2 \sin bt) \quad \text{with } a = -\frac{k}{2m}, \, b = \frac{1}{2m}\sqrt{4mk - c^2}.$$

EXAMPLE 2   *RLC circuit with forced oscillations*

There is a beautiful analogy between mechanical systems and electrical circuits. It comes from comparing $mu'' + cu' + ku = 0$ with the equation for the current:

$$L\frac{dI}{dt} + RI + \frac{1}{C}\int I dt = \text{applied voltage} = V \sin \omega t.$$

To make the comparison better we differentiate both sides, and write u instead of I for the current:

$$Lu'' + Ru' + \frac{1}{C}u = V\omega \cos \omega t. \tag{16}$$

On the left side the analogies are

> inductance $L \leftrightarrow$ mass m
>
> resistance $R \leftrightarrow$ damping constant c
>
> $1/$(capacitance C)$\leftrightarrow$spring constant k

On the right side something is new. We have a forcing term $f = V\omega \cos \omega t$, and what it forces is oscillation at the *driving frequency* ω. Our goal is to solve equation (16) for the current. At the same time we are finding the displacement of a spring with driving force $f(t)$. It is *forced* oscillation and not the free oscillation we had with $f = 0$.

Experimentally these analogies are useful. The circuit in Fig. 6.2 is easy to assemble and the current is easy to measure. A mechanical model is more difficult to prepare. Mathematically the analogies allow us to solve both problems at once. To readers who are economists or biologists or chemists, I apologize and offer a challenge: Find a fresh example of $mu'' + cu' + ku = f$.

The presence of f demands a new idea. We have not previously solved second-order equations with a forcing term, and there are at least two approaches. One of them—the method of *variation of parameters*—produces a solution comparable to $\int e^{a(t-s)} f(s) ds$ found earlier for first-order equations. The exponent became $h(t) - h(s)$ for problems with variable coefficients, and there would be something similar here. Unfortunately it is complicated. The second approach—the method of *undetermined coefficients*—is much more special. It prefers to have constant coefficients m, c, k, L, R, C on the left side, and it accepts e^{at} or $\cos \omega t$ or $\sin \omega t$ or t^n on the right side. Since this fits the equation (16) we are given to solve—and such problems are fortunately very common in applications—we choose this simpler method.

Method of Undetermined Coefficients

When the right side is a sine or a cosine, the idea of the method is to look for u as a combination of both:

$$u = a \cos \omega t + b \sin \omega t. \tag{17}$$

The *undetermined coefficients* are a and b. They become determined when we substitute u into the differential equation (16):

$$\left(-L\omega^2 a + R\omega b + \frac{1}{C} a \right) \cos \omega t + \left(-L\omega^2 b - R\omega a + \frac{1}{C} b \right) \sin \omega t = V\omega \cos \omega t.$$

This holds if the cosine terms and the sine terms both agree:

$$-L\omega^2 a + R\omega b + \frac{1}{C} a = V\omega$$

$$-L\omega^2 b - R\omega a + \frac{1}{C} b = 0. \tag{18}$$

Those are linear equations in a and b and their solution is

$$a = \frac{LV\omega(\omega_0^2 - \omega^2)}{L^2(\omega_0^2 - \omega^2)^2 + \omega^2 R^2}, \quad b = \frac{V\omega^2 R}{L^2(\omega_0^2 - \omega^2)^2 + \omega^2 R^2}. \tag{19}$$

We now have the solution (17). It is a **particular solution** to (16); it matches the right hand side but not the initial conditions. The ratio $1/LC$ has been replaced for convenience by ω_0^2. This is the frequency of free oscillation when there is no resistance and no damping, $Lu'' + u/C = 0$. In the mechanical case the undamped equation is $mu'' + ku = 0$ and ω_0^2 is k/m. These equations have solutions $\cos \omega_0 t$ and $\sin \omega_0 t$, and we will try to show that everything depends on the separation between the *driving* frequency ω and the *natural* frequency ω_0.

Undamped forced oscillations: resistance $R = 0$ or damping constant $c = 0$

Without damping the sine coefficient b is zero; it contains the factor $R = 0$. The cosine coefficient a is also simplified by the disappearance of R:

$$u = a \cos \omega t = \frac{V\omega}{L(\omega_0^2 - \omega^2)} \cos \omega t. \tag{20}$$

From that formula you see the main point: *As ω approaches ω_0 there is resonance*. The denominator in this solution becomes small and the amplitude goes to infinity. In fact the solution breaks down completely at $\omega = \omega_0$; we deal with that frequency later.

If the initial conditions are $u(0) = 0$ and $u'(0) = 0$, then the formula $u = a \cos \omega t$ needs another term:

$$u = a(\cos \omega t - \cos \omega_0 t). \tag{21}$$

At $t = 0$ the cosines cancel; $1 - 1 = 0$ so that $u = 0$. The derivatives involve sines so that $u' = 0$. This function u is still a solution of the equation $Lu'' - u/C = f$ because it has the form

$$u = u_{\text{particular}} + u_{\text{homogeneous}}. \tag{22}$$

The particular solution is $a \cos \omega t$, from the method of undetermined coefficients. It oscillates at the driving frequency ω and matches the driving force f. The homogeneous part $-a \cos \omega_0 t$ satisfies the equation with $f = 0$ on the right side. It oscillates at the natural frequency $\omega_0 = \sqrt{1/LC}$ and adjusts for the initial conditions.

The solution of any linear equation is the sum of a particular solution like (21) and a homogeneous solution.† The two constants in the homogeneous solution can match any initial values of u and u'.

How does the solution behave? It has the form $u = a(\cos \omega t - \cos \omega_0 t)$, and near $t = 0$ the cosines are close. At later times the cosines begin to separate, and at a much later time they actually become 1 and -1. At that time their difference is 2 and the solution has a maximum:

$$u_{max} = 2a = \frac{2V\omega}{L(\omega_0^2 - \omega^2)}.$$

For ω close to ω_0 this is close to resonance. The solution does not blow up but it goes far beyond the amplitude $V\omega$ of the driving force. It is like pushing a swing or a rocking chair at almost the right frequency but not quite. Sometimes the push is so much in the wrong direction that it momentarily stops the swing. Sometimes it is so much in the right direction that it sends the swing to amplitude $2a$. In a radio signal the large oscillations are heard as "beats." The difference of cosines can be written as

$$u = 2a \sin \frac{\omega_0 + \omega}{2} t \sin \frac{\omega_0 - \omega}{2} t,$$

which contains the two frequencies in Fig. 6.4—the fast frequency $\frac{1}{2}(\omega_0 + \omega)$ inside the envelope, and the slow frequency $\frac{1}{2}(\omega_0 - \omega)$ of the envelope itself.

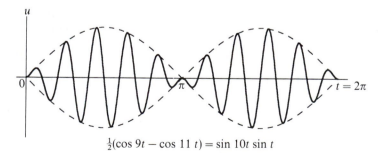

$$\tfrac{1}{2}(\cos 9t - \cos 11 t) = \sin 10t \sin t$$

Fig. 6.4. Undamped forced oscillations: $\omega = 9$ near $\omega_0 = 11$.

---

† The same was true for $Ax = b$. We could add to a particular x any vector from the nullspace.

Damped forced oscillations: resistance $R > 0$

With damping the particular solution is still $a \cos \omega t + b \sin \omega t$. It is a steady oscillation driven by f. However the homogeneous solution is different. It was found earlier from the two exponents λ_1 and λ_2:

$$u_{\text{homogeneous}} = c_1 e^{\lambda_1 t} + c_2 e^{\lambda_2 t} \to 0 \text{ as } t \to \infty.$$

The real parts of λ_1 and λ_2 are negative, as long as there is damping. Energy is dissipated as heat—electrical energy is lost in the resistor and mechanical energy in the dashpot. Therefore **at large times the initial conditions are washed out**. The conditions at $t = 0$ determine c_1 and c_2, but the exponentials $e^{\lambda t}$ become so small that only the particular solution remains. That solution continues because the force never stops:

$$u = a \cos \omega t + b \sin \omega t \quad \text{or} \quad u = A \cos(\omega t - \theta). \tag{23}$$

This is **steady oscillation** at the driving frequency ω.

We should explain this second form of u. Any combination of $\cos \omega t$ and $\sin \omega t$ is an ordinary oscillation. The coefficients a and b determine its amplitude A and its phase angle θ, but they do not affect the frequency. For example $\cos \omega t + \sin \omega t$ is the sum of two oscillations that are out of phase by $90°$. The sine function goes up as the cosine starts down. If you add them, the peak amplitude is reached at $45°$— where $\sin = \cos = 1/\sqrt{2}$ and $\sin + \cos = 2/\sqrt{2}$. In that case $A = 2/\sqrt{2}$ and $\theta = 45°$.

In general A is $\sqrt{a^2 + b^2}$. This peak is reached at the time when $\cos \omega t = a/A$ and $\sin \omega t = b/A$. That value of ωt is the phase angle θ, and the amplitude at that time is verified to be

$$a \cos \omega t + b \sin \omega t = \frac{a^2}{A} + \frac{b^2}{A} = A.$$

In our case the coefficients a and b in (19) lead to

$$A = \frac{V\omega}{\sqrt{L^2(\omega_0^2 - \omega^2)^2 + \omega^2 R^2}}. \tag{24}$$

That is an awkward expression but you can see its main features:

if $R \to 0$, A approaches the undamped amplitude in (20)

if $\omega \to \omega_0$, A approaches V/R—only resistance is left

if $R \to 0$ and also $\omega \to \omega_0$, A approaches infinity

The last is called **practical resonance**, when A almost explodes. As long as there is damping the amplitude is finite, but when A is very large the system may be destroyed. The Tacoma Narrows Bridge went down in 1940 when the oscillations from vortices produced by the wind resonated with the natural frequency of the bridge. There was plenty of warning before it went:

It is incredible that the bridge galloped for months with drivers coming from everywhere. Then something gave way, the amplitude went to 28 feet, and resonance took over. I am proud to say that the last man on the bridge was a professor. (He assured everyone of its stability.) After taking measurements he followed the yellow line to safety, while the local bank removed its sign saying "as safe as the Tacoma Bridge."

After collapse the governor of Washington promised to replace the bridge: "We are going to build the exact same bridge, exactly as before."

I had almost forgotten, but RLC circuits appeared already in Section 2.3. They went into the framework of $A^T C A$, except that the material constant was a complex number. The resistance R in Ohm's law was changed to the *impedance*:

$$Z = R + i\left(\omega L - \frac{1}{\omega C}\right).$$

We were solving the same problem of steady oscillations—when the transients are damped out. Therefore there must be a connection between Z and the new formula (24) for the amplitude. In fact the connection had better be extremely close:

$$A = \frac{V}{|Z|}. \tag{25}$$

The magnitude of Z determines the amplitude of the current, and its angle θ in the complex plane is the phase angle found above. Z is real only at the natural frequency, where $\omega^2 = 1/LC$ brings us back to R. It is an exercise to show that (25) is the same as (24).

We summarize the solution of second-order equations, with constant coefficients m, c, k taken from mechanics:

6B   The equation $mu'' + cu' + ku = F \cos \omega t$ leads to four important oscillations, at the driving frequency ω and the natural frequency $\omega_0 = \sqrt{k/m}$:

Undamped and free:	$c = F = 0$	$u^{(1)} = d_1 \cos \omega_0 t + d_2 \sin \omega_0 t$
Damped and free:	$F = 0$	$u^{(2)} = c_1 e^{\lambda_1 t} + c_2 e^{\lambda_2 t} \to 0$
Undamped and forced:	$c = 0$	$u^{(3)} = a \cos \omega t + u^{(1)}$
Damped and forced:		$u^{(4)} = A \cos(\omega t - \theta) + u^{(2)}$

As ω approaches ω_0 there is resonance in $u^{(3)}$ and practical resonance in $u^{(4)}$.

The only case not solved is $\omega = \omega_0$, at resonance itself.

The Solution at Resonance

For pure resonance the method of undetermined coefficients needs a new idea.

The equation is undamped:

$$mu'' + ku = F \cos \sqrt{\frac{k}{m}} \, t. \tag{26}$$

The driving frequency on the right equals the natural frequency on the left. The electrical analogue would be $Lu'' + u/C = V\omega_0 \cos \omega_0 t$. The same terms $\cos \omega_0 t$ and $\sin \omega_0 t$ appear in the homogeneous solution and in the particular solution used thus far, so a new particular solution is required. It has an extra factor t:

$$u_{\text{particular}} = t(a \cos \omega_0 t + b \sin \omega_0 t), \ \omega_0 = \sqrt{\frac{k}{m}}.$$

Substituting into (26) produces $a = 0$ and

$$u = \frac{F}{2m\omega_0} t \sin \omega_0 t. \tag{27}$$

There is oscillation at the only frequency in the problem, but with an amplitude that keeps growing. The dotted lines in Fig. 6.5 show the effect of the factor t.

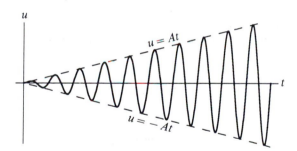

Fig. 6.5. The solution $At \sin \omega_0 t$ at resonance.

Remark   The *method of undetermined coefficients* also applies when a polynomial is on the right side of the equation. In that case the particular solution is a polynomial of the same degree:

$$\text{for } mu'' + cu' + ku = 1 + t^2 \quad \text{try } u = A + Bt + Ct^2. \tag{28}$$

When the right side is an exponential the rule is similar:

$$\text{for } mu'' + cu' + ku = Fe^{\lambda t} \quad \text{try } u = Ae^{\lambda t}. \tag{29}$$

The value of the method is its simplicity. If it fails the modification is also simple— as at resonance multiply u by an extra factor t. The most extreme case may need t^2. There is one exercise without the extra t and one that requires it.

Nonlinear Oscillations

The picture of linear oscillations is pretty clear. When they are undamped, as in $mu'' + ku = 0$, they are periodic. When they are damped, as in $mu'' + cu' + ku = 0$, they lose energy and die out. We did not say so, but if $c < 0$ then the solution grows. That is negative damping.

What if the equation is *not linear*? We do not expect to solve it exactly, even if it is the simplest equation for a pendulum:

$$\boxed{\frac{d^2\theta}{dt^2} + \sin \theta = 0.} \tag{30}$$

However we do expect the thing to oscillate. If θ stays small the equation is close to $\theta'' + \theta = 0$, which is linear with $\theta = a \cos t + b \sin t$. When θ is larger we cannot pretend that $\sin \theta = \theta$; the force that brings the pendulum back is nonlinear. The overall picture should still show undamped oscillations, periodic as before but with a period different from 2π. Here are three facts that can be confirmed even for nonlinear oscillations:

1. The total energy $E = \frac{1}{2}(\theta')^2 + 1 - \cos \theta$ is constant throughout the motion
2. The pendulum goes up to the point θ_{max} where $1 - \cos \theta_{max} = E$, and then drops back toward $\theta = 0$
3. The time for one swing back and forth (the period) is

$$T = 4 \int_0^{\theta_{max}} \frac{d\theta}{\sqrt{2E - 2(1 - \cos \theta)}}. \tag{31}$$

The energy, amplitude, and period are all available, even if the complete solution is not. Each of those properties can be verified in one or two lines:

Energy: The derivative dE/dt is $\theta'(\theta'' + \sin \theta)$. From equation (30) this is zero and E is constant

Amplitude: At the top of its motion, where the pendulum reverses direction and momentarily has velocity $\theta' = 0$, all the energy is potential: $E = 1 - \cos \theta_{max}$. When $E = 2$ the pendulum reaches 180° and swings through a full circle. When the energy is larger than 2 it goes over the top and keeps turning

Period: The square root of $(\theta')^2 = 2E - 2(1 - \cos \theta)$ is

$$\frac{d\theta}{\sqrt{2E - 2(1 - \cos \theta)}} = dt.$$

Integrating from 0 to θ_{max} is a quarter-period.

These three steps are possible for other oscillations, linear or nonlinear. In the linear case $\theta'' + \theta = 0$ leads to constant energy by the following trick: **Multiply by θ' and integrate**:

$$\theta'(\theta'' + \theta) = 0 \text{ gives } \tfrac{1}{2}(\theta')^2 + \tfrac{1}{2}\theta^2 = \text{constant} = E.$$

The amplitude is the point where θ' is zero: $\theta_{max} = \sqrt{2E}$. The period is

$$T = 4 \int_0^{\sqrt{2E}} \frac{d\theta}{\sqrt{2E - \theta^2}} = 2\pi.$$

In this linear case the energy and amplitude don't affect the period. Normally they do, and undamped oscillations follow the example set by the pendulum; we change θ to u.

6C For the nonlinear oscillations $u'' + V'(u) = 0$, the energy E and the amplitude u_{max} and the period T are given by

$$E = \tfrac{1}{2}(u')^2 + V(u) = V(u_{max}) \quad \text{and} \quad T = 4 \int_0^{u_{max}} \frac{du}{\sqrt{2(E - V(u))}}$$

When $u'' + V'(u) = 0$ is multiplied by u' both terms can be integrated: $V'u'$ is $(dV/du)(du/dt)$ and its integral is V:

$$\tfrac{1}{2}(u')^2 + V(u) = E. \tag{32}$$

This energy is kinetic plus potential. When the kinetic part is zero the oscillation is at full amplitude: $E = V(u_{max})$. With $V(u) = V(-u)$ the tension and compression are symmetric—otherwise u_{min} is not the negative of u_{max}, which causes no difficulty.

EXAMPLE $\qquad\qquad\qquad\qquad u'' + u + u^3 = 0$

The energy is $E = \tfrac{1}{2}(u')^2 + \tfrac{1}{2}u^2 + \tfrac{1}{4}u^4$. The equation (Duffing's equation) was multiplied by u' and integrated. The amplitude is reached where $u' = 0$ and

$$\tfrac{1}{2}u_{max}^2 + \tfrac{1}{4}u_{max}^4 = E \quad \text{or} \quad u_{max} = (\sqrt{1 + 4E} - 1)^{1/2}.$$

The integral that gives the period leads to "elliptic functions," which are well known but not actually known.

This spring is *hard* compared to a linear spring. Its restoring force $u + u^3$ grows faster than the linear force from Hooke's law (Fig. 6.6). The opposite case $u - u^3$ gives a soft spring, whose force is below Hooke's law. If we push far enough the spring may become so soft, at the extreme right of the figure, that greater displacement takes *less* force. Such a spring is unstable and stretches like a stick of gum that has reached the point of no return. The pendulum behaves like a soft spring, with $V'(u) = \sin u$ and with a period greater than 2π. Other springs lead to a fascinating problem of *phase transition*, when $V'(u)$ increases again after decreasing. Then there are two stable phases surrounding the unstable one, and the choice becomes hard to predict.

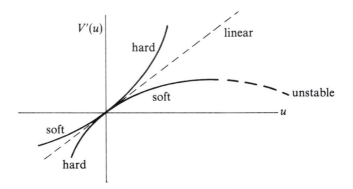

Fig. 6.6. Hard, soft, and linear springs.

The nonlinear picture is incomplete until damping is allowed. That makes an exact solution more difficult; we lose the integration that took us to E. The energy is not constant. But there is a beautiful method to study the oscillation, called the *phase plane*, which goes far beyond springs and electrical circuits. We will describe it after testing the equation for stability.

EXERCISES

6.1.1 Solve $u' + u = e^{2t}$, $u' + u = e^{i\omega t}$, $u' + u = e^{-t}$, and $u' + u = 1$ all with $u_0 = 5$. Which solutions go to a steady state u_∞?

6.1.2 If $u' + 2u = (\text{delta function at } t = 1) + c(\text{delta function at } t = 4)$, find the solution from equations (4–5). What choice of c will turn the solution off, so that $u = 0$ for $t > 4$?

6.1.3 Solve $du/dt = u^{1-k}$ with $u_0 = 1$, $k \neq 0$, by separating $u^{k-1}du$ from dt and integrating. When does u blow up if $k < 0$? Which of $u' = u^3$ and $u' = 1/u^3$ can be solved with $u_0 = 0$?

6.1.4 Solve $u' - u \cos t = 1$ with $u_0 = 4$.

6.1.5 Find the general solution to the separable equations

(a) $u' = -tu$ (b) $u' = -u/t$ (c) $uu' = \tfrac{1}{2}\cos t$

6.1.6 The example $u' - u/t = 3t$ started from $u_0 = 0$ at $t = 1$. What is the integrating factor $e^{-h(t)}$? Multiply the equation by that factor, express the left side as an exact derivative, and integrate from $t = 1$ to find u.

6.1.7   Change signs and solve $u' + u/t = 3t$ with $u(1) = 0$.

6.1.8   Suppose a rumor starts with one person and spreads according to $u' = u(N - u)$. Find $u(t)$ for this logistic equation. At what time T does the rumor reach half of the population $(u(T) = \frac{1}{2} N)$?

6.1.9   Show by differentiating $v = u/(a - bu)$ that if $u' = au - bu^2$ then $v' = av$. The nonlinear logistic equation is linearized by a change of variable.

6.1.10   Differentiate $u' = au - bu^2$ to show that $u'' = (au - bu^2)(a - 2bu)$. Where is the inflection point $u'' = 0$ in Fig. 6.1a, at which the curve changes from convex to concave?

6.1.11   Find the solution with arbitrary constants C and D to

(a) $u'' - 9u = 0$ (b) $u'' - 5u' + 4u = 0$ (c) $u'' + 2u' + 5u = 0$

6.1.12   Find an equation $u'' + pu' + qu = 0$ whose solutions are

(a) e^t, e^{-t} (b) $\sin 2t, \cos 2t$ (c) $1, t$ (d) $e^{-t} \sin t, e^{-t} \cos t$

6.1.13   (a)   What damping constants c in $\frac{1}{2} u'' + cu' + \frac{1}{2} u = 0$ produce overdamping, critical damping, underdamping, no damping, and negative damping?
   (b)   Find the exponents λ_1, λ_2 and solve with $u_0 = 2$ and $u_0' = -2c$. For which c does $u(t) \to 0$?

6.1.14   Find the undamped forced oscillation (21) for

(a) $u'' + u = \cos 2t$ (b) $u'' + 9u = \cos t$ (sketch u).

6.1.15   Solve with $u_0 = 2$, $u_0' = 0$ and find the steady oscillation (23):

(a) $u'' + 2u' = \cos \omega t$ (b) $u'' + 2u' + 2u = \sin \omega t$

6.1.16   What driving frequency ω will produce the largest amplitude A in equation (24)? For small R this is the "resonant frequency under damping."

6.1.17   Show that (25) is the same as (24), with $\omega_0^2 = 1/LC$.

6.1.18   (a)   Solve $u'' + u' + u = t^2$ by assuming $u = A + Bt + Ct^2$.
   (b)   If the right side is e^{-t}, find A in $u = Ae^{-t}$
   (c)   If the right side is $\cos t$, assume $u = A \cos t + B \sin t$.

Note: Exercise 18 uses the method of undetermined coefficients; 19 and 20 add a factor t when A or Ae^{-t} or $A \cos t$ would fail.

6.1.19   (a)   Solve $u'' + u' = t^2$ with $u = At + Bt^2 + Ct^3$
   (b)   Solve $u'' + u' = e^{-t}$ with $u = Ate^{-t}$.

6.1.20   For $u'' + u = \cos t$, show that $u = A \cos t + B \sin t$ fails to give a solution. This is resonance: solve with $u_0 = 0$ and $u_0' = 1$ and $u = C \cos t + D \sin t + At \cos t + Bt \sin t$.

6.1.21   Find the energy $E(u)$ for the equations

(a) $u'' + \frac{1}{2}e^u - \frac{1}{2}e^{-u} = 0$ (b) $u'' + u - u^3 + u^5 = 0$.

If $u_0 = 0$ and $u_0' = 1$, what equation gives the amplitude u_{max}? Are these springs hard or soft?

6.1.22   Suppose $a = 1$ but $b = -1$ in the logistic equation, giving cooperation instead of competition: $u' = u + u^2$. Solve for $u(t)$ if $u_0 = 1$. When does the population become infinite?

6.2 ■ STABILITY AND THE PHASE PLANE AND CHAOS

What facts are important to know about a system of differential equations? You may answer: the solution if possible. It is hard to argue with that. For the linear oscillators in the last section, it was important and possible to find a formula for $u(t)$. The same is true for 2 by 2 systems with constant coefficients. Beyond that it quickly becomes impossible to find the solution, except numerically. A numerical solution has its own importance but it is not everything.

What we hope for analytically is a qualitative picture of $u(t)$. It might oscillate, or decay, or grow, or drop suddenly from one value to another. The exact solution may be out of our reach, but its essential properties can still be established. In fact it would be better to speak about solutions, in the plural, since different initial values start the solution into different orbits. When the initial values are close the orbits are close—at least for a time. They oscillate or decay together, but for how long? In some sense it is not so totally important to compute one solution, in comparison with knowing about the whole field of solutions.

The underlying question is *stability*. If solutions come closer the system is stable. If they grow far apart, the system is unstable. Those are properties we can hope to predict, by localizing the problem and linearizing it. A linear problem with constant coefficients can be understood completely, and if it is clearly stable or clearly unstable then so is the nonlinear problem. As always the marginal case is the most difficult. When the linear problem is neutrally stable, the nonlinear problem can in theory do almost anything. That is strongly debated for weather prediction, where we simply do not know whether the flight of an airplane could change the weather two weeks later. If so, predictions beyond a certain time are not possible. The same question arises for our own history. Tolstoy thought that history flows from the actions of entire populations, not from a general or a tsar. But the individual life history of a single person seems to make a difference—or is it only that the population is ready?

What is certainly possible is to understand 2 by 2 systems, and we do that first.

Stability of $u' = \begin{bmatrix} a & b \\ c & d \end{bmatrix} u$

The solutions that count are the exponentials $u = e^{\lambda t}x$. The differential equation $u' = Au$ becomes $\lambda e^{\lambda t}x = Ae^{\lambda t}x$, so that λ is an *eigenvalue* of A. Therefore the question is whether λ is positive (producing growth) or negative (producing decay). More precisely, we ask whether the *real part* of λ is positive or negative or zero. For symmetric matrices that was decided by the tests for positive definiteness in Chapter 1. Now the matrix is not symmetric, but if it is 2 by 2 a decision cannot be difficult.

The eigenvalues come from the characteristic equation

$$\det \begin{bmatrix} a - \lambda & b \\ c & d - \lambda \end{bmatrix} = \lambda^2 - (a+d)\lambda + (ad - bc) = 0. \tag{1}$$

The roots λ_1 and λ_2 depend only on two numbers, the trace $a + d$ and the determinant $ad - bc$. In fact *the sum of the eigenvalues is the trace and the product is the determinant*. Therefore the stability decision in Fig. 6.7 is based entirely on those two numbers. The result is straightforward: *Stability requires a negative trace and a positive determinant*.

If both eigenvalues are negative, then $\lambda_1 + \lambda_2$ is negative and $\lambda_1 \lambda_2$ is positive. If the eigenvalues are complex numbers $x \pm iy$ and x is negative, the conclusion is the same:

$$\lambda_1 + \lambda_2 = 2x < 0 \quad \text{and} \quad \lambda_1 \lambda_2 = x^2 + y^2 > 0.$$

Thus only the second quadrant in Fig. 6.7 corresponds to stability. Its boundaries are neutrally stable, because crossing them gives instability. On the lower boundary one eigenvalue is zero (since $\lambda_1 \lambda_2 = 0$) and on the side both eigenvalues are imaginary (since $\lambda_1 + \lambda_2 = 2x = 0$). Those crossings are the two fundamental ways in which stability can be lost, in what is called bifurcation theory.

AN EXAMPLE FROM EACH QUADRANT: $\begin{bmatrix} 1 & 0 \\ 0 & 2 \end{bmatrix} \begin{bmatrix} -1 & 0 \\ 0 & -2 \end{bmatrix} \begin{bmatrix} 1 & 0 \\ 0 & -2 \end{bmatrix} \begin{bmatrix} -1 & 0 \\ 0 & 2 \end{bmatrix}$

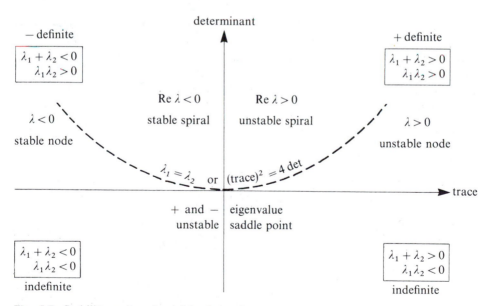

Fig. 6.7. Stability regions for a 2 by 2 matrix.

There is more to the figure than stability. It also indicates the dividing line (a parabola) between real eigenvalues and complex eigenvalues. That comes directly from the equation $\lambda^2 - (\text{trace})\lambda + (\text{det}) = 0$. The quadratic formula gives

$$\lambda = \tfrac{1}{2}[\text{trace} \pm \sqrt{(\text{trace})^2 - 4(\text{det})}]. \tag{2}$$

The roots are real and distinct if $(\text{trace})^2 > 4(\text{det})$. That is the region below the parabola. Above it the roots are complex, because they contain the square root of a negative number. On the dividing line the discriminant under the square root is zero, $(\text{trace})^2 - 4(\text{det}) = 0$, and the eigenvalues are equal: $\lambda_1 = \lambda_2 = \tfrac{1}{2}(\text{trace})$. Half the parabola is stable and half is unstable.

The labels positive definite, negative definite, and indefinite are included in the figure although they apply only to symmetric matrices. Since those have real eigenvalues they all lie below the parabola. We confirm that the square root is never imaginary when $b = c$:

$$(\text{trace})^2 - 4(\text{det}) = (a + d)^2 - 4(ad - b^2) = (a - d)^2 + 4b^2 > 0.$$

Complex eigenvalues are possible only when b and c have opposite signs and are sufficiently large.

A parabola could have entered the section on damping (but was never drawn). Overdamping is below it and underdamping is above it. The borderline of critical damping is the parabola $c^2 = 4km$. When we change $mu'' + cu' + ku = 0$ to a system $u' = Au$, the two parabolas will be seen as identical.

The other things to notice in Fig. 6.7 are the words **node, spiral**, and **saddle point**. Those describe what actually happens to the solution. For a 2 by 2 system the solution u is a point in the plane, and as time goes forward it traces out an orbit. For example its components might be $u_1 = Ae^t$ and $u_2 = Be^{3t}$. This is unstable with eigenvalues 1 and 3. The solution certainly goes to infinity, and the second component (with e^{3t}) goes faster than the first. The signs depend on A and B, but the orbit will look like one in Fig. 6.8a. Since it is an unstable node, the arrows on the orbits would be reversed to flow outwards.

We discuss each possibility in turn:

1. **Stable node.** Both eigenvalues are negative. An example is

$$u' = \begin{bmatrix} -1 & 0 \\ 0 & -3 \end{bmatrix} u \quad \text{with} \quad \begin{array}{l} u_1 = Ae^{-t} \\ u_2 = Be^{-3t} \end{array}$$

The solution $u(t)$ goes directly to zero as in Fig. 6.8a.

2. **Stable spiral.** The eigenvalues are $x \pm iy$ with x negative. For example

$$u' = \begin{bmatrix} -1 & 1 \\ -1 & -1 \end{bmatrix} u \quad \text{has } \lambda = -1 \pm i, \quad \begin{array}{l} u_1 = e^{-t}(C \cos t + D \sin t) \\ u_2 = e^{-t}(-C \sin t + D \cos t) \end{array}$$

The solution $u(t)$ spirals into the focal point at zero (Fig. 6.8b). It corresponds to underdamping, where e^{-t} pulls you toward equilibrium through smaller and smaller oscillations.

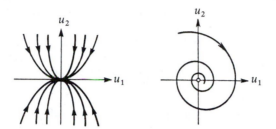

Fig. 6.8. A node and a spiral: arrows reversed when unstable.

3.   *Unstable saddle point.* The eigenvalues have opposite sign, as in

$$u' = \begin{bmatrix} 1 & 0 \\ 0 & -1 \end{bmatrix} u \quad \text{with} \quad \begin{matrix} u_1 = Ae^t \\ u_2 = Be^{-t} \end{matrix}$$

The first component goes to infinity as the second goes to zero (Fig. 6.9a). If the matrix reverses sign then so do the arrows, but it remains an unstable saddle point.

4.   *Neutrally stable center.* The eigenvalues are imaginary, as in

$$u' = \begin{bmatrix} 0 & 1 \\ -1 & 0 \end{bmatrix} u \quad \text{with} \quad \lambda = \pm i, \quad \begin{matrix} u_1 = C \cos t + D \sin t \\ u_2 = D \cos t - C \sin t \end{matrix}$$

The solution circles the origin (Fig. 6.9b), while the energy $u_1^2 + u_2^2 = C^2 + D^2$ stays constant. The numbers C and D determine which circle it is, and which is the starting point at $t = 0$. For the center in Exercise 6.2.4 it is $u_1^2 + \frac{1}{4}u_2^2$ that is constant and the circles become ellipses.

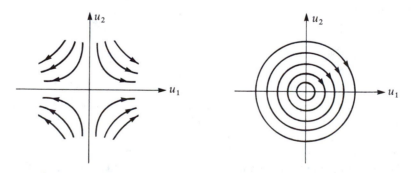

Fig. 6.9. A saddle point and a neutrally stable center.

A center is one of the borderline cases not marked on the main diagram of Fig. 6.7. It lies on the axis above zero. Physically it corresponds to an undamped spring that oscillates freely. That is a very special case, but in the applications it is extremely important. It represents *periodic motion*, and it is always present when $A^T = -A$—since skew-symmetric matrices have pure imaginary eigenvalues, and that is the requirement for a center.

Another marginal case lies on the parabola where the eigenvalues are equal. The two possibilities are in the figure below and in the exercises. If A is a multiple of the identity we have a ***star***; otherwise it is an ***improper node***.

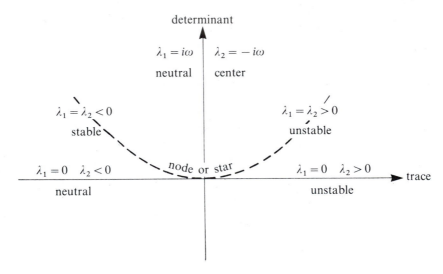

Fig. 6.10. Stability types at the region boundaries.

Stability for Nonlinear Systems

The equations in this section will not be linear; they will be linearized. They start in the form $u' = F(u)$. We continue to work with two equations in two unknowns, and a typical example is

$$u_1' = au_1 - bu_1u_2$$

$$u_2' = cu_1u_2 - du_2. \tag{3}$$

That is an important equation in population dynamics, where u_1 represents the population of the ***prey*** and u_2 is the population of the ***predator***. Because of the quadratic term u_1u_2, which is bad for the prey and good for the predator, the equation is nonlinear and an exact solution is not to be expected.

Our goal is not quantitative but qualitative—to get a clear idea of $u(t)$. All constants a,b,c,d are positive, and so are the initial conditions. Left alone, the prey would increase ($u'_1 = au_1$) and the predators would decrease ($u'_2 = -du_2$). The number of meetings between them is proportional to $u_1 u_2$, and each meeting gives the predator a chance to recoup. The model is similar to $u' = au - bu^2$ of Section 6.1, which had only one population—competing with itself. Here we have two populations, competing with each other according to $u'_1 = F_1$ and $u'_2 = F_2$.

The approach to linearized stability is straightforward:

1.   Look for **critical points** $u*$ where $F(u*) = 0$

2.   Compute the matrix $A = \begin{bmatrix} \partial F_1/\partial u_1 & \partial F_1/\partial u_2 \\ \partial F_2/\partial u_1 & \partial F_2/\partial u_2 \end{bmatrix}$ at $u = u*$

3.   Decide the stability or instability of A from its eigenvalues.

A critical point is also a *stagnation point*; since $u' = F(u*) = 0$ the solution does not move. It rests at equilibrium. The constant $u = u*$ satisfies the differential equation and the orbit of that solution is a single point. The question is whether nearby solutions come in toward $u*$ (stability), circle around it (neutral stability), or leave it (instability). Except for the neutral case, which is always more delicate, that can be decided by linearizing the equation around $u*$. We are looking at the equation locally, through a microscope that keeps the linear terms in $u - u*$ and throws away quadratic and higher terms. If u leaves $u*$ we cannot tell where it goes.

The predator-prey example has two points where $F(u*) = 0$:

$$\text{if} \quad au_1 = bu_1 u_2 \quad \text{and} \quad cu_1 u_2 = du_2 \quad \text{then}$$

$$\text{either} \quad u_1 = u_2 = 0 \quad \text{or} \quad u_1 = \frac{d}{c}, u_2 = \frac{a}{b}.$$

The derivatives of the two components of F go into A:

$$A = \begin{bmatrix} a - bu_2 & -bu_1 \\ cu_2 & cu_1 - d \end{bmatrix}.$$

At the critical point $u_1^* = u_2^* = 0$, this matrix is

$$A = \begin{bmatrix} a & 0 \\ 0 & -d \end{bmatrix}.$$

The eigenvalues a and $-d$ have opposite signs, so (0,0) is a saddle point. It is unstable. From a small start the prey increases because $a > 0$; we leave the origin.

At the other point $u_1^* = d/c$, $u_2^* = a/b$, the matrix is

$$A = \begin{bmatrix} 0 & -bd/c \\ ca/b & 0 \end{bmatrix}.$$

The trace is zero and the determinant is ad. This critical point is a **center**. The eigenvalues are $+i\sqrt{ad}$ and $-i\sqrt{ad}$. If the problem had been genuinely linear, not linearized, its solutions would oscillate with $\cos\sqrt{ad}\,t$ and $\sin\sqrt{ad}\,t$.

Remark   The oscillation is now around u^*, not around the origin. The expansion $F(u) = F(u^*) + A(u - u^*) + \cdots$ starts with $F(u^*) = 0$, so the equation can be written $(u - u^*)' = A(u - u^*) + \cdots$. Linearization throws away the three dots. You see why only critical points are tested, since at other points the leading term is some nonzero value $F(u^*)$ and the solution goes steadily along. It is like minimization; we are not interested unless the derivative is zero.

The center point $u_1^* = d/c$, $u_2^* = a/b$ is the key to the predator-prey equation. Even though nonlinear, the problem has periodic solutions. There is a cycle in which the prey grows, the predator takes over, and then they both drop back to their starting values. The solution repeats, and in the next pages we show how to draw this periodic orbit (Fig. 6.13a below) in the "phase plane." First we summarize the analysis of linearized stability:

6D   If $F(u^*) = 0$ then for initial values near u^* the nonlinear equation $u' = F(u)$ imitates the linearized equation with matrix $A = \partial F/\partial u$:

$$\text{if } A \text{ is unstable at } u^* \text{ then so is } u' = F(u)$$

$$\text{if } A \text{ is stable}\quad \text{at } u^* \text{ then so is } u' = F(u)$$

The nonlinear equation has spirals, nodes, and saddle points according to A. However if A has a center (neutral stability with imaginary eigenvalues) the stability of $u' = F(u)$ is undecided.

This linearization extends to n equations and an n by n matrix A. But the phase plane will be special to $n = 2$, where curves have no room to escape. If one curve spirals, so must all the curves inside it—and the analysis is much more complete. Only the case of a center is not settled by linearization.

EXAMPLE   The damped pendulum $\theta'' + c\theta' + \sin\theta = 0$

The extra term $c\theta'$ removes energy. Of course nonlinearity comes from $\sin\theta$. Prior to linearizing we write this equation as a first-order system $u' = F(u)$:

Reducing a second-order equation $\theta'' = G(\theta,\theta')$ to a first-order system: Let $u_1 = \theta$ be the first unknown and introduce the "velocity" $u_2 = \theta'$ as a second unknown. Then the first equation is $u_1' = u_2$ and the second is $u_2' = G(u_1,u_2)$.

For the equation of a damped pendulum, this produces

$$u'_1 = u_2$$
$$u'_2 = -\sin u_1 - cu_2. \tag{4}$$

The right side is $F(u)$. Its critical points and their stability come from the linearization steps 1–3:

(1)   We solve $F(u^*) = 0$. In the first component $u_2^* = 0$. Then $\sin u_1^* = 0$ and u_1^* can be any multiple of π. There is an infinite row of critical points $u^* = (0,0),(\pm\pi,0),$ $(\pm 2\pi,0), \ldots$. At even multiples of π the pendulum hangs down and at odd multiples it is stationary at the top.

(2)   The derivative matrix (the Jacobian matrix $\partial F/\partial u$) is

$$A = \begin{bmatrix} \partial F_1/\partial u_1 & \partial F_1/\partial u_2 \\ \partial F_2/\partial u_1 & \partial F_2/\partial u_2 \end{bmatrix} = \begin{bmatrix} 0 & 1 \\ -\cos u_1 & -c \end{bmatrix}.$$

At $u_1^* = 0$ the cosine is 1. At $u_1^* = \pm\pi$ it is -1. The matrix has one form for even multiples of π and another for odd multiples:

$$A = \begin{bmatrix} 0 & 1 \\ -1 & -c \end{bmatrix} \quad \text{or} \quad A = \begin{bmatrix} 0 & 1 \\ 1 & -c \end{bmatrix}$$

(3)   The trace of both matrices is $-c$. This is the diagonal sum, equal to the sum $\lambda_1 + \lambda_2$ of the eigenvalues. We are on the left side of the stability diagram in Fig. 6.7 if the damping constant c is positive.

The second matrix has determinant -1, giving a saddle point. All stationary points with the pendulum at the top are unstable. The first matrix has determinant $+1$, and the pendulum hanging down is stable. The exact type of stability depends on $(\text{trace})^2 - 4(\text{det})$, which is $c^2 - 4$:

$c^2 > 4$:   A stable *node*, below the parabola (overdamping)

$c^2 = 4$:   A stable *improper node*, on the parabola

$c^2 < 4$:   A stable *spiral*, above the parabola (underdamping)

$c = 0$:   A neutrally stable *center* (no damping)

$c = 0$ gives the equation $\theta'' + \sin\theta = 0$ of the last section, with steady oscillations. The energy was

$$\tfrac{1}{2}(\theta')^2 + 1 - \cos\theta = E = \text{constant}.$$

Now we want to draw this curve (which is closed) and the curves with damping (which are not closed) in the plane of θ and θ'. That is the celebrated phase plane.

The Phase Plane

In the phase plane we lose track of time. Like the orbit of a planet, the path is drawn without recording the time when it passes each point. However we have much more than just a path in space; it is a path in *phase space*, where the velocity is another coordinate. For one planet in 3-dimensional space the phase space is 6-dimensional, for N planets it is $6N$-dimensional, and for a 2 by 2 system it is a plane. Note that we could recover the time, if necessary, when the velocity is part of the picture; for a more general $u_1 - u_2$ plane, time is really lost. Note also that the equation $u' = F(u)$ is *autonomous*; F does not depend directly on t and the critical points $u*$ do not change with time. Those are the stagnation points at which $F(u*) = 0$, and linearization gives a local picture around each of them. The object of the phase plane is to achieve a global picture of the motion.

The picture is easy when the energy is constant. For the undamped pendulum it was $E = \frac{1}{2}(\theta')^2 + 1 - \cos\theta$, and the paths in the phase plane are the curves $E = \text{constant}$ (Fig. 6.11). If the solution starts on a curve it stays on that curve. You recognize the saddle points at odd multiples of π. Those are stationary but unstable, at the top of the pendulum where $u' = 0$ and $\cos\theta = -1$. The energy is exactly $E = 2$. Smaller energies give the closed curves around the centers. Those are the ordinary back and forth oscillations, around the neutral equilibrium at the bottom. In the $\theta - \theta'$ plane they are closed curves on which energy passes between kinetic and potential. A closed curve is a periodic motion.

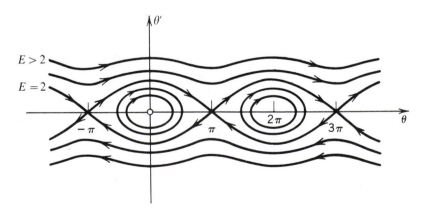

Fig. 6.11. The curves $E = \text{constant}$ for $\theta'' + \sin\theta = 0$: no damping.

For energies $E > 2$ the pendulum goes over the top. The curves continue on to $\pm 2\pi$ and never return to $\theta = 0$. Those are the upper and lower curves in Fig. 6.11, on which the velocity θ' is never zero. In that case the largest potential energy is 2 and some kinetic energy always remains. Note that where θ' is positive, above the horizontal axis, the arrows point to increasing θ.

Phase plane for the damped pendulum: $\theta'' + c\theta' + \sin\theta = 0$

If we multiply by θ' two terms can still be integrated but not the third:

$$\frac{d}{dt}[\tfrac{1}{2}(\theta')^2 + 1 - \cos\theta] = -c(\theta')^2. \qquad (5)$$

The energy is no longer constant. In fact *the energy is decreasing since the right side is negative*. This changes the picture in the $\theta - \theta'$ plane to curves that spiral in to equilibrium.

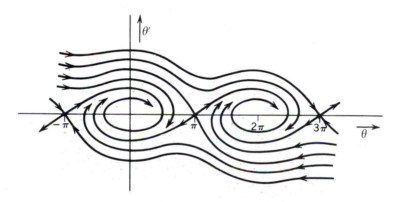

Fig. 6.12. Spiral points and stable equilibria: damped pendulum.

   The initial conditions pick out a point on this plane—a starting value of θ and θ'. As t increases, the points $\theta(t)$, $\theta'(t)$ trace out a curve in the figure. When the initial condition is near equilibrium, the curve spirals immediately toward a focal point. In the physical plane, the pendulum goes back and forth but with less and less energy. That is the normal behavior. But if the initial velocity is large enough, then even with damping the pendulum makes it over the top. It is on one of the upper curves which reach $\theta = 2\pi$ or even $\theta = 4\pi$ before spiralling to equilibrium. The lower curves do the same at a point $\theta = -2n\pi$, when their energy is exhausted. There are no centers.

   When the damping is greater, $c^2 \geq 4$, the picture changes again and the stationary points are *nodes*. Instead of going back and forth the pendulum swings back no more than once (Exercise 6.2.10). The spirals disappear and the curves go directly to equilibrium.

   We want to discuss the "phase portraits" of three other problems. The first two can be drawn, but not the last. It comes from one of the basic unsolved problems of differential equations; please don't miss it.

Periods, Limit Cycles, and Strange Attractors

1.   *The predator-prey problem.* The two equations were

$$u_1' = au_1 - bu_1u_2 \quad \text{and} \quad u_2' = cu_1u_2 - du_2.$$

The first step is to find a single equation in the $u_1 - u_2$ plane. Roughly speaking, time is eliminated by using the chain rule:

$$\frac{du_2}{du_1} = \frac{du_2}{dt}\frac{dt}{du_1} = \frac{du_2/dt}{du_1/dt} = \frac{cu_1u_2 - du_2}{au_1 - bu_1u_2}. \tag{6}$$

This equation tells *where* the unknowns go but not *when*. It describes the orbits in the phase plane. At $u_1 = d/c$ the numerator is zero, the slope is zero, and the curve is horizontal. At $u_2 = a/b$ the denominator is zero, the slope is infinite, and the curve is vertical. In between we can solve equation (6) by separating u_1 from u_2:

$$\frac{du_2}{du_1} = \frac{(cu_1 - d)u_2}{(a - bu_2)u_1} \quad \text{or} \quad du_2\left(\frac{a}{u_2} - b\right) = du_1\left(c - \frac{d}{u_1}\right)$$

Integration gives $a \log u_2 - bu_2 = cu_1 - d \log u_1 + \text{constant}$. That takes some effort to plot carefully, but it is roughly an ellipse (Fig. 6.13a). The curve is closed and the predator-prey populations are *cyclic*. The initial values give a point in the $u_1 - u_2$ plane and the populations follow the curve from that point. In this model they return to their initial values and start again—a phenomenon that is not easy to see in experiment.

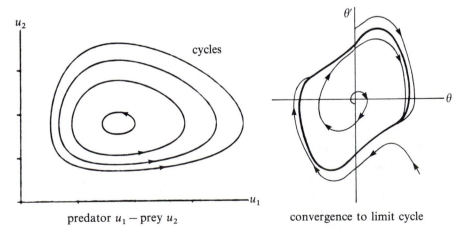

predator u_1 − prey u_2 convergence to limit cycle

Fig. 6.13. Phase planes for predator-prey and van der Pol ($c = 1$).

2.   *van der Pol's equation* $\theta'' + c(\theta^2 - 1)\theta' + \theta = 0$. This describes damped oscill-ations, but it is not certain that the damping is positive. That changes with the

factor $\theta^2 - 1$. For small θ the damping is negative and the amplitude grows. For large θ the damping is positive and the solution decays. This may sound periodic but it is not. When the solution leaves a very small or very large value, it has no way to get back. Instead all orbits spiral toward a *limit cycle* which is the unique periodic solution to van der Pol's equation.

This illustrates how qualitative information comes from the phase plane:

> If there is a region R without critical points and a solution whose orbit stays in R, then either it is periodic or it spirals toward a closed curve (periodic solution) in the phase plane.

That is the Poincaré-Bendixson theorem, only possible for a plane. Van der Pol's equation cannot be solved exactly for $u = (\theta, \theta')$, whether time is left in or removed:

$$
\begin{aligned}
u_1' &= u_2 \\
u_2' &= c(1 - u_1^2)u_2 - u_1
\end{aligned}
\quad \text{or} \quad
\frac{du_2}{du_1} = \frac{c(1 - u_1^2)u_2 - u_1}{u_2}.
\tag{7}
$$

However we can check that there are no critical points away from $(0,0)$ and that orbits stay in a ring R. Therefore they approach a limit cycle.

3. *Erratic oscillations from steady forcing*. Suppose we combine two things that have been studied separately: nonlinear oscillations and a periodic force $f(t)$. Remember that linear oscillations had a natural frequency, and the force had a driving frequency. Nonlinear oscillations are different. The natural frequency depends on the amplitude, which is constantly changed by the force. For example:

(1) You push a nonlinear swing: $\theta'' + \sin \theta = A \cos \omega t$
(2) You bounce an elastic basketball by moving your hand up and down at a fixed frequency (it doesn't work too well)
(3) You send a small satellite into the periodic force field between twin stars.

The motion can be crazy, or it can remain regular. It is the coexistence of almost periodic motion and stochastic motion (which looks random by any test, but is completely deterministic) that is most remarkable. The greatest detail will come in the quadratic model below, after two comments on problems in higher dimensions:

1. *Conservative systems:* For 3 bodies attracted to each other by gravity, it was argued for centuries whether all solutions are regular. Now we know that some are. They lie on "islands" in phase space surrounded by "seas" of stochastic motion. On a finer scale there are islands within those islands, and stochastic seas surround them too. This nesting never ends.

2. *Dissipative systems:* For $u' = Au$ all solutions approach $u = 0$. For van der Pol they approach a periodic orbit. In 1963 Lorenz discovered a new limit that is hard to describe—it is a "*strange attractor*." It has zero volume in phase space, and Hénon's attractor in Fig. 6.14a looks like a thick curve. On examination that curve is woven out of thinner curves, which on microscopic examination are woven out of even thinner curves. This infinite sequence of scales had been seen earlier in the

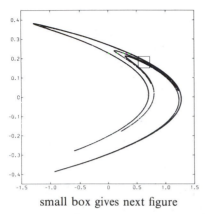

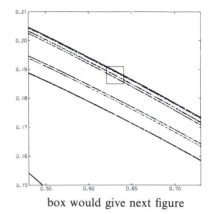

small box gives next figure   box would give next figure

Fig. 6.14. Hénon's strange attractor: limit set for $u_{n+1} = v_n - \frac{7}{5}u_n^2 + 1$, $v_{n+1} = \frac{3}{10}u_n$.

Cantor set—in which all middle thirds like $(\frac{1}{3}, \frac{2}{3})$, $(\frac{1}{9}, \frac{2}{9})$, $(\frac{7}{9}, \frac{8}{9})$, ... are removed from the interval $(0, 1)$. (Infinitely many points are left, and not just the ends of those removed intervals!) The striking feature is the *self-similarity*: From 0 to $\frac{1}{3}$ you see the same Cantor set, three times smaller. The middle thirds are still removed—starting with $(\frac{1}{9}, \frac{2}{9})$. From 0 to $\frac{1}{9}$ or from $\frac{2}{9}$ to $\frac{1}{3}$ the Cantor set is there again, scaled down by 9. Every section, when blown up, copies the larger picture. That is a typical cross-section of a strange attractor, invented a century ago and now actually encountered in differential equations.

We turn to the most famous example of chaos. It resembles the logistic equation $u' = au - bu^2$ solved on page 477, but it switches to a difference equation. Instead of a steady population growth, we are taking snapshots of an epidemic. It can oscillate (measles used to have a 2-year period), or it can settle down, or it can become chaotic—solutions from nearby starting points can diverge so quickly that there is no hope to predict the future.

The Quadratic Model of Chaos

Suppose the difference equation $u_{n+1} = F(u_n)$ starts at u_0. After one step, u_1 is $F(u_0)$. After two steps, u_2 is $F(F(u_0))$. After three steps, the notation $F(F(F(u_0)))$ is becoming hopeless but the idea is clear. Solving the equation is the same as *iterating* the function—applying it over and over in a feedback loop, where each output goes back as input. It is easy to execute, and ideal for experiments, even when there is no formula for the solution. We report here the unexpected results of experiments with a particular F.

The most interesting functions F are nonlinear. But we start with $u_{n+1} = au_n$, where every step is multiplication by a (and therefore u_n is $a^n u_0$). There are four possibilities and they depend on a:

(1)   If a is small then $u_n \to 0$ (*stability*)
(2)   If a is large then $|u_n| \to \infty$ (*instability*)
(3)   If $a = 1$ then $u_n = u_0$ (*steady state*)
(4)   If $a = -1$ then the sequence is $u_0, -u_0, u_0, -u_0, \ldots$ (***this has period*** 2).

The fourth is the most interesting. It occurs because $F(F(u))$ is $(-1)^2 u$. Two steps with $a = -1$ are equivalent to a double step with $a = +1$, giving periodicity.

The nonlinear iteration $u_{n+1} = a u_n^2$ contains something new but not much. It can blow up or decay or stay at the steady state $u = 1/a$. With $a = 1$ the sequence is u_0, u_0^2, u_0^4, u_0^8, ... and u_0 decides everything! That is the new nonlinear effect.

Now comes the **quadratic model.** It combines au with au^2, using a minus sign to balance those terms:

$$u_{n+1} = F(u_n) = au_n - au_n^2. \qquad (8)$$

The graph of F is in Fig. 6.15, between $u = 0$ and $u = 1$. Its greatest height is $F = \frac{1}{4}a$ at the middle of the interval. Therefore if the number a is not larger than 4, *the sequence u_n stays between 0 and 1.* It may approach a single point U, or it may approach a cycle of period 2 (jumping between U_1 and U_2), or it may do something new—but it does not leave the interval.

Take for example $a = 2$. *If* the u_n approach U, then

$$u_{n+1} = 2u_n - 2u_n^2 \quad \text{approaches} \quad U = 2U - 2U^2.$$

Therefore $2U^2 = U$ and the limit point U is either $\frac{1}{2}$ or 0. Those are stationary points; if the sequence starts there it stays there. The real question is what happens from other starting values u_0. The sequence u_n will or will not be attracted to $\frac{1}{2}$ or 0, according to the fundamental rule of stability:

A stationary point U is attractive if $|F'(U)| < 1.$

The reason can be seen in Fig. 6.15 on the left. The derivative of $2u - 2u^2$ is $2 - 4u$. At $U = 0$ this derivative is $F' = 2$; that point is repulsive. At $U = \frac{1}{2}$ the derivative is $F' = 0$; the point is extremely attractive. If the sequence starts at $u_0 = \frac{1}{4}$, *it goes up to the curve and along to the $45°$ line.* Distance up/distance along $\approx F'(u)$, so the process converges if $|F'| < 1$ and spreads outward if $|F'| > 1$. In this case it diverges from $U = 0$ and converges quickly to $U = \frac{1}{2}$.

Thus the derivative of F is the key. For each value of a, the stationary point satisfies $U = aU - aU^2$ or $U = (a-1)/a$. At that point the derivative of $au - au^2$ is

$$F' = a - 2aU = a - 2a\,\frac{a-1}{a} = 2 - a.$$

The stationary point is attractive *when a is between 1 and 3.* Below $a = 1$, the stationary point $U = 0$ is attractive. The derivative there is $F' = a$, which is stable. *What happens above $a = 3$?* Both stationary points are repulsive ($F' = 2 - a$ will be below -1) and the sequence cannot converge to either one.

Do not forget possibility 4, that the sequence alternates between a stable pair U_1 and U_2. That happens immediately after $a = 3$. To see it geometrically we graph the

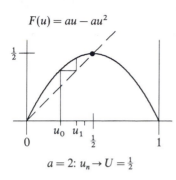

$F(u) = au - au^2$

$\frac{1}{2}$

0 u_0 u_1 $\frac{1}{2}$ 1

$a = 2:\ u_n \to U = \frac{1}{2}$

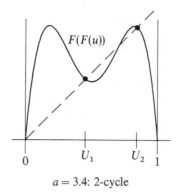

$F(F(u))$

0 U_1 U_2 1

$a = 3.4:$ 2-cycle

Fig. 6.15. The iterations of $u_{n+1} = au_n - au_n^2$.

function $G(u) = F(F(u))$ in the two-humped figure on the right. Iterating a quadratic gives the quartic $G = a(au - au^2) - a(au - au^2)^2$. It has the same two stationary points as F, at $U = 0$ and $U = (a - 1)/a$, but G *also has two more*. Its graph crosses the 45° line two more times, at U_1 and U_2. What is more, the slope at those new crossings satisfies $|G'| < 1$. They are attractive. The sequence $u_0, u_2 = G(u_0), u_4 = G(u_2), \dots$ is attracted to one of those points. The sequence $u_1, u_3 = G(u_1), u_5 = G(u_3), \dots$ is attracted to the other one. The whole sequence approaches a cycle of period 2.

In short, the system "bifurcates" as the number a passes 3. The stable point U splits into a stable pair U_1, U_2. It is even possible to show that G has the same slope at U_1 and U_2. But eventually, as a continues to increase toward 4, that slope will exceed 1 and *the 2-cycle will become unstable*. That happens near $a = 3.45$.

The next thing to appear is a stable 4-cycle. We will not graph $H(u) = G(G(u))$, which is a polynomial of degree 16. It exhibits a splitting of G, just as G split the stationary point of F. But since H is a 4-fold iterate of the original F, a stationary point for H is a cycle U_1, U_2, U_3, U_4 in the original sequence. Since it is stable, and everything else is repulsive, the sequence u_n approaches this 4-cycle. That is possibility 5, but of course it is not the last.

The situation seems clear. The 4-cycle will split into an 8-cycle and then into a 16-cycle. Each period 2^n has a "window of stability" through which the number a eventually passes. But there is a large cloud on the horizon. The bifurcation to an 8-cycle takes place rather soon. We are not going to make it to $a = 4$, because the stability intervals are rapidly getting smaller. The 2-cycle is stable from $a = 3$ to 3.45, but the window for the 4-cycle is less than $\frac{1}{4}$ as long. The later cycles survive for very short times, and the 128-cycle is possible but not at all easy to find. To look for it, the rule is that the distance between bifurcations shrinks geometrically:

The ratio between the length of one stability window and the next approaches a constant known as *Feigenbaum's number*, $\delta = 4.66920166 \dots$

That number kept turning up on the computer, for one function F after another, until there could be no doubt that it is universal. This was a sensational discovery about *period doubling*, that it is governed by the completely unexpected number

δ—which had never been seen. Now the same constant has been found in physical experiments, and in higher dimensions as well as one dimension. Experiments in Texas have reached as high as period 64. There is special interest in turbulence, where the mechanism for breakdown of smooth flow has resisted analysis for years. The small eddies were studied statistically, but we now know that a deterministic process can be so irregular that it can scarcely be distinguished from a random process. Nonlinear dynamics can be totally unstable and at the same time totally determined. All this is modeled by the humble parabola $au - au^2$, after the stable cycles of length 2^n have ended.

Because of δ, the stability intervals for the periods 2,4,8,16, ... end just before $a = 3.57$. What happens after that is partly chaotic and partly stable. Looking at the back cover of the book, the figure starts at the bottom with $a = 3.5$ (period 2) and moves up to $a = 4$. On each row the computer has plotted the values of u_{1001} to u_{2000}, omitting the first thousand points to let a stable period (or chaos) become established. The cover is dark where no points appeared; its brightness reflects the density of points. A few observations about that white figure:

(1)   The big window at period 3 is followed by periods 6, 12, 24, ... There is *period doubling* at the top of every window (including all the windows you cannot see). If one of them is blown up, it repeats the whole picture (*self-similarity*).

(2)   Period 4 appears again in the window near the top. That cycle is different—1324 vs. 1234—from the one at the bottom.

(3)   No odd periods come before the big wedge, in which no points were found.

(4)   The order in which periods appear is another universal phenomenon, not special to this quadratic.

We mention that the quadratic was transformed to $x_{n+1} = A - x_n^2$, to make the image clearer. This leaves unchanged the sequence of periods, as A increases, and it also leaves the more exotic alternative of chaos—which appears in two ways:

(1)   The points u_n can be attracted to a whole interval, in which they become dense. That happens at $a = 4$.

(2)   The points can be attracted to a Cantor set. That happens at $a = 3.5699$ ... immediately after the periods of length 2^n.

Those are not isolated exceptions; the probability that they will happen is positive. There is a good chance of hitting chaos with a random quadratic, but no certainty. I think it is the most impressive experiment this book can propose.

Julia Sets, Fractals, and the Mandelbrot Set

It often happens in mathematics that going from real numbers to complex numbers produces tremendous new insight. A one-dimensional problem moves into the complex plane, and the pictures can become amazingly beautiful. The exhibition "Frontiers of Chaos" has been showing computer-generated sets in art museums around the world, and we want to explain what they are.

The pictures come from our same quadratic! In the exhibition it is written $f(z)$ $= z^2 + c$, a minor change—but now all numbers are complex. The idea of *iteration* remains fundamental: Start from any $z_0 = x_0 + iy_0$ and compute $z_1 = z_0^2 + c$, $z_2 = z_1^2 + c$, and so on. The sequence $z_0, z_1, z_2, \ldots$ has three possibilities:

(1) z_n approaches a limit   (2) $|z_n| \to \infty$ (3) the other possibility.

When c is zero and the sequence is $z_0, z_0^2, z_0^4, \ldots$, those three cases are simple. Points inside the disk $|z_0| < 1$ go toward zero, points outside go toward infinity, and points on the circle $|z_0| = 1$ stay on the circle. The remarkable thing is what happens to the disk and the circle when c moves away from zero. You would expect these regions to change shape as c changes, and perhaps to bulge out a little. What actually happens is much more extreme. The boundary becomes totally irregular, so its length isn't even finite. Depending on c, the disk changes into "a fatty cloud, a skinny bush, the sparks after a firework, the shape of a rabbit, sea horse tails, ..." Figure 6.16a is an example. All these are *Julia sets*:

> The *filled-in Julia set* of f contains the points
> z_0 for which z_n stays bounded as $n \to \infty$.

This set is typical of a *fractal*, so named by Mandelbrot because its boundary (which is the Julia set itself) has *fractional dimension*. The circle is smooth, with dimension 1—but for $c \neq 0$ the boundary looks like an infinitely ragged coast line. To compute its dimension D we cover it with circles of small radius r; roughly r^{-D} circles are needed. (Example: The Cantor set has dimension $D = \log 2/\log 3$, between 0 and 1. It is covered by 2 intervals if their length is $\frac{1}{3}$, or 4 intervals of length $\frac{1}{9}$, or $2^k = r^{-D}$ intervals of length $r = (\frac{1}{3})^k$.) Of course many sets of fractional dimension are extremely strange; the special thing about fractals is that they are more or less self-similar. Nature produces such boundaries, with dimension between 1 and 2, and so did Julia.

There is another set M to explain. It comes by varying c instead of z_0. Each value of c leads to a Julia set in the z_0-plane, but the new set M is in the c-plane. It is somehow an index to all those Julia sets. It may be the most complicated figure ever to be studied, and partly understood, by humans. Still its definition is remarkably simple (all these fantastic sets are described by simple keys):

> M contains all numbers c for which the sequence
> $z_0 = 0, z_1 = c, z_{n+1} = z_n^2 + c, \ldots$ stays bounded.

This is the *Mandelbrot set*. A computer can draw it by testing many values of c. Another description, proved equivalent by Julia in 1919, is this: c is in M if its filled-in Julia set is connected. That definition applies to other functions $f(z)$—where the first definition started at $z_0 = 0$ because the derivative of $z^2 + c$ is zero at that point.

The Mandelbrot set begins with a big cardioid, whose points c give nice fractals. Attached to it are infinitely many disks, each with an infinity of smaller attached disks (and so on!). You also discover smaller cardioids off the real axis (with attached disks, and more disks, and more cardioids). They are all linked together

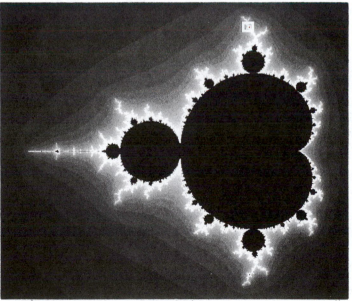

Fig. 6.16. A Julia set above, choosing c at the edge of the main cardioid in the Mandelbrot set below. The end of the antenna corresponds to the top of the back cover of the book (the real quadratic map). Courtesy Artmatrix (J. Hubbard–H. Smith, Cornell University) and The Beauty of Fractals (H.-O. Peitgen–P. Richter, Springer).

by thin filaments—the set M is connected. The branches follow a delicate combinatorial pattern. For values of c near the boundary of M, and especially down in "sea-horse valley" where a disk is attached, the Julia sets become wild and pleasurable. For c outside the Mandelbrot set, they break up into disconnected Cantor sets, or "Fatou dust".

We close with instructions for plotting a Julia set on a Macintosh (in 30 minutes?). Compute $x_1 = x_0^2 - y_0^2 - 1$ and $y_1 = 2x_0y_0$, then x_2 and y_2, stopping at x_{12} and y_{12}. Color the original x_0, y_0 white if $|x_{12}| > 2$ or $|y_{12}| > 2$. Do this for 129^2 points, with x_0 and y_0 equal to $-\frac{64}{32}$, $-\frac{63}{32}$, ..., $\frac{64}{32}$. For a new Julia set move c from -1. For the Mandelbrot set use a bigger machine.

EXERCISES

6.2.1   Solve the system $u_1' = -u_1, u_2' = -u_2$ and draw the paths of the point $u_1(t), u_2(t)$ starting from various initial values. This is the *stable star* produced by the matrix $A = -I$ with equal negative eigenvalues.

6.2.2   What types of critical points can $u' = Au$ have if
(1) A is symmetric positive definite
(2) A is symmetric negative definite
(3) A is skew-symmetric
(4) A is negative definite plus skew-symmetric (choose example).

6.2.3   Reduce $\theta'' + 2\theta' + \theta = 0$ to a system $u' = Au$ with $u_1 = \theta$ and $u_2 = \theta'$. A has equal negative eigenvalues but only one eigenvector, indicating a *stable improper node*. With $\theta = te^{-t}$ sketch the path of $(u_1, u_2) = (\theta, \theta')$ approaching the origin.

6.2.4   (a)   Solve $u_1' = -u_2$, $u_2' = 4u_1$ starting from $(1,0)$ to confirm that stability is neutral and the origin is a center.
(b)   Find the orbit by eliminating time, leaving $du_2/du_1 = -4u_1/u_2$, and show that the circles in Fig. 6.9b become ellipses.

6.2.5   For the skew-symmetric "cross product equation"

$$u' = \begin{bmatrix} 0 & c & -b \\ -c & 0 & a \\ b & -a & 0 \end{bmatrix} \begin{bmatrix} u_1 \\ u_2 \\ u_3 \end{bmatrix} = Au$$

(a)   write out u_1', u_2', u_3' and confirm that $u_1'u_1 + u_2'u_2 + u_3'u_3 = 0$
(b)   show that the energy $E = \frac{1}{2}(u_1^2 + u_2^2 + u_3^2)$ is constant
(c)   find the eigenvalues of A.
Since $u' = u \times w$ the solution rotates around the fixed vector $w = (a,b,c)$.

6.2.6   If $a = 2u_1$, $b = 3u_2$, $c = 4u_3$ the equation above becomes nonlinear:

$$u_1' = u_2u_3, \quad u_2' = -2u_1u_3, \quad u_3' = u_1u_2.$$

(a)   Show that u' is still perpendicular to u and $E = $ constant

(b)   Find the linearized matrix A at the stationary points $u^* = (1,0,0)$, $(0,1,0)$, and $(0,0,1)$ and decide if those points are stable, unstable, or neutral.

The equation describes the rotation of this book in the air. If you make it spin you will see which of its three axes is unstable; please catch the book.

6.2.7   Convert a 2 by 2 system $u' = Au$ into a single equation for u_1 by differentiating $u_1' = au_1 + bu_2$ and substituting $u_2' = cu_1 + du_2 = cu_1 + d(u_1' - au_1)/b$. How are the coefficients in the single equation connected to A?

6.2.8   Sketch the curves $E = $ constant in the phase planes for $\theta'' + \theta = 0$ and $u'' + u + u^3 = 0$, after multiplying by θ' and u' and integrating to find E.

6.2.9   If the linearized problem has a center then nonlinear stability is undecided, as in

$$\text{(a)} \quad \begin{aligned} u_1' &= u_2 + u_1(u_1^2 + u_2^2) \\ u_2' &= -u_1 + u_2(u_1^2 + u_2^2) \end{aligned} \qquad \text{(b)} \quad \begin{aligned} u_1' &= u_2 - u_1(u_1^2 + u_2^2) \\ u_2' &= -u_1 - u_2(u_1^2 + u_2^2) \end{aligned}$$

The right sides F are zero at $u^* = (0,0)$. Compute A at u^* and show it has a center. Then multiply the equations for u_1' by u_1 and the equations for u_2' by u_2, and add to find differential equations for $E = u_1^2 + u_2^2$. Show that (a) is unstable and (b) is stable; neither is neutrally stable.

6.2.10   If $c^2 > 4$ then Fig. 6.12 is wrong; the damped pendulum no longer spirals in to equilibrium. Identify the types of critical points and sketch the correct picture in the phase plane.

6.2.11   Solve the following equations and draw solution curves in the phase plane:

(a) $\theta'' + \theta = 0$ (c) $\theta''/\theta' = -\theta'/\theta$

(b) $\theta'' - \theta' = 0$ (d) $\theta'' + (\theta')^2 = 0$

6.2.12   With internal competition the predator-prey system might be

$$u_1' = u_1 - u_1^2 - bu_1u_2, \qquad u_2' = u_2 - u_2^2 + cu_1u_2$$

Find all equilibrium points and their stability (for $c < 1$ and $c > 1$). Which points make sense biologically?

6.2.13   According to Braun, reptiles, mammals, and plants on the island of Komodo have populations governed by

$$u_r' = -au_r - bu_ru_m + cu_ru_p$$
$$u_m' = -du_m + eu_ru_m$$
$$u_p' = fu_p - gu_p^2 - hu_ru_p.$$

Who is eating whom? Find all equilibrium solutions u^*.

6.2.14   If $u' = Au$ and A is skew-symmetric show that the energy $E = u^Tu$ is a constant. (The derivative of this inner product is $E' = u^Tu' + (u')^Tu$.)

6.2.15   If $u' = Au$ and $MA + A^T M$ is negative definite show that $E = u^T Mu$ is decreasing. This is the Liapounov approach—to find a symmetric positive definite M that yields decreasing energy and proves stability for nearby nonlinear equations.

6.2.16   (a)   Find the type of critical point at $u^* = (0,0)$ for

$$u_1' = u_1 + u_2 - u_1(u_1^2 + u_2^2), \quad u_2' = -u_1 + u_2 - u_2(u_1^2 + u_2^2).$$

(b)   Multiply the equations by u_1 and u_2 respectively and add to find an equation for $E = u_1^2 + u_2^2$.

(c)   Show that $E' = 0$ for the special value $E = 1$, and sketch orbits spiralling out and in to this limit cycle $u_1^2 + u_2^2 = 1$.

6.2.17   From their trace and determinant, at what times t do the following matrices change type (bifurcation)?

$$A_1 = \begin{bmatrix} 1 & -1 \\ t & -1 \end{bmatrix} \quad A_2 = \begin{bmatrix} 0 & 4-t \\ 1 & -2 \end{bmatrix} \quad A_3 = \begin{bmatrix} t & -1 \\ 1 & t \end{bmatrix}$$

6.2.18   Compute the solution to van der Pol's equation by finite differences and sketch the limit cycle for $c = \frac{1}{2}$.

6.2.19   (Epidemic theory). Suppose $u(t)$ people are healthy at time t and $v(t)$ are infected. If the latter become dead or otherwise immune at rate b and infection occurs at rate a, then $u' = -auv$, $v' = auv - bv$.

(a)   Show that $v' > 0$ if $u > b/a$, so the epidemic spreads.
(b)   Show that $v' < 0$ if $u < b/a$, so the epidemic slows down. (It never starts if $u_0 < b/a$.)
(c)   Show that $E = u + v - (b/a) \log u$ is constant during the epidemic.
(d)   What is v_{max} (when $u = b/a$) in terms of u_0?

6.2.20   For freely falling bodies with $u = \frac{1}{2}gt^2 + u_0't + u_0$, sketch the curves $(u(t), u'(t))$ in the phase plane starting from three different initial values.

6.2.21   Invent a real function F such that $F(F(x)) = -x$.

6.2.22   Differentiate $G(u) = F(F(u))$ by the chain rule, and show that the slope G' has the same value $F'(U_1)F'(U_2)$ at both points $u = U_1$ and $u = U_2$ of a 2-cycle—for which $F(U_1) = U_2$ and $F(U_2) = U_1$.

6.2.23   On a computer with sound, assign different notes to subintervals of $(0,1)$ so that you can hear the 2-cycles and 4-cycles of $u_{n+1} = au_n - au_n^2$.

6.2.24   Add periodic forcing to Duffing's equation, $u'' + u'/10 + u^3 = 12 \cos t$, and display the solutions at many multiples of $t = 2\pi$ in the $u - u'$ phase plane.

6.2.25   Change the coefficient from 1.4 to 1.3 in Hénon's 2 by 2 system, and iterate 500 times. The limit is believed to have period 7.

6.2.26   The Cantor set is left when the middle thirds like $(\frac{1}{3}, \frac{2}{3}), (\frac{1}{9}, \frac{2}{9}), (\frac{7}{9}, \frac{8}{9})$, ... are removed from $(0, 1)$. All numbers like $0.0200202 \ldots$ are still there—if they have no 1's when written in base 3. Where do the removed intervals $(\frac{1}{9}, \frac{2}{9})$ and $(\frac{7}{9}, \frac{8}{9})$ have a 1?

THE LAPLACE TRANSFORM AND THE z-TRANSFORM ■ 6.3

The Fourier transform is used for boundary-value problems. The Laplace transform is used for initial-value problems. They both work with exponentials, and they both turn differential equations into algebraic equations—but there is a difference. A function like e^{-x} is not suited for the Fourier transform; it is too large when x is negative. A function like e^{-t} is perfectly suited for the Laplace transform, because in an initial-value problem *the time is never negative*. The Laplace transform is one-sided. It looks only forward and prefers a solution that decays, where the Fourier transform looks both ways and prefers a solution that oscillates. That difference is felt everywhere:

Fourier	*Laplace*
whole line $-\infty < x < \infty$	half line $t \geq 0$
oscillations $e^{i\omega x}$	transients e^{at}
Re $i\omega = 0$	Re $a \leq 0$
input affects all values	input affects later values
ends at boundary conditions	starts with initial conditions

Despite those differences, the two transforms are in some deeper sense the same. They use the same functions (exponentials). They impose the same requirements (constant coefficients in the differential equation). And they succeed for the same reason, that those exponentials are eigenfunctions of d/dx and d/dt:

$$\frac{d}{dx} e^{ikx} = ike^{ikx} \quad \text{and} \quad \frac{d}{dt} e^{st} = se^{st}.$$

It is even possible to use two transforms in one problem. The solutions to $u_{xx} = -u$ oscillate in space, the solutions to $u_t = -u$ decay in time, and the solutions to the heat equation $u_t = u_{xx}$ do both. They are $e^{ikx}e^{-k^2 t}$. Each frequency k has its decay rate k^2. Superposition gives a complete solution, as long as there is no driving force $f(x)$ or $f(t)$ or $f(t,x)$. But the power of these transforms is that they also convert driving forces into exponentials, and for that we go directly to the definitions:

Fourier Transform	*Laplace Transform*
$\hat{f}(k) = \displaystyle\int_{-\infty}^{\infty} f(x)e^{-ikx}dx$	$F(s) = \displaystyle\int_{0}^{\infty} f(t)e^{-st}dt$

The imaginary number ik corresponds to the real number s. Otherwise there is no difference, when f is zero for negative x. The Laplace transform of e^{-at} is

$$\int_{0}^{\infty} e^{-at}e^{-st}dt = \left[\frac{e^{-(a+s)t}}{-(a+s)}\right]_{0}^{\infty} = \frac{1}{a+s}. \tag{1}$$

The Fourier transform of e^{-ax} is not defined, if x goes from $-\infty$ to ∞, but in any case that is not the same function. It should start at 0, not $-\infty$. The pulse is one-sided; f jumps to 1 and decays. The Fourier transform of that pulse was computed in Section 4.3, and it was $1/(a+ik)$. This confirms that $F(ik)$ equals $\hat{f}(k)$. When both are applicable, Fourier and Laplace produce the same transform—subject to the identification $ik = s$.

It follows that their inversion formulas also contain this change:

Inverse Fourier Transform	*Inverse Laplace Transform*
$$f(x) = \frac{1}{2\pi} \int_{k=-\infty}^{\infty} \hat{f}(k)e^{ikx}dk$$	$$f(t) = \frac{1}{2\pi i} \int_{s=-i\infty}^{i\infty} F(s)e^{st}ds$$

One integrates along the real line, combining harmonics. The other integrates along the imaginary axis; the integrand is real but not the path. Section 4.5 showed how it could automatically produce $f = 0$ for $t < 0$; the "residue method" found no poles. This Laplace inversion is more difficult than Fourier, and in practice it is seldom done (tables are easier), but in principle F is an analytic function over half of the complex plane and a contour integration can reproduce f.

What is really important is not the formulas but the operations. If we know that the transform of $f = e^{at}$ is $1/(s-a)$, and with $a = 0$ the transform of $f = 1$ is $1/s$, we can go directly to the applications. The only additional information, which is needed immediately in a differential equation, is how to transform a derivative. That looks easy, if we differentiate the inversion formulas to produce factors ik and s:

$$\text{Fourier:} \quad \frac{df}{dx} \to ik\hat{f}(k) \qquad \text{Laplace:} \quad \frac{df}{dt} \to sF(s).$$

Again ik corresponds to s. *But this rule for Laplace is wrong.* The transform of $f = 1$ is $1/s$. However the transform of $f' = 0$ is not s times $1/s$; it is zero. The problem is at the cutoff $t = 0$. The function did not always equal one; it had to jump there. It was a *step* function, not a constant function. Its true derivative is not $f' = 0$ but $f' = \delta$, and we account for the jump by going back to the definition and integrating by parts:

$$\int_0^\infty \frac{df}{dt}e^{-st}dt = \int_0^\infty f(t)(se^{-st})dt + [f(t)e^{-st}]_0^\infty. \tag{2}$$

The first term on the right is $sF(s)$, as expected. The other term is $-f(0)$, which we missed. The product $f(t)e^{-st}$ is required to be zero at $t = \infty$—it is to those s that the calculation applies—but at $t = 0$ it equals $f(0)$. Thus the troublesome function $f' = 0$ does have Laplace transform $s(1/s) - 1 = 0$, since $f(0) = 1$.

If the transform of $u(t)$ is $U(s)$, then

the transform of $\dfrac{du}{dt}$ is $sU(s) - u(0)$

the transform of $\dfrac{d^2u}{dt^2}$ is $s[sU(s) - u(0)] - u'(0)$

The rule continues in the same way for higher derivatives. It is stated for u instead of f because du/dt appears in differential equations. Capital letters will indicate Laplace transforms, and it is time for applications.

Solution of Differential Equations

EXAMPLE 1
$$\frac{du}{dt} - au = 0 \quad \text{with} \quad u_0 = 4$$

$$sU(s) - 4 - aU(s) = 0$$

$$U(s) = \frac{4}{s - a} \quad \text{or} \quad u(t) = 4e^{at}$$

That is only the familiar $u_0 e^{at}$, nothing new.

EXAMPLE 2
$$\frac{du}{dt} - au = e^{bt} \quad \text{with} \quad u_0 = 0$$

$$sU(s) - a\overset{\circ}{U}(s) = \frac{1}{s - b}$$

$$U(s) = \frac{1}{(s - a)(s - b)}$$

To find u we need another idea. Both factors $s - a$ and $s - b$ are in the denominator (we assume $a \neq b$) and we need the inverse transform. It is not $e^{at}e^{bt}$! There are at least two ways to find it—by splitting U into a sum (**partial fractions**) or by keeping U as a product (**convolution rule**). They are so important that we do both:

1.   Sum: $\dfrac{1}{(s - a)(s - b)} = \dfrac{C}{s - a} + \dfrac{D}{s - b}$ (3)

Multiply through by $s - a$ and then set $s = a$: $\dfrac{1}{a - b} = C$

Multiply through by $s - b$ and then set $s = b$: $\dfrac{1}{b - a} = D$

The two parts of U are the transforms of Ce^{at} and De^{bt}

Solution
$$u = Ce^{at} + De^{bt} = \frac{e^{at} - e^{bt}}{a - b}. \tag{4}$$

The solution is a combination of the natural mode e^{at} and the driving mode e^{bt}. In other words it is $u_{\text{homogeneous}} + u_{\text{particular}}$. The combination using C and D satisfies the initial condition $u(0) = 0$. We show below how this method of partial fractions is like the method of undetermined coefficients—it is a direct approach when the answer has a known form.

The second method keeps U as a product of two transforms:

2.   **Product:** $\dfrac{1}{(s - a)(s - b)} = G(s)F(s)$ where $g = e^{at}$ and $f = e^{bt}$

We need a formula for u, given that its transform is $U = GF$. That question was answered for Fourier transforms by the convolution of g and f, and the answer is the same for Laplace.

> Convolution rule: the convolution $u = g * f = \displaystyle\int_{s=0}^{t} g(t - s)f(s)\, ds$
> is transformed to $U = GF.$

This would look identical to the Fourier convolution if the integral went from $-\infty$ to ∞. In fact it is identical; the difference is that f and g are one-sided. Since $f(s) = 0$ for s below zero and $g(t - s) = 0$ for s above t, all contributions come between 0 and t. We therefore changed to those limits, but the rule and its proof are as before.

In our example f was e^{bt} and the response is

$$u(t) = \int_0^t e^{a(t-s)}e^{bs}ds = e^{at}\left[\frac{e^{(b-a)s}}{b-a}\right]_0^t = \frac{e^{at} - e^{bt}}{a - b}.$$

This convolution rule solves the general equation $u' - au = f(t)$. Taking transforms,

$$U = GF = \frac{1}{s - a}\,F \quad \text{and} \quad u = g * f = \int_0^t e^{a(t-s)}f(s)ds.$$

That is exactly the solution formula of Section 6.1.† Each input $f(s)$ is carried forward by the differential equation—which means it grows like e^{at} but only for the remaining time $t - s$. The solution at time t is the result of all earlier inputs $f(s)$. The inputs are a train of delta functions, or impulses of strength f, and the function $g = e^{at}$ that carries them along is the *fundamental solution*. It is the Green's function of the problem, or the response function. ***The convolution of g with f combines the responses at time t to impulses at all earlier times.***

---

† G was written before F, and g before f, to produce this formula.

EXAMPLE 3 $\dfrac{d^2u}{dt^2} - 5\dfrac{du}{dt} + 6u = 0$ with $u(0) = 1$ and $u'(0) = 9$

$$s[sU(s) - 1] - 9 - 5[sU(s) - 1] + 6U(s) = 0$$

$$(s^2 - 5s + 6)U(s) = s + 4.$$

We used the rule for transforming du/dt and also d^2u/dt^2. The transform U is a ratio of $s + 4$ to $s^2 - 5s + 6$, and after factoring the latter into $(s - 3)(s - 2)$ the problem is ripe for partial fractions:

$$\frac{s + 4}{(s - 3)(s - 2)} = \frac{C}{s - 3} + \frac{D}{s - 2}.$$

Multiply through by $s - 3$ and then set $s = 3$: $C = 7$

Multiply through by $s - 2$ and then set $s = 2$: $D = -6$

The two parts are the transforms of Ce^{3t} and De^{2t}

Solution $u = 7e^{3t} - 6e^{2t}.$

Remark   Before Laplace transforms the solution was found in two steps:

(1)   Substitute $u = e^{\lambda t}$ to find $\lambda^2 - 5\lambda + 6 = 0$ and $\lambda_1 = 3$, $\lambda_2 = 2$
(2)   Choose constants in $u = Ce^{3t} + De^{2t}$ to match u_0 and u_0'

The Laplace transforms must do the same things:

(1)   The poles of U correspond to the exponents in u
(2)   Partial fractions chooses C and D to match u_0 and u_0'

No magic is left. The transform $U(s)$ is described by its poles 3 and 2 (the denominator is zero and U is infinite) and by the residues C and D at those poles. If we put U into the Laplace inversion formula, the resulting u is again $7e^{3t} - 6e^{2t}$.

EXAMPLES 4–5 $\dfrac{du}{dt} - au = e^{at}$ and $\dfrac{d^2u}{dt^2} - 2a\dfrac{du}{dt} + a^2u = 0$

These are the cases we have been avoiding. In the first equation, the driving frequency equals the natural frequency; $b = a$ and we expect resonance. In the second, the exponents in $u = e^{\lambda t}$ satisfy $\lambda^2 - 2a\lambda + a^2 = 0$. Therefore $(\lambda - a)^2 = 0$ and the exponent $\lambda = a$ is a double root. In both cases a is repeated and the earlier solutions break down. We now see this happen in the Laplace transform, where a *double pole* replaces the two simple poles at a and b:

In the first equation, $sU - aU = \dfrac{1}{s-a}$, with $u_0 = 0$

In the second, $s[sU - u_0] - u_0' - 2a[sU - u_0] + a^2 U = 0$.

After the algebra of solving for U, we see the point:

$$U = \frac{1}{(s-a)^2} \quad \text{or} \quad U = \frac{(s-2a)u_0 + u_0'}{(s-a)^2}. \tag{5}$$

The factors $(s-a)$ and $(s-b)$ have come together into $(s-a)^2$, and at this moment we do not know what function u that comes from. But the first transform U is still a product, and the convolution rule still applies:

$$u = g * f = \int_0^t e^{a(t-s)} e^{as} ds = te^{at}. \tag{6}$$

That is the solution to $u' - au = e^{at}$. The resonance between natural frequency and driving frequency produced the extra factor t. This is exactly the limiting case as b approaches a in the old solution (4):

$$\frac{e^{at} - e^{bt}}{a-b} \rightarrow \frac{d}{da} e^{at} = te^{at}.$$

The derivative produced t. Notice also how that factor entered (6); e^{-as} cancelled e^{as} and left the integral of a constant, which was t.

This calculation suggests a general rule: The transform of $t^n e^{at}$ should have $(s-a)^{n+1}$ in the denominator. With $n = 0$ the transform is $1/(s-a)$, a simple pole. With $n = 1$ it is $1/(s-a)^2$, a double pole. For other n the constant $n!$ appears in the numerator. You can see that from an even more general rule, which shows the effect on any transform when u is multiplied by t:

> If $U(s)$ is the Laplace transform of $u(t)$,
> then $-dU/ds$ is the transform of $tu(t)$.

Proof The derivative of $U = \displaystyle\int_0^\infty u e^{-st} dt$ is $\dfrac{dU}{ds} = \displaystyle\int_0^\infty -tu e^{-st} dt$.

That is nearly symmetric with our earliest formula, in which it was u and not U that was differentiated:

$$u \leftrightarrow U \qquad \frac{du}{dt} \leftrightarrow sU - u(0) \qquad tu \leftrightarrow -\frac{dU}{ds}.$$

Differentiation corresponds to multiplication (that is why differential equations correspond to algebraic equations) but the symmetry is incomplete.

In our case differentiating $U = 1/(s-a)^n$ produces $-n/(s-a)^{n+1}$, with a pole that is one order higher. At $a = 0$ this yields the important transform

$$u = t^n \leftrightarrow U = \frac{n!}{s^{n+1}}.$$

We return to finish Example 5.

A double pole in $U = 1/(s-a)^2$ is now settled. The resonant solution is te^{at}. But the other example in (5) also had s in its numerator, and it provides a perfect chance to show how partial fractions are modified at a double pole:

$$U = \frac{(s-2a)u_0 + u_0'}{(s-a)^2} = \frac{C}{s-a} + \frac{D}{(s-a)^2}. \tag{7}$$

Again this is like the method of undetermined coefficients, or matching the solution to initial conditions. We *know* that for some constants C and D equation (7) is possible, just as we know that for some constants the solution is $u = Ce^{at} + Dte^{at}$. In fact, they are the same constants! To find them, multiply through by $(s-a)^2$ and then set $s = a: D = -au_0 + u_0'$. Remove D from the numerator of U leaving $(s-a)u_0/(s-a)^2$ or $C = u_0$. The two parts of U are the transforms of Ce^{at} and Dte^{at}.

Solution $\qquad\qquad\qquad u = u_0 e^{at} + [-au_0 + u_0']te^{at}. \tag{8}$

When $a = 0$ this is $u_0 + u_0't$, the solution to $u'' = 0$.

That completes the examples. We did not emphasize, as we should, that the exponents a and b are not necessarily real. They can be complex, and in particular they can be pure imaginary. That is the oscillating case, when it was correct to speak about frequencies. For real exponents it is natural decay versus a decaying source, or natural growth versus forced growth. In the imaginary case we have not only $e^{i\omega t}$ but its sine and cosine:

$$u = e^{i\omega t} \quad \leftrightarrow U = \frac{1}{s - i\omega} \qquad \text{(with } a = i\omega)$$

$$u = \cos \omega t \leftrightarrow U = \frac{s}{s^2 + \omega^2} \qquad \text{(real part)}$$

$$u = \sin \omega t \leftrightarrow U = \frac{\omega}{s^2 + \omega^2} \qquad \text{(imaginary part)}$$

Of course there are the free and forced and damped and undamped oscillations that we studied before, but perhaps you don't want to see them again.

Applications and Table of Transforms

The remark above, that inputs can be exponentials like $e^{i\omega t}$ and not necessarily like e^{-t}, implies a corresponding fact about s. *The transform variable s is by no means limited to real numbers.* It is a "*complex frequency*." The Laplace transform is studied throughout the complex plane. It is defined in at least half of that plane, to the right of a line Re $s = s_0$. When the real part is large enough, ue^{-st} goes to zero.

The table below gives a more complete list, including the transform of a delta function. We separate the table of specific transforms from the table of operational rules:

Table of Laplace Transforms

$u = \delta$	$\leftrightarrow$	$U = 1$
$u = 1$	$\leftrightarrow$	$U = 1/s$
$u = t^n$	$\leftrightarrow$	$U = n!/s^{n+1}$
$u = e^{at}$	$\leftrightarrow$	$U = 1/(s-a)$
$u = t^n e^{at}$	$\leftrightarrow$	$U = n!/(s-a)^{n+1}$
$u = \cos \omega t$	$\leftrightarrow$	$U = s/(s^2 + \omega^2)$
$u = \sin \omega t$	$\leftrightarrow$	$U = \omega/(s^2 + \omega^2)$
$u = e^{-at} \cos \omega t$	$\leftrightarrow$	$U = \dfrac{s+a}{(s+a)^2 + \omega^2}$
$u = e^{-at} \sin \omega t$	$\leftrightarrow$	$U = \dfrac{\omega}{(s+a)^2 + \omega^2}$

Table of Transform Rules

Derivative du/dt	$\leftrightarrow$	$sU(s) - u(0)$
2nd derivative d^2u/dt^2	$\leftrightarrow$	$s[sU(s) - u(0)] - u'(0)$
Integral $\int_0^t u$	$\leftrightarrow$	U/s
Multiplication tu	$\leftrightarrow$	$-dU/ds$
Delay (or shift) $u(t-T)$	$\leftrightarrow$	$Ue^{-sT}, \quad T > 0$
Transform shift $e^{at}u$	$\leftrightarrow$	$U(s-a)$
Scaling $u(kt)$	$\leftrightarrow$	$k^{-1}U(s/k)$
Convolution $g*f$	$\leftrightarrow$	$G(s)F(s)$

The delta function is easy. It picks out the value of e^{-st} at $t = 0$ and $\int \delta(t)e^{-st}dt = 1$. A delayed delta function $\delta(t - T)$ similarly picks out e^{-sT}. When other functions are delayed, they also have transforms multiplied by e^{-sT}:

$$\int_0^\infty u(t - T)e^{-st}dt = \int_{-T}^\infty u(x)e^{-s(x + T)}dx = U(s)e^{-sT}.$$

Notice why this requires $T > 0$. When the change to $x = t - T$ moves the lower limit below zero, nothing is new; one-sided functions have $u = 0$ from $-T$ to 0. In contrast, a value like $T = -2$ would start the integral at $x = +2$. Part of u would be chopped off, and what is left has an entirely different transform.

Perhaps the most important property should be mentioned separately:

$$Linearity \quad u + v \leftrightarrow U + V$$

Often $u_{particular}$ (at the driving frequency) is added to $u_{homogeneous}$ (at the natural frequency). Or u may come from one driving force and v from another. Linearity is a powerful property and a powerful limitation.

A final remark. Second order equations have two exponents,

$$mu'' + cu' + ku = f \leftrightarrow (ms^2 + cs + k)U = F.$$

The transform has $ms^2 + cs + k$ in its denominator. The roots of this quadratic are the poles of G, just as they are the exponents λ_1 and λ_2 in u. If those poles are not real, they are conjugate complex numbers $a \pm i\omega$. We have a choice, in applying partial fractions to U and in the final expression for u, whether to separate those roots or not. Take the harmonic motion $u'' + u = \delta$ that starts from $u_0 = u'_0 = 0$ with an impulse:

$$(s^2 + 1)U = 1 \quad \text{or} \quad U = \frac{1}{s^2 + 1}.$$

If we separate into linear factors, then

$$\frac{1}{s^2 + 1} = \frac{i/2}{s + i} - \frac{i/2}{s - i} \quad \text{and} \quad u = \frac{i}{2}e^{-it} - \frac{i}{2}e^{it}. \tag{10}$$

If we keep all calculations real, then

$$U = \frac{1}{s^2 + 1} \rightarrow u = \sin t \quad \text{from the table.}$$

The result is the same, and Exercise 6.3.4 is a further example.

The correspondence between poles and exponents is fundamental. It appears inside and outside of engineering. For the scattering matrix in physics the position of the poles is everything. In electrical networks it is the **system function** $G(s)$ whose poles indicate the natural frequencies. It contains total information about the input-output behavior of the system:

transform variable s: (output U) = (system function G)(input F)

time variable t: (output u) = (fundamental soln. g)∗(input f)

In network theory G is a "transfer function" if input and output are measured at different terminals. It is a "driving-point impedance" if input is current and output is voltage, at the same terminals. It is an "admittance" if it is output current divided by input voltage. In every case it comes by transforming the differential equation— or equivalently by substituting $u = G(s)e^{st}$ for the output, when the input is e^{st}.

The result is an algebraic equation in s, easier to manipulate than a differential equation in t. Of course even polynomials are not simple beyond degree 2 or 3. That is the range in which transform techniques are pushed all the way to analytical solutions. Beyond that point the calculation of partial fractions, or of eigenvalues in the matrix case, becomes inhuman. The real value is the insight into the design and analysis of larger systems.

We cite one example among many. It will be in the language of circuits, although it has analogies in acoustics, hydrodynamics, mechanics, biology, thermal conduction, ... It appears equally in economics, where the input of a new tax or invention has output in employment or productivity or income. Those outputs become the input in social sciences, where the system is even more delicate and depends strongly on the loop design that we now discuss. It is a **feedback loop**, intended to reduce distortion and control the system.

The loop is sketched in Fig. 6.17. The sign is chosen so that the input to the upper box is $F - BU$. The output from the box is

$$U = K(F - BU) \quad \text{or} \quad U = \frac{K}{1 + BK} F. \qquad (9)$$

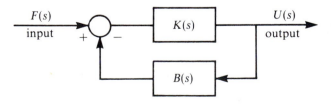

Fig. 6.17. A feedback loop.

In other words, the system function is $G = K/(1 + BK)$. That is the output-input ratio, and it is controlled by BK:

$$\text{if}\quad BK \ll 1 \quad \text{then}\quad G \approx K$$

$$\text{if}\quad BK \gg 1 \quad \text{then}\quad G \approx \frac{1}{B}.$$

The K block is an active element, like an amplifier. Standing alone it might be designed with gain $K = 10$. With feedback $B = 1/10$ we can add a pre-amplifier to have $K = 1000$, and the system function is hardly changed:

$$G = \frac{1000}{1 + \dfrac{1}{10}1000} \approx 9.9.$$

Furthermore the new system is insensitive to aging and degeneration in the amplifier. If K is reduced to 500, G slips only to 9.8. Would that a similar feedback were available to authors.

This feedback was negative, reducing the input to K and increasing the control. There is also *positive feedback*, in which the signal BK reinforces the input. The system function becomes $G = K/(1 - BK)$ and the minus sign changes the behavior:

$$\text{if}\quad BK \ll 1 \quad \text{then } G \approx K$$

$$\text{if}\quad BK \to 1 \quad \text{then } G \to \infty$$

That is desirable in detecting fluctuations (and encouraging friends) while negative feedback controls them. The point to emphasize is the simplicity of the picture in frequency space. A similar analysis in the time domain, with coupled differential equations, would be far less revealing.

To conclude this discussion we contrast active elements like amplifiers with the *passive RLC* elements studied previously. Those elements absorb power. Mathematically that is reflected in a stable differential equation, and in a "positive-real impedance:"

$$\operatorname{Re}\frac{U(s)}{F(s)} \geq 0 \quad \text{for}\quad \operatorname{Re} s \geq 0.$$

The poles of a passive system are on the left side of the imaginary axis. The marginal case is an LC circuit, which is lossless; there is no resistance and no damping. Its poles are *on* the imaginary axis and it is neutrally stable. For partial differential equations (distributed systems instead of lumped systems) that corresponds to hyperbolic equations, or wave equations. The dissipation of energy in a resistor or dashpot corresponds to parabolic equations, or diffusion equations.

Discrete Time and the z-Transform

For the Laplace transform time is continuous. It solves differential equations; it could solve integral equations. However it does not apply to discrete sequences $u_0, u_1, u_2, \ldots$. Those *digital signals* are no less important, and we need a transformation that can detect growth and decay factors in discrete time. Instead of exponentials e^{st} it will be built around powers z^n.

This new and closely parallel method is the *z-transform*. It solves difference equations rather than differential equations. The equations jump forward in time, like $u_{n+1} = A u_n$, rather than accumulating gradually like $u' = au$. In economics, it is a comparison of steady growth (continuous compounding) with interest paid at fixed times. Note that an 8% interest rate corresponds to $a = 0.08$ but in the difference equation it is $A = 1.08$. After a year the principal is multiplied by A, which is close to but less than $e^{0.08}$. In a "no growth" situation we have:

differential equation		difference equation
$a = 0$	$\leftrightarrow$	$A = 1$
$\text{Re } a = 0$	$\leftrightarrow$	$\lvert A \rvert = 1$

Stated more concisely, $e^0 = 1$ and $\lvert e^{i\theta} \rvert = 1$.

The simplest models—first-order difference and differential equations—are easy to solve:

$$u_{n+1} = A u_n \text{ leads to } u_n = A^n u_0; \quad u' = au \text{ leads to } u = e^{at} u_0.$$

Powers of A correspond to exponentials of a. The continuous variable $t \geq 0$ becomes the discrete variable $n \geq 0$. And the Laplace transform, made discrete, becomes the z-transform:

6E   The z-transform $U(z)$ of a discrete sequence u_n is

$$U(z) = \sum_{n=0}^{\infty} u_n z^{-n} \text{ in analogy with } U(s) = \int_0^{\infty} u(t) e^{-st} dt.$$

We emphasize the analogy by using the same notation U. The difference between t and n reflected in the difference between s and z.

EXAMPLE 1   The unit step sequence $u_n = 1$ for $n \geq 0$

The transform is $U = 1 + z^{-1} + z^{-2} + \cdots = \left(1 - \dfrac{1}{z}\right)^{-1} = \dfrac{z}{z-1}$.

EXAMPLE 2 The geometric sequence $u_n = A^n$

The transform is $U = 1 + A/z + A^2/z^2 + \cdots = \left(1 - \dfrac{A}{z}\right)^{-1} = \dfrac{z}{z - A}$.

Note that $U(z)$ has a pole at $z = A$. The poles in U correspond to the growth factors in the discrete sequence u_n, just as they did to the exponents in $u(t)$. The factor A can be a real number or a complex number; it is the *magnitude* of A that decides between growth or decay:

$$\text{Re } a > 0 \text{ corresponds to } |A| > 1: \text{ unstable growth}$$

$$\text{Re } a < 0 \text{ corresponds to } |A| < 1: \text{ stable decay}$$

Stability in a discrete system is reflected in a z-transform with its poles in the disk $|z| < 1$. Stability in a continuous system produces a Laplace transform with its poles in the half-plane Re $s < 0$.

EXAMPLE 3 The inverse transform of $U = \dfrac{z^2}{z^2 - \frac{1}{4}}$

The denominator is zero at $z = \pm\frac{1}{2}$; those are the poles. As with Laplace transforms, we can separate U into a product and use the convolution rule, or into a sum and use linearity. The latter sounds easier:

$$\text{Sum:} \qquad \frac{z^2}{z^2 - \frac{1}{4}} = \frac{Cz}{z - \frac{1}{2}} + \frac{Dz}{z + \frac{1}{2}}$$

Multiply through by $z - \frac{1}{2}$ and then set $z = \frac{1}{2}$: $C = \frac{1}{2}$

Multiply through by $z + \frac{1}{2}$ and then set $z = -\frac{1}{2}$: $D = \frac{1}{2}$

The two terms are the transforms of $C(\frac{1}{2})^n$ and $D(-\frac{1}{2})^n$

Solution $\qquad u_n = \frac{1}{2}[(\frac{1}{2})^n + (-\frac{1}{2})^n] = \{1, 0, \frac{1}{4}, 0, \frac{1}{16}, \ldots\}$

Of course that was partial fractions again. But convolution is important too:

$$\text{Product:} \qquad \frac{z^2}{z^2 - \frac{1}{4}} = \left(\frac{z}{z - \frac{1}{2}}\right)\left(\frac{z}{z + \frac{1}{2}}\right) \quad \text{or } u = v * w$$

Multiplication VW of transforms is convolution $v * w$ of sequences. The poles are $\frac{1}{2}$ and $-\frac{1}{2}$, and the sequences are $v_n = (\frac{1}{2})^n$, $w_n = (-\frac{1}{2})^n$. All we have to do is convolve them:

$$u_n = v_0 w_n + v_1 w_{n-1} + v_2 w_{n-2} + \cdots + v_n w_0$$

$$= (-\tfrac{1}{2})^n + (\tfrac{1}{2})(-\tfrac{1}{2})^{n-1} + (\tfrac{1}{2})^2(-\tfrac{1}{2})^{n-2} + \cdots + (\tfrac{1}{2})^n.$$

There are $n + 1$ terms in the sum, all the same except for alternating sign. If n is odd, they cancel to leave $u_n = 0$. If n is even, one term is left and $u_n = 1/2^n$. That agrees, by a miracle, with the result from the sum method.

With $z - A$ in the denominator the sequence has one growth factor A. With a quadratic in the denominator of U, this example had two growth factors $\frac{1}{2}$ and $-\frac{1}{2}$. When there are k poles there are k growth factors, and this corresponds to a difference equation of order k. That equation can be solved by z-transforms just as constant-coefficient differential equations were solved by Laplace transforms. The calculations can be awkward—that is why we appealed to a miracle—but in principle they are possible. For the example above it is not hard to guess the equation:

$$u_{n+2} - \tfrac{1}{4} u_n = 0 \quad \text{with} \quad u_0 = 1, \quad u_1 = 0. \tag{11}$$

For odd n the solution starts at zero ($u_1 = 0$) and stays zero. For even n the solution is multiplied at every double step by $\frac{1}{4}$. Thus $u_0 = 1, u_2 = \frac{1}{4}, u_4 = \frac{1}{16}, \ldots$, which is the predicted solution.

If that solution were not known, and (11) were presented as a typical second-order difference equation, there would be three ways to solve for u_n:

(1)   Substitute $u_n = A^n$ into the difference equation, find the two roots from $A^{n+2} = \frac{1}{4} A^n$ or $A^2 = \frac{1}{4}$, and match the initial conditions by $u_n = C(\frac{1}{2})^n + D(-\frac{1}{2})^n$.

(2)   Convert the second-order equation $u_{n+2} = \frac{1}{4} u_n$ into two first-order equations; $u_{n+1} = v_n$ and $v_{n+1} = \frac{1}{4} u_n$. This is a system

$$\begin{bmatrix} u \\ v \end{bmatrix}_{n+1} = \begin{bmatrix} 0 & 1 \\ \frac{1}{4} & 0 \end{bmatrix} \begin{bmatrix} u \\ v \end{bmatrix}_n.$$

The eigenvalues are $\frac{1}{2}$ and $-\frac{1}{2}$, controlling the powers.

Note   The analogous method introduced u' as unknown in reducing to a system of differential equations. There must be a "phase plane" for difference equations, but instead of traveling on curves the equation moves by jumps.

(3)   Take z-transforms of the difference equation, solve for $U(z)$, and transform back.

This third method requires the transform of the shifted sequence $v_n = u_{n+1}$, and the doubly-shifted $w_n = u_{n+2}$, just as the continuous case required the transforms of du/dt and d^2u/dt^2. The rules are sure to be similar:

$$V(z) = \sum_0^\infty \frac{v_n}{z^n} = \sum_0^\infty \frac{u_{n+1}}{z^n} = \sum_1^\infty \frac{u_k}{z^{k-1}} = z[U(z) - u_0]$$

$$W(z) = z[V(z) - v_0] = z^2 U(z) - z^2 u_0 - z u_1$$

With this we can transform the difference equation (11):

$$z^2 U(z) - z^2 - \tfrac{1}{4} U(z) = 0 \quad \text{or} \quad U = \frac{z^2}{z^2 - \tfrac{1}{4}}.$$

That is the transform we started with. Its inverse transform is the solution $1, 0, \tfrac{1}{4}, 0, \tfrac{1}{16}, \ldots$, which is now totally and excessively confirmed.

Transfer Functions: A Final Look

In all constant-coefficient problems we have a choice—to work in *state space* where the unknown is the state variable u, or to work in *transform space* with the unknown U. In state space there is a differential equation or a difference equation. In transform space, if everything is right, there is multiplication or division by polynomials. The link between them is still the convolution rule.

There is something new about initial-value problems, not seen in the doubly-infinite matrices of Section 4.2. It is *causality*. Time starts at zero, not at $-\infty$. Inputs at time t affect outputs at later times, not at earlier times. Stability is tested as time goes forward, not backward. Everything about the problem is one-sided, including the transforms themselves; Laplace transforms and z-transforms produce one-sided transfer functions. We begin with an example from economics and control theory, allowing what has been excluded up to now—a difference equation with a forcing term on the right hand side.

EXAMPLE The cobweb model for supply and demand

As the price increases, demand goes down and supply goes up:

$$d = d_0 - ap \quad \text{and} \quad s = s_0 + bp.$$

At equilibrium, demand equals supply (Fig. 6.18a) and the price satisfies

$$d_0 - ap = s_0 + bp, \text{ or } p = \frac{d_0 - s_0}{a + b}.$$

The question is whether this equilibrium will or will not be reached, if it takes a year for supply to react to price:

$$d_{n+1} = d_0 - ap_{n+1} \text{ (immediate)} \quad \text{but} \quad s_{n+1} = s_0 + bp_n \text{ (delayed)}.$$

We impose that the demand d_{n+1} still equals the supply s_{n+1}:

$$ap_{n+1} + bp_n = d_0 - s_0. \tag{12}$$

This can be solved by finding the transform $P(z)$. It can be solved more quickly by recognizing $(-b/a)^n$ as solution when the right side is zero; the homogeneous equation is $p_{n+1} = (-b/a)p_n$ and the growth factor is $-b/a$. Then the full solution is

$$p_{\text{homogeneous}} + p_{\text{particular}} = C\left(-\frac{b}{a}\right)^n + D.$$

The convergence of the price to its equilibrium value D **depends on the growth factor**; $(-b/a)^n$ goes to zero when $b < a$. Geometrically,

if $b < a$ the cobweb spirals in to equilibrium

if $b > a$ the cobweb spirals out to inflation-depression.

The cobweb starts with a price p_0. The supply at year 1 is $s_1 = s_0 + bp_0$, on the supply curve. Then the demand is d_1, at the same height on the demand curve. That produces a new price p_1 which leads to a new supply s_2. The spiral continues and if $b < a$ it converges; the producers have to be less sensitive to price changes than the consumers. If the producers are *more* sensitive, and $b > a$, the arrows in Fig. 6.18 go the other way toward disaster.

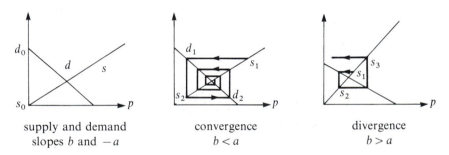

supply and demand   convergence   divergence
slopes b and $-a$ $b < a$ $b > a$

Fig. 6.18. The cobweb model for supply and demand.

EXAMPLE CONTINUED   Economy controlled from above

Suppose the price is set by the government, not by the hard laws of economic equilibrium. This becomes a problem in **control theory**, to set prices so that supply equals demand. The inputs are prices p_n, the outputs are differences $y_n = s_n - d_n$, the controls are b_1, b_2, and the "state equation" is

$$\begin{bmatrix} s \\ d \end{bmatrix}_{n+1} = A \begin{bmatrix} s \\ d \end{bmatrix}_n + \begin{bmatrix} b_1 \\ b_2 \end{bmatrix} p_n. \tag{13}$$

The government aims for $y = 0$ by *regulating* the price in an optimal way, instead of leaving the market to determine it. We give no detail, except to point to the z-transform of (13):

$$\begin{bmatrix} zS(z) \\ zD(z) \end{bmatrix} = A \begin{bmatrix} S(z) \\ D(z) \end{bmatrix} + bP(z), \quad \text{or} \quad \begin{bmatrix} S \\ D \end{bmatrix} = (zI - A)^{-1}bP.$$

The transform of y is then $Y = c(zI - A)^{-1}bP$, where $c = [1 \quad -1]$ because $y_n = s_n - d_n$. This function $H = c(zI - A)^{-1}b$ is the **transfer function** that connects input to output, and it is the key to an optimal control—but this example has already gone further than we intended.

Remark $H(z)$ is also the transform of the sequence y_n generated by a unit impulse $p_0 = 1$. In other words, **the transfer function is the z-transform of the impulse response**.

I want to leave control theory at that point and give a more direct explanation of transfer functions. It is easiest to do for matrices. They are the natural models for linear systems, because a matrix has linearity built into it. In matrix multiplication, the output depends linearly on the input. For initial-value problems the matrices are singly infinite—the input and output start at $n = 0$—and a typical example is

$$A = \begin{bmatrix} a & & & \\ b & a & & \\ 0 & b & a & \\ 0 & 0 & b & a \\ \cdot & \cdot & \cdot & \cdot & \cdot \end{bmatrix}.$$

We go immediately to the essential properties of A:

$$\textit{linearity:} \quad \text{automatic for matrices}$$

$$\textit{time-invariance:} \quad A \text{ has constant diagonals}$$

$$\textit{causality:} \quad A \text{ is lower triangular}$$

The system is time-invariant because the response to $u_0 = 1$ and the response to $u_k = 1$ are the same—except that the response is shifted in time:

$$\text{if } u = (1,0,0, \ldots) \text{ then } Au = (a,b,0,0, \ldots)$$

$$\text{if } u = (0,1,0, \ldots) \text{ then } Au = (0,a,b,0, \ldots).$$

A is a Toeplitz matrix, or discrete convolution matrix, or constant-diagonal matrix. Furthermore it is causal: The response Au does not start before u starts. It is time-invariance that makes $Au = f$ solvable by transform methods, and it is causality that picks out a one-sided transform. For differential equations and continuous convolutions (integral equations) it is the Laplace transform. Here it is the z-transform, for the difference equation $Au = f$:

$$
\begin{bmatrix}
a & & & \\
b & a & & \\
& b & a & \\
& & b & a \\
& & & . & .
\end{bmatrix}
\begin{bmatrix}
u_0 \\
u_1 \\
u_2 \\
u_3 \\
.
\end{bmatrix}
=
\begin{bmatrix}
f_0 \\
f_1 \\
f_2 \\
f_3 \\
.
\end{bmatrix}
\quad \text{is } au_n + bu_{n-1} = f_n. \tag{14}
$$

That is the equation derived earlier for prices. The triangular matrix carries us forward in time, and the right side f contains not only the forcing function but also the initial values. The first line of (14) is $au_0 = f_0$, specifying u_0. The next line gives u_1, and each succeeding line gives one more component of u.

This matrix has a further property, impossible to miss. It is a *band matrix*. The system $Au = f$ is a finite difference equation, and with bandwidth 1 it is a first-order equation. The effect will be that the transfer function is a first-order polynomial. This simplifies the algebra, but it does not decide between a good problem and a bad. That is the essential decision, and it depends on a fourth and final property—which A may or may not possess:

stability: A has a bounded inverse

In a well-posed problem, A^{-1} is time-invariant and causal and bounded. It is like A itself, except it is not banded. The equation $Au = f$ can be solved, and its solution decays if f decays. In a badly-posed problem, $u = A^{-1}f$ is unbounded even when $f = (1, 0, 0, \ldots)$ is a single impulse.

The property of stability depends on the numbers a and b. In our example it can be seen directly from the inverse:

$$
\begin{bmatrix}
a & & & \\
b & a & & \\
& b & a & \\
& & b & a \\
& & & . & .
\end{bmatrix}
\begin{bmatrix}
1/a & & & \\
-b/a^2 & 1/a & & \\
b^2/a^3 & -b/a^2 & 1/a & \\
-b^3/a^4 & b^2/a^3 & -b/a^2 & 1/a \\
. & & & . & .
\end{bmatrix}
= I. \tag{15}
$$

The diagonals of A^{-1} are constant. They are growing by the same factor $-b/a$ that appears in the solution of $au_n + bu_{n-1} = f_n$. There is stability if $|b/a| < 1$, neutral stability if $|b/a| = 1$, and instability if $|b/a| > 1$. We look now for a pattern that applies to all banded matrices A.

The first column of A^{-1} is the *impulse response*. When f is the vector $(1, 0, 0, \ldots)$, the multiplication $A^{-1}f$ produces that column. It contains complete information about A^{-1}, because of time-invariance. The other columns are shifted forward without change. Thus the computation of A^{-1} and the verdict on stability depend on finding that first column—which comes from the multiplication

$$\left(a + \frac{b}{z}\right)\left(\frac{1}{a} - \frac{b}{a^2 z} + \frac{b^2}{a^3 z} \cdots\right) = 1. \tag{16}$$

That is the key equation. The first factor reflects the entries in A; the second factor reflects A^{-1}. More than that, those factors are z-transforms:

$$au_n + bu_{n-1} = f_n \text{ transforms to } \left(a + \frac{b}{z}\right) U(z) = F(z)$$

$$\text{the solution transforms to } U = \left(a + \frac{b}{z}\right)^{-1} F$$

$$\text{the transfer function is } \left(a + \frac{b}{z}\right)^{-1} = \frac{1}{a} - \frac{b}{a^2 z} + \frac{b^2}{a^3 z} \cdots$$

$$\text{the solution is } u_n = \frac{1}{a} f_n - \frac{b}{a^2} f_{n-1} + \frac{b^2}{a^3} f_{n-1} + \frac{b^2}{a^3} f_{n-2} \cdots \text{ or } u = A^{-1}f.$$

To invert A we expand $(a + b/z)^{-1}$. Its pole is at the growth factor $z = -b/a$. Therefore stability of the difference equation and stability of the matrix both depend on the poles of the transfer function:

6F If the singly-infinite lower triangular banded matrix A has constant entries a_k on its diagonals, then the transfer function for A^{-1} is

$$\frac{1}{a_0 + a_1/z + \cdots + a_N/z^N} = g_0 + g_1/z + g_2/z^2 + \cdots \tag{17}$$

The numbers g_k lie on the diagonals of A^{-1}, and they decay to zero when A satisfies the *stability condition*:

the transfer function (17) must be bounded for $|z| \geq 1$.

All poles must lie inside the unit circle, and A^{-1} is lower triangular.

EXERCISES

6.3.1 Find the Laplace transform and its poles for

(a) $u = 1 + t$ (d) $u = \cos^2 t$

(b) $u = t \cos \omega t$ (e) $u = 1 - e^{-t}$

(c) $u = \cos(\omega t - \theta)$ (f) $u = te^{-t} \sin \omega t$

(g) $u = 1$ for $t \le 1$, $u = 0$ elsewhere   (i) $u =$ next integer above t

(h) $u = 2$ for $1 \le t \le 2$, $u = 0$ elsewhere   (j) $u = t\delta(t)$

6.3.2   Find the function $u(t)$ if $U(s)$ equals

(a) $\dfrac{1}{s - 2\pi i}$ (f) $\dfrac{s + 1}{s^2 + 1}$

(b) $\dfrac{1}{(s - 1)(s - 2)}$ (g) $\left(\dfrac{1}{s - 1}\right)\left(\dfrac{1}{s - 2}\right)$ by convolution

(c) $\dfrac{6}{s^2 + 3s}$ (h) e^{-s}

(d) $\dfrac{s}{s - 1}$ (i) $e^{-s}/(s - a)$

(e) $\dfrac{1}{s^2 + 2s + 2}$ (j) s

6.3.3   Solve $u'' + u = 0$ with initial values u_0 and u_0': (i) find the transform $U(s)$; (ii) express it as a combination of $s/(s^2 + 1)$ and $1/(s^2 + 1)$; (iii) find the inverse transform u from the table.

6.3.4   Solve $u'' + 2u' + 2u = \delta$ starting from $u_0 = 0$, $u_0' = 1$ by transforming to find U and then

(a)   finding the poles and linear factors as in equation (10)
(b)   searching directly in the table for u.

6.3.5   Solve the following initial-value problems by Laplace transform:

(a) $u' + u = e^{i\omega t}$, $u_0 = 8$
(b) $u' - i\omega u = \delta(t)$, $u_0 = 0$
(c) $u' + u = e^{-t}$, $u_0 = 2$
(d) $u'' + u = 6t$, $u_0 = u_0' = 0$
(e) $u'' - u = e^t$, $u_0 = u_0' = 0$
(f) $mu'' + cu' + ku = 0$, $u_0 = 1$, $u_0' = 0$

6.3.6   Show that a system function like $G = 1/(s^2 + s + 1)$ is positive-real: Re $G \ge 0$ when Re $s \ge 0$.

6.3.7   The Laplace transform of a *matrix* exponential e^{At} is

$$U(s) = \int_0^\infty e^{At}e^{-st}dt = [(A - sI)^{-1}e^{(A - sI)t}]_0^\infty = (sI - A)^{-1}.$$

(a)   Compute $e^{At} = I + At + A^2t^2/2! + \cdots$ for $A = 2I$ and $A = \dfrac{1}{2}\begin{bmatrix} 1 & 1 \\ 1 & 1 \end{bmatrix}$

(b)   Compute $U = (sI - A)^{-1}$ for those two matrices
(c)   Why do the poles of U agree with the eigenvalues of A?

Remark   The matrix exponential solves a first-order system $u' = Au$. The solution vector is $e^{At}u_0$ and its transform is $(sI - A)^{-1}u_0$. If A is diagonalized to give $A = S\Lambda S^{-1}$ with eigenvalues in Λ, then $e^{At} = Se^{\Lambda t}S^{-1}$ is an alternative to the sum $I + At + A^2t^2/2! + \cdots$.

For the next eight exercises as well as network applications in the text I am indebted to William M. Siebert's new book *Circuits, Signals, and Systems*, M.I.T. Press and McGraw-Hill, 1986.

6.3.8 If du/dt decays exponentially, use its transform to show that

(i) $sU(s) \to u(0)$ as $s \to \infty$

(ii) $sU(s) \to u(\infty)$ as $s \to 0$.

6.3.9 Transform Bessel's time-varying equation $tu'' + u' + tu = 0$ to find a first-order equation for U. By separating variables or by direct substitution find $U = C/\sqrt{1+s^2}$, the Laplace transform of the Bessel function $J_0(t)$.

6.3.10 Find the Laplace transforms of (a) a single arch of $u = \sin \pi t$ and (b) a single ramp $u = t$. First graph both functions, which are zero beyond $t = 1$.

6.3.11 Find the Laplace transforms of (a) the rectified sine wave $u = |\sin \pi t|$ and (b) the sawtooth function $S(t) =$ fractional part of t. This is the previous exercise extended periodically to all positive t; use the shift rule and the sum $1 + x + x^2 + \cdots = (1-x)^{-1}$.

6.3.12 Suppose your acceleration dv/dt on the highway depends on the velocity v^* of the car ahead: $v' = c(v^* - v)$.

(a) Find the ratio of Laplace transforms $V^*(s)/V(s)$
(b) If that car has $v^* = t$ find $v(t)$ starting from rest.

6.3.13 For a line of cars with velocities $v_0, v_1, v_2, \ldots$ let

$$v_n' = c[v_{n-1}(t-T) - v_n(t-T)], \quad v_0(t) = \cos \omega t.$$

(a) Substitute $v_n = A^n e^{i\omega t}$ to find the amplitude factor $A = (1 + i\omega e^{i\omega T}/c)^{-1}$ in steady oscillation.
(b) Show that $|A| < 1$ and the amplitudes are decreasing if $cT < \frac{1}{2}$.
(c) If $cT > \frac{1}{2}$ show that $|A| > 1$ for small frequencies ω. (Use $\sin \theta < \theta$.) Human reaction times are $T = 1$ or 2 seconds, and typical aggressiveness is $c = 0.4/\text{sec}$.—so danger is pretty close. Probably drivers have adjusted to be barely safe.

6.3.14 The *Pontryagin maximum principle* says that the optimal control is "bang-bang"—it takes on only the extreme values permitted by the constraints.

(a) With maximum acceleration A and maximum deceleration $-A$, how should you get from rest at $x = 0$ to rest at $x = 1$ in the minimum possible time?
(b) If the maximum braking is changed to $-B$, what is the optimal velocity $v(t)$ and the minimum time?

6.3.15 (a) Transform $u'' + 3u' + 2u = f(t)$ to find $U(s)$, and recover $u(t)$ if $u_0 = 1$, $u_0' = 2$, $f = 0$.
(b) Which constant control f will eliminate the e^{-t} decay and (with $u_0 = 1$, $u_0' = 2$) leave an e^{-2t} decay to the steady state $\frac{1}{2}f$?

6.3.16 Prepare a table of z-transforms for the four discrete functions 1, a^n, δ, n, and a second table of z-transform rules: forward shift to $u(n+1)$, delay to $u(n-1)$, multiplication by n, multiplication by a^n.

6.3.17   Solve the following difference equations by the z-transform:

(a) $u_{n+1} - 2u_n = 0$, $u_0 = 5$
(b) $u_{n+1} - u_n = 2^n$, $u_0 = 0$
(c) $u_{n+2} - 3u_{n+1} + 2u_n = 0$, $u_0 = 1, u_1 = 0$
(d) $u_{n+1} - nu_n - u_n = 0$, $u_0 = 1$

6.3.18   The eigenvalues of $\begin{bmatrix} a & b \\ c & d \end{bmatrix}$ satisfy

$$\lambda^2 - (a + d)\lambda + ad - bc = 0, \quad \text{or} \quad \lambda^2 - (\text{trace})\lambda + (\text{determinant}) = 0.$$

Show that if $|\lambda_1| < 1$ and $|\lambda_2| < 1$ then $|\text{trace}| < 2$ and $|\text{determinant}| < 1$. Draw the stability diagram for difference equations that corresponds to Fig. 6.7 for differential equations.

6.3.19   Show that $p_{n+1} - Ap_n = f_{n+1}$, with initial condition $p_0 = 0$, has the solution

$$p_n = \sum_{k=1}^{n} A^{n-k} f_k.$$ What is the analogous solution to $u' - au = f(t)$ from Section 6.1? Notice how f_k is carried through the remaining $n - k$ steps by A^{n-k}.

6.3.20   Suppose you start gambling with k chips, the house has $N - k$, and at each play you have a 5/11 chance of winning a chip. What is your probability u_k of breaking the bank before it breaks you? Certainly $u_0 = 0$—you have no chance with no chips—and $u_N = 1$.

(a)   Why does $u_k = \frac{5}{11} u_{k+1} + \frac{6}{11} u_{k-1}$?
(b)   Find λ in the general solution $u_k = C\lambda^k + D$
(c)   Choose C and D to match u_0 and u_N
(d)   If you start with $k = 100$ out of $N = 1000$ chips, your chance is approximately zero: $u_k \approx (5/6)^{900}$. Is it better to start with 1 superchip out of $N = 10$?

6.3.21   (Genetics) The frequency of a recessive gene in generations k and $k+1$ satisfies $u_{k+1} = u_k/(1 + u_k)$ if receiving two such genes prevents reproduction.

(a)   Verify that $u_k = u_0/(1 + ku_0)$ satisfies the equation
(b)   If the original frequency is $u_0 = \frac{1}{2}$, which generation has $u_k = \frac{1}{100}$?
(c)   Derive the solution in (a) by rewriting $u_{k+1} = u_k/(1 + u_k)$ as a simpler equation for $v_k = 1/u_k$.

6.3.22   A nonlinear difference equation $u_{n+1} = G(u_n)$ corresponds to a nonlinear differential equation $u' = F(u)$; the unknowns u are vectors.

(a)   Stationary points in $u' = F(u)$ satisfy $F(u^*) = 0$. What is the corresponding condition for u^* to be a steady state solution of the difference equation?
(b)   The derivative matrix $A = \partial F/\partial u$ gave linearized stability around u^* if its eigenvalues had $\text{Re } \lambda < 0$. What is the analogous condition on the matrix $B = \partial G/\partial u$ for linearized stability in the difference equation?
(c)   What are the stationary points $u^* = (v^*, w^*)$ and their stability for $v_{n+1} = w_n$, $w_{n+1} = \sin v_n$?

6.3.23   Transform the scalar control system

$$s_{k+1} = as_k + bp_k, \quad y_k = cs_k$$

to show that $Y(z) = [bc/(z - a)]P(z)$. What sequence has this transform $H = bc/(z - a)$? Compare with the responses y_k to a unit impulse $p_0 = 1$.

6.3.24   A system like (13) is *controllable* (or completely reachable) if with the right inputs it can reach any state (s,d) in finite time from any other state (s_0,d_0). Show that if b is an eigenvector of A the system is not controllable. When A is n by n, controllability requires the vectors $b, Ab, \ldots, A^{n-1}b$ to be independent.

6.3.25   Which 3-diagonal matrix has a bounded inverse according to 6F?

$$A_1 = \begin{bmatrix} 0 & & & \\ 2 & 0 & & \\ 1 & 2 & 0 & \\ 0 & \cdot & \cdot & \cdot \end{bmatrix}, \quad A_2 = \begin{bmatrix} 1 & & & \\ \frac{5}{2} & 1 & & \\ 1 & \frac{5}{2} & 1 & \\ 0 & \cdot & \cdot & \cdot \end{bmatrix}, \quad A_3 = \begin{bmatrix} 1 & & & \\ 1 & 1 & & \\ \frac{1}{4} & 1 & 1 & \\ 0 & \cdot & \cdot & \cdot \end{bmatrix}.$$

6.4 ■ THE HEAT EQUATION vs. THE WAVE EQUATION

This section is about initial-value problems for partial differential equations. That subject is enormous. It is probably larger, in a way that is hard to measure but still has scientific meaning, than the topic of any other section. Nevertheless the main points can come through clearly, because they are so extremely close to physical observations that everyone has made. By comparing a gas with a liquid, we compare the heat equation to the wave equation. The comparison will be made with three initial conditions—*steps, waves,* and *delta functions*.

   1.   Suppose the initial values are step functions. In the wave equation there is a wall of water, momentarily held in place. In the heat equation there is hot gas and cold gas, momentarily separated. At $t = 0$ the separations disappear and the flow begins. The mass and momentum and energy of the water move forward, and so does the hot gas. But there is a difference in the speed and the pattern of those two movements, reflecting the difference in the solutions to $u_t = u_x$ and $u_t = u_{xx}$.

   For the wave equation the speed is finite. At a time like $t = 2$, one region knows what happened at $t = 0$ and another region does not know. The hyperbolic equation $u_t = u_x$ sends out information about its initial condition, but those signals are not heard instantaneously. In contrast, the hot gas travels with infinite speed (mathematically if not physically) and the initial condition is felt immediately, at all points.

   The solutions themselves are very different. For the wave equation, the wall of water remains a wall. It travels down a pipe without breaking, when the equation is linear and contains no u_{xxx} that would disperse even a linear wave. The step function persists. For the heat equation, that function is gone in the first instant and can never be recovered. The diffusion of one gas into another is *irreversible*—we cannot imagine a situation at $t = -1$ from which diffusion would produce two separated gases at $t = 0$. For the wave equation there is no difficulty in going backward; the wall of water is just further back in the pipe, and it moves into place at $t = 0$. The backward heat equation is generally unsolvable because the forward heat equation smooths out all discontinuities.

   2.   As a second initial condition, let $u_0(x) = \sin kx$. In the equation $u_t = u_x$, this is a flood wave on a long river. Its wave number is k and the distance between peaks (the wavelength) is $2\pi/k$. As time goes forward the wave moves without changing shape, and the traveling wave solution is

$$u(t,x) = \sin k(x + t). \tag{1}$$

That is correct at $t = 0$, and its time derivative obviously agrees with its space derivative. Thus (1) solves $u_t = u_x$.

   We admit that an oscillation like $\sin kx$ is harder to imagine in the temperature of a gas, but starting from that initial condition the heat equation can also be solved:

$$u(t,x) = e^{-k^2 t} \sin kx. \tag{2}$$

The oscillation stays in place, while its amplitude goes down exponentially. A sine wave in the temperature is so beautifully balanced that diffusion does not destroy the wave. It does blur the extremes between hot and cold, since $e^{-k^2 t}$ is rapidly decreasing. Of course the average temperature, like the average height of a wave, remains fixed at zero.

The initial condition $\sin kx$ or $\cos kx$ may not be important in itself, but collectively this family of initial conditions is very valuable. Any function $u_0(x)$ is a combination of harmonic functions. That is Fourier's contribution to partial differential equations—to decompose the solution $u(t,x)$ into pure exponential solutions. Those solutions travel and decay at their own rate, depending on k. For a *convection-diffusion* equation like $u_t = u_x + u_{xx}$, the waves travel and decay at the same time; nevertheless they remain sine waves. Linearity allows all other initial conditions to be handled by superposition.

3.   There is a third initial condition, with a discontinuity even more exaggerated than a step function. It is $u_0(x) = \delta$, *a delta function at the origin*. In the heat equation, a finite amount of infinitely hot gas is released at the point $x = 0$ and the instant $t = 0$. It diffuses; in no time the gas is everywhere. The concentration and temperature remain highest at $x = 0$, but the delta function becomes scattered over the whole line. For some unlikely but highly probable reason it spreads into the bell-shaped *normal distribution*

$$u(t,x) = \frac{1}{2\sqrt{\pi t}} e^{-x^2/4t}. \tag{3}$$

Each molecule executes a continuous random walk, or Brownian motion, going out from $x = 0$. The net result is the distribution (3), to be derived below. It is the impulse response to $u_0 = \delta$—otherwise known as the fundamental solution to the heat equation. At time t its variance reaches $\sigma^2 = 2t$ as its graph becomes less steep. At all times the total mass is $\int u \, dx = 1$, and at $t = 0$ the mass is entirely concentrated at $x = 0$.

In the wave equation, $u_0 = \delta$ is an impulse of a different kind. If the step function was like a red light turning green, allowing a steady flow of traffic to pass through, a delta function represents the *derivative* of the density. Initially it is zero except at the light. Later it is zero except at the single point $x = t$, in front of the traffic. You see the simplicity of this model, but also its drawback—a true description is not given by the linear equation $u_t = u_x$. Cars do not start together; they spread out. When the front car reaches another light, or slows down for an accident, that sends a shock wave back through the line of traffic. This nonlinear model appears in Section 6.6, whereas here the initial function simply moves forward with speed equal to 1:

$$u(t,x) = \text{delta function at the point } x = t. \tag{4}$$

In the $x - t$ plane the signal travels along the straight line $x = t$. This is a *characteristic* line, and in all hyperbolic equations discontinuities are propagated along characteristics.

The equation $u_t = u_x$ is rather simple even as a model of linear wave equations. A better model is $u_{tt} = u_{xx}$, which allows two waves instead of one. When you hit a baseball, a wave goes to the end of the bat and another goes directly toward your hand. The first is reflected and the second is absorbed (by you). Those are the principal types of boundary conditions, for heat as well as for waves, and they are needed when the region in space is bounded. Similar waves appear when a rock is dropped into a pond (two dimensions) or a sound is emitted in the air (three dimensions). They are again the responses to an impulse $u_0 = \delta$, computable either directly or by superposition of harmonics. The reference to air allows us to emphasize that unlike the *diffusion* of a gas, the *dynamics* of a gas produces all the phenomena of waves and shocks that are associated with hyperbolic equations. For small disturbances of the air density this is acoustics—and the sound is governed by the linear wave equation.

To this general picture of diffusion vs. oscillation we now add detail.

The Heat Equation

We begin with the heat equation on the line from $x = -\infty$ to $x = +\infty$:

$$\boxed{\begin{array}{ll} u_t = u_{xx} & \text{for} \quad t > 0 \\ u = u_0(x) & \text{at} \quad t = 0 \end{array}} \tag{5}$$

The conductivity or diffusivity has been set at $c = 1$. With varying material (an inhomogeneous medium) this equation is $u_t = (c(x)u_x)_x$. *In the framework of the early chapters that would be* $u_t = -A^T C A u$. There is no longer equilibrium at every point, although heat is conserved. The new term u_t allows the temperature to change, but still in accordance with Kirchhoff's current law. Heat can enter or leave a region, **but only by going in or out through the boundary**. That is the conservation law—the model that stands behind the partial differential equation. When the region is the interval from x to $x + \Delta x$,

$$\text{heat inside} \qquad = \int_x^{x+\Delta x} u \, dx$$

$$\text{its rate of change} \qquad = \frac{\partial}{\partial t} \int_x^{x+\Delta x} u \, dx \approx \Delta x u_t$$

$$\text{flux passing in at } x \qquad = -\frac{\partial u}{\partial x}(x)$$

$$\text{flux passing in at } x + \Delta x = \frac{\partial u}{\partial x}(x + \Delta x)$$

Dividing by Δx, the rate of change balances the flux:

$$u_t = \frac{1}{\Delta x}\left[\frac{\partial u}{\partial x}(x + \Delta x) - \frac{\partial u}{\partial x}(x)\right] \to u_{xx}. \tag{6}$$

If there is also a heat source this becomes $u_t = u_{xx} + f$. As $t \to \infty$ the diffusion is complete and the time-dependent equation returns to the steady-state equilibrium equation: $u_{xx} + f = 0$ or $A^T C A u = f$.

Our goal is to solve the heat equation $u_t = u_{xx}$. For the right initial conditions that is easy:

$$\text{if } u_0 = \sin kx \text{ or } \cos kx \text{ or } e^{ikx} \text{ then } u = e^{-k^2 t} u_0. \tag{7}$$

The two x derivatives multiply u by $-k^2$, and so does the t derivative. Therefore $u_t = u_{xx}$ as required. These special solutions could have been found by separation of variables, starting with $u = A(t)B(x)$ and requiring A to be an eigenfunction of $\partial/\partial t$ and B to be an eigenfunction of $\partial^2/\partial x^2$. There are pure exponential solutions for every frequency k, and they are enough to solve the heat equation.

6G   If the initial function is a combination of harmonics

$$u_0(x) = \frac{1}{2\pi} \int_{-\infty}^{\infty} e^{ikx} \hat{u}_0(k) dk,$$

then the solution to $u_t = u_{xx}$ is the combination

$$u(t,x) = \frac{1}{2\pi} \int_{-\infty}^{\infty} e^{-k^2 t} e^{ikx} \hat{u}_0(k) dk. \tag{8}$$

In other words, u is a superposition of the special solutions in (7). You can see why the backward heat equation is in trouble, because the factors $e^{-k^2 t}$ will be exponentially large if t is negative and the integral (8) has no meaning.

Equivalently we can transform both sides of $u_t = u_{xx}$. The Fourier transform of the left side is $\hat{u}_t$. The Fourier transform of the right side is $-k^2 \hat{u}$, since x derivatives contribute ik. That leaves an ordinary differential equation for each frequency k:

$$\hat{u}_t = -k^2 \hat{u} \text{ or } \hat{u}(t,k) = e^{-k^2 t} \hat{u}(0,k). \tag{9}$$

Again, the harmonic e^{ikx} decays with the factor $e^{-k^2 t}$.

As it stands the solution formula (8) requires two infinite integrals—first to transform $u_0(x)$ to $\hat{u}_0(k)$, followed by (8) itself. That is the normal outcome of the Fourier method, but here we can do better. When u_0 is a delta function at the origin, its transform is $\hat{u}_0 = 1$ for all frequencies. Substituting $\hat{u}_0 = 1$ into (8) leaves the inverse Fourier transform of $e^{-k^2 t}$, which is computed in Exercise 6.4.4 below:

$$u(t,x) = \frac{1}{2\sqrt{\pi t}} e^{-x^2/4t}. \tag{10}$$

This is the *fundamental solution*. It is the response to a unit impulse at $x = 0$, $t = 0$. If the impulse is at another point y, so that $u_0 = \delta(x - y)$, then $x - y$ replaces x in the response. If the impulse is at another time s, then $t - s$ replaces t. If the equation changes to $u_t = cu_{xx}$, then ct replaces t. If there are many impulses, one more assist from superposition (which means linearity) yields the solution. The answer for any u_0 comes from one integral instead of two:

6H   Every function is trivially a superposition of delta functions,

$$u_0(x) = \int_{-\infty}^{\infty} \delta(x - y)u_0(y)dy.$$

Therefore the solution to $u_t = u_{xx}$ that starts from u_0 is a superposition of responses to those delta functions:

$$u(t,x) = \int_{-\infty}^{\infty} \frac{1}{2\sqrt{\pi t}} e^{-(x-y)^2/4t} u_0(y)dy. \tag{11}$$

EXAMPLE 1   If $u_0(x) = 1$, the temperature stays at $u = 1$ for all t.

EXAMPLE 2 $u_0 = 1$ for $0 \le x \le 1$ and otherwise $u_0 = 0$

The integral (11) from 0 to 1 is an "error function," tabulated in many places but impossible to do in closed form.

EXAMPLE 3 $u_0(x) = e^{-x^2/4}$

The integral (11) becomes possible but there is a better idea. This u_0 is virtually the fundamental solution (10) at time $t = 1$. That means it is the response at $t = 0$ to an impulse that occurred at $t = -1$ (maybe I am solving the backward heat equation but not further back than that impulse!). Thus going forward from $e^{-x^2/4}$ to time t is the same as going forward from δ to time $t + 1$:

$$u(t,x) = \frac{1}{\sqrt{t+1}} e^{-x^2/4(t+1)}.$$

Probably the right way to look at (11) is as a convolution of u_0 with the fundamental solution. That is expected for other partial differential equations. But the eye tends to glaze over in the presence of infinite integrals, and it is important to pick out what is significant:

(1)   *The solution is infinitely smooth.* The Fourier coefficients are multiplied by $e^{-k^2 t}$, so they drop sharply as soon as t leaves zero.

(2)   **If $u_0 \geq 0$ then $u \geq 0$ for all time**. This is clearest in (11): Nothing in that formula will be negative.

(3)   The scaling matches x^2 with t. That is the parabola that makes the heat equation parabolic.

(4)   In two or three dimensions the equation is $u_t = u_{xx} + u_{yy}$ or $u_t = u_{xx} + u_{yy} + u_{zz}$. The x^2 in (11) changes to $x^2 + y^2$ or $x^2 + y^2 + z^2$, and $2\sqrt{\pi t}$ is squared or cubed.

That describes the heat equation without boundaries. For a semi-infinite problem, with temperature $u = 0$ at $x = 0$ (insulated boundary), the adjustment is easy: Extend u_0 to be an *odd* function, $u_0(-x) = -u_0(x)$, and keep the solution (11). It will continue as an odd function, vanishing at $x = 0$. If the boundary condition changes to $u_x = 0$ (absorbing boundary) there is a similar "mirror image." In this case u_0 is extended to negative x as an *even* function, the solution (11) remains even, and even functions have zero slope at $x = 0$.

Finally there are problems with two boundaries, at $x = 0$ and $x = 1$. The boundary conditions admit certain frequencies k and reject all others. For a bar whose ends are held at $0°$, the acceptable frequencies are multiples of π: the pure exponential solutions are $u_n = e^{-\pi^2 n^2 t} \sin \pi n x$.

EXAMPLE $u_t = u_{xx}$ with $u(t,0) = 0$, $u(t,1) = 0$, $u_0(x) = 1$

The initial function is expanded in a sine series:

$$u_0 = \frac{4}{\pi}\left[\sin \pi x + \frac{\sin 3\pi x}{3} + \frac{\sin 5\pi x}{5} + \cdots\right]$$

At later times each sine has its decay factor:

$$u = \frac{4}{\pi}\left[e^{-\pi^2 t}\sin \pi x + e^{-9\pi^2 t}\frac{\sin 3\pi x}{3} + \cdots\right] \tag{12}$$

The odd function u_0 is the right half of a square wave (Exercise 4.1.3). The solution u is the right half of a "rounded wave," with the corners taken off to satisfy $u = 0$ at the boundaries.

I regret these infinite series even if they do give the solution. Numerically they are not attractive; the spectral method switches to finite series and the Fast Fourier Transform. Analytically they have an important place, as a combination of pure exponential solutions, but it has been made too important in the past and some revision is due. Series solutions should not be made the centerpiece of a course on applied mathematics. The attractive part of the heat equation is an exact closed-form solution but not a series. It is the fundamental solution itself, with the exponential $e^{-x^2/4t}$ which shows how diffusion spreads from a point source. That is memorable.

The Wave Equation

There is no doubt which is the simplest hyperbolic equation. It is $u_t = u_x$. It is a single linear first-order equation, which cannot model all possible wave phenomena. However it does model the basic property of waves, that they travel with finite speed. This is completely different from Laplace's equation, where any change in the boundary condition is felt at every other point (but smoothed out). It is also different from the heat equation, where the change is felt at all points at all later times (again with exponential smoothing). In the wave equation, a point feels the change only when the signal reaches it. The result is a solution that is no smoother than the initial values and boundary values—and may be rougher when the waves focus (linear case) or when they break up into shocks (nonlinear case).

The starting point is the pure initial-value problem

$$u_t = u_x \text{ with } u_0(x) \text{ given for } -\infty < x < \infty. \tag{13}$$

This equation is satisfied by any function $F(x + t)$, depending on the combination $x + t$. Examples are $(x + t)^2$ and e^{x+t}; the x derivative and the t derivative both produce F'. This is like the analytic functions $f(x + iy)$ that automatically satisfied the Cauchy-Riemann equations $f_y = if_x$. Of course i has an important effect; one problem is hyperbolic and the other is elliptic. The line $x + t = c$ is real for the hyperbolic equation, and the function $F(x + t)$ keeps the same value $F(c)$ at all points on that line.

That value is known when $t = 0$, because $u_0(x)$ is given. The solution $F(x + t)$ must match u_0 when $t = 0$. In this simple case F is identical with u_0, and *the solution is*

$$\boxed{u(t,x) = u_0(x + t).}$$

Waves move to the left with speed 1. They are not distorted (Fig. 6.19), and a peak which was initially at x will reach the point $x - t$ at time t. All frequencies travel with the same speed.

EXAMPLE   The dam-break problem

The dam is at $x = 0$. The water levels are $u_0 = 0$ to the left of the dam and $u_0 = H$ to the right. At $t = 0$ the dam breaks, and the water moves left. At time t the solution is

$$u_0(x + t) = \begin{cases} 0 & \text{for } x + t < 0 \\ H & \text{for } x + t > 0 \end{cases}$$

In other words, an imaginary dam moves along the line $x + t = 0$.

Practical problems are almost certain to have boundaries. Suppose x goes from 0 to 1 instead of $-\infty$ to ∞. In equilibrium equations there was a boundary condition at each end, but for a first-order equation $u_t = u_x$ the situation is very different:

At the left boundary: The wave goes in and no boundary condition is wanted or required.
At the right boundary: The wave comes out and a boundary condition is needed. Otherwise the solution $u_0(x + t)$ will not be known along a line like $x + t = \frac{3}{2}$.

That line strikes the right side of Figure 6.19b and not the initial line. Therefore the problem must specify u along the boundary to the right. When the specified value is $u(t, 1) = g(t)$, the solution is

$$u(t,x) = \begin{cases} u_0(x + t) & \text{for} \quad x + t < 1 \\ g(x + t) & \text{for} \quad x + t > 1 \end{cases}$$

The signal comes out from the right boundary, travels across on a characteristic line $x + t = \text{constant}$, and disappears into the left boundary.

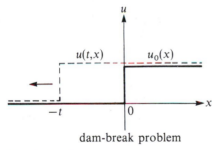

dam-break problem

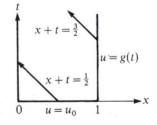

initial-boundary problem

Fig. 6.19. One-way waves from $u_t = u_x$.

That equation $u_t = u_x$ is a basic model for wave motion. To make it more general, the wave speed 1 can change to a function $c(x,t,u)$ that varies with position or time or with u itself. The new equation is $u_t = cu_x$, and there are at least four possibilities:

(1) $c = \text{constant}$: The solution is $F(x + ct)$ and signals travel with velocity c along the lines $x + ct = \text{constant}$. If $c < 0$, waves move to the right
(2) $c = c(t)$: The solution is $F(x + C(t))$, with $C' = c$. Signals travel on the space-time curves $x + C(t) = \text{constant}$
(3) $c = c(x)$: The solution is $F(D(x) + t)$, with $D' = 1/c$. Signals travel on the space-time curves $D(x) + t = \text{constant}$
(4) $c = c(u)$: The solution satisfies $u = u_0(x + c(u)t)$. The signal paths are back to straight lines, but the slope depends on u and those lines can intersect. That is what produces shock waves in the solution.

The last problem comes from a *nonlinear conservation law*. The equation $u = u_0(x + c(u)t)$ has to be solved for u, and the geometry of the lines—whether they intersect or fan out—determines the physical picture. We postpone that discussion until linear wave equations have gone further than $u_t = cu_x$.

The immediate problem is to go beyond a single equation of first order. Either the number of equations should increase or the order. It was the same for ordinary differential equations, where $u' = au$ could extend to a system $u' = Au$ or to a second-order equation like $u'' = -u$. The minus sign in the latter gave oscillating solutions $u = \cos t$ and $u = \sin t$. Similarly the negativity of the second derivative $\partial^2/\partial x^2$ will produce waves.

Second-order Wave Equation: $u_{tt} = c^2 u_{xx}$

We approach this like any other constant-coefficient equation, looking for exponential solutions. There are exponentials e^{ikx} in space, out of which all other functions can be built. There are exponentials $e^{i\omega t}$ in time, to give the growth factor or decay factor or in this case the oscillation frequency. **The frequency ω depends on the wave number k.** To find that dependence, substitute $u = e^{i\omega t}e^{ikx}$ into the wave equation:

$$-\omega^2 u = -k^2 c^2 u, \quad \text{or} \quad \omega^2 = c^2 k^2, \quad \text{or} \quad \omega = \pm ck.$$

Each wave e^{ikx} has two frequencies, consistent with the fact that this is a second-order equation. The solutions with $\omega = +ik$ and $\omega = -ik$ are

$$u = e^{ik(x + ct)} \quad \text{and} \quad u = e^{ik(x - ct)}. \tag{14}$$

Those are waves traveling at the same speed in opposite directions. The first solves $u_t = cu_x$ and the second solves $u_t = -cu_x$. Either will lead to $u_{tt} = c^2 u_{xx}$. The wave equation is effectively factored into

$$\left(\frac{\partial^2}{\partial t^2} - c^2 \frac{\partial^2}{\partial x^2}\right)u = \left(\frac{\partial}{\partial t} - c\frac{\partial}{\partial x}\right)\left(\frac{\partial}{\partial t} + c\frac{\partial}{\partial x}\right)u = 0, \tag{15}$$

and we have two families of solutions.

The general solution comes from combining exponentials. It is superposition when solutions are added, and it is the inverse Fourier transform when those solutions are harmonics. Normally that means an integral over all k in order to match initial conditions. In this case the result is special and no integral is needed: **The general solution to the wave equation is**

$$\boxed{u = F_1(x + ct) + F_2(x - ct).} \tag{16}$$

A superposition of the pure solutions $e^{ik(x + ct)}$ is a function $F_1(x + ct)$. Similarly the exponentials $e^{ik(x-ct)}$ combine to give $F_2(x-ct)$. Normally ω is not proportional to k, but here $\omega = ck$ and $\omega = -ck$ give the general solution in the remarkable form (16).†

That form has to fit the initial conditions. There are two conditions for a second-order equation, the initial position $u_0(x)$ and the initial velocity $u_t(0,x) = v_0(x)$. To match those with F_1 and F_2 requires

$$u_0 = F_1 + F_2 \quad \text{and} \quad v_0 = cF'_1 - cF'_2. \tag{17}$$

Case 1   Starting from rest ($v_0 = 0$) this gives $F_1 = F_2 = \tfrac{1}{2} u_0$:

$$u(t,x) = \tfrac{1}{2} u_0(x+ct) + \tfrac{1}{2} u_0(x-ct). \tag{18}$$

Case 2   Starting from equilibrium $u_0 = F_1 + F_2 = 0$ with velocity v_0, (17) becomes $v_0 = 2cF'_1$. Then $F_1 = -F_2$ is an integral of $v_0/2c$ and the solution is

$$u(t,x) = \frac{1}{2c} \int_0^{x+ct} v_0 - \frac{1}{2c} \int_0^{x-ct} v_0 = \frac{1}{2c} \int_{x-ct}^{x+ct} v_0(s)\,ds. \tag{19}$$

The complete solution is the sum of Case 1 and Case 2. It is d'Alembert's solution to the wave equation $u_{tt} = c^2 u_{xx}$, illustrated by four examples.

EXAMPLE 1
$$u_0 = \begin{cases} 1 & \text{for} \quad x > 0 \\ -1 & \text{for} \quad x < 0 \end{cases} \quad \text{and} \quad v_0 = 0$$

This step function splits in two pieces (Fig. 6.20b). Half the step goes left and half goes right. For large x, when the signal has not arrived from the origin, $x + ct$ and $x - ct$ are both positive. That is indicated by the dark point on the right, where $u = \tfrac{1}{2} + \tfrac{1}{2} = 1$ comes from the initial values $u_0 = 1$ at the ends of the dotted lines. At a smaller x the signal has arrived and gone past. That is the dark point near the center of the figures, where $u = \tfrac{1}{2} - \tfrac{1}{2} = 0$ is taken from $u_0 = 1$ and $u_0 = -1$ at the ends of *its* dotted lines. The "wall" that started at the origin passes exactly through the points $x = ct$ and $x = -ct$.

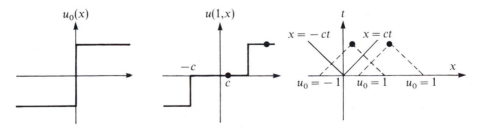

Fig. 6.20. The wave equation starting from a step function.

† F_1 and F_2 are "Riemann invariants," constant on the characteristics $x \pm ct = x_0$.

EXAMPLE 2 $u_0 = \delta$ and $v_0 = 0$

The impulse starts from the origin and splits into

$$u(t,x) = \tfrac{1}{2}\,\delta(x + ct) + \tfrac{1}{2}\,\delta(x - ct).$$

This is a sharp signal with two spikes—heard *only* at $x = \pm\, ct$.

EXAMPLE 3 $u_0 = 0$ and $v_0 = \begin{cases} 1 & \text{for} \quad x > 0 \\ -1 & \text{for} \quad x < 0 \end{cases}$

After substituting into (19), this step function in the velocity creates a ramp function in the displacement:

$$u(t,x) = \begin{cases} t & \text{for} \quad x > ct \\ x/c & \text{for} \quad |x| < ct \\ -t & \text{for} \quad x < -ct \end{cases}$$

The velocity $v = \partial u/\partial t$ is the same $-1,0,1$ function that appeared in Example 1. There it was u that split into two steps, here it is v. Both started from a single step function at time 0, and both satisfy the wave equation. This is an important observation, that when u satisfies a constant-coefficient equation like $u_{tt} = c^2 u_{xx}$ so do all its derivatives. Direct differentiation confirms that $u_{ttt} = c^2 u_{xxt}$, in other words that the velocity $v = u_t$ satisfies $v_{tt} = c^2 v_{xx}$.

EXAMPLE 4   It is quite possible that u_0 and v_0 are identically zero and all motion comes from the boundary conditions. If $u = g(t)$ at the endpoint $x = 0$, this signal moves to the right along the lines $x - ct = \text{constant}$. Without another boundary the waves go only one way, and the solution is $u = g\!\left(t - \dfrac{x}{c}\right)$. Where the combination $t - \dfrac{x}{c}$ is negative, no signal has arrived and $u = 0$.

Derivation of the Wave Equation

Some equations are harder to derive than to solve. That may be true for the wave equation, whose solution is so exceptional—but its derivation is not difficult either. Consider a string pulled tight (with tension T) between fixed ends at $x = 0$ and $x = 1$. Its height is u, its slope is $\partial u/\partial x$, and the upward pull of tension is $T\partial u/\partial x$. For the small piece of string in Fig. 6.21, this pull is different at x_1 and x_2. That difference is the net force, and it moves the string in accordance with $F = ma$. The mass m is the linear density ρ times the length $x_2 - x_1$:

$$\rho(x_2 - x_1)u_{tt} = T\left[\frac{\partial u}{\partial x}(x_2) - \frac{\partial u}{\partial x}(x_1)\right].$$

Dividing by $x_2 - x_1$ and allowing x_2 to approach x_1, the right side is the derivative of $\partial u/\partial x$. It is the second derivative of u: $\rho u_{tt} = Tu_{xx}$. That is the wave equation with $c^2 = T/\rho$.

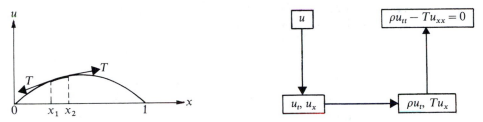

Fig. 6.21. A vibrating string: $u_{tt} = c^2 u_{xx}$ in the standard framework.

The figure also fits the wave equation into the standard framework. The gradient A is now in space-time, and produces both u_x and u_t. The constitutive matrix C is **not positive definite**: it is

$$C = \begin{bmatrix} T & 0 \\ 0 & -\rho \end{bmatrix}.$$

That is the difference between elliptic and hyperbolic equations, or between equilibrium problems and dynamic problems.† It is like relativity, in which the distance $dx^2 - c^2 dt^2$ is indefinite. The final equation is still a balance of forces, tension vs. inertia, and with an external force f per unit length the equation would be $\rho u_{tt} = Tu_{xx} + f$.

The wave equation is repeated for an elastic rod. Unlike a string, the displacement u is *along* the rod. The derivative $\partial u/\partial x$ is the stretching and not the slope. The elastic constant k takes the place of T, but the equation is still a balance of forces. The force on each section of the rod comes at the ends x_1 and x_2, and Newton's law $ma = F$ is

$$\rho(x_2 - x_1)u_{tt} = k\left[\frac{\partial u}{\partial x}(x_2) - \frac{\partial u}{\partial x}(x_1)\right].$$

Again this goes to $\rho u_{tt} = ku_{xx}$ as the section length goes to zero.

---

† Or else t is changed to it, so that $A^T A$ becomes $\partial^2/\partial x^2 - \partial^2/\partial t^2$ instead of the Laplacian. Then C can be positive definite.

For a chain of springs the equation would have been

$$\rho u_{tt} = \frac{k}{(\Delta x)^2} \begin{bmatrix} -2 & 1 & & & \\ 1 & -2 & 1 & & \\ & 1 & \cdot & 1 & \\ & & 1 & -2 \end{bmatrix} u.$$

This is the lumped system. As $\Delta x \to 0$ it becomes the distributed system. The ordinary differential equation $u'' = Au$ changes into the wave equation, as the second difference matrix A changes into the second derivative. The discrete problem becomes continuous—and in this case, with $u = F_1(x + ct) + F_2(x - ct)$, the continuous problem is easier to solve.

First-order Systems

After $u_{tt} = c^2 u_{xx}$ there are two natural alternatives. One is to take the wave equation into more space dimensions. The other is to stay in one dimension, with equations that illustrate more fully the problems that are actually met in applications. Those problems begin as first-order equations, for mass or momentum or displacement or pressure, and they are coupled into a *first-order system*. We could attempt elimination, and try to get back to the wave equation, but here we prefer to stay with a *vector unknown* $u = (u^1(t,x), \ldots, u^n(t,x))$:

$$\boxed{\frac{\partial u}{\partial t} = G \frac{\partial u}{\partial x}.} \tag{20}$$

G is an n by n matrix and initial values $u_0(x)$ are prescribed.

In physics and engineering, a linear equation generally deals with a small disturbance. Something from outside acts to change the equilibrium, but not by much:

   in acoustics it is a slowly moving body
   in aerodynamics it is a slender wing
   in elasticity it is a small load
   in electromagnetism it is a small source

Up to some transition point, the cause-effect relation is close to linear. In acoustics the sound speed is steady when pressure is nearly constant; otherwise $\partial p / \partial \rho$ produces nonlinearity. In elasticity Hooke's law holds until the geometry changes or the material begins to break down. In electromagnetism nonlinearity comes late, with relativistic and quantum effects. Prior to those transitions we have linear equations like (20). There is no damping to bring the system to a new equilibrium state.

The case to analyze is the one with a constant matrix G. Everything comes from its eigenvalues and eigenvectors. If $GU = \lambda U$, then

   the vector equation $u_t = Gu_x$ becomes a scalar equation $w_t = \lambda w_x$, when the solutions are related by $u = wU$.

To see why, substitute wU into the system $u_t = Gu_x$. The result is $w_t U = Gw_x U = \lambda w_x U$. The eigenvector U is fixed, so the scalar $\partial w/\partial t$ must agree with the scalar $\lambda \, \partial w/\partial x$.

The solution to $w_t = \lambda w_x$ is a wave traveling with speed λ. It has exactly the form discovered earlier: $w = w_0(x + \lambda t)$. All components of wU travel together, and we look for n solutions of this kind—one from each eigenvector of G. In other words, *the solution to a first-order system consists of n waves.*

There is a point not to be overlooked. This is a traveling wave only if λ is real. Therefore **the eigenvalues of G must be real**. Without that the equation is no longer hyperbolic. The solution can grow and the problem becomes improperly posed— like the backward heat equation. For the scalar equation c was assumed real, and now it must be assumed that the eigenvalues of G are real. That is certain to hold when G is symmetric and the problem is **symmetric hyperbolic**. It also holds in physical applications, where energy is constant.

Before the examples we summarize the main conclusion—that $u_t = Gu_x$ can be reduced to n scalar equations. That reduction is impossible in more space dimensions, but here it solves the problem:

6I The system $u_t = Gu_x$ is *hyperbolic* when G has n independent eigenvectors U_j, all with real eigenvalues. The solution is $u = w^1 U_1 + \cdots + w^n U_n$, where each w satisfies a scalar wave equation:

$$\frac{\partial w^j}{\partial t} = \lambda_j \frac{\partial w^j}{\partial x} \text{ with solution } w^j = w_0^j(x + \lambda_j t).$$

The initial values for w come from the initial values for the vector u: $u_0 = w_0^1 U_1 + \cdots + w_0^n U_n$. In effect, the change to the w's turns G into a diagonal matrix. **The problem has n signal speeds λ_j and it sends out n waves**. The w's are called Riemann invariants, because they are constant along the characteristic lines $x + \lambda_j t = x_0$. They also exist for nonlinear systems $u_t = G(u)u_x$—provided there are only two equations in the system.

EXAMPLE 1 The wave equation as a first-order system

Writing $u_{tt} = c^2 u_{xx}$ as a system, the unknowns can be cu_x and u_t:

$$\frac{\partial}{\partial t}(cu_x) = c\frac{\partial}{\partial x}(u_t)$$

$$\frac{\partial}{\partial t}(u_t) = c\frac{\partial}{\partial x}(cu_x). \tag{21}$$

The first is automatic; the second contains the wave equation. The matrix hidden on the right hand side is symmetric,

$$G = \begin{bmatrix} 0 & c \\ c & 0 \end{bmatrix} \text{ with eigenvalues } \lambda = \pm c.$$

Its eigenvectors are $U_1 = (1,1)$ and $U_2 = (-1,1)$, and the scalars in $u = w^1 U_1 + w^2 U_2$ satisfy

$$\frac{\partial w^1}{\partial t} = c \frac{\partial w^1}{\partial x}, \frac{\partial w^2}{\partial t} = -c \frac{\partial w^2}{\partial x}.$$

Thus w^1 is a function of $x + ct$, and is carried to the left with speed c. The other invariant w^2 is a function of $x - ct$; it travels to the right. Going back to u this is the same solution found earlier in (16)—with two waves and a solution $F_1(x + ct) + F_2(x - ct)$.

These diagonalizing w's are seldom used numerically, just as solutions of $u_t = Au$ are not always computed from eigenvectors of A. The reason lies in the boundary conditions. For the wave equation on a finite string, each end may have one of the conditions

$$u = 0 \text{ (fixed end)} \quad \text{or} \quad u_x = 0 \text{ (free end)}.$$

To know u at an endpoint is to know $\partial u / \partial t$, so the system prescribes either u_t or cu_x. One component is given because one wave enters from the boundary. (The other wave leaves.) In fact the prescription need not be zero; the ends can be moved. In any case these boundary conditions are more convenient for u_t or cu_x separately than for their combinations $w = \frac{1}{2}(u_t \pm cu_x)$.

EXAMPLE 2   Vibrating string

In the physical variables of Fig. 6.21 the equations are

$$\frac{\partial}{\partial t}(u_x) = \frac{\partial}{\partial x}(u_t)$$

$$\frac{\partial}{\partial t}(\rho u_t) = \frac{\partial}{\partial x}(T u_x). \tag{22}$$

The first is *conservation of slope*. The steepness of the string between x_1 and x_2 changes as "slope flows through the ends"—or more literally, as the ends move. The difference in velocities $u_t(x_1)$ and $u_t(x_2)$ changes the slope, and as $\Delta x \to 0$ that change becomes $(u_x)_t = (u_t)_x$.

The second equation is *conservation of momentum*. The momentum per unit length is ρu_t, and $\partial / \partial t(\rho u_t) = \partial / \partial x(T u_x)$ is Newton's law—the balance of inertial force with string force. The matrix in this 2 by 2 system is

$$G = \begin{bmatrix} 0 & 1 \\ T/\rho & 0 \end{bmatrix} = \begin{bmatrix} 0 & 1 \\ c^2 & 0 \end{bmatrix}.$$

The eigenvalues $\pm c$ show its equivalence to the wave equation.

EXAMPLE 3   Gas dynamics: Disturbance of uniform flow

The conserved quantities are **mass** and **momentum**. Those are nonlinear conservation laws, but for small disturbances the pressure changes are linearly proportional to the density changes: $p - p_0 = c^2(\rho - \rho_0)$. The leading terms in the two laws are

$$\frac{\partial \rho}{\partial t} = -\rho_0 \frac{\partial u}{\partial x}$$

$$\rho_0 \frac{\partial u}{\partial t} = -\frac{\partial p}{\partial x} = -c^2 \frac{\partial \rho}{\partial x}. \tag{23}$$

Dividing the second equation by ρ_0 the matrix is

$$G = \begin{bmatrix} 0 & -\rho_0 \\ -\dfrac{c^2}{\rho_0} & 0 \end{bmatrix}.$$

This is similar to the matrices in Examples 1 and 2, with the same eigenvalues $\pm c$. Eliminating ρ brings back the wave equation $u_{tt} = c^2 u_{xx}$ for the velocity of the gas, and now we move to more space dimensions.

Wave Equations in Higher Dimensions

The wave equation in three-dimensional space is

$$\boxed{\begin{array}{ll} u_{tt} = u_{xx} + u_{yy} + u_{zz} & \text{for } t > 0 \\ u = u_0 \text{ and } u_t = v_0 & \text{for } t = 0. \end{array}}$$

It sends waves in all directions. The solution is a superposition of those waves— which can be started from pure harmonics or from point sources δ. Those are the two supreme and successful ways to approach a constant-coefficient partial differential equation:

1. ***Exponential solutions***: $u = e^{i(\omega t + k_1 x + k_2 y + k_3 z)}$

2. ***Fundamental solution***: $u = $ response to the impulse $v_0 = \delta$

We expect the signal speed to be $c = 1$. What is less predictable is whether the signal will be sharp. Does the solution started by the delta function remain a pure impulse, or is there a "tail" following the wave front? In one case the solution is everywhere zero, except when the signal goes past. In the other case $u = 0$ until the signal arrives, but after that an oscillation remains. In one dimension the signal was sharp: $u = \frac{1}{2} \delta(x + ct) + \frac{1}{2} \delta(x - ct)$. We want to show that it is again sharp in three dimensions, allowing us to see clearly. In two dimensions it is not sharp. If you throw a stone in a pond, the wave front is followed by smaller waves.

1. (*Exponential solutions*) In three space dimensions there are three wave numbers k_1, k_2, k_3.† Those numbers multiply x, y, z while ω multiplies t, and to satisfy the wave equation there is a relation between them:

$$u = e^{i(\omega t + k_1 x + k_2 y + k_3 z)} \text{ requires } \omega^2 = k_1^2 + k_2^2 + k_3^2 = |k|^2.$$

That comes directly from substituting u into the equation. It gives two frequencies $\omega = \pm |k|$, and two plane waves. Combining $e^{i\omega t}$ and $e^{-i\omega t}$ produces sines and cosines, and the solution that starts from a harmonic (with $v_0 = 0$) is the one with a cosine:

$$u(t, x) = \cos |k| t \; e^{i(k_1 x + k_2 y + k_3 z)}. \tag{24}$$

That completes the first approach, except for superposition of these exponentials to allow any u_0. The weights are the three-dimensional Fourier transform $\hat{u}_0(k)$, which comes in the exercises.

2. (*Fundamental solution*) The impulse is a delta function in 3-space:

$$\delta = \delta(x)\delta(y)\delta(z) \text{ with the property } \int\int\int f(x,y,z)\delta = f(0,0,0).$$

Its Fourier transform is $\hat{\delta} = 1$ and it gives the same weight to all harmonics. Rather than computing that superposition we go back to the wave equation. The solution starting from $u_0 = 0$ and $v_0 = \delta$ will be symmetric around the origin, and with spherical symmetry the wave equation becomes

$$\frac{\partial^2 u}{\partial t^2} = \frac{\partial^2 u}{\partial r^2} + \frac{2}{r} \frac{\partial u}{\partial r}. \tag{25}$$

The Laplacian on the right normally has u_θ and u_ϕ terms in spherical coordinates (Section 3.4), but those disappear with symmetry and (25) can be written as

$$\frac{\partial^2}{\partial t^2} (ru) = \frac{\partial^2}{\partial r^2} (ru).$$

---

† Initial-value problems distinguish **frequencies** ω from **wave numbers** k.

In other words, ru **satisfies the 1-dimensional wave equation**. It is on the half-line $r \geq 0$, since distance from the origin is never negative. The initial condition $v_0 = \delta$ is concentrated at $r = 0$, and it can only move to the right. Therefore the solution is a function $F(r - t)$ which is zero except at the point $r = t$. Going back to 3 dimensions, u is zero except on the sphere $r^2 = x^2 + y^2 + z^2 = t^2$. Every point hears the signal only once, as the sphere passes by.

The response to δ is a signal on an expanding sphere—whose radius is t. The response to other initial functions is carried on a sphere from each point. This is "Huygens' principle," that the point (x, y, z) at time t sees the initial functions only on the sphere S of radius t around that point. The **domain of influence** of an initial point is a sphere at time t, and the **domain of dependence** of a later point is the sphere S at time zero. The solution is found by integrating over that spherical surface: if $u_0 = 0$ then

$$u = \frac{1}{4\pi t} \int\!\!\int_S v_0 \, dS. \tag{26}$$

A sharp impulse produces a sharp response, in three dimensions.

In two dimensions the signal is not sharp; it is felt indefinitely. The solution u does not return to zero for $t > r$. My own picture is to imagine a *point source* in two dimensions as a *line source* in the z-direction in three dimensions. The 3-dimensional solution is independent of z, so it satisfies $u_{tt} = u_{xx} + u_{yy}$. But in three dimensions, the spheres growing around a point continue to hit the line. They hit further and further away so the solution decays—but it is not zero. The wave front passes, but waves keep coming.

There is another property of the wave equation: **energy is conserved**. That seems natural without dissipation, but we need the right expression for the energy. It comes from reducing the second-order equation to a first-order system:

Let the unknown vector be (u_x, u_y, u_t). Then the trivial equations $u_{xt} = u_{tx}$ and $u_{yt} = u_{ty}$ join the wave equation $u_{tt} = u_{xx} + u_{yy}$ to give

$$\begin{bmatrix} u_x \\ u_y \\ u_t \end{bmatrix}_t = \begin{bmatrix} 0 & 0 & 1 \\ 0 & 0 & 0 \\ 1 & 0 & 0 \end{bmatrix} \begin{bmatrix} u_x \\ u_y \\ u_t \end{bmatrix}_x + \begin{bmatrix} 0 & 0 & 0 \\ 0 & 0 & 1 \\ 0 & 1 & 0 \end{bmatrix} \begin{bmatrix} u_x \\ u_y \\ u_t \end{bmatrix}_y. \tag{27}$$

These matrices are symmetric and (27) is *symmetric hyperbolic*.

The wave equation has become $U_t = GU_x + HU_y$, with symmetry in the nine entries of G and H. It is the two-dimensional analogue of the system $u_t = Gu_x$. It is probably true that all the conservative equations of mathematical physics can be stated in this form. Furthermore a symmetric hyperbolic equation shows immediately the energy that is conserved:

6J   The energy $E = \frac{1}{2} \displaystyle\int_{-\infty}^{\infty} \int_{-\infty}^{\infty} U^T U \, dx \, dy$ is constant.

Since $U^T U = u_x^2 + u_y^2 + u_t^2$, this energy is potential plus kinetic. To prove it is constant, differentiate $U^T U$ by the product rule:

$$\frac{\partial}{\partial t} U^T U = U^T U_t + U_t^T U = U^T G U_x + U^T H U_y + U_x^T G^T U + U_y^T H^T U.$$

Since G and H are symmetric, this is the x-derivative of $U^T G U$ plus the y-derivative of $U^T H U$. Integrating these exact derivatives over all space gives $U^T G U$ and $U^T H U$ at infinity, which is zero.† On the left side, the integral of $\partial/\partial t(U^T U)$ is therefore $dE/dt = 0$. The energy E remains constant.

The Relation of ω to k

It is time to leave the wave equation. It was typical but also special—special because of exact solutions and (sometimes) sharp signals, typical because the starting point was $e^{i(\omega t + kx)}$. Those exponentials are the key to every linear differential equation, and we want to apply them to a whole sequence of important problems. It is so simple to do, and so revealing, that this section cannot end without summarizing it and using it:

(1)   Substitute $u = e^{i(\omega t + kx)}$ into the equation
(2)   Find the relation between ω and k
(3)   Distinguish wave equations (real ω) from diffusion equations

(4)   If desired, compute solutions $\displaystyle\int \hat{u}_0(k) e^{i(\omega t + kx)} dk$ and in particular the fundamental solution with $\hat{u}_0 = 1$.

The crucial step is the relation of ω to k.

EXAMPLE 1   Convection-diffusion: $u_t = u_x + \dfrac{1}{P} u_{xx}$

Substituting $u = e^{i\omega t} e^{ikx}$, the derivatives produce $i\omega = ik + (ik)^2/P$. That is the $\omega - k$ relation. The real part of ω is k and the imaginary part is k^2/P. The equation is parabolic, because of that imaginary part, but *the solution travels downstream as it diffuses*. The fundamental solution, responding to $u_0 = \delta$, is

$$u(t,x) = \frac{\sqrt{P}}{2\sqrt{\pi t}} e^{-(x+t)^2 P/4t}. \tag{28}$$

---

† U drops to zero at infinity if its energy is finite.

It copies the solution for the heat equation, except that x is transported by the convection term to $x + t$. On the whole line this extra term is no problem, and changing to moving coordinates $X = x + t$ and $T = t/P$ recovers $u_T = u_{XX}$.

Convection-diffusion is a fundamental problem in chemical engineering. There u is the concentration of a solute, and one fluid is pushed by another. They mix as they flow down a pipe. Initially both are homogeneous; the *Péclet number P* (or *Pe*) controls the rate of mixing. It is a dimensionless ratio of convection to diffusion, and the limiting cases bring back pure convection or pure diffusion:

  $P \to 0$ (*perfect mixing*): The diffusion term overwhelms the convection term, and the fluids are instantly mixed.

  $P \to \infty$ (*perfect displacement*): The diffusion disappears and it leaves the wave equation $u_t = u_x$. One fluid is being pushed piston-like by the other, and the interface between them remains sharp.

In practice convection dominates over a short time, and diffusion over a long time. The definition of short and long is in the Péclet number, which is constructed from the actual velocity, length, and diffusivity: $P = VL/D$. The dimensionless form $u_t = u_x + u_{xx}/P$ comes from rescaling the true equation $U_T = VU_X + DU_{XX}$. The Péclet number is to mass transport what the Reynolds number is to momentum transport.

A few words about the physical problem. The flow velocity in a real pipe is not uniform. The flow is fastest in the center, and drops to zero at a no-slip boundary.† The velocity profile for "Poiseuille flow" is a parabola. Over a long time, however, a dissolved particle will sample all parts of the cross-section and the averaged concentration equation has constant coefficients. An interesting case is a flow whose average velocity is zero—for example in the heart. Blood is pumped back and forth, so the convection term disappears in the averaged equation—but the diffusion is much greater than the coefficient D in the unaveraged equation. The same is true for the flow of tides, which are major sources of diffusion.

We must comment on numerical difficulties when P is large. The equation is officially parabolic, because of u_{xx}, but practically it is hyperbolic. There may be boundary conditions $u(t,0) = A$ and $u(t,1) = B$, as there would be if time were absent and the equation were $u_x + u_{xx}/P = 0$. When P is small the convection term u_x is not troublesome—except that it spoils the symmetry of u_{xx}. If $P = 100$ the situation changes. The equation *tries to be hyperbolic*, and for $u_t = u_x$ there is only *one* boundary condition: The wave $u_0(x + t)$ moves left and we do not want to be told that $u(t,0) = A$. That contradicts the value $u(t,0) = u_0(t)$ from the arriving wave. The only possible resolution is a boundary layer near $x = 0$, where u has a fast transition between its two values. In that thin layer the neglected term $\frac{1}{100}u_{xx}$ becomes more important than u_x. Parabolicity is pushed to the wall, and u behaves hyperbolically elsewhere, but the boundary conditions are met.

Numerically the boundary layer is hard to resolve, and centered differences for u_x are dangerous.

---

† On the Mass. Turnpike it is the opposite; the outer lanes are both faster than the center. Is that true everywhere?

EXAMPLE 2   The dispersive equation $u_t = -u_{xxx}$

Substituting the exponential yields $\omega = k^3$. The frequency ω is real and there is no dissipation. However the k^3 term creates *dispersion*. It does not affect a single sine wave, which travels with velocity ω/k as before. This "phase velocity" is k^2; it varies with k. Therefore a sum of two waves, or a sum of n waves, **changes shape as it travels**. A concentrated pulse containing all wave numbers breaks up into a train of oscillations—with the highest wave numbers traveling fastest because of k^2.

In this example $\omega = k^3$ is called the **dispersion relation**. Its first purpose is to produce the solution itself:

$$u(t,x) = \frac{1}{2\pi} \int_{-\infty}^{\infty} e^{ik^3 t} e^{ikx} \hat{u}_0(k) dk. \tag{29}$$

At $t = 0$ this is u_0, the inverse transform of $\hat{u}_0$. At later times each harmonic e^{ikx} has its own factor $e^{i\omega t}$, and (29) combines those harmonics. For $u_0 = \delta$ and $\hat{u}_0 = 1$, that gives the fundamental solution. The integral with $e^{ik^3 t}$ is not elementary; it indicates a new scaling $x \sim t^{1/3}$. However the function with transform $e^{ik^3/3}$ is familiar in dispersion theory and especially in diffraction. It is known as the *Airy function*, and satisfies $Ai'' = xAi(x)$.

Now we take the harder step 4, to understand the equation and its solution. It is certainly a wave problem with real ω, not a diffusion problem. Since waves split up, it appears difficult to see far ahead in time—but the dispersion relation is the key. Watching a disturbance on the line $x + ct = 0$ in space-time, (29) is

$$u = \frac{1}{2\pi} \int e^{i(\omega - ck)t} \hat{u}_0 dk.$$

The main contribution to such an integral occurs at wave numbers where $\omega - ck$ is not changing. Otherwise a large t gives rapid oscillations that virtually cancel out—as they did for quantum theory in Section 3.6. This is the idea of *stationary phase*, that the important wave numbers at speed c satisfy

$$\frac{d}{dk}(\omega - ck) = 0 \quad \text{or} \quad \frac{d\omega}{dk} = c.$$

The quantity $d\omega/dk$ is the **group velocity**. In this example it is $3k^2$. It is different from the ratio ω/k, and considerably more subtle. I will describe a paradox and try to explain it:

> Suppose a stone is thrown into a deep pond. The longest waves spread out fastest, because ω is an increasing function of k. The crest of a wave moves at the velocity $c = \omega/k$. You would expect those crests to be found later at a distance ct from the center, but they are not there. If you try to watch the outside crests, you lose them! New crests come from behind, you follow them,

and they disappear too. The crests may be going at speed c, but the energy in a wave moves with its group velocity, which in deep water is $\frac{1}{2}c$.

A surfboarder must experience that effect, even in shallow water where the group velocity is higher. The surfboarder stays on the crest, so he travels with the phase velocity ω/k. Underneath him the wavelength is changing. Therefore his speed has to change and eventually he drops off the wave.

One way to understand group velocity is to look at two nearby waves. The first has wave number k and frequency ω; the second increases those by Δk and $\Delta\omega$. Their combined height is

$$e^{i\omega t}e^{ikx}[1 + e^{i\Delta\omega t}e^{i\Delta kx}].$$

This is a high frequency wave, but it is modulated by the term in brackets—which is like the low frequency envelope of Fig. 6.4. The sum in the brackets is 2 when

$$(\Delta\omega)t + (\Delta k)x = 0, \quad \text{or} \quad \frac{x}{t} = -\frac{\Delta\omega}{\Delta k} \approx -\frac{d\omega}{dk}.$$

This is reinforcement at the group velocity, and cancellation at all other velocities. The minus sign is from our use of e^{+ikx}; our velocities are to the left. In deep water the frequency ω is proportional to $\sqrt{k}$, $d\omega/dk$ is half of ω/k, and storms in the Pacific have been tracked for 2000 miles (at the group velocity!). In two dimensions the $\omega - k$ relation gives Kelvin's angle of $19.5°$ for the wedge behind a ship. I believe the speed of light was in error by a mile per second until the group velocity was put right.

EXAMPLE 3 The Klein-Gordon equation $u_{tt} = u_{xx} - u$

Substituting $e^{i\omega t}e^{ikx}$ gives $\omega^2 = k^2 + 1$. The frequencies are real (there are two square roots) and they depend nonlinearly on k. There is dispersion and no diffusion. Waves change shape, but less radically than in water because only the zero-order term $-u$ causes dispersion. At high wave numbers ω is close to k. The Klein-Gordon equation is hyperbolic, and from Exercise 6.4.21 it is also the telegraph equation. Its deeper applications are in quantum theory.

EXAMPLE 4 The beam equation $u_{tt} = -u_{xxxx}$

In this case the frequencies are $\omega^2 = k^4$, or $\omega = \pm k^2$. The group velocities are $\omega' = \pm 2k$, faster than the phase velocities $\omega/k = \pm k$. The waves come from bending a beam (or striking it). The equilibrium equation was $A^T C A u = u_{xxxx} = f$, and when the load depends on time the inertia term u_{tt} joins the equation. The motion is perpendicular to the beam and not parallel—as it was for a line of springs or a bar with $u_{tt} = u_{xx}$.

Notice that $u_{tt} = +u_{xxxx}$ has $\omega^2 = -k^4$; ω is not real. The equation has solutions $u = e^{k^2 t} e^{ikx}$ with an unbounded rate of growth k^2. Like the backward heat equation, this problem is totally ill-posed. Unlike the backward heat equation, we cannot reverse the time and cure it.

EXAMPLE 5   The Schrödinger equation $iu_t = u_{xx}$

The solutions $e^{i\omega t} e^{ikx}$ have $\omega = k^2$. Like the beam equation, the group velocity is $d\omega/dk = 2k$. Like other purely dispersive equations, ω is real. There is no dissipation and energy is conserved. For quantum mechanics that is a crucial point, because $|u|^2$ represents a *probability density*—associated with the chance of finding the particle at point x. For all times the total probability must be one, so we need $\int |u|^2 dx = 1$. That can be verified in frequency space, where $|e^{i\omega t}| = 1$ for real ω and the energy is fixed:

$$\int |e^{i\omega t} \hat{u}_0(k)|^2 dk = \int |\hat{u}_0(k)|^2 dk.$$

EXERCISES

6.4.1   Verify that the fundamental solution (10) satisfies $u_t = u_{xx}$.

6.4.2   At $t = 0$ a hot gas is on one side and a cold gas on the other: $u_0 = 1$ for $x > 0$, -1 for $x < 0$. Write down the solution (11) to $u_t = u_{xx}$, and with $z = y - x$ simplify to

$$u(t,x) = \frac{1}{2\sqrt{\pi t}} \int_{-x}^{x} e^{-z^2/4t} \, dz.$$

6.4.3   If $u(t,x)$ satisfies the one-dimensional heat equation, show that $U = u(t,x)u(t,y)$ satisfies the two-dimensional equation $U_t = U_{xx} + U_{yy}$. Why does this fail for the wave equation?

6.4.4   The fundamental solution with $\hat{u}_0 = 1$ is $u = \frac{1}{2\pi} \int e^{-k^2 t} e^{ikx} dk$. Then

$$\frac{du}{dx} = \frac{1}{2\pi} \int_{-\infty}^{\infty} e^{-k^2 t}(ik) e^{ikx} dk = -\frac{1}{4\pi t} \int_{-\infty}^{\infty} (e^{-k^2 t})(xe^{ikx}) dk = -\frac{xu}{2t}.$$

Solving $du/dx = -xu/2t$ leads to the correct response $u = ce^{-x^2/4t}$; c is decided at $x = 0$. *Problem*: Justify the step from one infinite integral to the other.

6.4.5   With $u = 0$ at the boundaries $x = 0$ and $x = \pi$, use the Fourier sine series for $u_0(x) = x$ in Section 4.1 to solve the heat equation.

6.4.6   For a bar with initial temperature $u_0 = 1$ and boundary conditions $u(t,0) = u(t,1) = 0$, how could u_0 be extended to an infinite bar $-\infty < x < \infty$ so as to keep the same solution to the heat equation?

6.4.7 (a)   If the ends of the bar are held at temperatures $u(t,0) = T_0$ and $u(t,1) = T_1$, what steady state $u_\infty(x)$ does the solution approach as $t \to \infty$?

(b)   Reduce $u_t = u_{xx}$ with initial value $u_0(x)$ and the boundary values T_0 and T_1 to the case of zero boundary values, by changing the unknown to $U = u - T_0 + (T_0 - T_1)x$. What is the equation and initial condition for U?

6.4.8   Substitute the Fourier formula $\hat{u} = \int u_0 e^{-ikx} dx$ into (8) and integrate with respect to k, using the same integral that led to (10). This is another derivation of the solution formula (11) for the heat equation.

6.4.9   Suppose the heat equation is fed by a constant source at the origin: $u_t = u_{xx} + \delta(x)$. Take Fourier transforms to find an ordinary differential equation for $\hat{u}(t,k)$, solve with zero initial conditions, and express $u(t,x)$ as an inverse Fourier transform.

6.4.10   Explain why the previous solution should agree with

$$u = \int_0^t \frac{1}{2\sqrt{\pi(t-s)}} e^{-x^2/4(t-s)}\, ds.$$

6.4.11 (a)   Solve $u_t = cu_x$ with $u_0(x) = \dfrac{1}{1+x^2}$

(b)   Solve $u_t = -u_x$ with $u(0,x) = 0$ for $x > 0$, $u(t,0) = 1$ for $t > 0$

(c)   Solve $u_{tt} = c^2 u_{xx}$ with initial conditions $u_0(x) = 0$ and $v_0(x) = 1$.

6.4.12   If $u_t = c(t,x)u_x$, the *characteristic curves* in the $x - t$ plane satisfy $dx/dt = -c(t,x)$. Along such a curve u is constant because

$$du = u_t dt + u_x dx = u_t dt + u_x(-c\,dt) = (u_t - cu_x)dt = 0.$$

(a)   Find the solution $X(t)$ of $dX/dt = e^t$ that has $X = x$ at $t = 0$

(b)   Verify that $u = u_0(X(t))$ solves $u_t = e^t u_x$

(c)   Do the same for the ordinary differential equation $dX/dt = X$ and the partial differential equation $u_t = xu_x$.

6.4.13   Solve the wave equation $u_{tt} = c^2 u_{xx}$ starting at rest $(v_0 = 0)$ from a square wave: $u_0 = 1$ for $|x| \le 1$, $u_0 = 0$ for $|x| > 1$. Graph the solution at $t = 2$.

6.4.14   How can $\sin x \cos t$ and $\cos x \sin t$ oscillate in place, when their sum $\sin(x + t)$ travels to the left?

6.4.15   For the wave equation $u_{tt} = c^2 u_{xx}$ starting from $u_0 = 0$ and $v_0 = \delta(x)$, find the solution given by (19). Sketch the response at $t = 1$ to striking a violin string at $x = 0$, $t = 0$.

6.4.16   Since ru satisfies the one-dimensional wave equation, the solution starting from a spherically symmetric $v_0(r)$ is given by (19):

$$ru = \tfrac{1}{2} \int_{r-t}^{r+t} sv_0(s)ds.$$

Solve the 3-dimensional wave equation if $u_0 = 0$ and $v_0 = e^{-r^2}$.

6.4.17   Find $u(r,t)$ from the formula above if $v_0 = 1$ in the solid sphere $|r| \leq R$ and $v_0 = 0$ outside. Show that u is discontinuous at $r = 0$, $t = 1/R$. The signals from the surface $r = R$ have focussed at the origin.

6.4.18   If the wave equation has $v_0 = 0$ then

$$u(t,x,y,z) = \frac{\partial}{\partial t} \left[\frac{1}{4\pi t} \int_S u_0 dS \right].$$

(a)   Find u in this 3-dimensional case if $u_0 = -1$ for $x < 0$, $u_0 = +1$ for $x > 0$. You will need the surface areas $2\pi t(t - x)$ and $2\pi t(t + x)$ where $u_0 = -1$ and $u_0 = +1$ on the sphere S of radius t around (x,y,z).

(b)   Why should the solution $u(x,y,z,t)$ agree with the 1-dimensional solution $u(x,t)$ in Fig. 6.19?

6.4.19   Verify that (28) satisfies the convection-diffusion equation, and by imitating (11) express the general solution u by superposition of this fundamental solution.

6.4.20   (a)   Show that the Klein-Gordon equation $u_{tt} = u_{xx} - u$ leads to $\frac{1}{2}(u_t^2 + u_x^2 + u^2)_t = (u_t u_x)_x$. Integrating over x with zero conditions at infinity, what energy E is constant in time?

(b)   Write Klein-Gordon as a symmetric system $U_t = \cdots$ with $U = (u_x, u_t, u)$.

6.4.21   Show that the telegraph equation $u_{tt} + 2au_t = u_{xx}$ becomes a Klein-Gordon equation with the change of variables $u = e^{-at}v$. (Another form of the equation is $w_{xt} = bw$.)

6.4.22   Convert the 3-dimensional wave equation from

$$u_{tt} = c^2(u_{xx} + u_{yy} + u_{zz}) \text{ to } U_t = GU_x + HU_y + KU_z,$$

with unknown vector $U = (cu_x, cu_y, cu_z, u_t)$ and symmetric matrices G, H, K. What energy is conserved by this system?

6.4.23   For a symmetric system $MU_t = GU_x + HU_y$ show that the energy—which is now the integral $\int\int U^T MU dx\, dy$—is still conserved.

6.4.24   Maxwell's equations in a vacuum reduce to $\partial E/\partial t = \text{curl } B$ and $\partial B/\partial t = -\text{curl } E$. Written as $U_t = GU_x + HU_y + KU_z$ with unknown vector $U = (E_1, E_2, E_3, B_1, B_2, B_3)$, show that G is symmetric.

6.4.25   Find the eigenvalues of $A = k_1 G + k_2 H + k_3 K$ above.

6.4.26   Find the $\omega - k$ relation for $u_t = u_x - u$ and $u_t = u_{xx} - u$, and write down the Fourier solutions $\int e^{i(\omega t + kx)} \hat{u}_0 dk$. Show that $e^{-t}F(x + t)$ solves $u_t = u_x - u$.

6.4.27   (a)   What is the $\omega - k$ relation for $u_t = cu_x$?

(b)   What is the fundamental solution—the response at time t to the initial value $u_0 = \delta(x)$?

(c)   How do you see this from the Fourier formula

$$u(t,x) = \frac{1}{2\pi} \int_{-\infty}^{\infty} e^{i\omega t} e^{ikx} \hat{u}_0 dk?$$

6.4.28 (a) Find the $\omega - k_1 - k_2 - k_3$ relation for the equation $u_t = c_1 u_x + c_2 u_y + c_3 u_z$, by substituting $u = e^{i\omega t} e^{ik_1 x} e^{ik_2 y} e^{ik_3 z}$.

(b) Verify that $u(t,x,y,z) = u_0(x + c_1 t, y + c_2 t, z + c_3 t)$ and describe what this solution does.

(c) The Fourier method (made obsolete by that explicit solution) is

$$u = \frac{1}{(2\pi)^3} \int\limits_{-\infty}^{\infty} \int \int e^{i(\omega t + k_1 x + k_2 y + k_3 z)} \hat{u}_0(k_1, k_2, k_3) dk_1 dk_2 dk_3.$$

If u_0 is a delta function at the origin, what is the response at time t?

6.4.29 What rescaling of X and T will change $U_T = V U_X + D U_{XX}$ into the dimensionless equation $u_t = u_x + u_{xx}/P$? The Péclet number is $P = VL/D$.

6.4.30 Find the $\omega - k$ relations and decide whether exponential solutions decay to zero, remain neutral (real ω), or grow:

$$u_{tt} = u_{xx} - u_{xxxx}, \quad u_t = u_{xx} + u_{xxx}, \quad u_{tt} = u_x.$$

What is the group velocity for the first equation?

6.4.31 Apply separation of variables to the equation $u_t = u_x$, by substituting $u = A(x)B(t)$ and then dividing by AB. If one side depends only on t and the other only on x, they must equal a constant k; what are A and B?

6.4.32 For the second-order wave equation $u_{tt} = u_{xx}$, the substitution of $u = A(x)B(t)$ will give second-order equations for A and B when the x and t variables are separated. From $B''/B = A''/A = -\omega^2$, find all solutions of the form $u = A(x)B(t)$.

6.4.33 For the heat equation $u_t = u_{xx}$, separation of variables leads to $u = e^{-\omega^2 t} \sin \omega x$. With the boundary conditions $u(t, 0) = 0$ and $u_x(t, 1) = 0$, what values of ω are allowed?

6.5 ■ DIFFERENCE METHODS FOR INITIAL-VALUE PROBLEMS

This section aims to give some guidance in the choice of difference methods. It is natural to start with ordinary differential equations. The true solution to $u' = f(t,u)$ evolves continuously in time, and we want to follow it by a discrete approximation. Both equations start from the initial value $u(0) = u_0$; the discrete one takes finite steps Δt and after n steps it reaches u_n. We hope and expect that this is close to the exact $u(n\Delta t)$, but a difference equation has a life of its own—it may or may not stay near the differential equation. There are at least three things to watch for:

(1) ***Accuracy***: the error $u(n\Delta t) - u_n$ should look like $C \Delta t^p$. To reduce this error we can either increase the order of accuracy p or decrease Δt

(2) ***Simplicity***: the step from u_n to u_{n+1} may be quick or slow, depending how often it makes use of the right side f in the differential equation

(3) ***Stability***: small errors are unavoidably introduced at every time step, and they become part of u_n—to grow or decay as *it* does and not as $u(t)$ does. If the errors grow uncontrollably (strong instability) then the difference equation may look accurate but it is useless.

To some extent these requirements are conflicting. The price of greater accuracy is likely to be paid in complexity—a method of "Runge-Kutta type" increases p by evaluating f more often at every step. I believe that the right balance is not extremely hard to find.† The more serious difficulty is with stability—to keep the difference equation as stable as the differential equation. That eliminates from the start some methods of exceptional accuracy. More important, it may force the time step Δt to be expensively small. We show by example how an otherwise stable method can become (weakly) unstable at a threshold value of Δt.

EXAMPLE 1 $\dfrac{du}{dt} = au$ is approximated by $\dfrac{u_{n+1} - u_n}{\Delta t} = au_n$

This is ***Euler's method***—the simplest of all difference equations, and probably the least accurate. Each new value u_{n+1} comes directly from $u_n + a\Delta t u_n$. The n steps from u_0 to u_n multiply n times by this growth factor $G = 1 + a\Delta t$. Thus

$$u_n = (1 + a\Delta t)^n u_0 \text{ is to be compared with the exact } u(n\Delta t) = e^{an\Delta t}u_0.$$

One way to make this comparison is to fix a point in time, say $t = n\Delta t$, and let $n \to \infty$ as $\Delta t \to 0$. This is the basic test of convergence:

$$\text{Does } u_n = \left(1 + \frac{at}{n}\right)^n u_0 \text{ approach } u(t) = e^{at}u_0?$$

---

† A rule of thumb is to choose p equal to the number of (decimal) digits required in u.

The answer is *yes*, as we show below. Euler's method is stable, in the sense that the discrete solution stays bounded by e^{Kt} or $e^{Kn\Delta t}$ for some constant K. However K may not be the same number as a; the growth in the difference equation could be faster than in the differential equation. The discrete equation cannot be *more* stable than the continuous problem, but it can certainly be less stable—especially at a finite value of Δt. In fact the difference approximation can grow when it should decay.

The growth factor (or amplification factor) $G = 1 + a\Delta t$ controls the discrete solution $u_n = G^n u_0$. If this factor has magnitude $|G| < 1$ we have decay. For positive values of a we must expect growth, and we see it: $1 + a\Delta t > 1$. The crucial test comes when a is negative and there should be decay: $1 + a\Delta t < 1$. The question is whether the amplification factor is pulled so far down by $a\Delta t$ that *it goes below* -1. To avoid that we need

$$\boxed{1 + a\Delta t \geq -1 \quad \text{or} \quad -a\Delta t \leq 2.} \tag{1}$$

This gives the threshold value of Δt. For small enough time steps (1) is satisfied, but when a is strongly negative—exactly the case of rapid decay in the true solution—we are compelled to keep Δt small. Therefore we go to an *implicit method* when $a \ll 0$.

EXAMPLE 2 $\dfrac{du}{dt} = au$ is approximated by $\dfrac{u_{n+1} - u_n}{\Delta t} = au_{n+1}$

This is ***backward Euler***—still only a first-order method ($p = 1$) but more stable than forward Euler. The difference is seen in the amplification factor:

$$u_{n+1} - a\Delta t u_{n+1} = u_n \quad \text{or} \quad u_{n+1} = \frac{1}{1 - a\Delta t} u_n = Gu_n.$$

For negative a this denominator is larger than 1. Therefore $|G| < 1$ and we have decay. Furthermore the same is true for any complex number in the left half-plane:

$$\boxed{\text{If} \quad \text{Re } a < 0 \text{ then } |G| < 1.} \tag{2}$$

That is called *absolute stability*, or *A-stability*. It is more demanding than the ordinary stability bound e^{Kt}. In fact the bound is now 1. The true solution $u = e^{at}u_0$ will decay when the real part of a is negative, and A-stability requires the same of the discrete solution.

For such stability there is a heavy price. It is not seen in this simple problem, where it was trivial to solve for u_{n+1}, but the difference is immediate for the nonlinear equations $u' = f(t,u)$ that we actually work with:

Forward: $u_{n+1} = u_n + \Delta t f(t_n, u_n)$ Backward: $u_{n+1} - \Delta t f(t_{n+1}, u_{n+1}) = u_n$

The first is *explicit*; it gives u_{n+1} directly. The cost is one substitution of known values $t_n = n\Delta t$ and u_n into f. The second is *implicit*; to find u_{n+1} we are required to solve a nonlinear equation. If we use Newton's method (probably modified as in Section 5.1, depending on the complexity of $\partial f/\partial u$) there will be repeated evaluations of f. This higher cost means that implicit methods are used only when necessary—but that necessity is not uncommon.

The first iteration in the implicit step can begin with u_n as a crude approximation to u_{n+1}. However a much better starting value is $u_n + \Delta t\, f(t_n, u_n)$. In other words Euler can be a predictor; we will see others.

EXAMPLE 3 $\dfrac{du}{dt} = au$ is approximated by $\dfrac{u_{n+1} - u_n}{\Delta t} = \dfrac{a}{2}(u_{n+1} + u_n)$

This is the *trapezoidal rule*. Like backward Euler, it is implicit and A-stable:

$$u_{n+1} = \frac{1 + \frac{1}{2}a\Delta t}{1 - \frac{1}{2}a\Delta t}\, u_n = Gu_n \quad \text{and} \quad |G| < 1 \quad \text{if} \quad \text{Re } a < 0.$$

For negative a the numerator is smaller than the denominator. The solution decays as it should. For imaginary a the numerator and denominator are complex and the magnitude is $|G| = 1$. This is a correct imitation of the true solution $e^{at}u_0$, which is conservative when a is imaginary: $|e^{at}| = 1$.

The trapezoidal method has another important advantage. It is more accurate than the other two methods. To test the accuracy we substitute the true solution $u(t)$ into the difference equation:

$$\frac{u(t + \Delta t) - u(t)}{\Delta t} = \frac{a}{2}(u(t + \Delta t) + u(t)) + \text{``local truncation error''}$$

After expanding $u(t + \Delta t)$ as $u(t) + \Delta t\, u'(t) + \cdots$ the leading terms are

$$\frac{\Delta t\, u' + \frac{1}{2}\Delta t^2 u'' + \frac{1}{6}\Delta t^3 u'''}{\Delta t} - \frac{a}{2}(2u + \Delta t\, u' + \frac{1}{2}\Delta t^2 u'') = O(\Delta t^2). \tag{3}$$

The accuracy is $p = 2$. Here u' cancelled au and $\frac{1}{2}\Delta t\, u''$ cancelled $\frac{1}{2}a\Delta t\, u'$, because of the original differential equation. The first of those cancellations is automatic for any consistent finite difference approximation. The cancellation of the Δt terms is not automatic, and would not have occurred in forward or backward Euler. Centering the method by using the average $\frac{1}{2}(u_{n+1} + u_n)$ has increased the accuracy. The exponent $p = 2$ remains correct in the nonlinear case $u' = f(t, u)$, where the right side of the difference equation is

$$\tfrac{1}{2}[f(t_{n+1}, u_{n+1}) + f(t_n, u_n)] = \tfrac{1}{2}[u'((n+1)\Delta t) + u'(n\Delta t)] = u' + \tfrac{1}{2}\Delta t\, u'' + \cdots$$

The cancellation still succeeds. ***There is no A-stable method with higher accuracy*** $p > 2$, and the constant $\frac{1}{6} - \frac{1}{4} = -\frac{1}{12}$ that multiplies the Δt^2 term in the truncation error (3) is known to be the smallest possible. That makes the trapezoidal method special, but we should not suggest that it is all-important. Other methods, explicit or implicit, may be completely satisfactory for the values of $a = \partial f / \partial u$ that appear in a given problem. If these methods are faster or more accurate or both, they will be preferred to the trapezoidal rule. We now turn in that direction, first to multistep methods and then briefly to Runge-Kutta methods.

More Accurate Approximations

At the beginning of this section I spoke about the choice of difference methods. Finite differences sounded only like a problem of *selection*, not *construction*. To a large extent that may be true, at least for ordinary differential equations. The standard codes can vary Δt and p as the calculation progresses, according to their own automatic internal estimates of the truncation error. Still we should present one example of constructing a difference equation to achieve a required order of accuracy:

$$\frac{u_{n+1} - u_n}{\Delta t} = \alpha f(t_n, u_n) + \beta f(t_{n-1}, u_{n-1}) \text{ is to have } p = 2.$$

This will be a two-step method, requiring values at time $t_{n-1} = (n-1)\Delta t$ as well as time t_n. Therefore it needs a special instruction to find u_1 (but that is not difficult). The method is explicit; u_{n+1} comes directly from previous values without involving f at the new time t_{n+1}. It will be essentially as fast as forward Euler, and now we make it more accurate:

$$\frac{u_{n+1} - u_n}{\Delta t} = u' + \tfrac{1}{2}\Delta t\, u'' + \cdots$$

is to match

$$\alpha f_n + \beta f_{n-1} = (\alpha + \beta)u' - \beta \Delta t\, u'' + \cdots$$

The coefficients agree if $\alpha + \beta = 1$ and $\beta = -\frac{1}{2}$, so that $\alpha = \frac{3}{2}$. That completes the construction of this ***multistep method***.

To test for stability we apply it to $u' = au$:

$$\frac{u_{n+1} - u_n}{\Delta t} = \frac{3}{2}au_n - \frac{1}{2}au_{n-1}. \tag{4}$$

Like Euler's method, the threshold of instability occurs when the growth factor reaches $G = -1$. Then $u_n = (-1)^n u_0$ satisfies the equation and does not decay.

Substituting into (4), this means that

$$\frac{(-1)^{n+1} - (-1)^n}{\Delta t} = \frac{3}{2} a(-1)^n - \frac{1}{2} a(-1)^{n-1} \quad \text{or} \quad a\Delta t = -1.$$

This value $a\Delta t = -1$ comes twice as quickly as the threshold $a\Delta t = -2$ for Euler—so there is a tighter limit on Δt.

It is useful to look at the factor G for other values of Δt, not only at the threshold. Since this is a two-step method it has *two* growth factors G_1 and G_2. They are determined so that $u_n = G^n u_0$ satisfies the difference equation.[†] Substituting into (4) and cancelling G^{n-1},

$$\frac{G^2 - G}{\Delta t} = \frac{3}{2} aG - \frac{1}{2} a.$$

The two roots of this quadratic are

$$G_1 = 1 + a\Delta t + \tfrac{1}{2} a^2 \Delta t^2 + \cdots \quad \text{and} \quad G_2 = \tfrac{1}{2} a\Delta t + \cdots$$

G_1 is close to $e^{a\Delta t}$, which amplifies the exact solution to the differential equation. The growth factor copies the exponential to second order because $p = 2$. The other root G_2 is small and causes no trouble. This is typical of all "Adams-Bashforth" multistep methods

$$\frac{u_{n+1} - u_n}{\Delta t} = c_1 f_n + c_2 f_{n-1} + \cdots + c_p f_{n-p+1}. \tag{5}$$

These are stable but not A-stable, with coefficients in the table below:

	c_1	c_2	c_3	c_4	stability threshold	error constant
$p=1$	1				-2	1/2
$p=2$	3/2	$-1/2$			-1	5/12
$p=3$	23/12	$-16/12$	5/12		$-6/11$	3/8
$p=4$	55/24	$-59/24$	37/24	$-9/24$	$-3/10$	251/720

The fourth-order method is often a good choice, although astronomers go above $p = 8$. The threshold is the lowest value of $a\Delta t$ for which $|G| \leq 1$; it gets progressively tighter. The error constant is the numerical coefficient of Δt^p in the truncation error. In many discussions the difference equation is multiplied through by Δt to

---

[†] There are two "power solutions" to the difference equation just as there are two exponential solutions to a second-order differential equation.

clear the denominator, and then the error constant is the coefficient of Δt^{p+1}. This is the *local* error, in which the other factor $d^{p+1}u/dt^{p+1}$ depends on the smoothness of u. Whether the local error is acceptable is a problem of step control; whether it is amplified by later steps is a problem of stability control.

A similar construction gives **implicit methods** of high accuracy. Compared with (5), they have an extra term $c_0 f_{n+1}$ at the new time level. Properly chosen, that adds one extra order of accuracy—as it did for the trapezoidal rule, which is the second method in the new table. Backward Euler is the first:

	c_0	c_1	c_2	c_3	stability threshold	error constant
$p=1$	1				$-\infty$	$-1/2$
$p=2$	1/2	1/2			$-\infty$	$-1/12$
$p=3$	5/12	8/12	$-1/12$		-6	$-1/24$
$p=4$	9/24	19/24	$-5/24$	$1/24$	-3	$-19/720$

Every row of both tables adds to 1, so that $u' = $ constant is solved at least to minimum accuracy.

You see that the error constants and the stability are all in favor of implicit methods. So is the user, except when solving for u_{n+1} becomes expensive. Then he thinks of a simple and successful trick:

P: Use the explicit formula to *predict* a new u^*_{n+1}
E: Use u^*_{n+1} to *evaluate* the right side $f^*_{n+1} = f(t_{n+1}, u^*_{n+1})$
C: Use f^*_{n+1} in the implicit formula to *correct* the new u_{n+1}

This is the **predictor-corrector** method, and it is best to keep the same accuracy (say $p = 4$) in both the explicit and implicit formulas. The stability is much improved if there is another E step to evaluate f_{n+1} with the corrected u_{n+1}. Then the sequence is *PECE*, followed by the next prediction u^*_{n+2}. Stability is even better if we continue the corrector to convergence: the CE steps are repeated until u_{n+1} no longer changes. It has reached its final value for the implicit formula. Frequently two or three corrections are enough, and this is faster than Newton's method in solving a single step of the implicit method.

In most codes the predictor-single corrector method is basic. There are two evaluations of f, but this extra calculation can be put to use. By comparing the predicted and corrected values of u_{n+1} the code can estimate the local truncation error as

$$\text{local error} = \frac{E}{E^* - E}(u_{n+1} - u^*_{n+1}),$$

where E^* and E are the error constants in the tables for the predictor and corrector respectively. It is this estimate that decides whether Δt should be decreased for greater accuracy or increased for greater speed (or kept as is).

For completeness here is a strongly unstable method, accurate but hopeless:

$$\frac{u_{n+1} - 4u_n + 3u_{n-1}}{\Delta t} = au_{n-1}. \tag{6}$$

The growth factors satisfy $G^2 - 4G + 3 = a\Delta t$. They are approximately $G_1 = 1$ and $G_2 = 3$. The first is genuine and desirable; G_1 stays near the exact exponential factor $e^{a\Delta t}$. But G_2 is no longer small, and it will destroy the calculation. There is always a spurious solution when a first-order equation is replaced by a second-order discretization, and it grows or decays according to G_2^n. In this case it grows much faster than any exponential, since 3^n far exceeds $e^{Kn\Delta t}$. This extreme instability would be caught immediately; for partial differential equations it can be more treacherous.

We summarize the main points:

6K For the model problem $u' = au$ the discrete equation determines

(1) the **accuracy**—by substituting the solution e^{at} of the differential equation
(2) the **stability**—by finding the solutions G^n of the difference equation.

With order of accuracy p, the growth factor G agrees with $e^{a\Delta t}$ through Δt^p. With stability, $|G| < 1$ in some range $C < a\Delta t < 0$. The constant C is the stability threshold, at which decaying solutions e^{at} begin to be approximated by non-decaying G^n.

A-stable methods have no threshold and $C = -\infty$. They are necessarily implicit—although implicit methods are not necessarily A-stable.

Multistep methods achieve high accuracy at minimal cost—only one or two evaluations of f at each step. However they are not perfect. If f is convenient and its evaluations are not expensive, **Runge-Kutta methods** are definitely competitive. They are compound one-step methods like

$$\frac{u_{n+1} - u_n}{\Delta t} = \frac{1}{2}[f(t_n, u_n) + f(t_{n+1}, u_n + \Delta t\, f(t_n, u_n))]. \tag{7}$$

You see the compounding. First there is the normal f_n, and then $u_n + \Delta t\, f_n$ appears *inside* the second evaluation. It partly imitates the trapezoidal rule, which uses $f(t_{n+1}, u_{n+1})$ and is implicit—while Runge-Kutta replaces u_{n+1} by Euler's prediction $u_n + \Delta t f_n$ and remains explicit. For the model equation $u' = au$ the result is

$$u_{n+1} = u_n + \frac{\Delta t}{2}[au_n + a(u_n + \Delta t a u_n)]$$

$$= (1 + a\Delta t + \tfrac{1}{2}a^2\Delta t^2)u_n = Gu_n. \tag{8}$$

This confirms the second-order accuracy, since G agrees with $e^{a\Delta t}$ through Δt^2. Of course there will be a stability threshold $a\Delta t = -C$, where the growth factor reaches $G = -1$ or $G = 1$ and we are on the margin of instability:

$$1 - C + \tfrac{1}{2}C^2 = -1 \quad \text{or} \quad 1 - C + \tfrac{1}{2}C^2 = +1.$$

Here is something new; the first equation has no real roots. G does not drop to -1. However it reaches $+1$ at the root $C = 2$ of the second equation. For larger Δt the discrete solution will grow while the true solution decays. We remark that the complete stability threshold is not a point but a closed curve in the complex plane—containing all the complex numbers $a\Delta t$ at which $|G| = 1$.

The discrete equation in (7) is known as *simplified* Runge-Kutta, with accuracy $p = 2$. The more famous version is compounded four times and achieves $p = 4$:

$$\frac{u_{n+1} - u_n}{\Delta t} = \frac{1}{3}(k_1 + 2k_2 + 2k_3 + k_4) \tag{9}$$

$$k_1 = \tfrac{1}{2} f(t_n, u_n) \qquad\qquad k_3 = \tfrac{1}{2} f(t_{n+1/2}, u_n + \Delta t k_2)$$
$$k_2 = \tfrac{1}{2} f(t_{n+1/2}, u_n + \Delta t k_1) \qquad k_4 = \tfrac{1}{2} f(t_{n+1}, u_n + 2\Delta t k_3)$$

As a one-step method no special instructions are necessary to start. Furthermore it is simple to change Δt; each step is independent of all information earlier than u_n. The growth factor is like (8) except that it reproduces $e^{a\Delta t}$ through $\frac{1}{24} a^4 \Delta t^4$. The error constant is the next coefficient $\frac{1}{120}$. Where the solution is flat (a is small) the step Δt can be large. Where the solution suddenly changes, the code reduces Δt with no effort. It is not convenient to alter p, and there is no absolute stability, but among highly accurate methods Runge-Kutta is especially easy to code and to run—probably the easiest there is.

To prove that the stability threshold is genuine we reproduce from the references the solution of $u' = -100u + 100 \sin t$. With $u(0) = 0$, Runge-Kutta gives

$$\boxed{\begin{aligned} u_{120} &= 0.151 \text{ with } \Delta t = \tfrac{3}{120} \\ u_{100} &= 670{,}000{,}000{,}000 \text{ with } \Delta t = \tfrac{3}{100} \end{aligned}} \tag{10}$$

One is the correct value of $u(n\Delta t) = u(3)$. The other seems to have stepped over the threshold, which is $a\Delta t = 2.78$.

Stiff Systems

It is much more common to have *systems* of differential equations than a single equation. The unknown u is a vector and so is the right side f. The model problem becomes $u' = Au$ and the eigenvalues λ_j of the square matrix A take the place of the single number a.

Fortunately, the finite difference formulas do not change. Euler's method gives the new vector $u_{n+1} = u_n + \Delta t f_n = (I + \Delta t A)u_n$ just as it gave the new scalar. *The growth factor is now a matrix:*

$$G = I + A\Delta t \text{ with eigenvalues } g_j = 1 + \lambda_j \Delta t.$$

At every step the discrete solution is multiplied by G. We can imagine that A and G are diagonalized, so that each eigenvector is growing at its own rate—g_j^n in the discrete equation and $e^{\lambda_j t}$ in the continuous equation. The differential equation is stable if all λ_j are negative, or more generally if their real parts are negative; then $e^{\lambda_j t} \to 0$. The difference equation is stable if all $|g_j| < 1$; then $g_j^n \to 0$. The trouble comes when the problem has different time scales:

Since Δt is the same for all components, its size may be controlled by the *most negative eigenvalue λ_j*—which corresponds to the fastest decay and dies out first in the true solution.

When the matrix A has eigenvalues of very different magnitudes, the system is called **stiff**. This is typical of chemical engineering problems and many others. There may be a large negative eigenvalue (strong damping of one component) or a large imaginary eigenvalue (rapid oscillation). The stability threshold forces Δt to be small, even when the real problem develops on a completely different time scale.

EXAMPLE $$\frac{du}{dt} = \begin{bmatrix} -3 & 1 \\ 0 & -100 \end{bmatrix} u \quad \text{or} \quad \begin{matrix} u_1' = -3u_1 + u_2 \\ u_2' = -100u_2 \end{matrix}$$

This is the model $u' = Au$, in which A has eigenvalues -3 and -100. If we use Euler's method the stability threshold is $a\Delta t = -2$; we need

$$3\Delta t < 2 \quad \text{and} \quad 100\Delta t < 2.$$

Stability requires $\Delta t < 2/100$ even though it is e^{-3t} that controls the true solution. In fact u_2 decays like e^{-100t} and for all practical purposes the solution is contained in u_1. But numerically we cannot ignore that very negative eigenvalue. The Runge-Kutta method would again blow up past 10^{10} for $\Delta t = 3/100$.

There is a distinction between single equations and systems that has not been made. For a single equation, the stability threshold is not unreasonable. We could not maintain accuracy if $a\Delta t$ were very large, so it is not only stability that is restrictive. Accuracy also limits the time step. For systems the situation is different. It is the part that decays like e^{-3t} that we want to compute, after e^{-100t} has virtually disappeared. In this case **accuracy is a restriction on $3\Delta t$ but stability is a restriction on** $100\Delta t$. The time step is 33 times smaller than necessary because of this stiff component.

The natural solution is to use an implicit method that keeps $|G| < 1$. For nonlinear equations, in which the derivative matrix $\partial f / \partial u$ decides the stiffness, a good code will provide a rough estimate of the eigenvalues. When Re λ becomes very negative it switches to an implicit approximation. And it may use a form of Newton's method, based on the same matrix $\partial f/\partial u$, to solve for u_{n+1}. Research into algorithms for stiff systems has become a dominant problem in this part of numerical analysis, and sources of software are discussed in the Appendix.

The Heat Equation

There is another important creator of stiff systems. They come from partial differential equations. When the space variable x is made discrete, a derivative like d^2/dx^2 is replaced by a finite difference matrix:

$$\frac{d^2u}{dx^2} \text{ becomes } \frac{1}{h^2}\begin{bmatrix} -2 & 1 & & \\ 1 & -2 & 1 & \\ & \cdot & \cdot & \cdot \\ & & 1 & -2 \end{bmatrix}\begin{bmatrix} u_1 \\ u_2 \\ \cdot \\ u_N \end{bmatrix} = Au.$$

The $1,-2,1$ matrix has negative eigenvalues, lying between -4 and 0. Those eigenvalues are all divided by the small number $h^2 = \Delta x^2$. Therefore the matrix A, which includes the factor $1/h^2$, has eigenvalues between $-4/h^2$ and 0. If the actual eigenvalues nearly reach these extremes—*which they do*—then the ratio of largest to smallest involves $1/h^2$. When the heat equation is approximated by

$$\boxed{\frac{u_{n+1} - u_n}{\Delta t} = Au} \tag{11}$$

we have produced an exceedingly stiff system. You see that the whole process was involuntary, and more or less inescapable. The derivative d^2/dx^2 has eigenvalues going all the way to $-\infty$, and its discrete approximations—by finite differences or finite elements—go as far down as the meshsize will allow. In this case it allows $-4/h^2$.

Equation (11) is the fundamental explicit method for the heat equation $u_t = u_{xx}$. It is Euler's method applied to $u' = Au$, after making the x-derivative discrete. We can find its stability threshold in three ways:

Fast way: The growth factor for Euler is $G = I + A\Delta t$ and the lowest eigenvalue is $1 - 4\Delta t/h^2$. This is above -1 if

$$\frac{4\Delta t}{h^2} < 2 \quad \text{or} \quad \frac{\Delta t}{h^2} < \frac{1}{2}. \tag{12}$$

Slow way: The difference equation for the approximation u_j^n at the point $t = n\Delta t$, $x = jh$ is

$$\frac{u_j^{n+1} - u_j^n}{\Delta t} = \frac{u_{j+1}^n - 2u_j^n + u_{j-1}^n}{h^2}.$$

The superscript is time, the subscript is space, and

$$u_j^{n+1} = \frac{\Delta t}{h^2} u_{j+1}^n + \left(1 - \frac{2\Delta t}{h^2}\right) u_j^n + \frac{\Delta t}{h^2} u_{j-1}^n. \tag{13}$$

Each new u_j^{n+1} is a weighted average of values at time level n, and the weight $1 - 2\Delta t/h^2$ is positive up to the same threshold as (12): $\Delta t/h^2 < \frac{1}{2}$. Positive weights are a guarantee of stability.

Right way: If at time 0 the starting value is an exponential $u_j^0 = e^{ijkh}$, then after one step (13) gives

$$u_j^1 = \frac{\Delta t}{h^2} e^{i(j+1)kh} + \left(1 - \frac{2\Delta t}{h^2}\right) e^{ijkh} + \frac{\Delta t}{h^2} e^{i(j-1)kh}$$

$$= \left(\frac{\Delta t}{h^2} e^{ikh} + 1 - \frac{2\Delta t}{h^2} + \frac{\Delta t}{h^2} e^{-ikh}\right) e^{ijkh}$$

$$= G(k)e^{ijkh}. \tag{14}$$

The expression in parentheses is the **growth factor for frequency** k. After n steps the exponential is multiplied by G^n; stability requires $|G| < 1$. At the frequency $e^{ikh} = -1$, the growth factor is

$$G = -\frac{\Delta t}{h^2} + 1 - \frac{2\Delta t}{h^2} - \frac{\Delta t}{h^2} = 1 - \frac{4\Delta t}{h^2}.$$

To have $|G| < 1$ we need $\dfrac{4\Delta t}{h^2} < 2$ or $\dfrac{\Delta t}{h^2} < \dfrac{1}{2}$.

You probably preferred the fast way, in which the extreme eigenvalue $-4/h^2$ led immediately to the threshold. But how are the eigenvalues to be found as the matrix gets large? They are certainly not computed in advance by any eigenvalue algorithm. *It is the Fourier method* (the right way) *that suggests the eigenvector and finds the eigenvalue*. The suggested eigenvector has components e^{ijkh}, and the suggested eigenvalue for a single time step is $G(k)$. The growth factor $G = 1 - 4\Delta t/h^2$ is the same as $1 + a\Delta t$. The Fourier method produced this factor automatically, and it also gave the frequency at which instability is most dangerous. (It was $k = \pi/h$, where $e^{ikh} = -1$.) This method of tracking exponentials, even if boundary conditions may prevent them from being the exact eigenvectors, continues to work when everything else gets too complicated. It will be developed more fully for the wave equation.

The threshold $\Delta t/h^2 < \frac{1}{2}$ was the first stability condition to be discovered for the heat equation. It was not noticed by the inventor of (11); without a computer he probably took only a few steps. However the rule is absolutely firm, and any choice larger than $\frac{1}{2}$ will produce oscillations. As h is reduced to improve accuracy in the x-direction, the time step must remain of order h^2. This is the dimensional balance between x and t in the heat equation $u_t = u_{xx}$, but its numerical consequences are very serious: parabolic equations usually require implicit methods to make high accuracy affordable.

An implicit method like backward Euler overcomes the stability restriction:

$$u' = Au \text{ becomes } u_{n+1} - \Delta t A u_{n+1} = u_n. \tag{15}$$

This is unconditionally stable; the only limitation on Δt is accuracy, not stability. The unknown is a vector, consisting of all the mesh values u_j^{n+1} at the new time level. The coefficient matrix $I - \Delta t A$ is positive definite. If we use finite elements in space, then $-A$ is the stiffness matrix K of Section 5.4. The mesh need not be regular. If we use finite differences, $-A$ can be tridiagonal in one space dimension and it can come from the 5-point scheme in two dimensions (for $u_t = u_{xx} + u_{yy}$). In every case the situation for a single time step is the same:

> *Each implicit step requires the solution of an equilibrium problem. Its positive definiteness is increased by the extra term $I/\Delta t$ from the time step. The initial-value problem becomes a sequence of boundary-value problems, one at each step.*

All the algorithms for the static problem can be employed here, with one addition. Since the equilibrium problem is solved many times, with right hand sides that come from the last known values u_n, we may store the triangular factors of $I - \Delta t A$ and use them at every step. This is natural for linear problems; in nonlinear problems the previous step will supply starting values for an iteration to reach u_{n+1}. However we must not underestimate the cost of implicit methods. On a large nonlinear problem, the squeeze between instability of explicit methods and complexity of implicit methods has made good solutions impossible. That is a critical deadlock, to be broken by the next generation of computers or numerical analysts.

Of course backward Euler is not the only implicit method. The trapezoidal rule is more accurate, and for heat equations it is called the ***Crank-Nicolson method***:

$$\frac{u_{n+1} - u_n}{\Delta t} = \frac{A}{2}(u_{n+1} + u_n). \tag{16}$$

Since this is A-stable for ordinary differential equations it is unconditionally stable for the heat equation. There are no restrictions on Δt or h^2, except for those that come from accuracy. We might also try analogues of Runge-Kutta or predictor-corrector, but discrete equations that are convenient for small problems can become unmanageable for the large systems that come from partial differential equations. Multistep methods are not popular. The order of accuracy $p = 2$ is generally acceptable—and frequently that is also the accuracy of the space discretization.

There is another aspect that is different for partial derivatives. It is not the equation $u_t = Au$, but rather the equation $u_t = u_{xx}$, that we ultimately want to solve. The first is a ***semi-discretization*** of the second; the space variables have been replaced by differences. This converted the heat equation to an ordinary differential equation $u_t = Au$, and we reached the final approximation in two steps—space was made discrete and then time. However the space step h and the time step Δt are linked. Both must go to zero. For explicit methods the ratio $r = \Delta t/h^2$ is constant. In other words, **the matrix A changes as Δt changes**. The problem $u_t = Au$ is not fixed, as it is for ordinary differential equations, but the matrix A (and the number of rows and columns and unknowns) varies with h.

This makes a difference in testing the accuracy. Euler's method is only first-order accurate for $u_t = Au$. However it may be possible to balance the error from discretizing time with the error from discretizing space. The space difference is in error by

$$u_{j+1} - 2u_j + u_{j-1} = (u + hu' + \tfrac{1}{2}h^2 u'' + \tfrac{1}{6}h^3 u''' + \tfrac{1}{24}h^4 u'''') - 2u$$

$$+ (u - hu' + \tfrac{1}{2}h^2 u'' - \tfrac{1}{6}h^3 u''' + \tfrac{1}{24}h^4 u'''')$$

$$= h^2 u'' + \tfrac{1}{12}h^4 u''''.$$

Dividing by h^2 this is $Au = u_{xx} + \tfrac{1}{12}h^2 u_{xxxx}$, ignoring higher order terms. The time difference in Euler's method is in error by

$$\frac{u_{n+1} - u_n}{\Delta t} = \frac{\Delta t\, u' + \tfrac{1}{2}\Delta t^2 u''}{\Delta t} = u_t + \tfrac{1}{2}\Delta t\, u_{tt}.$$

The method is consistent with $u_{xx} = u_t$. But if the next coefficient $\tfrac{1}{12}h^2$ is matched with the coefficient $\tfrac{1}{2}\Delta t$, the accuracy jumps to $p = 2$. (Note that $u_{xxxx} = u_{tt}$!) This matching gives the ratio $\Delta t/h^2 = \tfrac{1}{6}$, which is below $\tfrac{1}{2}$ and therefore stable as well as accurate.

Roughly speaking, Au was a lower order term in the ordinary differential equation, but in the partial differential equation it comes from u_{xx} and has equal importance with u_t. The factor $1/h^2$ makes stability more difficult to achieve, but it gives an extra chance for accuracy.

Convergence

Does the discrete solution $u_n = G^n u_0$ approach the true solution $u = e^{n\Delta t A}u_0$? The expected answer is yes. But there are two requirements for convergence, and one of them—*stability*—is by no means automatic. The other requirement is *consistency*—that the discrete problem approximates the right continuous problem. The fact that these two properties are sufficient for convergence, and that they are in some sense also *necessary* for convergence, is the **fundamental theorem of numerical analysis**:

6L (Lax equivalence theorem) The combination of consistency and stability is equivalent to convergence.

This equivalence requires careful definitions of the three properties— consistency, stability, and convergence. When u and G and A are scalars that is not difficult. For systems of equations, we need "norms" to measure the error vector $u - u_n$ and the matrix powers G^n. For partial differential equations the techniques go one step further; the proof that stability is necessary uses one of the basic theorems of functional analysis. Surprisingly, the innocent property of consistency is the hardest to define. But even in the scalar case—I should say *especially* in the scalar case—you will see how stability and consistency combine to control the error.

Lax proved the equivalence theorem in the context of initial-value problems. It is equally true for boundary-value problems, and for the approximation of functions, and for the approximation of integrals. It applies to every discretization, when the given problem $Lu = f$ is replaced by $L_h u_h = f_h$. We want to assume that the problems are close and prove that the solutions are close. That takes only a few lines when the equation $Lu = f$ is linear, and it is the essence of the fundamental theorem:

> Suppose f is changed to f_h and L is replaced by L_h. The requirements are
>
> consistency: $f_h \to f$ and $L_h u \to Lu$
>
> stability: the inverses L_h^{-1} remain uniformly bounded.
>
> Then as h goes to zero there is convergence:
>
> $$u - u_h = L_h^{-1}(L_h u - Lu) + L_h^{-1}(f - f_h) \to 0.$$

Consistency controls the quantities in parentheses, and stability controls the operators L_h^{-1} that act on them.

To apply this theorem to initial-value problems we have to make it concrete— to identify f_h and L_h and consistency and stability in each specific situation. Sometimes a direct approach is simpler, once we know what to look for. The crucial part is usually stability, that the approximate solutions remain within limits. If they stay bounded, they have nothing left to do but converge.

We prove this convergence directly for $u' = au$, without referring to the general theorem but by making the definitions specific. The parameter h becomes Δt, the approximations $u_h = L_h^{-1} f_h$ become the discrete solutions $G^n u_0$, and the requirements are

$$\text{stability:} \quad |G^n| \le e^{Kn\Delta t}$$

$$\text{consistency:} \quad |G - e^{a\Delta t}| \le C\Delta t^{p+1}.$$

Then the difference between $G^n u_0$ and the true solution $e^{na\Delta t} u_0$ converges to zero. It is a difference of powers—the powers of G and the powers of $H = e^{a\Delta t}$—and there is a "telescoping identity" that does the trick:

$$G^n - H^n = G^n - G^{n-1}H + G^{n-1}H - G^{n-2}H^2 + \cdots + GH^{n-1} - H^n$$

$$= G^{n-1}(G - H) + G^{n-2}(G - H)H + \cdots + (G - H)H^{n-1}. \tag{17}$$

Every term has a factor $G - H$, which is small by consistency. Every term has a power of G (possibly G^0) which is bounded by stability. And every term has a power of $H = e^{a\Delta t}$, which satisfies a similar bound because the continuous problem is well-posed. Since there are n terms in (17), or $t/\Delta t$ terms, the powers differ by

$$|G^n - H^n| \le \frac{t}{\Delta t} e^{Kt} C\Delta t^{p+1} = O(\Delta t^p). \tag{18}$$

Each part was necessary, and all parts together were sufficient, to achieve this order Δt^p. The telescoping sum in (17) is exactly how error accumulates in a difference equation. The powers H^k carry the true solution to time $k\Delta t$, $G - H$ produces the local truncation error at that time, and then the powers of G bring this truncation error forward *with the difference equation*. $G^{n-1}(G - H)$ reflects the earliest truncation error, committed at step 1, and $(G - H)H^{n-1}$ is the most recent error.

Ordinary stability permits the error bound (18) to grow like e^{Kt}. That is enough to prove convergence. It may not be enough for success in practice. *A*-stability, which has $K = 0$, gives an overwhelming improvement. For systems of equations an extra constant enters into stability and *A*-stability—they become $|G^n| \le De^{Kn\Delta t}$ and $|G^n| \le D$—and D also appears in (18). More important is the stability problem for partial differential equations, in which *every frequency has its own growth factor* $G(k)$. Each $G(k)$ is required to be stable, and more than that they must be uniformly stable; the same bound must apply to all frequencies.

We turn to that central problem, the Fourier analysis of difference equations in space as well as time.

The Stability of Partial Difference Equations

Stability for the heat equation was not difficult, provided Δt was very small. The second derivative in $u_t = u_{xx}$ is so strongly negative that the finite difference growth factor G seldom went above 1. The danger was at -1; to prevent a more negative G required a bound on $\Delta t/h^2$. In fact the negativity pushed us into implicit methods.

For the wave equation and other hyperbolic equations it will be different. In a model like $u_t = u_x$ the right side is neither negative nor positive. In a Fourier sense it is *pure imaginary*; the corresponding ordinary differential equation is $u_t = iku$. It is conservative; it neither grows nor decays. The differential equation is on the neutral borderline between stability and instability, and the difference equation has to fall the right way.

We begin with a single equation and a one-way wave:

$$\frac{\partial u}{\partial t} = c\frac{\partial u}{\partial x} \quad \text{with} \quad u(x,0) = u_0(x) \quad \text{and} \quad c > 0. \tag{19}$$

In a pure initial-value problem $u_0(x)$ is given on the whole line from $-\infty$ to ∞. There are no boundaries in the x-direction. The solution is required for all $t \ge 0$

and all x, and it is $u(t,x) = u_0(x + ct)$. The x derivative is u_0'; the t derivative is cu_0'. Thus (19) is satisfied, with the correct initial value. The graph of u is unchanged as time goes forward, except that *it is shifted to the left* (Fig. 6.22). The whole graph moves with velocity c. Starting from any point x^* at time 0, a signal travels along a straight line in the $x - t$ plane. The message is the value of u_0, and it is received at all points on the line $x + ct = x^*$. There is no damping and no distortion, and all oscillations $u_0 = e^{ikx}$ travel with the same speed.

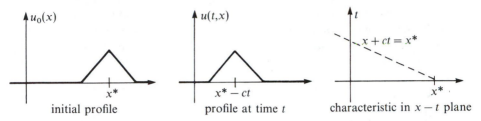

Fig. 6.22. The wave travels left with velocity c.

What is the corresponding difference equation? The first idea is

$$\frac{u(t + \Delta t, x) - u(t,x)}{\Delta t} = c\,\frac{u(t,x + h) - u(t,x)}{h}. \tag{20}$$

The t and x derivatives are replaced by forward differences. Since these are one-sided, the expected accuracy is $p = 1$. The question is whether (20) is stable:

Starting from a function like $u_0 = e^{ikx}$ at $t = 0$, does the difference equation imitate the differential equation or does it blow up?

In some sense those are the only possibilities. A stable equation will converge to the correct solution as Δt and h go to zero. An unstable equation will amplify some or all of the Fourier components of u_0, and the difference approximation will grow. This is not a question of absolute stability versus slow growth. It is a question of explosive growth, as the mesh is refined and the ratio of Δt to h is kept constant.

We test the difference equation for stability. Starting from $u_0 = e^{ikx}$ and $t = 0$, the first step of (20) produces

$$u(\Delta t, x) = e^{ikx} + \frac{c\Delta t}{h}\left[e^{ik(x + h)} - e^{ikx}\right].$$

The oscillation e^{ikx} is still there, multiplied by a **growth factor** G:

$$u(\Delta t, x) = \left[1 + \frac{c\Delta t}{h}(e^{ikh} - 1)\right]e^{ikx} = G(k)e^{ikx}. \tag{21}$$

G depends on the frequency k. At $k = 0$ it is $G = 1$; the constant function $e^{i0x} = 1$ is propagated without change (as it is in the differential equation). At the high frequency $k = \pi/h$, where $e^{ikh} = -1$, the growth factor is

$$G = 1 + \frac{c\Delta t}{h}(-1-1) = 1 - 2\frac{c\Delta t}{h}. \tag{22}$$

This is below 1; it may even be below -1. *On the mesh*, the function e^{ikx} looks like a sawtooth (Fig. 6.23). After n steps it still looks like a sawtooth. It may be reversed in sign, as we have drawn it, or the factor G^n may be positive. The second part of the figure shows a stable case, in which $|G| < 1$. The third part shows an unstable case, in which the amplification exceeds 1 (G drops below -1) and every step reverses the sign and increases the amplitude. After $t/\Delta t$ steps the initial e^{ikx} is multiplied by $G^{t/\Delta t}$, and is totally out of control.

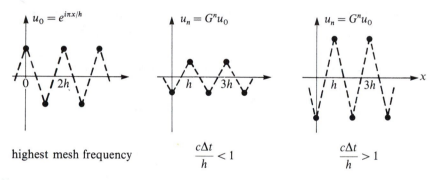

Fig. 6.23. Decay or growth of the highest frequency component.

The true solution $u = e^{ik(x+ct)}$ has constant amplitude. It also reverses sign, but it does so when e^{ikct} is -1, in other words when $kct = \pi$. At that time the sawtooth shifts left by h, which turns it upside down. Probably that is not the time at which the difference equation reverses sign. Even in the stable case, there is error in the phase as well as error in the amplitude.

Note that the still higher frequency $k = 2\pi/h$ brings us back, *on the mesh*, to a constant function. The initial function $e^{i2\pi x/h}$ is equal to 1 whenever x is a multiple of h. This is another example of **aliasing**, in which samples of a high frequency cannot be distinguished from samples of a lower frequency. Of course $k = 4\pi/h$ has the same alias, and $k = 3\pi/h$ has the alias π/h. That is the highest frequency which can be seen on the mesh.

For this discretization the highest frequency decides stability. That is not always so, but here the instability $|G| > 1$ strikes first for the growth factor in (22). It can be avoided by staying below the stability threshold of this difference equation:

$$G = 1 - 2\frac{c\Delta t}{h} \geq 1 \quad \text{provided} \quad \frac{c\Delta t}{h} \leq 1. \tag{23}$$

The ratio $r = c\Delta t/h$ is called the **Courant number**. It is dimensionless, since c has the dimensions of velocity. There are two ways to explain why r should not exceed 1:

6M The difference equation (20) is stable if $r = c\Delta t/h \leq 1$:

$$|G| = |1 - r + re^{ikh}| \leq 1 \text{ for all frequencies } k.$$

The equation is unstable if $r > 1$, for two reasons:

von Neumann: At the frequency where $e^{ikh} = -1$, the growth factor is $1 - 2r$, which is below -1

Courant-Friedrichs-Lewy: If $r > 1$ then the finite difference solution at t,x does not use information from u_0 near the correct point $x^* = x + ct$.

The von Neumann method is based on watching the Fourier components e^{ikx} and their growth factors. The Courant† method is based on comparing the signal speed c in the differential equation with the speed $h/\Delta t$ in this difference equation. If the approximation is too slow, so that $h/\Delta t < c$ (which means $r > 1$), then the signal from the point x^* at time 0 cannot reach other points when it should. The difference equation only moves it left by h in each time step, instead of by $c\Delta t$. The message about $u_0(x^*)$ arrives too late. This does not prove that the difference equation blows up; however blow-up occurs. It also does not prove convergence when $r \leq 1$, although convergence occurs. It does prove that convergence is generally impossible for $r > 1$, since the approximation computed at t,x depends on u_0 over the wrong interval. If the *domain of dependence* of the difference equation (the bold baseline in Fig. 6.24) does not include the source x^* of the signal, the discrete solution cannot know the value to which it should converge.

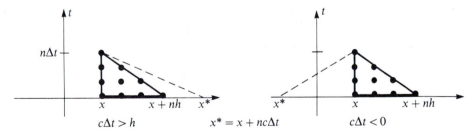

Fig. 6.24. Courant's proof of non-convergence for $r > 1$ and $r < 0$.

Courant's argument is even clearer if c is negative. Then $x^* = x + ct$ is to the *left* of x (Fig. 6.24b). The true solution at t,x is the initial value at x^*, while the difference equation uses values to the *right* of x. The discrete problem is looking for information in the wrong place. It is looking "downwind," whereas signals are

---

† More accurately Courant-Friedrichs-Lewy, here and below.

arriving with the wind. In this homely metaphor, it is ***upwind*** methods that succeed.

We emphasize that Courant's condition is not the whole story. Suppose the x-derivative in $u_t = cu_x$ is replaced by a ***centered difference***, which is more accurate than a one-sided difference:

$$\frac{u(t + \Delta t, x) - u(t,x)}{\Delta t} = c\,\frac{u(t, x + h) - u(t, x - h)}{2h}. \tag{24}$$

The triangles in Fig. 6.24 would now extend to the left as well as the right. The wind can come from either side. They are doubled in size, since all points between $x - nh$ and $x + nh$ contribute after n time steps to the approximation at x. Courant would still recognize that convergence is impossible if $r > 1$; x^* would remain too far to the right. But his argument would not recognize that ***this difference equation fails even if r is small***. The equation is unstable and the solution blows up, even when the dotted line to x^* stays *inside* the triangle. The difference equation uses the right information, but not in the right way.

The von Neumann test with $u_0 = e^{ikx}$ finds the difficulty immediately. After one step the discrete solution is

$$u(\Delta t, x) = u(0,x) + \frac{c\Delta t}{2h}\,[u(0, x + h) - u(0, x - h)]$$

$$= e^{ikx} + \frac{r}{2}\,[e^{ik(x + h)} - e^{ik(x - h)}].$$

The exponential e^{ikx} is multiplied by the growth factor

$$G(k) = 1 + \frac{r}{2}\,[e^{ikh} - e^{-ikh}] = 1 + ir\sin kh.$$

Even if r is small, this factor has $|G| > 1$. The magnitude is the square root of $1 + r^2\sin^2 kh$, always greater than 1. The next figure shows the growth factors for three difference equations: the one-sided case with $r < 1$ (*stable*), the one-sided case with $r > 1$ (*unstable* according to Courant), the centered case with $r < 1$ (*unstable* but not so recognized by Courant).

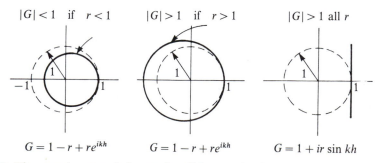

$$G = 1 - r + re^{ikh} \qquad\qquad G = 1 - r + re^{ikh} \qquad\qquad G = 1 + ir\sin kh$$

Fig. 6.25. The complex growth factors for all frequencies k.

Courant's condition is derived physically; von Neumann's is mathematical. The picture in physical space is not as complete as the picture in frequency space. Courant's is necessary for stability; von Neumann's is necessary and *sufficient*. They both apply to any difference equation that finds each new value u_j^{n+1}, at the point $x = jh$ on the time level $(n + 1)\Delta t$, from known values at time n—going L meshpoints to the left and R meshpoints to the right:

$$u_j^{n+1} = a_{-L}u_{j-L}^n + \cdots + a_0 u_j^n + \cdots + a_R u_{j+R}^n. \tag{25}$$

In the one-sided method $L = 0$ and $R = 1$. In the centered method $L = R = 1$. For the stability of any such method we have two tests:

> Courant's condition: $-Lh \le c\Delta t \le Rh$
>
> von Neumann's condition: $|G(k)| \le 1$ for all k.

The growth factor is $G = a_{-L}e^{-iLkh} + \cdots + a_0 + \cdots + a_R e^{iRkh}$. It multiplies e^{ikx} at each time step. It is a complex number, and von Neumann requires it to be in the unit disk: $|G| \le 1$. That implies the Courant condition (proof omitted). Each exponential grows like $G^n e^{ikx}$, and by Fourier analysis every function is a combination of exponentials:

$$\text{if} \quad u_0 = \int \hat{u}_0(k)e^{ikx}dk \quad \text{then} \quad u_n = \int G^n(k)\hat{u}_0(k)e^{ikx}dk. \tag{26}$$

With the von Neumann condition, u_n cannot grow (Exercise 6.5.21). It uses values of u_0 along the base of the triangle in Fig. 6.24, which now extends all the way from $x - Lnh$ to $x + Rnh$. As $\Delta t \to 0$, it somehow concentrates its attention on points near x^*.

EXAMPLE   A centered equation that is stable

The discrete equation (24) is totally unstable, but Friedrichs noticed that it was better to make no use of the center point t,x:

$$\frac{u(t + \Delta t,x) - \frac{1}{2}[u(t,x + h) + u(t,x - h)]}{\Delta t} = c\frac{u(t,x + h) - u(t,x - h)}{2h}. \tag{27}$$

Each new value is a weighted average of two old values:

$$u(t + \Delta t,x) = \left(\frac{1}{2} + \frac{r}{2}\right)u(t,x + h) + \left(\frac{1}{2} - \frac{r}{2}\right)u(t,x - h).$$

If u is an exponential e^{ikx}, each step multiplies it by

$$G(k) = \left(\frac{1}{2} + \frac{r}{2}\right) e^{ikh} + \left(\frac{1}{2} - \frac{r}{2}\right) e^{-ikh}.$$

The real part is $\cos kh$, the imaginary part is $r \sin kh$, and

$$|G|^2 = \cos^2 kh + r^2 \sin^2 kh \leq 1 \quad \text{if } -1 \leq r \leq 1. \tag{28}$$

This has all the stability that Courant allows. The triangles in Fig. 6.24 are again doubled in size, by including their mirror image to the left of x. Convergence is possible if x^* lies between $x - nh$ and $x + nh$, on the base of this doubled triangle. In other words, convergence is possible if the Courant number $r = c\Delta t/h$ is between -1 and 1. Since the growth factor is below 1 in (28), this convergence actually happens. Friedrichs' method allows c to be positive or negative—waves can move left or right, as long as their velocity c is not above the discrete velocity $h/\Delta t$.

With c allowed to have either sign, equation (27) is immediately useful for *systems*. It goes beyond the scalar model $u_t = cu_x$ to a vector equation $u_t = Cu_x$; C is a square matrix. For a wave equation (hyperbolic equation) C has real eigenvalues. By a change of variables it can be made symmetric or even diagonal. With c changed to C, and with all its eigenvalues satisfying $|\lambda| \leq h/\Delta t$, Friedrichs' method is stable and convergent.

EXAMPLE   The wave equation $u_{tt} = c^2 u_{xx}$

This is equivalent to a system of two first-order equations:

$$\begin{array}{c} v = u_t \\ w = cu_x \end{array} \qquad \begin{bmatrix} v \\ w \end{bmatrix}_t = \begin{bmatrix} 0 & c \\ c & 0 \end{bmatrix} \begin{bmatrix} v \\ w \end{bmatrix}_x = C \begin{bmatrix} v \\ w \end{bmatrix}_x. \tag{29}$$

The eigenvalues of this matrix are c and $-c$. There are two waves (or two signals), one moving left and the other moving right. The difference equation (27) handles them both, as long as the time step satisfies the Courant condition $c\Delta t/h \leq 1$.

The one-sided equation (20) will fail, since signals are coming from both sides. For the signal from the right it is an upwind method and it works. For the signal from the left it is unstable. One remedy is to diagonalize C and create an upwind scheme for both signals—by going to one side for $u_t = cu_x$ and to the other side for $u_t = -cu_x$. In practice, this extra effort is not made for linear equations but upwinding is important for the nonlinear conservation laws of the next section.

Finally we come to a more accurate and more desirable approximation. All the previous methods had $p = 1$. Each new value $u(t + \Delta t, x)$ was a combination of values from time level t, and the weights in that combination were always positive. If not, the methods failed. The one-sided method had weights r and $1 - r$, and Friedrichs' method had weights $\frac{1}{2}(1 + r)$ and $\frac{1}{2}(1 - r)$. To reach the higher accuracy $p = 2$ we have to allow negative weights, chosen so that $|G|$ is still below 1. The natural choice when $u_t = cu_x$ is the **Lax-Wendroff method**:

$$\frac{u(t + \Delta t, x) - u(t, x)}{\Delta t} = c \frac{u(t, x + h) - u(t, x - h)}{2h}$$

$$+ \frac{1}{2} c^2 \Delta t \frac{u(t, x + h) - 2u(t, x) + u(t, x - h)}{h^2}. \tag{30}$$

Without the c^2 correction, this is the unstable equation (24). **With that correction it becomes stable and also more accurate.** Courant's condition with one meshpoint on each side is $-h \le c\Delta t \le h$ or $-1 \le r \le 1$. This is generally only a necessary condition but here it is also sufficient:

6N   The Lax-Wendroff approximation to $u_t = cu_x$ is

$$u_j^{n+1} = (1 - r^2)u_j^n + \tfrac{1}{2}(r^2 + r)u_{j+1}^n + \tfrac{1}{2}(r^2 - r)u_{j-1}^n. \tag{31}$$

If $|r| \le 1$ then its growth factor satisfies the von Neumann condition:

$$|G| = |1 - r^2 + \tfrac{1}{2}(r^2 + r)e^{ikh} + \tfrac{1}{2}(r^2 - r)e^{-ikh}| \le 1. \tag{32}$$

Furthermore G differs from the true growth factor $e^{ikc\Delta t}$ by terms in Δt^3, so the order of accuracy is $p = 2$.

Equation (31) was copied from (30), in which you see the higher accuracy. Expanding the left side gives $u_t + \frac{1}{2}\Delta t\, u_{tt}$, as it did for ordinary differential equations. The centered difference on the right side approximates cu_x and the second difference approximates $\frac{1}{2}c^2\Delta t\, u_{xx}$. Since u_{tt} is the same as $c^2 u_{xx}$ (as a consequence of $u_t = cu_x$) these Δt terms agree and we have $p = 2$. The von Neumann condition (32) is an exercise in algebra.

Final Remarks

(1)   The second difference in Lax-Wendroff is an example of **dissipation**—the addition of a negative definite term to stabilize the discrete equation. Whenever $|G|$ is below 1 some dissipation is present. It can also be added as an artificial viscosity $-\varepsilon u_x^2$ that acts only when u_x is large.

(2)   All these methods have analogues for systems $u_t = Cu_x + Du_y + E(u_{xx} + u_{yy})$ in two space variables. There are also **splitting methods** that act alternately in the x and y directions—or alternately on the hyperbolic part (explicitly) and the parabolic part (implicity). This avoids a bound on $\Delta t/h^2$, for example in the Navier-Stokes equations.

(3)   In practice there are boundaries in the space directions; the problem is not infinite. For waves leaving through the boundary we look for **absorbing conditions**, to prevent a numerical reflection back into the domain.

(4)   The Fourier-von Neumann test $|G| \le 1$ applies **at each point** even if the material is not homogeneous. It is necessary and almost sufficient for stability, when the coefficients depend on x and t.

(5)   For systems $G(k)$ is a matrix and the stability test is $|G(k)^n| \le$ constant. This requires **each eigenvalue** of G to satisfy $|\lambda| \le 1$—or $|\lambda| \le 1 + K\Delta t$ since stability allows growth that is limited by e^{Kt}. The necessary and sufficient condition on G is a delicate problem in matrix theory.

(6)   Beyond the stability question, it is $G(k)$ that determines what happens to waves in the discrete model—how they **disperse**, how they are **damped**, and even **which direction they travel**. Beyond that, nonlinearity takes over.

EXERCISES

6.5.1   For $u' = -2u$ what is the largest Δt for which Euler's method is stable? What are the discrete solutions for $\Delta t = \frac{1}{2}$ and $\Delta t = 1$?

6.5.2   For $u' = -2u$ solve the backward Euler equation from $u_0 = 1$, with $\Delta t = \frac{1}{2}$ and $\Delta t = 1$. At $t = 5$ which is closer to the solution $u = e^{-2t} = e^{-10}$?

6.5.3   For $u' = -100u$ and $\Delta t = 1$, find the growth factors $G = (1 - a\Delta t)^{-1}$ and $G = (1 + \frac{1}{2} a\Delta t)(1 - \frac{1}{2} a\Delta t)^{-1}$ for backward Euler and the trapezoidal rule. Which solution oscillates with slow decay?

6.5.4   Motion around a circle (eigenvalues $\pm i$) is described by

$$u' = Ku = \begin{bmatrix} 0 & -1 \\ 1 & 0 \end{bmatrix} u \quad \text{with solution } u = \begin{bmatrix} \cos t \\ \sin t \end{bmatrix}.$$

Suppose we try to produce a circle on a video terminal from the approximations of Euler, the trapezoidal rule, and backward Euler:

(a) $u_{n+1} - u_n = \Delta t K u_n$,   or $u_{n+1} = (I + \Delta t K)u_n = G_1 u_n$
(b) $u_{n+1} - u_n = \frac{1}{2} \Delta t K(u_{n+1} + u_n)$,   or $u_{n+1} = (I - \frac{1}{2} \Delta t K)^{-1}(I + \frac{1}{2} \Delta t K)u_n = G_2 u_n$
(c) $u_{n+1} - u_n = \Delta t K u_{n+1}$,   or $u_{n+1} = (I - \Delta t K)^{-1} u_n = G_3 u_n$.

Find the eigenvalues of G_1, G_2, G_3 (depending on Δt) and decide from their absolute values whether u_n goes to zero, produces a circle, or leaves the screen.

6.5.5   (a)   For an explicit method the growth factor G is a polynomial in $a\Delta t$. Why is A-stability ($|G| < 1$ for all $a < 0$) not possible?

(b)   For an implicit method G has $1 - c_0 a\Delta t$ in the denominator. Show that the 3rd-order method in the implicit table is still not A-stable.

6.5.6   Find the growth factors G_1 and G_2 for the *leapfrog method* $u_{n+1} - u_{n-1} = 2a\Delta t u_n$ by solving $G^2 - 1 = 2a\Delta t G$. Show that one of the factors is below -1 if a is negative.

6.5.7   Choose the constants in $u_{n+1} - u_{n-1} = 2\Delta t(c_0 u_{n+1} + c_1 u_n + c_2 u_{n-1})$ to achieve 3rd-order accuracy in approximating the solution $u_n = e^{n\Delta t}$ of $u' = u$.

6.5.8   Write a brief code for the explicit and implicit methods in the two tables and show experimentally that the stability thresholds are correct. Compare accuracies for $u' = -u$ at $t = 1$.

6.5.9   The growth factor for 4th order Runge-Kutta, applied to $u' = au$ with $x = a\Delta t$, is $G = 1 + x + \dfrac{x^2}{2} + \dfrac{x^3}{6} + \dfrac{x^4}{24}$. Show that $G > 1$ for the example in equation (10) that grew to $670{,}000{,}000{,}000$ with $x = (-100)(3/100)$.

6.5.10   The solution to $u' = f(t)$, in which f is independent of u, is just the integral $u = \int f(t)dt$. Apply the Runge-Kutta method (9) and show that it reduces to *Simpson's rule* for numerical integration:

$$\int_{n\Delta t}^{(n+1)\Delta t} f(t)dt \approx \frac{\Delta t}{6} [f(t_n) + 4f(t_{n+1/2}) + f(t_{n+1})].$$

6.5.11   A single equation like $u' = -100u + 200$ is stiff for the initial value $u_0 = 2$. Find the general solution and choose the constant to match this initial condition. What is the largest time step for which Euler's method will be stable?

6.5.12   Second-order equations $u'' = f(t,u)$ are important in applications and can be approximated by $u_{n+1} - 2u_n + u_{n-1} = \Delta t^2 f_n$. For the model problem $u'' = -qu$ the two growth factors satisfy

$$G^2 - 2G + 1 = -q\Delta t^2 G.$$

Up to what threshold $q\Delta t^2 = C$ do the two roots satisfy $|G| = 1$? *Note*: At that limit the two roots are $G = -1$; beyond it one root drops below -1; within the threshold the roots give pure oscillations that correspond to the true solution of the differential equation.

6.5.13   A more accurate approximation to $u'' = f(u)$ is Numerov's method

$$u_{n+1} - 2u_n + u_{n-1} = \Delta t^2 \left(\tfrac{1}{12} f_{n+1} + \tfrac{10}{12} f_n + \tfrac{1}{12} f_{n-1}\right).$$

By expanding all terms in Taylor series around time n find the first power of Δt that does not cancel. What is the difference equation for the growth factors G in the model $u'' = -qu$?

6.5.14   Use the telescoping sum in (17) to show that

$$(1 + \Delta t)^n - e^{n\Delta t} \to 0 \quad \text{as} \quad \Delta t \to 0, \quad \text{with} \quad n\Delta t = t \text{ fixed.}$$

6.5.15   If $u_t = -u_x$ is approximated "downwind" by

$$\frac{u(t + \Delta t, x) - u(t,x)}{\Delta t} = -\frac{u(t, x + h) - u(t)}{h},$$

find the solutions $u = G^n e^{ikx}$ and confirm that $|G| > 1$.

6.5.16   What are the exponential solutions $u = e^{i\omega t}e^{ikx}$ to the "semidiscretization" $du/dt = [u(t, x + h) - u(t,x)]/h$, continuous in time and discrete in space? Is it stable?

6.5.17   For the equation $u_t = cu_x$, suppose each new computed value is a combination of the three previous values

$$u(t + \Delta t, x) = a_0 u(t, x) + a_1 u(t, x + h) + a_2 u(t, x + 2h).$$

What is the Courant condition for convergence? Draw a figure like 6.24.

6.5.18   Sketch the points $G = \cos kh + ir \sin kh$ inside the unit circle in the complex plane, with $r < 1$ in Friedrichs' method (25). The points should lie on the ellipse $x^2 + (y/r)^2 = 1$.

6.5.19   If u_{xx} is replaced by Au, with eigenvalues between $-4/h^2$ and 0, and $u_t = u_{xx}$ is replaced by an explicit-implicit mixture

$$\frac{u_{n+1} - u_n}{\Delta t} = \theta A u_{n+1} + (1 - \theta) A u_n,$$

find the stability threshold $\Delta t / h^2 \le (2 - 4\theta)^{-1}$ for $\theta < \frac{1}{2}$. Show that for $\theta \ge \frac{1}{2}$ the growth factor is always below 1; there is A-stability for Crank-Nicolson ($\theta = \frac{1}{2}$) and backward Euler ($\theta = 1$) and all methods in between.

6.5.20   The Lax-Wendroff growth factor is $G = 1 + ir \sin kh - r^2(1 - \cos kh)$, after simplifying (32). Compute $|G|^2$ from the squares of real and imaginary parts and verify the von Neumann condition $|G| \le 1$ if $r^2 \le 1$.

6.5.21   Show that $\int u_n^2(x)dx \le \int u_0^2(x)dx$ if all $|G(k)| \le 1$, by using the Plancherel equality 4.3.9 to convert to transform space: Why is $\int |\hat{u}_n(k)|^2 dk \le \int |\hat{u}_0(k)|^2 dk$?

6.5.22   Suppose A is the doubly-infinite constant-diagonal matrix with a_j on its jth diagonal. If these numbers come from the difference equation (25) then each time step is $u(t + \Delta t) = Au(t)$. Under what condition do the powers A^n remain bounded?

Difference methods for $u_t = cu_x$

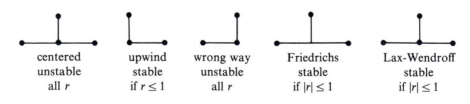

centered	upwind	wrong way	Friedrichs	Lax-Wendroff
unstable	stable	unstable	stable	stable
all r	if $r \le 1$	all r	if $|r| \le 1$	if $|r| \le 1$

NONLINEAR CONSERVATION LAWS ■ 6.6

Nature is nonlinear. That does not mean it is chaotic, although chaos sometimes occurs. It does not mean instability, although solutions can blow up. In some cases the nonlinear effects are *stabilizing*, and a differential equation that is forced to choose among many solutions will choose the best one. In other cases solutions remain smooth, and a change of variables makes the problem linear. That is an everyday event for ordinary differential equations, but it was an extraordinary event for partial differential equations. Solitary waves appeared on the computer, they were unmistakably regular, and the theory of solitons followed. Possibly this is the deepest contribution to mathematical physics the computer will ever make. It lies at the other extreme from the chaotic numerical experiments with $u_{n+1} = au_n - au_n^2$, but it was totally unexpected and beautiful.

The nonlinear problems that are stabilized or linearized will appear very innocent. They look close to the wave equation and diffusion equation and dispersive equation of Section 6.4, but you notice an extra u:

$$u_t + uu_x = 0 \qquad \text{(conservation law)}$$

$$u_t + uu_x = u_{xx} \qquad \text{(Burgers' equation)}$$

$$u_t + uu_x = -u_{xxx} \qquad \text{(KdV equation)}$$

The first is hyperbolic, and shock waves appear. The second is parabolic, but the diffusion term u_{xx} has to battle with the shocks. The third is the Korteweg-de Vries equation, for which certain solutions pass through each other and emerge unchanged—like a linear superposition of $\sin(x + t)$ and $\sin(x + 2t)$. This time the dispersion term u_{xxx} controls the shocks; it takes over when there is a rapid change in u. The real surprise was the superposition of solitons.

Our starting point is a nonlinear example from geometry. I am tempted to call it simple, and it is, even though it contains more mathematics than we can do. It begins with a curve C in the plane, and it moves that curve *perpendicular to itself*. If C is the x-axis, that line moves upwards (say with speed 1). If C is a circle, that circle expands. At every point where the curve has a tangent, the movement is perpendicular to that tangent. It is like a fire spreading forward. *At time t it reaches all unburned points which are a distance t from the original curve.* We could draw circles of radius t around the points of C, and the "envelope" of those circles would be the new curve C_t.†

Eventually we want an equation for C_t. It will not be linear, and to see that fact and its consequences is more important than the equation itself. Although not linear, it is absolutely controlled by a family of straight lines! Those are the rays going out at right angles to the curve C. When everything is smooth, each new

---

† C_{200} is like a 200-mile fishing limit around the coastline C.

curve C_t will be found at a distance t along those rays. C_t may be longer than C, as it was for expanding circles. It might be shorter, if time goes the other way and the circles contract. But the decisive effect of nonlinearity does not come *until the rays intersect*. From that time on, part of the original curve is completely lost from the problem—which can never happen for a linear equation.

Fig. 6.26 shows four of these possibilities. In region ① the front expands with no problems. Next to that, corner ② in C becomes a circular arc in C_t. The front contains all points at distance t, and an outward-pointing corner is rounded off. In region ③ the rays point inward. They intersect and *do not reach the front*. The construction of circles around C remains correct, but some circles fall short of the front and make no contribution. The fire would reach the same curve C_t whether it was burning at the start of those lost rays or not.

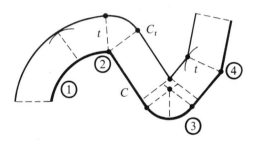

Fig. 6.26. The front C_t at a distance t from C.

The corner at point ④ in the figure is an extreme case. The rays near it intersect immediately and are lost inside the front. This inward-pointing corner is not smoothed out. In fact all the curves C_t have inward corners, which lie on a "shock path" that bisects the corner angle.

Note You can add to the figure by putting the fire above C_t, and watching the front move down. Everything is reversed—but not exactly! The fire will follow the rays that start from C_t and continue past C. At ① the front contracts. At ② it shrinks to a point. The circular arc above ② in C_t disappears into the corner in C. Carried below C that corner will persist and a shock will start—as it did going up at ④. On the other hand, the corner at ④ will now spread downward into a circular arc. In a word, the situations at ② and ④ are opposite. We pick out two of the many nonlinear effects:

(1) Part of C is lost, where the burned area is concave.

(2) Going forward and backward does *not* reproduce the original curve C (in particular at ④).

This is not like the irreversible heat equation, which smoothed everything in sight. Here $u_t = +uu_x$ is no less solvable than $u_t = -uu_x$, but there is loss of information and irreversibility both ways. The loss comes when corners are *created* by the nonlinear wave equation, and when they are *destroyed* by the linear heat equation.

Which partial differential equation will give rise to these moving fronts? That question has more than one answer. We first propose the wave equation $u_{tt} = u_{xx} + u_{yy}$, with initial values given along C in the x–y plane. Signals spread in circles around each point of C. The sharpest signal is heard *on the rays*. In x–y–t space those rays generate characteristic surfaces, like characteristic lines in one space dimension. Note that where rays intersect and fall short of the wave front—coming from the part of C that was "lost"—the values of u_0 still influence the solution u. They cannot disappear from a linear equation, but they arrive later and have less effect.

The front itself satisfies a nonlinear equation. For each t there are two ways to describe it, and an equation for each:

$$
\text{the curves } y = U(t,x) \quad \text{satisfy} \left(\frac{\partial U}{\partial t}\right)^2 - \left(\frac{\partial U}{\partial x}\right)^2 = 1
$$

$$
\text{the curves } S(y,x) = t \quad \text{satisfy} \left(\frac{\partial S}{\partial y}\right)^2 + \left(\frac{\partial S}{\partial x}\right)^2 = 1
$$

(1)

In the first, a front like C_t is the graph of $y = U(t,x)$. The function U gives the shape of the curve. In the second, its equation is $S(y,x) = t$. The two are equivalent, and this is clearest for circular fronts $x^2 + y^2 = t^2$. Looked at one way, the circle is $\sqrt{x^2 + y^2} = t$. The function $S = \sqrt{x^2 + y^2}$ solves the second differential equation. In the other form, the circle is the graph of $y = \sqrt{t^2 - x^2}$ and $U = \sqrt{t^2 - x^2}$ solves the first equation. The two descriptions are cross-sections of the same surface in x–y–t space. That surface is the cone $x^2 + y^2 = t^2$ containing the rays—with circles for the cross-sections $t = $ constant and hyperbolas for $y = $ constant.

The geometry of a moving front is more attractive than the equations (1), but we take this chance to mention that those are Hamilton-Jacobi equations. Their classical application is the one given here—to make surfaces out of rays. Their modern application is to optimal control theory, and in engineering texts they become Bellman equations—because they are the continuous counterpart of *dynamic programming*. The first one is

$$
\frac{\partial U}{\partial t} = H\left(\frac{\partial U}{\partial x}\right) = \sqrt{1 + (\partial U/\partial x)^2}.
$$

(2)

Note that the rays (straight lines) are shortest paths. They minimize the integral of $\sqrt{1 + (du/dx)^2}$. That leads to Hamilton's equations $du/dt = \partial H/\partial p$ and $dp/dt = -\partial H/\partial u$, by the least action principles of Section 3.6. They come from wave fronts and geometrical optics, not only for $u_{tt} = u_{xx} + u_{yy}$ but also for Schrödinger and Klein-Gordon and Maxwell. In each case we look for the *caustic* curve at which rays intersect and discontinuities develop.

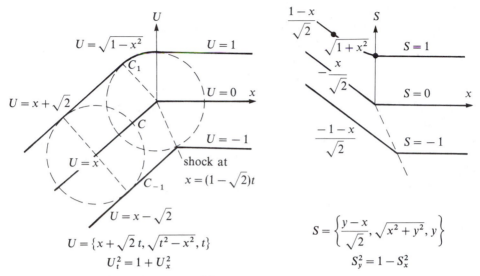

Fig. 6.27. A wave front that starts with a corner.

Because 3-dimensional surfaces are tricky (even when generated by straight lines) another example is shown in Fig. 6.27. The three parts of the cross-sections come from the planes $y = t$ and $y = x + \sqrt{2}\,t$, with the cone $x^2 + y^2 = t^2$ in between. I prefer the figure on the left, showing the curve C with an ordinary corner. Moving forward, that corner spreads into a circular arc as it did previously at point ②. Moving backward it remains a corner, as it did at point ④. One is a "rarefaction wave" and the other is a "shock wave."

The caustic is really the frontier between classical and modern. In the past, the "eikonal equation" (1) was solved up to the discontinuities and not beyond. Now the solution is continued further, even though the rays intersect. A decision is needed: which rays are kept and which are lost? In our geometric problem this was settled by Huygens' principle—to maintain the outer envelope of the circles around C. In conservation laws it will be the *entropy condition*. We have to prevent rays from crossing, even if entropy increases and shocks appear.

Conservation Laws

Suppose a material is distributed along a line. *The rate of change of the amount in a fixed interval equals the flux through the boundaries of the interval.* That is the conservation law. The total amount of material is conserved. If the system is in equilibrium, the divergence is zero. (In the continuous case div $w = 0$; in the discrete case $A^T y = 0$.) If the system is *not* in equilibrium, and the density u is changing because of the flux $F(u)$, conservation in the interval from x to $x + \Delta x$ is

$$\frac{d}{dt} \int_x^{x+\Delta x} u \, dx + [F(x + \Delta x) - F(x)] = 0. \tag{3}$$

As $\Delta x \to 0$ this integral law becomes a differential equation:

$$\frac{\partial u}{\partial t} + \frac{\partial F(u)}{\partial x} = 0. \tag{4}$$

It is a nonlinear equation when the flux is a nonlinear function of u. We are given the initial function $u_0(x)$. If the derivative of F is $dF/du = -c(u)$, the equation becomes

$$\frac{\partial u}{\partial t} = c(u)\frac{\partial u}{\partial x} \quad \text{with solution} \quad u = u_0(x + c(u)t). \tag{5}$$

The problem seems solved. The solution u is constant along every straight line $x + c(u)t = $ constant. Those are the characteristic lines on which the signals travel. But the slope of each line depends on $c(u)$, and therefore **the slope depends on the signal.** Some values of u_0 are carried faster than others, and when a fast signal overtakes a slower signal we have a problem:

> When rays intersect, **which value of u_0 is the correct u?**

EXAMPLE $\qquad\qquad F(u) = \frac{1}{2}u^2 \quad \text{and} \quad u_0(x) = \begin{cases} A \text{ for } x < 0 \\ B \text{ for } x > 0 \end{cases}$

This is the most frequently studied of all conservation laws:

$$\frac{\partial u}{\partial t} + \frac{\partial}{\partial x}(\tfrac{1}{2}u^2) = 0 \quad \text{or} \quad u_t + uu_x = 0. \tag{6}$$

The solution formula is $u = u_0(x - ut)$. Each signal travels along a straight line $x - ut = $ constant. The signal that starts at the point x^* (when $t = 0$) is the initial value $u_0(x^*)$; call it u^*. The solution keeps this value u^* along the line $x - u^*t = x^*$. The slope in the $x - t$ plane (Fig. 6.28) is $1/u^*$. In our example u^* is either A or B, according to whether the starting point x^* is to the left or right of zero.

In the first figure A is less than B. The initial function u_0 is increasing; the jump is positive. In that case, for the flux $F = \frac{1}{2}u^2$, the rays do not meet. The lines leaving from the left of the origin with slope $1/A$ do not catch those from the right. The signal speeds are A and B, and when $A < B$ the faster signal starts out ahead and stays ahead. The space has been filled in by a fan.

In the second figure A is greater than B. The initial function is decreasing, from 0 to -1, and the signal starting from $x > 0$ catches the signal from $x < 0$. It is easy to find the time of intersection:

$$x - At = x^* \text{ intersects } x - Bt = x^{**} \text{ at } t = \frac{x^{**} - x^*}{A - B}.$$

Then $u(t,x)$ is assigned *both* values A and B. When u_0 is decreasing, **the nonlinear equation $u_t = -uu_x$ has no solution** (in the ordinary sense) that extends to all $t > 0$. We have to settle for a "weak solution," after the signals overlap.

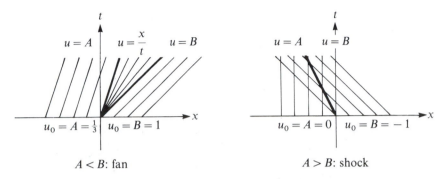

$A < B$: fan $\qquad\qquad\qquad\qquad A > B$: shock

Fig. 6.28. The lines $x - ut = \text{constant}$, spreading or intersecting.

That difficulty is also hidden in $u = u_0(x - ut)$. Several values of u may satisfy that formula. If $u_0(x) = x^2$, then $u = (x - ut)^2$ can have two solutions. Two rays have reached the same point x,t. For the more extreme case of a jump from A to B, there may be one solution, or two, or none.

This example indicates the problem but not the solution. The jump in u_0 makes the difficulty more than usually severe. Rays starting just to the left and right of $x = 0$ will either *intersect immediately* (if $A > B$) or *separate immediately* (if $A < B$). New input is needed, and it comes from going back to the conservation law. It leads to the following rule:

60 If the solution to $u_t + F(u)_x = 0$ has different values across a shock curve $x = X(t)$, then the jumps in u and $F(u)$ must satisfy

$$s[u] = [F]. \qquad (7)$$

The shock speed s is dX/dt. The jumps $[u]$ and $[F]$ are $u_{\text{right}} - u_{\text{left}}$ and $F(u_{\text{right}}) - F(u_{\text{left}})$.

To justify this rule $s[u] = [F]$, suppose $u_{\text{left}} = A$ and $u_{\text{right}} = B$. On any interval the conservation law states that

$$\frac{\partial}{\partial t} \int_{x_1}^{x_2} u \, dx = F(x_1) - F(x_2).$$

If the shock location $x = X(t)$ is between x_1 and x_2, this becomes

$$\frac{\partial}{\partial t} [(x_2 - X)B + (X - x_1)A] = F(A) - F(B).$$

With $\partial X/\partial t = s$ this is exactly $s(A - B) = F(A) - F(B)$ or $s[u] = [F]$.

That proof assumed constant values $u = A$ and $u = B$ on the two sides. Normally u on each side is smooth but not constant. In this case we move x_1 and x_2 close to the shock, and the correction terms from variations in u become arbitrarily small—in comparison with its sudden jump. The jump curve $X(t)$ is a *shock*, and $s[u] = [F]$ is known in gas dynamics as the Rankine-Hugoniot condition—or more simply the **jump condition**. In our example with $F(u) = \frac{1}{2}u^2$, the condition is

$$s(u_r - u_l) = \tfrac{1}{2}u_r^2 - \tfrac{1}{2}u_l^2, \quad \text{or} \quad s = \tfrac{1}{2}(u_r + u_l). \tag{8}$$

This means that $u_t + uu_x = 0$ has a shock speed s which is the *average* of the states $u_r = B$ and $u_l = A$, to the right and left of the shock. That jump condition picks out the correct solution when $A > B$:

> u should jump from A to B on the line $x = st = \frac{1}{2}(A + B)t$.

The shock curve is the straight line $x = X(t) = \frac{1}{2}(A + B)t$, and the jump condition is satisfied. Shocks are shown in Figs. 6.28 and 6.29, propagating with speed s (and slope $1/s$ because t is vertical).

For $A < B$ the rays do not intersect; they spread out. This produces the opposite difficulty, blank regions that are not reached by any ray. Instead of being overdetermined, u is underdetermined. There are at least two ways to connect $u_l = A$ and $u_r = B$, *both satisfying the jump condition.*

> Solution 1: u jumps from A to B at $x = \frac{1}{2}(A + B)t$
>
> Solution 2: u has no jump and increases linearly:

$$u = A \text{ for } x < At, \ u = \frac{x}{t} \text{ for } At < x < Bt, \ u = B \text{ for } x > Bt.$$

Both are solutions to $u_t + uu_x = 0$. The first jumps as before, although the rays no longer intersect. The second does not jump. Only one can be chosen, and it is Solution 2. The rule that eliminates Solution 1 is the **entropy condition**:

6P The characteristic lines must go *into* the shock curve as t increases, so those on the left go faster than the shock and those on the right go slower:

$$\left[\frac{dF}{du}\right]_{\text{left}} > s > \left[\frac{dF}{du}\right]_{\text{right}} \tag{9}$$

Our example has $F = \frac{1}{2}u^2$ and $dF/du = u$. With $u_{\text{left}} = A$ and $u_{\text{right}} = B$ and $A < B$, the entropy condition (9) would be violated by a jump. That condition allows all points to be connected back to the initial line, on rays that hit no shocks. (They hit only when going forward.) In compressible flow, (9) is the requirement that material crossing the shock must suffer an increase of entropy. Together, the jump condition (7) and the entropy condition (9) pick out a unique solution to $u_t + F(u)_x = 0$.

Examples and a Solution Formula

Figure 6.29 shows a more general u_0. There is still a fan at the left, in the blank triangle. In the center, where $u_0(x) = 1 - x$, the solution comes from $u = u_0(x - ut)$:

$$u = 1 - (x - ut) \text{ gives } u = \frac{1 - x}{1 - t}.$$

But this solution only applies in the small triangle before the shock starts. At $x = t = 1$, which is the caustic, the fast rays from the left with the signal $u = 1$ catch the standing rays where $u = 0$. The resulting shock speed is the average $\frac{1}{2}$.

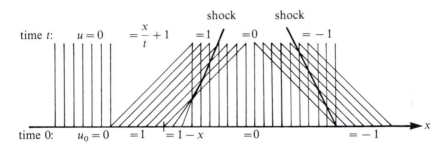

Fig. 6.29. Shocks and fans that eventually meet.

Overall this is a weak solution because it satisfies the weak form of the conservation law. As in Chapter 3, $u_t + F(u)_x = 0$ can be multiplied by a test function v and integrated by parts:

$$\int_0^\infty \int_{-\infty}^\infty [uv_t + F(u)v_x] \, dx \, dt = \int_{-\infty}^\infty u_0 v \, dx.$$

When shocks appear and the strong form is in trouble, this form survives.

EXAMPLE 1   A moving front that starts with a corner

The geometry of the moving front has all the symptoms of a conservation law! Some corners were smoothed out; others propagated without change; corners could appear in C_t when the original curve C was smooth. The outward corner, at point ② of Fig. 6.26, disappeared as the front developed. The slope discontinuity was smoothed. The inward corner at point ④ did not disappear; it moved along a "shock line" in $x - t$ space. The curve did not jump across this line, but *its slope did*. That suggests that the equation $U_t = \sqrt{1 + U_x^2}$ for the moving front should lead to a *conservation law for the slope* $u = \partial U / \partial x$. It does, if we differentiate both sides:

$$\frac{\partial}{\partial x} U_t = \frac{\partial}{\partial x} \sqrt{1 + U_x^2} \quad \text{or} \quad \frac{\partial u}{\partial t} = \frac{\partial}{\partial x} \sqrt{1 + u^2}. \tag{10}$$

Thus $F = -\sqrt{1 + u^2}$ in the conservation law for the slope.

At corner ②, where $F'(u)$ is increasing, the entropy condition prevents a shock. We have Solution 2 above, the one with no jump. The left side of (9) is zero, the right side is $1 / \sqrt{2}$, and a shock is not allowed. That is why the corner spreads forward into a circle. At corner ④ the slopes are $A = 0$ and $B = 1$, and the jump condition (7) gives the shock speed:

$$s[1 - 0] = [-\sqrt{2} + 1] \quad \text{with shock } X = st = (1 - \sqrt{2}) t.$$

You can see that shock path in Fig. 6.27. In this geometric illustration of a conservation law, which was developed by Sethian, the shock speed is not the average $\frac{1}{2}(A + B)$. The jump condition for $F = -\sqrt{1 + u^2}$ forces the shock to *bisect the angle* between intersecting rays. So does Huygens' construction of the wave front. There is a "lost region" of initial values, from which rays disappear into the shock.

Remark 1. Lax found a formula that solves both nonlinear equations

$$\frac{\partial U}{\partial t} + F\left(\frac{\partial U}{\partial x}\right) = 0 \quad \text{and} \quad \frac{\partial u}{\partial t} + \frac{\partial}{\partial x} F(u) = 0.$$

Remember that the conservation law on the right is the x-derivative of the Hamilton-Jacobi equation on the left: u is $\partial U / \partial x$. The initial values have the same relation $u_0 = \partial U_0 / \partial x$. The formula is amazingly simple:

$$\boxed{U(t,x) = \min_y \left[U_0(y) + t F^*\left(\frac{x - y}{t}\right) \right].} \tag{11}$$

The novelty is in F^*. It is the "conjugate function" of F, introduced in Section 3.6 for functions that have $F'' \geq 0$:

$$F^*(w) = \max [uw - F(u)].$$

The conservation law $u_t + u u_x = 0$ has $F = \frac{1}{2} u^2$, and its conjugate happens to be

$F^* = \frac{1}{2} w^2$. Therefore the law has the closed-form solution

$$u(t,x) = \frac{\partial}{\partial x} \min_{y} \left[\int_0^y u_0(x)\, dx + \frac{1}{2t} (x-y)^2 \right]. \tag{12}$$

For the Riemann problem, where u_0 jumps from A to B, this must yield a shock for $A > B$ and a fan $u = x/t$ for $A < B$.†

EXAMPLE 2 $u_0 = \delta(x)$ with total mass $\int \delta(x)dx = 1$

The integral of u_0 jumps from 0 to 1, so (12) is

$$u(t,x) = \frac{\partial}{\partial x} \min_{y} \left[\begin{matrix} 0 \; (y \le 0) \\ 1 \; (y > 0) \end{matrix} + \frac{1}{2t} (x-y)^2 \right].$$

If x is negative, the choice $y = x$ makes the expression in brackets zero. That is the minimum. If x is positive, the choice $y = x$ gives 1 inside the brackets and the choice $y = 0$ gives $x^2/2t$. The minimum is $x^2/2t$ up to $x = \sqrt{2t}$, and after that the minimum is 1:

$$u = \frac{\partial}{\partial x} \left\{ \begin{matrix} 0 \\ x^2/2t \\ 1 \end{matrix} \right\} = \left\{ \begin{matrix} 0 \\ x/t \\ 0 \end{matrix} \right\} \quad \begin{matrix} (x < 0) \\ (0 < x < \sqrt{2t}) \\ (x > \sqrt{2t}) \end{matrix}$$

The delta function spreads into a fan with a shock in front, where u drops from $u_{\text{left}} = \sqrt{2t}/t$ to $u_{\text{right}} = 0$. Everything seems right:

Conservation: $\displaystyle \int u\, dx = \int_0^{\sqrt{2t}} \frac{x}{t}\, dx = \left[\frac{1}{2} \frac{x^2}{t} \right]_0^{\sqrt{2t}} = 1$

Jump condition $s[u] = [F]$: $\displaystyle s = \frac{d}{dt} \sqrt{2t} = \frac{1}{\sqrt{2t}} = \frac{1}{2}(u_{\text{left}} + u_{\text{right}})$

Entropy condition for $F = \frac{1}{2} u^2$: $u_l > s > u_r$ or $\sqrt{\dfrac{2}{t}} > \dfrac{1}{\sqrt{2t}} > 0$

Remark 2.   You are not allowed to change variables in conservation laws, if shocks are present. The reason is this: The new jump condition alters the problem. If you innocently multiply $u_t + uu_x = 0$ by $2u$, to give

$$2uu_t + 2u^2 u_x = 0 \quad \text{or} \quad \frac{\partial}{\partial t}(u^2) + \frac{\partial}{\partial x}\left(\frac{2}{3} u^3 \right) = 0,$$

---

† Hopf found (12), Lax found (11), and Osher went beyond $F'' \ge 0$.

the weak solutions (shock solutions) are changed. The jump condition for the new variable u^2 is not $s[u] = [\frac{1}{2} u^2]$ but $s[u^2] = [\frac{2}{3} u^3]$. When the original solution jumped from $A = 1$ to $B = 0$, the shock speed was the average $s = \frac{1}{2}$. When $v = u^2$ jumps from $A = 1$ to $B = 0$, the shock speed will be $s = \frac{2}{3}$.

The quantities to conserve are decided by the integral form, not the differential form, of the physical law. In gas dynamics this point is driven home by the entropy. Mass and momentum and energy are conserved. For smooth solutions those equations lead to $\partial S/\partial t = 0$; the entropy is also constant. But at a shock $\partial S/\partial t$ is positive.

EXAMPLE 3 A neat application of conservation laws is to traffic flow. The density of cars is $u(t,x)$, and the flow is F—the rate at which cars go past point x at time t. On a freeway we assume *conservation of cars*. Each car's speed is decided by the local density—drivers respond only and immediately to the car ahead. Then F depends on the density u. It is zero when $u = 0$, and it is zero again at the bumper-to-bumper density ($u = 225$ cars per mile). The maximum flow is observed to be $F = 1600$ cars per hour, at a density near $u = 80$. In other words, the greatest movement on a freeway is (seriously) at about 20 miles per hour.

For this problem F is not $\frac{1}{2} u^2$. In fact F'' is negative rather than positive; velocity decreases as density increases. This reverses the shock direction in comparison with $u_t + u u_x = 0$, and the shock goes back down the line of traffic. That is what happens in Boston. The drivers at the back are going too fast and they enter the shock. They come out of it going slower (temporarily). You recognize the entropy condition: speed behind > speed ahead. When this inequality is reversed no shock will form, and the cars in front pull away smoothly.

Both solutions occur if there is a traffic light. When it goes red, a shock starts backward. When it goes green, a fan starts forward. The critical question is whether the shock gets through while the light is green; otherwise the traffic crawls. The model becomes more realistic by accounting for drivers who look ahead—their speed depend on the density gradient $\partial u/\partial x$ as well as the density. That leads to diffusion of the shock, and to Burgers' equation with a term proportional to u_{xx}.

Burgers' Equation

The equation has a term that prevents shocks:

$$u_t + u u_x = v u_{xx}. \tag{13}$$

The viscosity (the Greek letter is "gnu") has been copied from the equations for fluid flow. Burgers wanted a simple model for those nonlinear equations, in which the term $\Sigma\, u_i \partial u_j/\partial x_i$ of Section 3.5 made analytical solutions virtually impossible. Here the problem is not only solved; it is linearized. That allows a complete study of the limiting case $v \to 0$, in which (13) approaches the conservation law

$u_t + uu_x = 0$—which is also known as the "inviscid Burgers' equation." It is a singular limit, because the highest derivative disappears and solutions that are kept smooth by the u_{xx} term suddenly develop shocks.

To linearize (13) we introduce U again:

$$U_t + \tfrac{1}{2} U_x^2 = vU_{xx} \quad \text{if} \quad u = \partial U/\partial x. \tag{14}$$

Differentiating with respect to x recovers equation (13) for u. In the new form, the substitution $U = -2v \log w$ turns (14) into the heat equation. The nonlinearity is gone, *the equation is $w_t = vw_{xx}$*, and that problem can be solved in closed form. It gives the exact solutions U and u to nonlinear partial differential equations, which is rather exceptional.

Reversing $U = -2v \log w$, the initial function w_0 is $e^{-U_0/2v}$. The heat equation diffuses this w_0 in accordance with its fundamental solution. The responses to all starting values go into a superposition, and

$$U(t,x) = -2v \log w = -2v \log \left[\frac{1}{2\sqrt{\pi vt}} \int_{-\infty}^{\infty} e^{-U_0(y)/2v} e^{-(x-y)^2/4vt} \, dy \right]. \tag{15}$$

The fundamental solution has vt in place of t, because the equation is $w_t = vw_{xx}$.

The important limiting cases are $v \to 0$ (vanishing viscosity) and $t \to \infty$ (large time). We look at the first because it fits with the shock problem. The integral in (15) has the form

$$\int_{-\infty}^{\infty} e^{-B/2v} dy \quad \text{with} \quad B = U_0(y) + \frac{1}{2t}(x - y)^2.$$

What happens as v goes to zero? The exponential $e^{-B/2v}$ concentrates on the place where $-B/2v$ is as large as possible; B itself is a minimum. By a standard asymptotic method called "steepest descent," the bracketed quantity in (15) approaches $ce^{-B_{\min}/2v}$. The factor c is bounded. Taking logarithms and multiplying by $-2v$, (15) becomes

$$\lim_{v \to 0} U(t,x) = B_{\min} = \min_y \left[U_0(y) + \frac{1}{2t}(x - y)^2 \right].$$

That is exactly Hopf's formula (12) for the equation $U_t + \tfrac{1}{2} U_x^2 = 0$. The zero viscosity limit brings back the earlier problem, and a final step to $u = \partial U/\partial x$ brings back the model equation $u_t + uu_x = 0$.

So what, you may say. The solution to $u_t + uu_x = vu_{xx}$ should approach the solution to $u_t + uu_x = 0$, or something is very wrong. When solutions are smooth I cannot disagree. But remember that $u_t + uu_x = 0$ had *no* solution that was smooth, then a possible *infinity* of solutions with jumps, and finally *one* solution with increasing entropy. It took physics and mathematics to discover the jump condition and the entropy condition. They picked out the one correct weak solution. Now we have another method that picks out that same solution automatically: *Add the term* vu_{xx} *and let* $v \to 0$.

The tough question is the extension to more than one conservation law. That is not completed in 1985. Despite outstanding work, nonlinear hyperbolic systems are not fully understood. The example that has pushed the theory is *compressible flow*. For gas dynamics in one dimension there are three conserved quantities: mass, momentum, and energy. The differential equations contain their densities as the three components of u: the mass density ρ, the momentum density ρv ($v = $ velocity), and the energy density $\rho e + \frac{1}{2} \rho v^2$ (internal plus kinetic). The system $u_t + F(u)_x = 0$ is

$$\rho_t + (\rho v)_x = 0$$

$$(\rho v)_t + (\rho v^2 + p)_x = 0 \tag{16}$$

$$(\rho e + \tfrac{1}{2} \rho v^2)_t + (v[\rho e + \tfrac{1}{2} \rho v^2 + p])_x = 0.$$

The internal energy e, density ρ, and pressure p are related by an equation of state, reducing the unknowns to ρ, p, and v. We mention again that velocity is frequently denoted by u. In three space dimensions there are 3 momentum densities, from the 3 components of velocity; altogether there are 5 equations and 5 unknowns. The derivative $\partial / \partial x$ is replaced by the divergence, for example in $(\rho v_1)_x + (\rho v_2)_y + (\rho v_3)_z$. Conservation laws are equations written in "*divergence form*," and their discrete approximations should preserve that form.

The solution of (16) is slightly outside the scope of this book. Ordinarily it is approximated by a difference equation, suggesting a third way to pick out the correct weak solution. The discretization can create its own viscosity, no more artificial than Burgers'. When Δt and Δx go to zero, Friedrichs' method is conjectured to converge—and the limiting solution should satisfy the jump condition and the entropy condition.

Remark At weak shocks where the jump is small, the increase in entropy is only of order $\Delta S = (\text{jump})^3$. It is reasonable to take $S = $ constant, which replaces conservation of energy and reduces the three nonlinear equations to two. Written in Lagrangian form (moving with the particles, instead of staying in one place as Euler did) the conservation of mass and momentum are

$$\left(\frac{1}{\rho}\right)_t = v_x \quad \text{and} \quad v_t = -p_x.$$

Energy is not conserved, and its decrease becomes the entropy condition. In steady supersonic flow x and y replace x and t.

The KdV Equation

Like most good applied mathematics, the observation preceded the equation. John Scott Russell wrote down how it happened to him in 1834:

"I was observing the motion of a boat which was rapidly drawn along a narrow channel by a pair of horses, when the boat suddenly stopped—not so the mass of water in the channel which it had put in motion; it accumulated round the prow of the vessel in a state of violent agitation, then suddenly leaving it behind, rolled forward with great velocity, assuming the form of a large solitary elevation, a rounded, smooth and well-defined heap of water, which continued its course along the channel apparently without change of form or diminution of speed. I followed it on horseback, and overtook it still rolling on at a rate of some eight or nine miles an hour, preserving its original figure some thirty feet long and a foot to a foot and a half in height. Its height gradually diminished, and after a chase of one or two miles I lost it in the windings of the channel."

He knew he had seen something extraordinary. "The first day I saw it was the happiest day of my life." The wave was above the level of the canal, and he found by experiment that its height was proportional to its speed. In fact he mastered the trick of creating a solitary wave, which is not so easy to do—a recent conference of scientists failed. Their motorboat worked well in practice, but when it broke down they had no horses. Nevertheless they knew what shape that "soliton" should have, from the wave forms found by Boussinesq and Rayleigh. Finally Korteweg and deVries wrote down the equation for a one-way wave in shallow water:

$$\boxed{u_t + 6uu_x + u_{xxx} = 0.}$$

(17)

The coefficient 6 is decided by the scaling for u, and it makes the later formula (20) a little simpler. The important thing is the step to a solitary wave. A wave is a solution $u = F(x - ct)$ that depends only on the combination $x - ct$. Substituting into (17), the shape F must satisfy

$$-cF' + 6FF' + F''' = 0.$$

Integrating once, with zero boundary conditions at infinity,

$$-cF + 3F^2 + F'' = 0.$$

(18)

This allows another integration, after the usual trick for nonlinear oscillations: Multiply first by F'. Then each term is a derivative, and

$$-\tfrac{1}{2}cF^2 + F^3 + \tfrac{1}{2}(F')^2 = 0.$$

That is the separable equation $dF/(cF^2 - 2F^3)^{1/2} = dt$, and its solution can be found in a table of integrals (or possibly on MACSYMA?):

$$F(x - ct) = \tfrac{1}{2}c \ \text{sech}^2 \ (\tfrac{1}{2}\sqrt{c}\ (x - ct)).$$

(19)

The function sech x, the hyperbolic secant, is the reciprocal of cosh x, the hyperbolic cosine $\frac{1}{2}(e^x + e^{-x})$. Therefore sech x goes exponentially to zero for large x. It is symmetric around the origin, and of course sech^2 is always positive. The wave has a single hump whose height is $\frac{1}{2}c$. It travels to the right with speed c, proportional to its height.

This soliton must be stable, or Scott Russell could not have watched it so long. Linearized analysis of (18) gives neutral stability (a center at $(0,0)$) since there is no diffusion. A shock is prevented by the dispersion u_{xxx}.

So far everything is classical and old. Over the years there were more solutions to the KdV equation, but nothing to make it special. Then came a numerical experiment at Los Alamos with 64 springs, which surprised everyone by failing. The energy was expected to distribute itself equally among all modes of oscillation, because the springs were nonlinear. It started to do so, by ordinary nonlinear interactions, but after a time the energy went back into the lowest mode. The problem was discrete, but its continuous analogue happened to be KdV.

After some years Kruskal and Zabusky got curious and tried again. They made the problem periodic and started with a sine wave at $t = 0$. The uu_x term began to steepen the wave, moving it toward a shock (the dashed line in Fig. 6.30). Then the wiggles in that line separated into a chain of solitary pulses, similar to sech^2 except periodic. The higher pulses go faster and overtake the smaller ones. Everything looks nonlinear and normal, especially during the complicated interaction, but two things happened:

(1)   After the fast wave passed the slow wave, they reappeared *looking the same*.
(2)   After many cycles (but not too many) there was a *near recurrence*: the sine wave at $t = 0$ almost reappeared at $t = T$.

It was like a sum of exponentials $e^{i\omega t}e^{ikx}$, with ω's that are not commensurate. Such a sum is quasiperiodic; at certain times the numbers ωt differ almost exactly by 2π, and the initial profile nearly recurs. It is the opposite of a strange attractor, and it suggests that the nonlinear equation is integrable.

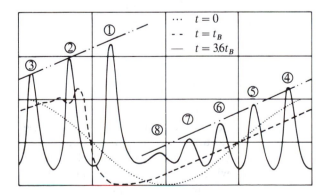

Fig. 6.30. A train of solitons appears (and later disappears).

If so, many quantities should be conserved. The KdV equation is

$$u_t + (3u^2 + u_{xx})_x = 0 \quad \text{which gives} \quad \int_{-\infty}^{\infty} u\,dx = \text{constant}.$$

That is conservation of mass. Conservation of energy comes from

$$u(u_t + 6uu_x + u_{xxx}) = 0 \quad \text{which gives} \quad \int_{-\infty}^{\infty} \tfrac{1}{2} u^2 dx = \text{constant}.$$

Whitham discovered that $\int (2u^3 - u_x^2)dx$ was also conserved. This started a search, which got stuck for a year on an algebraic mistake. Finally an infinite family of conserved quantities was found. It became less surprising that "multisoliton" solutions could exist, recovering their shape after nonlinear interaction, but the KdV equation was still not solved.

There is a change of variables that helps. It is not unlike the $u = -2vw_x/w$ that linearized Burgers' equation, and here it produces the Schrödinger eigenvalue problem:

$$w_{xx} + uw = \lambda(t)w. \tag{20}$$

That is remarkable—the original unknown u has become Schrödinger's potential function—but it is still not linear: u multiplies w. The key idea is on the next page, to *watch the eigenvalues*. While u is changing with time from (17), and Schrödinger's equation is changing, λ **stays fixed**:

$$\frac{d\lambda}{dt} = 0. \tag{21}$$

When $u(t,x)$ satisfies KdV, the eigenvalues of (20) do not change. It turns out that the crucial information about w (its reaction at $x = +\infty$ to radiation e^{ikx} at $x = -\infty$) evolves in a simple linear way. That information is called the "scattering data," and it is enough to recover u. Thus the final linearization is much more subtle than for Burgers' equation:

(1)   From $u_0(x)$ find the eigenvalues λ and the scattering data
(2)   Solve linear equations to find this data at time t
(3)   Recover $u(t,x)$ by "inverse scattering."

The Fourier transform also took three steps: Transform at time 0, solve $\hat{u}_t = ik^3\hat{u}$ to time t, and invert to find u. Fourier separates $u_t = -u_{xxx}$ into fragments that develop linearly; the steps 1–3 do the same for KdV. The difference is that the Inverse Scattering Transform, which recovers the shape of an obstacle from what it does to incoming waves, is not linear. It has opened up a small new world of exceptional equations, including the following three:

the Sine-Gordon equation $u_{tt} - u_{xx} + \sin u = 0$

the cubic Schrödinger equation $iu_t + u_{xx} + |u|^2 u = 0$

the Toda lattice: $\dfrac{d^2 u_i}{dt^2} = e^{u_i - u_{i-1}} - e^{u_{i+1} - u_i}$

The soliton has been joined by the instanton, from the Yang-Mills equations of field theory. However the deepest question is still to be answered: Which equations can be linearized? Which Hamiltonian problems are integrable? There are a great many clues, but something is still missing.

Lax Equations

After that historical essay we return to one central point: the constancy of the eigenvalues λ. That is something which could also happen in matrix equations. It almost happened once, when the matrix $Q(t)$ satisfied $Q' = BQ$ with a skew-symmetric B. Starting from $Q(0) = I$, the solution $Q = e^{Bt}$ is an orthogonal matrix:

$$Q^T Q = e^{B^T t} e^{Bt} = e^{-Bt} e^{Bt} = I.$$

This is the key to conservation of energy; Q is a rotation. Its eigenvalues lie on the unit circle. They have constant absolute value 1, but they are not actually constant. That requires a change to the **Lax equation**

$$\boxed{L' = BL - LB.} \tag{22}$$

Now the solution is $L(t) = e^{Bt} L_0 e^{-Bt} = Q L_0 Q^T$. If the initial matrix L_0 is symmetric, so is $L(t)$. Furthermore *the eigenvalues of L are fixed*:

6Q The eigenvalues of $L = Q L_0 Q^T$ equal the eigenvalues of L_0, because $Q^T = Q^{-1}$. The symmetric matrix $L(t)$ remains similar to L_0 for all time.

Chapter 1 diagonalized any symmetric matrix: $L_0 = Q_* \Lambda Q_*^T$. Then $L(t) = (QQ_*)\Lambda(QQ_*)^T$ is also diagonalized, and its eigenvalue matrix is also Λ.

Lax carried this idea from matrix equations to differential equations. For the Korteweg-deVries equation he found the pair

$$Lw = \left(\frac{\partial^2}{\partial x^2} + u\right)w, \quad Bw = -\left(4\frac{\partial^3}{\partial x^3} + 3u\frac{\partial}{\partial x} + 3\frac{\partial}{\partial x}u\right)w. \tag{23}$$

The equation $L' = BL - LB$ is KdV. Therefore the eigenvalues of L are fixed. Since $Lw = \lambda w$ is Schrödinger's equation (20), this is the crucial fact (21) that led to the solution. It turns out that the Sine-Gordon equation also fits into Lax's form, and so do all the other equations that have so far been solved. Perhaps there is a Lax pair for all integrable equations.

EXERCISES

6.6.1 Suppose the coastline C is a $45°$ sawtooth with slopes alternately $+1$ and -1 on the intervals between integers. What shape is the "fishing limit" C_t for large t?

6.6.2 If C has the V-shape of the absolute value $y = |x|$, what is the curve C_t for positive t and for negative t?

6.6.3 Find the slope of the tangent line and also of the perpendicular line to the parabola $y = -\frac{1}{2}x^2$. Sketch this curve C and find the point at a distance $t = 1$ from (x,y) along the perpendicular line.

6.6.4 Why are the rays perpendicular to the wave front? Use Huygens' construction of the front from circles of radius t around the points on C; the rays are the radii of those circles that reach the front.

6.6.5 For $u_t + uu_x = 0$ the characteristic lines are $x - u^*t = x^*$ and $x - u^{**}t = x^{**}$, starting from x^* with speed $u^* = u_0(x^*)$ and from x^{**} with speed $u^{**} = u_0(x^{**})$. At what time t do they intersect? If u_0 is decreasing then t is positive and shocks are certain to form.

6.6.6 Differentiate $u = u_0(x + c(u)t)$ by the chain rule to show

$$u_t = cu_0'/(1 - c'u_0't) \quad \text{and} \quad u_x = u_0'/(1 - c'u_0't) \quad \text{and} \quad u_t = cu_x.$$

6.6.7 A 2-dimensional conservation law is "rate of change = flux in":

$$\frac{d}{dt} \iint_S u \, dx \, dy = - \int_C F \cdot n \, ds.$$

What theorem changes the right side to $-\iint \text{div } F \, dx \, dy$? Then the law in differential form is $u_t + \partial F_1(u)/\partial x_1 + \partial F_2(u)/\partial x_2 = 0$.

6.6.8 For slowly varying oscillations $u = A(t,x)e^{i\alpha(t,x)}$, the phase α is close to $\omega t - kx + \alpha_0$. From $\alpha_t = \omega$ and $\alpha_x = -k$, derive the conservation law $k_t + \omega_x = 0$. Show that the wave number is constant $(dk = 0)$ on the line $x = \dfrac{d\omega}{dk} t$. Moving on that line at the group velocity keeps the observer even with wave number k.

6.6.9 Using formula (12) solve $u_t + uu_x = 0$ starting from $u_0 = -\delta(x)$. Compare with the example that follows the formula.

6.6.10 Show that the minimum in formula (12) occurs where $u_0(y) = (x - y)/t$, if the y-derivative is set to zero. At that point

$$u = \frac{\partial}{\partial x} \left[\int_0^y u_0(x)dx + \frac{(x - y)^2}{2t} \right] = u_0(y)\frac{dy}{dx} + \frac{(x - y)}{t}\left(1 - \frac{dy}{dx}\right) = \frac{x - y}{t}.$$

From $u_0(y) = (x - y)/t = u$ deduce $u = u_0(x - ut)$, which shows that Hopf's formula gives a correct answer. With shocks allowed it gives *the* correct answer.

6.6.11 Solve $u_t + uu_x = 0$ by (12) if u_0 jumps from A to B:

$$u(t,x) = \frac{\partial}{\partial x} \min_y \left[\begin{matrix} -Ay(y<0) \\ By(y>0) \end{matrix} + \frac{1}{2t}(x - y)^2 \right].$$

For $A > B$ compare with the shock along $x = \frac{1}{2}(A + B)t$ and for $A < B$ compare with the fan $u = x/t$.

6.6.12 Show that $U = -2v \log w$ turns the equation $U_t + \frac{1}{2}U_x^2 = vU_{xx}$ into the heat equation $w_t = vw_{xx}$. What is the relation between w and the original unknown u?

6.6.13 Derive from (15) the exact solution to Burgers' equation:

$$u = \frac{\displaystyle\int_{-\infty}^{\infty} \frac{x - y}{t} e^{-B/2v}\, dy}{\displaystyle\int_{-\infty}^{\infty} e^{-B/2v}\, dy}.$$

6.6.14 Construct a traveling wave solution $u = W(x - t)$ to Burgers' equation $u_t + uu_x = u_{xx}$ with $W(-\infty) = 2$ and $W(\infty) = 0$, in 3 steps:
  (a) Substituting this u, find an ordinary differential equation for W
  (b) Integrate once to reach $W' = \frac{1}{2}W^2 - W$
  (c) Integrate $\dfrac{dW}{W - \frac{1}{2}W^2} = \dfrac{dW}{W} + \dfrac{dW}{2 - W} = -dx$ to find $u = \dfrac{2e^{t-x}}{1 + e^{t-x}}$.

6.6.15 If $u = W(x - t)$ solves Burgers' equation with $v = 1$, show that $u = W\left(\dfrac{x - t}{v}\right)$ solves $u_t + uu_x = vu_{xx}$. Does $v \to 0$ produce a shock with speed $1 = $ average of 2 and 0?

$$\lim_{v \to 0} \frac{2e^{(t-x)/v}}{1 + e^{(t-x)/v}} \overset{?}{=} \begin{cases} 2 & \text{for } x < t \\ 0 & \text{for } x > t \end{cases}$$

6.6.16 Verify that $L' = BL - LB$ is the KdV equation, for the Lax pair in (23).

6.6.17 Solve $L' = BL - LB$ with $B = \begin{bmatrix} 0 & 1 \\ -1 & 0 \end{bmatrix}$ and $L_0 = \begin{bmatrix} 1 & 0 \\ 0 & -1 \end{bmatrix}$. The eigenvalues of L should be constant.

6.6.18 Find a solitary wave solution $u = F(x - ct)$ to the modified KdV equation $u_t + 6u^2u_x + u_{xxx} = 0$. Substitute for u and integrate three times with zero conditions at $\pm\infty$.

6.6.19 (a) If $u_0(x) = x$ show that $u = u_0(x - ut)$ gives $u = x/(t + 1)$.
  (b) Find the solution and the blow-up time if $u_0(x) = -x$.
  (c) Sketch the lines $x + x_0t = x_0$ in an $x - t$ plane for various x_0.

6.6.20 What happens after the shock lines meet in Fig. 6.29? What happens when the shock hits the fan?

CHAPTER 6 IN OUTLINE: INITIAL-VALUE PROBLEMS

7

NETWORK FLOWS AND COMBINATORICS

This chapter turns the book in a new direction. Matrices still appear, but they seldom multiply one another; they are used to record information, not to operate on it. The information is about a *network*, or a *graph*, and the questions to be answered are like this: What is the ***shortest path*** from one node to another, or what is the ***maximal flow*** through the network, or what is the ***minimum cost*** of a given flow? The facts we need are the lengths of the edges or their capacities or their costs. In many cases the information is even more combinatorial: Is there an edge or not? The recording matrix just contains 0's and 1's, and the mathematics is clean and elegant—but not trivial.

Much of this mathematics is comparatively recent. It applies to computer science and operations research more than to the classical areas of mechanics and physics. This brings a shift of emphasis, away from eigenvalues and differential equations and even Gaussian elimination, and toward ideas of a different kind. The best illustration is to study some specific problems, and find algorithms to solve them. We have chosen three problems in graph theory.

The first was mentioned much earlier in Exercise 1.6.1, and now its time has come:

> ***Problem 1*** Decide whether or not a given graph is ***connected***. Is there a path from every node to every other node?

The graph consists of a set of nodes, also known as vertices, and a set of (undirected) edges. Given the list of edges, our goal is to find a reasonable way to test for connectedness. I think that looking at one pair of nodes, and searching for a path between them, and then beginning again with another pair, is not reasonable. That idea can certainly be made into a correct algorithm—but it will be too slow.

It is better to start from one node and build outwards. That is not the only possibility but it is successful. If the graph is connected we reach every node; if not, the construction should find all nodes connected to the first one and then stop. As we add edges there is no reason to close a loop; the loop cannot take us to a new node. Therefore what we build is at every step a *tree*.

> *A tree is a connected graph with no loops.*

Two trees are illustrated in Fig. 7.1. The underlying graphs are the same; the edges not used in the trees appear as dotted lines. Either one is enough to answer our question: This graph is certainly connected. Note that the constructions could be stopped at any point (say after node 4) and they would still yield trees. The proof of connectedness comes when all nodes are included and we reach a *spanning tree*:

> A spanning tree is a tree that connects all n nodes of the graph.
> It has $n - 1$ edges.

The two trees in the figure are spanning trees, with five edges connecting the six nodes.

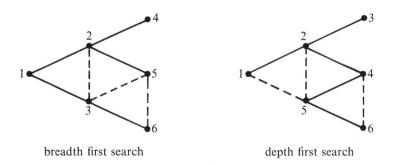

breadth first search   depth first search

Fig. 7.1 Two spanning trees for the same graph.

The real goal of the construction is to find a spanning tree, if it exists. But the algorithm of "building outwards" is still not decided. There are many spanning trees, and we want to propose two different systems of construction. They start from an arbitrarily chosen node 1.

> **Breadth first search**: Look for *every* new node that can be reached on an edge from the current node. Nodes are visited in the same order that they are found.

This algorithm was used for the spanning tree in part (a). From node 1 we found nodes 2 and 3. From node 2 we found 4 and 5 (rejecting the edge from 2 to 3, which closes a loop). Since node 3 was the next earliest, we visited it and found an edge to node 6—which completed the spanning tree.

The alternative is to keep going as far as possible along a single path, and turn back only if necessary:

> **Depth first search**: Look for *one* new node that can be reached on an edge from the current node. If a new node is found, it is the next to be visited; if it is not found, the search returns to the previous node.

This was used in Fig. 7.1(b). Node 1 led to 2 which led to 3. From 3 a new node was impossible; we therefore returned to 2. In general we might return all the way to node 1; if no suitable edge can be found then the graph is not connected (and the algorithm stops). In this case 2 led to 4, 4 led to 5, and 5 led to 6. The spanning tree has five edges, as it must, but its final shape is quite different.

I don't know which algorithm you prefer. (There is another one in Exercise 7.1.10.) They are both easy to code, except when special efficiency is required; then the data structures can become very intricate. The ideas of breadth and depth are in competition throughout the whole theory of combinatorial optimization. Neither one can be totally victorious, as long as there is a graph for which the other is better. When the spanning tree is unique, both methods must find it. Breadth first search (BFS) will finish immediately if all edges come out of node 1, like the spokes of a wheel. That graph would be a disaster for depth first search (DFS), which will backtrack every time to node 1. On the other hand DFS wastes no time when the graph is one long chain (provided node 1 is not in the middle).

We note that a good code should, for a graph that is not connected, find a spanning tree for each connected component of the graph.

Labeling and Scanning

There is one other point about a good code. When it finishes, the user should be able to travel easily along the tree. The main point is for him to know, at every node i, the edge that led to it. Of course he can search the list of edges in the tree, but that should not be necessary. It is much better to **label** node i with the name or the number of its "parent"—the node just before it, which comes first as we travel back to node 1. Then it is easy to backtrack along the tree, one step at a time, by reading the labels.

The label on node i will be $i - 1$ for a single long chain. But when several edges lead out from the same node, those new nodes are all given the same label. A node that is a dead end will never be used as a label. Our figures are not yet labeled—but they should be, or the backtracking needed in depth first search becomes ridiculously awkward.

There is one other operation that has a widely accepted name. It is the basic step of looking out from a given node—in this case to find all new nodes connected to it (BFS) or only one new node (DFS). This is called **scanning** the given node. Most network algorithms repeat this fundamental pattern:

> *scan node i* (to test its neighbors for some property)
> *assign label i* (to some subset of those neighbors)
> *continue* (according to the rules of the algorithm)

Eventually all possible nodes are labeled, and the algorithm takes action.

Minimum Spanning Trees

In the applications of graph theory there is another ingredient. A *length* is assigned to every edge. In that case the goal is not just to find a spanning tree, but to find the shortest spanning tree. That brings us to the second problem of this section:

> **Problem 2**   Find the **minimum spanning tree**, for which the sum of the edge lengths is as small as possible.

We may be given 50 cities that are to be connected by cables. There is a length (or a cost) associated with each pair of cities, and the best cable will connect 49 pairs—to form a minimal spanning tree. In this problem all edges are allowed, but in general that is not so. When the graph has no edge between nodes i and j we can assign an infinite length $l_{ij} = \infty$ (or an extremely large value, in an actual code). Then the algorithm can be written as if all edges are present, and the shortest spanning tree will have finite length exactly when the graph is connected.

Properly speaking, we should use the word *network* when a length or cost or capacity or resistance is assigned to each edge.

The minimum length problem will be attacked by starting with an example and looking for an algorithm. Fig. 7.2(a) shows the set of finite edge lengths, and a tree is growing in Fig. 7.2(b). It uses short edges; we hope it is the start of a minimum spanning tree. That remains to be proved but it is true. What is not so clear is the edge to be added next. Should it be the shortest edge incident to the existing tree (length 7)? Or should we choose the shortest edge remaining in the network (length 3)? Two or more trees could grow at once—a set of trees is called a *forest*—or we can stay with a single tree. It is not even clear that a correct and permanent decision

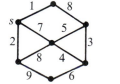 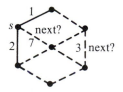

Fig. 7.2. A network and a minimum spanning tree.

can be made at this point. Most algorithms try a first guess and then improve it, but this problem is more special.

It is also not so clear whether we can start from an arbitrary source node s, or whether the first edge should be the shortest in the network. In fact there are two quite different algorithms:

(1) (Single tree: Prim's algorithm) Start from any node s and build a tree by repetition of the following rule: Add the shortest edge that is incident to the existing tree.

This rule chooses the edge of length 7, then the edge of length 4, then 3, and then 6. Notice that the last step skips the edge of length 5, which would close a loop. An edge is "incident" on a tree provided one end touches the tree but the other end does not. Prim's algorithm has constructed a spanning tree of total length $1 + 2 + 7 + 4 + 3 + 6 = 22$. It looks minimal.

(2) (Minimum forests: Kruskal's algorithm) Add edges in increasing order of length, rejecting any edges that complete a loop.

This algorithm choose edges in the order 1, 2, 3, 4, 6 (rejecting 5), 7. It produces the same tree, which is the only solution for this network: the example has a unique minimum spanning tree. If the length that is now 6 is changed to 9, then two different trees are equally short. Prim's algorithm will not use that new edge; Kruskal's algorithm might use it, depending on the way it resolves a tie in the ranking of edge lengths. In fact much of the work in Kruskal's method goes into sorting the edges in order of length. That requires $O(m \log n)$ steps in general, but it is faster if the edges are small integers. After this pre-processing step Kruskal maintains node lists for the trees, in order to exclude loops.

We intend to prove that both algorithms always work. They are examples of a general approach known as the ***greedy algorithm: Do the best thing at every step***. In other words, ignore all difficulties that might come later and make the optimal choice now.† For this model problem the greedy algorithm does succeed, and while forests are growing we will propose a third and more extreme possibility:

(3) (General case of the greedy algorithm) Start trees at all n nodes, and repeat the following step $n - 1$ times: Select a tree and add the minimum-cost edge that is incident to it.

Of course the actual steps depend on the selection order of the trees. Prim selects the same tree every time. Kruskal makes the selection based on edge lengths. Boruvka (who seems to have been first) goes through *all* the existing trees at each sweep—this produces trees of different sizes for the next sweep. And Tarjan, who is the world champion in network algorithms, moves a tree to the end of the queue after its minimum-cost edge has been added. Despite its old-fashioned appearance, algorithm 1 (Prim's) can be made quite efficient for dense graphs; it is worse than

---

† We avoid the temptation to mention politicians.

Kruskal's for large m and n only when sorting of edge lengths is especially easy. Tarjan's "round-robin" principle is currently best in this asymptotic sense, and can always finish in less than $Cm \log \log n$ steps.

We still have to prove that the greedy algorithm works. It starts safely with n separate trees. At each step, it chooses a tree T and adds a minimum-cost edge e. The goal is to show that this decision can never miss the best final tree.

7A   At every step of the greedy algorithm, the current trees and the new edge e are part of a minimum spanning tree. Therefore at the end the algorithm produces an optimal tree.

For the proof imagine the opposite. Suppose that at some step the existing trees (including T) are part of a minimum spanning tree S, but the new edge e (incident to T) is not part of S. Certainly S must contain a path which leaves T for the new node on e; otherwise S is not spanning. Let e' be the first edge on that path. It is not shorter than e, which was the shortest edge incident to T. Then replacing e' by e can only make S shorter.

You need to be certain that replacing e' by e leaves a spanning tree. There is still a path from every node to every other node; the bridge at e goes in or out of T. There are still $n - 1$ edges. No loop was created. Therefore we do not lose by requiring the edge e to be included, and this step is safe—completing the proof.

Does the stingy algorithm work? Remove the longest edge, keeping the graph connected.

Shortest Paths from the Source

Our third problem is one of the most basic in graph theory—or rather in network theory, since lengths $l_{ij} \geq 0$ are assigned to the edges. In this problem all paths start from one particular node; call it the source and identify it with the letter s. Then it is the length of the ***paths from the source***, not the total length of the tree, that we are required to minimize.

> ***Problem 3***   For each node in the graph, find the ***shortest path*** that connects it to the source node s.

This shortest path from s may not go along the minimum spanning tree. In Fig. 7.2, the best path going east has length $1 + 8$ and leaves the minimum tree. That tree was best for the whole graph, but not particularly best for the node s. In fact there will be a new tree, the "shortest path tree," in which s plays a special role.

How do we find the shortest paths? One method is to enumerate all possibilities; that is more or less insane. The competition among algorithms is based on their complexity, and the winner is the one that takes the fewest steps. This means either comparing worst cases or testing the algorithms on typical examples. Frequently the worst cases are not hard to identify, and we can determine how the number of operations increases with the number of nodes.

For shortest paths the key idea came from the Dutch mathematician Dijkstra.

(The first syllable is pronounced as in "dike.") His algorithm is not complicated. The object is to build out from s and to get one node permanently settled at each step. At step 1 we find the node nearest to s; it is s itself (with distance $d_1 = 0$). After k steps we know the distances $d_1, \ldots, d_k$ to the k nearest nodes. Suppose those nodes are numbered $1, \ldots, k$ as they are found, with the source as node 1. Then the problem is to find the next nearest node:

> **Dijkstra's algorithm**: Minimize $d_i + l_{ij}$.

Here i represents a node that is already settled and j is a new node. For a minimum spanning tree it was only the length l_{ij} that was minimized; that gave the length of the new edge. Now, because all distances are measured from s, the first part d_i (from the source to i) is also included. The best path will jump in a single step from a settled node to a new node—because if j were several steps away, the intermediate nodes would be closer to s.

The algorithm **minimizes $d_i + l_{ij}$ over all old nodes $i = 1, \ldots, k$ and all new nodes $j = k + 1, \ldots, n$**. That gives k times $n - k$ different possibilities. For a small graph they are easy to list, as in the following example:

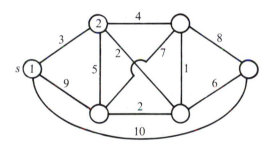

Fig. 7.3. A network for the shortest path problem.

Nodes 1 and 2 are already settled. The source is always node 1, at distance $d_1 = 0$, and the nearest node is at distance $d_2 = 3$. To decide on the *next* nearest node we look at all of the $k(n - k) = 2(6 - 2) = 8$ possible values of $d_i + l_{ij}$:

from node $i = 1$: $0 + 9$ $0 + 10$ $0 + \infty$ $0 + \infty$

from node $i = 2$: $3 + 5$ $3 + \infty$ $3 + 4$ $3 + 2$

The winner is $3 + 2 = 5$. Therefore the third node to be settled is Atlanta, in the southeast corner. Its distance from s is $d_3 = 5$. The fourth node will come in a similar way, by minimizing over $k(n - k) = 3(6 - 3) = 9$ values of $d_i + l_{ij}$:

from node $i = 1$: $0 + 9$ $0 + 10$ $0 + \infty$

from node $i = 2$: $3 + 5$ $3 + \infty$ $3 + 4$

from node $i = 3$: $5 + 2$ $5 + 6$ $5 + 1$

The shortest of these is $d_4 = 5 + 1 = 6$, going to Boston through Atlanta. That node and the final two nodes are shown in the shortest path tree. In practice the algorithm labels each node to indicate its ancestor.

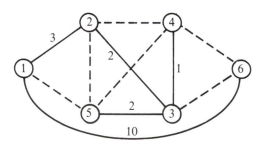

Fig. 7.4. The shortest path tree from node 1.

You may object that these distances are crazy. It should be shorter to go directly between nodes 2 and 4 than to travel through node 3. In other words the triangle inequality should hold: $l_{ij} \leq l_{ik} + l_{kj}$. However in that case all the interest in our problem is destroyed; every shortest path comes directly from s. And if you consider travel *times* instead of distances, the triangle inequality is not so sacred. The airlines violate that inequality every day, and $l_{ij} = \infty$ is extremely common—it means that there is no direct flight from i to j.

One application is the routing of messages through the computer network ARPANET. The vertices are computers, the edges are communication lines, and the l_{ij} are delay times. In practice Dijkstra's algorithm finds the best paths almost instantaneously; the delay times can be checked every few seconds and the paths are updated. We are assuming that the capacity of the lines is not a constraint (later in the chapter it will be). We look for the shortest path in the network— corresponding to a straight line in the plane, or a great circle on a sphere, or a helix on a cylinder—and the flow follows that path. In classical physics it is a geodesic, travelled by light. In relativity it is a world line. In automobile racing it is the only way to win—provided that "shortest" still refers, as it does for light passing from water to air, to time and not distance.

I hope you will look one more time at the algorithm. As it stands, it is not optimal. There is *too much repetition* in our tables of $d_i + l_{ij}$, and repetition is a sign of weakness. In fact the second table only removed a column from the first table and added a row—because one more node had been settled. It is a mistake to recompute $d_i + l_{ij}$ for all the unchanged pairs i and j, and Dijkstra saw the remedy: maintain a list of the best distances $D_j = \min(d_i + l_{ij})$ as they are discovered. After step k this distance to node j will change only if $d_k + l_{kj}$ gives an improvement, by using the path through the newest vertex k. Otherwise the current D_j is still the best we know. Therefore each step *updates* the distances D_j and then chooses the next nearest node:

$$\boxed{\begin{aligned} D_j &\leftarrow \min\left\{D_j, d_k + l_{kj}\right\} \\ d_{k+1} &\leftarrow \min\left\{D_j : j > k\right\}. \end{aligned}} \tag{1}$$

Instead of tabulating all possibilities $d_i + l_{ij}$, we are keeping a record of the smallest value in each column (namely D_j) and updating it as each new row is added.

Without improvement the original algorithm required

$$\sum_{k=1}^{n} k(n-k) = \frac{n^3 - n}{6} \quad \text{values of } d_i + l_{ij}.$$

That is the sum over all n steps, using standard formulas for the summation of k and k^2. For the streamlined version (1), this count is reduced from $O(n^3)$ to $O(n^2)$.† Therefore the updating in (1) is the way to execute Dijkstra's algorithm.

EXAMPLE In the first table above, the best distances from nodes 1 and 2 to the four unsettled nodes were $D_j = 8,10,7,5$. Therefore d_3 was 5 and Atlanta was node 3. Then the updated distances allow paths through Atlanta; the new D_j are

$$\min\{8,5 + 2\} = 7, \quad \min\{10,5 + 6\} = 10, \quad \min\{7,5 + 1\} = 6.$$

These are the shortest in the columns of the second table; its third row is compared with the best results already known from its previous rows. Therefore $d_4 = \min\{7,10,6\} = 6$ and the next step updates the other two distances 7 and 10.

Dynamic Programming

These shortest path algorithms are absolutely typical of dynamic programming. The philosophy is this: Whatever final edge ij is chosen to reach node j, *the path to node i should be optimal.* Of course that applies to every node! Whatever node is chosen before i, the path to that node is also optimal. It seems extremely obvious, but we emphasize the effects of this recursive optimization:

(1) While finding the shortest path to one node it finds the shortest paths to many nodes.

(2) It considers all paths but avoids trying them all. For $n = 20$ it uses less than 300 instead of more than a million additions.

(3) It works backwards equally well.

Frequently dynamic programming does go backward. It is a ***multistage decision process***, where the goal is known. Unlike the greedy algorithm, it is not true that at

---

† The notation $O(n^2)$ is pronounced "Oh of n^2" and means that some multiple Cn^2 is an upper bound. It gives the rate of growth with n, and ignores C. Quantities that are $O(n)$ have linear growth; $O(n^a)$ is polynomial; $O(\alpha^n)$ is exponential.

every point you should pick the shortest edge. The choice depends on that immediate step, but it also depends on how far there is to go *after* that step.

Here is an example: Pick a point on line A and a point on line B. The points are to give the shortest path from source to sink. On line A, point 1 is closest. However starting from 3, point 1 is a loser. The distance from 3 to the sink is

$$\textit{the minimum of } \left\{\begin{matrix} \text{distance from 3 to a point on line } A \\ + \text{ distance from that point to the sink} \end{matrix}\right\}.$$

This is like the $d_i + l_{ij}$ of Dijkstra; point 2 is minimizing. At the final stage, *all* these points are losers. The last decision is based on distances from s to line B plus line B to the sink (known from the previous family of decisions).

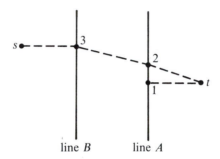

This is pretty basic stuff, arriving at a straight line, but comparing it to minimizing $\int \sqrt{1+(u')^2}\, dx$ brings out an unexpected point:

In the calculus of variations, the unknown was the *path from s to t*
In dynamic programming, the unknown is the *distance from each x to t*

The calculus of variations never considered distances from other points. It tested u against $u + v$ on the whole interval. Dynamic programming looks at the problem differently. It breaks the optimization into smaller steps and finds the minimum distances from all intermediate points x. There is always the crucial recursive decision

$$x - t \text{ distance} = \min_{y} (x - y \text{ distance} + y - t \text{ distance}).$$

That is Bellman's equation of dynamic programming in its simplest form.

Atlas of Shortest Paths

We mention two extensions of the shortest path problem. The first is to allow the possibility of negative lengths; one or more edges may have $l_{ij} < 0$. In this case a loop may suddenly become profitable. The distance around it could be negative.

That sends the shortest distances to minus infinity, by repeating the loop infinitely often (see Exercise 7.1.16). It is also possible to have $l_{ij} \neq l_{ji}$; Dijkstra's algorithm is unchanged.

The second extension asks for the shortest path between *every pair* of nodes. The special source has disappeared. It is like constructing a road atlas, and one possibility is to repeat Dijkstra's algorithm n times; each node in turn becomes the source. There is a simpler method proposed by Warshall (and by Floyd). They keep a matrix of shortest distances d_{ij} from i to j, starting from the direct distances l_{ij}—possibly infinite or negative. The key idea is first to admit paths like i to 1 to j that go through node 1, in case they are shorter. Using these paths, they next allow $m = 2$ as another possible intermediate node, and so on. Where Dijkstra introduces a new destination at every step, Warshall introduces routes through a new node. At least the algorithm looks simple:

$$\text{Initially } d_{ij} = l_{ij}$$

$$\text{FOR } m = 1, \ldots, n$$

$$\text{FOR } i = 1, \ldots, n$$

$$\text{FOR } j = 1, \ldots, n$$

$$d_{ij} \leftarrow \min \{d_{ij}, d_{im} + d_{mj}\}.$$

At the end all nodes and all paths have been allowed, and the shortest distances are the d_{ij}. Somehow this search requires only n^3 steps, even though the number of paths is exponential.

We cannot resist suggesting one more algorithm. The shortest paths can be found with pieces of string. Cut them to the lengths given by the network and attach them to the nodes. Then pick up node i in one hand and node j in the other, and pull until some path between them becomes taut. That is the optimal path.

Looking back at this section, I realize that it was entirely algorithmic. It simply computed spanning trees and shortest paths. By contrast, the rest of the chapter offers two answers to each question—an algorithm that computes the minimum or maximum, and also an optimality condition that identifies a proposed answer as correct. We will not find the optimum in a single sweep, and we have to know when to stop. That will be the link between theory and computation—the optimality condition acts as a "stopping test" to tell the algorithm it has finished.

EXERCISES

7.1.1   Find a graph whose edges cannot be drawn in a plane unless they are allowed to cross.

7.1.2   Continue by BFS and then by DFS to number the remaining nodes and find a spanning tree (if one exists) for the following graphs:

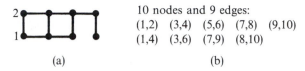

10 nodes and 9 edges:
(1,2) (3,4) (5,6) (7,8) (9,10)
(1,4) (3,6) (7,9) (8,10)

(a)   (b)

7.1.3   In ordinary language, describe a code that executes a depth first search for a spanning tree.

7.1.4   The adjacency matrix A of a graph has $a_{ij} = 1$ if there is an edge connecting i to j; otherwise $a_{ij} = 0$.
   (a)   Interpret A^2: what does it mean if the ij entry of A^2 is positive?
   (b)   Here is another test for connectedness: All entries of A^{n-1} should be positive. Is that test correct?

7.1.5   Does the adjacency matrix have rank $n - 1$ exactly when the graph is connected?

7.1.6   Why does every spanning tree have $n - 1$ edges if the graph has n nodes?

7.1.7   Does the following algorithm produce a minimum spanning tree whenever the graph is connected? Start with all edges and look for a loop. Remove the longest edge in that loop. Continue until reaching a tree.

7.1.8   Is it possible to have two minimum spanning trees with no edges in common?

7.1.9   What is the minimum spanning tree for the network in Fig. 7.3?

7.1.10   To look for a spanning tree without worrying about lengths, apply Kruskal's method to any list of edges: Keep edge i unless it forms a loop with edges already kept. List the edges in Fig. 7.1 and use this algorithm. If the graph is not connected, what will be the situation after the algorithm goes through the edge list?

7.1.11   How can Prim's algorithm be organized to avoid looking every time at the edges (of length 8 and 9 in the example) which were rejected in an early step? Are they always rejected permanently?

7.1.12   Use the previous exercise to describe an on-line method that finds minimum spanning trees as new edges are announced.

7.1.13   Show that if an edge e is longer than any other edge in a given loop, then it cannot be in a minimum spanning tree T.

7.1.14   If the source s is moved to Boston, so that the first edge in the shortest-path tree of Fig. 7.4 is Boston-Atlanta, list the 8 possible values for $d_i + l_{ij}$ at the next step. Find d_3 and sketch the new shortest path tree.

7.1.15   In the streamlined form of Dijkstra's algorithm, what will be the two updated values of D_j at the next step in the example—after the step that decides $d_4 = 6$?

7.1.16   Construct two networks with edge lengths allowed to be negative, one in which shortest distances from s are still positive and another in which the distance around some loop is negative. (In currency exchange this corresponds to "arbitrage," a paradise in which exchanging dollars for yen for francs for pounds for dollars produces a profit.)

7.1.17   Find the shortest path tree rooted at the node s in Figure 7.2.

7.1.18 Describe in words the steps of a code for the streamlined form of Dijkstra's algorithm.

7.1.19 For the six-node network in Fig. 7.3 write down the original 6 by 6 matrix with entries l_{ij} and then update with the step $m = 1$ of Warshall's algorithm: Allow paths from i to 1 to j as well as from i to j, and find the best distances d_{ij}.

7.1.20 Apply Dijkstra's algorithm to find the shortest distances d_i from node 1 to all other nodes in the networks below:

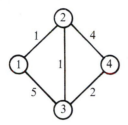
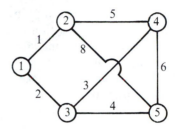

7.1.21 The bottleneck path problem looks for the paths from s that minimize the length of the *longest edge* rather than the total length. Construct the tree containing these paths for the graph in Fig. 7.3.

7.2 ■ THE MARRIAGE PROBLEM

You may think that marriage is outside the scope of applied mathematics. That is true. Once you are married, mathematics is of no possible use. But there is a superficial model for a simpler problem—getting married in the first place—and it leads directly to a fundamental question in combinatorics.

The model is easy to visualize as a matrix of zeros and ones. Imagine n rows corresponding to girls $g_1, \ldots, g_n$ and n columns corresponding to boys $b_1, \ldots, b_n$. Then $a_{ij} = 1$ if g_i and b_j are willing to marry, and $a_{ij} = 0$ if not. (No shades of emotion are allowed.) The question is: *How many marriages are possible?* This is also known as the simple assignment problem, the maximum transversal problem, and matching theory. A full set of n marriages (if it is possible) is a *complete matching*.

We begin with an example:

$$A = \begin{bmatrix} 1 & 1 & 1 & 1 \\ 0 & 0 & 0 & 1 \\ & 1 & 0 & 1 \\ & 1 & 0 & 0 \end{bmatrix}.$$

Depending on the missing entries, three or four marriages are possible. If either of the blanks is a 1 there is a complete matching, for instance in

$$A = \begin{bmatrix} 1 & 1 & \boxed{1} & 1 \\ 0 & 0 & 0 & \boxed{1} \\ 0 & \boxed{1} & 0 & 1 \\ \boxed{1} & 1 & 0 & 0 \end{bmatrix}.$$

The matching amounts to a permutation matrix—no row or column is represented twice, since bigamy is not allowed. It has also been called a transversal, or a nonzero diagonal, since after row exchanges the marriages fill the main diagonal.

Now suppose both blank entries are zero:

$$A = \begin{bmatrix} 1 & 1 & 1 & 1 \\ 0 & 0 & 0 & 1 \\ 0 & 1 & 0 & 1 \\ 0 & 1 & 0 & 0 \end{bmatrix}.$$

Then a set of four marriages is impossible. That is easy to check by trial and error, but it is essential to have a clear statement of the reason: There is a submatrix of zeros, leaving *three girls* g_2, g_3, g_4 who among them *like only two boys*. The girls in the last three rows cannot all be satisfied:

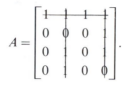

$$A = \begin{bmatrix} 1 & 1 & 1 & 1 \\ 0 & 0 & 0 & 1 \\ 0 & 1 & 0 & 1 \\ 0 & 1 & 0 & 0 \end{bmatrix}.$$

From the point of view of the columns (the boys), there is a set of two boys (1 and 3) who can marry only one girl. A complete matching is made impossible by this submatrix of zeros.

Eventually these submatrices will answer the question in general. A full set of marriages can be arranged if and only if every set of girls likes at least an equal number of boys. Equivalently *no zero submatrices can be too large*. Before the theory we look once more at the example, to express in different words what it implies:

(1) (For chess) If rooks can be placed on any of the squares containing 1's, with the condition that no rook can take any other rook, then the maximum number of rooks is 3.

(2) (For duality) The maximum number of marriages (3) equals the minimum number of horizontal or vertical lines that cover all the 1's.

(3) (For algebra) Every matrix with these zeros (any numbers can replace the 1's) is automatically singular. Its determinant is zero.

This last point depends on rows 2,3,4; if the zeros remain where they are, those rows must be linearly dependent. The same is true of columns 1 and 3; one is a multiple of the other. And an equivalent test comes from the determinant of A. It is a sum of $4! = 24$ terms like $a_{12}a_{24}a_{33}a_{41}$, each term coming with plus or minus sign from one of the possible matchings. All of these terms are zero in the example, because at least one of the a's is zero. Therefore the determinant is zero. Such a matrix is "structurally singular," since its structure of zeros means that it can never be invertible.

The simplest structurally singular matrix has a complete row of zeros, which makes the determinant zero regardless of the other entries. The row is a 1 by n submatrix of zeros, and it prevents marriage for that girl.

Now we come to the general result: *An r by s submatrix of zeros makes a matching impossible if $r + s > n$.* Because of the zeros, the r girls can marry only boys outside the submatrix, and there are only $n - s$ such boys. Since $r > n - s$, this group of r girls likes too few boys and cannot all be married.† In the example with $n = 4$, r is 3 and s is 2; three girls liked only $4 - 2$ boys and 2 boys liked only $4 - 3$ girls. Thus the necessary condition, due to Philip Hall, can be expressed in three

---

† The situation is symmetric; the s boys whose columns are in the submatrix can marry only the $n - r$ girls whose rows are not. There are too few girls for that group of boys, since $s > n - r$.

completely equivalent ways:

> (1)   There is no r by s submatrix of zeros for which $r + s > n$
> (2)   Every set of r girls, $1 \le r \le n$, likes at least r boys
> (3)   Every set of s boys, $1 \le s \le n$, likes at least s girls.

We check condition 2 in the example. First, every girl does like at least one boy. Second, every set of two girls does like at least two boys; so far the test is passed. But the three girls who like only two boys violate Hall's condition. Conditions 1 and 3 must also be violated, since they are equivalent to 2.

If Hall's condition fails, a complete matching is impossible. That is clear. What is remarkable is that the converse is also true. If Hall's condition is satisfied, we can show that a complete matching *must be possible.* The proof is by induction, assuming the conclusion to be correct for matrices of order less than n and deducing that Hall's condition is sufficient also for order n. That conclusion is easy to check if $n = 1$; Hall's condition means that the one girl likes the one boy, and their marriage is a complete matching for $n = 1$. (Equivalently, there is no 1 by 1 zero submatrix.) The harder step is to go from order less than n to order n. It is more subtle than many proofs by induction, and you have to watch closely to see the point.

7B   Hall's condition, in any of its equivalent forms, is both necessary and sufficient for a complete matching to be possible.

The condition is certainly necessary. To prove it is sufficient, we assume it is satisfied and show how to construct a complete matching. There are two cases:

Case 1.   For every set of r girls, $r < n$, there are *more than* r acceptable boys. This should be the easier case. There is enough freedom for the first girl to marry any boy she likes; then Hall's condition holds for the remaining $n - 1$ girls and $n - 1$ boys. Every r of these girls liked more than r boys. One of those boys may have been the one who married the first girl—but without him there are still at least r boys. Thus Hall's condition is satisfied by the remaining girls and boys, and by the induction hypothesis they can be completely matched.

The zero submatrices in Case 1 all have $r + s$ *less than* n. After the first marriage they still have $r + s \le n - 1$, which is Hall's condition on the remaining matrix of order $n - 1$; it guarantees a matching. Case 2 will be tighter.

Case 2.   Some set of r girls, $r < n$, likes *exactly* r boys. Within this smaller set Hall's condition still holds (it applied to every set of girls). Therefore, by induction, r marriages can be arranged in this set.

To complete the proof we verify that for the remaining subsets of $n - r$ girls and $n - r$ boys, Hall's condition also holds. Consider any R girls in this remaining subset. They certainly like at least R boys, but our problem is to show that they like at least

R of the remaining *unmarried* boys. The trick is to consider these *R* girls together with the original *r*. By Hall's condition, this group likes at least $r + R$ boys. Therefore the *R* girls, among themselves, must like at least *R* of the unmarried boys (since the *r* girls liked only the *r* boys they married). This is Hall's condition for the remaining girls and boys, and again (since $n - r$ is smaller than *n*) they can be married by induction.

In the language of matrices, Case 2 refers to the possibility

$$A = \begin{bmatrix} B & C \\ 0 & D \end{bmatrix} \begin{matrix} n-r \\ r \end{matrix} .$$
$$ n-r \quad r$$

The last *r* girls like only the last *r* boys. The zero submatrix is *r* by $n - r$, the largest that still allows a complete matching. Hall's condition for *A* implies Hall's condition for *D* and *B*—that was the key point established above—and then the complete matchings in *D* and *B* combine into a complete matching for *A*.

That completes a short and elegant proof. However it is not a constructive proof. As it stands, all 2^n subsets of the girls would have to be checked just to verify Hall's condition. It is natural to look for a fast algorithm to find the matching that is guaranteed to be present. But first comes the other crucial step of the theory—to consider the general case, when a complete matching may be impossible, and ask how many marriages can still be achieved.

A large zero submatrix is now allowed, and the answer comes from a completely different direction—which is typical of *duality*. In Chapter 2 duality occurred when $-P = Q$; the quantities $-P$ and Q were quadratics. Here the maximum of one quantity again equals the minimum of another, even though quadratics are gone. The problem is one of counting instead of calculus, but the key to everything is still duality.

Duality in Combinatorics

7C The *maximum number of marriages* equals the *minimum number of lines* (rows or columns or a combination of both) that contain all the 1's in the matrix.

In the example with three marriages, three lines covered the 1's. When there is a complete matching the maximum and minimum both equal *n*. In fact we need *n* lines to cover the particular 1's in the matching. You could not cover the 1's in the identity matrix, or in any permutation matrix, with fewer than *n* lines—since no two 1's can be covered by the same line.

This argument applies in general: If we can achieve *N* marriages, we need *N* lines to cover those 1's, and therefore we need at least *N* lines to cover all the 1's. That inequality is *weak duality*: *minimum number of lines* ≥ *maximum number of marriages.* It is the easy half of any duality theorem. The problem is to push the maximum up and the minimum down until they meet.

Suppose the minimum number of lines is M. The goal is to find M marriages. All the 1's in the matrix are covered by p rows and q columns, with $p + q = M$. For convenience we reorder the matrix so that the p rows come first and the q columns come last:

$$A = \begin{bmatrix} B & C \\ 0 & D \end{bmatrix} \begin{matrix} p \\ m - p \end{matrix}$$
$$\begin{matrix} n - q & q \end{matrix}$$

Since the p rows and q columns cover A, the lower left corner must be zero as shown. When this zero submatrix is big, a complete matching is impossible—but our goal is to find a *maximum* matching, with M marriages. If we can prove that B and D both satisfy Hall's condition, then there are p marriages within B and q more in D. The first p girls can be matched to p of the first boys, and the last q boys can be matched to q of the last girls. That gives $p + q = M$ marriages, as we wished to show.

The key is to check Hall's condition for B. *Suppose r of the first girls like only k* of the first boys, with $k < r$. Then the 1's in A could be covered by those k columns, and the other $p - r$ first rows, and the last q columns. But $k + p - r + q = k - r + M$ is fewer lines than the minimum number M, a contradiction. So each subset of the first p girls does like enough boys; p marriages are possible in B. Similarly Hall's condition holds for D, giving q more marriages and completing the proof.†

With this duality theorem we can tell when a set of marriages is as large as possible: There must be a corresponding set of lines covering all the 1's. Then we know we can stop.

Bipartite Graphs: The Same Theorem

The duality theorem for marriages and lines is identical to a duality theorem for "bipartite graphs." Historically it is hard to say which came first; the link could not be closer. Hall's condition will guarantee a matching between the two parts of the graph. When it fails duality will still say how large a matching is possible. This provides an ideal opportunity to introduce more of the language of graph theory.

Remember that a graph consists of a set V of vertices and a set E of edges between them. In general it does not contain all possible edges. It is a ***directed graph*** (or *digraph*) if each edge is given a direction; the edge starts at a vertex i and ends at a vertex j. We don't allow two edges to do that, but there could also be an edge from j to i. In the marriage problem the vertices are the n girls and n boys, with a directed edge for every 1 in the matrix (Fig. 7.5). Thus the edge ij is present exactly when girl i is willing to marry boy j. The result is a special directed graph, called a ***bipartite graph***: the set of vertices divides into two parts, the girls and the boys, and all edges

---

† B and D can be rectangular. Exercise 7.2.5 shows that Hall's test still applies to the girls in B and the boys in D.

go from one part to the other. There are no edges within the set of girls or within the set of boys.

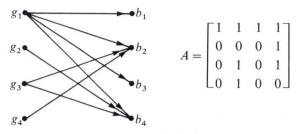

Fig. 7.5. A bipartite graph and the corresponding matrix.

Converting from matrices to graphs, the definitions become:

a ***matching*** is a set of *edges* that have no vertices in common; no girl and no boy can appear twice. The matching may or may not be complete.

a ***cover*** is a set of *vertices* that meets every edge. The vertices correspond to rows (girls) and columns (boys), in other words to lines of the matrix. A set of lines covering all the 1's becomes a set of vertices that meets every edge.

Figure 7.6 shows a maximum matching—3 heavy edges in the graph and 3 circled 1's in the matrix. It also shows a minimum cover—3 squared vertices in the graph (they meet every edge) and the corresponding 3 lines in the matrix. This equality of maximum and minimum is an instance of the duality theorem for matrices. Translated to bipartite graphs, it is called the König-Egervary theorem:

7D The maximum number of edges in a matching equals the minimum number of vertices in a cover.

There is no need to repeat the proof! It is the same as 7C. We mention only that the result is equally true whether or not the graph has the same number of vertices (girls and boys) in its two parts. If the maximum number of independent edges is N, and if we pick correctly N vertices (choosing one end or the other from each edge in an optimal way) then all other edges in the graph will meet these N vertices.

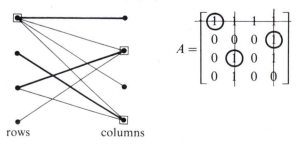

Fig. 7.6. A maximum matching and a minimum cover.

Doubly Stochastic Matrices

There is a neat application of the marriage theorem to *doubly stochastic matrices* B—which have all $b_{ij} \geq 0$, and the sums along every row and every column equal to one. Examples are the identity matrix, or any permutation matrix, or all $b_{ij} = 1/n$. This last matrix is a combination of n different permutation matrices, each multiplied by $1/n$. Birkhoff discovered that a similar idea works whenever B is doubly stochastic:

7E   Every doubly stochastic matrix is a combination $B = w_1 P_1 + \cdots + w_k P_k$ of permutation matrices P_i, with positive weights w_i that add to one.

EXAMPLE

$$\begin{bmatrix} \frac{2}{3} & \frac{1}{3} & 0 \\ \frac{1}{3} & 0 & \frac{2}{3} \\ 0 & \frac{2}{3} & \frac{1}{3} \end{bmatrix} = \frac{2}{3}\begin{bmatrix} 1 & 0 & 0 \\ 0 & 0 & 1 \\ 0 & 1 & 0 \end{bmatrix} + \frac{1}{3}\begin{bmatrix} 0 & 1 & 0 \\ 1 & 0 & 0 \\ 0 & 0 & 1 \end{bmatrix}.$$

The key is to see that any zero submatrix of B, with r rows and s columns, must have $r + s \leq n$. *Proof*: Outside the submatrix of zeros will be entries in the r rows adding up to r (a sum of 1 along each row). Also outside, and not overlapping, will be entries in the s columns adding up to s. But the total of all entries in B is n, so $r + s \leq n$.

Thus Hall's condition is satisfied. There must exist a complete matching—a permutation matrix P_1 which has its n nonzeros only where B is nonzero. Let w_1 be the smallest of these particular n nonzero entries of B. Then $B - w_1 P_1$ is still nonnegative, and has at least one more zero than B; progress is being made. The new row and column sums are all $1 - w_1$; if $w_1 = 1$ we are finished ($B = P_1$), and otherwise we divide by $1 - w_1$ to get a second doubly stochastic matrix B'. Since B' has at least one more zero than B, repeating the argument eventually finishes the proof.

EXERCISES

7.2.1   Suppose A is the 4 by 4 matrix with $a_{ij} = 1$ on the diagonals just above and just below the main diagonal. Is a matching possible, and is it unique? Draw the corresponding graph with eight vertices.

7.2.2   If A is the 5 by 5 matrix with $a_{ij} = 1$ on the diagonals above and below the main diagonal, show that a complete matching is impossible by finding
    (i)   a set of girls who do not like enough boys
    (ii)   a set of boys who do not like enough girls
    (iii)   an r by s zero submatrix with $r + s > n$
    (iv)   four lines which cover all the 1's in A.

7.2.3 Let A_n be the n by n tridiagonal matrix with $a_{ij} = 1$ if $j = i - 1$, i, or $i + 1$. Apart from these three diagonals, $a_{ij} = 0$. For the cases $n = 1,2,3,4$, count the number of complete matchings.

7.2.4 For the same tridiagonal matrix A_n, let F_n be the number of complete matchings:
    (i) The first girl could marry the first boy. Explain why F_{n-1} matchings start with this marriage.
    (ii) The first girl could marry the second boy, and the second girl the first boy. Explain why F_{n-2} matchings start with these marriages.
    (iii) The first girl could marry the second boy, and the second girl the third boy. Show that no matching can start with these marriages.
Thus $F_n = F_{n-1} + F_{n-2}$, and the F_n are the *Fibonacci numbers* 1,2,3,5,8,13, ...

7.2.5 If A is a rectangular matrix, with m girls and n boys and $m < n$, and if every set of r girls likes at least r boys, show from Hall's theorem 7B that m marriages are possible. (*Hint:* Create $n - m$ more girls who will marry anyone.)

7.2.6 (a) Suppose there are exactly two 1's in each row and column; every girl likes two boys and every boy likes two girls. Show that Hall's condition is satisfied and a complete matching is possible. (How many lines are needed to cover the $2n$ 1's?)
    (b) After subtracting the permutation matrix P_1 that gives the matching, show that what is left is another permutation matrix P_2 and construct an example of $A = P_1 + P_2$ with $n = 4$.
    (c) Suppose there are two *or more* 1's in each row and column. Show by an example (I think $n = 5$) that a complete matching may be impossible.

7.2.7 Suppose the zero-one matrix A has the form $A = \begin{bmatrix} R & 0 \\ T & S \end{bmatrix}$, with square blocks R and S of order r and s. If a complete matching is possible, show that T makes no contribution: none of the last s girls can marry any of the first r boys.

7.2.8 Suppose P is a 20 by 20 permutation matrix, subdivided into a 10 by 10 array of small blocks (each 2 by 2). This produces a 10 by 10 matrix A, with $a_{ij} = 0$ if the corresponding block in P is all zero and $a_{ij} = 1$ if the block in P contains any ones. Show that A allows a complete matching with 10 marriages.

7.2.9 Verify that a doubly stochastic matrix B satisfies Hall's condition in the equivalent form: No set of k rows has all its nonzero entries in only $k - 1$ different columns. Add up the entries in such a submatrix, first by rows and then by columns, to show that it cannot exist.

7.2.10 Write this doubly stochastic matrix B as a combination of permutation matrices:

$$B = \begin{bmatrix} .4 & .3 & .3 \\ .6 & .4 & 0 \\ 0 & .3 & .7 \end{bmatrix}.$$

7.2.11 For *infinite sets* Hall's test can fail: If the first girl likes all boys, and then girl k likes only boy $k - 1$, $k = 2,3,4, \ldots$, show that
    (i) every set of r girls likes at least r boys, but
    (ii) there is no complete matching.
If each girl likes only a finite number of boys, Hall's theorem becomes true again.

7.2.12   The *continuous* marriage problem also fails. If there is a girl at every point $0 \le x \le 1$, and she likes the two boys at $y = \frac{1}{2}x$ and $y = \frac{1}{2}(x + 1)$, show that
   (i)   Any open sets of girls—say all the girls in the interval $a_1 < x < b_1$—likes a set of boys in intervals of total length $(b_1 - a_1)$ but
   (ii)   A complete matching is impossible.

7.2.13   Suppose there are n suns and n moons, and every set of moons is within a million miles of at least an equal number of suns. Show from Hall's theorem that the suns and moons can be paired off, each moon within a million miles of its own sun.

7.2.14   How many planes does an airline need to make all n of its scheduled flights? Set $a_{ij} = 1$ if flight j starts after flight i ends; otherwise $a_{ij} = 0$ (and in particular $a_{ii} = 0$). Suppose the largest matching contains M marriages; the 1's in the matrix can be covered by M lines.
   (a)   Show that all flights can be made with $n - M$ planes.
   (b)   Show that fewer than $n - M$ planes are not sufficient.
(The M lines in the cover come from at most M different flights, leaving at least $n - M$ flights; explain why no plane can do two of them.)

7.2.15   (recommended) If a 7 by 7 matrix has 15 1's, prove that it allows more than two marriages.

MATCHING ALGORITHMS ■ 7.3

Hall guaranteed a complete matching whenever his condition is satisfied: The n by n matrix should have no r by s submatrix of zeros, with $r + s > n$. Every set of r girls should like at least r boys. In this case the matrix possesses a full set of n nonzero entries, with each row and column represented once; that is the complete matching. By reordering the rows of the matrix those nonzeros will appear on the main diagonal, which is then "zero-free."

This section asks how such a matching can be found, when the matrix is given. Up to now our argument was completely nonconstructive—or rather, if the last section's proof by induction were converted directly into an algorithm, it would take forever. Much better and simpler algorithms are known, and they give a beautiful contrast between *depth first search* and *breadth first search*. The police must make a similar choice every day, in following up a crime.† The result is that if A has m nonzero entries, and a matching is possible, it can be found in nm operations. For a sparse matrix ($m \ll n^2$) this is comparatively inexpensive, but there are faster algorithms and the competition to find the fastest is still going on. The winners so far are Hopcroft and Karp, who can finish in $Cn^{1/2}m$ steps.

A successful algorithm should also confirm the duality theorem 7C. If the maximum number of marriages is N (possibly less than n) it should find N lines that cover the 1's in the matrix. If a bipartite graph has at most N independent edges, it should discover N vertices that meet all edges. More than that, the matching algorithm should introduce the key ideas that come later, and even imitate as far as possible the great *simplex algorithm*:

(1) It starts from a position that is acceptable but probably not optimal
(2) At every stage it improves the position by a single move
(3) When it stops, it has solved not only the primal problem (maximum number of marriages) but also the dual (minimum number of lines).

In that sense we also get a new and more constructive approach to the theory. I think the specific problem is less fundamental than others in the book and less important than the underlying principles.

The algorithms start by choosing some set M of acceptable marriages. Probably it is not the largest possible. A reasonable construction of M is to go down the rows, and choose in each row the first 1 that is in a column not already used. It is important to see that this may not produce the best matching. For the matrix

$$A = \begin{bmatrix} 1 & 1 \\ 1 & 0 \end{bmatrix} \text{ it chooses } \begin{bmatrix} ① & 1 \\ 1 & 0 \end{bmatrix}.$$

The second row has no 1 in an unused column, so our initial matching M has only

---

† We do not pursue this remark.

one marriage—even though two are possible. Unlike the greedy algorithm, the wrong choice for the first girl can get us temporarily stuck.

Every later stage begins with a matching and has to decide whether it is best possible or can be improved. In this case it looks at girl 2. She can marry boy 1, if we annul his marriage to girl 1. Therefore the algorithm looks next at girl 1. She can marry boy 2, who is not already married, and this produces a ***reassignment chain***. That is a chain of the form girl-boy-girl-·····-boy (no one can appear twice) with three properties:

(1)   the first girl and last boy are outside M; neither one is already married

(2)   the first, third, fifth, ... links in the chain are possible marriages but are not in M

(3)   the second, fourth, ... links are already in M. In the example, the reassignment chain is girl 2-boy 1-girl 1-boy 2. Neither girl 2 or boy 2 is in M, whose only marriage was boy 1 to girl 1—the second link in the chain.

Whenever there is a reassignment chain, there is an easy way to reach a matching with one more marriage than before:

Annul the marriages in links 2, 4, ... ***and add the marriages in links*** 1, 3, 5, ...

This is legal by properties 1–3, and it creates one more marriage than it annuls. In the example it leaves us with girl 2-boy 1 and girl 1-boy 2. In that case the new M is optimal; in general we start again to look for a reassignment chain, and a further improvement.

The goal of every algorithm is to find such a chain. *If there is no reassignment chain, then M is best possible.* This is a subtle point, since we might imagine that some totally different matching allowed more marriages, while trying to improve M leads to a dead end. It is a fundamental fact in this whole area of mathematics that such a thing never happens. Either M is optimal, or it allows a step by step improvement—a sequence of reassignment chains or "augmenting paths," adding one marriage each time—that takes us to an optimum. This is verified in Exercise 7.3.8.

We begin with a ***depth first search*** (DFS). It starts with a girl not in the existing M, as every reassignment chain must. In her row it chooses the first column with a 1—the first acceptable boy. If he is also outside M, we have found a chain with only one link, and we add that marriage immediately to M. This was too easy, and normally it will not happen. Probably the boy is already married in M, to a girl who will be next in our chain. In her row we again choose the first column with a 1 (but not the first boy in the chain); that gives the second boy. If he is unmarried in M, we have a reassignment chain with three links; otherwise his marriage identifies the third girl, and the chain continues.

Eventually, we reach either an unmarried boy (who completes the chain) or a girl in whose row all the 1's have been used. The acceptable boys are already in the chain, and the construction cannot go forward. In that case we must ***backtrack*** to the previous girl in the chain, and look in her row for a different 1. If we find one not already used, we start forward again. But if all her 1's are in the chain, or if we tried them and were unable to complete a reassignment, then we backtrack again. We may get all the way back to the very first girl, and find no successful chain. Then a new search begins from the next girl who is not in M. If M is not optimal, we eventually complete a chain. If M is optimal, the search finally fails and the algorithm stops.

EXAMPLE 2   In the 5 by 5 matrix of Fig. 7.7a, the original matching contains three marriages (circled). Its reassignment chain starts with the first unmarried girl, row 4, and chooses boy 1. It annuls his marriage to girl 1, and looks in her row for the next acceptable boy; he is boy 2. From there, the chain alternates between uncircled and circled entries and ends at an unmarried boy: $g_4 - b_1 - g_1 - b_2 - g_2 - b_4$. This reassignment gives one more marriage, and there are four circles in the second matrix.

The two circles in the chain are erased—those marriages are annulled—and the three squares become circles in the new M.

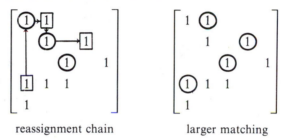

reassignment chain   larger matching

Fig. 7.7. Depth first search: one step.

The next chain enters at g_5, but after $g_5 - b_1 - g_4 - b_2 - g_1$ it halts. There is no boy to continue the chain from g_1, since in the first row the only candidate b_1 is already in the chain (Fig. 7.8). Therefore the algorithm backtracks (dotted line) to the girl g_4, and begins again from there; it chooses the second possible boy in her row. He is b_3, and the chain is completed to $g_5 - b_1 - g_4 - b_3 - g_3 - b_5$. Annulling two marriages and creating three gives the final assignment. In this case we reach the complete matching in Fig. 7.8.

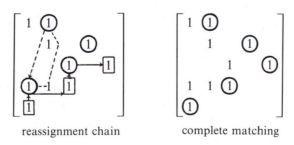

reassignment chain   complete matching

Fig. 7.8. Depth first search: final step.

Note the main drawback of depth first search. It may start down a long and possibly wrong track, and take a long time to fail, when there is a shorter chain that succeeds. The algorithm can miss an easy solution. On the other hand, looking at all possible short chains is not automatically successful. It is the same dilemma you face in buying a car: Do you choose one model, and one dealer, and one salesman, or do you try to check them all? A compromise seems optimal, but we have to make it into a precise algorithm.

The other extreme is a **breadth first search** (BFS). The difference is this:

DFS: Each step starts from the row or column that was visited *most recently*
BFS: Each step starts from the row or column that was visited *least recently*.

It is easy to describe the first step of BFS. For the girls not in M, find *all* acceptable boys. Then check to see whether those boys are already matched in M. If not we have a new assignment and M is improved. Otherwise they lead to a second set of girls, and we start again. In other words, we pursue many reassignment chains at once.

As it stands, this is not totally efficient. Nevertheless the idea has merit; the best algorithms seem to combine BFS and DFS. One compromise by Duff introduces a one-step BFS into the previously described DFS: as each girl is reached, we look for a boy who can immediately finish the chain (he is not in M). Rather than accepting without question the first unused column with a 1, all columns are checked for a "cheap assignment" that would end the step. If there is none, the depth first search continues. Another version, discovered by Hopcroft and Karp, makes the worst case even shorter. It is the leader at this moment. But it is a little complicated to describe and the underlying BFS ideas are clearest in the following simple algorithm.

We use labels to keep track of the steps. When a row is scanned and unused 1's are found, their columns are labeled by the row number. Similarly if a column is scanned, its number is the label assigned to a row.

Breadth First Search
 1. Given a matching M, label with (*) all girls that are not matched.
 2. (*Label boys*) For the first newly labeled girl g_i, find all acceptable boys not yet labeled and label them (g_i). Repeat for all other newly labeled girls.
 3. (*Label girls*) For each newly labeled boy b_j, look for the girl to whom he is matched by M and label her (b_j). If any boy is unmatched, *stop* with a completed chain; otherwise return to step 2.

If we meet a boy in step 3 who is not matched in M, a reassignment chain has been found. He is the end of the chain. His label takes us to the previous girl, whose label takes us one more step back. Eventually we come to a girl labeled (*). She is the start of this successful chain.

The alternative is that on some entrance to step 2, all acceptable boys are already labeled. Then step 2 adds no new labels. In this case M itself is optimal; there is no reassignment chain and the algorithm stops. To see how this works, we consider two examples. They can use either graphs or matrices.

EXAMPLE 3A The edges in the bipartite graph give the possible marriages, corresponding to 1's in the matrix. The dark lines in Fig. 7.9 are the three edges in the current matching.

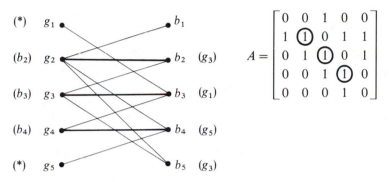

$$A = \begin{bmatrix} 0 & 0 & 1 & 0 & 0 \\ 1 & ① & 0 & 1 & 1 \\ 0 & 1 & ① & 0 & 1 \\ 0 & 0 & 1 & ① & 0 \\ 0 & 0 & 0 & 1 & 0 \end{bmatrix}$$

Fig. 7.9. Initial matching with three marriages.

The labels are found by the algorithm, which starts with the two unmatched girls:

(1) Label g_1 and g_5 with (*).
(2) Label b_3 with (g_1) and b_4 with (g_5).
(3) Label g_3 with (b_3) and g_4 with (b_4).
(2) Label b_2 and b_5 with (g_3).
(3) Label g_2 with (b_2); b_5 is unmatched and completes a chain.

Backtracking to find the chain, b_5 came from g_3, g_3 from b_3, and b_3 from g_1. We annul $g_3 - b_3$ and create the other two marriages to reach a new M, with which we start again. This is example 3B, beginning from the marriages on the four dark edges:

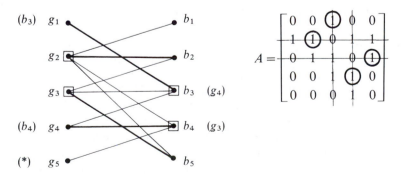

$$A = \begin{bmatrix} 0 & 0 & ① & 0 & 0 \\ 1 & ① & 0 & 1 & 1 \\ 0 & 1 & 1 & 0 & ① \\ 0 & 0 & 1 & ① & 0 \\ 0 & 0 & 0 & 1 & 0 \end{bmatrix}$$

Fig. 7.10. Final matching with four marriages.

EXAMPLE 3B   The steps are

(1)   Label g_5 with (*).
(2)   Label b_4 with (g_5).   (not in M)
(3)   Label g_4 with (b_4).   (in M)
(2)   Label b_3 with (g_4).   (not in M)
(3)   Label g_1 with (b_3).   (in M)
(2)   No unlabeled boys are connected to g_1; stop the algorithm.

Algorithmic Duality

It remains to prove that this breadth first search always works; it finds the largest possible matching. At the same time it solves the dual problem. It finds the minimum number of lines that cover all the 1's—or the smallest set of vertices meeting every edge. In this example four marriages were maximal and four vertices are minimal. They are the *unlabeled girls* (g_2 and g_3) and the *labeled boys* (b_3 and b_4). Every edge in the graph touches at least one of these marked vertices. (They were marked by extra squares after the algorithm stopped.) Similarly every 1 in the matrix lies on at least one of the four lines: row 2, row 3, column 3, and column 4. Note the 3 by 3 zero submatrix.

> **7F**   If the algorithm stops without finishing a chain then M is a maximal matching. The set C of unlabeled girls and labeled boys meets every edge in the graph, and no smaller set of vertices can do so. C gives a minimal cover of all the 1's in the matrix.

Proof   Suppose there is an edge not touching C. Then it must go from a labeled g to an unlabeled b. If the edge is not in M, b would have been labeled after g was (step 2). If the edge is in M, then g got its label from b (step 3). One way or the other, b must have been labeled and this nontouching edge cannot exist. The contradiction implies that every edge touches C.

Now we show that C is minimal and M is maximal. First of all, C is not larger than M. Reason: Each vertex in C meets an edge in M. Every unlabeled girl certainly meets an edge in M; otherwise she would have the label (*). A labeled boy meets an edge in M, or the algorithm would have stopped with him and produced a new chain.

Since no vertex can ever meet two edges of a matching (those edges are always independent) a cover C can never be smaller than a matching M. Therefore our two sets are the same size, which completes 7F.

This gives a constructive proof of the duality in 7D. The algorithm actually finds C and M. It seems to do so with at most $m + n$ operations at each stage. The steps numbered 2 look along the rows of the matrix for the m 1's. The steps numbered 3 are quicker, since for each boy they just find the edge already in M. Since there may be up to n stages, each producing a reassignment chain and one new marriage, the complexity is $(m + n)n$—and a slight reorganization gives mn. It is this which Hopcroft and Karp were able to defeat.

Block Triangularization of A

We want to show, in a completely different situation, how a complete matching can be useful. It picks out a permutation that will make the main diagonal free of zeros. For any matrix A_0 we put 1's where the entries are not zero, look for a matching with n nonzeros, and then reorder rows to put those nonzeros on the diagonal. The new matrix is $A = PA_0$, where P is the permutation matrix that reorders the rows.

This is the first step in a bigger problem: to find a further permutation Q that will make the matrix block triangular. If there is such a Q, it can bring enormous savings in the solution of a linear system. And if this can be done, the combination of P and Q will do it.

The matching part is complete once the diagonal is nonzero. Then with Q there will be one important difference; it is applied to the columns as well as the rows. It takes A to $A_1 = QAQ^T$, and we try for

$$QAQ^T = \begin{bmatrix} A_{11} & & & \\ A_{21} & A_{22} & & \\ \cdot & \cdot & \cdot & \\ \cdot & \cdot & \cdot & \cdot \\ A_{p1} & \cdot & \cdot & \cdot & A_{pp} \end{bmatrix}. \tag{1}$$

The diagonal blocks are square and may have different sizes; the other blocks are rectangular. Of course this form is always possible with a single block, $p = 1$, when the block is A itself. It is $p > 1$ that gives a genuine triangle of blocks.

In the extreme case the blocks are 1 by 1. Then QAQ^T is actually triangular, and a linear system is solved from top to bottom. In a less extreme case the order is the same but A_{11} is a small matrix. It determines the first few unknowns and we move down to the next block. The smaller the blocks, the smaller the cost. Note that $p = 2$ gives

$$QAQ^T = \begin{bmatrix} A_{11} & 0 \\ A_{21} & A_{22} \end{bmatrix}.$$

If A_{11} is r by r and A_{22} is s by s, then certainly $r + s = n$. Therefore the zero submatrix is exactly as large as Hall's condition allows. If it were bigger, the original A had no complete matching (and was not even invertible). If all zero submatrices are smaller, $r + s < n$, the block triangular attempt will fail. It fails on symmetric matrices, unless they become block diagonal. But many applications produce matrices with a hidden order, on which the attempt succeeds.

There are two ways to visualize the change from A to QAQ^T. In matrix terms the rows and columns of A have been rearranged. The equations $Ax = b$ have been multiplied by Q and the components of x have been renumbered. The new unknown is Qx and the new system is $(QAQ^T)(Qx) = Qb$, which is the same as $Ax = b$. That is certainly simple, but the other way to see this change is the one we prefer; it uses not matrices but graphs. There will be a graph for A and another graph for the matrix QAQ^T. The essential point is that those two graphs are practically the same. Only the numbering of the nodes is different, and the search for Q can be completely understood by looking at the graph.

Graphs *G* and Adjacency Matrices *A*

The previous section introduced a graph with $2n$ nodes, one for each row and column of A. Whenever $a_{ij} \neq 0$ there was an edge leaving the node for row i and entering the node for column j. It was a "bipartite graph," since the nodes were in two groups and all edges went from one group to the other. Now we change to a smaller graph G with only n nodes. *The graph has an edge from node i to node j whenever a_{ij} is not zero.* It is again a directed graph; there is an edge from j to i only if a_{ji} is also nonzero. For convenience, Fig. 7.11 omits the edges ii that come from the diagonal entries and go from each node to itself.

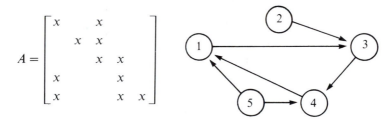

$$A = \begin{bmatrix} x & & x & & \\ & x & x & & \\ & & x & x & \\ x & & & x & \\ x & & & x & x \end{bmatrix}$$

Fig. 7.11. The directed graph G corresponding to the matrix A.

If we reverse the numbers on nodes 1 and 2, there is a change in the matrix. Since rows 1 and 2 give the edges that go out from those nodes, those rows are reversed. And since columns 1 and 2 give the edges that go into those nodes, they are also reversed. The matrix A is taken to QAQ^T, where Q exchanges rows 1 and 2 but the graph looks the same. Therefore the possibility of a triangular form must be revealed by the graph.

Which graphs are associated with block triangular matrices? The block A_{11} comes from a set of nodes from which *no edges lead out.* All other entries in those top rows are zero, and the block triangular form is begun. Next comes a second set of nodes, from which edges may go back to the first set by A_{21}. Within this second set the connections are given by A_{22}, but no edges go out to any remaining nodes. Continuing, there are p sets of nodes with no edges leaving one set and entering a later set.

The other question is, how do we know when a block like A_{11} cannot be split into smaller blocks? The test is to see whether its nodes are *strongly connected: There is a path from any node to any other node* in the block. Then edges come out from every set of nodes, and a splitting is impossible. If the entire graph G is strongly connected, the matrix A cannot be decomposed; it remains in one large block. Thus our first goal is to find a strongly connected set of nodes, say N_i, with no edges going out to the other nodes. Then we look among those other nodes for a second set N_2, again strongly connected within itself, with no edges going out except possibly back to N_1. This continues to N_3, and eventually to the last set N_p.

Consider the following algorithm to find N_1. Start from any node and go along any path through the graph. The path corresponds to entries $a_{ij} \neq 0$, and eventually we either encounter the same node twice or we reach a node that has no edges leaving it. In the first case we have found a *cycle*, or loop: there is a sequence of edges $i_1i_2, i_2i_3, \ldots, i_ki_1$ which ends where it begins, at the node i_1 we encountered twice. This gives a cycle of nonzeros in the matrix, which must go into the same block. Therefore we collapse them into a single "composite node" and start the algorithm again. When it finds a node (or composite node) that has no edges leaving it, that node becomes the first set N_1 and it gives the first block A_{11} in the triangular form. It is deleted and the algorithm continues.

This is the Sargent-Westerberg algorithm to find Q. It can be coded so that every edge in the graph is inspected only once. It is simple, but because recording composites and then composites of composites requires overhead, it is not the last word. An improvement proposed by Tarjan maintains a stack of nodes with a pointer, and is actually faster (but not simpler to describe). It has been developed by Duff and Reid into a Fortran code with operation count $C(n + m)$—which is less than the matching algorithm that comes first, to make the diagonal free of zeros.

EXAMPLE A has a zero-free diagonal but is not yet block-triangular (each x is a nonzero entry):

$$A = \begin{bmatrix} x & 0 & x & 0 \\ x & x & 0 & x \\ x & 0 & x & 0 \\ 0 & 0 & x & x \end{bmatrix} \quad \text{with graph } G:$$

Node 1 is connected to node 3, which is connected back to 1; they become a composite node. No edges leave this node, so it gives the first block A_{11}, and the path is now empty. Starting again, node 2 is connected to node 4, which is connected to no new nodes (it only goes back to 3). Therefore the block A_{22} contains node 4. Finally A_{33} contains node 2. The permutation Q that sends row $1 \to$ row 1, row $3 \to$ row 2, row $4 \to$ row 3, and row $2 \to$ row 4 is

$$Q = \begin{bmatrix} 1 & 0 & 0 & 0 \\ 0 & 0 & 1 & 0 \\ 0 & 0 & 0 & 1 \\ 0 & 1 & 0 & 0 \end{bmatrix}, \quad \text{and} \quad QAQ^T = \left[\begin{array}{cc|c|c} x & x & 0 & 0 \\ x & x & 0 & 0 \\ \hline 0 & x & x & 0 \\ \hline x & 0 & x & x \end{array} \right]$$

is block triangular.

EXERCISES

7.3.1   Apply a depth first search to improve the original matching with three marriages in example 3A. You can trace the reassignment chain on the graph or on the matrix.

7.3.2   Apply a depth first search to the matching with four marriages in example 3B. Show that no successful chain starts from g_5.

7.3.3   Use either DFS or BFS to improve the circled matchings:

$$A = \begin{bmatrix} ① & & & \\ & ① & 1 & \\ 1 & & & \\ & 1 & & \end{bmatrix} \qquad B = \begin{bmatrix} ① & 1 & 1 & 1 \\ 1 & ① & & \\ 1 & & & \\ 1 & & & \end{bmatrix}$$

Show that three marriages are optimal by covering the 1's in A and B with three lines. Which zero submatrices of A and B prevent the fourth marriage?

7.3.4   Draw the bipartite graphs corresponding to the previous matrices A and B. Indicate three possible marriages and three vertices which cover all edges.

7.3.5   Apply the breadth first search algorithm to the 5 by 5 depth first example, starting from the 3 original marriages in Fig. 7.7(a) and ending at a complete matching.

7.3.6   How many complete matchings are there for an n by n matrix with all $a_{ij} = 1$?

7.3.7   Can you describe a 4 by 4 matrix that allows exactly 4 complete matchings?

7.3.8   Suppose M is not optimal; there is a different matching M' with more marriages. To show that there must be a reassignment chain which improves M, start with a girl who is married in M' but not in M. That marriage is the first link in a chain, which alternates between marriages in M' (and not M) and marriages in M (and not M'). Give an example of all the following possibilities:

   (1)   After 2 or 4 or 6 … links we return to a girl who is not married in M'. This chain fails, but we can discard all its marriages (an equal number in M' and M) and start again with a smaller problem.

   (2)   After 4 or 6 or 8 … links we return to a girl already in the chain. This chain also fails (it has cycled) but again we discard its equal number of marriages from M and M' and start over.

   (3)   After 1 or 3 or 5 … links we reach a boy not married in M. Then M can be improved by this chain, as we wanted to show.

7.3.9   (a)   Suppose an Np by Np matrix is block diagonal, with p blocks of order N. What replaces the usual operation count of $(Np)^3/3$ elimination steps?

   (b)   If the matrix is only block triangular instead of block diagonal, what is the operation count?

7.3.10   Draw the connections between four nodes that arise from a 4 by 4 tridiagonal matrix, and show that no block triangular form is possible.

7.3.11   Renumber the nodes in the following graphs to make the associated matrices block triangular. Indicate by x's the nonzero entries in the original matrices A (adding x's on the main diagonal) and in the block triangular QAQ^T.

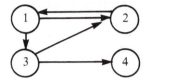 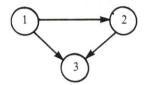

7.3.12 Find a triangular form with 3 blocks for the matrix in Fig. 7.11. Redraw the graph G with the nodes numbered to match this triangular form.

7.4 ■ MAXIMAL FLOW IN A NETWORK

Network flow is one of the major successes of operations research. Commercially it is certainly a success; it has saved millions of dollars by designing and using networks in an optimal way. In fact that saving was achieved in a single application of network analysis to gas pipelines in the Gulf of Mexico (*Scientific American*, July 1970). The nodes are at the gas fields and the delivery point. The network costs depend on the pipe lengths and pipe diameters, which determine the price of compressors. There is an optimization problem, much too complicated to be solved by trial and error, whose solution has systematically improved pipeline designs by more than 25%.

Mathematically, the problem of network flow lies between combinatorics and linear programming. We can apply flow theory to the marriage problem, and we can apply the general principles of linear programming to network flow. It has a middle position and a character of its own. The flow travels through a network of vertices and edges, but there is not necessarily a bipartite graph—with boys lined up against girls. Any pair of nodes can be connected, and the underlying problem is not speed, or cost, or matching; it is *capacity*. There is a classical maximization problem of linear programming, with large incidence matrices to specify the nodal connections and with inequality constraints from the capacities. For such a network we prove the great duality principle—the *max flow-min cut theorem*—by an algorithm that also computes the maximal flow and minimal cut.

A network starts with a directed graph—a set of nodes connected by some (or all) of the possible edges. Loops from a node to itself are ruled out, and each edge is directed by an arrow. We draw only one edge connecting i and j, but it may permit flow *in either direction*. That flow depends on the capacities, which complete the definition of the network. Every edge has a capacity c_{ij} in the forward direction and a capacity c_{ji} in the backward direction (which may be the same as c_{ij}, or different, or zero). These capacities are nonnegative integers, and they impose bounds on the flow. We write x_{ij} for the amount of flow from i to j, so that always $x_{ji} = -x_{ij}$. Then on every edge the flow is required to satisfy

$$x_{ij} \le c_{ij} \quad \text{and} \quad x_{ji} \le c_{ji}. \tag{1}$$

Defining $c_{ij} = c_{ji} = 0$ when there is no edge, this constraint applies everywhere. In a network of roads or communication lines, the capacities depend on the separate pieces of the network—and the problem is *to find the capacity of the network* as a whole.

The maximal flow problem identifies two special nodes, the source s and the sink t. At all other nodes the flow in equals the flow out, as in Kirchhoff's current law:

$$\sum_{i=1}^{n} x_{ij} - \sum_{k=1}^{n} x_{jk} = 0 \quad \text{for every node } j \ne s,t. \tag{2}$$

Then *the problem is*, subject to (1) and (2), *to maximize the flow from s to t.*

You can find that maximum in the following example, which shows the capacities (in both directions) on each edge:

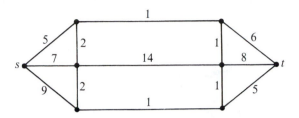

There are two ways to start. Either you look for a way to maximize the flow, by an intelligent choice of paths, or you look for an upper limit imposed by the capacities. The capacities give the bound $5 + 7 + 9 = 21$ on the flow that can leave the source. That limit cannot be reached, since there is a tighter bound $6 + 8 + 5 = 19$ at the sink. Can a flow of 19 get through the network? It has difficulty at the center, where the horizontal edges have a total capacity of $1 + 14 + 1 = 16$. The flow must cross those edges. They are an example of a *cut* in the network—a separation of the nodes into two groups S and T, one containing the source s and the other containing t. The *capacity of the cut* is the total capacity on edges crossing from S to T, and the flow cannot exceed this capacity. The cut at the source had capacity 21; it separated the source from all other nodes. The cut at the sink had capacity 19. So far, 16 is an upper bound on the flow.

In fact, 16 is not possible either. One method is trial and error; 7 can go straight from the source, and we can start with 3 upwards and 3 downwards. But there is a cut that does not even let through 13. Its capacity is $1 + 1 + 8 + 1 + 1 = 12$, and it is the dotted line in the figure below. S contains four nodes plus the source, and T contains two nodes and the sink. The figure also defines a flow of 12: each pair is (c_{ij}, x_{ij}), the capacity and then the actual flow on the edge.

There are two absolutely crucial properties of this cut. *Every edge across it is filled to capacity, and every node to its left* (in S) *can be reached with more flow*. It is a cut of the smallest possible capacity—a *minimal cut*—since if there existed a cut of smaller capacity this flow of 12 would have been impossible.

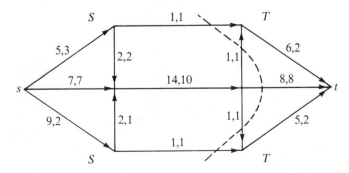

Fig. 7.12. Maximal flow (total 12) and minimal cut (capacity 12).

The minimal cut is not necessarily unique. If the edge of capacity 14 is reduced to 10, a second cut of capacity 12 goes down the center. Similarly, the maximal flow is not necessarily unique. One more unit could go down from the source and one less unit across, changing the figure to 9,3 and 7,6 (and then 2,1 to 2,2). What *is* uniquely determined is the value $v = 12$ of the maximal flow, and the capacity of the minimal cut. The beautiful result of duality is that these two numbers are equal.

7G (Max flow-min cut theorem) The value of the maximal flow equals the capacity of the minimal cut.

We give a nonconstructive proof and then a constructive proof.

First imagine we have found a maximal flow (which certainly exists). Construct a cut as follows: S contains the source and every node that can be reached from it with additional flow. To such a node there is an "*augmenting path*" from s, along which flow can be added without exceeding any capacities. If such a path does not exist, the node goes into the complementary set T. Note that *the sink is in T*; if it could be reached by an augmenting path the flow would not be maximal.

To prove the theorem, look at the edges from S to T. Each of those edges must be filled to capacity, or the path could have been continued across to a node in T. Therefore the capacity of this cut (the sum of individual capacities) equals the total flow (the sum of individual flows). Since no cut could have smaller capacity, the theorem is proved—but the maximal flow was not actually found.

Here is a special case. Suppose the edges have capacity 1 in both directions. The minimal cut is the one with fewest edges across it—its capacity is the number of edges from S to T. The maximal flow can be sent on disjoint paths—the paths have no edges in common and one unit flows on each path. The duality expressed by max flow = min cut becomes:

> The maximum number of disjoint paths from source to sink equals the minimum number of edges across a cut.

In other words, the maximum number of *paths* with no edge in common equals the minimum number of *edges* whose removal disconnects s from t.

The Labeling Algorithm

The constructive proof of 7G discovers the largest flow and the smallest cut. The idea is to look for augmenting paths that reach the sink. If one is found, the flow is increased. Eventually the flow becomes maximal. This takes longer than the first proof, but it brings out one point that was obscure above: *An augmenting path need not always add flow in the direction already taken.* It is perfectly legal for the path to go in the opposite direction! In other words, it might **cancel** one unit of the existing flow. We illustrate with the flow in the figure on the left. All forward capacities are $c_{ij} = 1$ (from left to right) and backward capacities are $c_{ji} = 0$:

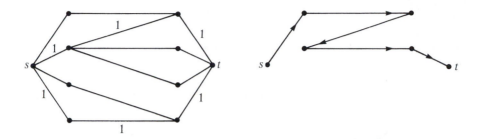

Which nodes can be reached by more flow? The two unused nodes at the left can be reached directly from the source, and two more can be reached from them. At first that seems to be all. However the sketch on the right shows another augmenting path that is legitimate. It sends one unit in the direction of *its* arrows, which means canceling the existing flow on the backward leg. The result is to reach more nodes, and even to reach the sink t. Therefore the flow can be increased by sending one unit along this path, to give the picture below:

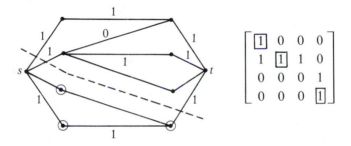

$$\begin{bmatrix} \boxed{1} & 0 & 0 & 0 \\ 1 & \boxed{1} & 1 & 0 \\ 0 & 0 & 0 & 1 \\ 0 & 0 & 0 & \boxed{1} \end{bmatrix}$$

From that flow we start again. The three circled nodes can be reached with more flow—one directly from s, and a second directly from the first. The third is reached by going *back* from the second, canceling the flow along the bottom. But the sink cannot be reached, so the flow is maximal and the dotted line is a minimal cut (with capacity 3).

That example is identical to a marriage problem. The source is connected to all girls, and the sink to all boys. The six middle edges give the six 1's in A. The matrix shows the maximal matching, which is also the maximal flow. Notice the 2 by 3 matrix of zeros in the lower corner, corresponding precisely to the cut in the network. Girls 3 and 4 have no edge to boys $1, 2, 3$.

Now we complete the constructive proof. The algorithm is a systematic way to increase a given flow x_{ij} by searching for an augmenting path. When the algorithm stops the flow is maximal.

The first step is to *scan* the source s—to find all nodes that can be reached directly from s by additional flow. If node j can be reached in this way, *label* it. The label has two pieces of information about the additional flow: the node it came from (in this case s) and the amount of new flow. This amount will be the difference $c_{sj} - x_{sj}$ between the capacity and the existing flow. Next we scan each labeled node in turn, to see if other nodes can be reached (and therefore labeled) from them. Once a node is labeled it waits to be scanned. Eventually the sink gets a label and an augmenting path has been discovered, or every labeled node has been scanned and no new labels are possible.

To be absolutely systematic, s itself receives the label ($,\infty$). It came from nowhere, and there is no bound on the flow out from it. The algorithm starts with all flows $x_{ij} = 0$.

Labeling Algorithm
  (1)   Scan s and label nodes that can be reached with additional flow.
  (2)   Choose the node i that was labeled earliest but is not yet scanned. For each edge from i to an unlabeled node j, determine whether $x_{ij} < c_{ij}$. If so, label j with (i,d_j), where the positive number d_j is the additional flow that can reach j through i:

$$d_j = \text{smaller of } d_i \text{ and } c_{ij} - x_{ij}.$$

  (3)   If the sink receives a label, increase the flow and return to 1. If the sink is not yet labeled, return to 2. If all labeled nodes have been scanned without reaching the sink, STOP.

The increase in flow is decided by the label on the sink. A label (k,d_t) means that d_t units of flow can be added, coming from node k. The complete path is found by backtracking to node k, then (according to the label on k) to the previous node, and finally back to the source. With this new and larger flow, and $x_{ji} = -x_{ij}$ as always, the algorithm begins again. When all labeled nodes are scanned without reaching the sink, the algorithm stops.

EXAMPLE   Suppose the capacities are zero against the arrows; the edges go one way. The numbers on each edge give first, the capacity forward, and second, the existing flow (which totals 5). We want to add more flow.

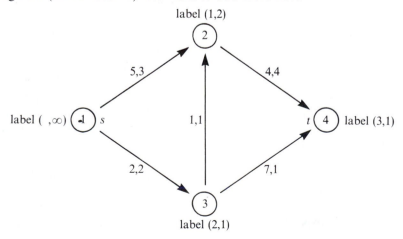

Scanning the source produces a label (1,2) on node 2, to show that node 1 can send 2 more units. Node 3 is not labeled from the source; that edge is saturated. Now scan the newly labeled node 2. Node 4 is not labeled from 2, but node 3 is—its label is (2,1) because node 2 can send 1 unit by *canceling the present flow*. From our rule, d_3 is 1:

$$d_3 = \text{smaller of } d_2 \text{ and } c_{23} - x_{23} = 0 - (-1)$$

$$= \text{smaller of 2 and 1.}$$

Finally scanning node 3, the sink at node 4 is labeled (3,1); one more unit can come from 3. It looks as if six units are possible, but $d_3 = 1$ means that only one unit reached node 3; d_4 is the smaller of 1 and 6.

The sink is labeled, so an added flow $d_4 = 1$ can go through. Its "augmenting path" is found by backtracking from the sink; 4 was labeled from 3, which was labeled from 2, which was labeled from 1. That flow is added to the figure:

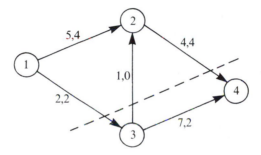

The flow of 6 is proved to be maximal by applying the algorithm. The sink is labeled (,∞), as always, and node 2 is labeled (1,1). No other nodes can be labeled. Therefore no more flow is possible, and the minimal cut separates the labeled nodes 1 and 2 from the unlabeled 3 and 4. Its capacity (in the forward direction from source toward sink!) is 6.

Remark 1 If the capacities are integers then the additional flows are always integers. Therefore there is an integral maximal flow.

Remark 2 With irrational capacities and a pathologically bad choice of augmenting paths, an infinite number of steps is possible. The graph below allows a very poor choice:

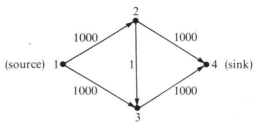

The paths could alternate between 1-2-3-4 and 1-3-2-4 with an increase of only one unit at each step. It would take 2000 steps to reach the correct maximal flow of 2000. However our algorithm of "first labeled—first scanned" reaches that flow in 2 steps. It always converges in polynomial time, since each augmenting path is as short as possible. An alternative is to maximize the increase in flow.

Applications

Minimum cuts turn up in unexpected places. One is the problem of choosing which courses to take, if the good courses have bad courses as prerequisites. This is made quantitative by rating the courses:

> with the winner $w_1 = 7$ come the losers $l_1 = 2$ and $l_2 = 6$
>
> with the winner $w_2 = 5$ comes the loser $l_2 = 6$
>
> with the winner $w_3 = 3$ come the losers $l_2 = 6$ and $l_3 = 4$

Each single choice is a net loser; 7 is below $2 + 6$. However a group of winners may come out ahead if they share the same losers.

There are many more businesslike problems equivalent to this one. An airline might decide what routes to fly, comparing income on each route with the cost of ground operations in each airport—but that cost can be shared by all routes that use the airport. Similarly the cost of a warehouse or a sales office is balanced by the set of products it can store or sell. In each case it is the sharing of different subsets that creates the optimization problem.

The key is to construct the right network. The source is connected to the winners by edges of capacity 7,5,3. The losers are connected to the sink with capacities 2,6,4. Between a winner and its required losers there are edges of capacity $+\infty$. They will never be separated by a minimal cut, because of that infinite capacity, so they are kept together in the solution. Between other winners and losers there is no edge; w_1 is not connected to l_3 and the cut is free to separate them. All edges go in the forward direction only:

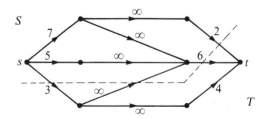

Fig. 7.13. The winner-loser network.

The dotted line is the minimal cut for this example. Its forward capacity from S to T is 11. To see what that means we compare it with the cut that separates s from the rest of the network, and the cut that separates the network from the sink:

cut at source (choose no w's or l's):   capacity $7 + 5 + 3 = 15$

cut at sink (choose all w's and l's):   capacity $2 + 6 + 4 = 12$

minimal cut (choose w_1, w_2, l_1, l_2):   capacity $3 + 6 + 2 = 11.$

The first takes no courses at all (with profit 0). The second does everything (with profit 3, when losses are subtracted from wins). The third case wins 7 and 5 while losing 2 and 6 (with profit 4). A minimal cut is a maximal profit because

$$\text{profit} = 15 - \text{capacity of cut}.$$

This application is less interested in the maximal flow. There must be a competitor in the dual problem offering to take those bad courses at a price.

VLSI Networks

A chip has an array of gates that need to be wired in pairs. Those pairs may not be close; wires may go everywhere. Since space is at a premium, the object is to keep the maximum "channel width" as small as possible. That width is the number of wires running side by side, across the boundary of a small square, and in the following designs the maximum widths are 3 and 2:

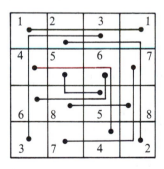

Vertical and horizontal wires do not actually cross because they are laid down at separate levels. The problem could allow many bends in a path or only one—we choose the latter, and search for the best routing of the connections.

This is a problem of ***multi-commodity flow***. The shipment from 4 to 4 is separate from the shipment from 7 to 7. It is also an *integer program*, because wires are not split and fractions are not allowed. In single commodity flows those fractions were not needed. Here the width in the 2 by 2 array could be reduced to one, if each pair could be connected by two wires of width $\frac{1}{2}$. The integer problem has a different minimum from the fractional problem. It is also much harder to find. The multi-commodity integer problem is NP-complete—all known algorithms require times that increase exponentially with the number of pairs, as discussed at the end of the chapter. In contrast an ordinary linear program is now known to take only polynomial time (although the simplex method, which remains best overall, can be exponential).

Integer problems can be attacked by "heuristic" methods that quickly get near the optimum—probably near enough for the applications. In the VLSI problem a probabilistic method has been suggested:

> Solve the linear program allowing fractions; x_i and $1 - x_i$ go on the two L-shaped routes connecting pair i. Then flip a coin that has probability x_i of heads and $1 - x_i$ of tails to decide on a single route.

With high probability the result is near the optimum. (The closer you require it to be, the lower the probability.) By contrast, choosing the integer solution that is closest to the fractional solution—rounding off, based on whether or not $x_i > \frac{1}{2}$—yields a routing whose width is no more than twice the optimum. In the 2 by 2 array, the width was two when the optimum with fractions was one.

EXERCISES

7.4.1   Suppose a graph with source s, sink t, and four other nodes has capacity 1 between every pair of nodes. (There are 15 pairs.) What is a maximal flow and a minimal cut?

7.4.2   In Fig. 7.12 suppose you could add one more edge of capacity 100 from the source to any node except t. Which edge would allow the greatest increase in flow, and by how much?

7.4.3   In Fig. 7.12 add 3 to every capacity. Find by inspection the minimal cut and the maximal flow.

7.4.4   For the flow shown in Fig. 7.12, apply the labeling algorithm and assign labels until the algorithm stops.

7.4.5   For the 2 by 2 marriage problem that admits the matchings g_1-b_2, g_2-b_1, and g_2-b_2, draw the network with infinite capacities on these edges. What should be the capacities on $s-g_1$, $s-g_2$, b_1-t, and b_2-t? Find a maximal flow and a minimal cut.

7.4.6   Does the labeling algorithm correspond to depth first search or breadth first search? Describe a labeling algorithm that corresponds to the opposite method of search.

7.4.7   Draw a graph with nodes 0,1,2,3,4 and capacity $|i - j|$ between node i and node j, and find the minimal cut and maximal flow from 0 to 4.

7.4.8   How quickly can the number of operations in the labeling algorithm grow with n (the number of nodes) and m (the number of edges)?

7.4.9   Suppose there are *node capacities* as well as edge capacities: The flow entering (and therefore leaving) node j cannot exceed c_j. Reduce this to the standard problem by replacing each node j by two nodes j' and j'', connected by an edge of capacity c_j. How should the rest of the network be changed?

7.4.10   The maximum number of paths from s to t that have no other nodes in common equals the minimum number of nodes whose removal disconnects s from t.
   (a)   Illustrate this equality on 3 graphs including the one below.
   (b)   Replace each node by two nodes j' and j'', as above, connected by an edge of capacity 1. If all other edges in your 3 graphs have infinite capacity, draw the minimal cuts and maximal flows.

7.4.11   The maximum number of paths from s to t with no *edge* in common equals the minimum number of *edges* whose removal disconnects s from t. Verify this equality on the graph above and connect it to the max flow-min cut theorem.

7.4.12   What is the maximal flow from the corner $s = (0, \ldots, 0)$ to the corner $t = (1, \ldots, 1)$ of an n-dimensional cube? Its edges have capacity 1 and connect pairs of points that differ in one component.

7.4.13   Explain how the bottleneck path problem—to find the path from s to t whose longest edge is as short as possible—is related to finding the augmenting path that allows the largest increase in flow.

7.4.14   (a)   Complete the "bottleneck analogy" of max flow-min cut:

   the maximum that can be sent on a *single path* from s to t equals the minimum for all cuts of _____.

   (b)   Prove weak duality: this maximum $\leq$ your minimum.

7.4.15   Suppose a directed graph has no loops. Show by backtracking that some node has no edge going into it.

7.4.16   (a)   Draw a directed graph with six nodes, ten edges, and no loops, and number the nodes so that all edges go from a lower to a higher number.
   (b)   Show that this is always possible. Assign 1 to a node with no edge going into it (Exercise 15) and explain step 2.

7.4.17   A scheduling graph has edge lengths l_{ij} equal to the minimum time between starting activity i and starting activity j. The edge is directed from i to j, and the problem is to find the *longest* path from s to t.
   (a)   Why does the longest path give the minimum time to complete the project, and why would a loop be fatal?
   (b)   Give an algorithm like Dijkstra's for the longest path.
   (c)   Construct a house-building example in which node $s =$ foundation, $2 =$ frame, $3 =$ wiring, $4 =$ plumbing, $5 =$ roofing, $\ldots, t =$ moving in.

7.4.18   In the winner-loser example which is the first flow sent from s to t by the labeling algorithm? When a flow of 11 has been reached, draw the labeled network at which the algorithm stops.

7.4.19 If five winners all have $w_i = 4$ and five losers have $l_i = 3$, draw the network when winner i is required to accept losers $i - 1$ and $i + 1$. Infinite capacity edges go from w_2 to l_1 and l_3, for example, and at the ends w_1 is connected only to l_2 and w_5 only to l_4. Find the minimal cut and the maximal profit.

7.4.20 Is there a routing in the 4 by 4 VLSI network that has no channel widths of 3? If so, draw another network whose minimum width is 3.

7.4.21 Suppose the gates are in a circle and each pair is connected by going around one way or the other. Estimate how large the minimum width might be for N pairs, and propose an algorithm that comes near to the optimal routing.

7.4.22 Construct a $2n$ by $2n$ VLSI network for which the minimum width is at least n. Can you find a 6 by 6 network with minimum width 4? *Theorem:* The minimum is never above $n + 2$ (in this case 5).

7.4.23 In *time-dependent* flow with nodes a,b,c,d, suppose 4 messages can travel every second between every pair of nodes. In two seconds how many messages can start from a and reach d? Draw a network with 3 copies of a,b,c,d (at times 0,1,2), choose edges and capacities, and add a source connected to a,a,a and a sink with infinite capacity from d,d,d.

THE TRANSPORTATION PROBLEM ■ 7.5

The transportation problem is just about the oldest of all linear programs. Two hundred years ago the French mathematician Monge studied the problem of moving earth from a volume V to an equal volume V'. The goal was to build a fort. From the simple fact that all movement is along straight lines and no lines should cross, he constructed a beautiful theory. Unfortunately, it was forgotten. One hundred years later it was studied again, and forgotten again. Then about 1941 a much more practical analysis was begun—it was discrete instead of continuous, with vectors instead of functions and linear algebra instead of calculus—and it has become a central problem in applied optimization.

It differs from maximal flow by imposing *costs instead of capacities* on the edges of the network. These edges connect m suppliers to n markets, with cost C_{ij} for each unit sent from supplier i to market j. We are given the amounts $a_1, \ldots, a_m$ to be sent from the suppliers and the demands $b_1, \ldots, b_n$ to be received at the markets. The problem is to find the cheapest transportation plan.

The unknowns are the shipments $x_{ij} \geq 0$, sent from i to j. Their cost is $z = \Sigma\Sigma\, C_{ij}x_{ij}$, the sum of the individual costs on all edges. The total shipment from supplier i is constrained to be

$$x_{i1} + x_{i2} + \cdots + x_{in} = a_1. \tag{1}$$

The total shipment into market j must be

$$x_{1j} + x_{2j} + \cdots + x_{mj} = b_j. \tag{2}$$

We assume that supply equals demand, $\Sigma\, a_i = \Sigma\, b_j$, so the m supply equations (1) are compatible with the n demand equations (2). Then the transportation problem is:

> **Minimize** $z = \displaystyle\sum_i \sum_j C_{ij}x_{ij}$ **subject to** (1), (2), **and** $x_{ij} \geq 0$.

This is not a small problem; it has mn unknowns x_{ij} and $m + n$ equality constraints. (It is true that one of these equality constraints is dependent on the other $m + n - 1$. The shipment into the last market must have the correct value b_n if the right amounts $a_1, \ldots, a_m$ were sent and the right amounts $b_1, \ldots, b_{n-1}$ were received; total supply equals total demand.) The solution will come from the simplex method of linear programming, but the problem is so important that everything has been specialized and refined to fit the transportation model. Therefore this problem is a perfect introduction to the general simplex method in Chapter 8.

We illustrate the underlying ideas with the following example:

Supplies	Demands		Costs		
$a_1 = 4$	$b_1 = 2$		5	3	1
$a_2 = 7$	$b_2 = 3$	$C_{ij} =$			
	$b_3 = 6$		8	4	3

There are six unknowns $x_{ij} \geq 0$ and $m + n - 1 = 4$ independent constraints.

First, before the algorithms, I have to write about the dual problem. It is a beautiful example of cutthroat competition. The customer is in the middle, looking for the cheapest routes. On one side are the costs C_{ij} from the shipper; they give the primal problem already described. On the other side is a toll collector offering free travel on *his* routes, except for tolls at the beginning and the end. He collects u_i for each unit sent out from supplier i, and v_j for each unit received at market j. In case all shipments go through him, he collects the amount

$$z' = u_1 a_1 + \cdots + u_m a_m + v_1 b_1 + \cdots + v_n b_n.$$

You can think of him as a parcel service, like Federal Express, and he is trying to maximize z'.

If the tolls are too high, the customer will stay with the original shipper (the post office). On each route, he compares their price C_{ij} with the alternative—to pay tolls at each end, or pickup and delivery costs, in the amount $u_i + v_j$. Federal Express can capture the business in a free market only if

$$u_i + v_j \leq C_{ij} \quad \text{for each } i \text{ and } j. \tag{3}$$

Subject to these constraints, Federal Express will choose the u_i and v_j so as to maximize their income z'. This is the **dual problem**:

> **Maximize z' subject to $u_i + v_j \leq C_{ij}$.**

It is natural to try for equality in every constraint, $u_i + v_j = C_{ij}$, which would certainly give a maximum. In general these mn equalities cannot all be achieved. There are only $m + n$ tolls to be chosen, and normally there is no solution to mn equations in $m + n$ unknowns. In fact, comparing any allowed tolls and any choice of shipments x_{ij} through the post office, the parcel service is at least as cheap: $z' \leq z$ because Federal Express is cheaper on every route. More formally, the constraints combine to give

$$z' = u_1 a_1 + \cdots + u_m a_m + v_1 b_1 + \cdots + v_n b_n$$
$$= u_1(x_{11} + \cdots + x_{1n}) + \cdots + v_1(x_{11} + \cdots + x_{m1}) + \cdots + v_n(x_{1n} + \cdots + x_{mn})$$
$$= (u_1 + v_1)x_{11} + (u_1 + v_2)x_{12} + \cdots + (u_m + v_n)x_{mn}$$
$$\leq C_1 x_{11} + C_{12}x_{12} + \cdots + C_{mn}x_{mn} = z. \tag{4}$$

This is no accident. It is **weak duality**, and it means that the largest z' cannot exceed the smallest z; Federal Express will not collect more than the post office. The important question is whether they can collect the same amount—in other words, whether z' can equal z. That will require the best choice of tolls in the dual, to maximize z', and the best choice of shipments x_{ij} in the primal, to minimize z. It is strong duality, and it is true.

We will see it in the example. Notice that the inequality $z' \leq z$ becomes $z' = z$ precisely when every term $(u_i + v_j)x_{ij}$ equals the corresponding $C_{ij}x_{ij}$. That was the only place in (4) where inequality might enter, so the key to duality is to achieve $(u_i + v_j)x_{ij} = C_{ij}x_{ij}$. Since always $u_i + v_j \leq C_{ij}$, equality can happen in only two ways: for every i and j,

$$\text{either} \quad u_i + v_j = C_{ij} \quad \text{or} \quad x_{ij} = 0. \tag{5}$$

This is the "optimality condition" or "complementary slackness condition," and it is the test for duality. If there are u_i, v_j, and x_{ij} for which it holds, then for those tolls and those shipments we have $z' = z$. No other toll collection could be larger, and no other transportation plan could be cheaper.

We have come back to the idea of achieving $u_i + v_j = C_{ij}$, but not on all of the mn routes—only on *those which are actually used*. In short, the parcel service has outsmarted the post office. Their tolls are lower on every route, or at least not higher, but Federal Express reaches the same income $z' = z$ by adjusting their tolls to be optimal on the routes that the customer will want to take. If the tolls are low where $x_{ij} = 0$, it doesn't matter.

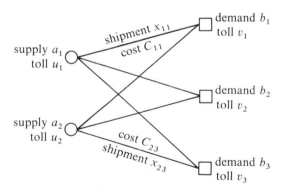

Fig. 7.14. The transportation problem (shipments) and its dual (tolls).

Transportation Algorithm

In the example with $m + n - 1 = 4$, we eventually have to find the four routes that are actually used. The first step is to satisfy the demands $b_1 = 2, b_2 = 3, b_3 = 6$ from the supplies $a_1 = 4, a_2 = 7$. If the x_{ij} are placed into a matrix, the *row sums* equal the a_i and the *column sums* are the b_j. To start, we look for any four shipments x_{ij} that satisfy these constraints: we begin with x_{11}.

$a \backslash b$	2	3	6
4	x_{11}	x_{12}	x_{13}
7	x_{21}	x_{22}	x_{23}

$a \backslash b$	2	3	6
4	2	x_{12}	x_{13}
7	0	x_{22}	x_{23}

In the northwest corner either $x_{11} = a_1$ (to use up the first supply) or $x_{11} = b_1$ (to fill the first demand). We choose the smaller: $x_{11} = b_1 = 2$. That leaves supplies of 2 and 7, and demands of 3 and 6. The new northwest corner is x_{12}, and by the same rule it is the smaller of the supply 2 and the demand 3. Therefore $x_{12} = 2$, which exhausts the first supply and forces $x_{13} = 0$:

$$
\begin{array}{c|ccc}
a \backslash b & 2 & 3 & 6 \\
\hline
4 & 2 & 2 & 0 \\
7 & 0 & x_{22} & x_{23}
\end{array}
\qquad
\begin{array}{c|ccc}
a \backslash b & 2 & 3 & 6 \\
\hline
4 & 2 & 2 & 0 \\
7 & 0 & 1 & 6
\end{array}
= \text{shipping matrix } x_{ij}
$$

The remaining demands are 1 and 6, the remaining supply is 7, and the matrix is completed with a total of four shipments. Of course **they may not be the cheapest four**. At costs 5, 3, 4, 3 the total is $z = \Sigma\Sigma\, C_{ij}x_{ij} = 5 \cdot 2 + 3 \cdot 2 + 4 \cdot 1 + 3 \cdot 6 = 38$.

This "northwest corner rule" settles a row or a column at each step. If $x_{11} = a_1$ the rest of row 1 is zero, exhausting the first supply; if $x_{11} = b_1$ the rest of column 1 is zero, and the first column is finished. There are $m + n$ rows and columns, so there ought to be $m + n$ steps. But when there is only one row and one column to go, in the lower right corner, a single shipment (in this case $x_{23} = 6$) satisfies them both. Therefore $m + n - 1$ shipments are always sufficient.

Next, the parcel service chooses its five prices u_1, u_2, v_1, v_2, v_3. Its immediate goal is to have $u_i + v_j = C_{ij}$ on the four routes that are tentatively chosen by the customer. The C_{ij} that have to be matched have circles in the cost matrix C at the same places where $x_{ij} > 0$ in the shipping matrix:

$$
C = \begin{array}{|ccc|}
\hline
\text{⑤} & \text{③} & 1 \\
8 & \text{④} & \text{③} \\
\hline
\end{array}
\quad \text{leads to} \quad
\begin{array}{cc}
 & \begin{array}{ccc} v_1 = 5 & v_2 = 3 & v_3 = 2 \end{array} \\
\begin{array}{c} u_1 = 0 \\ u_2 = 1 \end{array} & \begin{array}{|ccc|} \hline 5 & 3 & \\ & 4 & 3 \\ \hline \end{array}
\end{array}
$$

With $u_1 = 0$, the other four tolls were easy to find. They were all positive but that is not necessary; we could increase all the u_i and decrease all the v_j by the same amount without changing the income to the parcel service. (It would just collect more at pickup and less at delivery, so we can arbitrarily "ground" $u_1 = 0$.) The tolls 0, 1, 5, 3, 2 match the four circled costs.

Now comes the moment of truth: Do we have $u_i + v_j \le C_{ij}$ on *all* routes? In this case we do not:

$$
\begin{array}{|ccc|}
\hline
5 & 3 & 1 \\
8 & 4 & 3 \\
\hline
\end{array}
\quad \text{is compared with} \quad
\begin{array}{cc}
 & \begin{array}{ccc} 5 & 3 & 2 \end{array} \\
\begin{array}{c} 0 \\ 1 \end{array} & \begin{array}{|ccc|} \hline 5 & 3 & 2 \\ 6 & 4 & 3 \\ \hline \end{array}
\end{array}
$$
$$
\qquad\quad C_{ij} \qquad\qquad\qquad\qquad\qquad\qquad u_i + v_j
$$

At the end of the first row $u_1 + v_3 = 2$ exceeds $C_{13} = 1$. This only means that our first attempt was not optimal. The customer will prefer the post office on the route from 1 to 3, and if he sends an amount s on that route we have to adjust the shipping matrix. Its entry x_{13} is now s. The last column sum is still the demand

$b_3 = 6$, so the shipment x_{23} must drop to $6 - s$. This increases x_{22} by s, which decreases x_{12} by s; the chain reaction stops there. Now the customer is using five

$$
\begin{array}{c|ccc}
a\backslash b & 2 & 3 & 6 \\
\hline
4 & 2 & 2-s & s \\
7 & 0 & 1+s & 6-s
\end{array}
= \text{new } x_{ij} =
\begin{array}{c|ccc}
a\backslash b & 2 & 3 & 6 \\
\hline
4 & 2 & 0 & 2 \\
7 & 0 & 3 & 4
\end{array}
\; \text{at } s = 2.
$$

routes, but with the right choice of s he will be back to four. All shipments must satisfy $x_{ij} \geq 0$, and one of the shipments drops to zero at $s = 2$. It is the first to disappear as s increases, and it leaves a new matrix with four shipments. The cheaper route is now in use.

Federal Express resets prices so that $u_i + v_j = C_{ij}$ on the new set of four circles— and the shipper uses those four routes:

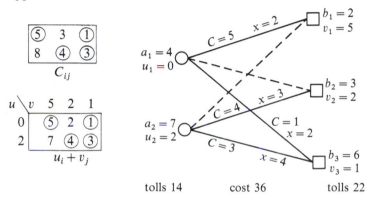

$$
z = \sum\sum C_{ij}x_{ij} = 5 \cdot 2 + 1 \cdot 2 + 4 \cdot 3 + 3 \cdot 4 = 36,
$$

and the income to Federal Express is

$$
z' = a_1 u_1 + a_2 u_2 + b_1 v_1 + b_2 v_2 + b_3 v_3 = 4 \cdot 0 + 7 \cdot 2 + 2 \cdot 5 + 3 \cdot 2 + 6 \cdot 1 = 36.
$$

Therefore *duality is achieved*. For any shipping plan and any acceptable prices we have $z' \leq z$; that was weak duality. For the optimal shipments and prices we have $z' = z$. The minimum transportation cost is 36, and the algorithm is easy to summarize.

Algorithm for the transportation problem
   (1)   Satisfy the demands b_j from the supplies a_i with at most $m + n - 1$ shipments $x_{ij} > 0$. The northwest corner rule gives one systematic starting choice for the x_{ij}.

(2) Choose prices u_i and v_j, with $u_1 = 0$, so that $u_i + v_j = C_{ij}$ on the routes used by the shipments.

(3) Test whether $u_i + v_j \leq C_{ij}$ on all routes. If this holds then the shipments and prices are optimal. Otherwise find the route ij with the largest overcharge $u_i + v_j - C_{ij}$.

(4) Ship an amount s on this route and adjust shipments on the earlier routes to $x_{ij} \pm s$ as necessary to maintain supplies a_i and demands b_j.

(5) Choose s so that a shipment $x_{ij} - s$ drops to zero (one old route drops out in favor of the new one) and the other shipments remain nonnegative. With this value of s in the updated matrix of shipments x_{ij}, return to step 2.

We have to mention one possibility. There might be *degeneracy*, in which the supplies can meet the demands with fewer than $m + n - 1$ shipments. Suppose $a_1 = a_2 = b_1 = b_2 = 1$. Then two shipments will be enough; the x_{ij} can come from the identity matrix, $x_{11} = x_{22} = 1$. This may not be optimal, since those routes may be more expensive than necessary. But the algorithm needs an extra instruction if too few routes are used, since step 2 will give too few conditions to determine the u_i and v_j. The classical remedy (not necessarily the optimal one) is to add some small fraction ε to each of the m supplies and then to add $m\varepsilon$ to the last demand.

The algorithm itself is a special form of the simplex method. It is rewritten so as to be more efficient for the transportation problem, but it contains the key idea that makes the simplex method work: From a first guess at the shipments x_{ij}, move step by step to the optimal choice. Each step opens one route and closes another. One of the x_{ij} that was zero becomes positive, and one that was positive becomes zero. After a finite number of steps—the cost is improved at every step—the simplex method arrives at the optimum.

The next chapter will begin by describing for any linear program how this idea succeeds. To complete this chapter we describe two more optimizations that are on everyone's list of fundamental network problems.

Optimal Assignments

> Given n applicants for n jobs, suppose C_{ij} measures the value of applicant i in job j. The optimal assignment maximizes the total value.

This is an extension of the marriage problem (or simple assignment problem). There every C_{ij} was 0 or 1; each marriage had value 1, and we maximized the number of marriages. A complete matching has total value n. Now the marriages have different values (even negative values) and the problem is more delicate. We maximize the *sum* of n values C_{ij}, and those values can come from any of the $n!$ assignments (or permutations) of the n applicants. Nevertheless the simpler problem is still the key—from any proposed assignment the marriage problem can decide if an improvement is possible.

At the same time, optimal assignments are a special case of shipments: we ship one applicant to each job. All supplies are $a_i = 1$ and all demands are $b_j = 1$. The cost of the shipment is still C_{ij}, but now we look for the *largest* instead of the smallest total cost. It will be simple to reverse signs in the transportation problem and find the optimal assignment.

These two approaches, one through the marriage problem and the other through the transportation problem, are in some crazy sense dual. Recall that the transportation algorithm works at every step with shipments x_{ij} that satisfy supplies and demands. The algorithm is repeated in the example below. What was not satisfied until reaching the optimum was the condition $u_i + v_j \le C_{ij}$. Any violation of this inequality pointed to a way to rearrange the shipments and reduce the cost.

In the marriage approach, it will be the u_i and v_j that are acceptable at every step and the shipments that are inadequate. An applicant is sent to a job only on routes for which equality holds, $u_i + v_j = C_{ij}$. For the first set of prices, there may be too few of these routes and some applicants will not be assigned. By adjusting the prices we create enough routes with $u_i + v_j = C_{ij}$ to complete the assignment.

Remember the underlying problem: To choose n numbers, one from each row and column of C, so that their sum is a maximum.

EXAMPLE Find optimal assignments for the matrix

$$C = \begin{bmatrix} 5 & 4 & 0 \\ 9 & 7 & 3 \\ 3 & 2 & 5 \end{bmatrix}.$$

First we apply the transportation algorithm. It begins with a possible assignment like $x_{11} = x_{22} = x_{33} = 1$. (This is always the choice of the northwest corner rule.) Since that uses only 3 routes instead of $m + n - 1 - 5$, we include two others with no flow: say $x_{12} = x_{23} = 0$. Then the prices that yield $u_i + v_j = C_{ij}$ on these five circled routes (with $u_1 = 0$) are easy to find:

$$u_i + v_j = \quad \begin{array}{c|ccc} u \backslash v & 5 & 4 & 0 \\ \hline 0 & ⑤ & ④ & 0 \\ 3 & 8 & ⑦ & ③ \\ 5 & 10 & 9 & ⑤ \end{array}$$

On the other routes we want $u_i + v_j \ge C_{ij}$, with the inequality in this direction because the assignment problem is a maximization. For the 2,1 entry this fails: $3 + 5$ is less than $C_{21} = 9$. Therefore applicant 2 is reassigned to job 1. To maintain a matching, applicant 1 will now be assigned to job 2. A new set of prices is

$$u_i + v_j = \quad \begin{array}{c|ccc} u \backslash v & 6 & 4 & 0 \\ \hline 0 & 6 & ④ & 0 \\ 3 & ⑨ & ⑦ & ③ \\ 5 & 11 & 9 & ⑤ \end{array} \quad \text{compared with } C = \begin{bmatrix} 5 & 4 & 0 \\ 9 & 7 & 3 \\ 3 & 2 & 5 \end{bmatrix}.$$

These prices give $u_i + v_j \geq C_{ij}$ in all entries, so the assignment that picks out the ratings 9,4,5 is optimal. Applicants 1,2,3 should go to jobs 2,1,3.

This method is imperfect because of degeneracy. Three routes are enough for an assignment, but two extra routes were needed to fix the prices. The optimal u_i and v_j are not unique—and although the defect could be fixed, it is worth looking at another method.

The opposite approach to optimal assignments is through the marriage problem. In this case we maintain prices satisfying $u_i + v_j \geq C_{ij}$. As a first guess, let all u_i be zero and let v_j be the largest cost in column j. That certainly achieves $u_i + v_j \geq C_{ij}$, but assignments are permitted only on routes for which equality holds. The circled

$$
\begin{array}{c|ccc}
u \backslash v & 9 & 7 & 5 \\
\hline
0 & 9 & 7 & 5 \\
0 & ⑨ & ⑦ & 5 \\
0 & 9 & 7 & ⑤
\end{array}
\quad \text{compared with } C = \quad
\begin{array}{|ccc|}
\hline
5 & 4 & 0 \\
9 & 7 & 3 \\
3 & 2 & 5 \\
\hline
\end{array}
$$

routes with $u_i + v_j = C_{ij}$ do not allow a complete matching. Therefore the prices u_i and v_j are not yet optimal.

We look for the rows and columns that prevented a matching. By Hall's condition r rows and s columns must be responsible, with $r + s > n$. In this case the first row and all three columns form a 1 by 3 submatrix without circles. We lower the prices v_j on these columns by 3, to produce a new circle (around the entry that equals 4). At the same time we *raise* prices by 3 on the remaining rows 2 and 3. This leaves the previous circles unchanged. The new equality $u_i + v_j = C_{ij}$ is circled in the first row, and we look again for a complete matching.

$$
u_i + v_j = \quad
\begin{array}{c|ccc}
u \backslash v & 6 & 4 & 2 \\
\hline
0 & 5 & ④ & 0 \\
3 & ⑨ & ⑦ & 3 \\
3 & 3 & 2 & ⑤
\end{array}
$$

It is there. Assigning rows 1,2,3 to columns 2,1,3, the total value is $4 + 9 + 5 = 18$ and it is maximal. The total price is also $0 + 3 + 3 + 6 + 4 + 2 = 18$, and it is minimal. It was inevitable that these two totals would be equal. Assignments were made only if $u_i + v_j = C_{ij}$, and summing both sides over the three assignments automatically produces $18 = 18$.

Remark 1 Among other assignment algorithms, one good method starts with the best single assignment—the largest entry in the matrix. Copying the reassignment chains of the marriage problem, it systematically finds optimal assignments for 2, then 3, and finally all n applicants.

Remark 2 The **bottleneck assignment problem** is different. Instead of maximizing the sum of n ratings, it makes the lowest rating as large as possible. It tries to avoid a serious mismatch. For the matrix

$$A = \begin{bmatrix} 2 & 4 \\ 1 & 2 \end{bmatrix},$$

the optimal assignment chooses 4 and 1 and the bottleneck assignment chooses 2 and 2. Both problems are included in linear programming, unlike the sums of squares in Chapter 2.

Exercise 7.5.17 tests for the best bottleneck assignment via the marriage problem.

Minimum Cost Flow

Our final topic combines two questions in network flows: it *maximizes the flow at minimum cost*. Both capacities and costs are present; the maximal flow problem is mixed with the transportation problem. The model is becoming more realistic. You will see how the following algorithm also brings in the shortest path problem.

Start from the zero flow. At each step add flow from source to sink along the cheapest path that is still open.

How can we find that cheapest path? At the start the length of an edge is its cost; $l_{ij} = c_{ij}$ unless the edge has no capacity. The cheapest path is the shortest path, and Dijkstra's algorithm will find it. At later steps l_{ij} becomes $+\infty$ if no capacity remains. In the direction from j to i the length becomes negative, as the following example shows. A simple change will rescue Dijkstra's algorithm from any difficulties with $l_{ji} < 0$.

EXAMPLE The network shows costs; all capacities are $c_{ij} = 1$ so the maximal flow is 2. The cheapest path costs $1 + 3 + 4 + 7 = 15$ and the first unit is sent. In that

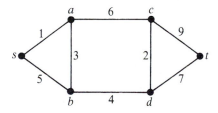

direction the four edges are filled; no capacity remains and their lengths become $l_{sa} = l_{ab} = l_{bd} = l_{dt} = \infty$. In the opposite direction their lengths become $-1, -3, -4, -7$; reversing a flow will recover its cost since a zero flow is free. The shortest (cheapest) path now costs $5 - 3 + 6 + 9 = 17$, and the other unit of flow goes along that path. The total for the minimum cost flow is $15 + 17 = 32$.

The key to this method is the following fact. If f is a minimum cost flow based on the original lengths and f' is a minimum cost flow based on the new lengths, then

$f+f'$ has minimum cost on the original network. The volume of flow can be increased by this rule until the flow is maximal—and is still at minimum cost.

In my experience minimum cost flow is the network problem with the most far-ranging applications. It is flexible enough to be realistic and simple enough to be solved quickly. 5,000 nodes and 600,000 edges have been handled in ten minutes, as a prototype for the merging of U.S. Treasury tapes—which will have 50,000 nodes and 60,000,000 edges. That could only be possible for network problems. They are governed by incidence matrices, whose special properties this book has emphasized. We present a typical application, which used a code developed in Austin by ARC to optimize the production and distribution of several models of Buicks:

> There are factories in Detroit, Los Angeles, ... and markets in Chicago, Houston, ... Each has an upper bound *and a lower bound* on its supply or demand. There are transportation costs and capacities between the cities, and production costs at the factories. The network has intermediate nodes represent-ing three models of Buicks—it is a ***transshipment problem*** because those nodes are neither sources or sinks. The maximal flow algorithm for capacities will decide if the lower and upper bounds can be satisfied. Then there is a minimum cost problem to compete with Toyota.

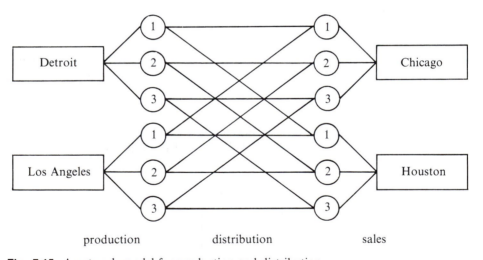

Fig. 7.15. A network model for production and distribution.

This concludes our list of network optimizations. The subject is a beautiful mixture of computer science and graph theory and data structure techniques. On the one hand these problems are excellent examples of duality and the simplex method; some will reappear in the next chapter. (We even invent a wedding game.) On the other hand the algorithms are efficient and simple, and the best ones compute exactly the relevant information.

All algorithms in this chapter finished in "*polynomial time*." Thus the problems are in the most manageable class, designated by *P*. The number of steps grows with the size of the problem, but the rate of growth is polynomial and not exponential—n^{α} and not $e^{\alpha n}$. Many other problems are suspected to be exponential, but that conjecture is not proved. They include

(1) Integer programming: linear programs with integer variables
(2) Traveling salesman: to find the shortest tour of a network
(3) Three color problem: to decide if three colors are enough for a given graph. (Nodes connected by an edge must have different colors.)

For any graph, four colors were proved to be enough (if you accept a computer proof). The nodes represent states on a map, with an edge when two states share a boundary. Some maps need only three colors but there is no fast way to decide.

These problems are typical of the class NP. A proposed answer can be checked for correctness in polynomial time. However it may require polynomially many lucky guesses to discover the solution in the first place—the letter N stands for "nondeterministic," allowing the algorithm to branch. The precise definition is technical but these problems are absolutely central—and they also belong to the subclass called **NP-complete**. If a polynomial-time algorithm could be found for one of them, it would yield a similar algorithm for all problems in NP (including factoring an integer, on whose difficulty modern cryptography depends). At this writing, most experts do not believe that such an algorithm exists.† That is the outstanding problem in complexity theory, **to decide whether or not $P = NP$.** A proof of the inequality $P \neq NP$, with a Turing machine as the model for the computer, will be sensational.

EXERCISES

7.5.1 For what costs C_{13} would the original "northwest corner" prices and shipments have been optimal for the example in the text?

7.5.2 If the cost C_{21} had been 4 instead of 8, why would the algorithm choose this first as the new route? Find the s matrix in step 4, the value of s in step 5, and the prices in step 2. Is this an optimal solution?

7.5.3 Suppose the supplies and demands are $a_1 = 8$, $a_2 = 5$, $b_1 = 1$, $b_2 = 6$, $b_3 = 6$. Find (in the northwest) four shipments that meet these conditions.

7.5.4 If the cost matrix is $\begin{bmatrix} 9 & 3 & 6 \\ 3 & 1 & 2 \end{bmatrix}$, find the prices that give $u_i + v_j = C_{ij}$ on the four routes of the previous exercise. Is $u_i + v_j \leq C_{ij}$ on all routes? If not, which new route should be introduced?

---

† A polynomial algorithm was found for linear programming (the simplex method is excellent but exponential) but somehow that always seemed easier than integer programming—it is not on the list of NP-complete problems.

7.5.5 Complete the algorithm with this cost matrix, by finding the right s to give a new shipping matrix (steps 4–5) and the right prices to match the new routes (steps 2–3). What is the minimal cost?

7.5.6 Solve the 3 by 3 transportation problem for optimal x_{ij}, u_i, v_j, and z with

$$a_1 = 1 \qquad b_1 = 3 \qquad \qquad \begin{bmatrix} 4 & 5 & 1 \\ 4 & 4 & 6 \\ 6 & 4 & 5 \end{bmatrix}.$$
$$a_2 = 5 \qquad b_2 = 4 \qquad C_{ij} = $$
$$a_3 = 7 \qquad b_3 = 6 \qquad $$

7.5.7 If the cost matrix is the 4 by 4 identity matrix, $C = I$, and if all $a_i = b_j = 1$, what is the minimal transportation cost?

7.5.8 Suppose the supplies $a_1 = 4$, $a_2 = 8$ combine to exceed the demands $b_1 = 2$, $b_2 = 3$, $b_3 = 6$. Introduce a new demand $b_4 = 1$ (at the dump) with zero cost on all routes to it: $C_{14} = C_{24} = 0$. What is the optimal solution with costs as in the text?

7.5.9 If all supplies and demands are $a_i = b_j = 1$, and if all C_{ij} are 0 or 1, explain why the transportation problem becomes a matching problem. Are we trying to match the ones in C or the zeros? What condition guarantees a shipping plan with zero cost?

7.5.10 The northwest corner rule finds an initial shipment but it ignores costs; it could be far from optimal. The minimum matrix method attempts a better guess by starting with the smallest cost C_{ij} instead of C_{11}. On this route it sends the maximum possible; x_{ij} is the smaller of a_i and b_j. Thus either the supply a_i is exhausted or the demand b_j is filled; a row or column of x_{ij} is settled. Then the route of next lowest cost is used as fully as possible, and so on. Find the x_{ij} in this way for the example of this section and compare the (improved?) cost with the northwest corner cost $z = 38$.

7.5.11 Starting with the smallest C_{ij} can lead to expensive shipments if it ends by using the largest C_{ij} too heavily. An alternative is to find in each row and column the difference d between the two smallest costs:

5	3	1	$d = 2$
8	4	3	$d = 1$.
$d = 3$	$d = 1$	$d = 2$	

Vogel's method chooses the smallest d and then the smallest C_{ij} in that row or column. If we break the tie with $C_{12} = 3$, it sends as much as possible over that route: $a_1 = 4$ and $b_2 = 3$ so $x_{12} = 3$. The demand b_2 is now filled and the second column is removed. Find the new row differences and column differences d, and show that the next shipment should be $x_{13} = 1$. Complete the Vogel approximation x_{ij} and find its cost.

7.5.12 How is the coefficient matrix for the $m + n$ equations (1) and (2) related to the incidence matrix of the graph connecting suppliers to markets?

7.5.13 In the optimal assignment example the initial shipments were

$$x_{ij} = \begin{bmatrix} 1 & 0 & \\ & 1 & 0 \\ & & 1 \end{bmatrix}.$$

Solving by the transportation method introduced a shipment x_{21}. With $x_{21} = s$, adjust the existing shipments x_{ij} to keep all row sums equal to $a_i = 1$ and all column sums equal to $b_j = 1$. Beyond what value of s does an x_{ij} drop below zero? What is the shipping matrix at this limiting value of s?

7.5.14 In applying the marriage problem to optimal assignments, the text chose row 1 and columns 1,2,3 as responsible for failure in the first guess. Equally responsible are rows 1 and 3 with columns 1 and 2; they form a 2 by 2 submatrix without circles. Reduce prices on these columns and raise them on the remaining row 2 so as to produce $u_i + v_j = C_{ij}$ for some entry in this submatrix. The final assignment is the same, but with what new set of optimal prices?

7.5.15 Suppose C_{ij} is the largest entry in its row and also in its column, as with $C_{33} = 5$ in the example. Show that this C_{ij} will always be in the optimal assignment.

7.5.16 If we reduce the prices v_j on s columns and raise each u_i an equal amount on $n - r$ rows, when is the sum of all prices reduced?

7.5.17 In the bottleneck assignment problem suppose we have a solution—a matching in which the smallest entry m is as large as possible. Circle all entries of the matrix that exceed m. Is a complete matching possible among the circled entries?

7.5.18 Solve the optimal and bottleneck assignment problems by trial and error for

$$A = \begin{bmatrix} 2 & 1 & 5 \\ 8 & 6 & 7 \\ 3 & 1 & 8 \end{bmatrix}.$$

7.5.19 Describe a network that converts the optimal assignment problem into the minimum cost flow problem.

7.5.20 If the text example of minimum cost flow is given an edge from b to c with cost 1, find the minimum cost to send the maximal flow.

CHAPTER 7 IN OUTLINE: NETWORK FLOWS AND COMBINATORICS

7.1 **Spanning Trees and Shortest Paths**—depth first vs. breadth first
Labeling and Scanning—backtracking along a tree
Minimum Spanning Trees—variations on the greedy algorithm
Shortest Paths from the Source—Dijkstra's algorithm
Dynamic Programming—each piece of the path is optimal
Atlas of Shortest Paths—minimum distance between every pair

7.2 **The Marriage Problem**—Hall's test for a complete matching
Duality in Combinatorics—marriages and lines: maximum = minimum
Bipartite Graphs: The Same Theorem—duality for edges and vertices
Doubly Stochastic Matrices—row sums = column sums = 1

7.3 **Matching Algorithms**—reassignment chains by DFS and BFS
Algorithmic Duality—construction of maximum and minimum

8

OPTIMIZATION

Here is a typical problem in linear programming. We are given

  (i)   an m by n matrix A
  (ii) a column vector b with m components
  (iii) a row vector c with n components.

Suppose that $m < n$ and that $Ax = b$ has infinitely many solutions. Linear programming adds an extra condition on x: it looks at *nonnegative solutions* of $Ax = b$. No component of x can be negative. Such a vector is called *feasible*: it satisfies $Ax = b$ and $x \geq 0$. These vectors make up the "feasible set" within n-dimensional space. Then the fundamental problem of linear programming is this: Among all vectors x in the feasible set, find the one that minimizes the inner product $cx = c_1 x_1 + \ldots + c_n x_n$. This is the **cost function**, or *objective function*, and its minimization we call problem (P):

(P)
$$\boxed{\textit{Minimize } cx \textit{ subject to } Ax = b \textit{ and } x \geq 0.}$$

It is this problem, in all its variations and disguises and extensions, which has created a new subject within applied mathematics.

   A linear program like (P) allows one great simplification, and we mention it immediately. It applies above all to the usual and desirable case, in which there is only one vector x that gives the minimum. That is the vector we want, and it has the following property: $n - m$ **of its components are zero.** If we knew those components in advance, the other m components could come directly from the m equations $Ax = b$. Since the zeros are not known, the chief task of every algorithm is to discover where they belong. This leads to the key idea underlying the *simplex method*:

> *Stage 1.* Find a vector x that has $n - m$ zero components, with $Ax = b$ and $x \geq 0$. It may not be the one that minimizes cx.
>
> *Stage 2.* Allow one of the zero components to become positive and force one of the positive components to become zero. The choice of these "entering" and "leaving" components is governed by two requirements: The cost cx should be reduced, while keeping $Ax = b$ and $x \geq 0$.

After a sequence of these pivot steps, exchanging one zero component for another and reducing the cost cx, the minimum must be reached. There are only a finite number of possible positions for the $n - m$ zeros among n components—so eventually the right one appears and the algorithm stops.

You should reread that paragraph. It has given away the main idea of the calculation, on the second page of this new chapter. Nevertheless some important things are held in reserve. The simplex method is a great algorithm but it is not the whole story. It is very much like Gaussian elimination: The system $Ax = b$ can be studied through the details of row operations, but a different and deeper understanding is also possible. A second approach is more geometric, to see the intersections of planes in n dimensions (now including the planes $x_j = 0$). A third approach is more algebraic. For $Ax = b$ it involves nullspaces and column spaces, to decide if there is a unique solution x, and for linear programming it is more subtle; Lagrange multipliers and duality will appear. Each of these contributes an essential part, and we take them in the natural order:

(1) Geometrical, by describing the feasible set (in the rest of this section)
(2) Computational, with details of the simplex method (Section 8.2)
(3) Algebraic, through the theory of duality (Section 8.3).

Section 8.4 applies linear programming to network problems and economics and game theory. And in Section 8.5 we reach nonlinear optimization, minimizing a general cost function $C(x)$ subject to constraints. These two sections are nearly independent of the previous two. They finish what was begun in the early chapters, with the minimization of quadratics.

This chapter also completes the last of the three fundamental areas of applied mathematics: *static problems*, *dynamic problems*, and *optimization*. There is no doubt that optimization requires the most finesse. It leads to the *best* design, while the others analyze the *given* design. In statics and dynamics the materials and the shapes and the equations were given; now we have to find the equations before solving them. But the ideas are central to so many applications—old and new— that they belong in this book.

Extreme Points of Feasible Sets

An example will reveal almost everything:

Minimize $5x_1 + 4x_2 + 8x_3$ subject to $x_1 + x_2 + x_3 = 1, \quad x_1, x_2, x_3 \geq 0.$

There is one equality constraint on the three unknowns; $m = 1$ and $n = 3$. The cost vector is $c = [5 \quad 4 \quad 8]$. The m by n matrix is $A = [1 \quad 1 \quad 1]$, and b is the short vector $[1]$ on the right side. The feasible set contains all x that satisfy $Ax = b$ and also $x \geq 0$. Since $x_1 + x_2 + x_3 = 1$ describes an ordinary plane in three dimensions, and $x \geq 0$ cuts that plane off as it leaves the "positive cone," the feasible set is a triangle (Fig. 8.1). Its vertices are at the points $P = (1,0,0)$, $Q = (0,1,0)$, and $R = (0,0,1)$. These are the *extreme points* of the feasible set.

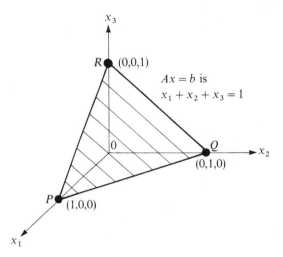

Fig. 8.1. A feasible set: the triangle PQR.

Within this triangle we have to minimize $cx = 5x_1 + 4x_2 + 8x_3$. That is easy to do, because the inside of the triangle is of no importance. The minimum occurs at one of the corners P, Q, R, and not in the interior. To decide which corner is optimal we can try all three: the winner is Q. The optimal x is $(0,1,0)$, and the "value" of the linear program is $cx = 4$. This is the minimum cost.

As expected, the minimum occurs at a point where $n - m = 2$ of the components are zero. This is the usual (nondegenerate) case. The m equations $Ax = b$ combine with $n - m$ equations $x_j = 0$, and those n planes intersect at a corner of the feasible set. It is a point with m positive components and $n - m$ zeros.

Geometrically, it is easy to see why the minimum is attained at a corner. All vectors x with *zero cost* would lie on the plane $5x_1 + 4x_2 + 8x_3 = 0$, or $cx = 0$, passing through the origin. None of these vectors are feasible, since the plane misses the feasible set. As the cost increases above zero, the plane moves toward the triangle. When the cost reaches 4, the plane $5x_1 + 4x_2 + 8x_3 = 4$ first makes contact. It has remained parallel to the original plane $5x_1 + 4x_2 + 8x_3 = 0$, and it touches the feasible set at the point Q. Therefore Q is the point with minimum cost.

It is clear that these parallel planes, or any family of parallel planes, will meet the triangle first at one of its corners. If the cost had been $5x_1 + 6x_2 + 8x_3$, the optimal corner would have been $(1,0,0)$ and the minimum cost would have been 5. The new cost planes are tilted a little, and they first touch the triangle at P.

This picture of first contact with the feasible set is very clear in two dimensions. The constraint $x_1 + 2x_2 = 4$ produces a line as feasible set. That line is chopped off by the constraints $x_1 \geq 0$ and $x_2 \geq 0$, leaving the segment PQ in Fig. 8.2. The figure also shows what happens if the constraint is an inequality $x_1 + 2x_2 \geq 4$; the feasible set becomes the whole shaded region above the line segment. Finally we see the parallel lines from the cost $cx = x_1 + x_2$ approaching this feasible set. The line $x_1 + x_2 = 0$ of zero cost misses the set. So does $x_1 + x_2 = 1$. The line that touches is $x_1 + x_2 = 2$, and the point of contact is P. The optimal solution is $P = (0,2)$, or $x_1 = 0$, $x_2 = 2$—which satisfies the constraint and minimizes the cost.

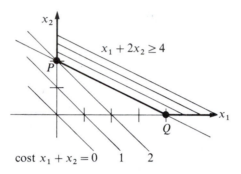

Fig. 8.2. The feasible set touched by $cx = x_1 + x_2 = 2$.

In general the feasible set lies in n-dimensional space and so do the cost planes $cx = $ constant.

We want to point out four other possibilities in linear programming. For the first two we refer to the original triangle in Fig. 8.1. For the third and fourth the feasible sets look very different.

(1)   If we look for the **maximum** of cx, instead of the minimum, we find it at the point R. Again the extreme value occurs at a corner of the feasible set, where $cx = [5 \ 4 \ 8] \ [0 \ 0 \ 1]^T = 8$. In the case of a maximum the cost planes $cx = $ constant move in from infinity instead of out; they first touch the triangle at R.

(2)   The cost could be **negative**. We could easily have $c = [5 \ -3 \ 8]$ and the minimum of $cx = 5x_1 - 3x_2 + 8x_3$ is at $Q = (0,1,0)$. There the cost is negative: $cx = -3$.

(3)   The feasible set could be **unbounded** even with equality constraints. If the constraint $Ax = b$ is $x_1 + x_2 - x_3 = 1$, then there are feasible points far out in the region $x \geq 0$; x_3 can be arbitrarily large, say 999, and $x_1 = x_2 = 500$. The points $P = (1,0,0)$ and $Q = (0,1,0)$ are still corners of the feasible set but R has gone out to infinity. Therefore $cx = 5x_1 + 4x_2 + 8x_3$ can be arbitrarily large, and the maximum value is $+\infty$. The minimum is still 4.

(4)   The feasible set can be **empty**, if $Ax = b$ is incompatible with $x \geq 0$. Example: $x_1 + x_2 + x_3 = -1$, which has no solution that meets the requirement $x_1 \geq 0$, $x_2 \geq 0$, $x_3 \geq 0$.

In the last case there is no problem to solve.

Those observations lead directly to the three alternatives in linear programming. Every problem, with constraints $Ax = b$ or $Ax \leq b$ or $Ax \geq b$, fits into one of these categories:

(1)   The minimum or maximum is reached at a corner of the feasible set (which may be bounded or unbounded)

(2)   The minimum is $-\infty$ (or the maximum is $+\infty$); in this case the feasible set must be unbounded

(3)   The feasible set is empty (no vector satisfies all the constraints).

To solve a linear program is to decide among these three alternatives and, in the case of possibility (1), to find a corner where the minimum or maximum is reached.

Inequality Constraints

If the constraint had been $x_1 + x_2 + x_3 \leq 1$, with inequality instead of equality, the feasible set would have been quite different. In addition to the triangle PQR, a whole pyramid of points going down to the origin $x_1 = x_2 = x_3 = 0$ becomes feasible. The constraint $x_1 + x_2 + x_3 = 1$ produced a plane, but an inequality $x_1 + x_2 + x_3 \leq 1$ produces the entire ***half-space*** on or below that plane. The feasible set is the intersection of this half-space with the positive cone $x \geq 0$. It is the tetrahedron in Fig. 8.1, with the old vertices $P, Q,$ and R, and a new vertex at the origin. The minimum or maximum of cx will be reached at one of the four points $P, Q, R,$ and O.

Inequality constraints are very common. If there are m inequalities $Ax \leq b$, each one produces its own half-space. The nonnegativity constraints add the further half-spaces $x_1 \geq 0, \ldots, x_n \geq 0$. Since these inequalities are all in force at the same time, the feasible set with inequality constraints turns into an *intersection of $m + n$ half-spaces*. It is a polyhedron—a polygon in n dimensions—containing the points that lie in all half-spaces. Those points satisfy every constraint, including non-negativity.

The feasible set has one important property: It is *convex*. It can look like the first set in Fig. 8.3, but it cannot look like the second set. There can never be an "internal corner" like Q. Equivalently, if two points like P and R satisfy all the constraints, then every point on the line between them also satisfies the constraints—and the second picture must be wrong. In n dimensions the idea is the same: if P and R lie in each of the required half-spaces, then the line segment between them is also in each half-space and therefore in the feasible set.

Algebraically, the points between P and R are the *convex combinations* $(1 - c)P + cR$—the linear combinations with coefficients that are nonnegative and add to one. In other words, the line segment from y to z contains all weighted averages $x = (1 - c)y + cz$; the number c ranges from zero (at y) to one (at z). Each half-space is determined by an inequality like $x_1 + x_2 + x_3 \leq 1$, or more generally by $ax \leq b$ for some row vector a. Then if y and z are in this halfspace, so is every point

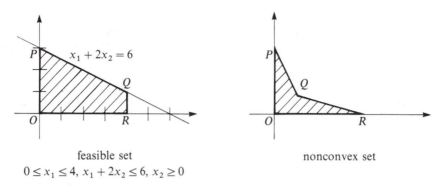

feasible set   nonconvex set
$$0 \leq x_1 \leq 4, \ x_1 + 2x_2 \leq 6, \ x_2 \geq 0$$

Fig. 8.3. A feasible set with inequalities, and a nonconvex set.

between them. If $ay \leq b$ and $az \leq b$, then also

$$ax = (1 - c)ay + caz \leq (1 - c)b + cb = b.$$

Thus halfspaces are convex, and so are the intersections of half-spaces.

There is one more thing about inequality constraints. They are not so convenient for computations. Therefore we turn them into equalities by introducing **slack variables**; $x_1 + x_2 + x_3 \leq 1$ becomes $x_1 + x_2 + x_3 + x_4 = 1$. The new variable x_4 is still nonnegative—it takes up the "slack" between $x_1 + x_2 + x_3$ and 1—and its cost is $c_4 = 0$. If the original problem asked us to minimize or maximize $5x_1 + 4x_2 + 8x_3$, then the new problem does the same; the new c is $[5 \quad 4 \quad 8 \quad 0]$. Only the feasible set is changed, from the tetrahedron $PQRO$ in 3-dimensional space to some set in 4-dimensional space.

I will be very happy if you can visualize that 4-dimensional set. It is the intersection of the plane $x_1 + x_2 + x_3 + x_4 = 1$, coming from the constraint $Ax = b$, with the positive cone $x_1 \geq 0, x_2 \geq 0, x_3 \geq 0, x_4 \geq 0$. The corners of this set seem to be at $P = (1,0,0,0)$, $Q = (0,1,0,0)$, $R = (0,0,1,0)$ and $S = (0,0,0,1)$. Somehow the set should be flat, because it lies on a plane. But a plane in four dimensions involves only one constraint, and must be three-dimensional—which doesn't seem flat to us. That is our fault. The feasible set is the tetrahedron $PQRS$, with no 4-dimensional volume. It corresponds exactly to the original tetrahedron $PQRO$ in three dimensions; lifted into four dimensions it lies on a plane. The origin $(0,0,0)$ needs the slack value $x_4 = 1$ to make $x_1 + x_2 + x_3 + x_4 = 1$, and that turns it into the fourth vertex S.

Once the inequalities are removed and we reach $Ax = b$ with $x \geq 0$, there is no distinction between the original vertices P,Q,R and the new vertex S. A quantity like $cx = 5x_1 + 4x_2 + 8x_3 + 0x_4$ is minimized at S (where it equals zero). It is maximized at R (where it equals 8). In general, a maximum problem with m inequality constraints looks like

Maximize CX subject to $AX \leq b, \ X \geq 0.$

Then a vector W of m slack variables changes the inequality constraints into the equation

$$AX + W = b, \quad \text{or} \quad [A \quad I] \begin{bmatrix} X \\ W \end{bmatrix} = b. \tag{1}$$

We rename this long matrix A, and call this long vector x. The new cost vector is $c = [C \quad 0]$, since the slack variables have no cost, and the problem is back to equality constraints:

$$\text{Maximize } cx \text{ subject to } Ax = b, \, x \geq 0.$$

We note one more possibility. The inequalities could go in the opposite direction, for example in the important problem

$$\text{Minimize } CX \text{ subject to } AX \geq b, \, X \geq 0.$$

In this case the slack variables are *subtracted*. The inequality $x_1 + x_2 \geq 1$ is changed to $x_1 + x_2 - x_3 = 1$ by introducing $x_3 \geq 0$. Similarly $-I$ replaces I in (1), to give $AX - W = b$, and we have again recovered the original form:

$$\text{Minimize } cx \text{ subject to } Ax = b, \, x \geq 0.$$

It is this problem, whether it comes directly from equality constraints like those in Kirchhoff's current law, or indirectly from inequality constraints and slack variables, that we solve in the next section by the simplex method.

EXERCISES

8.1.1 Find the vertices of the feasible set $x_1 + 2x_2 + 3x_3 = 6$, $x_1 \geq 0$, $x_2 \geq 0$, $x_3 \geq 0$. Maximize and minimize $cx = 2x_1 + 7x_2 + 5x_3$ over this set.

8.1.2 The two equations $x_1 + x_2 + x_3 = 4$ and $x_1 - x_2 = 2$ determine a line. Where does it meet the plane $x_1 = 0$? Where does it meet $x_2 = 0$, and $x_3 = 0$? What are the vertices of the feasible set, when these are the equations $Ax = b$? Note that the vertices have $n - m \, (= 3 - 2)$ zero components.

8.1.3 For what values of a does $x_1 + ax_2 = -1$, $x_1 \geq 0$, $x_2 \geq 0$, give an empty feasible set? For what values of a is the set unbounded?

8.1.4 If $cx = x_1 + x_2 + 3x_3$ is minimized subject to $x_1 + x_2 + x_3 = 1$, $x \geq 0$, show that *two* corners of the feasible set give the same minimum $cx = 1$. Find all solutions x, and explain geometrically how more than one solution is possible.

8.1.5 Starting from four points P, Q, R, O, consider the set of all their convex combinations (weighted averages) $x = c_1 P + c_2 Q + c_3 R + c_4 O$ with $c_i \geq 0$ and $c_1 + c_2 + c_3 + c_4 = 1$. This is the **convex hull** of the given points—the smallest convex set that contains them. What is the convex hull of the points P, Q, R, O in Figs. 8.3a and 8.3b?

8.1.6   Sketch the feasible set for $x_1 + 2x_2 \geq 6, 2x_1 + x_2 \geq 6, x_1 \geq 0, x_2 \geq 0$. What points lie at the corners?

8.1.7   On the preceding feasible set, minimize the cost $cx = x_1 + x_2$. Draw the line $x_1 + x_2$ that first touches the feasible set. How about the cost functions $3x_1 + x_2$ and $x_1 - x_2$?

8.1.8   Show that the feasible set for $2x_1 + 5x_2 \leq 3, -3x_1 + 8x_2 \leq -5, x_1 \geq 0, x_2 \geq 0$ is empty.

8.1.9   Show that the maximum cost $cx = x_1 + x_2$ is unbounded, if the constraints are $x_1 \geq 0$, $x_2 \geq 0$, $-3x_1 + 2x_2 \leq -1$, $x_1 - x_2 \leq 2$.

8.1.10   What inequality constraints will give a unit square as feasible set in two dimensions? When slack variables are introduced to give equality constraints $Ax = b$, what are A and b?

THE SIMPLEX METHOD AND KARMARKAR'S METHOD ■ 8.2

The simplex method does for linear programs what Gaussian elimination does for linear equations. It produces the answer in a finite time. The number of steps is not completely fixed, as it is for elimination, because we cannot tell in advance how many corners of the feasible set the method will try. This number can in theory be very large, but in practice that rarely happens; it seems clear that the average number of operations grows only polynomially in m and n, although the worst case is exponential. There have been years of experience with the simplex method, and it has survived and flourished. It must be good.

Nevertheless it is not the only method. It has variations that succeed on special problems, like the network flows of Chapter 7. And a few years ago there appeared a completely different algorithm with a remarkable property: The number of operations never exceeds a fixed polynomial bound, even in the worst case. This was "*Khachian's algorithm*," which keeps the unknown solution—the vector that minimizes or maximizes cx—inside a series of ellipsoids that are shrinking to a point. That point is the solution, and the possibility of computing it in a new and faster way was given tremendous publicity. (More than the proof of the four color theorem; probably less than Fermat's last theorem will receive, if $x^n + y^n = z^n$ is ever proved impossible for $n > 2$.) There were telephone calls to George Dantzig, the inventor of the simplex method, who said "wait and see." And up to now, it has not come close to the efficiency of the older method.

Recently another idea appeared, again with extraordinary publicity. (These problems are of more than theoretical importance.) The new algorithm is called "*Karmarkar's method*," and it also operates inside the feasible set—while the simplex method stays on the edges. The solution is found in polynomial time, and furthermore the method was claimed to be *fifty times faster* in actual experiment than an established simplex code. That is an astonishing claim. I don't think there is independent evidence to support it. The situation is quite exceptional in mathematical research, that a new algorithm is being developed for such a basic problem and access to it is so limited. The underlying idea is attractive, and rich with possibilities; it stands or falls on the central questions of numerical linear algebra. Experience both inside and outside AT&T Bell Labs has led to substantial agreement on two points:

(1)   The new method uses significantly fewer iterations than the simplex method.

(2)   The speed of an iteration step is strongly dependent on the structure of the constraint matrix A.

The time-consuming part of each step is a least squares problem with a diagonal weighting matrix D. It is governed by the matrix AD^2A^T. That is our framework again, with A and A^T exchanged as expected. The success of Karmarkar's method will depend on the fast solution of this least squares problem—by elimination (LU) or orthogonalization (QR) or a preconditioned iterative method like conjugate gradients. The choice will be controlled by A, and it is being investigated as I write.

I have just come from Karmarkar's report on test problems proposed by Dantzig and others. It was taped by PBS; you may see it as this book appears. His reports are still sensational, particularly for large problems. It is far from established that the new method is generally better, but it might be superior in some serious applications. I believe the method will be important. The key ideas fit beautifully with what has come before, and they are described at the end of this section.

The Simplex Method

First place still goes—without doubt—to the simplex method. Our goal is to make that method clear, and we start immediately. Suppose we have found a corner (or vertex, or extreme point) of the feasible set. It is a point that satisfies $Ax = b$ and $x \geq 0$, and it does so with only m nonzero components . It is the meeting point of the m planes $Ax = b$ and the $n - m$ planes $x_{m+1} = 0, ..., x_n = 0$ (if we number the zero components last). Probably it is not the solution, where the cost is a minimum; but it is a start. The main idea of the simplex method is to move from this vertex to a neighboring one at which the cost is lower. After a finite number of steps—there are only a finite number of corners of the feasible set—the cost is reduced as far as possible and the current vertex is optimal.†

What does it mean to be a neighbor of the given vertex? We imagine the feasible set as a complicated (but convex!) polyhedron with many edges and faces, and we are at one of the corners. Its neighboring corners are the vertices that can be reached by traveling along a single edge. The simplex method carries us on a path around the feasible set, one edge at a time, until we reach the solution.

Algebraically, this step from one vertex to the next is simple to describe. One of the zero components of x is allowed to become positive; it increases along the edge. The other zero components of x remain zero, and the equations $Ax = b$ remain in force—so that the original nonzero components of x have to adjust to the presence of a new one. As they adjust, some of these components may decrease toward zero. When a component of x reaches zero, we stop; we have come to the end of the edge, and found a neighboring vertex. It will again have $n - m$ zero components (the new zero replaces the old one). It is still feasible, because $Ax = b$ was maintained and no component of x became negative. Thus a simplex step is really an *exchange step*, in which a zero component of x enters the action and a positive component leaves (it becomes zero).

Since each component of x is associated with a column of A—it multiplies that column when forming Ax—the exchange can equally well be expressed in terms of the columns. At the start of the step, m columns are "in the basis." Those are the columns corresponding to positive components of x. At the end of the step, one column has left the basis and a new one has entered. Solving $Ax = b$ means expressing b as a combination of the basic columns; those columns form a *square*

---

† Experiments suggest that the average number of simplex steps, or corners to be visited, is about $3m/2$. At this moment the theory applies to variants of the method.

matrix B. The other columns are not used since they multiply the zero components of x.

There remains an important decision: Which edge to choose? Starting from a given vertex there are $n - m$ zero components that might be allowed to increase, and therefore $n - m$ edges to select from. We choose an edge along which the cost drops as rapidly as possible. The cost vector c must enter into that choice, and there must be some matrix algebra. It is better to begin with an example.

EXAMPLE   Minimize the cost $cx = 3x_1 + x_2 + 9x_3 + x_4$, subject to $x \geq 0$ and these two equations $Ax = b$:

$$\begin{array}{l} x_1 \quad\ + 2x_3 + x_4 = 4 \\ \\ x_2 +\ \ x_3 - x_4 = 2. \end{array} \tag{1}$$

The starting corner is $x = (4, 2, 0, 0)$, with $m = 2$ positive entries and $n - m = 2$ zeros. The first two columns of A are basic, and conveniently they form the identity matrix: the basis matrix is $B = I$. The question is whether to increase x_3 or x_4, or stop. If x_3 is increased (keeping $x_4 = 0$) then equations (1) require

$$x_1 = 4 - 2x_3$$
$$x_2 = 2 - x_3$$
$$cx = 3(4 - 2x_3) + (2 - x_3) + 9x_3 = 14 + 2x_3.$$

The cost increases with x_3, because $2x_3$ becomes positive. That coefficient 2 is called the **reduced cost**, and its sign is decisive. The edge that makes x_3 positive is useless; the cost goes the wrong way. On the other hand an increase in x_4 (keeping $x_3 = 0$) means from (1) that

$$x_1 = 4 - x_4$$
$$x_2 = 2 + x_4$$
$$cx = 3(4 - x_4) + (2 + x_4) + x_4 = 14 - x_4.$$

It pays to bring in x_4, since there is a minus sign in the cost cx. The reduced cost is -1, and this edge decreases the total cost.

When x_4 enters, either x_1 or x_2 must leave. That is decided by looking again at

$$\begin{array}{l} x_1 = 4 - x_4 \\ \\ x_2 = 2 + x_4. \end{array} \tag{2}$$

As x_4 increases, it is only x_1 that moves toward zero. It reaches $x_1 = 0$ when $x_4 = 4$. That signals the end of the edge. The new vertex has $x_1 = 0$ (and still $x_3 = 0$), with

$x_4 = 4$ and now $x_2 = 6$. Its cost is $cx = 3x_1 + x_2 + 9x_3 + x_4 = 10$, smaller than the original cost (which was 14).

To complete this simplex step—or rather to prepare for the next step—we look at the columns of A. The fourth and second columns are now basic; they form the 2 by 2 basis matrix $B = \begin{bmatrix} 1 & 0 \\ -1 & 1 \end{bmatrix}$. It is no longer the identity, but it can be made so by adding the first of equations (1) to the second:

$$\begin{aligned} x_1 \quad + 2x_3 + x_4 &= 4 \\ x_1 + x_2 + 3x_3 \quad &= 6. \end{aligned} \tag{3}$$

We have effectively multiplied through by B^{-1}, leaving the identity matrix in the basic columns 4 and 2. This separates out x_4 and x_2, in the same way that x_1 and x_2 were distinguished in (1). It will make the next decision clear.

The current corner is $x = (0, 6, 0, 4)$, and I think it is optimal. To verify that, the simplex method computes the new reduced costs.† If x_1 were introduced, forcing $x_2 = 6 - x_1$ and $x_4 = 4 - x_1$ and keeping $x_3 = 0$, then

$$cx = 3x_1 + x_2 + 9x_3 + x_4 = (3 - 1 - 1)x_1 + 10. \tag{4}$$

The coefficient of x_1 is $r_1 = 1$, a positive reduced cost, so x_1 should not be allowed to grow. That could have been predicted, since if x_1 increases we go back along the edge we arrived on. But it is certainly possible, in a larger problem, that a leaving variable could reenter several steps later and be active in the optimal solution.

The other reduced cost is r_3, the coefficient of x_3. The constraints require $x_2 = 6 - 3x_3$ and $x_4 = 4 - 2x_3$, with $x_1 = 0$, and

$$cx = 3x_1 + x_2 + 9x_3 + x_4 = (9 - 1 \cdot 3 - 1 \cdot 2)x_3 + 10. \tag{5}$$

The coefficient of x_3 is $r_3 = 4$, again positive. Since no reduced cost is negative, no edge gives an improvement. **The current corner $x = (0, 6, 0, 4)$ is optimal.**

Remark In this calculation, the numbers $r_1 = 1$ and $r_3 = 4$ were computed separately. Matrix algebra will soon give an explicit formula; in this case it is

$$[r_1 \quad r_3] = [c_1 \quad c_3] - [c_2 \quad c_4] \begin{bmatrix} 1 & 2 \\ 1 & 3 \end{bmatrix}$$

$$= [3 \quad 9] - [1 \quad 1] \begin{bmatrix} 1 & 2 \\ 1 & 3 \end{bmatrix} = [1 \quad 4].$$

The direct costs $c_1 = 3$ and $c_3 = 9$ are *reduced* by a term that accounts for c_2 and c_4. The matrix came from columns 1 and 3 of the constraints in equation (3).

---

† In economics this is called **pricing out** those variables.

It was noticed early in the history of linear programming that the cost coefficients could form a new row at the bottom of the matrix A and elimination could be applied to this row too. The bigger matrix is called a ***tableau***, and it contains all information about the linear program. At the start, when x_1 and x_2 were basic, the tableau is formed from A, b, c, and zero:

$$T_0 = \left[\begin{array}{c|c} A & b \\ \hline c & 0 \end{array} \right] = \left[\begin{array}{cccc|c} 1 & 0 & 2 & 1 & 4 \\ 0 & 1 & 1 & -1 & 2 \\ \hline 3 & 1 & 9 & 1 & 0 \end{array} \right].$$

Subtracting row 2 and also 3 times row 1 from the last row completes the elimination, and yields the new tableau

$$T_1 = \left[\begin{array}{cccc|c} 1 & 0 & 2 & 1 & 4 \\ 0 & 1 & 1 & -1 & 2 \\ \hline 0 & 0 & 2 & -1 & -14 \end{array} \right].$$

We recognize the last row. It contains the reduced costs $r_3 = 2$ and $r_4 = -1$ found earlier at the original corner. It also shows the original cost $cx = 14$.[†] That row tells us what to do: bring in x_4 because r_4 is negative. It also decides whether x_1 or x_2 should leave, by the following trick. Take the ratios of 4 and 2 in the last column to 1 and -1 in the entering column:

$$\frac{4}{1} = 4, \quad \frac{2}{-1} = -2.$$

Those ratios give the limit on x_4, when the end of the edge is reached. In this case we can go as far as $x_4 = 4$, when x_1 touches zero and must leave. The negative ratio can be ignored. As in (2), it means that x_2 increases as x_4 enters; x_2 never drops to zero.

The first step ends at $x_1 = 0$ and a new elimination prepares for the next step. It distinguishes the active unknowns x_2 and x_4 by clearing out their columns. Adding the first row of T_1 to the second and third gives (as before, but now in a tableau) the identity matrix in columns 4 and 2:

$$T_2 = \left[\begin{array}{cccc|c} 1 & 0 & 2 & 1 & 4 \\ 1 & 1 & 3 & 0 & 6 \\ \hline 1 & 0 & 4 & 0 & -10 \end{array} \right].$$

---

† With a minus sign, which is part of the magic of the tableau.

This is the situation at the new (and optimal) corner. The last column shows $x_4 = 4$ and $x_2 = 6$, at a cost of 10. The reduced costs are $r_1 = 1$ and $r_3 = 4$, appearing in the last row. Neither one is negative so the corner is optimal and the simplex method stops. *The solution is $x = (0,6,0,4)$.*

The Algebra of a Simplex Step

From this example we can write down the pieces of a simplex step, for a system of any size. It was the same for Gaussian elimination; after a 2 by 2 example the whole pattern could be put into matrix notation. Here A is rectangular, and the vector x at a corner of the feasible set is separated into m nonzero components (call that part x_B) and $n - m$ zeros. The columns of A that correspond to the nonzero components form a square matrix B. If x has its nonzero components first, then the columns of B are the first m columns of A. The other columns—the nonbasic columns in the rest of A—form a matrix N. Similarly the entries in the cost vector c are separated into c_B and c_N. Then the constraint $Ax = b$ is

$$Ax = [B \quad N] \begin{bmatrix} x_B \\ 0 \end{bmatrix} = b, \quad \text{or} \quad x_B = B^{-1}b. \tag{6}$$

This is a genuine corner (it is feasible) provided $x_B \geq 0$. Its cost is

$$cx = [c_B \quad c_N] \begin{bmatrix} x_B \\ 0 \end{bmatrix} = c_B B^{-1} b.$$

The question is where to go next, after leaving this corner. It requires some matrix algebra.

The decision is made easy by elimination on A, which reduces the square part B to the identity matrix. In matrix notation this multiplies $Ax = b$ by B^{-1}:

$$[I \quad B^{-1}N] \begin{bmatrix} x_B \\ 0 \end{bmatrix} = B^{-1}b. \tag{7}$$

What happens if the zero components of x increase to some values x_N? The nonzero components x_B must drop by $B^{-1}Nx_N$ to maintain equality in (7). That changes the cost to

$$cx = c_B(x_B - B^{-1}Nx_N) + c_N x_N.$$

Rearranging this into

$$cx = (c_N - c_B B^{-1}N)x_N + c_B x_B, \tag{8}$$

we can see whether the cost goes up or down as x_N increases. Everything depends on the signs of the vector r in parentheses:

$$r = c_N - c_B B^{-1} N. \tag{9}$$

This vector contains the **reduced costs**. If $r \geq 0$ then the current corner is optimal. The product $r x_N$ in (8) cannot be negative since $x \geq 0$, so the best decision is to keep $x_N = 0$ and stop. On the other hand, suppose a component of r is negative. Then the cost is reduced by increasing that component of x_N. The simplex method chooses one entering variable—normally the one with the most negative component of r— and it allows that component x_i to increase from zero.

After r is computed and the entering x_i is chosen, there is one more question. Which component x_j should leave? It will be the first to reach zero as x_i increases. To find it, let v be the column of $B^{-1} N$ that multiplies x_i. Then $Ax = b$, which is equation (7), becomes

$$x_B + v x_i = B^{-1} b. \tag{10}$$

The kth component of x_B will drop to zero when the kth components of $v x_i$ and $B^{-1} b$ are equal. This happens when x_i grows to

$$x_i = \frac{k\text{th component of } B^{-1} b}{k\text{th component of } v}. \tag{11}$$

The smallest of these ratios determines how large x_i can become. If the jth ratio is the smallest, the leaving variable will be x_j. At the new corner x_i has become positive and x_j has become zero.

Note   In the example the entering variable was $x_i = x_4$ and equation (10) appeared as equation (2):

$$x_1 + x_4 = 4$$

$$x_2 - x_4 = 2.$$

The vector v has components 1 and -1, and the ratios are $4/1$ and $2/-1$. The smallest positive ratio is $4/1$, so the leaving variable is x_1. As before, we do not worry about the ratio that is negative! It means that x_2 will *never* drop to zero as x_4 increases. If there are no positive ratios then x_i can be increased forever; the edge is infinitely long and the cost can be decreased all the way to $-\infty$.

All this matrix algebra can be collected into a few lines.

A Step of the Simplex Method

(1)   Compute $r = c_N - c_B B^{-1} N$.

(2)   If $r \geq 0$ *stop*; the current x is optimal. Otherwise find the most negative component r_i, and let the corresponding x_i increase from zero—it is the ***entering variable***. Let v be the corresponding column of $B^{-1} N$.

(3)   Compute the ratios in (11), admitting only positive components of v. If the jth ratio is the smallest, then x_j is the ***leaving variable***.

(4)   The new corner satisfies $Ax = b$ with x_i now positive and x_j now zero. Compute this corner and by row operations in the tableau (or otherwise) prepare for the next simplex step.

The Organization of a Simplex Step

We have achieved a transition from the geometry of the simplex method to the algebra—from the visualization of corners and edges to their actual calculation. The vector r and the ratios in (11) became decisive. Now it remains to organize this calculation in an efficient and numerically stable way. We are at the heart of the method, and there are at least three ways to proceed:

(a)   By updating the whole tableau, as above

(b)   By recording only the matrices that multiply each tableau to give the next one

(c)   By keeping the triangular factors L and U of B and updating them at each step.

In the early days of the simplex method, the tableau was all important. It was also a little mysterious. Without computers every problem had to be small, and like our example it could be written out in full. (In those days you could not assume that everyone knew linear algebra.) But when the problems got larger and matrix notation made the simplex step clear, it was noticed that a slight change would save an enormous amount of time.

The inefficiency is this, that when the negative r_i indicates the entering variable x_i, all work on the other variables and their columns has been wasted. Those variables are fixed at zero throughout the step. It is excessive to do a complete elimination in advance, reducing B to the identity and reaching a tableau like T_2. This tableau makes the step simple, but it is really the vector r that should come first. Therefore we think through one more time the essential part of the simplex method.

At each step, some set of m components of x is basic; all other variables are zero. The corresponding m columns of A form a square submatrix B, the "basis matrix" that controls the step. The rest of the matrix is N. What elimination did was to change B into the identity matrix, and to change N into $B^{-1} N$, simplifying the remaining calculations but doing more than was necessary. If we knew B^{-1}, or if we knew the factorization $B = LU$, that would be enough. The formula for r is

$$r = c_N - c_B B^{-1} N,$$

and we are crazy to compute the matrix $B^{-1}N$ and multiply by c_B. Any rational person would compute the vector $z = c_B B^{-1}$, and multiply by N. In other words he would solve $zB = c_B$, an ordinary linear equation, rather than do the elimination on A that led to $B^{-1}N$.†

In fact there are three linear systems to be solved at each step. They all share the same coefficient matrix B:

   (i) $Bx_B = b$ gives the nonzero part of x
   (ii) $zB = c_B$ gives z and the reduced costs $r = c_N - zN$
  (iii) $Bv = y$ gives v and then the ratios (y is the entering column of N so that v is the column of $B^{-1}N$).

The *revised simplex method* solves these three equations and never computes $B^{-1}N$. What it needs is the current B^{-1}.

That raises a natural question in matrix algebra. If a column leaves B and is replaced by a column of N, what is the change in B^{-1}? The step starts with a basis matrix B_{old} and ends with a basis matrix B_{new}. One column is changed—say the last column of B_{old} is replaced by the entering column y taken from N. The vector v mentioned above is $B_{old}^{-1} y$. We claim that this vector v leads to B_{new}^{-1}, if we multiply by the right matrix:

$$B_{new}^{-1} = \begin{bmatrix} 1 & & v_1 \\ & \ddots & \vdots \\ & 1 & \\ & & v_m \end{bmatrix}^{-1} B_{old}^{-1}. \tag{12}$$

To prove it, invert both sides. B_{new} will agree with B_{old} except in the last column. On the right that will be B_{old} times v, which is y, and this is the correct last column of B_{new}.

Thus by storing the original B^{-1}, and also the vectors v at each simplex step, we know everything about the current B^{-1}. This is called the "product form of the inverse." Of course we also know which variables have entered and left the basis, so B^{-1} can be computed from scratch whenever the product has too many terms.

It is equally possible to update the triangular factors of B, going from $L_{old} U_{old}$ to $L_{new} U_{new}$. In case roundoff error is dangerous, this is preferable. But in every case the revised method is faster than the old-fashioned tableau.

---

† It is like computing ABx; you should do the matrix-vector products Bx and $A(Bx)$, never the matrix multiplication AB. Exercise 8.2.13 gives the right order of multiplication in other triple products ABC.

Remark on degeneracy. Up to now x was assumed to have exactly $n - m$ zero components. But it could certainly happen that there is an extra zero. In fact that does happen in our example, at the other corner $x = (0,0,2,0)$ of the feasible set. This satisfies $Ax = b$, equation (1), and it could have been the starting corner for the simplex method. We might have chosen columns 3 and 4 as basic, with $x_1 = x_2 = 0$, and then discovered that $x_4 = 0$ in the solution. This means that an extra plane happens to go through the point x; it is even more of a corner than expected. Its cost is $cx = 3x_1 + x_2 + 9x_3 + x_4 = 18$, and we still look for an edge along which this cost will decrease.

In this problem the simplex method easily finds that edge. But it is conceivable, if the numbers were chosen differently, that the leaving variable might be x_4. It is supposed to drop to zero, but it is already zero. The new corner will be the same as the old one; the solution is still $x = (0,0,2,0)$, and we can hardly tell that x_4 has left and x_1 or x_2 has entered. Furthermore, with a very exceptional choice of coefficients, this might happen forever! In theory the method could cycle, going from one basis to the next without ever leaving the corner. It is possible to watch for this cycling, but apparently it is so rare—Dantzig has seen only one case in his lifetime†—that the important codes do nothing about it. The codes do make computer-related adjustments (scaling, pivot tolerances, perturbation of degeneracies, ...) in the simplex method and might therefore produce "computer cycling"—but again that is not a serious danger.

It is *not at all rare* for a corner to be degenerate, with extra zeros in the vector x. The usual simplex steps eventually choose an entering x_i that really enters, with $x_i > 0$, and then move away from the degenerate corner.

Remark on Phase I. We have described how the simplex method goes from one corner to the next. That is "Phase II," the most important part. But the method needs a corner at which to begin, and finding that corner is not completely trivial. We could allow any m components of x to be nonzero and solve $Ax = b$, but some of those components might turn out to be negative. We would have to try again, and to compute all of the $n!/m!(n - m)!$ possibilities would be disastrous. Instead there is a neat trick to find a starting corner, known as "Phase I" of the simplex method. We illustrate it on our example.

Phase I invents a new problem. It creates $m = 2$ new variables x_5, x_6 and inserts them into $Ax = b$ (after reversing signs in any equation where the right side was negative):

$$
\begin{aligned}
x_1 \quad &+ 2x_3 + x_4 + x_5 &&= 4 \\
x_2 + \; &x_3 - x_4 \quad\;\; + x_6 &&= 2.
\end{aligned}
$$

It also invents the new cost $cx = x_5 + x_6$, so that $c = [0 \quad 0 \quad 0 \quad 0 \quad 1 \quad 1]$. Then it solves this new problem by the simplex method. The minimum is zero where $x_5 = x_6 = 0$, and when the simplex method finds that point it has completed Phase I.

---

† Which equals the lifetime of the simplex method.

The first four components of x will be a corner to start from on the original problem.

You may ask, where do we start on the invented problem? Fortunately it does not need its own Phase I; it begins immediately at $x_5 = 4$ and $x_6 = 2$. The construction itself suggests the starting corner $x = (0,0,0,0,4,2)$. Applying the simplex method with the invented cost and with $B = I$ in the last two columns— here B is on the *right* of N—we have

$$r = c_N - c_B B^{-1}N = [0 \ \ 0 \ \ 0 \ \ 0] - [1 \ \ 1] \begin{bmatrix} 1 & 0 & 2 & 1 \\ 0 & 1 & 1 & -1 \end{bmatrix}$$

$$= [-1 \ \ -1 \ \ -3 \ \ 0].$$

The third component is the most negative, so x_3 is the entering variable. Whether x_5 or x_6 should leave depends on the constraints

$$2x_3 + x_5 \qquad = 4$$

$$x_3 \qquad + x_6 = 2.$$

The ratios are $4/2$ and $2/1$—they are accidentally equal—so the entering variable can increase to $x_3 = 2$. The new corner using x_3 is $x = (0,0,2,0,0,0)$. By a degenerate accident, x_5 and x_6 fell to zero in the same step; normally Phase I takes more work. But the result is what we want, a corner with $x_5 = x_6 = 0$ in the invented problem— and thus a corner $(0,0,2,0)$ from which Phase II can begin on the original problem.

Karmarkar's Method

We turn to one of the most sensational developments in numerical optimization. It has implications for nonlinear programming, but we concentrate on the linear case—where an important fraction of scientific computing takes place. In one sense, Karmarkar† proposes to treat the minimization of cx in a more standard way—by moving within the feasible set, subject to the constraints $Ax = b$ and $x \geq 0$, toward the minimum. At the same time he has introduced a change of variables that transforms the problem in an original way. In its simplest form it is a *rescaling algorithm*; after describing that form we can add his nonlinear transformation. In approaching the optimal x he determines which of its components are nonzero, and then there is a jump to a corner of the feasible set—where the solution lies.

The starting point is the following question. From an initial guess x^0, which direction and which stepsize will take us toward the optimum? There is a natural answer:

(1)   Move in the direction that decreases cx as quickly as possible
(2)   Stop at the boundary of the feasible set.

---

† The accent is on the first syllable.

With modifications this is what we do. A step Δx changes the cost by $c\Delta x$. If Δx is in the direction of $-c$—which the constraints are not likely to allow—the cost goes down most quickly. For other Δx the decrease is $|c||\Delta x| \cos \theta$, proportional to the step size and to the cosine of the angle between Δx and $-c$. Therefore a simplified first step of Karmarkar's method is

(1*)   Choose the direction that makes $\cos \theta$ as large as possible. It is the *projection* of $-c$ onto the nullspace of A.

(2*)   Stop close to, but *a little before*, the boundary beyond which a constraint is violated.

A typical coefficient in 2* is $\alpha = 0.98$. The largest allowed step is multiplied by α, so that no inequality constraint becomes tight. Otherwise the next step is restricted and the tight constraint (which might not be active at the optimal solution) is hard to loosen. The value $\alpha = 0.25$ was originally published by Karmarkar and applied to his transformed problem; it was chosen to prove convergence in polynomial time and not to achieve convergence in real time.

The important calculation, on which everything rests, is 1*. Since x^0 satisfies the constraint $Ax^0 = b$ and the next approximation must satisfy $Ax^1 = b$, the difference satisfies $A\Delta x = 0$. Each step must lie in the nullspace of A, which is parallel to the feasible set. Projecting $-c$ onto that nullspace gives, among allowed directions, the one of steepest descent—maximizing $\cos \theta$ by minimizing the angle between Δx and $-c$.† The formula for projection onto the nullspace of A is pure linear algebra:

$$P = I - A^T(AA^T)^{-1}A. \tag{13}$$

A vector in the nullspace is not moved by P; if $Az = 0$ then $Pz = z$. Furthermore P takes every vector into that nullspace, because $AP = A - (AA^T)(AA^T)^{-1}A = 0$. Thus $APc = 0$, and Pc is the projection of c onto the nullspace.

In computations, formula (13) is not taken literally. The inverse matrix $(AA^T)^{-1}$ is never formed. Instead we find $y = (AA^T)^{-1}Ac$ by solving a linear equation—the normal equation—and then y is multiplied by the final A^T:

$$\boxed{(AA^T)y = Ac \quad \text{and} \quad Pc = c - A^Ty.} \tag{14}$$

The move Δx is a multiple of $-Pc$. **The cost is reduced** because $c^T\Delta x$ is that same multiple of $-c^TPc$, which cannot be positive. Algebraically, $-P$ is negative semi-definite; geometrically, the angle θ is at most $90°$.

EXAMPLE   If the constraint is $x_1 + x_2 + x_3 = 1$, the feasible set is the triangle PQR in Fig. 8.4. The cost $cx = 5x_1 + 4x_2 + 8x_3$ is minimized at a corner, and that corner is $Q = (0,1,0)$. It is the x_2 component that has the lowest cost, namely 4, so the

---

†Henceforth c is a column vector and the cost is c^Tx.

optimal corner has all its weight in that component. The initial guess will be $x^0 = (\frac{1}{3},\frac{1}{3},\frac{1}{3})$, at the center of the feasible set, and the first step projects c onto the plane of the triangle:

$$A = [1 \quad 1 \quad 1] \quad \text{so} \quad AA^Ty = Ac \quad \text{becomes} \quad 3y = 17$$

$$Pc = c - A^Ty = \begin{bmatrix} 5 \\ 4 \\ 8 \end{bmatrix} - \frac{17}{3}\begin{bmatrix} 1 \\ 1 \\ 1 \end{bmatrix} = \frac{1}{3}\begin{bmatrix} -2 \\ -5 \\ 7 \end{bmatrix}.$$

This vector should be in the nullspace of $A = [1 \quad 1 \quad 1]$, and it is; the components of Pc add to zero. Moving from x^0 in the opposite direction $-Pc$ gives

$$x^0 - sPc = \frac{1}{3}\begin{bmatrix} 1 + 2s \\ 1 + 5s \\ 1 - 7s \end{bmatrix}.$$

The largest possible step before a component becomes negative is $s = \frac{1}{7}$. That brings the third component $1 - 7s$ to zero; we are at the boundary point B in Fig. 8.4. Its coordinates, with $s = \frac{1}{7}$, are $(\frac{3}{7},\frac{4}{7},0)$. The constraint $x_3 \geq 0$ has become tight, and in this example that is good. The constraint is also tight at the optimal corner $Q = (0,1,0)$ and we could move there in one more step. For safety, however, the steplength s is reduced by $\alpha = 0.98$:

$$s = 0.98 \left(\tfrac{1}{7}\right) = 0.14 \quad \text{and} \quad x^1 = x^0 - 0.14Pc = \frac{1}{3}\begin{bmatrix} 1.28 \\ 1.70 \\ 0.02 \end{bmatrix}.$$

That is the new x^1 from which the next step starts.

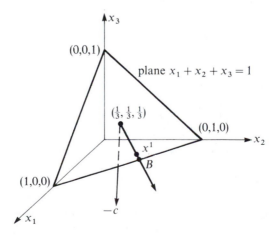

Fig. 8.4. A step of the rescaling algorithm.

The second step is still to be decided. It needs a new idea, since to move again in the same direction $-Pc$ is useless. Karmarkar's suggestion is to transform x^1 back to the central position $(\frac{1}{3}, \frac{1}{3}, \frac{1}{3})$. That change of variables will alter the linear programming problem, and then we project the new c onto the nullspace of the new A.

Karmarkar's change of variables was nonlinear (a projective transformation discussed below). The problem can be kept linear, however, by a simple rescaling. The components of x^1, and the components of every vector x, are divided by the numbers 1.28, 1.70, and 0.02. If those numbers are put into a diagonal matrix D, then *the rescaling multiplies all vectors by* D^{-1}:

$$D = \begin{bmatrix} 1.28 & & \\ & 1.70 & \\ & & 0.02 \end{bmatrix} \quad \text{and} \quad D^{-1}x^1 = \begin{bmatrix} \frac{1}{3} \\ \frac{1}{3} \\ \frac{1}{3} \end{bmatrix}.$$

We are back at the center of the feasible set! However the set itself has changed, and so has the cost vector. The rescaling from x to $X = D^{-1}x$ has three effects:

the constraint $Ax = b$, which is $ADD^{-1}x = b$, becomes $ADX = b$
the cost c^Tx, which is $c^TDD^{-1}x$, becomes $(D^Tc)^TX$
the constraint $x \geq 0$ becomes $X \geq 0$.

Since D is diagonal these changes are easy. The columns of $A = \begin{bmatrix} 1 & 1 & 1 \end{bmatrix}$ and the entries of $c = (5,4,8)$ are multiplied by the scaling numbers:

$$\boxed{AD = \begin{bmatrix} 1.28 & 1.70 & 0.02 \end{bmatrix} \quad \text{and} \quad D^Tc = (6.4, 6.8, 0.16).}$$

The matrix AD replaces A in the projection formula (13), and we apply it to D^Tc instead of c. That gives the direction of movement from the center point $(\frac{1}{3}, \frac{1}{3}, \frac{1}{3})$:

$$\Delta X = -s[I - D^TA^T(ADD^TA^T)^{-1}AD]D^Tc. \tag{15}$$

Now you have the whole algorithm, except for starting and stopping. It can start from any point x^0 inside the feasible set. At each step the current guess x^k is rescaled to $e = (1,1, \ldots, 1)$—it is simpler to omit factors like $\frac{1}{3}$—and then the change Δx is computed from a projection in the rescaled variables. The output from the step is x^{k+1}:

Rescaling Algorithm

---

(1) Construct a diagonal matrix D from the components of x^k, so that $D^{-1}x^k$ $= (1,1, \ldots, 1) = e$
(2) Apply the projection in equation (15)
(3) Determine s so that $e + \Delta X$ has a zero component
(4) Multiply that correction ΔX by a factor α $(0.9 \leq \alpha < 1)$
(5) The new vector is $x^{k+1} = D(e + \alpha\Delta X) = x^k + D\alpha\Delta X.$

All the work is in step 2. It is a *weighted projection*. In the language of Section 2.5 the weighting matrix is $W = D^T$. In our framework the matrix C is DD^T (or D^2 since these matrices are diagonal). As always, the inverse of ADD^TA^T is not computed. Instead the weighted normal equations are solved for y:

$$(ADD^TA^T)y = ADD^Tc. \tag{16}$$

This is the heart of the problem—it is weighted least squares—and we comment on three points before describing Karmarkar's nonlinear change of variables.

1. *Numerical linear algebra.* The normal way to compute y is by Gaussian elimination. That certainly succeeds if the dimension is not large. It is also effective when ADD^TA^T is sparse—provided the fill-in of zeros during elimination is not excessive. Perhaps the results so far reported by Karmarkar fall into this category. Note that one dense column in A fills up AA^T; that column could be dealt with privately, but 10 or 20 dense columns become more difficult. In the sparse case, a reordering to reduce fill-in is a good investment, since the minimal degree algorithm is not expensive.

An alternative is to factor D^TA^T into QR, by a stable orthogonalization of the columns. That leads to $ADD^TA^T = R^TR$, so that R agrees with the upper triangular factor in elimination—but again there is the difficulty of fill-in. A more sophisticated approach uses a preconditioning matrix M, followed by conjugate gradients. The choice of M is crucial, and the special feature of rescaling is that A stays the same; only the diagonal matrix D is changed. At each step there are n scaling numbers (where the simplex method looked for one lowest reduced cost). It is reasonable to hope that the computations can use the simplicity of D, especially since an exact projection is not necessary.

Degeneracy is an interesting question. It is bad for the simplex method, which is confused by extra zero components and takes time to find an edge that leaves the corner. On the other hand degeneracy is most common in network problems, on which simplex steps are extremely fast. Its effect on Karmarkar's method is not yet clear.

2. *Dual variables.* It is a beautiful and remarkable fact that the vector y in the middle of the projection represents the dual variables. The minimum cost can be estimated from y^Tb as well as c^Tx; in fact y^Tb is probably better. Furthermore the largest entries in $r = c - A^Ty$ indicate which components of x are likely to be zero (since $x^Tr = 0$ at the solution) and allow a final jump to that optimal corner. The jump must be made as soon as possible if the simplex method is to be beaten.

The connections come directly from matrix algebra. Equation (16) gives $y = (ADD^TA^T)^{-1}ADD^Tc$. Then (15) gives $\Delta x = D\Delta X$ as a multiple of $z = DD^Tc - DD^TA^Ty$. That is the same as DD^Tr. Apart from scaling, *the step is in the direction of the reduced cost r*. It is like the simplex method, except it uses all components of r. In fact the earlier reduced cost $r = c_N - c_BB^{-1}N$ coincides exactly with the r given here, when both are computed at a corner of the feasible set—where x and

D have $n - m$ zeros. The simplex method had the stopping test $r \geq 0$. The rescaling algorithm stops when that is approximately satisfied and it jumps to x^*.

I want to emphasize that the projection formula is familiar. At each step the problem is: Find the vector z that is closest to c but still in the nullspace of A. It is weighted least squares, with the constraint $Az = 0$. According to Chapter 2 it leads to the equilibrium equations

$$\begin{bmatrix} (DD^T)^{-1} & A^T \\ A & 0 \end{bmatrix} \begin{bmatrix} z \\ y \end{bmatrix} = \begin{bmatrix} c \\ 0 \end{bmatrix} \tag{17}$$

The primal variable is the projection z, giving the step direction. The dual variable is y. And formulas (15) and (16) for z and y come directly from these equilibrium equations.

3. *Starting values.* By adding the extra column $b - Ae$ to a given matrix A, the vector e of 1's is made feasible. The larger matrix is renamed A, so that $Ae = b$ and the algorithm can start at $x^0 = e$. When the new variable is given a high cost, it is driven toward zero and the solution of the enlarged problem is also optimal for the original problem. Thus Phase I goes simultaneously with Phase II, and uses one artificial variable where the simplex method used m. If the artificial variable is not driven to zero then the original feasible set was empty.

We come finally to several ingenious ideas put forward by Karmarkar. They go beyond the rescaling $X = D^{-1}x$, which preserves a linear cost function. Karmarkar intends to replace that rescaling by a nonlinear transformation

$$\bar{x} = \frac{nD^{-1}x}{e^T D^{-1}x} \quad \text{with} \quad e^T = \begin{bmatrix} 1 & 1 & \cdots & 1 \end{bmatrix}. \tag{18}$$

This is a projective transformation (not a projection) that has several effects. First, the product $e^T\bar{x}$ is always n. All points lie on the plane $\bar{x}_1 + \bar{x}_2 + \cdots + \bar{x}_n = n$. Whether x is large or small is made irrelevant by the quotient in (18). In particular, the vector x whose components match the diagonal of D is taken to the center point $\bar{x} = e$—since every component of $D^{-1}x$ is 1.

This transformation is suited to special problems and special cost functions, so Karmarkar changes both. He puts the constraints into the form

$$Ax = 0, \ e^T x = n, \ x \geq 0. \tag{19}$$

He also requires that $Ae = 0$, so the starting point $x^0 = e$ is feasible. The algorithm is more efficient when the unknown minimum cost is zero; therefore it modifies the problem as it goes forward. Furthermore, the cost $c^T x$ is replaced by the nonlinear and nonconvex function

$$f(x) = \sum_{1}^{n} \log\left(\frac{c^T x}{x_j}\right). \tag{20}$$

Since $f(x)$ differs from $f(\bar{x})$ by a constant term $\log \det D$, the projective transformation (18) preserves the form of this "potential function" f. Such functions were proposed earlier, but the conclusions of Karmarkar's analysis are new.

The algorithm is guaranteed to reduce f by a constant at every step. Then $c^T x$ goes to zero at exponential speed, which produces convergence in polynomial time (polynomial in the number of bits in the data). The original cost $c^T x$ does not always decrease; but when it increases, the other factor $\Sigma \log x_j$ in (20) must also increase and the algorithm gains altitude for later steps—by moving away from the coordinate planes $x_j = 0$. That might be an advantage over simple rescaling and projection, which decreases $c^T x$ but can suffer from the chronic disease of steepest descent (the bends). Near convergence, steepest descent can go back and forth in short and frustrating steps—across a valley in Fig. 5.6 and possibly between constraint planes in linear programming. Where $c^T x$ drops to zero, the potential function aims for $f(x^*) = -\infty$. It may come sooner to a point where the optimal corner x^* can be identified.

The algorithm starts from $x^0 = e$. At that center point, steepest descent for f has the same direction as the projection Pc computed earlier. This is still the main computational step. An extra row e^T is included with A because it enters the constraint (19); its corresponding diagonal entry in D is always 1. The movement Δx in the direction $-Pc$ can again go to within a factor $\alpha \approx 0.98$ of the boundary, or we can minimize the potential function f along that search direction. The new point $x^1 = x^0 + \Delta x$ gives the entries in the next scaling matrix D, and (18) returns that point to $\bar{x} = e$ for the next step. Each successive step uses a rescaled matrix and cost, and the computed $\bar{y}$ again represents the dual variable.

Todd and Burrell have improved the algorithm originally published in volume 4 of *Combinatorica* (1984), which should not be coded as is. Many others, and especially Karmarkar himself, are active in the analysis—of the projections P, of the starting method (an iterative equivalent of Phase I), and of the cost function and its convergence. That is entirely normal for a new algorithm. Whether linear programming will ever return to normal it is too early to say.

EXERCISES

8.2.1 Suppose the cost is $cx = x_1 + x_2 + x_3$ and the constraint is $x_1 + 2x_2 + 3x_3 = 12$ with $x \geq 0$. Start from the corner $x = (12, 0, 0)$ and take a simplex step:

(a) If x_2 goes up from zero with $x_3 = 0$ and $x_1 + 2x_2 = 12$, find the cost in terms of x_2. Is the cost decreasing as x_2 enters?

(b) If x_3 is increased from zero with $x_2 = 0$ and $x_1 + 3x_3 = 12$, find the cost in terms of x_3. What is the reduced cost r_3—the coefficient of x_3? Why should x_3 be the entering variable?

(c) The new corner is $(0, 0, 4)$ when x_3 enters and x_1 leaves. Show that increasing x_2 from zero with $x_1 = 0$ and $2x_2 + 3x_3 = 12$ increases the cost; the corner $(0, 0, 4)$ is optimal.

8.2.2 With cost vector $c = [0 \quad 7 \quad 9 \quad 0]$ and with constraints $x_1 + x_2 + x_3 = 6$, $x_2 + 2x_3 + x_4 = 1$, $x_i \geq 0$, explain directly why $x = (6, 0, 0, 1)$ is optimal.

8.2.3   Write down the 3 by 5 tableau T for the previous question. How do you see that the stopping test is passed?

8.2.4   For the same problem start the simplex method from the wrong corner $x = (5,1,0,0)$. Use elimination on T to reach a tableau T' in which columns 1 and 2 come directly from the identity matrix. What is r, the vector of reduced costs in the last row? What is the cost at the corner $(5,1,0,0)$? Should x_3 or x_4 be the entering variable?

8.2.5   Instead of the tableau split A and c into

$$A = \begin{bmatrix} 1 & 1 & \vdots & 1 & 0 \\ 0 & 1 & \vdots & 2 & 1 \end{bmatrix} = [B \ \ N] \quad \text{and} \quad c = [0 \ \ 7 \ \ \vdots \ \ 9 \ \ 0] = [c_B \ \ c_N].$$

Compute the reduced costs $r = c_N - c_B B^{-1} N$. Should x_3 or x_4 be the entering variable?

8.2.6   If x_4 enters the basis then x_1 or x_2 must leave. Keeping $x_3 = 0$ and $Ax = b$, show that $x_1 = 5 + x_4$ and $x_2 = 1 - x_4$. Which of these is first to reach zero as x_4 increases? What are x_1, x_2, x_3, x_4 at this new corner?

8.2.7   The leaving variable can also be decided by the ratios $(B^{-1}b)_j / v_j$ in (11). What is $B^{-1}b$, the nonzero part x_B of the current corner? What is $v = B^{-1}y$? The column $y = [0 \ \ 1]^T$ for the entering variable x_4 is the fourth column of A. What are the ratios, if we allow only those with positive v_j?

8.2.8   In equation (12) what is the inverse of

$$\begin{bmatrix} 1 & & & v_1 \\ & \ddots & & \vdots \\ & & 1 & \\ & & & v_m \end{bmatrix} ?$$

8.2.9   If we wanted to maximize instead of minimize the cost cx (still with $Ax = b$ and $x \geq 0$), what would be the stopping test on r and what rules would choose the entering variable x_i and the leaving variable x_j?

8.2.10   Apply Phase I of the simplex method to find a corner of the feasible set $x_1 - x_2 - x_3 = 6$, $x \geq 0$. With only one equation, a single new variable is invented and started at $x_4 = 6$. How many corners does the feasible set actually have?

8.2.11   Starting from $x = (6,0,0)$ to minimize $cx = x_2 - x_3$ subject to $x_1 - x_2 - x_3 = 6$ and $x \geq 0$, show that $r_3 < 0$ and the third variable should enter. But then show that x_1 never drops to zero as x_3 increases, and the minimum cost is $-\infty$. To be specific, what feasible x has cost below -1000?

8.2.12   In the product ABx, when both matrices are n by n, how many multiplications are needed to form Bx and then $A(Bx)$? How many multiplications enter the usual matrix product AB?

8.2.13   (a)   Why does ordinary multiplication of an m by n matrix times an n by p matrix lead to mnp individual multiplications?
  (b)   To compute ABC, where A is m by n, B is n by p, and C is p by q, how many

multiplications are used in the order $(AB)C$? How many are used in $A(BC)$, when B is multiplied first by C?

(c)   If B is square $(n = p)$ which order is better?

(d)   Which order is better in general?

8.2.14   In the example of Karmarkar's algorithm with $A = [1 \quad 1 \quad 1]$, what is the projection matrix P? Verify that it is positive semidefinite and that $AP = 0$.

8.2.15   With a calculator take the second step of the rescaling algorithm, starting from $x^1 = \frac{1}{3}(1.28, 1.70, 0.02)$. Compute the change in cost.

8.2.16   With the constraint $x_1 + 2x_2 = 1$ as well as $x_1 + x_2 + x_3 = 1$, what direction Δx does the rescaling algorithm take with cost vector $c = (5,4,8)$? What is the optimal solution?

8.2.17   If c is perpendicular to the nullspace of A, show that the step in Karmarkar's algorithm is $\Delta x = 0$ and any feasible solution is optimal.

8.2.18   Verify that the equilibrium equations (17) give the correct y and z.

8.2.19   With cost $cx = 8x_1 + 2x_2 + 7x_3$ and constraint $2x_1 + 2x_2 + x_3 = 5$, project $c = (8, 2, 7)$ onto the nullspace of $A = [2 \quad 2 \quad 1]$:

$$\text{solve } AA^T y = Ac \text{ and compute } Pc = c - A^T y.$$

Starting from $x^0 = (1, 1, 1)$, show that $x^0 - \frac{1}{4} Pc$ reaches the boundary point $(\frac{1}{2}, 2, 0)$. Which of the vertices $(\frac{5}{2}, 0, 0)$, $(0, \frac{5}{2}, 0)$, $(0, 0, 5)$ do you suspect gives minimum cost?

8.2.20   If Karmarkar's method stops in the previous exercise at $x^1 = (0.52, 1.96, 0.04)$, construct a diagonal matrix D from those numbers and solve

$$AD^2 A^T y = AD^2 c \quad \text{and} \quad PDc = Dc - DA^T y.$$

Show that $s = \frac{1}{3}$ brings the first component near zero, when $sPDc$ is subtracted from $(1, 1, 1)$. Multiplying by D, the second step ends near $(0.02, 2.46, 0.04)$.

8.3 ■ DUALITY IN LINEAR PROGRAMMING

The dual of a linear program is another linear program. It is a maximization, if the original problem (the primal) was a minimization. It uses the same matrix A and the same vectors b and c, but the feasible set is completely different; it is a set in m-dimensional space R^m rather than in R^n. The unknown vector y has m components. Certainly the dual problem could be solved by a direct application of the simplex method, like any other linear program. But it is much more special than that: When the simplex method solves the primal, it discovers at the same moment the solution to the dual.

The two problems stand in a unique relation to each other (and the dual to the dual brings back the primal). The question is whether to start by trying to explain the details of that relationship. It is beautiful but subtle. The alternative is to write down the dual immediately, without any special explanation or confusion, and then gradually establish its connection to the primal. I think that is the right thing to do, beginning with a primal problem that has equality constraints:

> (**Primal**)   *Minimize* cx *subject to* $Ax = b$ *and* $x \geq 0$.

That is the problem we already know. In the dual, the unknown will be a *row vector* $y = [y_1 \ \cdots \ y_m]$. It is subject to n constraints, and they are inequalities rather than equations:

> (**Dual**)   *Maximize* yb *subject to* $yA \leq c$.

The vector c has moved from the cost to the constraint. The right side of $Ax = b$ has gone into the cost. Thus b and c are reversed, and there is no requirement $y \geq 0$. Finally the matrix A has been transposed. That is hidden in $yA \leq c$, but it is revealed in $A^T y^T \leq c^T$ and in the tableau for the dual problem:

$$T_{\text{dual}} = \left[\begin{array}{c:c} A^T & c^T \\ \hdashline b^T & 0 \end{array}\right].$$

We use row vectors exactly to avoid all these transposes.

A simple example with $A = [1 \ \ 1]$ is

(Primal)   Minimize $5x_1 + 4x_2$ subject to $x_1 + x_2 = 1$, $x \geq 0$

(Dual)   Maximize y_1 subject to $y_1 \leq 5$, $y_1 \leq 4$.

The minimum is 4, attained at $x_1 = 0$, $x_2 = 1$. The maximum is also 4, attained at $y_1 = 4$. In this case the dual is easier (they are both pretty easy) because it has only one unknown.

You can already guess the main theorem of the whole subject: *The minimum in one problem equals the maximum in the other.* That is not at all obvious, but somehow the necessary y must exist. It is exactly like the max flow–min cut theorem of Chapter 7, or the largest matching–smallest cover theorem for marriages, or the equality between minimum transportation cost and maximum income to Federal Express. In fact those are special cases of this present duality theorem, which we now want to prove.

Remember that in those cases it was simple to verify "maximum ≤ minimum." The same is true here. This one-sided result, the easy half of every duality theorem, is still called *weak duality*:

8A If x and y are any feasible vectors in the minimum and maximum problems, then $yb \le cx$. Therefore the maximum of yb cannot exceed the minimum of cx. If $yb = cx$ then those vectors x and y are optimal.

For the proof we write down the feasibility conditions on x and y:

$$Ax = b, \ x \ge 0, \ yA \le c.$$

Multiply the first equation by y and the last by x:

$$yAx = yb \quad \text{and} \quad yAx \le cx. \tag{1}$$

The left sides are identical, so $yb \le cx$. The cost in the dual is never above the cost in the primal; that is all there is to weak duality.

If some choice of x and y gives $yb = cx$, then those vectors x and y must be optimal. No y could make yb larger than this number cx, and no x could give a cost below yb. The simple inequality $yb \le cx$, which holds for all feasible vectors, is remarkably powerful.

Notice how the constraint $x \ge 0$ played its part. Without it we could not know that the inequality sign in $yA \le c$ is still there in $yAx \le cx$. (If $x \le 0$ the sign would be reversed.) And notice also the consequences of $yb \le cx$ in case one of the problems is unbounded. *If the minimum equals $-\infty$ or the maximum equals $+\infty$, then the feasible set in the other problem must be empty.* Otherwise, if there were a feasible y, all possible costs cx would be above yb. The cost could not go down to $-\infty$. Similarly, a feasible x would keep every yb below cx; the maximum of yb could not be $+\infty$. This leaves three degenerate cases, and they are all possible:

> (1) Both feasible sets are empty
> (2) The minimum in the primal is $-\infty$ and there is no feasible y
> (3) The maximum in the dual is $+\infty$ and there is no feasible x.

Of course it leaves out the most important case, when the minimum and maximum are finite and we want to prove them equal.

It is easy to see what must happen if $yb = yAx$ is equal to cx. It does not follow that yA equals c—but something special has occurred. In the inner product of $yA \leq c$ with $x \geq 0$, the inequality signs have disappeared. In other words $(c - yA)x$ is zero, although both of the vectors in this inner product are nonnegative. That can happen in only one way: *x must be zero in every component where $c - yA$ is positive, and vice versa.* This condition ties together the solution x to the primal and the solution y to the dual. It is the "optimality condition," or the "complementary slackness condition," or the "Kuhn-Tucker condition."

8B   The feasible vectors x and y give duality if and only if $(c - yA)x = 0$. For each $j = 1, ..., n$, optimality requires

$$\textit{either} \quad x_j = 0 \quad \textit{or} \quad (yA)_j = c_j. \tag{2}$$

If this condition holds then $cx = yAx$, or $cx = yb$, which makes x and y optimal. If it fails then the inner product $(c - yA)x$ will be positive, and cx will be larger than $yAx = yb$. The duality theorem will say that this x and y are not optimal.

The test (2), by which we recognize an optimal x, involves y. That seems unusual; the primal is apparently not self-contained. We may have found the best x, but to *know* that we have found it requires something more—the solution to the dual. This is very different from an ordinary unconstrained minimization (Chapter 1) where the derivative was just set to zero. It is much closer to Chapter 2, where the constraint $A^T y = f$ led to Lagrange multipliers and a dual problem. The optimality condition (2) is illustrated by the example of the previous section:

(Primal)   *Minimize $cx = 3x_1 + x_2 + 9x_3 + x_4$ subject to $x \geq 0$ and*

$$\begin{bmatrix} 1 & 0 & 2 & 1 \\ 0 & 1 & 1 & -1 \end{bmatrix} \begin{bmatrix} x_1 \\ x_2 \\ x_3 \\ x_4 \end{bmatrix} = \begin{bmatrix} 4 \\ 2 \end{bmatrix}.$$

The minimum cost was $cx = 10$, at $x = (0,6,0,4)$. It was found by the simplex method.

(Dual)   *Maximize $yb = 4y_1 + 2y_2$ subject to*

$$[y_1 \quad y_2] \begin{bmatrix} 1 & 0 & 2 & 1 \\ 0 & 1 & 1 & -1 \end{bmatrix} \leq [3 \quad 1 \quad 9 \quad 1].$$

The maximum is $yb = 10$, at $y = [2 \quad 1]$.

To show that $y = [2 \quad 1]$ is feasible we check $yA \leq c$:

$$y_1 \leq 3, \; y_2 \leq 1, \; 2y_1 + y_2 \leq 9, \; y_1 - y_2 \leq 1.$$

In fact the first and third become strict inequalities, $2 < 3$ and $5 < 9$, and the second and fourth become equations: $c - yA$ is $[1 \quad 0 \quad 4 \quad 0]$. This is exactly the right pattern to complement $x = (0, 6, 0, 4)$. Both vectors are nonnegative and their inner product is zero. The optimality condition (2) is satisfied and $yb = 10 = cx$.

It remains to prove that $yb = cx$ will always occur, and to understand y.

8C   Duality theorem. If there is an optimal x in the primal problem then there is an optimal y in the dual, and **the minimum value of cx equals the maximum value of yb.** Otherwise one or both of the feasible sets must be empty, and the three degenerate possibilities were listed earlier.

The proof is short and constructive. It depends on the simplex method, which reaches the optimal x. At that point $n - m$ components of x are zero. The other m components form the nonzero part (call it x_B) and the corresponding m columns of A form a square submatrix B. To satisfy $Ax = b$ the nonzero part is $x_B = B^{-1}b$. The cost at this vertex, and the stopping test which it must pass, are

$$cx = c_B B^{-1}b \quad \text{and} \quad r = c_N - c_B B^{-1}N \geq 0. \tag{3}$$

Here N comes from the nonbasic $n - m$ columns of A—those not in B—and $c = [c_B \quad c_N]$ is the corresponding splitting of the cost vector.

To prove duality we need to recognize the optimal y. It is $y = c_B B^{-1}$. That choice certainly satisfies

$$yb = c_B B^{-1}b = cx.$$

Furthermore $yA \leq c$ follows from $r \geq 0$, so y is feasible:

$$yA = c_B B^{-1}[B \quad N] = [c_B \quad c_B B^{-1}N] \leq [c_B \quad c_N] = c.$$

Since x and y give the same value $yb = cx$, and both are feasible, they must be optimal. The maximum has met the minimum, and duality is proved.

In the example above, $y = [2 \quad 1]$ agrees with $c_B B^{-1}$:

$$[2 \quad 1] \begin{bmatrix} 0 & 1 \\ 1 & -1 \end{bmatrix} = [1 \quad 1] \quad \text{or} \quad yB = c_B.$$

The columns that were basic for x—the second and fourth columns of A—are also the key to y. The object of the simplex method was to pick out those two columns. Since they produce B and $y = c_B B^{-1}$, the simplex method has also solved the dual.

Remark 1   As it stands, the duality theorem begins with the primal and goes to the dual. But the dual of the dual problem is the primal. Therefore the theorem could be written in a completely symmetric way:

If either the primal or the dual has an optimal vector then so does the other, and $yb = $ maximum $ = $ minimum $ = cx$.

Of course the simplex method can be applied directly to the dual. It reaches the same answer by a different route, along the edges of the feasible set $yA \leq c$.

Remark 2   If the original problem has $Ax \geq b$—an inequality constraint in place of an equation $Ax = b$—then the dual has the extra condition $y \geq 0$. Both x and y are nonnegative, which contributes to the symmetry:

(P)	Minimize cx subject to $Ax \geq b$ and $x \geq 0$
(D)	Maximize yb subject to $yA \leq c$ and $y \geq 0$.

Again the minimum equals the maximum, because this pair can be reduced to the previous one. By introducing slack variables x_s into (P), with zero cost,

$$Ax - x_s = b \quad \text{or} \quad [A \quad -I] \begin{bmatrix} x \\ x_s \end{bmatrix} = b.$$

Then the dual uses this long matrix and its constraint becomes

$$y[A \quad -I] \leq [c \quad 0], \quad \text{or} \quad yA \leq c \quad \text{and} \quad y \geq 0.$$

This agrees with (D), which is therefore the dual to (P).

Interpretation of Duality

So far the theory has been very formal. Given a linear program, its dual was produced out of thin air. It is true that the connections between the two were extremely close, much too close to be an accident. But there was no hint to suggest where the dual came from, and no way to give it a reasonable interpretation.

For that we go back to one of the most basic optimization problems, the shortest path through a network. It was certainly not written as a linear program, so we need to show first that it could have been. The original question was simple:

For the network in Fig. 8.5 what is the shortest path from source to sink?

We can think of the nodes s, t, u, v as computers; lengths are the costs of sending a message between them. Then the problem becomes:

What is the cheapest way to send a message from s to t?

The fractions of the message sent from s to v and from u to t should be zero, $x_{sv} = x_{ut} = 0$. The fractions sent along the other three edges should be one:

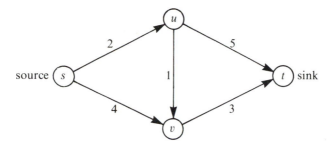

Fig. 8.5. Edge lengths = costs.

$x_{su} = x_{uv} = x_{vt} = 1$. The whole message goes along this shortest path. The minimal cost (or the shortest distance from s to t) is $2 + 1 + 3 = 6$. In this language we see the underlying linear program:

Minimize the cost $cx = 2x_{su} + 4x_{sv} + x_{uv} + 5x_{ut} + 3x_{vt}$ *subject to* $x \geq 0$ *and*

$$x_{su} - x_{uv} - x_{ut} = 0 \qquad \text{(message not lost at } u)$$

$$x_{sv} + x_{uv} - x_{vt} = 0 \qquad \text{(message not lost at } v)$$

$$x_{ut} + x_{vt} = 1 \qquad \text{(message reaches } t)$$

Adding these equations gives $x_{su} + x_{sv} = 1$; the message goes out from s. The equations are completely analogous to Kirchhoff's current law; it is *conservation of message*. The coefficient matrix is like the incidence matrices of Chapters 1 and 2, if we write out the three constraints:

$$Ax = \begin{bmatrix} 1 & 0 & -1 & -1 & 0 \\ 0 & 1 & 1 & 0 & -1 \\ 0 & 0 & 0 & 1 & 1 \end{bmatrix} \begin{bmatrix} x_{su} \\ x_{sv} \\ x_{uv} \\ x_{ut} \\ x_{vt} \end{bmatrix} = \begin{bmatrix} 0 \\ 0 \\ 1 \end{bmatrix} = b.$$

There is a column for each edge and a row for each node—except that the row $[1 \ 1 \ 0 \ 0 \ 0]$ for the source node was not included. It is the sum of the other rows, and the source is the "grounded node."

The equations are totally familiar, with one difference. In the early chapters the constraint would have been $A^T y = f$. **The notation has been reversed.** To match Chapter 2, where the current was y, with Chapter 7, where the flows and shipments were x, you need the exchanges

$$\boxed{x \leftrightarrow y, \ b \leftrightarrow f, \ A \leftrightarrow A^T.} \tag{4}$$

The vector x now contains the *edge* variables. It gives the flow. The vector b has become the current source, sending one message from s to t. And the node-edge incidence matrix A^T in Kirchhoff's current law is now A. If we complete the framework of Chapter 2, but reverse the notation, it looks like Fig. 8.6.

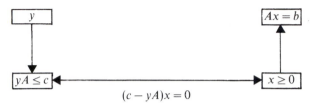

Fig. 8.6. The framework reversed for linear programming.

The right side of the figure is clear. It is the left side, involving the dual, that is still unexplained. The dual variable is y, and it is called a "price" instead of a potential—but only because linear programs are more likely to come from economics than electronics. The price drives the flow, and what is amazing is the parallel to Chapter 2.

The relation between primal and dual is really a battle. We can describe it in words. In the primal, the cost on each edge is set by the telephone company. In the dual, the prices are set by a competitor (we will call him *Sprint*). He does not reveal direct information about the edges; in fact the customer has no idea how the message is sent. It leaves one node and turns up somehow at the next. What Sprint does announce is the price of the message at each node; he chooses y_s, y_u, y_v, and y_t. He will buy it at this price at any node (for example the source) and sell it back at any other node (for example the sink). The difference $y_t - y_s$ is his income for carrying the message. ***Sprint maximizes that income***. Since he is working only with price differences, he might as well fix the source price at $y_s = 0$. (The source is grounded.) Then the problem is to maximize the price y_t at the sink.

To stay competitive, Sprint cannot charge more than AT&T on any edge. Unless the prices at the nodes satisfy

$$y_u - y_s \le 2$$
$$y_v - y_s \le 4$$
$$y_v - y_u \le 1 \qquad (5)$$
$$y_t - y_u \le 5$$
$$y_t - y_v \le 3$$

across the five edges, the customer will go partly or completely to AT&T. With $y_s = 0$, that is exactly the dual problem: *Maximize y_t subject to $yA \le c$, or*

$$[y_u \quad y_v \quad y_t] \begin{bmatrix} 1 & 0 & -1 & -1 & 0 \\ 0 & 1 & 1 & 0 & -1 \\ 0 & 0 & 0 & 1 & 1 \end{bmatrix} \le [2 \quad 4 \quad 1 \quad 5 \quad 3].$$

Sprint maximizes, the customer minimizes, and the result is a deadlock. The lowest cost on AT&T matches the highest income available to Sprint. That is duality, and we check the details:

Basic columns of A [submatrix B]: The first, third, and fifth columns of A correspond to edges on the optimal path $s - u - v - t$. These columns enter

$$B = \begin{bmatrix} 1 & -1 & 0 \\ 0 & 1 & -1 \\ 0 & 0 & 1 \end{bmatrix},$$ (6)

which is the basis at the end of the simplex method. (It may have started with three other columns.) The optimal y and the nonzero part $x_B = (x_1, x_3, x_5)$ of the optimal x satisfy

$$Bx_B = b, \text{ or } \begin{bmatrix} 1 & -1 & 0 \\ 0 & 1 & -1 \\ 0 & 0 & 1 \end{bmatrix} \begin{bmatrix} x_1 \\ x_3 \\ x_5 \end{bmatrix} = \begin{bmatrix} 0 \\ 0 \\ 1 \end{bmatrix}, \text{ or } \begin{bmatrix} x_1 \\ x_3 \\ x_5 \end{bmatrix} = \begin{bmatrix} 1 \\ 1 \\ 1 \end{bmatrix} \quad \begin{array}{l} \text{(1 message} \\ \text{is sent on} \\ \text{edges 1,3,5)} \end{array}$$

$$yB = c_B, \text{ or } \begin{bmatrix} y_u & y_v & y_t \end{bmatrix} \begin{bmatrix} 1 & -1 & 0 \\ 0 & 1 & -1 \\ 0 & 0 & 1 \end{bmatrix} = \begin{bmatrix} 2 & 1 & 3 \end{bmatrix}, \text{ or } y = \begin{bmatrix} 2 & 3 & 6 \end{bmatrix}.$$

The prices set by Sprint are $y = 2$ at u, $y = 3$ at v, $y = 6$ at t. To buy the message back at the sink costs 6.

Duality [$yb = cx$]: The minimum cost on AT&T, which is the shortest distance $2 + 1 + 3$, equals the maximum income to Sprint: $cx = yb$. The shortest path has length 6, and the highest price Sprint can charge is 6.

Optimality [$(c - yA)x = 0$]: The edges on the shortest path have $x_j > 0$. On those edges Sprint's price difference in (5) equals AT&T's charge: $(yA)_j = c_j$. The inner product $(c - yA)x$ is zero, which again guarantees duality.

   You may have only a very moderate interest in this optimality condition $(c - yA)x = 0$, which is the same as $cx = yb$. I thought it was just one more expression of duality, until I realized that it is the new form of Ohm's law. Of course the elements in the network are not ordinary resistors, with Ohm's law $V = IR$, or we would be back to the quadratics of Chapter 2. Instead the elements are **ideal diodes**. A diode provides no resistance in one direction and infinite resistance in the other. It reduces to a short circuit for positive current and to an open circuit for negative voltage. The figure shows a graph of the voltage-current curve and also the standard symbol for a diode.

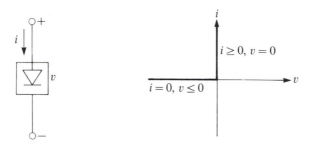

Fig. 8.7. The voltage-current rule $i \geq 0$, $v \leq 0$, $iv = 0$ for a diode.

Remark   Diodes can remove the backward half of every sine curve in alternating current. That is useful in rectifiers. A second diode can put back that half with the sign reversed. This "full-wave rectifier" leaves a current like $|\sin t|$, always in the same direction and sending direct current to a power supply. We mention that real diodes, made of semiconducting materials and doped by impurities, will not give exactly the $v - i$ characteristics in Fig. 8.7. The infinite resistance breaks down when v is sufficiently negative. Prior to breakdown diodes can be nearly ideal, and their development led to the most important active device in electronic circuits—the *transistor*. It has three terminals instead of two, and a weak signal at one source controls the flow from its second and more powerful source. The weak signal is amplified—the transistor controls an op-amp—and part of its behavior imitates a diode.

Of course the diode is nonlinear. The curve in Fig. 8.7 is only *piecewise* linear, with two pieces. An ordinary resistor has voltage drop proportional to current. Instead of this linearity, the diode resistance jumps between $R = 0$ and $R = \infty$. The voltage difference is $c - yA = 0$ whenever $x \neq 0$, and the current is $x = 0$ whenever $c - yA \neq 0$. Therefore $(c - yA)x = 0$, and this law completes the pattern for a linear program (Fig. 8.6). The *reduced costs* $c - yA$ are completely analogous to the *voltage drops* $b - Ax$. It is only Ohm's law that has changed, and the notation.†

We close with three comments about the shortest path problem:

(a)   The price y at each node is the length of the shortest path from the source. This gives a special way to solve the dual, quite different from the simplex method. We can price one node at a time, starting with $y_s = 0$ at the source and finding the next nearest node at every step. That is exactly Dijkstra's algorithm from Section 7.1.

(b)   The simplex method is a little surprising. A corner of the feasible set is a flow pattern x, satisfying Kirchhoff's current law $Ax = b$ with at most 3 nonzeros. In other words, the flow uses at most 3 edges. Suppose we try the edges of length 2, 4, 5 in Fig. 8.5 as a starting point (initial corner) for the simplex method. You see that the path does not actually use all three; the cost is $2 + 5$ and the flow from s to v is

---

† Actually the diode should be put in backwards, since we want voltage drops $c - yA \geq 0$ and it has $v \leq 0$ in Fig. 8.7.

zero. In other words, the corner is *degenerate*: it doesn't need three nonzeros. The first simplex step removes the unused edge of length 4 in favor of the edge of length 1, but the new path will not use this edge either. The cost is still $2 + 5$; it is *not improved* by this step and the new corner is still degenerate. Nevertheless we continue. The next step brings in the edge of length 3, throws out the edge of length 5, finds the minimal cost $2 + 1 + 3$.

Examples like this are extremely common in applications—with degeneracy of corners, fewer than m nonzeros, and success of the method.

(c) If combinatorial problems turn into linear programs, as the shortest path problem just did, there must be an explanation for the fact that the solutions are always integers. A linear program could easily lead to fractions, even if A, b, c contain integers. That is already true for a linear system $Ax = b$. Somehow it doesn't happen in combinatorics.

The answer lies in the properties of incidence matrices A and their submatrices B. The optimal x and y come from linear equations $yB = c_B$ and $Bx_B = b$. If the determinant of B is 1 or -1 then by Cramer's rule (much maligned in Section 1.1) the solutions x and y are integral whenever the right sides are. The determinant is in the denominator, and if it is ± 1 there are no fractions. Therefore we prove:

> All nonsingular submatrices of an incidence matrix have det $B = 1$ or -1.

This is the "totally unimodular" property of incidence matrices. All entries are 0, except for at most a single $+1$ and a single -1 in each column.† If every column has both, the columns add to zero and B is singular. If a column is all zero, B is again singular. If any column has *one* nonzero, either $+1$ or -1, the determinant is found from a smaller matrix (in which the row and column of that one nonzero are removed). This reduction eventually leads to a single number, either 1 or -1, as in the matrix displayed in (6) above. The first row and column are removed, then the second, and finally det $B = 1$.

Every combinatorial problem in Chapter 7 can be rewritten as a linear program—and the matrix A is an incidence matrix, or almost. In maximum flow it comes from Kirchhoff's current law (plus capacity constraints $x_i \leq c_i$ that do not spoil det $B = \pm 1$). In the transportation problem, the right sign changes will produce A. Therefore these linear programs are automatically integer programs. We can look for the optimal x_i among real numbers, or fractions, or integers—and it will always be an integer.

EXERCISES

8.3.1 In the AT&T-Sprint example in the text, show that there is equality in $yA \leq c$ (equation (5)) exactly on the edges that carry the message.

---

† They were rows in Chapter 2.

8.3.2   For any network the lengths of the shortest paths from the source give optimal prices for Sprint. Show that these prices satisfy "price difference ≤ edge length" on each edge, so they are feasible in the dual problem.

8.3.3   Number all nodes on the network of Fig. 7.3 and find their shortest distances from s (those are optimal prices for Sprint). On which edges does price difference = edge length?

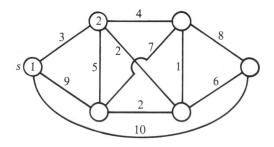

8.3.4   (i)   If a new edge is added to Fig. 8.5 with very negative cost (-6 from v to u) show that the minimum distance from source to sink is now $-\infty$. There is not just a path with zero cost; you can describe a path with cost below -1000.
   (ii)   If the simplex method starts at the current optimum of length $2 + 1 + 3$, it will introduce the new edge and allow $x_6 = x_{vu}$ to grow. What are the three equations $Ax = b$ in six unknowns, and what happens to the nonzero x_1, x_3, x_5 as x_6 is increased?

8.3.5   The dual problems in Remark 2 were

   (P)   Minimize cx subject to $Ax \geq b$ and $x \geq 0$

   (D)   Maximize yb subject to $yA \leq c$ and $y \geq 0$.

   (a)   Prove that always $yb \leq cx$ (weak duality). How does the new condition $y \geq 0$ enter the proof?
   (b)   Find the minimizing x and maximizing y without the simplex method, if

$$A = [1 \quad 3 \quad 4], \quad b = [12], \quad c = [2 \quad 4 \quad 5].$$

8.3.6   Construct a 1 by 1 example in which the feasible set $Ax \geq b$, $x \geq 0$ is empty and the dual problem is unbounded.

8.3.7   Construct a 1 by 1 example in which both feasible sets are empty: no x or y can satisfy $Ax \geq b$, $x \geq 0$, or $yA \leq c$, $y \geq 0$.

8.3.8   Write down the dual of the following problem: Maximize $x_1 + x_2 + x_3$ subject to $2x_1 + x_2 \leq 4$, $x_3 \leq 6$. What are the solutions to this problem and the dual?

8.3.9   Write as a linear program the transportation example of Section 7.5, with supplies 4 and 7, demands 2, 3, and 6, and cost matrix

$$C = \begin{bmatrix} 5 & 3 & 1 \\ 8 & 4 & 3 \end{bmatrix}.$$

What are A, b, and c, and what is the dual linear program?

8.3.10 Suppose A is the identity matrix and the vectors b and c are positive. Explain why $x = b$ is optimal in the minimum problem with constraints $Ax \geq b$, $x \geq 0$, and $y = c$ is optimal in the dual.

8.3.11 Show that the vectors $x = (1,1,1,0)$ and $y = (1,1,0,1)$ are feasible in the standard dual problems, with

$$A = \begin{bmatrix} 0 & 0 & 1 & 0 \\ 0 & 1 & 0 & 0 \\ 1 & 1 & 1 & 1 \\ 1 & 0 & 0 & 1 \end{bmatrix}, \quad b = \begin{bmatrix} 1 \\ 1 \\ 1 \\ 1 \end{bmatrix}, \quad c = \begin{bmatrix} 1 \\ 1 \\ 1 \\ 3 \end{bmatrix}.$$

Then, after computing cx and yb, explain how you know they are optimal.

8.3.12 Verify that the vectors in the previous exercise satisfy the complementary slackness conditions (2), and find the one slack inequality in both the primal and the dual.

8.3.13 The edges in the following network allow flow in only one direction, with the given capacities:

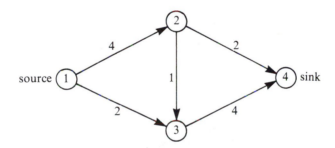

Write the maximal flow problem as a linear program, with 5 flow variables and 5 slack variables for the capacities. What is the solution?

8.3.14 Write down the dual using A, b, and c of the previous exercise, and find an optimal y. It gives a "price" at each node. Connect the price differences to the minimum cut in the network.

8.3.15 If $L(x,y)$ is any function of x and y, and $\bar{x}$ and $\bar{y}$ are any particular points, explain why $\min_x L(x,\bar{y}) \leq \max_y L(\bar{x},y)$. The inequality remains true if the left side is maximized over $\bar{y}$ and the right side is minimized over $\bar{x}$. It becomes maximin $\leq$ minimax—which is weak duality.

8.4 ■ SADDLE POINTS (MINIMAX) AND GAME THEORY

I begin this new section thinking about the example in the last one. It used shortest paths to illustrate duality—the maximum income to Sprint matched the minimum to the telephone company. It also did more. The shortest path problem went into the same framework that dominated the first half of the book. The new variables were prices instead of potentials, but that is not significant. The problem came from a network and the matrices A and A^T were incidence matrices. The resistors changed to diodes and Ohm's law was nonlinear, but the basic pattern of applied mathematics was still there. I would like to see if that pattern continues.

If it does continue there must be a reason. Therefore we return to the original problem, which minimized cost subject to constraints. In Chapter 2 the cost $Q(y)$ was a quadratic, and the constraints were $A^T y = f$. Now the cost cx is linear, but the constraints include inequalities; we have $x \geq 0$ as well as $Ax = b$ (or $Ax \geq b$). In the quadratic case it was Lagrange multipliers that built the constraints into the function L and turned out to be the dual variables. The same should be true in linear programming, except that inequalities in the constraints will lead to inequalities on the multipliers.

To see how that happens, we take the first step. The dual variables multiply $Ax - b$ and turn the minimum into a minimax:

$$\min_{\substack{Ax=b \\ x \geq 0}} cx = \min_{x \geq 0} \max_{y} \left[cx - y(Ax - b) \right]. \tag{1}$$

Here y is the vector of Lagrange multipliers, enforcing the constraint $Ax = b$. If that constraint fails, the maximum over y will be $+\infty$ (choose $-y$ to be a large multiple of $Ax - b$). This value $+\infty$ will certainly not be the minimum. Therefore the minimum occurs at a point where $Ax = b$, and we are back to the problem on the left side of equation (1).

Suppose we can interchange *max* and *min*, to reach

$$\max_{y} \min_{x \geq 0} \left[(c - yA)x + yb \right]. \tag{2}$$

The minimization comes first, admitting all vectors $x \geq 0$. Now the undesirable answer is $-\infty$; if this is the minimum it is useless when we maximize. We do get that undesirable minimum unless $c - yA \geq 0$. (If any component of $c - yA$ is negative, take the corresponding component of x to be extremely positive and the product approaches $-\infty$.) Therefore the only good vectors y are those with $c - yA \geq 0$, in which case the best x is one with $(c - yA)x = 0$. The minimum is the untouched term yb, and we reach the maximization

$$\boxed{\max_{c - yA \geq 0} yb.} \tag{3}$$

You recognize that as the dual problem. In three lines, Lagrange arrives at the right answer. He also finds the optimality test $(c - yA)x = 0$.

8D In linear programming as elsewhere, duality is equivalent to finding a *saddle point*. The pair of vectors (x, y) maximizes L with respect to y and minimizes with respect to x. A saddle point solves both the primal and the dual, and it gives a "minimax theorem"

$$\min_x \max_y L(x, y) = \max_y \min_x L(x, y).$$

The Lagrangian in linear programming is $L = cx - yAx + yb$.

Inequality Constraints and Competitive Equilibrium

If the original problem had inequalities $Ax \geq b$ instead of the equation $Ax = b$, there is a sign constraint on y:

$$\min_{\substack{Ax \geq b \\ x \geq 0}} cx = \min_{x \geq 0} \max_{y \geq 0} [cx - y(Ax - b)]. \tag{4}$$

Every inequality on x leads to an inequality on y. The maximum is now $+\infty$ when $Ax \geq b$ fails; previously it was $+\infty$ when $Ax = b$ failed. (That is the effect of $y \geq 0$.) Duality hinges again on interchanging *max* and *min*, so that (4) is equal to

$$\max_{y \geq 0} \min_{x \geq 0} [(c - yA)x + yb] = \max_{\substack{yA \leq c \\ y \geq 0}} yb. \tag{5}$$

This is the correct dual problem: *Maximize yb subject to $y \geq 0$ and $yA \leq c$.* But now there are two optimality conditions or complementary slackness conditions instead of one. In the minimization (5) we need $(c - yA)x = 0$ as before. In the maximization (4) we need $y(Ax - b) = 0$—which was automatic and therefore unnoticed for the equality constraint $Ax = b$.

This produces a startling change in the basic framework (Fig. 8.8), where a fourth connection suddenly appears—in addition to the exchanges of y for x, c for b, and A^T for A.

The difference is in Kirchhoff's current law. In the past, conservation of current gave $Ax = b$. Now there is only an inequality $Ax \geq b$; a one-way ground is attached to every node, drawing off whatever current is needed to recover Kirchhoff's law. It is *one-way*, because it never supplies current; we cannot have a negative component of $Ax - b$. It is a *ground*, because it acts only when the potential is $y_j = 0$. If $y_j > 0$ then the node is above ground and Kirchhoff's law goes back to $(Ax - b)_j = 0$. In every case $y_j(Ax - b)_j = 0$, and this one-way ground (like an ideal diode) uses no power. In mechanical terms it is a supporting wall, which pushes back if you lean

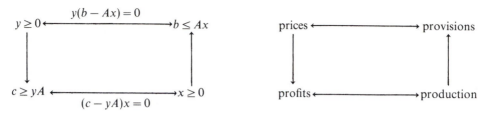

Fig. 8.8. The framework with inequality constraints; competitive equilibrium.

on it but not if you don't. The wall never pulls, and it pushes only when the displacement is zero.†

You see how networks lead to linear programs. Nevertheless, I cannot leave the impression that the great applications of linear programming come from diodes. That is not true. Networks provide a beautiful model when A happens to be an incidence matrix. But the framework in Fig. 8.8 applies to *any* matrix A, and the subject of linear programming is more often associated with operations research and management.

We can describe one central theme; you will recognize the application. It starts with supplies $b_1, \ldots, b_m$ of m raw materials (including labor and capital). It is based on a **production model** in economics. An amount a_{ij} of material i is consumed in manufacturing product j. The supplies are not exceeded if $Ax \le b$; x gives the amounts produced. If the value of product j is c_j, the firm will choose production levels x_j so as to maximize the total value of their products:

$$\text{Maximize } cx = c_1 x_1 + \cdots + c_n x_n \text{ subject to } Ax \le b, x \ge 0.$$

This is the standard maximum problem; the simplex method will find the solution.

Economic theory goes one step further, with far-reaching results. It asks about the *prices* of the raw materials. If they are set at $y_1, \ldots, y_m$, it is easy to compute the profit to the firm. The income is cx, as above, and the expenses are yAx, the cost of the supplies. Therefore the profit, income minus expense, is

$$cx - yAx = (c - yA)x.$$

What do the *suppliers* do? If this profit is positive they increase their prices. They can get away with it, since the producer will keep going as long as he makes a profit. At some point the prices would be too high and the producer would be out of business; the supplier doesn't want that. An equilibrium occurs just at the point where the profit is zero! The producer stays alive, hoping for a better day, but his activity level is zero for any product whose cost exceeds its value:

$$\boxed{\text{if } (c - yA)_j < 0 \text{ then } x_j = 0.} \tag{6}$$

---

† The mechanical analogue of a diode might be a rigid cable. It has no resistance to compression and infinite resistance to tension.

There is also another condition, if the economy is really competitive. Whenever supply exceeds demand, the price of a material will drop. If the material is not fully used at the lower price, then that price drops further. At equilibrium the price will actually reach zero, if supply still exceeds demand:

$$\text{if } (Ax - b)_j < 0 \text{ then } y_j = 0. \tag{7}$$

In this case the material becomes a "free good."

You see that the final prices are exactly the solution to the dual problem. The equilibrium rules (6) and (7) are the slackness conditions on prices $y \geq 0$ and production levels $x \geq 0$. We interpret them as follows:

8E In economic competition there is equilibrium when

(1) The producer makes zero profit: $(c - yA)x = 0$
(2) Leftover supplies are worthless: $y(b - Ax) = 0$.

Those are the optimality conditions when the producer maximizes cx subject to $Ax \leq b$, $x \geq 0$, and the supplier maximizes yb subject to $yA \geq c$, $y \geq 0$.

This model is fundamental to the theory of *competitive equilibrium*. That lies at the heart of mathematical economics, and it looks not at one firm but at the whole economy. Everybody is producing, consuming, supplying, and surviving (at zero profit). The prices of finished products and raw materials are set by the economy itself. There may be whole firms whose optimal activity levels are zero, so they close up shop. Similarly, there may be suppliers whose goods turn out to be worthless (that is, free). They also have troubles. The model need not be linear; there are economies of scale available to a big company and nonlinear programming will enter. But the conclusion from the mathematical model is quite remarkable:

> At a competitive equilibrium all firms behave as if they belonged to a single giant company (the whole economy) which maximizes the value of its output.

This might be the best argument (if it is true) for free enterprise: small-scale competition is automatically converted into large-scale cooperation. The firms that cooperate by going out of business may think there is a flaw in the model.

Sensitivity Analysis

Up to now the dual solution has not contributed directly to the primal problem. It was interesting in itself, it appeared at the same time as the primal solution, and it had a natural interpretation in terms of "shadow prices." We want to show that it also answers an important question about the original problem.

That question is: How much does the minimum cost change if the constraints are changed? This is the problem of *marginal cost*, the sensitivity of cx to a change in b.

The marginal cost is crucial to any decision taken on the basis of linear programming. It is seldom that an economist or an executive can replace his whole model, even if he doesn't like the answer it gives. The most he can change is b or A or c, to make cost go down or the profit go up. Geometrically, he can move the feasible set a little (by changing b or A), and he can tilt the planes that come to meet it (by changing c). Provided those movements are small, the planes will touch the feasible set first at the same (but slightly shifted) corner.

In other words, the corner which was optimal remains optimal. The same m columns of A go into the basis matrix B, and the nonzero components of x remain nonzero. This applies only to small changes; if the feasible set is turned significantly, or the planes are rotated because of a large change in c, then the solution can jump to a new corner. It will be possible to say when such a jump can occur.† But the key information is in the reaction to small changes.

8F A change Δb in the constraints produces a change $y\Delta b$ in the minimum cost; the dual solution y gives the *derivative of the cost with respect to b*. Similarly a shift from c to $c + \Delta c$ alters the minimum cost by $(\Delta c)x$.

In the first case, when b is changed, the optimal prices $y = c_B B^{-1}$ do not move. They do not depend on b. The old cost was yb—by duality this equals cx at the saddle point— and the new cost is $y(b + \Delta b)$. The change in cost is therefore the difference $y\Delta b$. It is exactly proportional to Δb—until a large change shifts the solution to a new corner and a new choice of B. The ratios $\Delta\,\text{cost}/\Delta b$, or the derivatives $d\,\text{cost}/db$, are the prices y. This is the final interpretation of the Lagrange multipliers—y gives *the change in solution divided by the change in data*.

When c changes, it will be x that stays the same. The feasible set does not move. Therefore the cost goes to $(c + \Delta c)x$. And if the entries of A are varied, altering B^{-1} by a small matrix ΔB^{-1}, we write the cost as $c_B B^{-1}b$ and recognize its change as $c_B \Delta B^{-1}b$.

Of course a jump to a new corner must occur whenever $x \geq 0$ or $r \geq 0$ breaks down at the old one.

Game Theory

The competition between primal and dual is clearest of all in game theory. I totally admit that it is hard to imagine edges competing with nodes, or currents with potentials, or flows with cuts. In a two-person game, the battle is out in the open. Player X has n different strategies and player Y has m. The payoffs for the mn possible combinations are entered in a matrix A, with positive payoffs going from Y to X; a negative a_{ij} is a payment to Y. Both players know the matrix but not the

---

†and even to follow the jump. There we would leave sensitivity analysis and begin parametric programming.

opponent's choice of strategy. The problem is to find the best overall plan for each player.

An example from football will clarify the rules. Suppose X is the offensive coach, choosing between a run and a pass. Y is the defensive coach trying to outguess X.[†] Suppose the defense concentrates against a run. Then if X does run, he gains only 2 yards; if he passes he gains 8. Those numbers go into the first row of A, corresponding to the first strategy of Y. The second strategy is to concentrate against a pass. In this case X gains 6 yards if he runs, and loses 6 if he passes. Therefore the payoff matrix (the same on every play) is

$$X$$

$$A = \begin{bmatrix} 2 & 8 \\ 6 & -6 \end{bmatrix} Y \qquad \begin{matrix} \text{defense against run} \\ \text{defense against pass} \end{matrix}$$

$$\text{run pass}$$

You see that no single strategy is perfect. If X does the same thing every time, Y defends against that and does well. (It is true that X can gain 2 on every play, but that is not enough to keep the ball.) Similarly if Y chooses always the same defense, X will do the opposite and win. Therefore both players must use a *mixed strategy*. The choice at every play must be absolutely independent of the previous plays, since any pattern can be recognized by the opponent. The game will be played over and over again, so surprise is not significant. The best thing a coach can do is just to flip a coin—but it should be a biased coin. The offense should run with probability x_1 and pass with the remaining probability $x_2 = 1 - x_1$, but the right probabilities are not $x_1 = x_2 = \frac{1}{2}$. If Y defends against a run, the expected gain from the first row of A is

$$2x_1 + 8x_2, \text{ or } 2x_1 + 8(1 - x_1) = 8 - 6x_1. \tag{8}$$

This is the average result, with payoffs 2 and 8 weighted by their probabilities. Next we find the expected gain if Y defends against a pass (the second row of A):

$$6x_1 - 6x_2, \text{ or } 6x_1 - 6(1 - x_1) = -6 + 12x_1. \tag{9}$$

Y will certainly choose the smaller of (8) and (9). X should make that smaller gain as big as possible; he *maximizes the minimum*. In this case $8 - 6x_1$ goes down as $-6 + 12x_1$ goes up, and the best strategy is to make them equal:

$$8 - 6x_1 = -6 + 12x_1, \quad \text{or} \quad 14 = 18x_1, \text{ or } x_1 = \tfrac{7}{9}.$$

[†] If you prefer linear algebra to football, the game is like reviewing for a test. You are Y, defending against proofs or computations. The professor is X.

The offense should run 7/9 of the time. If it does, then regardless of the defense the average gain will be

$$8 - 6(\tfrac{7}{9}) = \tfrac{72}{9} - \tfrac{42}{9} = \tfrac{30}{9} = \tfrac{10}{3}.$$

On the average X will gain 10 yards in 3 plays.

What about Y? There is nothing he can do if X follows this mixed strategy. But still Y must do the right thing, or X will change to take advantage of him. Therefore Y chooses his rows with probabilities y_1 and $y_2 = 1 - y_1$, yielding on the average $2y_1 + 6(1 - y_1)$ to a run and $8y_1 - 6(1 - y_1)$ to a pass. The best value of y_1 will *minimize the maximum* of these two gains, and in this case it makes them equal:

$$2y_1 + 6(1 - y_1) = 8y_1 - 6(1 - y_1).$$

That happens at $y_1 = \tfrac{2}{3}$. This strategy not only gives equality, it brings back the same value 10/3 that was found earlier by X:

$$2(\tfrac{2}{3}) + 6(\tfrac{1}{3}) = 8(\tfrac{2}{3}) - 6(\tfrac{1}{3}) = \tfrac{10}{3}.$$

Y should defend against a run 2/3 of the time, which is less frequently than X actually runs. With this defensive strategy by Y, X cannot gain more than 10/3.

We have found a **saddle point**. It is a pair of strategies, one for X and one for Y, with the following property: **Neither player can do better by making a change**. The saddle point is a maximum for X against the strategy of Y, and a minimum for Y against the strategy of X. Therefore the game is "solved," and its value is 10/3.†

The fundamental result of game theory—the *minimax theorem*—is that a saddle point always exists. This is the theorem discovered by John von Neumann, who at mid-century was the greatest mathematician in the world. (Probably Hilbert was the greatest at the turn of the century, and Gauss the greatest ever.) The proof is not as easy as the example, and the original reasoning was complicated. But if we describe a game in the right way, you will see the connection with linear programming. That gives a quick proof of the minimax theorem, and we can go on to a new game.

The Minimax Theorem

Player X is free to decide on any mixed strategy $x = (x_1, \ldots, x_n)$. The probabilities x_j are nonnegative, and they add to one. As in the example, they are the frequencies of his n pure strategies (the n columns of A). At each turn of the game he calls on a random device, constructed to pick column j with probability x_j. Then if

---

† You appreciate the brilliant realism of this number.

Y plays row 1, the payoff a_{1j} occurs with probability x_j, and the average payoff is

$$a_{11}x_1 + a_{12}x_2 + \cdots + a_{1n}x_n = \text{first component of } Ax.$$

If Y plays row i, the average payoff is the ith component of Ax. Y will concentrate his strategy on the row that gives the smallest component of Ax. Therefore the task of X is to make this smallest component (call it v) as large as possible. If b is the column vector of m 1's, his problem is:

Choose x to maximize v, subject to $Ax \geq vb$, $x_j \geq 0$, $x_1 + \cdots + x_n = 1$.

X wants to keep Ax above $(v, v, \ldots, v)$, with v as large as possible.

The task of Y is similar but opposite. He chooses frequencies $y = (y_1, \ldots, y_m)$, also with $y_i \geq 0$ and $\Sigma y_i = 1$, for the *rows*. Against column j he loses a_{ij} with probability y_i, and on the average he loses $y_1 a_{1j} + y_2 a_{2j} + \cdots + y_m a_{mj}$. His loss is the jth component of yA. X will concentrate on the column that gives the largest component of yA (call it w). Therefore Y wants to make w as small as possible. If c is the row vector of n 1's, the goal of Y is:

Choose y to minimize w, subject to $yA \leq wc$, $y_i \geq 0$, $y_1 + \cdots + y_m = 1$.

Y wants to keep yA below $(w, w, \ldots, w)$, with the smallest possible w.

Notice that $y_1 + \cdots + y_m = 1$ is exactly $yb = 1$. In the problem facing X the equation was $xc = 1$. The same b and c appear in both problems; they must be near to duality. It helps to suppose that all payoffs are positive, by adding the same large number N to every a_{ij}. This has no effect on the choice of strategy—Y just pays an extra amount N at every turn—and it guarantees that Ax and yA and v and w are all positive.

Now a small trick produces the right linear programs. Divide x by the positive number v. Since the components of x add to 1, the components of $\bar{x} = x/v$ add to $1/v$—which means $c\bar{x} = 1/v$, where $c = (1, 1, \ldots, 1)$. Therefore X makes v large by making $c\bar{x}$ small, and his new task is:

Choose $\bar{x}$ to minimize $c\bar{x}$, subject to $A\bar{x} \geq b$ and $\bar{x} \geq 0$.

This is a standard linear program. And since minimizing $c\bar{x} = 1/v$ is the same as maximizing v, it is the right problem for X. Similarly $\bar{y} = y/w$ has components that add to $\bar{y}b = 1/w$. Y wants w to be small and $1/w$ to be large. Therefore his problem is now a maximization:

Choose $\bar{y}$ to maximize $\bar{y}b$, subject to $\bar{y}A \leq c$ and $\bar{y} \geq 0$.

The fundamental result of game theory—the minimax theorem—is in our hands. These two linear programs are dual. The minimum of $c\bar{x} = 1/v$ equals the maximum of $\bar{y}b = 1/w$. Therefore w is the same as v. If we multiply the optimal $\bar{x}$

and $\bar{y}$ by v to give $x^* = v\bar{x}$ and $y^* = v\bar{y}$, we are back to mixed strategies; the components of x^* and y^* add to 1. Furthermore

$$Ax^* \geq vb = \begin{bmatrix} v \\ v \\ \vdots \\ v \end{bmatrix} \text{ and } y^*A \leq vc = [v \quad v \quad \cdots \quad v]. \tag{10}$$

X cannot be prevented from gaining v, if he uses the mixed strategy x^*. Y cannot be forced to pay more than v, if he uses y^*. *The game is solved*, and its value is v.

8G (*Minimax theorem*) For any m by n matrix A there are mixed strategies x^* and y^* that satisfy (10). They guarantee a payoff of at least v and at most v, respectively. For all other strategies y and x,

$$yAx^* \geq yvb = v \text{ and } y^*Ax \leq vcx = v. \tag{11}$$

Therefore the pair x^*, y^* is a saddle point from which neither player wants to move, and the value of the game is $v = y^*Ax^*$:

$$y^*Ax \leq y^*Ax^* \leq yAx^* \text{ for all } x \text{ and } y. \tag{12}$$

Equation (11) came directly from (10), since $yb = cx = 1$; the frequencies in any strategy add to 1. And if N was added to every entry of A, when for convenience we made every payoff positive, it can be subtracted again from all sides of all equations. For the original payoff matrix A, the strategies x^* and y^* are still optimal and the value is $v - N$.

Escape and the Wedding Game

This completes the minimax theorem, and it also suggests how to play the game: Solve the equivalent linear program. In general that means the simplex method. In games like *poker*, trial and error have produced solutions that are nearly optimal. (Two-person poker fits our theory; games with several players are more complicated, because of alliances in which one group gangs up on the others.) Even two-person poker is extremely difficult, because of the number of strategies. There are so many cards and so many decisions that m and n are enormous. Nevertheless the ingredient that makes poker different from chess—namely bluffing—is what fits it into game theory. If you bluff too seldom or too often you lose, and in between is the optimal strategy.†

---

† Ankeny has written a good book on poker (*Poker Strategy*, Basic Books, 1981). I guess the theory of bluffing applies also in war—but that game is played only once.

In this section we introduce two more games, one old and one new. The old one is *Escape*, and its payoff matrix is simple; it is the identity matrix $A = I$. Player X chooses one of the n columns and Y chooses one of the n rows. If they match, X catches Y and wins $1; they meet on the main diagonal of the identity matrix. Otherwise, if the row and column are different, Y escapes and the (off-diagonal) payoff is $a_{ij} = 0$. The game is in favor of X, but less so as n increases.

It is easy to guess the optimal strategies x^* and y^*. X and Y should play their columns and rows with equal frequencies $1/n$. With this random strategy for X, he expects to catch Y at least 1 time out of n. Whatever Y does, that cannot be avoided. On the other hand, the strategy y^* means that Y cannot be caught more often. Therefore this is a saddle point, where (with $A = I$) we can see v and w:

$$Ax^* = \begin{bmatrix} 1/n \\ \vdots \\ 1/n \end{bmatrix} \text{ and } y^*A = [1/n \cdots 1/n].$$

The game is solved. Its value is $v = y^*Ax^* = 1/n$, which does go down as n increases. There are more rows in which Y can hide.

As in the football game, this example actually has $Ax^* = vb$ (not just $Ax^* \geq vb$) and $y^*A = vc$ (not just $y^*A \leq vc$). If this were always true, two-person games would be easy to solve; they would lead directly to linear equations. Usually the situation is like linear programming, in which some equalities and some inequalities hold at the optimum. Some columns are played with *zero frequency* $x_j = 0$. This is true in football if a third strategy of the offense is "fumble," with payoff -10 or even -1. That column will be out of the optimal strategy.

The situation is identical with complementary slackness and competitive equilibrium. If the jth component of y^*A is *below* v, the optimal strategy for X has $x_j^* = 0$. X omits column j because it fails to win v. Similarly Y omits row i if $(Ax^*)_i$ is above v. Within the payoff matrix A is hiding a square submatrix B (a basis matrix!) for which equalities do hold: $Bx^* = vb$ and $y^*B = vc$. B looks like a smaller game, but it is the real game inside A. The other components of x^* and y^* are zero—those columns and rows are never used—and the game based on A is solved once we know B.

Finally we present a new and more remarkable game. It comes from the marriage problem. Suppose there are 10 girls and 8 boys, and out of the 80 possible pairs only 33 marriages are allowed. Player Y chooses one of these 33 marriages. At the same time X chooses one of the 18 people who might be involved. Thus $m = 33$ and $n = 18$. If X guesses right and picks someone in the marriage chosen by Y, then X wins $1; otherwise the payoff is zero. This is the **Wedding Game**.

It has other interpretations. Suppose the U.S. has 10 agents, and a secret is to be passed to one of Canada's 8 agents—when there are 33 pairs of agents who recognize each other. X is a spy who can follow any one of the 18 agents; if he guesses right he wins. What strategy should the spy use, and what strategy should the agents use, and what is the chance of success?

The answer must depend on the 33 by 18 matrix A. It is like an incidence matrix, with a row for every marriage and a column for every person. Each row has two nonzeros, in the two columns that are involved in the marriage. Both entries are now $+1$, since the payoff is always 1. Thus A has 66 1's and a lot of zeros; the game looks complicated. Player X chooses columns, or agents to follow, with frequencies $x_1, ..., x_{18}$. Player Y chooses marriages, or pairs of agents, with frequencies $y_1, ..., y_{33}$. It is like Escape, but more delicate.

We want to propose a strategy for each player. Suppose the maximum number of marriages (without bigamy) is 7. From such a set of 7 marriages, Y will choose with equal frequencies $\frac{1}{7}$. This strategy uses only those 7 rows; the remaining rows will have frequencies $y_i^* = 0$. By duality in the marriage problem (Section 7.2) there must also exist 7 people who cover all possible marriages. (None of the other 11 can marry one another.) From one such set of 7 people, X will choose with equal frequencies $\frac{1}{7}$. He uses only those 7 columns, and the others have $x_j^* = 0$. Our goal is to show that those strategies are optimal. Here are the main points:

(1)   These 7 rows and columns are just a permutation of the identity matrix. On that submatrix we are back to the game of Escape—where equal frequencies $y_i^* = x_j^* = \frac{1}{7}$ are known to be optimal.

(2)   Every column of the payoff matrix A has at most one 1 in the 7 rows chosen by Y, since no person can be in two of those 7 marriages. Therefore X has no motive for changing to other columns; no column can win more than $\frac{1}{7}$ of the time.

(3)   Every row of A has at least one 1 in the 7 columns chosen by X, since that group of 7 people is involved in every marriage. Therefore Y has no motive for changing to other rows; no row can avoid losing at least $\frac{1}{7}$ of the time, when X stays with this strategy.

Inside the wedding game, the real game is Escape.

EXAMPLE   Suppose the matrix that lists the possible marriages is

$$
M = \begin{array}{c} \\ \begin{bmatrix} b_1 & b_2 & b_3 \\ 1 & 0 & 1 \\ 1 & 0 & 0 \end{bmatrix} \end{array} \begin{array}{l} \\ \text{girl 1} \\ \text{girl 2} \end{array}.
$$

There are five people and three possible marriages, so the payoff matrix in the wedding game will be 3 by 5:

$$
A = \begin{bmatrix} 1 & 0 & 1 & 0 & 0 \\ 1 & 0 & 0 & 0 & 1 \\ 0 & 1 & 1 & 0 & 0 \end{bmatrix} \begin{array}{l} \text{marriage 1} \\ \text{marriage 2*} \\ \text{marriage 3*} \end{array}
$$
$$
\quad\; g_1^* \;\; g_2 \;\; b_1^* \;\; b_2 \;\; b_3
$$

The maximum number of marriages is 2 (the second and third marriages). There are 2 people whose lines cover all marriages (girl 1 and boy 1). Therefore Y can play those two rows and X can play those two columns, all with frequency $\frac{1}{2}$. The rows and columns are marked by asterisks, and *the submatrix they form is the identity.* The average payoff to X is $\frac{1}{2}$. Y will not play row 1, since that would lose him \$1 every time. X has no reason to play other columns, since no column has more than a single 1 in the two rows being played by Y. Therefore these strategies give a saddle point and the game is solved.

The marriage problem must hold the record for versatility. It is a combinatorial problem, a graph theory problem, a linear program, and a two-person game. Combinatorially, it maximizes the number of marriages and minimizes the number of lines to cover the 1's. On a bipartite graph, it maximizes the number of disjoint edges and minimizes the number of vertices that touch every edge. It is written as a linear program in Exercise 8.4.8. And it is equivalent to the wedding game.

EXERCISES

8.4.1 With payoff matrix $A = \begin{bmatrix} 5 & 0 \\ 0 & 2 \end{bmatrix}$, X should choose columns 1 and 2 with frequencies that satisfy $5x_1 = 2x_2$. Find x_1 and $x_2 = 1 - x_1$, and show that X wins $v = \frac{10}{7}$ against both strategies (both rows) of Y. What should X and Y do, and what is the value v, if the payoff 5 is changed to -5?

8.4.2 With payoff matrix $A = \begin{bmatrix} 1 & 4 \\ 3 & 2 \end{bmatrix}$, find the best strategy for X (as in (8) and (9) for the football game) and the best strategy for Y.

8.4.3 A game is completely symmetric (and fair, with value $v = 0$) if its payoff matrix is *skew-symmetric*: $A^T = -A$. Whatever X wins with column i against row j, he loses with column j against row i. Find optimal strategies for

$$A = \begin{bmatrix} 0 & 1 & 0 \\ -1 & 0 & -1 \\ 0 & 1 & 0 \end{bmatrix} \quad \text{and} \quad A = \begin{bmatrix} 0 & 1 & -1 \\ -1 & 0 & 1 \\ 1 & -1 & 0 \end{bmatrix}.$$

8.4.4 If a_{ij} is the largest entry in its row and the smallest in its column, why will X always choose column j and Y always choose row i? Show that the previous exercise had such an entry. (The football game did not, and therefore needed a mixed strategy.)

8.4.5 For the wedding game example in the text, what other pairs of columns could X choose in an optimal strategy?

8.4.6 The linear programs associated with the wedding game example are

$$\text{Minimize } \bar{x}_1 + \bar{x}_2 + \bar{x}_3 + \bar{x}_4 + \bar{x}_5 \text{ subject to } \bar{x} \geq 0, \begin{array}{c} \bar{x}_1 + \bar{x}_3 \geq 1 \\ \bar{x}_1 + \bar{x}_5 \geq 1 \\ \bar{x}_2 + \bar{x}_3 \geq 1 \end{array}$$

$$\bar{y}_1 + \bar{y}_2 \le 1$$
$$\bar{y}_3 \le 1$$
Maximize $\bar{y}_1 + \bar{y}_2 + \bar{y}_3$ subject to $\bar{y} \ge 0$, $\qquad \bar{y}_1 + \bar{y}_3 \le 1$
$$\bar{y}_2 \le 1$$

Show that maximum = minimum = 2. This number is $1/v$, where $v = \frac{1}{2}$ is the value of the game.

8.4.7   Suppose B is a $0-1$ matrix in which every row contains at least a single 1 and no column contains more. Show that all rows and columns contain exactly one 1, so B is a permutation of the identity matrix.

8.4.8   For the marriage matrix

$$M = \begin{bmatrix} 1 & 1 & 1 \\ 1 & 0 & 0 \\ 1 & 0 & 0 \end{bmatrix},$$

how many marriages are possible? Write out the 6 by 5 wedding game as a linear program and find an optimal strategy for X and Y. On the graph associated with M, show a maximal set of independent edges (marriages) and a cover of all edges by a minimal set of nodes.

8.4.9   If $Ax^* = vb$ and $y^*A = wc$, show directly from these equations (and $y^*b = cx^* = 1$) that $v = w$. The frequencies in x^* and y^* add up to 1, when b and c are vectors of 1's. Under what additional condition do x^* and y^* solve the game with payoff matrix A?

8.4.10   Compute the minimax of yAx when A is the identity matrix. Maximize, and then minimize the maximum, by computing

$$F(y_1, y_2) = \max_{\substack{x_i \ge 0 \\ x_1 + x_2 = 1}} (x_1 y_1 + x_2 y_2) \quad \text{and} \quad \min_{\substack{y_i \ge 0 \\ y_1 + y_2 = 1}} F(y_1, y_2).$$

8.4.11   If Y writes down a prime number while X guesses whether it is odd or even (winning or losing $10), which player would you choose to be? What is the best strategy for X?

8.4.12   Suppose grapes, labor, and capital go into the production of grape juice and wine, in the amounts a_{ij}:

$$A = \begin{bmatrix} 3 & 1 \\ 1 & 2 \\ 1 & 3 \end{bmatrix} \quad \begin{matrix} \text{grapes} \\ \text{labor} \\ \text{capital.} \end{matrix}$$

juice wine

Let the juice and wine have values $c_1 = 2$ and $c_2 = 3$, and the inputs of grapes, labor, and capital be limited to $b_1 = b_2 = b_3 = 8$. Maximize cx subject to $Ax \le b$ and $x \ge 0$ (by inspection rather than the simplex method) to find the best production levels x_1 and x_2. Which input is not completely used?

8.4.13   Write down the dual to the previous exercise and find the optimal prices y_1, y_2, y_3 of grapes, labor, and capital. Which one is a free good? If the availability of grapes or labor or capital goes from 8 to $8 + \delta$, what are the three corresponding changes in the value of total output?

8.4.14 If the value of wine changes to $c_2 = 6$, show that all production should be concentrated on wine. Find the optimal production levels x and prices y, with $Ax \leq b$ and $yA \geq c$, and verify (6) and (7).

8.4.15 At what level $\bar{c}_2$ for the value of wine will all production be switched to wine? Draw the feasible set $Ax \leq b$, $x \geq 0$ in the plane, and sketch the cost lines touching this set—first the line $2x_1 + 3x_2 = $ constant and then the line $2x_1 + 6x_2 = $ constant. How will the switch-over line $2x_1 + \bar{c}_2 x_2 = $ constant touch the feasible set?

8.5 ■ OPTIMIZATION AND NONLINEAR PROGRAMMING

An optimal solution is an admissible solution of minimum cost. That is easy to say, but it contains a lot. It means, first, that there is a *cost function* $C(x)$ to be minimized. It also implies some *admissibility constraints* on x: the flow out must equal the flow in, or capacities cannot be exceeded, or the structure must withstand the load. If x represents a set of prices, or a set of shipments, or a set of probabilities, they cannot be negative. The constraints probably prevent the cost C from reaching its absolute minimum, and the problem is one of *constrained optimization*.

Without constraints, the minimization of $C(x)$ is an ordinary problem in calculus. When x is a vector we need multivariable calculus, and when it is a function we need the calculus of variations, but these just give extensions of the same basic idea: The derivative C' should be zero at the minimum. That is the grandfather of all optimality conditions! The constraints seem to destroy this simple rule, and the chief goal of the theory is to save as much of it as possible.

Thanks to Lagrange, it is almost entirely saved. We look first at the case of linear constraints, then at nonlinear constraints, and finally at inequality constraints. In each case there is a test on the derivative C' at the optimal point x^*, but it applies *only in the directions permitted by the constraints*. In the other directions we cannot move from x^* without leaving the admissible set. This is clear from the geometry, and we will rely more than usual on insight and examples. At the end, however, it is mathematics that produces the critical quantities—the Lagrange multipliers y. They convert the constrained minimization of $C(x)$ into an unconstrained minimization of $L(x) = C(x) + y^T(A(x) - b)$. A chief object of this section is to find and understand these magic numbers y.

Before we start, it is worth thinking for a moment about inequality constraints. Suppose there are m of them, linear or nonlinear, and they are written as $A_i(x) \le b_i$. At the optimal x^* they fall into two groups—those for which there is equality $A_i(x^*) = b_i$ and a "tight" constraint, and those for which there is strict inequality $A_i(x^*) < b_i$. The first constraints are *active* at x^*, the others are *inactive*. As far as inactive constraints are concerned, the test for a minimum hardly notices them. They will be satisfied by any x near x^*, and their Lagrange multipliers y_i will be zero. It is the active constraints that need Lagrange multipliers. When $A_i(x^*) - b_i$ is zero, y_i is almost certainly nonzero; it measures the force of the constraint, preventing $A_i(x)$ from exceeding b_i as it would like to.

One or the other of these quantities, either y_i or $A_i(x^*) - b_i$, is always zero. In the case of equality constraints, $A_i(x^*) - b_i$ is zero because that is the constraint. In every case the product $y_i(A_i(x^*) - b_i)$ is zero, and the sum of all m products—which is the inner product of y with $A(x^*) - b$—is also zero:

$$\boxed{y(A(x^*) - b) = 0.} \tag{1}$$

This is the *complementarity condition*, a nonlinear copy of the previous section's

$y(Ax - b) = 0$. The zeros in y (we should really write y^*) are in complementary positions to the zeros in $A(x^*) - b$. When b_i is large, and the ith material is oversupplied, the constraint $A_i(x) \le b_i$ is not really restrictive. The material becomes a free good, and its price y_i drops to zero. At an equilibrium point x^*, each Lagrange multiplier y_i tells the real price of its constraint—by revealing how much a small change in b_i would affect the minimum cost.†

You might say that (1) must automatically hold, if the constrained minimum of C and the unconstrained minimum of $L = C + y(Ax - b)$ are the same. That is true! The y_i have the remarkable property that the minima of C and L are attained *at the same point* x^*. Of course the y_i are not known—they might be regarded as the fundamental unknowns, if the final unconstrained problem is easy—and it cannot be predicted in advance which of them are nonzero. We do not know which constraints are active, until the whole problem is solved. Just as in the simplex method, the correct y_i emerge at the same time as the correct x_j.

Numerically, the simplex method is a very limited model for solving nonlinear programs. For *quadratic programming*—when C is a quadratic and the constraints are linear equations and inequalities—you can see what will happen. Moving along an edge of the feasible set, the cost looks like a parabola. It is decreasing at the start of the edge or we would not move. If it is still decreasing at the end, we stop there. But unlike the linear case, a parabola may start down and later go up; in that case we stop at its minimum, and take the next step from there. This requires only small changes in the simplex method, and quadratic programming is not excessively hard—but the minimum may occur *inside* the feasible set. The solution is not always at a corner.

For a general nonlinear program there are many possibilities, and no algorithm is the clear winner. We will choose a direction d_k in which to move from the current guess x_k. We may conduct a line search in that direction—to find the new $x_{k+1} = x_k + s d_k$ that minimizes the cost $C(x)$ while remaining admissible (or close to it). This search is one-dimensional, with a scalar unknown—the step size s. The direction d_k is the critical choice. Frequently the gradient of C is projected onto the subspace of directions permitted by the active constraints (as Karmarkar did). It is like Newton's method, linearizing near the current x_k and venturing a step on the basis of C'. In fact Newton's method becomes a quadratic program at each step; the cost and the active constraints are decided at x_k. That may be the best. At the end of the step new constraints will be active, just as one component became nonzero and another became zero in each simplex step. But nonlinear constraints bring extra difficulties, and a full discussion is hardly possible.

We might remark on the choice between minimizing C and solving an equation like $C' + yA' = 0$. With inequality constraints most algorithms choose the minimization; with equality constraints $L' = C' + yA' = 0$ becomes reasonable. In structural optimization there was a war between these two camps, recently settled by a

---

† When b_i goes up by δ, more vectors x become admissible and the minimum cost goes *down* by $y_i \delta$. If b_i is already large and $y_i = 0$, the change by δ makes no difference. In each case y_i measures the sensitivity to b_i.

compromise—in which the multipliers y are improved at the same time as the primal unknowns x. Duality won again.

Conditions for a Constrained Minimum

Imagine that the cost function $C(x)$ has a bowl-shaped graph. If it comes from a positive definite matrix, $C(x) = \frac{1}{2}x^T M x$, then the bowl is perfect. Its cross-sections $C = $ constant, called "level curves" or "level surfaces," are ellipsoids. In general C will not be exactly a quadratic and the graph will be more uneven; its level curves are sketched in Fig. 8.9. The inner curves come from the low values of C, near the bottom of the bowl. The cost increases as we move out and up the bowl. The problem is to find the lowest point that satisfies the constraints, and we proceed in four steps.

1. To begin, let there be **one equality constraint** and let it be linear: $a_1 x_1 + \cdots + a_n x_n = b$. This is $A(x) = b$. It gives a vertical plane that slices through the graph of C. The constrained problem looks for the lowest point of their intersection, the minimum of $C(x)$ subject to $A(x) = b$. The idea is to look down from above on the graph of C, and watch what happens at that lowest point x^*—marked P in Fig. 8.9. The level curve through P is *tangent* to the cutting plane $Ax = b$.†

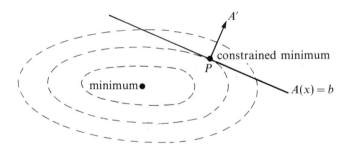

Fig. 8.9. Constraint $A(x) = b$ tangent to level curve at the solution x^*.

When two surfaces $A(x) = b$ and $C(x) = $ constant are tangent, their perpendicular directions are the same. One perpendicular comes from the vector $a = (a_1, \ldots, a_n)$, and the other comes from the gradient $C' = (\partial C/\partial x_1, \ldots, \partial C/\partial x_n)$. Since these vectors are in the same direction, the gradient must be a multiple of a:

$$\boxed{C' + ya = 0 \text{ for some multiplying factor } y.} \qquad (2)$$

That is the key to constrained optimization. The partial derivatives of C may not be zero at the point P, but **the derivatives of $C + yA$ are zero**. There are $n + 1$

---

† The level curve cannot pass through the plane, or there would be an even lower level curve inside it still touching the plane. Since the curve just touches at P, it must be tangent.

unknowns $x_1, \ldots, x_n, y$, and $n+1$ equations

$$\frac{\partial C}{\partial x_1} + ya_1 = 0, \ldots, \frac{\partial C}{\partial x_n} + ya_n = 0 \text{ and } A(x) = b. \tag{3}$$

EXAMPLE 1   *Minimize* $C = \frac{1}{4}(x_1^4 + x_2^4)$ *subject to* $x_1 + 8x_2 = 34$.
The $n+1 = 3$ equations are

$$\frac{\partial C}{\partial x_1} + ya_1 = 0, \quad \text{or} \quad x_1^3 + y = 0$$

$$\frac{\partial C}{\partial x_2} + ya_2 = 0, \quad \text{or} \quad x_2^3 + 8y = 0$$

$$A(x) = b, \quad \text{or} \quad x_1 + 8x_2 = 34.$$

To solve them, $x_1 = -y^{1/3}$ and $x_2 = -2y^{1/3}$ give

$$x_1 + 8x_2 = -17y^{1/3} = 34, \quad \text{or} \quad y^{1/3} = -2, \quad \text{or} \quad y = -8.$$

The point P has $x_1 = 2$, $x_2 = 4$, and the minimum is $C = \frac{1}{4}(16 + 256) = 68$.

2. The step to *one nonlinear constraint* is easy. The surface $A(x) = b$ becomes curved instead of flat, but it is still tangent to the level surface of C at the point $x^* = P$. Therefore the two perpendicular vectors still go in the same direction. One is C', as before, and the other is A'. Previously, $A(x) = a_1 x_1 + \cdots + a_n x_n$ was linear and its gradient was always $A' = (a_1, \ldots, a_n)$. Now A' varies from point to point, as C' does, but what matters is the situation at P—where they are in the same direction: $C' + yA' = 0$ for some multiplier y. In the next example the constraint surface is a circle $x_1^2 + x_2^2 = 1$.

EXAMPLE 2   *Minimize* $C = ax_1^2 + 2bx_1 x_2 + cx_2^2$ *subject to* $x_1^2 + x_2^2 = 1$.
The $n+1$ equations are

$$\frac{\partial C}{\partial x_1} + y\frac{\partial A}{\partial x_1} = 0, \quad \text{or} \quad 2(ax_1 + bx_2 + yx_1) = 0$$

$$\frac{\partial C}{\partial x_2} + y\frac{\partial A}{\partial x_2} = 0, \quad \text{or} \quad 2(bx_1 + cx_2 + yx_2) = 0$$

$$A(x) = b, \quad \text{or} \quad x_1^2 + x_2^2 = 1.$$

Cancelling the factor 2, the first equations are

$$\begin{aligned} ax_1 + bx_2 &= -yx_1 \\ bx_1 + cx_2 &= -yx_2 \end{aligned} \quad \text{or} \quad \begin{bmatrix} a & b \\ b & c \end{bmatrix} \begin{bmatrix} x_1 \\ x_2 \end{bmatrix} = -y \begin{bmatrix} x_1 \\ x_2 \end{bmatrix}.$$

Thus the optimal $x = (x_1, x_2)$ is an *eigenvector* of this matrix M; we have $Mx = -yx$. The minimum value of C is exactly the *smallest eigenvalue*, since

$$C = [x_1 \quad x_2]\begin{bmatrix} a & b \\ b & c \end{bmatrix}\begin{bmatrix} x_1 \\ x_2 \end{bmatrix} = -y[x_1 \quad x_2]\begin{bmatrix} x_1 \\ x_2 \end{bmatrix} = -y.$$

Geometrically, the level curve of C is an ellipse that touches the circle $x_1^2 + x_2^2 = 1$ at the ends of its longest axis—and that axis points along the eigenvector. It is like minimizing the Rayleigh quotient

$$\frac{ax_1^2 + 2bx_1 x_2 + cx_2^2}{x_1^2 + x_2^2} = \frac{x^T M x}{x^T x}.$$

For any symmetric matrix, of any size, the minimum of $C(x) = x^T M x$ subject to $x^T x = 1$ is the smallest eigenvalue of M.

3. The next step is to admit **two or more constraints**. Separately they are $A_i(x) = b_i$, and collectively (in a vector equation) they are $A(x) = b$. If they are linear, the gradients A_i' are the rows of a fixed matrix. If they are nonlinear, the A_i' depend on x. In either case, the minimizing point P must satisfy $A(x) = b$. If we move away from P, staying on the surface $A(x) = b$, the cost $C(x)$ must not decrease; otherwise P would not be minimal. Therefore the vector C', which is orthogonal at P to the surface $C = $ constant, must also be orthogonal to the surface $A(x) = b$.

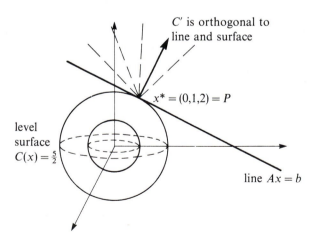

Fig. 8.10. The derivative C' at x^* is orthogonal to $A(x) = b$.

When there was one constraint, this orthogonality determined the direction of C'. With more constraints the surface $A(x) = b$ will be lower dimensional, like a line in three dimensions, and many directions are orthogonal to it. The gradient C' goes in one of those directions (Fig. 8.10), so

$$C' + y_1 A'_1 + \cdots + y_m A'_m = 0 \text{ for some multiplying factors } y_i. \tag{4}$$

That is subtle but important. Think of the linear case, with constraint $Ax = 0$. The admissible x are in the nullspace of A. Each row of A is orthogonal to the nullspace, and C' can be any combination of those rows. That is equation (4), which has $n + m$ unknowns $x_1, \ldots, x_n, y_1, \ldots, y_m$. There are also $n + m$ equations, the n given by (4) and the m constraints. They are combined by saying that **all partial derivatives of** $L = C + y(Ax - b)$ **are zero**:

$$\frac{\partial L}{\partial x_j} = 0, j = 1, \ldots, n, \quad \text{and} \quad \frac{\partial L}{\partial y_i} = 0, i = 1, \ldots, m. \tag{5}$$

The last m equations are the constraints $A(x) = b$.

EXAMPLE 3 *Minimize* $C = \frac{1}{2}x^T H x - x^T f$ *subject to* $Ax = b$.
The gradient C' is $Hx - f$ and the $n + m$ equations are

$$\begin{array}{ll} \partial L/\partial x_j = 0: & Hx + A^T y^T = f \\ \partial L/\partial y_i = 0: & Ax \qquad\quad = b \end{array} \quad \text{or} \quad \begin{bmatrix} H & A^T \\ A & 0 \end{bmatrix}\begin{bmatrix} x \\ y^T \end{bmatrix} = \begin{bmatrix} f \\ b \end{bmatrix}. \tag{6}$$

These are the all-important equations from Chapter 2, with the notations reversed: $x \leftrightarrow y$, $b \leftrightarrow f$, $A \leftrightarrow A^T$. We denoted the matrix by H, since C is now cost, and y has become a row vector: $y_1 A'_1 + \cdots + y_m A'_m$ is $A^T y^T$. But the underlying problem is identical with the one in Chapter 2, to minimize a quadratic with linear constraints.

Figure 8.10 has $C = \frac{1}{2}(x_1^2 + x_2^2 + x_3^2)$, with spheres as level surfaces. The rows from the constraint are $A'_1 = [1 \quad 0 \quad 0]$ and $A'_2 = [0 \quad 1 \quad 2]$. At the solution C' is A'_2 and the optimality condition (4) is satisfied.

4. The final step is to allow **inequality constraints**. They may be active or inactive—either they alter the minimum or they don't. Both possibilities appear in the simplest problem, *to minimize the number* x^2 *subject to* $x \le b$. If b is positive, the minimum of x^2 is at $x = 0$. It is the absolute minimum and whether $b = 10$ or $b = 1000$ makes no difference. But if b is negative, and $x = 0$ is inadmissible because it violates $x \le b$, the minimum is changed. It occurs where x *equals* b. The constraint becomes active, and it changes the minimum of x^2 from 0 to b^2.

How is this reflected in the Lagrange multiplier y? When the constraint is active, the derivatives of $x^2 + y(x - b)$ are

$$\frac{\partial L}{\partial x} = 2x + y = 0$$

$$\frac{\partial L}{\partial y} = x - b = 0.$$

Thus $y = -2b$.† When the constraint is inactive, the multiplier y is zero and the minimum is zero (at $x = 0$).

The pattern is the same when there are n unknowns and m constraints, and it is the fundamental condition for optimality:

8H   (Kuhn-Tucker optimality conditions) The minimum of $C(x_1, \ldots, x_n)$ subject to $A_i(x) \leq b_i$ occurs where

$$\frac{\partial C}{\partial x_j} + y_1 \frac{\partial A_1}{\partial x_j} + \cdots + y_m \frac{\partial A_m}{\partial x_j} = 0, j = 1, \ldots, n \tag{7}$$

with y and x also subject to

$$y_i \geq 0, A_i(x) \leq b_i, y_i(A_i(x) - b_i) = 0, i = 1, \ldots, m. \tag{8}$$

There are $n + m$ equations, but we cannot predict in (8) whether $y_i = 0$ or $A_i(x) = b_i$. Either the constraint is active or the multiplier is zero; the right equation is not known in advance.

In a full-scale treatment of optimization we would have to discuss the extra hypotheses that make this literally true. First, the functions C and A_i have been assumed smooth. If the graph of C has a corner, there is a whole family of "derivatives" and any one is acceptable in (7). Second, the vectors A_i' should be independent at the minimizing point or the y_i are not well determined. Third, and most important, the functions C and A_i should be **convex** (see below). Convexity is the prime requirement in proving that there is a constrained minimum. Without it the solution to (7–8) can be a saddle point, or a maximum, or fail to exist. With a strengthened form of convexity, the minimization succeeds.

EXAMPLE 4   (Linear programming)   *Minimize* $C = c_1 x_1 + \cdots + c_n x_n$ *subject to* $Ax \leq b$.
The equations (7) in vector notation are $c + yA = 0$, and (8) is complementary slackness:

$$y \geq 0, Ax \leq b, y(Ax - b) = 0. \tag{9}$$

They are the optimality conditions connecting the primal to the dual, when there is no sign constraint on x.

This is an example in which C is convex but not *strictly* convex—its second derivatives are zero. The zero matrix is positive semidefinite but certainly not positive definite. Therefore the minimum may fail to exist. It is $-\infty$, if we minimize $2x$ subject to $x \leq 4$.

In the final example C is strictly convex; the matrix H is to be positive definite.

---

† y is the derivative of the minimum value $x^2 = b^2$, but with opposite sign.

EXAMPLE 5 (Quadratic programming) *Minimize* $C = \frac{1}{2}x^T H x$ *subject to* $x_1 \le b_1, \ldots, x_n \le b_n$.

There are n constraints, n multipliers y_i, and $n + n$ equations:

$$(7) \text{ becomes } Hx + y^T = 0 \tag{10}$$

$$(8) \text{ becomes } y \ge 0, \ x \le b, \ y(x - b) = 0. \tag{11}$$

For $n = 1$ we are back to the two possibilities $x = b$ (active constraint) or $y = 0$ (inactive constraint). For $n = 2$ it is reasonable to test all four possibilities. As n increases, the number of combinations climbs to 2^n; each constraint can be active or inactive. For large n a good algorithm finds the right combination without trying them all.

Convex Functions

We need to know which functions $C(x)$ fit naturally into these minimum problems. They will not be the only functions that can be minimized, but they are the best ones. They were described earlier in terms of a "bowl-shaped graph"— which was intuitively correct but not overwhelmingly precise. The right description is in the definition of a ***convex function***, which extends one of the basic ideas of calculus—that the second derivative satisfies $f'' \ge 0$ at a minimum. The first requirement is $f' = 0$, and in our constrained problems that became $L' = C' + yA' = 0$. Without this *first-order condition*, a point is not even a candidate for a minimum. But for points which survive that test, there has to be a *second-order condition* (involving the second derivatives of L) to distinguish between minima and maxima and saddle points. A convex function will pass that second-order test, and a strictly convex function will pass with something to spare.

A convex function is defined in the same way as a convex set:

A set E is convex if the line segment between any two of its points stays within the set.

A function F is convex if the line segment between any two points of its graph lies on or above the graph.

If all line segments go strictly inside the set E, or strictly above the graph of F, then the set or the function is *strictly convex*. There are no flat segments on the boundary of the set or on the graph. A function $F(x) = \text{constant}$, or a linear function $F(x) = a^T x$, or a feasible set in linear programming, is convex but not strictly convex.

There are three ways to test for convexity. The first two come directly from the definition and the third, which extends $f'' \ge 0$, is provided by calculus.

(1) The set of points on or above the graph of f should be a convex set. This set is called the "epigraph."

(2) At every point $x = cx_1 + (1 - c)x_2$ between x_1 and x_2, the value of $F(x)$ must not be above the straight line value $cF(x_1) + (1 - c)F(x_2)$:

$$F(cx_1 + (1-c)x_2) \le cF(x_1) + (1-c)F(x_2) \quad \text{for } 0 \le c \le 1. \tag{12}$$

(3)   The matrix H of second derivatives of F, $H_{ij} = \partial^2 F/\partial x_i \partial x_j$, must be positive semidefinite at every point.

The link between test 2 and test 3 comes from Taylor series expansions:

$$F(x) = F(x_1) + (x-x_1)^T \nabla F(x_1) + \tfrac{1}{2}(x-x_1)^T H(x-x_1) + \text{cubic terms}.$$

Expanding both sides of (12), the quadratic terms produce $(x_2 - x_1)^T H(x_2 - x_1) \ge 0$. This is the positive semidefiniteness of H. It means that all eigenvalues of H are ≥ 0, all pivots are ≥ 0, and all symmetrically placed subdeterminants are ≥ 0. If H is positive definite, these numbers are strictly positive, (12) is true with strict inequalities, and F is strictly convex.

EXAMPLE   The third test confirms the convexity of $F_1 = x^4 + y^6$ and it refutes the convexity of $F_2 = x^2 y^2$. The function F_2 is not convex even though x^2 and y^2 are separately convex. The matrices H, called second gradients or Hessians, are easy to find from the second derivatives of F_1 and F_2:

$$H_1 = \begin{bmatrix} 12x^2 & 0 \\ 0 & 30y^4 \end{bmatrix} \quad \text{and} \quad H_2 = \begin{bmatrix} 2y^2 & 4xy \\ 4xy & 2x^2 \end{bmatrix}.$$

The determinant of H_2 is negative so one of its eigenvalues must be negative. The graph of F_2 is a parabola in the x and y directions, but overall it cannot be convex. $F_2 = x^2 y^2$ is zero at $x = 0$, $y = 2$ and also at $x = 2$, $y = 0$, but between them it is positive. It goes *above* the line segment connecting those points, and all tests for convexity must fail.

This definiteness condition on the second derivatives is completely successful when F is smooth, but it is crucial to recognize that there are nonsmooth possibilities:

(a)   A convex function like the absolute value $|x|$ has a corner, where the second derivative becomes a delta-function. The graph resembles the letter V. In n dimensions it turns into the length function $F(x) = \|x\|$, whose graph is the cone in Fig. 8.11. The set E above the graph is pointed but still convex, and condition (12) becomes the triangle inequality

$$\|cx_1 + (1-c)x_2\| \le \|cx_1\| + \|(1-c)x_2\|.$$

(b)   A convex function may be **infinite** at some points. Condition (12) keeps it finite on the line between x_1 and x_2, if it is finite at those two points. Therefore the set on which F is finite must be convex. One particular function is important: $F(x) = 0$ when x is in the convex set S, and $F(x) = +\infty$ when x is not in S. It is known as the "indicator function" $I(x)$ of the set S, and it is extremely useful for constraints:

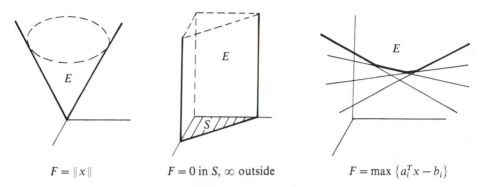

$$F = \|x\| \qquad\qquad F = 0 \text{ in } S, \infty \text{ outside} \qquad\qquad F = \max\{a_i^T x - b_i\}$$

Fig. 8.11. Convex sets E above the graphs of convex functions.

To minimize $C(x)$ subject to x in S, we minimize $C(x) + I(x)$ for all x. The points not in S lead immediately to $+\infty$, so effectively it is a minimum of $C(x)$ over the set S.

The sum of convex functions is automatically convex. So is the maximum of two or more convex functions, and the third graph in Fig. 8.11 is an example. With an infinite number of planes, the graph of the maximum can be curved (but still convex!). We could even reach the other two graphs in the figure, by choosing the right planes. In fact every convex function is the maximum of a family of linear functions—and to understand that we go back to convex sets.

The great property of a convex set E is that through every boundary point there is at least one **supporting hyperplane**. This is a plane that touches the boundary of E, keeping the rest of the set on one side. We assume the boundary is included in E; E is a "closed" convex set. Where the boundary is smooth, the only supporting plane is the one tangent to E. Where the boundary is pointed, at a corner, there are a lot of supporting planes through the boundary point. If we know all these planes, we can reconstruct E.† It will be the intersection of all the halfspaces cut out by the supporting planes.

Moving from convex sets to convex functions, these supporting planes become *tangent planes* to the graph. Their slopes are the *derivatives* of the function. It is the existence of these derivatives—the fact that there is at least one tangent plane at every point of the graph (and more than one, at the vertex of the cone)—that will produce a saddle point. The planes are exactly what is needed for a general duality theorem, bringing together all the specific cases proved earlier in this book.

Convexity and Duality

Suppose the cost $C(x)$ and the constraints $A_i(x)$ are all convex functions. If they are linear, we have linear programming. If C is quadratic, we have quadratic programming. In general we have nonlinear programming, and we are ready to show how it is transformed by Lagrange multipliers.

---

† This is basic to duality: A convex set can be described by saying which points the set contains, or by saying which half-spaces contain the set.

The theory is based on one brilliant idea (I don't know who it came to). That idea is to look beyond the particular values $b = (b_1, \dots, b_m)$ in the constraints of the given problem, and to admit all vectors b. Our one problem is embedded in a whole family of problems, and their minimum values produce a *minimum function* that varies with b:

$$M(b) = \text{minimum value of } C(x) \text{ subject to } A(x) \le b. \qquad (13)$$

To distinguish our particular b we denote it by b^*; the specific problem is to find $M(b^*)$. We denote by x^* the minimizing point in that problem (if it exists). Thus $A_i(x^*) \le b_i^*$ for each $i = 1, \dots, m$, and the minimum cost is $M(b^*) = C(x^*)$.

Now enters the hypothesis that C and the A_i are convex functions. It follows that the minimum value $M(b)$ is not only decreasing as b increases (because more candidates x are admitted). *It is also a convex function of b.* The example with cost $C = x^2$ and constraint $x \le b$ is sketched in Fig. 8.12—its minimum $M(b)$ equals b^2 for negative b and zero for positive b. The fact that M is convex when C and A are convex is verified in the exercises. This convexity allows the general theory to make its contribution: At any point like b^*, the graph of M has a supporting tangent plane. The plane has height $M(b^*) = C(x^*)$ at the point b^*, and at no point does the plane go above the graph of M.

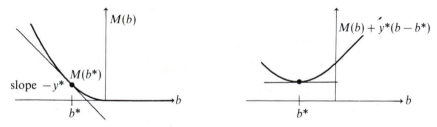

Fig. 8.12. $M(b) = $ minimum of x^2 with $x \le b$.

If you tilt your head to make the plane horizontal, the whole curve has its minimum at b^*. That is the point of Fig. 8.12b; we produce an *unconstrained minimum* by adding a linear term that comes from the plane. If the slope of the plane is $-y^*$, the linear term to add is $y^*(b - b^*)$. It equals zero at b^*, but for larger b it raises the curve and for smaller b it lowers it. Then the value $C(x^*)$ at $b = b^*$ becomes an absolute minimum:

$$C(x^*) = \min_b \; \min_{A(x) \le b} \; [C(x) + y^*(b - b^*)]. \qquad (14)$$

This y^* is the right Lagrange multiplier for the original problem. It is nonnegative, as expected with inequality constraints. Its components satisfy $y_i^* \ge 0$, because $-y^*$ is the slope of a decreasing function $M(b)$. That slope is the *sensitivity of the* *minimum with respect to b*:

$$-y_i^* = \frac{\partial M}{\partial b_i} \text{ at } b = b^*. \tag{15}$$

The Lagrange multiplier—which will be the solution to the dual problem, when that appears—is the *marginal cost*: y^* gives the change in the minimum as the constraints are changed. The whole theory of sensitivity comes from the tangent plane to $M(b)$.

To go from sensitivity to duality, and to show that y^* is the correct multiplier, the technical step is to verify from (14) that also

$$C(x^*) = \min_x [C(x) + y^*(A(x) - b^*)]. \tag{16}$$

We do it quickly. For any x in (16), choose the same x in (14) and choose $b = A(x)$. Then the two expressions agree, and since (14) allows other choices we have $(16) \geq (14)$. On the other hand, for any b and x the requirement $A(x) \leq b$ makes $(16) \leq (14)$, remembering $y^* \geq 0$. Therefore the two are equal.

At the point x^* something special must happen. The term $y^*(A(x^*) - b^*)$ could not be negative, or equation (16) would be ridiculous. Since $y^* \geq 0$ and $A(x^*) \leq b^*$, the only alternative is the *complementarity condition*:

$$\text{for each } i, \text{ either } y_i^* = 0 \text{ or } A_i(x^*) = b^*. \tag{17}$$

Then the inner product $y^*(A(x^*) - b^*)$ is zero, which is the Kuhn-Tucker condition. We have reached the main result of Lagrange duality:

8I If the cost $C(x)$ and the constraint functions $A_i(x)$ are strictly convex, then

$$\min_{A(x) \leq b^*} C(x) = \max_{y \geq 0} \min_x [C(x) + y(A(x) - b^*)]. \tag{18}$$

The constrained minimization splits into an unconstrained minimization of L with a parameter y, followed by a maximization (the dual problem) over y. The optimal x^* in the primal problem on the left and the optimal y^* in the dual problem on the right are related by the Kuhn-Tucker conditions (7) and (8).

Proof At $y = y^*$ the two minima in (18) agree; that is $(14) = (16)$. For other $y \geq 0$ the right side could not be larger than the left. This is weak duality, which is always easy:

$$\text{if } y \geq 0 \text{ and } A(x) \leq b^*, \quad \text{then } C(x) \geq C(x) + y(A(x) - b^*).$$

Therefore (18) is correct, the minimum equals the maximum, and duality holds. Our simplest example will illustrate it best.

EXAMPLE   *Minimize $C = x^2$ subject to $x \le b$*
The unconstrained minimum of $L = x^2 + y(x - b)$ comes first:

$$L' = 2x + y = 0 \text{ at } x = -\tfrac{1}{2}y, \text{ so the minimum is } L = -\tfrac{1}{4}y^2 - by.$$

Then the maximum over $y \ge 0$ (the dual problem) is

$$\max(-\tfrac{1}{4}y^2 - by) = \begin{cases} b^2 & \text{at} & y = -2b, & \text{if } b \le 0 \\ 0 & \text{at} & y = 0, & \text{if } b \ge 0 \end{cases}$$

This is the right side of (18); the minimum of C is b^2 or zero.†

Conjugate Convex Functions

Behind this analysis of duality lies a beautiful piece of geometry. We have hinted at it, twice at least, and the book will be incomplete until we say what it is. It brings together the applications, and then we are finished.

The geometry starts with a convex function, for example $F(x) = x^2$. The question is: Which straight lines lie below this parabola? The graph of $yx - d$ is a line with slope y, and the line is below the parabola if $yx - d \le x^2$ for all x. That is true if the depth d is large enough:

$$d \ge yx - x^2 \text{ for all } x. \tag{19}$$

The right side is largest where its derivative is zero. At that point $y - 2x = 0$, or $x = \tfrac{1}{2}y$, and the requirement becomes $d \ge \tfrac{1}{4}y^2$. The line will just touch the parabola, as in Fig. 8.13, if $d = \tfrac{1}{4}y^2$.

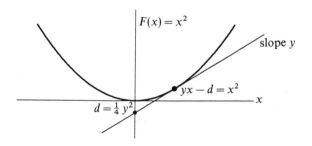

Fig. 8.13. Tangent line under $F(x)$, at depth $d = F^*(y)$.

---

†I always doubted that duality could make a problem easier (here it doesn't). But for a continuous maximum flow problem with capacity $|v| \le 1$ in the unit square, duality shows that div $v = 2 + \sqrt{\pi}$ is possible. There is a prize of 10,000 yen for v.

Now consider all these touching lines, with different slopes y. Their envelope is the parabola! We get back to x^2 by looking always for the highest line:

$$\max_{y} \left[yx - \frac{y^2}{4} \right] = x^2. \tag{20}$$

The maximum is at $y = 2x$, and at that point (20) is $2x^2 - x^2 = x^2$.

This duality between convex functions and tangent lines extends far beyond parabolas. The functions depend on x and the tangents depend on y. It is usual to write $F^*(y)$ rather than $d(y)$, to emphasize the parallel between F and F^*. This *conjugate function* F^* looks so ordinary and innocent; it is the constant term d that raises or lowers the line until it touches the parabola. However its construction is at the center of convex analysis, and the steps that succeeded for a parabola in (19) and (20) will succeed for every F:

8J   Suppose $F(x)$ is a convex function. For each slope y let

$$d = F^*(y) = \max_{x} [yx - F(x)]. \tag{21}$$

This conjugate function F^* is also convex, and for every x and y it satisfies $F^* \geq yx - F$, or $F \geq yx - F^*$. Then the maximum over the tangent lines $yx - d$ brings back F:

$$F(x) = \max_{y} [yx - F^*(y)]. \tag{22}$$

Since (22) repeats the operation in (21), the conjugate of F^* is $F^{**} = F$. In other words, the dual of the dual is the primal.

The step from F to F^* is the *Legendre-Fenchel transform*—named after the mathematician who used it in physics and the one who saw its possibilities as mathematics. Legendre concentrated on smooth functions; Fenchel allowed corners, and jumps to infinity. For smooth F the maximum occurs where the derivative of $yx - F(x)$ is zero: $y = F'(x)$. For the parabola this was $y = 2x$ and it gave $x = \frac{1}{2}y$. Notice that we are looking for x! The equation $y = F'(x)$ has to be solved—the function F' has to be "inverted" to find $x = (F')^{-1}(y)$—and this is the subtle point in the calculation. Fortunately F' is an increasing function (since F was convex). Then transforming from F^* back to $F = F^{**}$ reverses this process. The derivative in (22) is zero at $x = (F^*)'(y)$; for the parabola this was $x = \frac{1}{2}y$. We are looking for y and we rediscover $y = 2x$. Let me try to put that relationship into words:

There is a pairing between slopes y and points x. The tangent with slope y touches the graph of F at the corresponding x. For each pair that means

$$F(x) + F^*(y) = xy, \quad \text{or} \quad F^* = xF'(x) - F(x). \tag{23}$$

Furthermore the derivatives $G = F'$ and $H = (F^*)'$ are inverse to each other: $H(G(x)) = x$ and $G(H(y)) = y$.

In the example F was x^2, F^* was $\frac{1}{4}y^2$, and the slope paired with x was $y = 2x$. You can check that (23) is correct. The last statement is easy, since G was multiplication by 2 and H was multiplication by $\frac{1}{2}$.

In Section 3.6 the transformation from F to F^* took the Lagrangian to the Hamiltonian. The pairing was between velocity v and momentum $p = mv$; the kinetic energy was given equally by $F = \frac{1}{2}mv^2$ and $F^* = \frac{1}{2}p^2/m$. Other pairs are fundamental in science and engineering. In statics the pairing is between strain and stress; F and F^* are strain energy and complementary energy. In thermodynamics one pair is pressure and volume, and the Gibbs free energy and the Helmholtz free energy. For electrical networks there is potential difference and current. In n dimensions the tangent lines become tangent planes, but the geometry holds on.

Applications: Minimum Norms

As we end this part of the book, devoted to optimization and duality, it is amazing to see how many applications come from a single source. That source was present at the beginning, in the first example of Chapter 2—the distance to a line. In n-dimensional space, it would become the distance to a flat surface $Ax = b$. And if different norms (measures of distance) are allowed, including the familiar $\|x\|^2 = x_1^2 + \cdots + x_n^2$ along with others, then the applications begin to appear. They all minimize that distance, subject to some form $Ax = b$ of Kirchhoff's current law, which permits a brief review of the whole subject:

(1)   *Transportation problem*: Minimize the shipment cost $C_1x_1 + \cdots + C_nx_n$
(2)   *Resistive network*: Minimize the heat dissipation $x_1^2 R_1 + \cdots + x_n^2 R_n$
(3)   *Maximal flow* (stated differently): Minimize the maximum of $|x_1|/c_1, \ldots, |x_n|/c_n$.

If the costs and resistances and capacities are all 1, the distances are

$$\|x\|_1 = |x_1| + \cdots + |x_n|; \quad \|x\|_2 = (x_1^2 + \cdots + x_n^2)^{1/2}; \quad \|x\|_\infty = \max|x_i|.$$

The first and third are the "l^1 norm" and "l^∞ norm" of x. They are associated with linear programming. If one of them appears in the primal, the other will appear in the dual. In between these two, and dual to itself, is the ordinary Euclidean length $\|x\|_2$. This is the "l^2 norm" of x. It is squared in the electrical problem and in all of Chapter 2, and it leads to quadratic programming.

I realize now that all these problems lead back to the first example of duality in this book: *The minimum distance to a line equals the maximum distance to planes through that line*. It is true that the line has become a higher-dimensional surface; it is the graph of $Ax = b$. Originally I was thinking of an ordinary line, with 2 equations in 3-dimensional space, but one virtue of algebra is its freedom from that limitation. The dual problems are associated with Federal Express, and Sprint, and a minimum cut. Perhaps even the marriage problem fits into this framework; it must. So the whole of duality theory in these applications has come to depend on one final calculation, the distance from a point to a plane.

That is the place to stop. I am sorry to put down my pen; by now I am writing to a friend. I very much hope that you will master applied mathematics, and like it, and use it. Thank you for reading so far.

EXERCISES

8.5.1 Minimize $C = \frac{1}{4}(x_1^4 + x_2^2)$ subject to $x_1 + x_2 = 3$. Show that the multiplier is $y = -1$.

8.5.2 Minimize $C = x_1^2 + 2x_1 x_2 + 5x_2^2$ subject to $x_1^2 + x_2^2 = 1$. Which matrix enters with its eigenvalues?

8.5.3 Minimize $C = \frac{1}{2}(x_1^2 + x_2^2 + x_3^2)$ subject to $x_1 + x_2 + x_3 = 8$ and $x_1 + 2x_2 - x_3 = 10$.

8.5.4 Minimize $C = x^2 - 2x$ without any constraint, and then subject to $x \leq b$. For which b is this constraint active, moving the minimum to $x = b$? Graph the function $M(b)$ that gives the minimum value of $x^2 - 2x$ for each b. (Its slope is $-y$.)

8.5.5 For the constraint $x_1 + 8x_2 = b$ in Example 1 in the text, we could solve for x_1 and substitute into C:

$$C = \frac{1}{4}(x_1^4 + x_2^4) = \frac{1}{4}((b - 8x_2)^4 + x_2^4).$$

Show that the minimum value is $M(b) = (b/17)^3$, and that the sensitivity $-M'$ agrees with the Lagrange multiplier in the text when $b = 34$.

8.5.6 Starting from the quadratic program

$$\text{Minimize } C = \frac{1}{2}X^T HX - X^T b \text{ subject to } X \geq 0,$$

change variables by $X = b - x$ and reach Example 5 in the text. What are the optimality conditions (10–11) in terms of X? What are the optimal x and X and y, if

$$H = \begin{bmatrix} 2 & -1 \\ -1 & 2 \end{bmatrix} \quad \text{and} \quad b = \begin{bmatrix} -1 \\ -4 \end{bmatrix}?$$

8.5.7 Find the second derivative matrix H and test the following functions for convexity:

$$F_1 = x^2 + y^2 + x^2 y^2, \quad F_2 = e^x e^y, \quad F_3 = \frac{1}{2}(ae^{2x} + 2be^{x+y} + ce^{2y}).$$

8.5.8 Show that $F(x) = x^T Hx$ passes the convexity test (12) if H is positive semidefinite (i.e. $x^T Hx \geq 0$ for all x) by simplifying and rearranging the required inequality

$$(cx_1 + (1 - c)x_2)^T H(cx_1 + (1 - c)x_2) \leq cx_1^T Hx_1 + (1 - c)x_2^T Hx_2.$$

8.5.9 If S is a convex set, which of the following functions are convex?

(a) The "characteristic function" $C(x)$, which equals 1 for x in S and zero otherwise
(b) The "support function" $h(x)$, which equals the maximum of $y^T x$ for y in S
(c) The "distance function" $d(x)$, which equals the distance from x to S.

8.5.10   Show that the intersection of two convex sets is a convex set and the maximum of two convex functions is a convex function: $F(x) = \max\left(F_1(x), F_2(x)\right)$ is convex.

8.5.11   If $F(x)$ is a convex function, show that the set S containing all x with $F(x) \leq 0$ is a convex set. (It may be empty.)

8.5.12 $M(b)$, the minimum of $C(x)$ subject to $A(x) \leq b$, is a convex function if C and A are convex. Explain the two inequality steps in the proof:

$$
\begin{aligned}
M(cb_1 + (1-c)b_2) &= \min\left\{C(x) : A(x) \leq cb_1 + (1-c)b_2\right\} \\
&\leq \min\left\{C(x) : x = cx_1 + (1-c)x_2, A(x_1) \leq b_1, A(x_2) \leq b_2\right\} \\
&\leq c\min\left\{C(x_1) : A(x_1) \leq b_1\right\} + (1-c)\min\left\{C(x_2) : A(x_2) \leq b_2\right\} \\
&= cM(b_1) + (1-c)M(b_2).
\end{aligned}
$$

8.5.13   For $L = x^2 - y^2$, find the minimum with respect to x—the depth of a "valley" in the y-direction. Find also the maximum of L with respect to y—the height of a "mountain range" in the x-direction. Why is the top of the valley below the bottom of the range (weak duality)? Where is the saddle point?

8.5.14   Explain why the maximum over y is zero if $A(x) \leq b^*$ and $+\infty$ otherwise, which leads to

$$
\min_{A(x) \leq b^*} C(x) = \min_x \max_{y \geq 0} \left[C(x) + y(A(x) - b^*)\right].
$$

Comparing with (18), duality becomes an exchange of *min* and *max*.

8.5.15   Test the duality equation (18) by minimizing $C(x) = |x|$ subject to $x \leq -1$. Note when the minimum of $|x| + yx + y$ is $-\infty$.

8.5.16   If $F(x) = x^4$ find its convex conjugate $F^*(y) = \max(xy - F)$. Then compute the conjugate of F^*.

8.5.17   Find the conjugate functions $F^*(y)$ of (a) $F(x) = e^x$ (b) $F(x) = 0$ for $x \leq 0$, $F(x) = +\infty$ for $x > 0$. In part (a) show that the pairs x and y satisfy (23), and that the derivatives $G = F'$ and $H = (F^*)'$ are inverses.

8.5.18   Let $F(x) = |x|$, the absolute value function with a corner at $x = 0$. Show from (21) that the depth $d = F^*(y)$ is zero for all slopes between $-1 \leq y \leq 1$. For steeper slopes $d = F^* = \infty$, or the straight line will not stay below F. Show that (22) gives back $F^{**} = F$.

8.5.19   The Schwarz inequality is $z^T x \leq \|z\|\ \|x\|$ (since $\cos \theta \leq 1$). Prove the analogous inequality linking the l^1 and l^∞ norms: $z_1 x_1 + \cdots + z_m x_m \leq (\max |z_k|)(|x_1| + \cdots + |x_m|)$.

8.5.20   Why is the distance d from source to sink not less than their separation s, if every pair of nodes is connected by a string whose length is the length of the edge in the network? The string may be slack or taut.

8.5.21   Why is the shortest distance equal to the maximum separation of source from sink?

APPENDIX: SOFTWARE
FOR SCIENTIFIC
COMPUTING

This appendix is a first step toward obtaining good software at moderate cost (normally under $1000, frequently under $100, sometimes free). I hope the listing will be useful; it can hardly be complete. Most codes are in FORTRAN and have been principally used on mainframes, but they are quickly being transferred to personal computers. We begin with two important sources of software for scientific computing in universities, and with subroutine libraries specifically intended for the IBM PC:

(1) IMSL: MATH/PC LIBRARY, STAT/PC LIBRARY, and SFUN LIBRARY   7500 Bellaire Blvd., Houston TX 77036; (800)222-4675

(2) NAG: PC50 LIBRARY (soon NAG Library for Scientific Workstations) 1101 31st Street, Downers Grove IL 60515; (312)971-2337
256 Banbury Road, Oxford, UK; (865)511245

Those libraries start with the basic operations of numerical linear algebra—the LU factorization for linear systems, the orthogonalization represented by $A = QR$, and the corresponding QR algorithm for the eigenvalue problem. The singular value decomposition and Fast Fourier Transform are also included. The libraries contain linear programming, nonlinear optimization, curve fitting, quadrature, special functions (Bessel...), and difference methods for ordinary differential equations. Thus they match the presentation in this book, up to the point at which a PC is limited by size—roughly speaking, up to partial differential equations.

For smaller pieces of software, and particularly for codes that are in the public domain, we mention two central sources. For each one we give a representative listing. The first provides magnetic tapes at a moderate charge for handling:

(3)   IMSL Distribution Services (address above)
      BSPLINE: interpolation and approximation
      EISPACK: eigenvalue problems
      GRAPHPAK: graphical displays
      ITPACK: iterative methods
      LINPACK: numerical linear algebra
      LLSQ: linear least squares
      MINPACK: nonlinear equations and least squares
      QUADPACK: numerical integration
      TOEPLITZ: constant-diagonal systems
      TOMS: Transactions on Mathematical Software (ACM Collected Algorithms)

The second source, called *"netlib"*, is new. It is a quick service to provide specific routines, not libraries. It is entirely by electronic mail sent over ARPANET, CSNET, Telenet, or Phone Net. Since no human looks at the request, it is possible (and probably faster) to get software in the middle of the night. The one-line message "send index" wakes up a server and a listing is sent by return electronic mail. There is no consulting and no guarantee, but in compensation the response is efficient, the cost is minimal, and the most up-to-date version is always available. The addresses are not recognized by the post office:

(4)   *netlib@anl-mcs* (ARPANET/CSNET) or *research!netlib* (UNIX network)
      EISPACK,   ITPACK,   LINPACK,   MINPACK,   QUADPACK,
      TOEPLITZ, TOMS
      PPPACK: piecewise polynomials including splines
      FFTPACK: Fast Fourier Transforms including sine and cosine
      FISHPACK: Poisson's equation using FFTPACK
      BLAS: Basic Linear Algebra Subprograms
      SCPACK: Schwarz-Christoffel conformal mapping
      FNLIB: special functions...and others to come

Beyond that there is a very wide range of codes. The tremendous success of EISPACK established that first class software could be written and would be used. We point to several sources:

(5)   PC/MATLAB for linear algebra commands in a fully-developed inter-active format: Mathworks, 158 Woodland Street, Sherborn MA 01770
(6)   ELLPACK for linear boundary-value problems on general 2-D and rectangular 3-D regions: Computer Science Department, Purdue University IN 47907
(7)   SPARSPAK for sparse systems of linear equations: Research Institute, University of Waterloo, Canada

(8)   MINOS for linear and nonlinear programming and SYMMLQ for indefinite systems: Software Distribution Center, Stanford CA 94305

(9)   ODEPACK and DEPAC for ordinary differential equations, FUNPACK for special functions: National Energy Software Center, Argonne IL 60439

(10)   Harwell Subroutine Library, especially for sparse systems—linear, nonlinear, eigenvalues, optimization, ODE's: AERE Harwell, Didcot, UK

(11)   Algorithms for Network Programming, subroutine listings for the algorithms of Chapter 7: textbook by Kennington and Helgason

(12)   SEQS and NLLSQ for nonlinear problems on Apple computers: CET, Box 2029, Norman OK 73070

(13)   LINDO and GINO for linear, integer, discrete and general optimization: Scientific Press, 540 University Avenue, Palo Alto CA 94301

(14)   PLTMG for linear finite elements with grid refinement: R. Bank, Math. Dept., UCSD CA 92093

(15)   LINPACK...in EEC countries: NEA Data Bank, B.P. 9, F-91191 Gif-sur-Yvette, France

(16)   DASSL and DSS/2 for differential/algebraic systems: W. Schiesser, Lehigh University, Bethlehem PA 18015

(17)   MacSimplex for teaching the simplex algorithm: A. Philpott, Engineering School, University of Auckland, New Zealand

(18)   GAUSS Program Language for PC computing: Aptech, 1914 N.34th, Seattle WA 98103

(19)   Alligator Transforms for FFT: Box 11386, Costa Mesa CA 92627

(20)   Dynamical Systems for ODE's, phase plane, fractals: Box 35241, Tucson AR 85740

(21)   FractalMagic from Sintar Software, Box 3746, Bellevue WA 98009

(22)   BCSLIB and VectorPak from PC's to supercomputers, Boeing Box 24346, 7W-05, Seattle WA 98124

Statistics and graphics codes are included in the IMSL and NAG libraries, and in more specialized packages. Linear programming is similar. Other libraries (PORT, SLATEC, YALE, Scientific Desk) are being established or reorganized. For finite elements the commercial codes are especially dominant: ABAQUS, ADINA, ANSYS, ASKA, GTSTRUDL, MARC, NASTRAN, PROBE (*hp*-version), and others. Those are major systems! We close with references to documentations and to a wider collection of codes, noting again the book *Numerical Recipes*:

(a) LINPACK User's Guide (J. Dongarra...), SIAM

(b) EISPACK (B. Garbow...), Lecture Notes in Computer Science 6 and 51, Springer

(c) QUADPACK (R. Piessens) and ELLPACK (J. Rice), Springer

(d) Handbook for Automatic Computation (J. Wilkinson, C. Reinsch), Springer

(e) Numerical Methods: IMSL Reference Edition (J. Rice), McGraw-Hill

(f) Software Catalog 1985: Science and Engineering, Elsevier

(g) Sources and Development of Mathematical Software (W. Cowell), Prentice-Hall

REFERENCES AND ACKNOWLEDGEMENTS

G. Bierman, Factorization Methods for Discrete Estimation, Academic (1977)

R. Blahut, Fast Algorithms for Digital Signal Processing, Addison-Wesley (1985)

V. Chvátal, Linear Programming, Freeman (1983)

P. Collet and J.-P. Eckmann, Iterated Maps on the Unit Interval as Dynamical Systems, Birkhäuser Boston (1980)

R. Devaney, Introduction to Chaotic Dynamical Systems, Cummings (1986)

I. Duff, A. Erisman, and J. Reid, Direct Methods for Sparse Matrices, Oxford (1986)

R. Fletcher, Practical Methods of Optimization, John Wiley (1980)

D. Gale, The Theory of Linear Economic Models, McGraw-Hill (1960)

A. George and J. Liu, Computer Solution of Large Sparse Positive Definite Systems, Prentice-Hall (1981)

G. Golub and C. Van Loan, Matrix Computations, Johns Hopkins Press (1983)

S. Goldstein, Lectures on Fluid Mechanics, Interscience (1960)

P. Henrici, Essentials of Numerical Analysis, John Wiley (1982)

F. B. Hildebrand, Advanced Calculus for Applications, Prentice-Hall (1962)

T. Hughes and J. Marsden, A Short Course in Fluid Mechanics, Publish or Perish (1976)

H. Kocak, Differential Equations through Computer Experiments, Springer (1986)

E. Kreyszig, Advanced Engineering Mathematics, 5th ed., John Wiley (1983)

J. D. Lambert, Computational Methods in Ordinary Differential Equations, John Wiley (1973)

E. Lawler, Combinatorial Optimization, Holt, Rinehart and Winston (1976)

D. Luenberger, Optimization by Vector Space Methods, John Wiley (1967)

A. Newell, Solitons in Mathematics and Physics, SIAM (1985)

J. Norwood, Intermediate Classical Mechanics, Prentice-Hall (1979)

A. Oppenheim (ed.), Applications of Digital Signal Processing, Prentice-Hall (1978)

C. Papadimitriou and K. Steiglitz, Combinatorial Optimization, Prentice-Hall (1981)

B. N. Parlett, The Symmetric Eigenvalue Problem, Prentice-Hall (1980)

H. O. Peitgen and P. H. Richter, The Beauty of Fractals, Springer (1986)

W. Press, B. Flannery, S. Teukolsky, W. Vetterling, Numerical Recipes, Cambridge (1986)

B. N. Parlett, The Symmetric Eigenvalue Problem, Prentice-Hall (1980)

R. Richtmyer and K. W. Morton, Difference Methods for Initial-Value Problems, Wiley-Interscience (1967)

E. B. Saff and A. D. Snider, Fundamentals of Complex Analysis, Prentice-Hall (1976)

S. Senturia and B. Wedlock, Electronic Circuits and Applications, John Wiley (1975)

W. M. Siebert, Circuits, Signals, and Systems, M.I.T. Press and McGraw-Hill (1985)

J. G. Simmonds, A Brief on Tensor Analysis, Springer-Verlag (1982)

G. Strang, Linear Algebra and Its Applications, 2nd ed., Academic Press (1980)

G. Strang and G. Fix, An Analysis of the Finite Element Method, Prentice-Hall (1973); (now published by Wellesley-Cambridge Press)

R. Tarjan, Data Structures and Network Algorithms, SIAM (1983)

R. Voigt, D. Gottlieb, and M. Y. Husseini (eds.), Spectral Methods for Partial Differential Equations, SIAM (1984)

G. B. Whitham, Linear and Nonlinear Waves, John Wiley (1974)

O. C. Zienkiewicz, The Finite Element Method, McGraw-Hill (1977)

More than those references, I want to acknowledge the help that has come from friends. It is a special pleasure to thank Sophia Koulouras for typing this book. Together with Mike Michaud of Unicorn Production Services and David Macdonald of M.I.T. Graphic Arts, she made it fun to publish as well as to .ite. The book relies on the line drawings by Cyndy Patrick and some remarkable figures from the computer—more than ten were prepared by Nick Trefethen, who created SCPACK for conformal mapping. Three other favorites (the delta function 4.2, the Gibbs phenomenon 4.3, and the Bessel function 4.5) were gifts of Alar Toomre. The beautiful picture of chaos (Fig. 6.16) was contributed by Jim Yorke. It arrived four feet long, from the walls of his office. Some of its patterns are still a mystery.

I am enormously grateful—the word is hardly adequate—for advice about the text itself. May I put on record my debt to twenty-two people: Howard Brenner, Iain Duff, Cathy Durso, Hermann Flaschka, Bill Gear, Gene Golub, Eric Grosse, John Guckenheimer, Robert Kotiuga, Ed Lorenz, Dick MacCamy, Andy Majda, Ignacio Mas, Marc Meketon, Lisa Newton, Nick Papageorgiou, Andy Philpott, Jim Simmonds, Mike Todd, Nick Trefethen, George Verghese, and Grae Worster (for figures too). If the book is not perfect, we accept collective responsibility. In truth, I owe more than I can say.

My gratitude goes above all to my wife Jill. The years of writing this book have been among the happiest of my life.

Notes of acknowledgement: to Claude Poirier and Ghyslaine McClure, who have kindly undertaken a French translation ; to the National Science Foundation and the Army Research Office, for support when needed ; to John Wiley for permission to reproduce Figs. 6.11–12 from the book by Kreyszig, to Sam Stearns for Fig. 4.1, and to Tom Hughes for Fig. 3.11 ; and especially to M.I.T., whose students it is still inspiring to teach. ·

SOLUTIONS TO
SELECTED EXERCISES

CHAPTER 1   SYMMETRIC LINEAR SYSTEMS

1.2.1 $(x_1, x_2, x_3) = (3, -2, 1)$ **1.2.4** (a) $q = 12$ (d) $b_2 = 2$

1.2.9 $B = \begin{bmatrix} 1 & -1 \\ 1 & -1 \end{bmatrix}$, $A = C = \begin{bmatrix} 0 & -1 \\ 1 & 0 \end{bmatrix}$, $D = \begin{bmatrix} 0 & 1 \\ 1 & 0 \end{bmatrix}$ **1.2.10** $A^{-1} = \begin{bmatrix} 3 & 2 & 1 \\ 2 & 2 & 1 \\ 1 & 1 & 1 \end{bmatrix}$

1.2.12-13 $3, -2, 1$

1.3.1 $LDL^T = \begin{bmatrix} 1 & 0 \\ -1 & 1 \end{bmatrix}\begin{bmatrix} 3 & \\ & 2 \end{bmatrix}\begin{bmatrix} 1 & -1 \\ 0 & 1 \end{bmatrix} = \begin{bmatrix} 1 \\ -1 \end{bmatrix}3[1 \quad -1] + \begin{bmatrix} 0 \\ 1 \end{bmatrix}2[0 \quad 1] \rightarrow 3(x_1 - x_2)^2$
$$+ 2x_2^2$$

1.3.2, 1.3.6 No   **1.3.8** $c > 2$; positive definite if $c > 1$

1.3.10 Eliminate upwards on A, last row first   **1.3.11** F_1: local minimum;
F_2: maximum

1.3.15 $\bar{L} = \begin{bmatrix} 2 & 0 \\ 6 & 3 \end{bmatrix}$ **1.3.20** $A^T A$ is increased by vv^T

1.4.2 $Ax = b$; $A^T Ax = A^T b$; $A^T Ax = A^T b$ **1.4.4** $C = 0$, $D = 1$

1.4.7 $5; -1, 6; 0, 0, 3$ **1.4.8** $C = 2$, $C + E = 2$, $C + D = 1$, $C + D + E = 5$

1.4.9 $m < n$ leads to dependent columns

1.4.12 $K = \begin{bmatrix} c_1 + c_2 & -c_2 & 0 \\ -c_2 & c_2 + c_3 & -c_3 \\ 0 & -c_3 & c_3 \end{bmatrix}$ has $K^{-1} > 0$

1.5.3 $\lambda = 3$, $x = \begin{bmatrix} c \\ c \\ c \end{bmatrix}$; $\lambda = 0$, $x = \begin{bmatrix} c \\ d \\ -c-d \end{bmatrix}$ **1.5.5** $\dfrac{6}{5}e^{5t}\begin{bmatrix} 2 \\ 1 \end{bmatrix} - \dfrac{12}{5}e^{-5t}\begin{bmatrix} 1 \\ -2 \end{bmatrix}$

1.5.8 $u = c_1\begin{bmatrix} 1 \\ 3 \\ 4 \end{bmatrix} + c_2e^{2t}\begin{bmatrix} 1 \\ 1 \\ -2 \end{bmatrix} + c_3e^{4t}\begin{bmatrix} 1 \\ -1 \\ 0 \end{bmatrix}$; the note refers to $\dfrac{d^2u}{dt^2} + Au = 0$

1.5.11   trace AB = trace $BA = \Sigma\Sigma\, a_{ij}b_{ji}$ **1.5.12**   Choose $\lambda = 0$

1.5.14 $v_1 = e^{3t} + te^{3t}$, $v_2 = e^{3t}$ **1.5.20** $(\lambda^2 M + \lambda F + K)x = 0$

1.5.23 $A^{1/2} = \begin{bmatrix} 3 & -1 \\ -1 & 3 \end{bmatrix}$ **1.5.24, 25** $\lambda = \frac{1}{2}(3 \pm \sqrt{3}) = \omega^2$

1.6.3   4 independent rows, 2 from 1–2–6, 2 from 3–4–5; $(1, 1, 0, 0, 0, -1)$ and $(0, 0, 1, 1, -1, 0)$

1.6.4 A has rank 7; $Ax = 0$ has p solutions, $A^Ty = 0$ has $m - n + p$

1.6.5   Along a path from j to k, x is the same at all nodes

1.6.7 Ax is in $\mathscr{R}(A)$, y is in $\mathscr{N}(A^T)$

1.6.10 A is orthogonal, invertible, permutation, diagonalizable; $1, -1, i, -i$

1.6.15 $y^Tb = y^TAx = (A^Ty)^Tx = 0$ **1.6.16** $r = n < m; r = m < n; r < m, r < n; r = m = n$

1.6.19 T, T, T, T, F **1.6.21** $\lambda = 0, 0; J = \begin{bmatrix} 0 & 1 \\ 0 & 0 \end{bmatrix}$

1.6.23 $\lambda_1\lambda_2 = 6$; $\lambda = 3 \pm \sqrt{3}$ **1.6.26** $QR = \begin{bmatrix} .6 & -8 \\ .8 & .6 \end{bmatrix}\begin{bmatrix} 5 & 4 \\ 0 & 3 \end{bmatrix}$

1.6.27 $Q = \dfrac{1}{\sqrt{5}}\begin{bmatrix} 2 & -1 \\ 1 & 2 \end{bmatrix}$, $B = \sqrt{5}\begin{bmatrix} 2 & 1 \\ 1 & 2 \end{bmatrix}$ **1.6.29-31** $\lambda = 1, 0; 1, 1, 0; 1, -1$

1.6.28 $\dfrac{1}{\sqrt{10}}\begin{bmatrix} 3 & 1 \\ -1 & 3 \end{bmatrix}\begin{bmatrix} \sqrt{5} & \\ & 3\sqrt{5} \end{bmatrix}\dfrac{1}{\sqrt{2}}\begin{bmatrix} 1 & -1 \\ 1 & 1 \end{bmatrix}$

1.6.34 $c = (w^Tv - 1)^{-1}$; no inverse if $w^Tv = 1$

1.6.35 $c = (w^TA^{-1}v - 1)^{-1}$; with $v = w = (1, 0, 0, 0)$, $qr^T/(s - 1)$ is subtracted from A^{-1}

CHAPTER 2   EQUILIBRIUM EQUATIONS AND MINIMUM PRINCIPLES

2.1.3 $x = (-1, -1, -1)$, $y = (0, 0, 0, -1, -1, -1)$ **2.1.4** $x = (-1, -1, -1)$, $y = 0$

2.1.10 $(D - A^TCA)x = f$, $(-A_1^TC_1A_1 - A_2^TC_2A_2)x = f$ **2.1.11** $A^TCAx = A^TCb$

2.2.1 $y = (2, 6)$, $P = 2x^2 + 8x$ **2.2.3** $Q = \dfrac{1}{4}, \dfrac{7}{3}, 3, -\infty$ **2.2.4** $y_1 = 4y_2$

2.2.9   The points are in the planes, so distance to point $\geq$ distance to plane

2.2.13 $Ax = Pb = (5, 2, -1)$ **2.2.15** $\lambda = 1, 0$; column space $= S$; rank $= n$; det $= 0$ if $n < m$

2.3.1 $\begin{bmatrix} R_1 & 0 & -1 \\ 0 & R_2 & -1 \\ -1 & -1 & 0 \end{bmatrix}\begin{bmatrix} I_1 \\ I_2 \\ x_1 \end{bmatrix} = \begin{bmatrix} 0 \\ 0 \\ -I \end{bmatrix}$ **2.3.2** $V/I = 1$ **2.3.4** $R = 3/5$

2.3.8 $A_1x_1 - A_2x_2 = b_1$, $A_2x_1 + A_1x_2 = b_2$ **2.3.12**   pressure

2.3.10   Euler's formula is # loops − # edges + # vertices = 1 (or see page 74)

2.4.2 $K = \dfrac{1}{4} \begin{bmatrix} 7 & -4 & -\sqrt{3} & 0 \\ -4 & 7 & 0 & \sqrt{3} \\ -\sqrt{3} & 0 & 5 & 0 \\ 0 & \sqrt{3} & 0 & 5 \end{bmatrix}$ **2.4.3** $y = f_H^1(1/\sqrt{3}, 0, 0, -2/\sqrt{3})$

2.4.4   solution if $f_H^1 = -f_H^2$ **2.4.9**   mechanism even with 2 bars and $m = n$

2.5.1   limits $\frac{1}{2}(b_1 + b_2)$, b_3 **2.5.2-3** $P = (\sigma_1^{-2} + \cdots + \sigma_m^{-2})^{-1}$,
$\hat{x} = P(b_1\sigma_1^{-2} + \cdots + b_m\sigma_m^{-2})$

2.5.5 $\hat{x} = 20/7$, $P = 5/7$ **2.5.6** $\sigma^2 = 1$ **2.5.11** $\sigma^2 = 35/6$

2.5.12 $\binom{n}{x}p^x(1-p)^{n-x}$ **2.5.15-16** $\sigma_1 = 1/\sqrt{5}$, $\sigma_2 = \sqrt{2}/\sqrt{5}$; see **2.5.2-3**

2.5.18 $1/P_3 = 1/(A^T A)_{44}^{-1} = 21/13$

CHAPTER 3   EQUILIBRIUM IN THE CONTINUOUS CASE

3.1.1 $u = \dfrac{1}{6c}[1 - (1-x)^3]$ **3.1.4** $\int f \, dx = 0$ **3.1.5** $1 - x + ex - e^x$

3.1.6 $u = -\frac{1}{2}fx^2 + fx$ for $x < \frac{1}{2}$, $-\frac{1}{4}fx^2 + \frac{1}{2}fx + \frac{3}{16}f$ for $x > \frac{1}{2}$

3.1.7 $W = x^5/240 - x^6/720 - x^3/144 + x/240$

3.1.10 $d_1 e^{1/\sqrt{c}} + d_2 e^{-1/\sqrt{c}} + 1 = 0$ leads to $d_2 = (e^{1/\sqrt{c}} - 1)/(e^{-1/\sqrt{c}} - e^{1/\sqrt{c}}) \approx -1$

3.1.12 $V = \frac{1}{6}(\frac{1}{2} - x)^3 - \frac{1}{24}(\frac{1}{2} - x)$

3.1.13 $A = -B = (e^{1/c} - 1)^{-1} \to 0$; $u \to x$ keeping $u(0) = 0$

3.1.15 $u = x$ or $u = \frac{1}{2}$; u' jumps by -1 **3.1.16** $u = \frac{1}{2}x$ or $u = \frac{1}{2}(1-x)$

3.2.1 $\int u''v''dx$ **3.2.2** $\frac{1}{6}(x^3 - x)$ **3.2.3** $\frac{1}{2}x^2 - \frac{1}{6}$

3.2.4 $u = \frac{1}{6}x^3 - \frac{1}{6}x$, $w = \frac{1}{2}x^2 - \frac{1}{2}$ **3.2.7-8** $\int \frac{c}{2}(u')^2 + w'u + \frac{w^2}{2c} = \int \frac{c}{2}\left(u' - \frac{w}{c}\right)^2 \geq 0$

3.2.10 $u = x^2(x-1)^2/24$ **3.2.16** $u = 4x$

3.2.18 $2s_0 + s_1 = (3u_1 - 3u_0)/h$ **3.3.1** $s = \sin 4\theta$

3.3.3 $(d_1 - d_6) + (d_2 - d_9 + d_7 - d_1) + () + () = d_3 - d_9 + d_8 - d_5$

3.3.8   (c) $s = \theta$ **3.3.9**   (a) $y = -e^{-x} + C$ (c) $y^2 - x^2 = C$ **3.3.16** $u = 3xy$

3.3.17 $u = x + x/(x^2 + y^2)$ **3.3.19** $-u_{xx} - u_{yy} - f = 0$, n·grad $u = 0$

3.4.1   arc lengths $2t$, $\sqrt{3t^2}$ **3.4.2-3** $(z, z, x + y) = $ grad $(xz + yz)$; $\int F \cdot dr = 2$

3.4.6 $(ds/dt)^2 = 1 + t^2$ **3.4.7**   perpendicular + parallel   **3.4.8** $w = (0, 0, 1)$

3.4.9 $v = (1, 0, y)$ **3.4.12** $S_2 = (5yz, 0, 2xy) + $ any grad u

3.4.16   (a) $u = $ constant; $u = 0$

3.4.17   (b)   div $f = $ div $v = $ div grad u is Poisson's equation for u

3.4.21   curl $(x, 0, 0) = 0$ **3.4.23** -1; 0 **3.4.24** $u = 1$, $w = (\phi_x, \phi_y)$

3.4.27   (d)   0   (f)   0   **3.4.28** $f_{uu} + f_{vv} + (u^2 + v^2)f_{ww} = 0$

3.4.29   (c) r^2 is now $x^2 + y^2 + z^2$ **3.4.30**   compare with **3.3.17**   **3.5.1** $Au = 0$

3.5.2 (d)   and **3.5.12** $\begin{bmatrix} 0 & 0 & \frac{1}{2}\partial u_3/\partial x \\ 0 & 0 & \frac{1}{2}\partial u_3/\partial y \\ - & - & 0 \end{bmatrix}, \sigma = 2\mu e, f = -2\mu \begin{bmatrix} 0 \\ 0 \\ \nabla^2 u_3 \end{bmatrix}$

3.5.5 $v = \lambda/(2\lambda + 2\mu)$ **3.5.8** $\dfrac{1}{4\mu}\sum \sigma_{ij}^2 - \dfrac{\lambda}{4\mu + 6\lambda}(\text{tr } \sigma)^2$

3.5.13   The first unit eigenvector has 2 degrees of freedom; the second has one

3.5.15 $4xy$ and $(x, y, 0)$; 0 and $(x, 0, z)$ **3.5.17**   Bernoulli

3.5.18 $p/\rho = \frac{1}{2}\omega^2 x^2 + \frac{1}{2}\omega^2 y^2 - gz$ **3.5.20**   center jet

3.5.23   In column j, $\dfrac{\text{div } \sigma}{\mu} = \dfrac{\partial}{\partial x_1}\left(\dfrac{\partial v_1}{\partial x_j} + \dfrac{\partial v_j}{\partial x_1}\right) + \cdots = \dfrac{\partial}{\partial x_j}(\text{div } v) + \nabla^2 v_j = \nabla^2 v_j$

3.6.2 $\int u'v' \, dx = 0$ for all v with $v(0) = v(1) = 0$; $-u'' = 0$

3.6.3   (b) $0 = 0$ (c) $(-2x^2 u')' = 0$ **3.6.7**   path goes to the left of 0

3.6.5 $u'/x\sqrt{1 + (u')^2} = \text{constant}$; $u = 1 - \sqrt{1 - x^2}$

3.6.9 $M = 1/m$; $l^2 < 1 + (b - a)^2$ is not achievable

3.6.11-12 $-\log u - 1 + m_1 + m_2 x + m_3 x^2 = 0$ **3.6.13**   helix

3.6.14 $\int \frac{1}{2}(u')^2 - \cos u$

3.6.15   (a)   biharmonic   (b) $(yu_x)_x + u_{yy} = 0$ (d) $-u_{xx} - u_{yy} = \lambda u$

CHAPTER 4   ANALYTICAL METHODS

4.1.1   (b) $\dfrac{2}{\pi} - \dfrac{4}{\pi}\sum\dfrac{\cos 2nx}{4n^2 - 1}$ (c) $\pi^2/3 + \Sigma(-1)^k 2k^{-2} \cos kx$

4.1.2 $\dfrac{4}{\pi}\left(\sin x + \dfrac{\sin 3x}{3} + \dfrac{\sin 5x}{5} + \cdots\right)$

4.1.3 $2\delta(x) - 2\delta(x + \pi) = \dfrac{2}{\pi}\sum (\cos kx - \cos k(x + \pi)) = \dfrac{2}{\pi}\sum 2\cos kx$ (odd k only: $\cos k(x + \pi) = -\cos kx$)

4.1.6 $\dfrac{4}{\pi}\left(r \sin \theta + \dfrac{r^3 \sin 3\theta}{3} + \cdots\right)$ **4.1.10** $\frac{1}{2}, \frac{3}{4}\cos x$

4.1.12 $\dfrac{1}{\pi} + \dfrac{1}{2}\sin x - \dfrac{2}{\pi}\sum\dfrac{\cos 2kx}{4k^2 - 1}$ **4.1.14** $\|u\|^2 = \|f\|^2 = \frac{4}{3}$, $\|v\|^2 = \|g\|^2 = \frac{9}{8}$, $u^T v = \frac{6}{5}$

4.1.16 $\frac{1}{2}$ (compare **4.1.6**)   **4.1.18**   (b) $\dfrac{1 - r^2}{2\pi(1 + r^2 - 2r \sin \theta)}$

4.1.21   error is orthogonal to P_0, P_1, P_2 **4.1.24**   all $c_{jk} = 1/4\pi^2$; only $c_{j0} = 1/4\pi^2$

4.1.29 $x^2 - y^2$, $x^2 - z^2$, xy, xz, yz **4.1.31** $p = r$, $q = -n^2/r$, $w = r$

4.1.36 $J_0(\sqrt{\lambda_1}) = 0 = J_0(\sqrt{\lambda_2})$ **4.2.1** $c = (1, 0, 0, 0)$, $(\frac{1}{2}, 0, \frac{1}{2}, 0)$, $(0, 1)$

4.2.2 $(4, 0, 0, 0)$, $(1, -1, 1, -1)$ **4.2.4** $(11, 11, 14)$ **4.2.9**   use $w^2 + w + 1 = 0$

4.2.10 $(3, 6, 9, 12, 9, 6, 3) = 3702963$; each component consists of n products, each $\leqslant 81$

4.2.11 $\lambda = 0, 2, 4, 2$

4.2.12 $x_1 = (1, 0, -1, 0)$, $x_2 = (0, 1, 0, -1)$, Fourier eigenvectors are $x_1 \pm ix_2$

4.2.17 3, 10, 8; $a * u = u$ **4.2.18** $a = l\overline{l} = \frac{1}{2}e^{-it} + \frac{5}{4} + \frac{1}{2}e^{it}$; L^{-1} has diagonals 1, $-\frac{1}{2}, \frac{1}{4}, \cdots$

4.2.20 eigenvector has $u_k = e^{-ikt}$ **4.2.22** A^{-1} has diagonals $-\frac{2}{3}, \frac{5}{3}, -\frac{2}{3}$

4.3.1 $1/(a - ik)$ **4.3.2** (b) $\pi\delta - 1/ik$ (c) $\hat{f} = 2\pi$ for $0 < k < 1$ from (5)

4.3.3 (b) $1/(\pi + \pi x^2)$ **4.3.4** π; $\pi/2a^3$ **4.3.6** $W_x = W_k = 1/\sqrt{2}$

4.3.9 delta function at 0; $-a/(a + ik) + 1 = ik/(a + ik)$

4.3.11 $(1/ik - ik)\hat{u} = 1$; $u = \frac{-1}{2}$ (odd pulse)

4.3.15 $\int e^{-ikx}\delta' dx = -(e^{-ikx})'(0) = ik$; $-\int ke^{ikx}dx = +2\pi i\delta' = \hat{x}$ **4.3.23** $g = \delta - \frac{8}{6}e^{-3|x|}$

4.3.27 $\hat{G} = 1/(k^2 + a^2)^2$; $G = e^{-a|x|} * e^{-a|x|}/4a^2$ **4.3.28** δ

4.4.2 (d) $-(\pi/2)(\log \pi/2 + i\pi/2)$ **4.4.3** (c) $|z_1 z_2| = 1$ (d) $|z_1 + z_2| \leqslant 2$

4.4.4 (a) 1 (c) 1 (e) e^3 **4.4.7** no; no; yes; yes   **4.4.8.** $u = x^2$ has $u_{xx} + u_{yy} = 2$

4.4.10 (b) $ru_r = s_\theta$, $-rs_r = u_\theta$ **4.4.11** circles reach to $z = 1$

4.4.13 angle b; radius e^c **4.4.17** $x > 1$ and $0 < x < 1$ go to $w > 1$; iy goes to 2 copies of iY

4.4.18 $(4X/5)^2 + (4Y/3)^2 = 2X^2 - 2Y^2 = 1$ **4.4.20** $-u$; $F = -if$

4.4.21 $u = (x + 1)(x_0 - 1)/2$ or $(x - 1)(x_0 + 1)/2$; $\int Ghdx_0$

4.4.23 $u = Y = 4x^3y - 4xy^3$ **4.5.2** (a) $\pi i/2, -3\pi i/2$ (b) πi (c) 0 (e) $6\pi i$

4.5.7 (a) (c) Res $= \pm\frac{1}{4}^+$ (e) Res $(z = 2n\pi i) = -1$ by l'Hôpital

4.5.8 (a) $-\pi i$ (c) $2\pi i$ **4.5.9** (a) $3\pi/4$ (b) $2\pi/\sqrt{a^2 - 1}$

4.5.10 (a) $\pi/2$ (c) $\pi/\sqrt{2}$

4.5.11 (b) poles at $n\pi$ (d) branch pt. at 1   (f) (h) branch pt. at 0. Also (d) (f) (h) branch pt. at ∞, (b) (h) ess. sing. at ∞

4.5.13 (a) $\sin t$ (b) $\cos t$ (c) $\frac{1}{2}t^2e^{-at}$

CHAPTER 5   NUMERICAL METHODS

5.1.1 x in diagonal, subdiagonal, whole last column of U **5.1.4** six -1's; $w = N^2$

5.1.5 pt. 1 is at most $N - 1$ steps from pt. N^2; numbering between must jump by $\dfrac{N^2 - 1}{N - 1}$

5.1.6 $P = \frac{1}{2}x_1^2 + x_1 \sin x_2 + \frac{1}{2}x_2^2$ **5.1.8** $c = 0, d = 3$

5.1.10–11 (c) $|1 - a| < 1$; each step multiplies error by $|f'(x*)|$

5.1.12 $x^1 = (2, 8)17/66$ **5.2.1** $c = (b_1, \ldots, b_n)$

5.2.2 $p = 7a/25$ **5.2.3** (2) $b = (1, 1, \ldots, 1)$ **5.2.5** $p = (0, 0, 1, 1)$

5.2.10 $P = QQ^T$ **5.2.12** $\tan\theta = -c/a$

5.2.17 $Q_{31} = Q_{41} = Q_{42} = 0$ **5.2.18** $Hy = HHx = x$

5.2.21 $3H_1 = \begin{bmatrix} 2 & 1 & 2 \\ 1 & 2 & -2 \\ 2 & -2 & -1 \end{bmatrix}$ **5.2.22** $Hu = -u$

5.2.24 $5H = \begin{bmatrix} 5 & 0 & 0 \\ 0 & 3 & 4 \\ 0 & 4 & -3 \end{bmatrix}$ **5.3.1** $\lambda^2 = bc/ad$ **5.3.2** $\lambda = 0, bc/ad$

5.3.3 $B = B^3 = B^5 \ldots$ **5.3.4** Jacobi converges, Gauss-Seidel diverges!

5.3.6 $\omega \to 2$ **5.3.7** (x, x) and $(x, -x)$

5.3.12 Fibonacci numbers $\approx (1 + \sqrt{5}, 2)$

5.3.17 $T_{n+1} = 2xT_n - T_{n-1}$ **5.3.18** $q_2 = (0, 0, 1), q_3 = (0, 1, 0)$

5.3.19 $\beta = -d^T Ar/d^T Ad$ **5.3.21** $a_1 = 1 + c^2, b_1 = cd, a_2 = 1 + d^2$

5.3.22 $x_1 = (1/6, 1/6)$ or $(5/54, 5/6)$

5.3.23 $c + d\lambda_i = \lambda_i^{-1}$ gives $c = \lambda_1^{-1} + \lambda_2^{-1}, d = -\lambda_1^{-1}\lambda_2^{-1}$

5.3.24 $x = (1, 1)$ and $(1, -1)$ **5.4.2** $U = 4y(1 - x)$ in lower triangle, $4x(1 - y)$ above

5.4.4 $U = (1 - x)(1 - y)(1 - 2x - 2y)$ **5.4.5** 24; 14; no neighbors outside its square

5.4.6 $B_0 = \dfrac{1}{4}\begin{bmatrix} -1 & 1 & 1 & -1 \\ -1 & -1 & 1 & 1 \end{bmatrix}$ (corners in order $(-1, -1)(-1, 1)(1, 1)(1, -1)$)

5.4.19 $K^* = 2/h + 4h/6 = 13/3, F = 1/2, U_1 = 3/26 = .115$;
$u = 1 - e^x/(e + 1) - e^{1-x}/(e + 1) = 1.113$

5.5.1 $F_4^4 = 16I$ **5.5.5** $(20, -4, -4, -4)$; $(0, -4i, 0, 4i)$

5.5.6 from (2) and 5.5.5, $y = (20, -4 - 4iw_8, -4, -4 + 4iw_8^3, 20, -4 + 4iw_8, -4, -4 - 4iw_8^3)$

5.5.10 $\lambda = d - w^j - w^{-j}, w = (1 + \sqrt{3}i)/2$

5.5.13 T_n^2 has diagonals 1, -4, 6, -4, 1 approximating u''''

5.5.16 $u = f/20\pi^2$; $u = f/(4 - 2\cos 2\pi h - 2\cos 4\pi h)$

5.5.19 (1 or θ) (1 or $\log r$), $(\cos n\theta$ or $\sin n\theta$) $(r^n$ or $r^{-n})$

CHAPTER 6   INITIAL-VALUE PROBLEMS

6.1.1 (b) $\dfrac{1}{1 + i\omega} e^{i\omega t} + \dfrac{4 + 5i\omega}{1 + i\omega} e^{-t}$ (c) $te^{-t} + 5e^{-t}$

6.1.2 $u = e^{2(1-t)}$ for $1 < t < 4$; $c = -e^{-6}$

6.1.3 $u = (1 + kt)^{1/k}$ blows up at $t = -1/k$; $u = (4t)^{1/4}$ solves $u' = 1/u^3$

6.1.4 $u = \int_0^t e^{\sin t - \sin s}ds + 4e^{\sin t}$

6.1.5 (b) $u = C/t$ (c) $u = (\sin t + C)^{1/2}$ **6.1.7** $u = t^2 - t^{-1}$

6.1.8 $T = N^{-1} \log (N - 1)$

6.1.11 (a) $u = Ce^{-3t} + De^{3t}$ (c) $u = Ce^{-t}\cos 2t + De^{-t}\sin 2t$

6.1.12 (c) $u'' = 0$ (d) $u'' + 2u' + 2u = 0$

6.1.13 $c > 1, c = 1, 0 < c < 1, c = 0, c < 0; \lambda = -c \pm \sqrt{c^2 - 1}$

6.1.14 (a) $-\frac{1}{3}\cos 2t$ **6.1.15** (a) $(2 \sin \omega t - \omega \cos \omega t)/(4\omega + \omega^3)$

6.1.16 $\omega = \omega_0$ **6.1.18** (a) $t^2 - 2t$

6.1.20 $u = \sin t + \frac{1}{2}t \sin t$

6.1.21 (a) $E = \frac{1}{2}(u')^2 + \frac{1}{2}e^u + \frac{1}{2}e^{-u}$; $e^U + e^{-U} = 3$, $U = u_{max}$ **6.1.22** $u = (2e^{-t} - 1)^{-1}$

6.2.2 (4) stable node or spiral   **6.2.4** $u_2^2 + 4u_1^2 = C$

6.2.5 $\lambda = \pm (a^2 + b^2 + c^2)^{1/2}i$, 0

6.2.6 A has eigenvalues $-1, 0, 1$ at $u^* = (0, 1, 0)$

6.2.7 $u_1'' = (\text{trace})u_1' + (\text{determinant})$ **6.2.9** $E' = \pm 2E^2$, $E = (E_0^{-1} \mp 2t)^{-1}$

6.2.11 (b) $45°$ lines   (c) rays

6.2.12 $u^* = (0, 1)$ stable for $b > 1$, $u^* = \left(\dfrac{1-b}{1+bc}, \dfrac{1+c}{1+bc}\right)$ stable for $b < 1$, $c > -1$, $bc > -1$

6.2.16 (a) unstable spiral

6.2.17 $t = 1$, center to saddle; $t = 5$, spiral to node; $t = 4$, node to saddle; $t = 0$, stable to unstable spiral

6.3.1 (b) $(s^2 - \omega^2)/(s^2 + \omega^2)^2$ (d) $(s^2 + 2)/s(s^2 + 4)$
(g) $(1 - e^{-s})/s$ (i) $1/s(1 - e^{-s})$

6.3.2 (b) (g) $e^{2t} - e^t$ (d) $\delta + e^t$ (f) $\cos t + \sin t$ (j) δ'

6.3.4 $2e^{-t}\sin t$ **6.3.5** (b) $e^{i\omega t}$ (d) $6t - 6\sin t$ (e) $\frac{1}{4}e^{-t} + \frac{1}{2}te^t - \frac{1}{4}e^t$

6.3.7 $e^{At} = I + A(e^t - 1)$ **6.3.9** $(1 + s^2)U' + sU = 0$

6.3.10 (b) $-e^{-s}/s - e^{-s}/s^2 + 1/s^2$

6.3.11 (b) multiply by $1 + e^{-s} + e^{-2s} + \cdots$.

6.3.12 $v = t + c^{-1}(e^{-ct} - 1)$

6.3.14 (a) brake at $x = \frac{1}{2}$ (b) $v = At$ and $d = \frac{1}{2}At^2$
so $t = \sqrt{2d/A}$ = time to brake at $x = d$; time to rest is
$\sqrt{2d/A} + \sqrt{2(1 - d)/B}$; minimum time if $d = B/(A + B)$.

6.3.15 $f = 4$ **6.3.16** $n \to \sum nz^{-n} = z/(z - 1)^2$

6.3.17 (b) $2^n - 1$ **6.3.20** $\lambda = 6/5$; 1 superchip is better

6.3.21 $k = 98$; $v_{k+1} = 1 + v_k$ **6.3.22** $u^* = G(u^*)$; $|\lambda| < 1$

6.3.23 cba^{n-1} **6.3.24** (c) bounded transfer function $z^2/(z + \frac{1}{2})^2$

6.4.5 $u = 2(e^{-t}\sin x - \frac{1}{2}e^{-2t}\sin 2x + \cdots)$

6.4.6 alternate between 1 and -1 **6.4.7** $u_\infty = T_0 + (T_1 - T_0)x$

6.4.9 $\hat{u} = (1 - e^{-k^2 t})/k^2$ **6.4.11** (a) $1/(1 + (x + ct)^2)$ (b) $u = 1$ for $t > x$

6.4.12 $X = x + e^t - 1$ **6.4.15** $u = 1/2c$ if $|x| < ct$

6.4.16 $u = (e^{-(r-t)^2} - e^{-(r+t)^2})/4r$ **6.4.20** $E = \frac{1}{2}(u_t^2 + u_x^2 + u^2)$

6.4.24 $G = \begin{bmatrix} 0 & g \\ g^T & 0 \end{bmatrix}$ with $g = \begin{bmatrix} 0 & 0 & 0 \\ 0 & 0 & -1 \\ 0 & 1 & 0 \end{bmatrix}$ **6.4.22–23** Compare 6J

6.4.27 $\omega = ck$; $\delta(x - ct)$ **6.4.29** $X = xL$, $T = tL/V$

6.4.30 neutral; decay; grow   **6.5.1** $\Delta t = 1$; 0; $(-1)^n$

6.5.2 2^{-10} is closer than 3^{-5} **6.5.3** $G = 1/101$, $-49/51$

6.5.4 $|\lambda| > 1$; $|\lambda| = 1$; $|\lambda| < 1$ **6.5.6** $G = a\Delta t - \sqrt{1 + a^2\Delta t^2}$ is below -1

6.5.7 $c_0 = c_2 = \frac{1}{2}$, $c_1 = 0$ **6.5.11** $u(t) = 2$; $\Delta t \le 0.02$ **6.5.12** $C = 4$

6.5.15 $G = 1 + r - re^{ikh}$ **6.5.16** Yes   **6.5.17** $0 \le r \le 2$

6.5.20 $|G|^2 = 1 - r^2(1-r^2)(1 - \cos kh)^2$ **6.5.22** $|\sum a_j e^{ijt}| \le 1$

6.6.1 arcs of circles   **6.6.2** $x + \sqrt{2t}$ except $-\sqrt{t^2 - x^2}$ if $2|x| < -t$

6.6.3 tangent slope $-x$; perpendicular slope $1/x$; point $(x + x/\sqrt{1 + x^2}$,

$y + \sqrt{1 + x^2})$

6.6.9 $u = x/t$ for $-\sqrt{2t} < x < 0$, $u = 0$ elsewhere

6.6.11 $U = Ax - \frac{1}{2}tA^2$ or $Bx - \frac{1}{2}tB^2$ and $u = U_x$

6.6.17 $L = \begin{bmatrix} \cos 2t & -\sin 2t \\ -\sin 2t & -\cos 2t \end{bmatrix}$

6.6.18 $(F')^2 = cF^2 - F^4$; sech $(x - t)$ when $c = 1$ **6.6.19** (b) $u = \dfrac{x}{t-1}$

CHAPTER 7   NETWORK FLOWS AND COMBINATORICS

7.1.4 $A_{ij}^2 > 0$ if a 2-edge path connects i to j; not unless we require $A_{ii} = 1$

7.1.5 No   **7.1.8** Yes

7.1.11 List rejected edges in order of length, ready for next selection

7.1.12 If the new edge closes a loop, discard the longest edge in that loop

7.1.13 If e is in T, deleting it leaves 2 smaller trees. The loop containing e has another edge connecting those trees. It is not in T (or T would have a loop) but substituting it for e reduces T—so T is not minimal.

7.1.14 4, 7, 8, ∞; 3, 3, 7, ∞

7.1.15 $D_5 = \min(7, 6 + 7)$, $D_6 = \min(11, 6 + 8)$ allow paths through node 4

7.2.6 (a) covering $2n$ 1's needs at least n lines, leaving at most n for a zero submatrix

7.2.8 If A has an r by s zero submatrix, P would have a $2r$ by $2s$ zero submatrix. But P is a matching so $2r + 2s \le 20$. Then $r + s \le 10$ and A satisfies Hall's condition.

7.2.12 The two boys at $\frac{1}{2}x$ and $\frac{1}{2}(x + 1)$ like only one girl.

7.2.14 If M matches i to j, a plane can do both flights (and it can do k next if j is matched to k). If flights i and j are not involved in the cover then $a_{ij} = a_{ji} = 0$ and no plane can do both flights.

7.2.15 It takes at least 3 lines to cover 15 1's.

7.3.6 $n!$ **7.3.7** block diagonal with 1's filling 2 by 2 blocks

7.3.9 $\frac{1}{3}pN^3$; add $\frac{1}{2}p^2N^2$ for back substitution

7.3.11 aim for *lower* triangular   **7.4.1** flow $= 5$ **7.4.3** flow $= 25$

7.4.6 BFS   **7.4.7** flow $= 10$

7.4.9 capacities into $j' =$ capacities into j, capacities out of $j'' =$ capacities out of j

7.4.11 give each edge capacity 1

7.4.14 the largest capacity crossing the cut

7.4.16 (b) Assign number 2 to a node with no edge into it, except from node 1

7.5.1 $C_{13} \geq 2$

7.5.2 $u_2 + v_1 - C_{12} = 3$ is largest overcharge; $s = 1$; $u = 0, -1$; $v = 5, 3, 4$

7.5.4 $u = 0, -4$; $v = 9, 3, 6$; introduce $x_{21} = 1$ **7.5.5** $u = 0, -4$; $v = 7, 3, 6$; cost 41

7.5.12 Reversing signs in (2), each column has 1 and -1

7.5.13 $1 = x_{21} = x_{12} = x_{33}$ **7.5.18** optimal $8 + 1 + 8$; bottleneck 3, 6, 5

7.5.20 cost $15 + 16 = 31$

CHAPTER 8   OPTIMIZATION

8.1.1 $(6, 0, 0)$ $(0, 3, 0)$ $(0, 0, 2)$ **8.1.3** $a \geq 0$; $a < 0$ **8.1.6** $(6, 0)$ $(2, 2)$ $(0, 6)$

8.1.7 min $= 4$; 6; $-\infty$ **8.2.1** $12 - x_2$ is decreasing; $12 - 2x_3$ decreases faster

8.2.4–5 $r = 0, 0, -5, -7$; cost 7; x_4 **8.2.6** 6, 0, 0, 1

8.2.8 last column $-v_1/v_m, \ldots, 1/v_m$ **8.2.9** $r \leq 0$

8.2.13 $mnp + mpq$ vs. $npq + mnq$ **8.2.14** $P_{ii} = \frac{2}{3}$, $P_{ij} = -\frac{1}{3}$

8.2.16 direction $(34, 22, 46)$

8.3.2 (shortest distance to j) is not greater than (shortest distance to i) $+ l_{ij}$

8.3.4 (ii) x_1, x_3, x_5 remain positive and the edge is infinitely long

8.3.5 (a) $y \geq 0$ multiplies $Ax - b \geq 0$ to give $yAx \geq yb$
(b) $x = (0, 0, 3)$; $y = 5/4$

8.3.9 A is 5 by 6   **8.3.13** A can be 3 by 10 (KCL at 3 nodes); minimax $= 5$

8.3.14 price difference only across min cut

8.3.15 min $L(x, \bar{y}) \leq L(\bar{x}, \bar{y}) \leq$ max $L(\bar{x}, y)$ **8.4.2** $x_1 = \frac{1}{2}$; $y_1 = \frac{1}{4}$

8.4.3 $X = (0, 1, 0) = Y$; $X = (\frac{1}{3}, \frac{1}{3}, \frac{1}{3}) = Y$

8.4.8 X follows g_1 or b_1, and Y chooses g_1–b_2 or g_2–b_1 with probabilities $\frac{1}{2}$

8.4.9 $x^* \geq 0$, $y^* \geq 0$ **8.4.10** $F = \max(y_1, y_2)$; $\frac{1}{2}$

8.4.12 $x_1 = x_2 = 2$; labor not completely used   **8.4.13** $y = (\frac{3}{8}, 0, \frac{7}{8})$

8.4.14 $x = (0, \frac{8}{3})$, $y = (0, 0, 2)$ **8.4.15** $\bar{c}_2 = 6$ **8.5.1** $x = (1, 2)$

8.5.2 $\begin{bmatrix} 1 & 1 \\ 1 & 5 \end{bmatrix}$ **8.5.4** active if $b < 1$

8.5.7 no; yes but not strictly convex; yes if $a \geq 0$, $ac \geq b^2$

8.5.9 no; yes; yes   **8.5.16** $3(y/4)^{4/3}$

8.5.17 $F^* = y \log y - y$; $F^* = 0$ if $y \geq 0$, ∞ if $y < 0$

TABLE OF SPECIAL FUNCTIONS: DIFFERENTIAL EQUATIONS, WEIGHTED ORTHOGONALITY, RECURSION FORMULA

Bessel Function $J_p(x)$: $\qquad x^2 J_p'' + x J_p' + (x^2 - p^2) J_p = 0$

$$\int_0^1 x J_p(w_m x) J_p(w_n x) dx = 0 \quad \text{if} \quad J_p(w_m) = J_p(w_n) = 0$$

$$x J_{p+1} = 2p J_p - x J_{p-1} \qquad J_p(x) = \frac{\Gamma(p+1)}{2^p} \sum_{k=0}^{\infty} \frac{(-1)^k (x/2)^{2k+p}}{k! \Gamma(k+p+1)}$$

Legendre Function $P_n(x)$: $\qquad (1-x^2) P_n'' - 2x P_n' + n(n+1) P_n = 0$

$$\int_1^1 P_m(x) P_n(x) dx = 0 \qquad (n+1) P_{n+1} = (2n+1) x P_n - n P_{n-1}$$

$$P_n(x) = \sum_{k=0}^{[n/2]} (-1)^k \binom{-\frac{1}{2}}{n-k} \binom{n-k}{k} (2x)^{n-2k}$$

Chebyshev Polynomial $T_n(x) = \cos n\theta$, with $x = \cos\theta$:

$$(1-x^2) T_n'' - x T_n' + n^2 T_n = 0$$

$$\int_{-1}^1 (1-x^2)^{-1/2} T_m(x) T_n(x) dx = 0 \qquad T_{n+1} = 2x T_n - T_{n-1}$$

$$T_n(x) = \frac{n}{2} \sum_{k=0}^{[n/2]} \frac{(-1)^k (n-k-1)!}{k!(n-2k)!} (2x)^{n-2k}$$

Laguerre Polynomial $L_n(x)$: $\qquad x L_{n+1}'' + (1-x) L_n' + n L_n = 0$

$$\int_0^{\infty} e^{-x} L_m(x) L_n(x) dx = 0 \qquad (n+1) L_{n+1} = (2n+1-x) L_n - n L_{n-1}$$

$$L_n(x) = \sum_{k=0}^n \frac{(-1)^k n!}{(k!)^2 (n-k)!} x^k$$

Hermite Polynomial $H_n(x)$: $\qquad H_n'' - 2x H_n' + 2n H_n = 0$

$$\int_{-\infty}^{\infty} e^{-x^2} H_m(x) H_n(x) dx = 0 \qquad H_{n+1} = 2x H_n - 2n H_{n-1}$$

$$H_n(x) = \sum_{k=1}^{[n/2]} \frac{(-1)^k n!}{k!(n-2k)!} (2x)^{n-2k}$$

Gamma Function $\Gamma(n)$: $\qquad \Gamma(n+1) = n\Gamma(n)$ or $\Gamma(n+1) = n!$

$$\Gamma(n) = \int_0^{\infty} e^{-x} x^{n-1} dx \text{ gives } \Gamma(1) = 0! = 1, \ \Gamma(\tfrac{1}{2}) = \sqrt{\pi}$$

$$\Gamma(n+1) = n! \approx \sqrt{2\pi n} \left(\frac{n}{e}\right)^n \quad \text{(Stirling's formula for large } n\text{)}$$

Binomial Symbol $\dbinom{n}{m} = \dfrac{n!}{m!(n-m)!} = \dfrac{\Gamma(n+1)}{\Gamma(m+1)\Gamma(n-m+1)}$

Integer Part Symbol $[x] = $ largest integer not greater than x

INDEX